FREE RADICALS

FREE RADICALS

Biology and Detection by Spin Trapping

GERALD M. ROSEN

BRADLEY E. BRITIGAN

HOWARD J. HALPERN

SOVITJ POU

New York   Oxford

Oxford University Press

1999

Oxford University Press

Oxford   New York
Athens   Auckland   Bangkok   Bogotá   Buenos Aires   Calcutta
Cape Town   Chennai   Dar es Salaam   Delhi   Florence   Hong Kong   Istanbul
Karachi   Kuala Lumpur   Madrid   Melbourne   Mexico City   Mumbai
Nairobi   Paris   São Paulo   Singapore   Taipei   Tokyo   Toronto   Warsaw

and associated companies in
Berlin   Ibadan

Library of Congress Cataloging-in-Publication Data

Free radicals : biology and detection by spin trapping / by Gerald M.
    Rosen ... [et al.].
        p.   cm.
    Includes bibliographical references and index.
    ISBN 0-19-509505-7
    1. Free radicals (Chemistry)—Physiological effect.   2. Free
radicals (Chemistry)—Analysis.   3. Electron paramagnetic resonance.
I. Rosen, Gerald M., 1945–
QP527.F727   1999
572'.3—dc21   98-38929

9 8 7 6 5 4 3 2 1

Printed in the United States of America
on acid-free paper

Preface

In 1954, Gerschman and co-authors proposed that free radicals were the causitive agent responsible for oxygen poisoning and ionizing radiation. Surprisingly, the implication of this hypothesis to aerobic organisms—oxygen-centered free radicals might be by-products of cellular metabolism—seemed remote at that time, as more than a decade would elapse before biological sources of superoxide would be inextricably linked to an enzyme—superoxide dismutase—that had evolved to scavenge this free radical. Now as we approach the thirtieth anniversary of the discovery of superoxide dismutase by Joe McCord and Irwin Fridovich, the importance of free radicals in biology is no longer controversial. These reactive species are common intermediates produced during cell metabolism where these reactive species play an essential role in the control of many physiological functions from cell signaling to host immune response.

Spin trapping traces its origin to a series of publications in the late 1960s when several laboratories reported the addition of free radicals to either nitrosoalkanes or nitrones. The resulting nitroxide—a spin trapped adduct—was found to exhibit remarkable stability at ambient temperature, far exceeding the lifetime of the parent free radical. Serendipitously, spin trapping was born at a time when it had become apparent other methods of identifying free radicals were so limiting that future research into free radicals in biology was problematic.

In the intervening years, spin trapping has been instrumental, for example, in defining how enzymes shunt electrons to their terminal acceptor, allowing insightful discoveries to be made. One only needs to appreciate that spin trapping was the method of choice in discovering the mechanism by which nitric oxide synthase secretes nitric oxide and superoxide to realize the contribution this analytical tool has made to our knowledge of free radicals in biology and chemistry.

Most recently, spin trapping and low frequency EPR spectroscopy have allowed us to peer inside a living animal and observe the generation of free radicals in real time and at their site of formation. Clearly, future advances in physiology and pathology of free radicals will be intimately linked to progress in this technology.

Like many scientists who have employed spin trapping in their research, we have entered the field of free radicals from diverse backgrounds. As such, we have long been cognizant of the lack of a single source that provides scientists of various disciplines, from biology to chemistry to phsyics, with the background and information they may need to logically apply the strength of spin trapping to their particular research. The spin trapping literature is replete with technically excellent data that has been misinterpreted due to an under appreciation of the limitations and complexity of the chemistry involved in this methodology. Similarly, there are myriad studies that inappropriately extrapolate spin trapping findings to highly complex biological systems. We, therefore, set out with the goal of writing a text that could serve as a practical guide to spin trapping that scientists, regardless of their expertise, research focus and experience with this technique, would find current and helpful. This book approaches the topics with the belief that progress is most often tied to an understanding and appreciation of past discoveries.

Any endeavor of this magnitude is never accomplished without the efforts of many fellow scientists who kindly supplied us with their papers, even keeping us abreast of their latest findings. The research of many graduate students, post-doctoral fellows and colleagues is evident throughout this text. Special thanks go to Drs. Eli Finkelstein, Elmer Rauckman, Michelle Kloss, Dan Hassett, Jimmy Turner, Les Ramos, Garry Buettner, Oyebode Olakanmi, Michael McCormick, Pei Tsai, and Myron Cohen for their many years of support, camaraderie and devotion as we explored the role of free radicals in biology.

We are grateful to our friends, who played an active role in developing this book, and through their suggestions and constructive criticism, helped produce this vast compendium. Thanks go to Dr. Barry Gittlen, whose accurate translation of a Hebrew text made, In the Beginning, meaningful. To Dr. Ann Kinney, whose lifetime of research can be described as the first three minutes of creation, we appreciate your labors in explaining the early events that led to the formation of our solar system and in particular Earth. To Dr. Howard Steinman, for his devotion to unraveling the mysteries of superoxide dismutase; we cherish the time you spent guiding us through the intricacies of this enzyme. And a special thanks go to Dr. Irwin Fridovich, whose love of science can only be matched by his support of these authors. More than a quarter of a century ago, Irwin listened to a young scientist (GMR) as he proposed the "crazy" notion that superoxide, secreted by the enzyme FAD-containing monooxygenase, was responsible for the one-electron oxidation of hydroxylamines to nitroxides. If his theory were correct, then another monooxygenase would be added to the growing list of enzymes that released superoxide during enzymic cycling. Irwin was, surprisingly, delighted at such a possibility and provided guidance and superoxide dismutase to verify the experimental observations. Over the years you have continued to be a sounding board for many new and sometimes provocative ideas. We will always be indebted for such attentiveness.

Dr. William Moore, late of the Brigham and Womens Hospital, Boston and Drs. Michael Bowman, Beverly Teicher, Miroslav Peric, and Mr. Gene Barth were crucial to the early development of low frequency EPR spectroscopy and imaging in our group. A community of scientists involved in *in vivo* EPR spectroscopy and imaging, too numerous to individually thank, have contributed to the vision expressed in this book. We would like to thank Dr. Yelena Galtseva for her assistance with the figures in chapter 7.

Thanks, of course, go to our editors and staff at Oxford University Press. We initially proposed to Oxford a modest size handbook on spin trapping. With time, such a project grew into the current text. During the years it took to create this large compendium, our editor, Kirk Jensen continued to support us, clearing the many hurdles that came our way. We are indebted to Lisa Stallings, who kept us on an even keel when too many obstacles seemed to be drowning us. We likewise appreciate the skills of the staff at Keyword Publishing Services, especially Maureen Allen, for making this book a beautiful reality.

Finally, as with almost all individual successes, this work is a testament to the enduring support and love of our families.

Baltimore	Gerald M. Rosen
Iowa City	Bradley E. Britigan
Chicago	Howard J. Halpern
Baltimore	Sovitj Pou

Contents

FREE RADICALS

1

In the Beginning

In the beginning of God's creation of the heaven and
the earth: the earth was without form and void, darkness was on the face of the deep, and a wind from God
vibrated over the surface of the waters. And God said,
let there be light: and there was light. And God saw
the light, that it was good: and God divided the light
from the darkness ... And God said, Let the waters
swarm abundantly with moving creatures that have
life.

Genesis 1: 1–7

This biblical account of creation, a theme common to the cultural life of many
societies over the past several millennia (Westerman, 1974), foretells the birth of
life on our planet. From the available evidence we have today, this epic began
about 15 billion years ago, when, we are told, a massive explosion of incompre-
hensible power, referred to as the "Big Bang," created everything that was, is, and
will be (Hawking, 1988; Peebles, *et al.*, 1994). From this singular event, the evolution of life commenced with the formation of hydrogen and, to a lesser extent,
helium. Eventually, these gases coalesced under the force of their own gravity into
stars. The more massive ones created, through fusion, the elements of carbon,
nitrogen, and oxygen (Mitchell, *et al.*, 1987). With time, increased density due to
the formation of the heavier elements caused the core of these stars to collapse. The
release of this gravitational energy blew off the outer regions, erupting in a super-
nova, during which heavier atoms, up to iron, were produced in the process. Debris
from these explosions were sent throughout the universe, enriching interstellar
space with matter for the next generation of stars (Joseph, 1994). This cascade
led, about 5 billion years ago, to the formation of our sun, leading to the creation
of, and, with time, the diversity of many life forms from the prebiotic assembly of
proteins to single-cell organisms to the evolution of humans with the capacity and
curiosity to explore the mysteries of the universe.[1]

Evolution of Life

^{13}C-Isotope labeling studies suggest microbial life appears to have begun about 3.8
billion years ago (Schidlowski, 1988; Mojzsis, *et al.*, 1996), relatively soon on the

cosmic time scale after Earth's formation.[2] Life started in a world hostile to most organisms that presently inhabit this planet (Figure 1.1). Ultraviolet light continually rained upon its surface, since O_2, so essential in shielding our planet from this electromagnetic radiation, was absent in Earth's primitive atmosphere. Yet this environment was ideal for the development of prebiotic life. Although little is known of Earth's early existence, evidence has suggested that the fundamental building blocks of life were absent at its dawn. Thus, a readily available energy source was essential for syntheses of even the simplest of amino acids. Although there is much speculation as to the nature of the driving force for this prebiotic engine, evidence exists that the photochemical redox cycling of Fe^{2+}/Fe^{3+} could have provided the necessary high-energy reducing equivalents for formation of H_2,

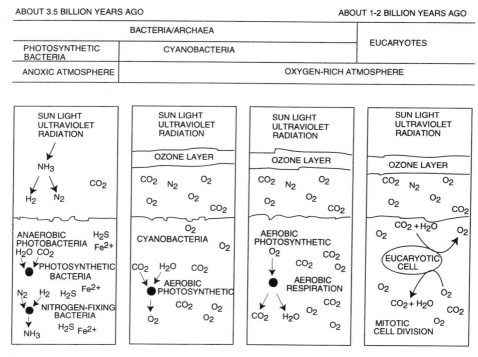

Figure 1.1 The evolution of life, from anaerobic to aerobic organisms, is seen through changes in the planet's atmosphere as a function of time. In this model, life appeared within the first billion years after formation of the planet. The earliest cells accumulated nutrients, synthesized from abiotic chemical reactions, through passive and active transport across primitive biomembranes. The inefficiency and limitations of these gathering processes forced the development of the first phototropes in which H_2S and thiol esters were the source of reducing equivalents. Even after the evolution of bacteria with the capability of water-splitting photosythesis to produce O_2, accumulation of this gas in the atmosphere required an additional billion years to reach measurable levels, however. Thereafter, rapid efflux took a surprisingly short period of time to reach current levels. (Adapted from Schopf, 1978.)

CH_4 and other important ingredients for the assembly of amino acids in this abiotic world (Braterman, *et al.*, 1983; Borowska and Mauzerall, 1987; Huber and Wächtershäuser, 1997). Later, this process could have given way to more sophisticated energy systems, such as thioesters and pyrophosphates; the latter became a forerunner for adenosine triphosphate (ATP). Thus, these energy sources could have been an additional route to the prebiotic syntheses of carboxylic acids and other basic cellular building blocks.

Let us assume that life began with the formation of primitive cells. Then, how did these fundamental units arise from the primeval environment of pre-Cambrian Earth? Since there is no fossil evidence of the earliest protocells, it is difficult to chronicle their evolution from abiotic chemistry. Nevertheless, scientists believe that this seminal series of events commenced several hundred million years after Earth's crust cooled and solidified. Many biologists subscribe to the theory that abiotic chemical syntheses led to the formation of amino acids and nucleotides, which, with time, resulted in the ordered assembly of proteins and nucleic acids. The next step—creation of an outer membrane encapsulating the genetic information to undergo metabolism and self-replication—must have been an enormously complex problem to solve. One only needs to consider the difficulty of transporting simple molecules and ions, let alone macromolecules, across liposomes of various composition, in order to appreciate the beauty of this evolutionary process. How this may have occurred remains a mystery.[3] Nevertheless, there are abundant theories as to its origin. One attractive hypothesis (Blobel, 1980) suggests that the evolution of biological membranes required the integration of proteins into the outer portion of lipid vesicles. As such, the building of enzymes involved in cellular metabolism, for example, as well as other macromolecules essential to support early life, must have proceeded on the surface of the vesicle (Blobel, 1980). Invagination led to the evolution of what are referred to as "inside-out cells" (Blobel, 1980; Cavalier-Smith, 1987a). Evidence in support of this theory comes, in part, from the observation that characteristics of the "inside-out cell"— two membranes, a lipoprotein outer membrane surrounding a plasma membrane with attached ribosomes and chromosomes (Cavalier-Smith, 1987a)—could have accounted for the development of early Gram-negative bacteria. Approximately 3.5 billion years ago, it appears that phylogenetic branching from these primitive ancestral cells to the first of the three domains, bacteria, appeared to have begun (Woese, 1987; Woese, *et al.*, 1990).

Little is known of the bioenergetics of these early organisms, even though there are numerous schools of thought on this subject. Most scientists agree, however, that primitive life developed in a reducing environment. One theory (DeDuve, 1991) advocates the importance of harnessing the energy of ultraviolet (UV) light to transport electrons, required for syntheses of essential amino acids and phosphorylation reactions. The efficiency of the UV-driven Fe^{2+}/Fe^{3+} redox cycling system certainly supports this hypothesis. Nevertheless, until free radical scavengers evolved to afford protection against cytotoxicity associated with UV-generated free radicals, growth and diversity of these early organisms must have been limited.

The nutritional demands of early cell life dictated a need for readily available and continual sources of energy. What were they? The universality of ATP attests to its appearance in early bacterial evolution and its continued conservation throughout the various domains (Woese, *et al.*, 1990). What drove the ATP engine? There is serious debate as to whether abiotic chemistry provided sufficient, as well as a broad spectrum of, organic molecules required for sustaining cell function. Clearly, in the beginning, this abiotic broth contained nutrients essential to drive primitive energy retrieval systems. For example, intracellular transport of glucose would have allowed glycolysis to phosphorylate adenosine diphosphate (ADP) to produce ATP, using nicotinamide adenine dinucleotide (NAD^+) as an electron acceptor. Transfer of two electrons as a hydride to pyruvate, giving lactic acid, would allow the cycling of $NADH/NAD^+$ to continue generating energy in the form of ATP. In some microorganisms, pyruvate is converted to ethanol. Thus, fermentation provided both a source of energy, ATP, and reducing equivalences, NADH (although $NADP^+$ could equally have substituted, giving NADPH), to conduct biosynthesis, essential for anaerobic metabolism. However, with the rapid proliferation of these primitive cells, abiotic chemical syntheses could not possibly meet the demands of this burgeoning process. Therefore, ecological pressures must have forced the bacterial population to seek alternative sources of ATP and the reducing equivalents NADH and NADPH.

This stress certainly led to the development of other photochemical reaction centers. With a continual increase in atmospheric levels of CO_2, as the result of ethanol fermentation, the high-energy UV-Fe^{2+}/Fe^{3+} cycle became less attractive and bacteria then evolved mechanisms of capturing the energy associated with the longer wavelengths of the visible spectrum. These early phototropes probably used the energy from one quantum of absorbed light to generate reducing equivalents in the form of NADPH. The ease of oxidizing H_2S, and some organic acids, as compared with H_2O, supports the efficiency of this early photosystem, which probably had characteristics similar to photosystem I of green sulfur bacteria:

$$H_2S \rightarrow S + 2H^+ + 2e^- \qquad E_0' = -280\text{mV}$$
$$H_2O \rightarrow \tfrac{1}{2}O_2 + 2H^+ + 2e^- \qquad E_0' = +820\text{mV}$$

The evolution of bacteria capable of using H_2O as an electron source in CO_2 reduction had far-reaching implications for both the growth and diversity of later organisms. With this evolutional advance, microbes could grow in an environment that was not limited by the accessibility of reducing equivalents. Yet, to overcome the poor oxidizing potential of H_2O, these new photosynthetic bacteria had to develop an additional photoacceptor—photosystem II, which in series with photosystem I provided the necessary electrochemical energy required for NADPH formation. In addition, these phototropes had to evolve a mechanism to "trap" electrons within the photosystem, thereby preventing release of the free radicals, HO· and $O_2 \cdot^-$, and H_2O_2—toxic intermediates that would have threatened the survival of these species:

$$2H_2O + 2NADP^+ + h\nu \rightarrow O_2 + 2NADPH + 2H^+$$

Initially, most of the O_2 produced by these photosynthetic bacteria was consumed by the formation of oxides of various metal ions. This chemical reaction led to the precipitation of ferric and ferrous oxides, resulting in the deposit of banded-iron throughout the surface of our planet (Braterman, *et al.*, 1983). Thus, by about 2 billion years ago, O_2 levels increased to such an extent that the UV-Fe^{2+}/Fe^{3+} era had virtually ended, while H_2O/O_2 photosystems emerged as the dominant source of energy (Figure 1.1). Over the next 500 million years, O_2 concentration slowly increased to present-day levels. At some point, however, the rapid accumulation of O_2 must have presented an enormous barrier to the survival of bacterial life, whose evolution had advanced in an anaerobic world. Evidence in support of this thesis comes, in part, from studies with obligate anaerobes such as *Porphysomonas gingivalis* (Nakayama, 1994). When this bacterium is grown in an aerobic environment, life is intimately linked to the presence of the enzyme, iron-superoxide dismutase (FeSOD), which scavenges the free radical $O_2\cdot^-$. Mutant strains of *P. gingivalis*, which are deficient in this enzyme, do not survive in this O_2-rich atmosphere.

Besides developing enzyme systems capable of neutralizing the toxicity associated with the presence of O_2, these early aerobic bacteria had to convert existing electron transport chains to accommodate O_2 as the new electron acceptor. Like most evolutionary processes, there was undoubtedly an intermediate step in which these microbes, similar to present-day purple bacteria, employed components of the electron transport chain to harness energy from sunlight, converting H_2O into O_2. With minor reshuffling of the electron flow, these bacteria reversed the cycle, using O_2 as an electron acceptor for respiration. With time, this dual system became less efficient, forcing some bacteria to lose their photosynthetic capability. In those organisms, the respiratory chain became the primary source of energy. About 1 billion years ago, with the dawn of eucarya (Woese, *et al.*, 1990), the primitive respiratory systems of more ancestral organisms gave way to the sophisticated engine of the mitochondrion.

It would be, at this point, remiss not to discuss, albeit briefly, the source of the mitochondrion. One very popular theory (Margulis, 1981) is that mitochondria were aerobic bacteria that were incorporated into early eucaryotic cells. There is, of course, much debate and controversy (Cavalier-Smith, 1987b) as to how this endosymbiotic relationship—"the formation of permanent associations between organisms of different species"—(Margulis, 1981) took place, whether through phagocytoses or endocytoses. In either case, overwhelming evidence supports the hypothesis that the mitochondrion was an unique organism, which, even today, retains many of its ancestral features. For example, mitochondria, like bacteria, multiply through fission. Similarly, this organelle contains DNA, ribosomes, and enzymes to synthesize proteins encoded in their DNA, as well as MnSOD to protect itself from the toxic effects of $O_2\cdot^-$ leakage during respiration. Yet the majority of the genetic information has been passed to the host, as a significant amount of mitochrondrial protein is transported from the cytosol into this organelle.

The advantage of aerobic versus anaerobic metabolism can be reflected in the amount of energy produced. For example, the driving force behind aerobic catabolism is the enormous gain in energy (686 kcal/mol generated aerobically versus about 20 kcal/mol produced anaerobically) by degradation of glucose to CO_2 and H_2O, resulting in the formation of 36 molecules of ATP:

$$C_6H_{12}O_6 + 6O_2 + 36P_i + 36ADP \rightarrow 6CO_2 + 6H_2O + 36ATP$$

Contrast this aerobic catabolism to anaerobic glycolysis in which some of the energy from glucose metabolism is converted to form 2 molecules of ATP and lactate, dictating the need for sequential steps for CO_2 production:

$$C_6H_{12}O_6 + 2ADP + 2NAD^+ + 2P_i \rightarrow 2C_3H_6O_3 + 2ATP + 2NADH$$

The enormous advantage offered by aerobic metabolism was, to some degree, offset by the need to expend large amounts of energy in the defense against O_2 toxicity.

Concluding Thoughts

During the past 2 billion years of evolutional history, organisms that have had to survive in a world hostile to aerobic life have developed families of enzymes capable of regulating the flux of $O_2 \cdot^-$ and products derived from this free radical. In some situations, sophisticated repair mechanisms have corrected the damage elicited by these reactive species (Demple and Halbrook, 1983; Demple, 1991; Farr and Kogoma, 1991), while in others, elaborate and frequently overlapping defense mechanisms have assured the continuum of the species. Thus, our life in a world surrounded by O_2 must be viewed with great trepidation. We depend upon its presence for survival, yet we must be on guard against untoward reactions elicited by this gas. In the next chapter, we will begin to explore the mysteries of this oxygen paradox.

References

Blobel, G. (1980). Intracellular protein topogenesis. *Proc. Natl. Acad. Sci. USA* **77**: 1496–1500.

Borowska, Z.K., and Mauzerall, D.C. (1987). Efficient near ultraviolet light induced formation of hydrogen by ferrous hydroxide. *Orig. Life* **17**: 251–259.

Braterman, P.S., Cairns-Smith, A.G., and Sloper, R.W. (1983). Photo-oxidation of hydrated Fe^{2+}—Significance for banded iron formations. *Nature* **303**: 163–164.

Cavalier-Smith, T. (1987a). The origin of eukaryote and archaebacterial cells. *Ann. N.Y. Acad. Sci.* **503**: 17–54.

Cavalier-Smith, T. (1987b). The simultaneous symbiotic origin of mitochondria, chloroplasts and microbodes. *Ann. N.Y. Acad. Sci.* **503**: 55–71.

DeDuve, C. (1991). *Blueprint for a Cell. The Nature and Origin of Life*, pp. 123–145, Neil Patterson Publishers and Carolina Biological Supply Company, Burlington, NC.

Demple, B. (1991). Regulation of bacterial oxidative stress genes. *Annu. Rev. Genet.* **25**: 315–337.

Demple, B., and Halbrook, B. (1983). Inducible repair of oxidative DNA damage in *Escherichia coli*. *Nature* **302**: 466–468.

Farr, S.B., and Kogoma, T. (1991). Oxidative stress response in *Escherichia coli* and *Salmonella typhimurium*. *Microbiol. Rev.* **55**: 561–585.

Hawking, S.W. (1988). *A Brief History of Time—From the Big Bang to Black Holes*. Bantam Books, Toronto, Canada.

Huber, C., and Wächtershäuser, G. (1997). Activated acetic acid by carbon fixation on (Fe, Ni)S under primordial conditions. *Science* **276**: 245–247.

Joseph, R.D. (1994). Dust in the distance. *Nature* **370**: 325.

Margulis, L. (1981). *Symbosis in Cell Evolution: Life and Its Environment on the Early Earth*, pp. 1–14, W.H. Freeman, San Francisco.

Mitchell, D.L., Lin, R.P., Anderson, K.A., Carlson, C.W., Curtis, D.W., Korth, A., Reme, H., Sauvaud, J.A., d'Utson, C., and Mendis, D.A. (1987). Evidence for chain molecules enriched in carbon, hydrogen and oxygen in comet Halley. *Science* **237**: 626–628.

Mojzsis, S.J., Arrhenius, G., McKeegan, K.D., Harrison, T.M., Nutman, A.P., and Friend, C.R.L. (1996). Evidence for life on Earth before 3,800 million years ago. *Nature* **384**: 55–59.

Nakayama, K. (1994). Rapid viability loss on exposure to air in a superoxide dismutase-deficient mutant of *Porphysomonas gingivalis*. *J. Bacteriol.* **176**: 1939–1943.

Peebles, P.J.E., Schramm, D.N., Turner, E.L., and Kron, R.G. (1994). The evolution of the universe. *Sci. Am.* **271**(4): 53–57.

Schidlowski, M. (1988). A 3,8000-million year isotope record of life from carbon in sedimentary rocks. *Nature* **333**: 313–318.

Schopf, J.W. (1978). The evolution of the earliest cells. *Sci. Am.* **239**(4): 110–138.

Westerman, C. (1974). *Creation*, Fortress Press, Philadelphia.

Woese, C.R. (1987). Bacterial evolution. *Micobiol. Rev.* **51**: 221–271.

Woese, C.R., Kandler, O., and Wheelis, M.L. (1990). Towards a natural system of organisms: Proposal for the domains Archae, Bacteria and Eucarya. *Proc. Natl. Acad. Sci. USA* **87**: 4576–4579.

2

The Oxygen Paradox

Continued exposure to O_2 might burn out the candle
of life too quickly.

Priestley, 1775

This prophetic observation by Priestley[1] set the stage for one of the great adventures of modern science—solving the mystery of why O_2, so crucial a fuel for the metabolic machinery of life, is yet so toxic to all creatures inhabiting this planet. Along the path of inquiry, many roads were taken. None were more important than the discovery of X-rays by Röntgen[2] in 1895, and the discoveries of the radioactive properties of uranium by Becquerel in 1896 and radium by Marie and Pierre Curie in 1898. Within a few short years, it was recognized that the radiation emitted from the Crookes tube and from specific isotopes of various elements had profound effects on H_2O. However, it was not until 1944 that this reaction was found to generate $HO\cdot$ (Weiss, 1944):

$$H_2O + h\nu \rightarrow H_2O^+\cdot + e^-(\rightarrow e_{aq})$$

$$H_2O^+\cdot + H_2O \rightarrow HO\cdot + H_3O^+$$

Even though the biological consequences of O_2 were recognized soon after its discovery, toxicity associated with this gas was not apparent until the middle of the next century (Bert, 1943; Bean, 1945; Gilbert, 1981). Even then, pathogenic effects were thought to be limited to unique and nonphysiologic exposures to O_2. For example, Behnke, et al. (1936) and Behnke (1940) reported adverse effects of increased O_2 concentrations at atmospheric pressures. Similarly, untoward reactions, such as convulsions and pulmonary dysfunction, after exposure to O_2 at higher pressures, were well documented by the early part of the twentieth century (Shilling and Adams, 1933; Bert, 1943; Bean, 1945).

Soon thereafter, evidence began to accumulate, suggesting that the injurious effects of radiation paralleled those noted for O_2 poisoning. Typical of the reports of the time was a study by Conger and Fairchild (1952), in which chromosomal aberrations found in *Tradescantia* microspores after exposure to elevated concentrations of O_2 were identical to the effects induced by ionizing radiation. Despite the apparent association between the pathogenicity of O_2 and radiation exposure, the connection between these two lines of research might have been lost for years had not an inquisitive young scientist come along at this auspicious time.[3] In an article published in *Science*, free radicals were proposed to be the connecting thread, uniting the effects of ionizing radiation with those associated with O_2 poisoning (Gerschman, *et al.*, 1954). At first, the impact of this hypothesis and its implication—free radicals may be intimately linked to the chemistry of life processes—seemed remote to biological scientists of that era. Yet, within a few decades, free radicals, which had become recognized as common intermediates in cellular metabolism, were found to play an essential role in the control of many physiologic functions, including those linked to host immune response (Malech and Nauseef, 1997).

Chemistry of Dioxygen

Dioxygen in its ground state, $^3\Sigma_g{}^-$, has two unpaired electrons with identical spins, thereby fulfilling the requirements of the Pauli exclusion principle (Pauling, 1960; Fee and Valentine, 1977; Hill, 1978, 1979). Since a change in the spin state, which is slow relative to the lifetime of a collisional complex, is required for ionic reactions to proceed, univalent reduction of O_2 is preferred over divalent pathways. Thus, the chemistry of O_2, except for the excited singlet states $^1\Delta_g$ and $^1\Sigma_g{}^+$, is favored by, but not limited to, free radical reactions (Figure 2.1).

The progressive four-electron reduction of O_2 results in the sequential formation of $O_2{}^{.-}$, H_2O_2, $HO\cdot$, and H_2O (Hill, 1979; Figure 2.2). Based on the oxidation–reduction potentials illustrated in Table 2.1, all intermediates are oxidants compared with H_2O (Fee and Valentine, 1977; White 1991). While $O_2{}^{.-}$ is the least powerful oxidant, $HO\cdot$ demonstrates remarkable strength.

As illustrated in Figure 2.2, these intermediates exhibit varying degrees of acidity. While O_2 is neutral throughout a broad pH range, $O_2{}^{.-}$ is a weak acid with a pK_a of 4.8 (Behar, *et al.*, 1970; Bielski, 1978; Bielski, *et al.*, 1985). Hydrogen peroxide with pK_as of 11.65 (Morgan, *et al.*, 1988; Czapski and Bielski, 1993) and ≈ 16 (Fee and Valentine, 1977) remains uncharged at physiologic pH.

The primary reaction of $O_2{}^{.-}$ in aqueous media is its disproportionation to H_2O_2, the rate of which is dependent upon the pH of the buffer, as depicted in Figure 2.3:

$$HO_2\cdot + HO_2\cdot \rightarrow H_2O_2 + O_2$$

$$HO_2\cdot + O_2{}^{.-} + H_2O \rightarrow H_2O_2 + O_2 + HO^-$$

$$O_2{}^{.-} + O_2{}^{.-} + 2H^+ \rightarrow H_2O_2 + O_2$$

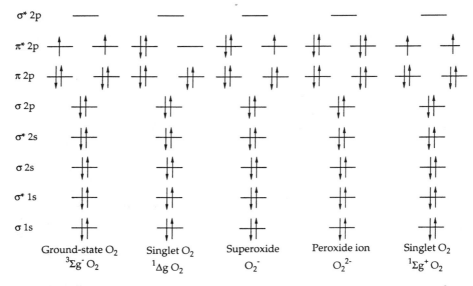

Figure 2.1 Molecular orbital depiction of ground-state O_2, singlet O_2, $O_2\cdot^-$, and O_2^{2-}.

From the curve we show in Figure 2.3, at pHs below 2, the rate of dismutation is constant. Then, it rapidly increases, leveling off at the pK_a of $HO_2\cdot/O_2\cdot^-$. Above pH 4.8, the line precipitiously decreases such that at pH values exceeding the physiologic range, the rate of $O_2\cdot^-$ disproportionation is exceptionally slow. Rate constants at selected pHs are presented in Table 2.2.

Enzymatic Sources of Superoxide

Early in its adaption to aerobic metabolism, microbial life had to evolve enzymatic pathways to regulate production and cellular effects of $O_2\cdot^-$. The most straight-

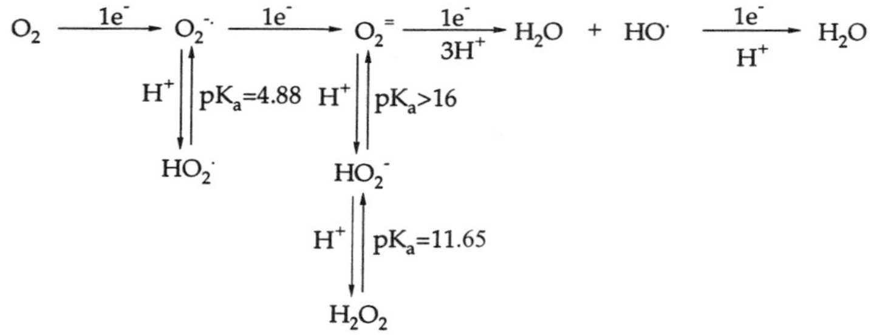

$$O_2 \xrightarrow{1e^-} O_2\cdot^- \xrightarrow{1e^-} O_2^= \xrightarrow[3H^+]{1e^-} H_2O + HO\cdot \xrightarrow[H^+]{1e^-} H_2O$$

$H^+ \big\Vert \, pK_a=4.88 \quad H^+ \big\Vert \, pK_a>16$

$HO_2\cdot \qquad\qquad HO_2^-$

$H^+ \big\Vert \, pK_a=11.65$

H_2O_2

Figure 2.2 Four-electron reduction of dioxygen to water. (Adapted from Fee and Valentine, 1977.)

Table 2.1 Electrochemical Potentials for the Reduction of Oxygen to Water

Reaction	Oxidation–Reduction Potential	
$O_2 + e^- \leftrightarrows O_2 \cdot^-$	$E^\circ = -0.33 \, V^a$	$E' = +0.28 \, V^b$
$O_2 \cdot^- + e^- + 2H^+ \leftrightarrows H_2O_2$	$E^\circ = +087 \, V^a$	$E' = +0.63 \, V^b$
$H_2O_2 + e^- + H^+ \leftrightarrows HO \cdot + H_2O$	$E^\circ = +0.38 \, V^a$	$E' = +0.73 \, V^b$
$HO \cdot + e^- + H^+ \leftrightarrows H_2O$	$E^\circ = +2.33 \, V^a$	$E' = +1.56 \, V^b$

[a]The introduction of a single electron to each species under standard conditions, pH 7.0, 25°C, and 1 M of reactants is presented (Data taken from Fee and Valentine, 1977).

[b]Midpoint potentials are calculated at physiological concentrations of the specific intermediate (Data taken from White, 1991).

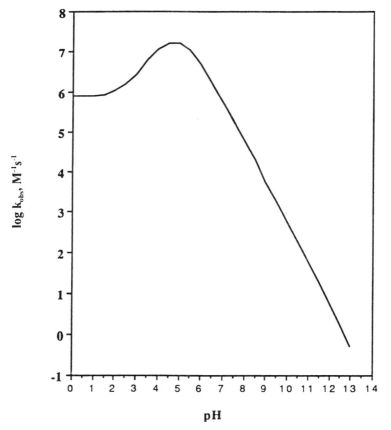

Figure 2.3 Rate constant (k_{obs} M^{-1} s^{-1}) for HO$_2$·/O$_2$·$^-$ dismutation to H$_2$O$_2$ as a function of pH. (Taken from Bielski and Cabelli, 1991.)

Table 2.2 Rate Constants for the Self-Dismutation of Superoxide[a]

Reaction	pH	Rate Constant $(M^{-1} s^{-1})$
$HO_2 \cdot + HO_2 \cdot \rightarrow H_2O_2 + O_2$	≤ 2	8.5×10^5
$HO_2 \cdot + O_2 \cdot^- + H^+ \rightarrow H_2O_2 + O_2$	4.8	2.5×10^7
$O_2 \cdot^- + O_2 \cdot^- + 2H^+ \rightarrow H_2O_2 + O_2$	7.4	3.0×10^5
$O_2 \cdot^- + O_2 \cdot^- + 2H^+ \rightarrow H_2O_2 + O_2$	12	6

[a]Data taken from Bielski and Allen, 1977; Bielski, *et al*, 1985; and Bielski and Cabelli, 1991.

forward approach would have been enzymic designs that precluded the generation of this free radical. Accomplishing this task dictated the development of several critical steps during enzymatic oxidative metabolism. First, O_2 had to be incorporated into the active site of the protein, whether as part of a coordinated metal complex or attached to a flavin. Second, once O_2 was "immobilized" within this enzymic pocket, a series of rapid electron transfers and intramolecular rearrangements would have to take place at rates faster than $O_2 \cdot^-$ could be released from the enzyme. The mitochondrial cytochrome c oxidase, which catalyzes the tetravalent reduction of O_2 to H_2O without the appearance of detectable levels of $O_2 \cdot^-$, illustrates the complexity of this problem.

Cytochrome c oxidase (Complex IV) is the terminal enzyme of the respiratory electron transport chain. This oxidase uses O_2 as an electron acceptor to transport protons into the intermembrane space of the mitochondrion (Babcock and Wikstrom, 1992; Varotsis and Babcock, 1995). The heart of the O_2 reductive process is the bimetal ion complex, composed of a heme iron from the polypeptide chain of cytochrome a_3 in close proximity to the histidine-bound Cu_B (Figure 2.4). The initial step in the formation of H_2O is the sequential one-electron reduction of the oxidized complex, giving $Fe_{a3}^{2+}-Cu_B^{1+}$. Once completed, O_2 binds to Fe_{a3}^{2+}, producing the peroxy intermediate, $Fe_{a3}^{3+} -O_2^{2-} - Cu_B^{2+}$. Addition of a third electron, which is followed by protonation, results in the formation of the hydroperoxy complex, $Fe_{a3}^{3+} -O^{1-} -OH-Cu_B^{1+}$. Subsequent transfer of the fourth electron followed by rearrangement yields two molecules of H_2O and the oxidized cytochrome c oxidase. At several steps along the way, $O_2 \cdot^-$ could arise if key intermediates fall apart. This apparently does not occur, since no measurable flux of this free radical has been observed with current highly sensitive detection systems.

Even though free radicals, including $O_2 \cdot^-$, were hypothesized to be the causitive agent responsible for injuries associated with ionizing radiation and O_2 poisoning (Gerschman, *et al.*, 1954), more than a decade would elapse before the elusive search for biologic sources of $O_2 \cdot^-$ would become intimately linked to an enzyme designed to eliminate this free radical (McCord and Fridovich, 1969). The story begins with the observation that sulfite oxidation by xanthine oxidase was dependent upon the formation of an O_2-centered free radical (Fridovich and Handler, 1958, 1961). At that time, $O_2 \cdot^-$ was thought to be tightly associated with xanthine oxidase, even though the nature of this "enzyme-bound radical" remained unclear

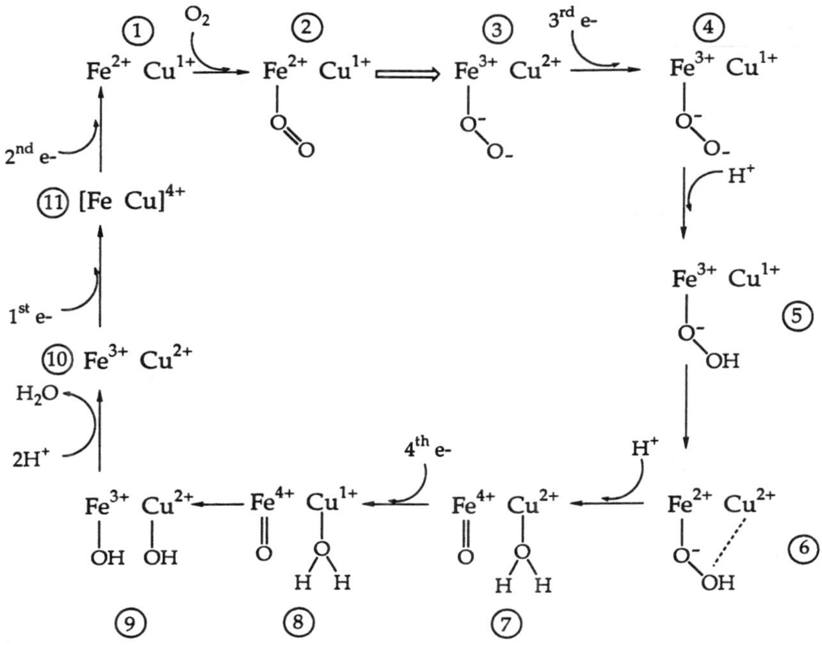

Figure 2.4 Four-electron reduction of O_2 to H_2O by cytochrome c oxidase. (Taken from Babcock and Wikstrom, 1992.)

(Fridovich and Handler, 1961). Soon thereafter, $O_2\cdot^-$ would be associated with other enzymes, such as aldehyde oxidase (Fridovich and Handler, 1961; Greenlee, et al., 1962; Rajagopalan and Handler, 1964) and dihydroorotate oxidase (Greenlee, et al., 1962; Miller and Massey, 1965)—its presence proposed to be linked to the active site of the enzyme.

As the 1960s drew to a close, two splendid discoveries[4] would forever change our understanding of aerobic metabolism, and usher in the biologic era of free radicals. The first finding pointed to the fact that $O_2\cdot^-$ was not bound to xanthine oxidase, but, rather, was released into the solution surrounding the enzyme where this free radical could initiate a variety of one-electron reductions or oxidations (McCord and Fridovich, 1968). The second observation described the catalytic action of a Cu/Zn-containing enzyme toward $O_2\cdot^-$ (McCord and Fridovich, 1969). Since then, this enzyme, which became known as superoxide dismutase (SOD), would play a pivotal role in defining the ubiquitous nature of $O_2\cdot^-$.

In the intervening years, additional enzymes have been shown to "leak" $O_2\cdot^-$ during their enzymic cycling, including aldehyde oxidase (Arneson, 1970), the cytochrome P-450 system, both the reductase and hemoprotein (Kuthan, et al., 1978; Kuthan and Ullrich, 1982), FAD-containing monooxygenase (Rauckman, et al., 1979), neutrophilic NADPH-dependent oxidase (Babior, 1978) and nitric oxide synthase (Pou, et al., 1992). Furthermore, several sites along the respiratory chain have been reported to generate $O_2\cdot^-$, such as NADPH–ubiquinone oxidoreductase, Complex I, and ubiquinol–cytochrome c oxidoreductase, Complex III (Boveris and Cadenas, 1982). In fact, it has been suggested that up to 2% of O_2 reduction by cells other than leukocytes occurs by univalent pathways (Boveris, *et al.*, 1972; Boveris and Chance, 1973; Chance, *et al.*, 1979).

Leukocyte Production of Superoxide and Other Free Radicals

In the quiescent state, polymorphonuclear neutrophils (PMN) are metabolically subdued. Yet, upon encountering a microorganism, these cells become stimulated, resulting in sequestration of the invading microbe into an enclosed vacuole, known as a phagosome. As the pathogen is being transported into this vesicle, the leukocyte increases its uptake of O_2 concomitant with the secretion of $O_2 \cdot^-$ (Babior, *et al.*, 1973; Bellavite, 1988; Segal, 1989). Even though this finding is now part of the well-accepted literature of phagocyte biology, the discovery that free radicals play a vital role in the microbicidal activity of these cells spanned a nearly 40-year period of time.

As early as 1933, it was recognized that O_2 consumption significantly increased during engulfment of bacteria (Baldridge and Gerard, 1933). At first, it was thought to be associated with increased respiration. However, it soon became apparent that this was not the case (Sbarra and Karnovsky, 1959; Cohn and Morse, 1960), but, rather, enhanced O_2 uptake was more closely linked with production of H_2O_2 (Iyer, *et al.*, 1961; Paul and Sbarra, 1968). This finding suggested the unusual ability of phagocytic cells to divalently reduce O_2 to H_2O_2 (Paul and Sbarra, 1968). However, with the discovery that in other biologic systems $O_2 \cdot^-$ can be generated during oxidative metabolism, it was soon realized that stimulated neutrophils secreted this free radical (Babior, *et al.*, 1973). The detection of neutrophil-derived H_2O_2 was, therefore, the result of the rapid disproportionation of $O_2 \cdot^-$ (Fee and Valentine, 1977; Babior, 1978) and not the direct generation of this peroxide (Table 2.2):

$$HO_2 \cdot + O_2 \cdot^- + H_2O \rightarrow H_2O_2 + O_2 + HO^-$$

In estimating the importance of the neutrophil "oxidative burst" to host immunity, one only needs to consider the clinical course of chronic granulomatous disease patients, who face repeated life-threatening infections primarily because of the inability of their phagocytic cells to generate $O_2 \cdot^-$ (Baehner and Nathan, 1967; Holmes, *et al.*, 1967; Tauber, *et al.*, 1983; Forrest, *et al.*, 1988, Erickson, *et al.*, 1992; Malech and Nauseef, 1997).

The neutrophilic enzyme responsible for $O_2 \cdot^-$ production is a unique NADPH-dependent oxidase (Bellavite, 1988; Segal, 1989; Chanock, *et al.*, 1994; Quinn, 1995; De Leo, *et al.*, 1996). In the resting cell, this oxidase is in an unassembled state, incapable of reducing O_2. Whereupon, after stimulation of specific cell surface receptors, three cytoplasmic proteins with molecular masses of 47 kDa ($p47^{phox}$), 67 kDa ($p67^{phox}$), and 22 kDa (rac-1, rac-2 or krev-1, related GTP proteins) translocate to the plasma membrane where they associate with the hemoprotein cytochrome b_{558}, consisting of a 91-kDa (gp91^{phox}) subunit and a 22-kDa (p22^{phox}) subunit (Figure 2.5) (Segal, 1989; Clark, $et\ al.$, 1990; Heyworth, $et\ al.$, 1991, 1994; Bokoch, $et\ al.$, 1994; Nanda, $et\ al.$, 1994; De Leo, $et\ al.$, 1996; Dorseuil, $et\ al.$, 1996; Freeman, $et\ al.$, 1996; Sumimoto, $et\ al.$, 1996). Recent evidence indicates that it is the gp91^{phox} subunit of this cytochrome that is the site of heme binding (Yu, $et\ al.$, 1998). Even though the exact pathway by which the various components assemble has not been fully described, the importance of phosphorylated p47^{phox} in the formation of the component oxidase has, nevertheless, been well documented (Segal, $et\ al.$, 1985; Hayakawa, $et\ al.$, 1986; Okamura, $et\ al.$, 1988; Heyworth and Badwey, 1990; Kleinberg, $et\ al.$, 1990; Steinbeck, $et\ al.$, 1991; Park, $et\ al.$, 1997; Inanami, $et\ al.$, 1998).

After cell activation, the SH3 domain of p47^{phox} binds to p22^{phox}, whereas p67^{phox} directly associates with p47^{phox} (de Mendez, $et\ al.$, 1994; Finan, $et\ al.$, 1994; Leto, $et\ al.$, 1994; Freeman, $et\ al.$, 1996; Sumimoto, $et\ al.$, 1996; de Mendez, $et\ al.$, 1997; Sathyamoorthy, $et\ al.$, 1997). What role p47^{phox} might play in the control of the NADPH-oxidase has remained unclear. However, recent studies suggest a down-regulatory function for this cytoplasmic-associated protein (Sathyamoorthy, $et\ al.$, 1997).

Independently, Rac is transported to binding sites on cytochrome b_{558} (Heyworth, $et\ al.$, 1994). A recent study suggests that even partial organization of the oxidase will, surprisingly, result in secretion of $O_2 \cdot^-$ (Koshkin, $et\ al.$, 1996). Once this complex of enzymes has been assembled, electrons flow from NADPH to cytochrome b_{558} by way of the flavin FAD, weakly bound to this terminal electron acceptor (Segal, $et\ al.$, 1983, 1985; Parkos, $et\ al.$, 1987; Segal, 1987; Nunoi, $et\ al.$, 1988; Volpp, $et\ al.$, 1988; Kleinberg, $et\ al.$, 1990; Rotrosen, $et\ al.$, 1990, 1992; Abo, $et\ al.$, 1991, 1992; Foroozan, $et\ al.$, 1992; Kreck, $et\ al.$, 1994; Uhlinger, $et\ al.$, 1994). Transfer of an electron from the low-spin state of cytochrome b_{558} (Fujii, $et\ al.$, 1995) to O_2 yields $O_2 \cdot^-$ (Isogai, $et\ al.$, 1995) (Figure 2.6). This elaborate scheme, requiring the transport of several cytoplasmic proteins to the cell membrane prior to the generation of $O_2 \cdot^-$, provides a safeguard against errant production of this free radical.

While there has been enormous interest in the assemblage of phagocytic NADPH oxidase, it has only been in recent years that evidence has emerged, suggesting that a similar mechanism of $O_2 \cdot^-$ production may be present in cells other than leukocytes. For instance, endothelial cells appear to express many of the components of the NADPH oxidase, previously thought to be solely the domain of the phagocytic cell, including gp91^{phox}, gp22^{phox}, p67^{phox} and p47^{phox} (Jones, $et\ al.$, 1996; Bayraktutan, $et\ al.$, 1998). Similar evidence has been observed in studies with vascular smooth muscle cells (Ushio-Fukai, $et\ al.$, 1996; Fukui, $et\ al.$, 1997; DeKeulenaer, $et\ al.$, 1998; Zafari, $et\ al.$, 1998) and glomerular mesangial cells (Jones, $et\ al.$, 1995). The magnitude of $O_2 \cdot^-$ secretion by these cells is far less than that found for activated phagocytes. Here, it is proposed to play a role in cell signaling events.

Since the pH of the phagosome is acidic, ranging from 5.8–6.1 for macrophages (Lukacs, $et\ al.$, 1990, 1991) to 6.5 for neutrophils (Grinstein and Furuya, 1988), secretion of $O_2 \cdot^-$ into the vacuole results in the accelerated formation of H_2O_2. This low-pH environment also enhances the reactivity of this free radical toward

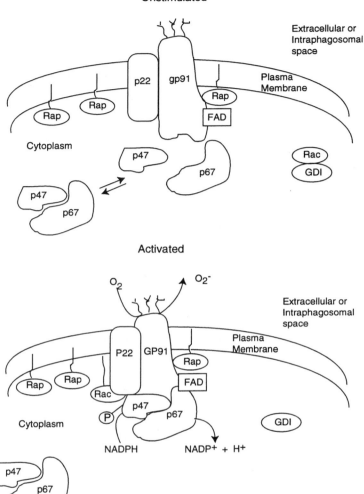

Figure 2.5 A current model for the assemblage of the neutrophil NADPH–oxidase. In the unstimulated state, the NADPH–oxidase exists as a cytosol component containing proteins: $p47^{phox}$, $p67^{phox}$ and Rac. A transmembrane hemoprotein, cytochrome b_{558}, (a dimer) consists of $gp91^{phox}$ and $p22^{phox}$. This cytochrome has binding sites for FAD and NADPH. Rap proteins control many important cellular functions, including the regulation of the NADPH–oxidase system, the modulation of the protein kinase C signal transduction pathway, and possibly the guidance of granule/vesicle trafficking. In the activated state, the cytosolic proteins translocate to the membrane and bind to cytochrome b_{558} in a complex — the nature of which is still under intense investigation. Binding of the various components renders the intact enzyme competent to transfer electrons from NADPH to O_2, resulting in the formation of $O_2 \cdot ^-$. (Taken from Quinn, 1995.)

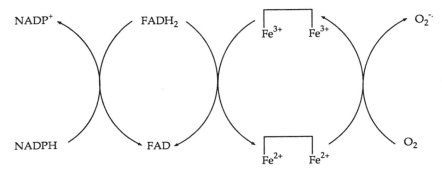

Figure 2.6 Proposed scheme for the generation $O_2 \cdot^-$ by cytochrome b_{558}. In the initial step, NADPH transfers $2e^-$, as a hydride, to FAD, forming $FADH_2$. Sequential one-electron reduction of the cytochrome b_{558}–Fe^{3+} to cytochrome b_{558}–Fe^{2+} allows the transfer of an electron to O_2, yielding $O_2 \cdot^-$.

many biologic targets. This latter point is appropriately illustrated by comparing the effect of pH on the rate of addition of $O_2 \cdot^-$ to the spin trap, 5,5-dimethyl-1-pyrroline-N-oxide (DMPO) (Figure 2.7). At the pK_a for $HO_2/O_2 \cdot^-$, the apparent rate constant, k_{app}, approaches $6.6 \times 10^3 \, M^{-1} \, s^{-1}$. As the pH increases toward the physiologic range, the rate constant drops to about $15 \, M^{-1} \, s^{-1}$ and even lower above pH 8 (Finkelstein, *et al.*, 1980a).

There is a school of thought that advocates that the primary purpose for the phagosomal secretion of $O_2 \cdot^-$ is to amplify the oxidative capacity of the vacuole by producing stronger oxidants, including hydrohalous acids (Albrich, *et al.*, 1981, 1986). To accomplish this, human neutrophils, eosinophils, and monocytes release from their cytoplasmic granules one of two distinct peroxidases, which have the ability to generate hypohalous acids (HOX, where X = Cl or Br) in the presence of H_2O_2:

$$H_2O_2 + HX \rightarrow HOX + H_2O$$

In neutrophils and monocytes, that enzyme is myeloperoxidase (MPO), which catalyzes the formation of hypochlorous acid from H_2O_2 and Cl^- (Klebanoff and

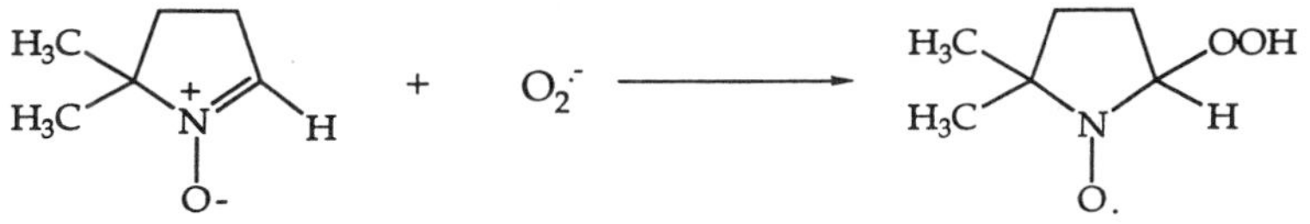

Figure 2.7 Reaction of $O_2 \cdot^-$ with the spin trap, 5,5,-dimethyl-1-pyrroline-N-oxide. The rate of this reaction is pH dependent, ranging from $6.6 \times 10^3 \, M^{-1} \, s^{-1}$ at the pK_a for $HO_2/O_2 \cdot^-$ to $10 \, M^{-1} \, s^{-1}$ at pH > 8.

Hamon, 1972; Rosen and Klebanoff, 1979a; Rosen, *et al.*, 1990). It has historically been thought that tissue macrophages, which are derived from peripheral blood monocytes, do not possess MPO and therefore do not contribute to the biology of MPO-mediated events. This hypothesis has been based mostly on evidence that when placed in culture, monocytes will differentiate into macrophages (monocyte-derived macrophages) and this process is associated with a loss of MPO in these cells (Johnston, *et al.*, 1977). However, recent work has suggested that some tissue macrophages may possess MPO (Daugherty, *et al.*, 1994).

In eosinophils, it is the antigenically distinct protein, eosinophil peroxidase, which mediates this reaction (Hurst and Barrette, 1989). Although both enzymes employ various halides, they differ in their substrate specificity. For example, myeloperoxidase preferentially uses Cl^- over Br^-, whereas the opposite is true for eosinophil peroxidase (Weiss, *et al.*, 1986; Hurst and Barrette, 1989; Mayeno, *et al.*, 1989). The ability of both enzymes to mediate bactericidal activity appears to be the result of the cationic nature of the proteins (Selvaraj, *et al.*, 1978; Ramsey, *et al.*, 1982; Miyaski, *et al.*, 1987). This allows them to stick to cell surfaces, thereby increasing the local concentration of hypohalous acid at the cell surface and promoting bactericidal activity (Selvaraj, *et al.*, 1978; Locksley, *et al.*, 1982; Ramsey, *et al.*, 1982; Miyaski, *et al.*, 1987; Britigan, *et al.*, 1996; Wright and Nelson, 1998).

Hypochlorous acid and other hypohalous acids are potent oxidants, whose microbicidal properties has long been recognized. Numerous studies over the years have confirmed that either myeloperoxidase or eosinophil peroxidase enhances $O_2\cdot^-$–mediated killing of various microorganisms (Hurst and Barrette, 1989; Weiss, 1989). The cytotoxic effects attributable to hypochlorous acid include oxidation and/or decarboxylation of membrane proteins, leading to increases in cell permeability (Albrich, *et al.*, 1986; Schraufstatter, *et al.*, 1990); oxidation of components of the bacterial respiratory chain (Albrich, *et al*, 1981; Rosen, *et al.*, 1987; Rakita, *et al.*, 1989; Schraufstatter, *et al.*, 1990; Hurst, *et al.*, 1991); inactivation of bacterial penicillin-binding proteins, leading to activation of autolysis similar to β-lactam antibiotics (Rakita and Rosen, 1991); membrane peroxidation (Winterbourn, *et al.*, 1992); formation of adenine nucleotide chloramine (Bernofsky, 1991); and generation of toxic aldehydes (Paul, *et al.*, 1970).

It appears that MPO-derived HOCl can likewise interfere directly with bacterial DNA replication by interacting with the hemimethylated DNA sequence of the bacterial chromosomal origin of replication, resulting in a loss of DNA synthesis and eventually bacterial viability (Rosen, *et al.*, 1990, 1998). Both HOCl and HOBr will react with *E. coli* phosphatidylethanolamine, the principal phospholipid in this organism's membrane, to form haloamines and halohydrins (Carr, *et al.*, 1998).

Several potential additional mechanisms for peroxidase-initiated damage unrelated to classical HOX-mediated oxidation of target molecules have come to light. MPO and other peroxidases in the presence of H_2O_2 will react with L-tyrosine to generate tyrosyl radical (TYR·), which can in turn react with free or protein-bound tyrosine molecules resulting in dityrosine cross linking (Heinecke, *et al.*, 1993a, 1993b; Winterbourn, *et al.*, 1997; McCormick, *et al.*, 1998a). This reaction is independent of the requirement for the presence of either halides or pseudohalides (Heinecke, *et al.*, 1993a). In the presence of L-tyrosine, PMA-stimulation of either neutrophils or monocyte-derived macrophages also results in the formation of dityrosine adducts (Heinecke, *et al.*, 1993a). Importantly, these adducts and cross-linked proteins have been found in association with human atherosclerotic plaques (Hazen and Heinecke, 1997; Leeuwenburgh, *et al.*, 1997):

$$H_2O_2 + \text{L-tyrosine} \xrightarrow{\text{MPO}} \text{TYR·}$$

$$\text{TYR·} + \text{TYR·} \longrightarrow \text{dityrosine}$$

$$\text{TYR·} + \text{TYR–protein} \longrightarrow \text{L-tyrosine} + \text{·TYR–protein}$$

$$\text{·TYR–protein} + \text{TYR–protein} \longrightarrow \text{protein dityrosine cross linking}$$

In the presence of $O_2\cdot^-$, TYR· will result in tyrosine peroxide (TYR-OOH) (Winterbourn, *et al.*, 1997). This reaction occurs at the expense of dityrosine formation and has the potential at physiologic concentrations of tyrosine to contribute to phagocyte-mediated cytotoxicity (Winterbourn, *et al.*, 1997):

$$\text{TYR·} + O_2\cdot^- \xrightarrow{H^+} \text{TYROOH}$$

MPO-mediated formation of HOCl can lead to chlorinated tyrosine by-products (Domigan, *et al.*, 1995; Kettle, 1996; Hazen and Heinecke, 1997). The mechanism of chlorotyrosine formation is currently unknown. However, substantial amounts of chlorotyrosine, as well as dityrosine, have been found in human atherosclerotic plaques (Hazen and Heinecke, 1997). Formation of chlorinated tyrosines appears to be intimately involved in the interaction of MPO-derived HOCl with tyrosine to form *p*-hydroxyphenylacetaldehyde (Hazen, *et al.*, 1996, 1998a). Essentially all α-amino acids can serve as substrates in this reaction (Hazen *et al.*, 1998a, 1998b). Recent data suggests that the process involves the initial formation of an unstable α-monochloramine, which in turn decomposes to the aldehyde (Hazen, *et al*, 1998a). Interestingly, HOBr, which would be the primary product of eosinophil peroxidase, leads to the generation of different products (Hazen, *et al.*, 1998a).

Formation of these aldehydes can in turn lead to covalent modification of lysine residues on proteins such as serum albumin (Hazen, *et al.*, 1997). The interaction of HOCl with lysine components of proteins directly leads to the formation of nitrogen-centered free radicals on the side chain amino groups, which can in turn react with other molecules such as ascorbate and glutathione (Hawkins and Davies, 1998).

Another recent development in the chemistry of MPO and MPO-derived oxidants has been the recognition that MPO, lactoperoxidase, horseradish peroxidase, and stimulated neutrophils will in the presence of NO_2^- and H_2O_2, result in the nitration of tyrosine, a process previously felt to be the exclusive domain of peroxynitrite (Klebanoff, 1993; Eiserich, *et al.*, 1996, 1998; van der Vliet, *et al*, 1997; Sampson, *et al.*, 1998). Recent evidence suggests that HOCl reacts with NO_2^- to form NO_2Cl (Eiserich, *et al.*, 1998). Nitrite will likewise react with compound II of MPO to form $\cdot NO_2$, resulting in regeneration of the active form of the enzyme. This free radical is then available to react with other biomolecules (Eiserich, *et al*, 1998). Taurine, another well-described target of HOCl (Weiss, *et al.*, 1983; Test, *et al.*, 1984; Thomas, *et al.*, 1985; Marquez and Dunford, 1994), will compete with NO_2^- for reaction with HOCl (Eiserich, *et al.*, 1998). Thus, detection of nitrosylated tyrosine can no longer be considered a definitive marker of peroxynitrite generation.

The search for neutrophil-derived HO· is a story filled with controversy. As recent as the late 1970s, it was well accepted that neutrophils, activated by either soluble or particulate stimuli, generated HO·, through a reaction now referred to as the metal ion-catalyzed Haber-Weiss reaction. Alternatively, it has been described as the superoxide-driven Fenton reaction[5] (Fenton, 1898; Haber and Weiss, 1934; Koppenol, 1993b; Lloyd, *et al.*, 1997):

$$O_2 \cdot^- + O_2 \cdot^- + 2H^+ \rightarrow H_2O_2 + O_2$$

$$O_2 \cdot^- + Fe^{3+} \rightarrow Fe^{2+} + O_2$$

$$Fe^{2+} + H_2O_2 \rightarrow HO \cdot + HO^- + Fe^{3+}$$

In support of phagocyte-derived HO·, the advocates of this hypothesis cited the studies of Tauber and Babior (1977) and Weiss, *et al.* (1978), in which ethylene formation from methional (or KMB) in the presence of these stimulated leukocytes appeared to confirm the formation of this free radical. However, within a few years, it became apparent that formation of ethylene in those studies was not specific for HO·, as it could arise from the action of alternative oxidants (Klebanoff and Rosen, 1978; Pryor and Tang, 1978).

As the 1980s began, other methods for identification of neutrophil-derived HO· were developed to determine whether phagocytic cells have the endogenous capacity to generate this free radical. Decarboxylation of benzoic acid (Sagone, *et al.*, 1980a, 1980b; Alexander, *et al.*, 1986), methane formation from dimethylsulfoxide (DMSO) (Repine, *et al*, 1979), and salicylate hydroxylation (Sagone and Husney, 1987), to name a few, were touted as the next generation of "specific" probes toward HO·. Unfortunately, many of these assays fell far short of expectations (Cohen, *et al.*, 1988a). In addition, the possible contribution of adventitious iron in the buffers used to characterize HO· was not appreciated and, therefore, an artifactual environment was created (Cohen, *et al.*, 1988a).

Undaunted, the search continued for the elusive neutrophil-derived HO·. This time, however, spin trapping became the method of choice, if for no other reason than the fact that the spin trap DMPO can discriminate between $O_2 \cdot^-$ and HO· (Finkelstein, *et al.*, 1980b). The first series of these studies (Green, *et al.*, 1979; Rosen and Klebanoff, 1979b) reported the spin trapping of HO·. This conclusion was based exclusively on electron paramagnetic resonance (EPR) spectral data, despite its dependence on SOD and its lack of inhibition by catalase. These findings suggested that HO· was generated by a mechanism that did not include H_2O_2 as a source for this free radical. More recent investigations (Bannister, *et al.*, 1982;

Finkelstein, *et al*, 1982; Britigan, *et al.*, 1986a; Rosen, *et al.*, 1988; Pou, *et al.*, 1989) demonstrated that Rosen and Klebanoff (1979b) actually spin trapped $O_2\cdot^-$. Rapid decomposition of the initial and unobserved spin trapped adduct, DMPO–OOH, to the recorded EPR spectrum of DMPO–OH by stimulated neutrophils led to a misinterpretation of their findings. Nevertheless, their novel approach to the detection of neutrophil-derived free radicals paved the way for later investigators to refine the technology. As the decade drew to a close, there was little evidence that neutrophils, in the absence of exogenous redox active metal ions capable of Haber-Weiss catalysis, produced $HO\cdot$.

We now know from experiments with a sensitive spin trapping system specific for $HO\cdot$ that human neutrophils and monocytes, but not macrophages, can produce this free radical (Ramos, *et al.*, 1992; Pou, *et al.*, 1994). This occurs, not by a metal ion-catalyzed Haber-Weiss reaction, but through a pathway that is dependent upon the presence of myeloperoxidase (MPO) (Bannister and Bannister, 1985; Ramos, *et al.*, 1992; Candeias, *et al.*, 1993):

$$O_2\cdot^- + O_2\cdot^- + 2H^+ \rightarrow H_2O_2 + O_2$$

$$H_2O_2 + Cl^- + MPO \rightarrow HOCl + H_2O$$

$$HOCl + O_2\cdot^- \rightarrow HO\cdot + Cl^- + O_2$$

The rate of reaction of hypochlorous acid with $O_2\cdot^-$ is pH dependent. Not surprisingly, its maximum, $k \approx 7.5 \times 10^6 \, M^{-1} \, s^{-1}$, is in the pH range of the phagosomal vacuole (Long and Bielski, 1980). Recent studies have found that eosinophil peroxidase (EPO) behaves in a fashion analogous to myeloperoxidase (McCormick, *et al.*, 1994):

$$O_2\cdot^- + O_2\cdot^- + 2H^+ \rightarrow H_2O_2 + O_2$$

$$H_2O_2 + Br^- + EPO \rightarrow HOBr + H_2O$$

$$HOBr + O_2\cdot^- \rightarrow HO\cdot + Br^- + O_2$$

Two separate and competing mechanisms by which stimulated neutrophils, monocytes, and eosinophils can generate $HO\cdot$ are the metal ion-independent (either MPO-dependent or EPO-dependent) and the metal ion-dependent (Haber-Weiss reaction) pathways. For the metal ion-dependent pathway, either the target cell or the microenvironment would provide the metal complex with the appropriate catalytic activity required to generate $HO\cdot$.

Superoxide Dismutase

Considering that $O_2\cdot^-$ is a consequence of aerobic metabolism, it is not surprising that much effort has been devoted toward understanding how cells protect themselves from the consequences of this free radical and products derived from this reactive species (Fridovich, 1997). Phylogenetic evidence (Schwartz and Dayhoff, 1978) points to the fact that SOD evolved early in Earth's microbial history, appearing in such obligate anaerobes as *Chromatium vinosum* (Kanematsu and Asada, 1978a), *Chlorobium thiosulfatophilum* (Kanematsu and Asada, 1978b) and *Desulfovibrio desulfuricans* (Hatchikian and Henry, 1977) at a time prior to the evolution of photosynthetic bacteria, including cyanobacterium. The presence of SOD must, therefore, suggest the accumulation of low levels of O_2 in the environment. Considering that these early organisms predated O_2-generating phototropes, how did these species come in contact with O_2? Even though there is considerable uncertainty, the source of O_2 must undoubtedly have resulted from abiotic organic syntheses and/or photolysis of H_2O (Asada, *et al.*, 1980).

The early presence of FeSOD and MnSOD among archae and bacteria and the independent evolution of Cu,ZnSOD and extracellular Cu,ZnSOD in eucarya attest to the importance of this enzyme family in sustaining life on a constantly changing planet (McCord, *et al.*, 1971, Steinman and Hill, 1973; Steinman, 1982a; Marklund, 1984a; Touati, 1992). Only within the past 15 years, however, with the advent of tools capable of manipulating genetic information, has this theory been adequately tested. In a series of simple, yet elegant studies, SOD-deficient mutants of *Escherichia coli* (Carlioz and Touati, 1986; Farr, *et al*, 1986; Natvig, *et al*, 1987; Touati, *et al.*, 1991), *Salmonella typhimurium* (Storz, *et al.*, 1987), *Streptococcus mutans* (Nakayama, 1992), *Porphysomonas gingivalis* (Nakayama, 1994), and *Saccharomyces cerevisiae* (Gralla and Valentine, 1991; Longo, *et al.*, 1996) have exhibited either retarded growth, decreased viability, or enhanced frequency of spontaneous mutation in O_2-containing media. The mechanism whereby the absence of SOD promotes oxidant-induced injury to *Escherichia coli* appears to be the result of an increase in $O_2 \cdot^-$-mediated reduction of iron from bacterial enzymes, such as aconitase (Keyer and Imlay, 1996). This, in turn, leads to an elevation in redox active iron within the organism and the resulting potential for enhanced formation of HO· (Keyer and Imlay, 1996; McCormick, *et al.*, 1998b). From these and other studies, there is little doubt that SOD is essential for life in an aerobic world.

Current data point to the fact that the most primitive of life forms contained only FeSOD, suggesting the appearance of MnSOD was linked to the presence of O_2-generating bacteria (Asada, *et al.*, 1980). Evidence in support of this theory comes, in part, from a series of experiments with *Escherichia coli* K12 (Hassan and Fridovich, 1977). This bacterium, when grown under strict anaerobic conditions, contains only FeSOD. Upon exposure to even the lowest of measurable O_2 levels, however, these microbes rapidly synthesize MnSOD. Based on these observations and other studies with this bacterium (Gregory and Fridovich, 1973; Fridovich, 1982; Hassan, 1988; Touati, 1988), FeSOD, encoded by the *sodB* gene (Sakamoto and Touati, 1984), is considered to be the constitutive isoform, whereas MnSOD, regulated by the *sodA* gene (Touati, 1983), is referred to as the inducible homolog of this enzyme family. In fact, distributional patterns among the archae and bacteria domains further support the separate roles for FeSOD and MnSOD. Anaerobes, anaerobically grown facultative anaerobes, and aerobic diazotrophs contain only FeSOD, whereas aerobes contain either MnSOD or FeSOD and MnSOD (Asada, *et al.*, 1980; Steinman, 1982a; Grace, 1990). Recent studies with *Escherichia coli* K12 suggest a theory as to why MnSOD is regulated by elevated concentrations of O_2 (Steinman, *et al.*, 1994). It has been found that MnSOD, but not FeSOD, is associated with DNA. Based on this observation, it has been proposed that MnSOD behaves as a "tethered antioxidant," responding to changes in $O_2 \cdot^-$ levels (Steinman, *et al.*, 1994). As such, MnSOD would offer increased protection to DNA from oxidative events and in some cases suppress carcinogenesis (Oberley and Buettner, 1979; Oberley and Oberley, 1988; Church, *et al.*, 1993).

Even though FeSOD and MnSOD dominate the archae and bacteria domains, there are examples of these isoforms appearing within the eucarya family, even in organisms where Cu,ZnSOD is present (Henry and Hall, 1977). On the other hand, Cu,ZnSOD appears to be primarily limited to eucarya (Asada, *et al.*, 1980; Joshi and Dennis, 1993), although there are a few exceptions to this general rule (Puget and Michelson, 1974; Steinman, 1982b, 1985, 1987; Beck, *et al.*, 1990; Steinman and Ely, 1990; Kroll, *et al.*, 1991; Stabel, *et al.*, 1994). In the case of *Escherichia coli*, the Cu,ZnSOD is periplasmic and plays an important role in aerobic growth of this bacterium (Benov and Fridovich, 1994, 1996; Benov, *et al.*, 1995). In eucaryotic species that contain both Cu,ZnSOD and MnSOD, the Cu,Zn isoform is localized in the cytosol, whereas the Mn homolog, which can be induced through cytokine exposure (Wong and Goeddel, 1988; Chang, *et al.*, 1992; Warner, *et al.*, 1996), is limited to the mitochondrion. This unique distribution of two evolutionary separate but functionally identical enzymes adds further support for the endosymbiotic origin of the mitochondrium (Fridovich, 1974).

The Cu,ZnSODs are a well-conserved family of enzymes (Fridovich, 1978, 1995) that are dimeric in nature with a molecular weight ranging from 31 to 33 kDa/unit, and contain two atoms of copper and two atoms of zinc/mol (Steinman, 1982a). Contrast this uniformity to the diversity of the MnSOD and FeSOD families. These enzymes contain one or two atoms of Fe or Mn, are either dimeric or tetrameric in their native state, and have been found to have molecular weights ranging from 19 to 22 kDa/unit (Steinman, 1978, 1982a).

In each of the enzymes, iron, manganese, or copper acts as a catalyst for the SOD-mediated disproportionation of $O_2 \cdot^-$ (Fridovich, 1978). For illustrative purposes, we show here the reaction of $O_2 \cdot^-$ with Cu,ZnSOD:

$$SOD\text{–}Cu^{2+} + O_2 \cdot^- \xrightarrow{k_1} SOD\text{–}Cu^{1+} + O_2$$

$$SOD\text{–}Cu^{1+} + O_2 \cdot^- + 2H^+ \xrightarrow{k_2} SOD\text{–}Cu^{2+} + H_2O_2$$

In this scheme, $O_2 \cdot^-$ reacts with the copper ion. Of importance is the finding that the rate constants, $k_1 \approx k_2 \approx 2 \times 10^9 \, M^{-1} \, s^{-1}$, are independent of the oxidation state of the copper (Klug, *et al.*, 1972; Rotilio, *et al.*, 1972; Forman and Fridovich, 1973; Klug-Roth, *et al.*, 1973; Fielden, *et al.*, 1974; Fee and Bull, 1986; O'Neill, *et al.*, 1988; Cabelli, *et al.*, 1989; Banci, *et al.*, 1990; Bolann, *et al.*, 1992; Getzoff, *et al.*, 1992) and of the pH in the range between 5.3 and 9.5 (Rotilio, *et al.*, 1972; Ellerby, *et al.*, 1996). By altering electrostatic interactions through site-specific mutations that increase the positive charge of the enzyme (Koppenol, 1981;

Getzoff, *et al.*, 1983), the rate constant for the Cu,ZnSOD-catalyzed dismutation of $O_2\cdot^-$ can be markedly enhanced, approaching nearly $7 \times 10^9\,M^{-1}\,s^{-1}$ (Getzoff, *et al.*, 1992; Banci, *et al.*, 1993). This high rate constant and its low activation energy (Takahashi and Asada, 1982) make Cu,ZnSOD one of the fastest known enzymes. The other isoforms of SOD behave similarly, where the oxidation states of the corresponding metal ions are Mn^{3+}/Mn^{2+} and Fe^{3+}/Fe^{2+}. Here, however, the second-order rate constants for FeSOD and MnSOD are slightly smaller, in the range of $0.5–2 \times 10^9\,M^{-1}\,s^{-1}$ (Lavelle, *et al.*, 1977; McAdam, *et al*, 1977a, 1977b; Bull and Fee, 1985).

Since the dismutation of $O_2\cdot^-$ can occur spontaneously—in the absence of SOD—what evolutional urgency dictated the development of this enzyme? The answer can be found by examining the disproportionation reaction more closely. In the absence of SOD, the formation of H_2O_2 is dependent upon the activational energy required for the collision of two molecules of $HO_2 \cdot /O_2\cdot^-$ As the pH increases, the repulsive effects of two negatively charged ions retard the rate of disproportionation. Even so, at physiologic pH, the rate constant is quite reasonable at $3 \times 10^5\,M^{-1}\,s^{-1}$ (Table 2.2). Since the self-dismutation reaction is second order, the rate of $O_2\cdot^-$ disappearance is

$$-d[O_2\cdot^-]/dt = k[O_2\cdot^-]^2 \qquad [2.1]$$

The half-life of this reaction is, therefore, inversely proportional to the steady-state concentration of $O_2\cdot^-$ and the rate constant,

$$t_{1/2} = 1/[O_2\cdot^-]_{ss}k \qquad [2.2]$$

At a high steady-state concentration[6] of $O_2\cdot^-$ of 3 µM, as may be obtained with PMA-stimulated neutrophils (Britigan, *et al.*, 1986a; Suzaki, *et al.*, 1994), the half-life at physiologic pH is ≈ 1 s. However, at lower levels of $O_2\cdot^-$, ≈ 1 nM, as might be achieved by other cells (Oshino, *et al.*, 1973; Imlay and Fridovich, 1992), the half-life is increased by a factor of 3000 (Table 2.3). For discussion purposes, let us choose a steady-state concentration of 0.10 nM, then, the rate of $O_2\cdot^-$ disappearance, $-d[O_2\cdot^-]/dt = k[O_2\cdot^-]^2$ becomes $3 \times 10^{-15}\,M\,s^{-1}$ when $k = 3 \times 10^5\,M^{-1}\,s^{-1}$.

In the presence of Cu,ZnSOD, the dismutation reaction can be described by the Michaelis-Menten relationship between an enzyme and a substrate:

$$E + S \rightleftarrows ES \rightarrow E + P$$

Since the cellular $[O_2\cdot^-]_{ss}$ is $<<< K_m$[7], the rate-limiting step is the binding of $O_2\cdot^-$ to the active site of the enzyme (Bull and Fee, 1985; Fee and Bull, 1986). As such, the rate equation becomes

$$-d[O_2\cdot^-]/dt = 2(1/k_1 + 1/k_2)^{-1}[SOD][O_2\cdot^-] \qquad [2.3]$$

which reflects the individual rate constants, $k_1 \approx k_2 \times 10^9\,M^{-1}\,s^{-1}$ and the dimeric nature of this enzyme, Cu_2Zn_2,SOD. If we assume a cellular concentration of SOD, $\approx 10\,\mu M$, as is contained within erythrocytes (Michelson, *et al.*, 1977), and

Table 2.3 Estimated Half-Lives of Oxygen-Centered Free Radicals

Free Radical	Half-Life (at 37°C)
HO·	$0.3–1 \times 10^9$ s
RO·	1×10^6 s
ROO·	7 s
$O_2\cdot^-$	≈ 3333 s[a]
NO·	3–45 s[b], 3 min[b], 2 hr[b]

Sources: Pryor, 1986; Saran and Bors, 1989; Sies, *et al.*, 1992.

[a]This number is based on a $[O_2\cdot^-]_{ss}$ of ≈ 1 nM (Imlay and Fridovich, 1992), and a second-order dismutation rate constant at pH 7.4, $k \approx 3.0 \times 10^5$ M^{-1} s^{-1}. This half-life is unrealistically high in cells containing SOD. In most cases, the lifetime of $O_2\cdot^-$ is, therefore, limited by diffusion to this enzyme.

[b]The calculated half-title of NO· at an initial aqueous concentration of NO· and O_2 of 100 nM and 230 µM, respectively, is 2 hr (Kharitonov, *et al.*, 1994). Shorter liftimes in tissue preparations of 3–45 s have been reported (Griffith *et al.*, 1984; Rubanyi *et al.*, 1985; Ignarro, *et al.*, 1986; Palmer, *et al.*, 1987; Kelm and Schrader, 1990; Taha, *et al.*, 1992). Using one of the highest cellular levels measured for NO· of 4 µM (Malinski, *et al.*, 1993), a 3-min half-life was calculated for the autoxidation of NO· (Kharitonov, *et al.*, 1994).

a $O_2\cdot^-$ concentration of ≈ 0.10 nM, then $-d[O_2\cdot^-]/dt = 2(1/k_1 + 1/k_2)^{-1}[\text{SOD}]$ $[O_2\cdot^-] = 2 \times 10^{-6}$ M s^{-1}. Thus, the enzyme-catalyzed dismutation of $O_2\cdot^-$ at physiologic pH is $\cong 10^9$-fold greater than that for the spontaneous disproportionation. As the concentration of $O_2\cdot^-$ is further diminished, the importance of SOD in affording protection against errant free radical-mediated cytotoxicity becomes even more critical.

The dismutation of $O_2\cdot^-$ results in the formation of H_2O_2, which has been reported to inactivate Cu,ZnSOD (Steinman, 1982a; Horton, *et al.*, 1989). Low concentrations of H_2O_2 reduce the cupric (Cu^{2+}) to cuprous (Cu^{1+}), whereas higher levels mediate irreversible inactivation of this enzyme through loss of one histidine residue per subunit (Bray, *et al.*, 1974; Hodgson and Fridovich, 1975a). Similar observations have been noted for FeSOD (Asada, *et al.*, 1975), but not for MnSOD (Steinman, 1982a).

Cu,ZnSOD, but not MnSOD, acts as a peroxidase toward a number of substrates, including H_2O_2 (Hodgson and Fridovich, 1975a, 1975b). Recent data have proposed that HO· can be produced through the action of Cu,ZnSOD on H_2O_2 and released therein into the surrounding solution (Yim, *et al.*, 1990, 1993). If this mechanism were correct, then addition of Cu,ZnSOD into a reaction mixture containing the spin trap DMPO, H_2O_2 and ethanol (CH_3CH_2OH) should have resulted in formation of DMPO–CH(CH_3)OH at the expense of DMPO–OH (Finkelstein, *et al.*, 1982a):

$$HO \cdot + CH_3CH_2OH \rightarrow CH_3 \cdot CHOH$$

$$CH_3 \cdot CHOH + DMPO \rightarrow DMPO\text{–}CH(CH_3)OH$$

$$HO \cdot + DMPO \rightarrow DMPO\text{–}OH$$

Since this competition reaction was not observed, a more likely scenario is that Cu,ZnSOD, acting as a peroxidase towards DMPO, produced DMPO–OH through direct reaction with the peroxidic form of the enzyme (Hodgson and Fridovich, 1975a; Wiedau-Pazos, *et al.*, 1996):

$$SOD-Cu^{2+} + H_2O_2 \rightarrow SOD-Cu^{1+} + O_2 \cdot^- + 2H^+$$

$$SOD-Cu^{1+} + H_2O_2 \rightarrow SOD-Cu^{2+} - HO \cdot + HO^-$$

$$SOD-Cu^{2+} - HO \cdot + DMPO \rightarrow DMPO-OH + SOD-Cu^{2+}$$

Nevertheless, peroxidase activity of Cu,ZnSOD may explain, in part, why this enzyme enhances the oxidative damage of biomacromolecules in certain disease states (Sinet, 1982; Balazs and Brooksbank, 1985; Elroy-Stein, *et al.*, 1986; Elroy-Stein and Groner, 1988; Wiedau-Pazos, *et al.*, 1996). As an alternative explanation, studies using oxygen[17]-enriched H_2O strongly suggest that most of the DMPO–OH detected in the reaction of Cu,ZnSOD with H_2O_2 resulted from the addition of an oxygen atom from H_2O and not H_2O_2, as would have occurred if HO· were formed (Singh, *et al.*, 1998).

In the case of FeSOD, H_2O_2 inactivates the enzyme primarily through the loss of tryptophan and to a lesser extent through the disappearance of iron (Beyer and Fridovich, 1987). The significance of these reactions *in vivo* must be seriously questioned, since catalase, in archae and bacteria, and catalase and glutathione peroxidase, in eucarya, catalyze the reduction of H_2O_2 to H_2O (Cohen and Hochstein, 1963; Schonbaum and Chance, 1976; Flohé, 1982). Finally, the various isoforms of SOD are differentially affected after exposure to cyanide (Borders and Fridovich, 1985). While Cu,ZnSOD is inhibited, the catalytic activity of FeSOD and MnSOD is not markedly impaired by treatment with this compound. This selective inhibitory effect of cyanide is a unique analytic tool to distinguish between Cu,ZnSOD and its familial isozymes.

In 1982, a new homolog of Cu,ZnSOD was discovered to localize within extracellular fluids, such as plasma and lymph. However, its primary site has subsequently been found to be the interstitial space of various tissues (Marklund, 1984a, 1984b). Based on its localization, the term extracellular superoxide dismutase (EC-SOD) was coined to differentiate this glycoprotein enzyme from the intracellular Cu,ZnSOD isozyme (Marklund, 1982). Most EC-SODs exist as a tetramer, formed by two disulfide-linked dimers (Carlsson, *et al.*, 1996; Oury, *et al.*, 1996). Three isoforms of EC-SOD have been isolated (Marklund, 1990). Both EC-SOD A and EC-SOD B are distributed in extracellular fluid (Karlsson and Marklund, 1987, 1988a), whereas EC-SOD C exists in an equilibrium between the fluid phase and a heparin-binding site in the glycocalyx of endothelial cells (Hjalmarsson, *et al.*, 1987; Karlsson and Marklund, 1987, 1988a, 1988b, 1989; Karlsson, *et al.*, 1988; Sandstrom, *et al.*, 1993). Upon contact with heparin, EC-SOD C is released from its binding site and enters the plasma (Sandstrom, *et al.*, 1993).

This enzyme shares many of the same properties attributed to Cu,ZnSOD, including the same ratio of Cu,Zn/subunit, high second-order rate constant, $k \approx 1 \times 10^9 \, M^{-1} \, s^{-1}$, and inactivation by H_2O_2, cyanide, azide and diethyldithiocarbamate (Marklund, 1982, 1984a, 1985; Tibell, et al., 1987). Unlike the cytosolic Cu,ZnSOD, distribution of EC-SOD is limited to a few cell types (Marklund, 1990; Sandström, et al., 1993; Oury, et al., 1994; Folz, et al., 1997; Frederiks and Bosch, 1997). Nevertheless, its importance cannot be minimized, as mice lacking this enzyme are considerably more sensitive to hyperoxia than wild-type mice (Carlsson, et al., 1995). Finally, it is worth noting that specific cytokines differentially influence the various isozymes of SOD. Whereas MnSOD is induced by TNF-α, IL-1α, and IFN-γ (Masuda, et al., 1988; Wong and Goeddel, 1988; Harris, et al., 1991), Cu,ZnSOD is unchanged and EC-SOD is variably affected (Marklund, 1992).

We would be remiss not to mention the ever-increasing number of diseases attributed to the overproduction of $O_2 \cdot^-$ (Cross, et al., 1987). A few of the well-characterized ones, such as ischemia/reperfusion injury (Granger, et al., 1981; Korthuis, et al., 1992), diabetes (Loven and Oberley, 1982; Tesfamariam, 1994), inflammation (Biemond, et al., 1988; Ward and Mulligan, 1992), aging[8] (Harman, 1956; Sohal and Weindruch, 1996; Beckman and Ames, 1998); age-related macular degeneration (Gottsch, et al., 1990; Seddon, et al., 1994), atherosclerosis (Steinberg, et al., 1989; Böger, et al., 1995), and Alzheimer's disease (Hensley, et al., 1994; Mattson, 1994; Busciglio and Yankner, 1995; Smith, et al., 1995) appear to arise when free radical production overwhelms cellular defense mechanisms. Likewise, $O_2 \cdot^-$ has been linked to amyotrophic lateral sclerosis, (Bowling, et al., 1993; Deng, et al., 1993; Rosen, et al., 1993; Gurney, et al., 1994), even though one recent study suggests the contrary (Wong, et al., 1995). Despite these infrequent events, SOD offers enormous protection against the toxic effects associated with O_2 reduction.

Catalase

Catalase is an enzyme that catalyzes the dismutation of H_2O_2 to O_2 and H_2O (Schonbaum and Chance, 1976). The importance of this enzyme can be inferred by its presence in most aerobic organisms (von Ossowski, et al., 1993) and by studies linking this enzyme to Escherichia coli viability. Mutants lacking catalase exhibited remarkable sensitivity to the toxic effects of H_2O_2 (Loewen, 1984; Rocha, et al., 1996). Similar observations have been reported in other bacteria (Bol and Yasbin, 1990), as well as eucaryotic organisms. For example, mutants of Drosophila melanogaster, which lack catalase, have shorter lifespans than control flies (Griswold, et al., 1993), while those mutants with overexpressed Cu,ZnSOD and catalase have a signficantly prolonged life expectancy (Sohal, et al., 1995). Enhancement of catalase activity has been observed after exposure of Escherichia coli to a continued source of H_2O_2 (Yoshpe-Purer, et al., 1977; Hassan and Fridovich, 1978; Loewen, 1984; Loewen, et al., 1985).

In its most common form, catalase is a heme-containing protein comprising four subunits (≈ 60-kDa/subunit) with one protoheme IX bound/subunit

(Deisseroth and Dounce, 1970; Schonbaum and Chance, 1976; Fita and Rossman, 1985). There are, of course, variations on this theme, such as a hexameric structure of HPII in *Escherichia coli* K12 and in *Bacillus subtilis* (Loewen and Switala, 1986, 1988) and manganese-containing isozymes from *Lactobacillus plantarum* and *Thermoleophilum album* (Kono and Fridovich, 1983; Allgood and Perry, 1986). Nevertheless, despite these variants, the catalases are a phylogenetically well-conserved group of proteins, thereby affirming the importance of regulating H_2O_2 in order to sustain life in an aerobic world.

The mechanism for the dismutation of H_2O_2 by catalase has been extensively explored and appears to involve similar redox transformations, independent of the source of the enzyme (Dounce, 1983) (Figure 2.8). In the resting state, the ferricatalase (catalase–Fe^{3+}) is oxidized by H_2O_2, yielding H_2O and Compound I (catalase–Fe^{5+}). Compound I is then reduced back to catalase–Fe^{3+} by H_2O_2, giving O_2:

$$H_2O_2 + E-Fe^{3+} \rightarrow 2H_2O + E-Fe^{5+}$$

$$E-Fe^{5+} + H_2O_2 \rightarrow E-Fe^{3+} + 2H^+ + O_2$$

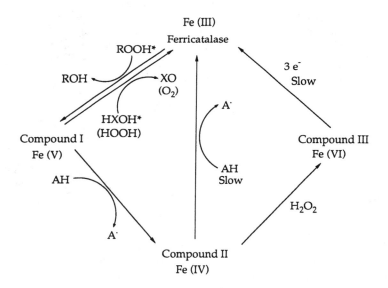

* Dominant pathways

Figure 2.8 Proposed redox cycling of catalase where Compound I, Compound II, and Compound III are enzyme–peroxide complexes. The oxidation state of the iron is +5, +4, and +6, respectively. XHOH is a two-electron donor; AH is a one-electron donor and ROOH is a hydroperoxide where R = H, alkyl, or acyl. (Taken from Schonbaum and Chance, 1976.)

Even though H_2O_2 is shown above, other hydroperoxides, ROOHs, where R = alkyl or acyl groups, are also substrates for this enzyme (Chance, *et al.*, 1952; Schonbaum and Chance, 1976). In this case, the corresponding alcohol, ROH, is generated along with H_2O and O_2. However, it should be noted that the low affinity of even the simplest of hydroperoxides makes this reaction an unlikely cytoprotective pathway. Compound I tends to show peroxidic activity. As such, this intermediate accepts electrons sequentially from a variety of one-electron donors, giving Compound II (catalase–Fe^{4+}) prior to returning to catalase–Fe^{3+}. One-electron reductants include hydrazoic, formic, and nitrous acids, small aliphatic alcohols (such as ethanol), and *N*-hydroxylamine (Schonbaum and Chance, 1976). The latter reaction can result in localized formation of nitric oxide (Craven, *et al.*, 1979). Considering that the estimated *in vivo* steady-state concentration of H_2O_2 is $\approx 1\,nM$ (Oshino, *et al.*, 1973), catalase undoubtedly displays catalatic and peroxidic activities.

In *Escherichia coli*, two unique forms of catalase, known as catalase HP-I and catalase HP-II, have been isolated and characterized (Gregory and Fridovich, 1974; Claiborne and Fridovich, 1979; Claiborne, *et al.*, 1979; Loewen and Switala, 1986). Catalase HP-I, encoded by *katG*, comprises four equal subunits, containing two molecules of protoheme IX/molecule of enzyme (Claiborne and Fridovich, 1979; Triggs-Raine, *et al.*, 1988). This unique enzyme demonstrates both catalatic and peroxidic activities (Claiborne and Fridovich, 1979). The peroxidative properties of catalase HP-I can be observed even at relatively high concentrations of H_2O_2. Despite this *in vitro* peroxidase activity, evidence seems to suggest that catalase HP-I primarily functions in *Escherichia coli* K12 as a catalase. Catalase HP-II, encoded by *katE*, exhibits only catalatic activity (Loewen and Switala, 1986; von Ossowski, *et al.*, 1991). Catalase HP-I is considered to be the constitutive isoform, present in *Escherichia coli* grown even under anaerobic conditions, whereas HP-II is the inducible isozyme (Claiborne, *et al.*, 1979). More recently, it has been found that other bacteria contain enzymes that exhibit catalatic activities similar to catalase HP-I and catalase HP-II of *Escherichia coli* (Switala, *et al.*, 1990).

The dismutation reaction for catalase HP-I, although considerably slower than either catalase HP-II at $k \approx 4 \times 10^7\,M^{-1}\,s^{-1}$ (von Ossowski, 1993) or horse erythrocyte catalase at $k \approx 1.7 \times 10^7\,M^{-1}\,s^{-1}$ (Chance, *et al.*, 1952), is still quite reasonable at $k \approx 4 \times 10^6\,M^{-1}\,s^{-1}$ (Claiborne and Fridovich, 1979) and parallels one calculated for the catalase of *Neurospora crassa*, $k = 4.6 \times 10^6\,M^{-1}\,s^{-1}$ (Jacob and Orme-Johnson, 1979).

Thermoleophilum album (Allgood and Perry, 1986), *Lactobacillus plantarum* (Kono and Fridovich, 1983; Beyer and Fridovich, 1985), and *Mycobacteria smegmatis* (Magliozzo and Marcinkeviciene, 1997) are species of bacterium, where manganese is at the catalytic site of their respective catalases. As such, these enzymes, like their catalase–Fe^{3+} counterpart, exhibit both catalatic and peroxidic activity. For these manganese-containing enzymes, the active site contains two manganese/subunit (Penner-Hahn, 1992). From low-temperature EPR spectroscopic studies (Khangulov, *et al.*, 1990), and X-ray absorption and X-ray absorption near-edge structure spectroscopies (Waldo, *et al.*, 1991; Waldo and

Penner-Hahn, 1995), it appears that the catalytic site is a binuclear manganese cluster in which one electron flows from each of the manganese ions to H_2O_2:

$$Mn^{2+}/Mn^{2+} \Leftrightarrow Mn^{3+}/Mn^{3+}$$

As in the case of the iron-based catalase from other bacteria, the rate constant for manganese catalase dismutation of H_2O_2 from *Lactobacillus plantarum* is quite acceptable: $k \approx 1.6 \times 10^6 \, M^{-1} \, s^{-1}$ (Kono and Fridovich, 1983).

In eucarya, catalase is localized within microbodies known as peroxisomes, self-replicating membrane-bound organelles that, unlike mitochondria, are without genomes. These organelles contain enzymes that use O_2 in their degradation of fatty acids and amino acids. This metabolic process frequently results in the formation of H_2O_2, which is dismutated by catalase to H_2O and O_2.

Peroxidases

Even though catalase displays catalatic and peroxidic activity, the major peroxidic enzyme in mammals and invertebrates, but not in bacteria or plants, is the selenium-containing glutathione peroxidase (Mills, 1957; Flohé, *et al.*, 1973; Rotruck, *et al.*, 1973; Flohé, 1982). This enzyme catalyzes the reduction of H_2O_2 and long-chain alkyl and branched hydroperoxides to H_2O and the corresponding alcohols (Cohen and Hochstein, 1963; Flohé, *et al*, 1973; Rotruck, *et al.*, 1973; Forstrom, *et al.*, 1979; Wendel, 1980; Flohé, 1982). For hydroperoxides derived from biomembrane lipids, early studies suggested that an initial cleavage by phospholipases was an essential step prior to reduction by this peroxidase (Grossmann and Wendel, 1983; Sevanian, *et al.*, 1983). However, this was soon found not to be the case with the discovery of a selenium-containing peroxidase with substrate specificity toward peroxidized phospholipids (Ursini, *et al.*, 1985; Thomas, *et al.*, 1990; Schuckelt, *et al.*, 1991). More recently, a third distinct selenoenzyme found in plasma has been isolated and characterized (Takahashi and Cohen, 1986; Broderick, *et al.*, 1987; Maddipati and Marnett, 1987; Takahashi, *et al.*, 1987, 1990; Avissar, *et al.*, 1989; Esworthy, *et al.*, 1993; Yamamoto and Takahashi, 1993). This unique glycosylated glutathione peroxidase, which is secreted into plasma from hepatocytes (Avissar, *et al.*, 1989, 1994), appears to use either the thioredoxin or the glutaredoxin systems as electron donors (Björnstedt, *et al.*, 1994) in place of glutathione (GSH), whose plasma levels (< 0.4 µM, Wendel and Cikryt, 1980) are well below the K_m for this enzyme (Maddipati and Marnett, 1987; Takahashi, *et al.*, 1987).

Cellular glutathione peroxidase is widely distributed in most tissues, localizing primarily in the cytosol and to a lesser extent in the mitochondria, as well as being the primary peroxidase in erythrocytes (Wahlländer, *et al.*, 1979). This enzyme is surprisingly rather uniform throughout, with a molecular weight of ≈ 84–88 kDa and comprising four identical subunits with one atom of selenium/unit (Flohé, *et al.*, 1973; Wendel, 1980; Ursini, *et al.*, 1985). Similarly, plasma glutathione peroxidase is composed of four identical subunits with one atom of selenium/unit. With glycosylation, it is not unexpected that its molecular weight would exceed cellular glutathione peroxidase, approaching 100 kDa (Broderick, *et al.*, 1987; Maddipati and Marnett, 1987; Takahashi, *et al.*, 1987; Esworthy, *et al.*, 1993; Ren, *et al.*, 1997). Phospholipid glutathione peroxidase, which is distributed in the cytosol and partly membrane-bound, is a monomer with one atom of selenium and a molecular weight of ≈ 20 kDa (Ursini, *et al.*, 1982; Thomas, *et al.*, 1990).

Although thiols are easily oxidized by hydroperoxides to the corresponding disulfides (Capozzi and Modena, 1974), the rate of this reaction is slow and dependent upon the pK_a and nature of the thiol:

$$2RSH + H_2O_2 \rightarrow RSSR + 2H_2O$$

In contrast, glutathione peroxidase-catalyzed reduction of hydroperoxides is considerably faster with rate constants approaching that of catalase toward these substrates. For example, the rate constant for glutathione peroxidase-mediated reduction of H_2O_2 is 0.6–$1.8 \times 10^8 \, M^{-1} \, s^{-1}$, depending on experimental conditions (Flohé, *et al.*, 1972), whereas the reduction of organic hydroperoxidase is somewhat slower. Even so, the rate constants are very reasonable at $k \approx 3 \times 10^7 \, M^{-1} \, s^{-1}$ for ethyl hydroperoxide and $k \approx 7.5 \times 10^6 \, M^{-1} \, s^{-1}$ for *tert*-butyl hydroperoxide (Günzler, *et al.*, 1972).

Unlike plant peroxidases, cellular glutathione peroxidase does not accept electrons from a wide variety of donors (Mills, 1957). Although most thiols will bind to the enzyme and transfer electrons to selenium, glutathione, because of its large concentration in most cells and its high specificity toward this enzyme, is the preferred physiologic substrate (Flohé, *et al.*, 1971).

From extensive studies with purified cellular glutathione peroxidase, the first step involves the reduction of the hydroperoxide by the resting enzyme, giving the corresponding alcohol (Flohé and Günzler, 1974; Flohé, 1982) (Figure 2.9). This reaction occurs at an unusual selenocysteine amino acid residue, which appears to be intimately linked to the catalytic site of the enzyme (Forstrom, *et al.*, 1978; Ren, *et al.*, 1997):

$$E\text{–}CySe^- + H^+ + ROOH \rightarrow E\text{–}CysSeOH + ROH$$

Based on kinetic studies with a wide variety of hydroperoxides, this reduction involves a simple transition-state reaction, rather than a specific substrate–enzyme complex of the Michaelis-Menten type (Flohé, *et al.*, 1972). Regeneration of the enzyme involves two sequential steps. Initially, the selenenic acid intermediate reacts with GSH, giving a selenosulfide:

$$E\text{–}CysSeOH + GSH \rightarrow E\text{–}CysSe\text{–}SG + H_2O$$

Once this complex is formed, addition of another molecule of GSH leads to the reduced enzyme and oxidized glutathione (GSSG):

$$E\text{–}CysSe\text{–}SG + GSH \rightarrow E\text{–}CySe^- + GSSG + H^+$$

Although cellular and phospholipid glutathione peroxidases play a critical role in protecting cells from peroxidative damage (Cohen and Hochstein, 1963; Sies

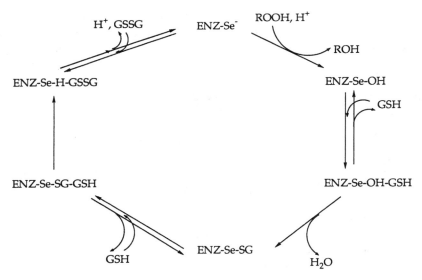

Figure 2.9 Proposed mechanism for the reduction of hydroperoxides by glutathione per-oxidase, where R = alkyl or H. (Taken from Flohé, 1982.)

and Summer, 1975; Burk, *et al.*, 1978), their ability to regulate cellular levels of a diversity of hydroperoxides and minimize oxidative damage is dependent upon the availability of reduced glutathione. To maintain cellular levels of this important thiol, the flavoprotein, NADPH-glutathione reductase catalyzes the reduction of GSSG to GSH (Rall and Lehninger, 1952; Black, 1963; Staal, *et al.*, 1969). Thus, removal of hydroperoxides is intimately linked to glutathione peroxidase and glutathione reductase and the integrated flow of electrons between these two enzymes (Pinto and Bartley, 1969) (Figure 2.10):

$$GSSG + NADPH \xrightarrow{\text{glutathione reductase}} 2\ GSH + NADP^+$$

As recent as the mid-1970s, a number of literature reports suggested that an enzyme other than the selenium-containing glutathione peroxidases could exhibit peroxidic activity (Lawrence and Burk, 1976, 1978; Prohaska and Ganther, 1977; Prohaska, 1980; Tappel, *et al.*, 1982; Carmagnol, *et al.*, 1983). This new peroxidic enzyme was considerably smaller than the well-described selenium-containing per-oxidases, having a molecular weight of ≈ 39 kDa (Nakamura, *et al.*, 1974). More striking, however, was the finding that this enzyme lacked the selenocysteine amino acid residue at its active site, thought to be so essential for peroxidic activity (Lawrence and Burk, 1976). Furthermore, this enzyme was found to exhibit remarkable selectivity toward only organic hydroperoxides—H_2O_2 was not a sub-strate for this enzyme (Lawrence and Burk, 1978). This nonselenium peroxidase was soon discovered to be a member of the glutathione-*S*-transferase family of enzymes, classically associated with conjugation reactions (Prohaska and Ganther,

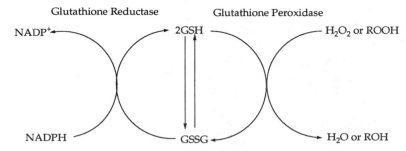

Figure 2.10 Interrelationship between glutathione peroxidase and glutathione reductase. In this scheme, glutathione peroxidase reduces hydroperoxides to alcohols and oxidized glutathione, GSSG. Glutathione reductase reduces GSSG to reduced glutathione, GSH, using NADPH as a source of electrons. (Taken from Pinto and Bartley, 1969.)

1977; Prohaska, 1980; Tappel, *et al.*, 1982). Like its selenium counterpart, this glutathione-*S*-transferase is predominately located in the cytosol of the cell (Sies and Moss, 1978; Wahlländer, *et al.*, 1979). The catalytic activity of this unique isozyme promotes the nucleophilic attack of GSH toward an organic hydroperoxide, leading to the sulfenic acid of glutathione. Following an S_N2 addition by another molecule of GSH, the resulting GSSG is then recycled to GSH by NADPH–glutathione reductase (Prohaska, 1980).

$$ROOH + GSH \xrightarrow{\text{glutathione-}S\text{-transferase}} GSOH + ROH$$

$$GSH + GSOH \rightarrow GSSG + H_2O$$

$$GSSG + NADPH \xrightarrow{\text{glutathione reductase}} 2GSH + NADP^+$$

What separate and overlapping functions do glutathione peroxidase and glutathione-*S*-transferase play in regulating peroxidative damage? Although there are no clear studies that will allow us to assign a specific degree of importance to each enzyme, the finding that glutathione-*S*-transferase activity is markedly enhanced in selenium-deficient rats suggests an adaptive role for this enzyme (Burk and Lane, 1979; Flohé, 1982). Finally, it is important to note that the location of cellular glutathione peroxidase and SOD—in the cytosol of eucaryotic cells—suggests an intimate association between these enzymes, as they regulate cellular levels of $O_2 \cdot^-$ and H_2O_2, thereby preventing or, in some cases, limiting cellular formation of HO· (Ceballos, *et al.*, 1988).

Lipid Peroxidation

Initiating events and common pathways for cellular injury have long been a puzzle to scientists. The situation is exceedingly complex, because of the diversity of subcellular reactions, leading to irreversible changes in the morphology of the cell. This process culminates in cell death, whether through apoptosis or cell necrosis. The role of O_2 reduction products in these events has been inferred based on the protective role that SOD (Imlay and Fridovich, 1992; Nakayama, 1994), catalase (Loewen, 1984), and glutathione peroxidase (Flohé, 1982) play in maintaining cell viability.

One of the accepted hypotheses to explain the molecular mechanism of cell death is lipid peroxidation. This theory postulates that peroxidative decomposition of structural lipids in membranes is the link between localized events involved in cell metabolism and the subsequent emergence of impaired cellular function (Mead, 1976; Girotti, 1985; Emerit and Chaudiere, 1989; Janero, 1990). Support for this thesis comes, in part, from the finding that the normal structure and function of lipoprotein membranes depends on the entirety of structural lipids. Alterations in phospholipid integrity as the result of a free radical cascade have been shown to impede the normal functioning of many diverse membrane-associated enzyme systems, such as glucose-6-phosphatase (Benedetti, et al., 1977) and cytochrome P-450 (Strobel, et al., 1970; Wills, 1971). In addition to oxidative alteration of the structure of complex lipids, free radical-mediated events may also affect other cellular components, such as proteins, carbohydrate-containing compounds, and nucleic acids (Emerit and Chaudiere, 1989).

The injury produced by $O_2 \cdot^-$ and/or $HO\cdot$ results not only from these primary reactive species, but also from secondary free radicals that are produced as a consequence of the initiating event with other cellular targets. These reactions may continue as cyclic chain processes further amplify the initial insult. This mechanism of cell death has been given the elusive and encompassing term "lipid peroxidation."

Lipid peroxidation is a free radical chain reaction that has separate initiation, propagation, and termination steps (Figure 2.11) (Mead, 1976; Halliwell and Gutteridge, 1984; Porter, 1984; Girotti, 1985; Buettner, 1993). Initiation occurs when a free radical abstracts a hydrogen atom from a polyunsaturated fatty acid (LH), producing a free radical of the original fatty acid (L·):

$$LH + HO\cdot \rightarrow L\cdot + H_2O$$

The preferential attack by the initiating free radical (e.g., $HO\cdot$) on allylic hydrogens is due to their lower bond dissociation energy (Figure 2.11). Resonance stabilization of L· guarantees a family of other secondary free radicals that can react with O_2 at near-diffusion-controlled rates.

The nature of this initiating free radical remains somewhat in doubt and is dependent upon experimental conditions. There is, nevertheless, serious debate as to whether $O_2 \cdot^-$, per se, participates in lipid peroxidation. Some argue, for instance, that since this free radical is too weak an oxidant, it is not possible to initiate this oxidation (Halliwell and Gutteridge, 1985; Afanas'ev, 1989). In contrast, at lower pH—as might be found in some organelles, such as lysosomes—the rate constant for the reaction of $HO_2\cdot$ with polyunsaturated fatty acids, such as linoleic acid, is $\leq 300\,M^{-1}\,s^{-1}$, suggesting the feasiblity of this initiation reaction

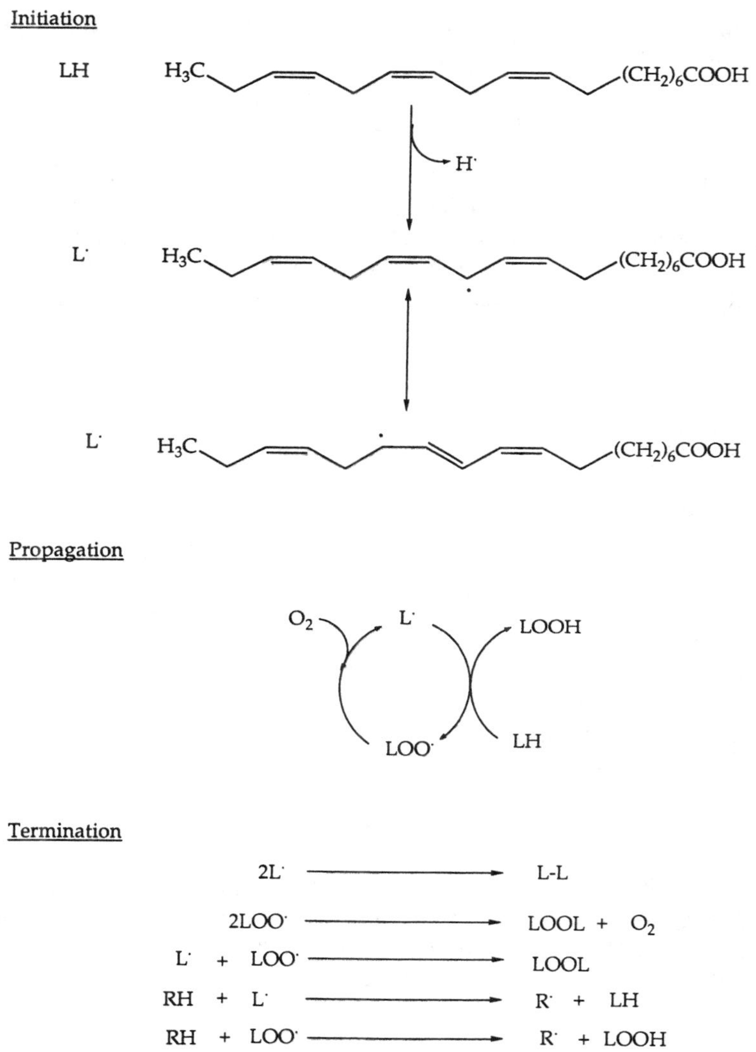

Initiation

LH

H$_3$C—≡—≡—≡—(CH$_2$)$_6$COOH

↓ → H·

L·

H$_3$C—≡—≡—≡·—(CH$_2$)$_6$COOH

↓

L·

H$_3$C—≡—·—≡—(CH$_2$)$_6$COOH

Propagation

O$_2$ ⟶ L· ⟶ LOOH

LOO· ⟶ LH

Termination

$$2L· \longrightarrow L\text{-}L$$

$$2LOO· \longrightarrow LOOL + O_2$$

$$L· + LOO· \longrightarrow LOOL$$

$$RH + L· \longrightarrow R· + LH$$

$$RH + LOO· \longrightarrow R· + LOOH$$

Figure 2.11 Proposed mechanism to account for the free radical-mediated peroxidation of linoleic acid. In the initiation step, a free radical abstracts a hydrogen atom from a poly-unsaturated fatty acid (LH), producing a free radical of the original fatty acid (L·). Delocalization can result in a number of lipid free radicals. Propagation of the free radical event is accompanied by the addition of O$_2$ to the lipid free radical. Termination of the chain reaction occurs with quenching of the propagating free radical through a number of different pathways. (Taken from Girotti, 1985).

(Gebicki and Bielski, 1981; Thomas, *et al.*, 1982, 1984; Sutherland and Gebicki, 1984; Aikens and Dix, 1991):

$$LH + HO_2 \cdot \rightarrow L \cdot + H_2O_2$$

Despite the rather academic question as to the role of $O_2 \cdot^-$ in mediating lipid peroxidation[9], there is little disagreement as to the ability of HO· to initiate this reaction, independent of the mechanism for its formation, even though most of the studies propose a metal ion-catalyzed Haber-Weiss reaction as its source (Fong, *et al.*, 1976; Girotti, 1985; Goldstein, *et al.*, 1993; Koppenol, 1993a; Rosen, *et al.*, 1995):

$$O_2 \cdot^- + O_2 \cdot^- + 2H^+ \rightarrow H_2O_2 + O_2$$

$$O_2 \cdot^- + Fe^{3+} \rightarrow Fe^{2+} + O_2$$

$$Fe^{2+} + H_2O_2 \rightarrow HO \cdot + HO^- + Fe^{3+}$$

In fact, the reaction of HO· with polyunsaturated fatty acids is considerably slower than other reactions with this free radical. For example, Barber and Thomas (1978) have found the rate constant for this reaction with artificial lecithin bilayers to be $5 \times 10^8 \, M^{-1} \, s^{-1}$. Other models may result in considerably higher rate constants.

Once a lipid dienyl free radical, L·, is formed, resonance stabilization forms conjugated dienes and trienes that react with O_2 at near-diffusion-controlled rates to generate a conjugated peroxyl radical (LOO·) (Aust and Svingen, 1982):

$$L \cdot + O_2 \rightarrow LOO \cdot$$

Formation of LOO· is subsequently followed by various chain-propagating reactions, including abstraction of a hydrogen atom from a neighboring polyunsaturated fatty acid, L'H. This results in the generation of the lipid hydroperoxide, LOOH, and a new lipid dienyl free radical, L'·:

$$LOO \cdot + L'H \rightarrow LOOH + L' \cdot$$

The hydroperoxide will not directly reinitiate chain reactions, although in the presence of redox active metal ions, peroxyl and alkoyl radicals can continue this free radical cascade:

$$Fe^{3+} + LOOH \rightarrow Fe^{2+} + LOO \cdot$$

$$Fe^{2+} + LOOH \rightarrow LO \cdot + Fe^{3+}$$

In some situations, the peroxidic action of glutathione peroxidase or glutathione-*S*-transferase can terminate the reaction by eliminating the hydroperoxide, thereby retarding further peroxidation of membranes. In other cases, however, autocatalytic destruction of the lipid structure proceeds through cross-linking polymerization and breakage of fatty acid chains, resulting in a loss of membrane integrity (Roubal and Tappel, 1966; Roubal, 1971; Witting, 1980).

Termination of the chain reaction occurs with quenching of the propagating free radical through pathways that include, for instance, the self-reaction of two lipid dienyl free radicals or two peroxyl radicals (Figure 2.11):

$$L \cdot + L \cdot \rightarrow L–L$$

$$L \cdot + LOO \cdot \rightarrow LOOL$$

$$L \cdot + LO \cdot \rightarrow LOL$$

$$LO \cdot + LO \cdot \rightarrow LOOL$$

$$LOO \cdot + LOO \cdot \rightarrow LOOL + O_2$$

Alternatively, scavengers can, by interrupting the free radical cascade, end the peroxidative process (Liebler, 1993). Although there are numerous examples of free radical scavengers, including spin traps and spin labels (Bolli, et al., 1989; DeGraff, et al., 1992), one of the most important families of lipid-soluble antioxidants is the tocopherols (α-, β-, γ- and δ-tocopherols), which collectively exhibit vitamin E activity (Witting, 1980; Cadenas, et al., 1995; Wagner, et al., 1996). The most active of these is α-tocopherol (α-TOH) (Diplock, 1985). Its reaction with peroxyl radicals is both thermodynamically and kinetically favored, giving α-TO$\cdot$ (Burton and Ingold, 1981; Burton, et al., 1983; Bowry and Ingold, 1995). This secondary free radical is so stable that α-TO$\cdot$ will not promote the propagation of other free radical events (Doba, et al., 1984; Burton and Ingold, 1989; Buettner, 1993). Recently, water-soluble quaternary ammonium analogs of α-tocopherol have been developed that extend the versatility of this important family of antioxidants to aqueous environments (Bolkenius, et al., 1991).

$$\alpha–TOH + ROO \cdot \rightarrow \alpha–TO \cdot + ROOH$$

Because vitamin E is in such low concentrations in biomembranes, it is difficult to make definitive conclusions as to its cellular location (Perly, et al., 1985; Burton and Ingold, 1989). Nevertheless, from studies by Csallany and Draper (1960), it appears that the largest concentration of this antioxidant is in the membranes of the endoplasmic reticulum and mitochondria and to a lesser extent in those of other organelles.

There are several unique features of α-tocopherol that enhance its physiologic function (Figure 2.12). The phytyl tail, with its high lipophilicity, is imbedded in the interior of the membrane, whereas the chromanol head is on the bilayer surface. As such, vitamin E bobs up and down in the membrane similar to a cork in water. This suggests a dual purpose: protecting membranes from internal propagated free radicals and complimenting hydrophilic cytoplasmic scavengers. Illustrating the role of α-tocopherol in protecting plasma membranes from lipid peroxidation is the demonstration that cells grown in the presence of supplemental

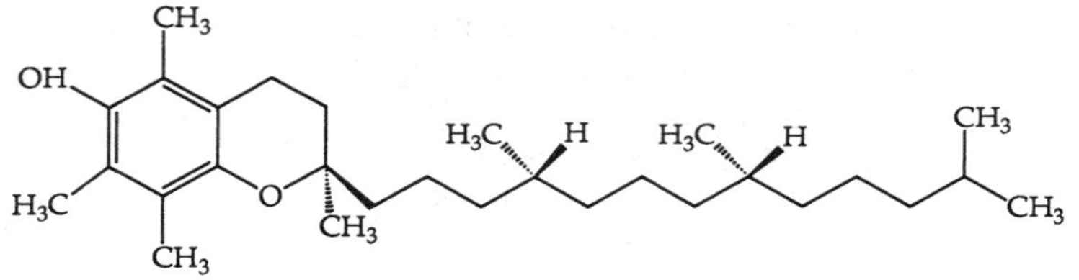

2R,4'R,8'R-α-Tocopherol (RRR-α-TOH)

Figure 2.12 The most active member of one of the most important families of lipid-soluble antioxidants is α-tocopherol (α-TOH). Its reaction with peroxyl radical gives the stable free radical α-TO·, which will not promote the propagation of other free radical events. The phytyl tail, with its high lipophilicity, is imbedded in the interior of the membrane, whereas the chromanol head is on the bilayer surface.

α-tocopherol for 24 hr exhibited less formation of lipid radicals, as detected by spin trapping, relative to control cells when both cells were exposed to oxidative stress (Wagner, *et al.*, 1996).

Like vitamin E, vitamin C plays an important role in cellular redox reactions (Swartz and Dodd, 1981; Machlin and Bendich, 1987; Niki, 1987). Because ascorbic acid is located in the cytoplasm of the cell, in which it exists primarily as the negatively charged lactone, with a pK_a of 4.17 (Rose and Bode, 1993) (Figure 2.13), it complements the reductive ability of glutathione toward free radicals, including $O_2\cdot^-$ (Nishikimi, 1975; Machlin and Bendich, 1987; Munday and Winterbourn, 1989). In this reaction, ascorbic acid (AH_2) is oxidized by $O_2\cdot^-$ in two sequential steps to dehydroascorbic acid (A) (Figure 2.13). The intermediate in the redox cycle is the ascorbyl free radical (AH·). The exceptional stability of AH· is due to resonance structures that contribute to its low reactivity.

Ascorbic Acid   Ascorbyl Radical   Dehydroascorbic Acid

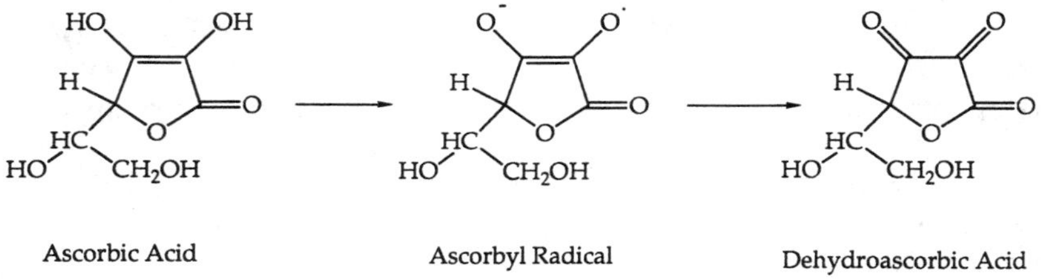

Figure 2.13 The most important hydrophilic antioxidant is ascorbic acid. In this scheme, ascorbic acid is oxidized in two sequential steps to dehydroascorbic acid. The intermediate in the redox cycle is the ascorbyl free radical, which is exceptionally stable, due, in part, to resonance structures that contribute to its low reactivity. One of the functions of ascorbic acid is to recycle vitamin E by reducing α-TO·.

$$O_2 \cdot^- + AH_2 + H^+ \rightarrow H_2O_2 + AH\cdot$$

$$O_2 \cdot^- + AH\cdot + H^+ \rightarrow H_2O_2 + A$$

In addition, ascorbate can enhance the antioxidant properties of α-tocopherol by recycling this vitamin (Tappel, et al., 1961; Liebler, et al., 1989; Van Den Berg, et al., 1989; Bisby and Parker, 1995; May, et al., 1996):

$$\alpha-TO\cdot + AH_2 \rightarrow \alpha-TOH + AH\cdot$$

Given its stability and the presence of ascorbate as a natural antioxidant, EPR-based measurements have become increasingly popular as a means to quantitate oxidative stress occurring *in vivo* in biological systems, through detection of ascorbyl radical in either an extracellular milieu, such as serum or in an isolated tissue preparation (Pietri, et al.,1990; Buettner and Jurkiewicz, 1994; Sharma, et al., 1994; Kihara, et al., 1995; Galley, et al, 1996; Nakagawa, et al., 1997). However, as carefully pointed out by Roginsky and Stegmann, factors other than oxidative stress can dramatically alter steady-state levels of ascorbyl radical and must be taken into account during the interpretation of data derived using this approach (Roginsky and Stegmann, 1994).

The specific compartmentalization of vitamins C and E has led some to suggest a synergy between α-tocopherol and ascorbic acid (Packer, et al., 1979; Doba, et al., 1985; Burton and Ingold, 1989; Chan, 1993). More recent studies, however, seem to support an additive, rather than a synergistic, effect between these anti-oxidants, pointing to a more independent role for each in the regulation of free radicals than previously theorized (Buettner, 1993; Pryor, et al., 1993; Glascott, et al., 1996).

Redox Properties of Iron—A Source of Free Radicals in Biological Systems

A number of free radical reactions of biologic relevance are dependent on the presence of iron in a redox active state. Of particular importance is the role of this metal ion in the formation of HO· via the Fenton or metal ion-catalyzed Haber-Weiss reaction (Haber and Weiss, 1934) and the resulting biologic consequences. Key to the participation of iron is its ability to redox cycle, where it serves as a means of reducing O_2 and its reduction products to more reactive species:

$$Fe^{2+} + O_2 \rightarrow O_2\cdot^- + Fe^{3+}$$

$$O_2\cdot^- + Fe^{3+} \rightarrow Fe^{2+} + O_2$$

$$Fe^{2+} + H_2O_2 \rightarrow HO\cdot + HO^- + Fe^{3+}$$

Under aerobic conditions, iron exists predominantly in the oxidized ferric (Fe^{3+}) state. Ferric ion is relatively insoluble in neutral solutions, as it rapidly precipitates out as ferric hydroxide. Ferrous ion (Fe^{2+}) is more soluble than Fe^{3+}. Under aerobic conditions in neutral solutions, however, Fe^{2+} undergoes rapid oxidation to Fe^{3+} unless a reducing agent, such as ascorbate, is available. Given the above, most experiments related to the role of iron in free radical chemistry are performed using iron bound to various chelating agents in order to maintain the metal in a soluble state. Not suprisingly, the ability of the iron to participate in free radical reactions varies considerably, depending on the nature of the chelating agent to which it is complexed. For example, Fe^{3+} bound to either ethylenediamine-tetracetate (EDTA) or nitrilotriacetate (NTA) is highly redox active, whereas iron bound to other chelating agents, such as deferoxamine (desferral) or diethylene-triaminepentaacetate (DTPA) has very low or no reactivity (Halliwell and Gutteridge, 1981; Sutton, 1985; Singh and Hider, 1988; Hamazaki, et al., 1989; Gutteridge, 1990; Smith, et al., 1990; Yamazaki, et al., 1990). This low redox activity with DTPA and deferoxamine appears to be due, to a great extent, to the slow rate at which Fe^{3+} can be reduced to Fe^{2+} when complexed to these compounds (Butler and Halliwell, 1982; Buettner, et al., 1983). Among the features of an iron chelate that have been suggested to contribute to its ability to participate in free radical reactions is whether the complex possesses an appropriate water coordination site on the Fe^{3+} complex (Graf, et al., 1984).

In addition to its role as a catalyst for the formation of HO· via Fenton reaction (Koppenol, 1993b), it has long been proposed that iron-dependent oxidative events may occur through the formation of a species in which iron exists as $[Fe^{5+} = O]^{3+}$ or Fe^{+4}-OH (Imlay, et al., 1988; Sutton and Winterbourn, 1989; Rush, et al., 1990). Like HO·, such species, would be expected to be highly oxidizing. Much of the evidence for the formation of $[Fe^{5+} = O]^{3+}$ or Fe^{+4}–OH in biologic systems comes from mechanistic studies with cytochrome P-450 in which such reactive intermediates, at the active site of this enzyme, appear to be the proximal species responsible for the oxidation of substrates (Augusto, et al., 1982; Guengerich, and Macdonald, 1984, White, 1991, 1994; Guengerich, et al., 1997) and from the specificity of oxidant scavengers (Sutton and Winterbourn, 1989). More recently, stoichiometric studies of spin adduct formation from the reaction of H_2O_2 and Fe^{2+} have provided additional experimental evidence for the potential formation of such species (Yamazaki and Piette, 1990; Rienecke, et al., 1994). However, there remains little definitive evidence for the formation and/or participation of such species in oxidant-mediated cell injury. The absence of such data may reflect more the lack of a definitive means of proving their existence rather than their lack of importance in biology.

The extracellular iron contained within multi-cellular organisms is chelated to iron-binding proteins (Bullen, et al., 1978; Halliwell and Gutteridge, 1986). In humans, the glycoprotein transferrin is the major chelator of iron in serum (Halliwell and Gutteridge, 1986). The related glycoprotein lactoferrin, and to a lesser extent transferrin, serve this function at mucosal surfaces and in milk (Halliwell and Gutteridge, 1986). Alternative iron-binding proteins with structural similarities to transferrin and lactoferrin are present in other animals, including ovotransferrin in egg white (De Jong, et al., 1990). Each protein can bind two molecules of Fe^{3+} by means of a separate N- and C-terminal high-affinity iron-binding site (Bates, et al., 1967; Aisen and Leibman, 1972; Aisen, et al., 1978;

Bluard-Deconinck, *et al.*, 1978; Evans and Williams, 1978; MacGillivray, *et al.*, 1983; Anderson, *et al.*, 1987). In contrast, these proteins have low affinity for Fe^{2+} (Aisen, *et al.*, 1978; Kojima and Bates, 1981). Optimal iron chelation by these proteins requires synergistic adsorption at their iron-binding sites of an associated anion, which is usually felt to be bicarbonate *in vivo* (Aisen, *et al.*, 1967; Schlabach and Bates, 1975; Harris, 1985; Foley and Bates, 1988; Eaton, *et al.*, 1990; Shongwe, *et al.*, 1992). The attachment of iron results in a conformational change in the protein structure in which the iron-binding sites become buried within the matrix (Aisen and Leibman, 1972; Anderson, *et al.*, 1990; Grossmann, *et al.*, 1992, 1993; Chung and Raymond, 1993; Gerstein, *et al.*, 1993; Nguyen, *et al.*, 1993; Ying, *et al.*, 1994). A trait common to these iron-binding proteins is their avidity toward Fe^{3+}, although, in some cases, this property may vary, allowing the distinctive nature of these proteins to be expressed (Aisen and Leibman, 1972; Halliwell, *et al.*, 1988b; Winterbourn and Malloy, 1988; Britigan, *et al.*, 1993). The effect of pH on the binding capacity of Fe^{3+} by transferrin and lactoferrin is illustrative of this point. Even though both proteins have a decreased capacity for Fe^{3+} as the pH of the environment decreases, transferrin releases Fe^{3+} at a higher pH than lactoferrin (Iacopetta and Morgan, 1983; Baldwin, 1980). Also, lactoferrin is more resistant than transferrin to digestion by some proteases (Brines and Brock, 1983). In addition to these differences between transferrin and lactoferrin, there also appears to be species-related variability between proteins of the same class (Spik, *et al.*, 1985, 1988, 1993; Coddeville, *et al.*, 1989, 1992; De Jong, *et al.*, 1990; Shimazaki, *et al.*, 1993).

From the standpoint of the free radical chemistry of iron bound to lactoferrin and transferrin, there has been a gradual evolution of thought on this subject. Initial reports in the early 1980s cited experimental data, suggesting that diferric lactoferrin and diferric transferrin were efficient catalysts of the metal ion-catalyzed Haber-Weiss reaction (Ambruso and Johnston 1981; Motohasi and Mori, 1983; Nakamura, 1990). It was thought that these studies provided a mechanism for explaining neutrophil-derived HO·, the existence of which was first proposed several years before (Tauber and Babior, 1977; Weiss, *et al.*, 1978; Repine, *et al.*, 1979; Rosen and Klebanoff, 1979a; Sagone, *et al.*, 1980a, 1980b; Sagone and Husney, 1987; Karnovsky and Badwey, 1983; Ambruso, *et al.*, 1984; Miyachi, *et al.*, 1987). Subsequent experiments by other laboratories, however, failed to confirm these initial observations (Gutteridge, *et al.*, 1981; Winterbourn, 1983; Baldwin, *et al.*, 1984; Britigan, *et al.*, 1986a, 1986b, 1987, 1989, 1990; Aruoma and Halliwell, 1987; Cohen, *et al.*, 1988a, 1988b, 1992; Pou, *et al.*, 1989; Shinmoto, *et al.*, 1992). These latter investigators offered a variety of explanations for the earlier findings, including the presence of iron bound to locations other than the "true" iron-binding sites on the protein. Thus, the initial observations of HO· production by diferric lactoferrin and diferric transferrin could be explained by overzealous iron-loading procedures, the failure to control for adventitious iron present in buffers, and the continued presence of other chelating agents, such as NTA, which were initially added as part of the iron-loading procedure. There now appears to be sufficient data to conclude that iron adherent to the iron-binding sites of either lactoferrin or transferrin lacks the capacity to generate HO· through the metal ion-catalyzed Haber-Weiss reaction. Despite this, proteolytic cleavage of transferrin, which can occur *in vivo* (Britigan, *et al.*, 1993), results in the formation of new and smaller iron complexes that can act as HO· catalysts (Britigan and Edeker, 1991; Miller and Britigan, 1995).

In addition to protein-bound iron, the extracellular iron pool contains a variety of low-molecular-weight nonprotein chelates of iron that have been detected in the serum of patients with hereditary hemochromatosis and leukemia, the synovial fluid of patients with arthritis, and the cerebrospinal fluid under both normal and pathologic conditions (Halliwell, *et al.*, 1988a; Buettner and Chamulitrat, 1990; Halliwell and Gutteridge, 1990). This low-molecular-weight iron pool is composed, at least in part, of iron complexes that, incidentally, are redox active. In fact, detection requires their reaction with bleomycin to oxidize thiobarbituric acid or for them to be acted upon by ascorbate to generate the ascorbyl radical (Gutteridge, *et al.*, 1985; Aruoma, *et al.*, 1988; Halliwell, *et al.*, 1988a; Buettner and Chamulitrat, 1990; Halliwell and Gutteridge, 1990).

The airways of smokers contain increased levels of iron (Cosgrove, *et al.*, 1985; Thompson, *et al.*, 1991) that is capable of acting as a HO· catalyst (Olakanmi, *et al.*, 1993). In hemochromatosis, low-molecular-weight iron appears to be in the form of iron citrate or potentially iron citrate acetate (Halliwell and Gutteridge, 1990), whereas its form in other conditions is not known. Evidence has also been obtained that iron bound to one of two classes of siderophores secreted *in vivo* by the pathogenic microorganism *Pseudomonas aeruginosa* is an efficient catalyst of the metal ion-catalyzed Haber-Weiss reaction (Coffman, *et al.*, 1990).

Despite the few examples cited above, there remains a high degree of uncertainty as to the exact compound to which iron is chelated under physiologic and pathologic conditions. From studies with a number of cell models, a picture has emerged in which low-molecular-weight iron complexes appear to enhance cellular susceptibility to damage initiated by H_2O_2 and other oxidants (Starke and Farber, 1985; Balla, *et al.*, 1992; Varani, *et al.*, 1992; Zager and Foerder, 1992; Miller and Britigan, 1995). Of particular interest is the correlation linking substantial cytotoxicity with increased hydrophobicity of the iron chelates (Balla, *et al.*, 1990; Britigan, *et al.*, 1992a, 1994). This presumably takes place through the greater capacity of these agents to place iron intracellularly or near the cell surface where site-specific production of HO· can mediate localized toxicity (Chevion, 1988).

In addition to extracellular iron chelates, in organisms that populate the archae, bacteria and eucarya domains, there exists a large pool of intracellular iron that may catalyze HO· generation (Gannon, *et al.*, 1987; Kvietys, *et al.*, 1989; Zarley, *et al.*, 1991; Britigan, *et al.*, 1992a), especially under conditions that promote large fluxes of $O_2\cdot^-$ (Liochev and Fridovich, 1994; Keyer, *et al.*, 1995). In these cases, $O_2\cdot^-$ behaves as a reductant toward the iron chelate, releasing Fe^{2+}, which can reduce H_2O_2 (Keyer and Imlay, 1996; McCormick, *et al.*, 1998b). Although the specific catalysts responsible have not been defined, possible candidates include ferritin, hemoglobin, and myoglobin; any of a number of iron-centered enzymes; and nonprotein iron complexes, such as Fe^{3+}–ADP, Fe^{3+}–citrate, and DNA-associated iron (Halliwell and Gutteridge, 1986; Imlay and Linn, 1988; Zarley, *et al.*, 1991; Britigan, *et al.*, 1992b; Rothman, *et al.*, 1992). Under conditions of cell stress, these intracellular iron chelates can be released into the extracellular space, thereby further contributing to cell injury. Evidence in support of this theory comes from the demonstration that iron associated with ferritin, hemoglobin, and myoglobin is capable of being reduced, of acting as a HO· catalyst, and of initiating lipid peroxidation (Weiss, 1982; Benatti, *et al.*, 1983; Sadrzadeh, *et al.*, 1984; Szebeni, *et al.*, 1984; Augusto, *et al.*, 1986; Gutteridge, 1986; O'Connell, *et al.*, 1986; Szebeni and Toth, 1986; Harel, *et al.*, 1988; Puppo and Halliwell, 1988; Reif, *et al.*, 1988; Harel and Kanner, 1989; Monteiro, *et al.*, 1989; Winterbourn, *et al.*, 1990, 1991; Abdalla, *et al.*, 1992; Reif, 1992; Ryan and Aust, 1992; Allen, *et al.*, 1994; Kukielka and Cederbaum, 1994; Mao, *et al.*, 1994). Likewise, microglial-secreted $O_2 \cdot^-$ has been reported to promote release of iron from ferritin. This increased "free" iron, resulting in sequelae of free radicals in response to this redox active metal ion, has been proposed to account for neurodegenerative disorders (Yoshida, *et al.*, 1995).

As discussed above, most theories as to how cells protect themselves against oxidant-mediated damage have focused on the role of SOD (Bannister, *et al.*, 1987; Hassett and Cohen, 1989), catalase (Heffner and Repine, 1989), and enzymes intimately involved in maintaining glutathione in a reduced state (Heffner and Repine, 1989). A less well-studied cellular antioxidant defense strategy is the regulation of iron available to redox cycle. This may, for example, involve heme oxygenase, which can convert redox active heme iron to biliverdin and Fe^{2+}, the latter of which can be inactivated through chelation (Stocker, 1990). Cellular levels of this enzyme can undergo modulation in response to various exogenous factors (Keyse and Tyrrell, 1989; Kim, *et al.*, 1989; Keyse, *et al.*, 1990; Kurata and Nakajima, 1990; Lin, *et al.*, 1990; Stocker, 1990; Eisenstein, *et al.*, 1991; Saunders, *et al.*, 1991; Smith, *et al.*, 1991; Mitani, *et al.*, 1992; Tyrrell and Basu-Modak, 1994; Nath, 1994; Vile, *et al.*, 1994). In addition to its role in intracellular iron metabolism, the protective effects of heme oxygenase toward oxidant stress may be through its role as a mediator of extracellular iron release. Evidence in support of this theory comes, in part, from mice deficient in either heme oxygenase 1 or heme oxygenase 2. These animals exhibit increased susceptibility to oxidative stress (Poss and Tonegawa, 1997a; Dennery, *et al.*, 1998). Further, mice lacking heme oxygenase 1 develop a profound anemia in association with low serum iron levels and an apparent inability to release iron from the liver and other sites of storage (Poss and Tonegawa, 1997b). The ability of heme oxygenase to mediate this process appears to be modified by cellular exposure to nitric oxide (NO·). Illustrative of this point is the finding that ·NO-mediated nitrosylation of cellular heme prevented iron release concomitant with decreased endothelial cell susceptibility to oxidant stress (Juckett, *et al.*, 1998).

Limiting iron availability also appears to be a strategy of host defense (Finkelstein, *et al.*, 1983). Acute infection leads to an increase in serum transferrin and lactoferrin. Not surprisingly, many pathogens are unable to utilize iron in such complexes (Finkelstein, *et al.*, 1983). There is also a shift in iron from serum to reticuloendothelial-derived macrophages.

Microbial pathogens utilize several distinct means to compete with the host for iron, among them siderophore production. These low-molecular-weight iron chelators, usually either hydroxamate or catechol derivatives (Neilands, 1981), compete with and/or remove Fe^{3+} from host iron-binding compounds (Neilands, 1981;

Otto, *et al.*, 1992). The organism then internalizes the siderophore-bound iron. *Escherichia coli* and other Enterobacteriacea produce enterochelin and aerobactin (Neilands, 1981, 1993; Williams and Carbonetti, 1986). *Pseudomonas aeruginosa* produces pyoverdin and pyochelin (Cox and Graham, 1979; Cox, 1980; Cox, *et al.*, 1981; Wendenbaum, *et al.*, 1983; Cox and Adams, 1985). The iron chelator desferrioxamine, which is used as a treatment for iron-overload (Halliwell, 1989; Brittenham, *et al.*, 1994), is a siderophore produced by the fungus *Streptomyces pilosus*. Not surprisingly, this chelate, which is a poor catalyst of the metal ion-catalyzed Haber-Weiss reaction, protects cells from HO·-mediated damage (Ward, *et al.*, 1983; Gannon, *et al.*, 1987; Halliwell, 1989; Kvietys, *et al.*, 1989; van Asbeck, *et al.*, 1989; Bolli, *et al.*, 1990; Williams, *et al.*, 1991; Zarley, *et al.*, 1991; Katoh, *et al.*, 1992; Rothman, *et al.*, 1992; Sullivan, *et al.*, 1992; Hershko, *et al.*, 1993; McBride, *et al.*, 1993; Le, *et al.*, 1994; Palmer, *et al.*, 1994). Similarly, the *P. aeruginosa* siderophore pyoverdin does not chelate iron in a catalytic form (Coffman, *et al.*, 1990; Morel, *et al.*, 1992, 1995). In contrast, iron bound to the other *P. aeruginosa* siderophore pyochelin catalyzes formation of HO·, while augmenting cell injury induced by $O_2 \cdot^-$ (Britigan, *et al.*, 1992a, 1994).

Glutathione and Related Thiols

Glutathione, a tripeptide (γ-glutamylcysteinglycine, GSH), is the major nonprotein thiol present in most species of the archaea, bacteria and eucarya domains, with levels in the millimolar range (Kosower and Kosower, 1978). Because of this rather high concentration, it is not surprising that this thiol is intimately involved in many biologic processes, besides those associated with glutathione peroxidase and glutathione-*S*-transferase (Smith, *et al.*, 1996). For instance, it is believed that GSH plays an integral part in protein synthesis by transporting small peptides from one portion of the cell to another, as well as enhancing protein disulfide bond formation through disulfide exchange reactions. As a strong nucleophile, GSH can react with a variety of structurally diverse but very reactive xenobiotics through either 1,2- or 1,4-addition to quinones or quinone imines (Hinson, *et al.*, 1982; Rosen, *et al.*, 1984, Cadenas and Ernster, 1990; March, 1992), to epoxides (Cadenas and Ernster, 1990), or to alkyl halides (Kosower and Kosower, 1976).

Besides nucleophilic addition reactions, GSH can act as a one-electron reductant, resulting, for example, in the termination of lipid peroxidation (Schöneich, 1995; Wardman and von Sonntag, 1995):

$$GSH + L \cdot \rightarrow GS \cdot + LH$$

The glutathionyl radical, GS·, can also be produced enzymatically as the result of peroxidase action on GSH (Ross, *et al.*, 1985; Eling, *et al.*, 1986; Harman, *et al.*, 1986; Schreiber, *et al.*, 1989; Maples, *et al.*, 1990; Mason and Rao, 1990; Stoyanovsky, *et al.*, 1996). Once this thiyl radical is generated, reactions involving GS· are numerous, depending on experimental conditions (Asmus, 1990; Koppenol, 1993a; Zhao, *et al.*, 1994). Formation of the disulfide through biradical reaction is a rapid process controlled by diffusion:

$$2GS \cdot \rightarrow GSSG$$

However, in a cell, GSSG formation via this pathway must compete with other targets that GS· may encounter. This free radical can, for instance, undergo an intramolecular transformation, leading to a carbon-centered free radical (Grierson, *et al.*, 1992). Addition of GS· to GS⁻ results in a strong reductant (Quintiliani, *et al.*, 1977; Buettner, 1993; Winterbourn, 1993), which, in the presence of O_2, leads to $O_2 \cdot^-$ production (Winterbourn and Metodiewa, 1994):

$$GS \cdot + GS^- \rightarrow GSSG \cdot^-$$

$$GSSG \cdot^- + O_2 \rightarrow GSSG + O_2 \cdot^-$$

Alternatively, GS· can react with O_2, producing glutathionyldioxyl radical (GSOO·), whose fate remains uncertain (Tamba, *et al.*, 1986; Mönig, *et al.*, 1987; Asmus, 1990; Buettner, 1993; Koppenol, 1993a):

$$GS \cdot + O_2 \rightleftarrows GSOO \cdot$$

Finally, GS· can react with either ascorbic acid (AH_2) or α-tocopherol (α-TOH), thereby enhancing the chain-breaking properties of these antioxidants (Winkler, *et al.*, 1994):

$$GS \cdot + AH_2 \rightarrow AH \cdot + GSH$$

$$\alpha\text{-TOH} + GS \cdot \rightarrow \alpha\text{-TO} \cdot + GSH$$

The reaction of $O_2 \cdot^-$ with GSH has received considerable attention in recent years, due, in part, to high cellular levels of this thiol (Winterbourn and Metodiewa, 1995). Surprisingly, estimates of the rate constant vary greatly from a high of $7-8 \times 10^5 \, M^{-1} \, s^{-1}$ (Asada and Kanematsu, 1976; Bielski, *et al.*, 1985; Dikalov, *et al.*, 1996) to a considerably slower reaction of $1 \times 10^3 \, M^{-1} \, s^{-1}$ (Winterbourn and Metodiewa, 1994). The source of such enormous differences in these rate constants must undoubtedly lie in methodological limitations. Given the most conservative of estimates, there is about a 10,000-fold difference in rate constants between SOD and GSH. One must wonder, therefore, how effective can GSH be in the cellular regulation of $O_2 \cdot^-$. Even with concentrations of GSH as high as 1 mM in some cells, considerably less than 10% of $O_2 \cdot^-$ could possibly be captured by this thiol, given that SOD levels are $\geq 1 \, \mu M$. Therefore, the biologic significance of this reaction must be seriously questioned.

These are but a few of the reactions GSH can undergo. It must be recognized, however, that specific reaction pathways are frequently dictated by localized cellular events and environmental conditions (Asmus, 1990). Because of this, it is not possible, a priori, to predict what role GSH or similar thiols may play in modulating intracellular free radical events.

Even though aerobes maintain high thiol levels, of which GSH is most ubiquitous, there are a few examples of evolutional diversity, resulting in the biosynthesis of alternative sulfhydryls. One such adaption can be found in the African trypano-some, *Trypanosoma brucei* (Fairlamb and Cerami, 1985; Fairlamb, *et al.*, 1985). Here, this protozoan has linked GSH to the polyamine spermidine to produce the spermidine thiol, bis(L-γ-glutamyl-L-cysteinyl-L-glycl)spermidine, commonly known as trypanothione (T[SH]$_2$) (Fairlamb and Cerami, 1992). With 2 mol of thiol/T[SH]$_2$, the reaction with a lipid radical L·, resulting in trypanothione di-sulfide (T[S]$_2$), should be considerably more efficient than would be GSH:

$$T[SH]_2 + 2L· \rightarrow T[S]_2 + 2LH$$

As in the case of GSH, a unique reductase, NADPH-trypanothione reductase, maintains high cellular levels of T[SH]$_2$ (Figure 2.14).

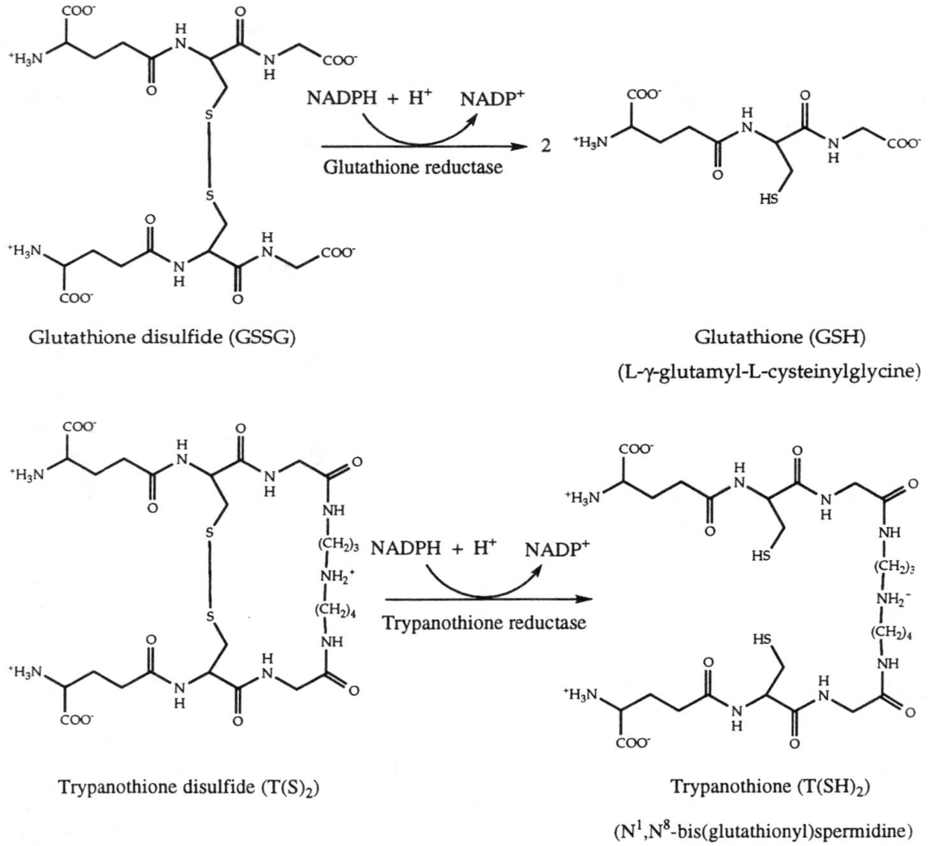

Figure 2.14 Reduction of glutathione disulfide (GSSG) and trypanothione disulfide (T[S]$_2$) to glutathione (GSH) and trypanothione (T[SH]$_2$) by their respective NADPH-dependent reductases. (Based on Fairlamb and Cerami, 1992.)

What environmental pressures forced the trypanosome family to forsake the GSH synthetic pathway in favor of T[SH]$_2$ are presently uncertain. The dissimilar net charge between these thiols, $+1$ for T[SH]$_2$ and -1 for GSH, may be an important factor in substrate discrimination and maintenance of cellular viability for this protozoan. Alternative explanations will undoubtedly be forthcoming.

Another low-molecular-weight thiol present in both eucaryotic and procaryotic cells for which there is increasing evidence for a role in defense against oxidant stress is the protein disulfide thioredoxin. It has been shown to be involved in a variety of cellular processes by modulating redox events. In conjunction with thioredoxin reductase, thioredoxin regulates oxidant-mediated signal transduction, cell growth, and proliferation (Nakamura, et al., 1997). Thioredoxin has also been shown to control levels of other antioxidant enzymes, including MnSOD (Das, et al., 1997), and protect against H$_2$O$_2$-mediated cytotoxicity (Takemoto, et al., 1998). A novel peroxidase that utilizes thioredoxin as a source of electrons for reduction and removal of H$_2$O$_2$, thioredoxin peroxidase has recently been described (Choi, et al., 1998; Kang, et al., 1998). Data to date suggest that it may play a novel role in the regulation of cellular response to oxidant stress and in the control of cell signaling events (Jin, et al., 1997).

Concluding Thoughts

Priestley's prophetic observations on the toxicity of O$_2$ ring clearer today than at any time in the intervening centuries since his discovery of this gas. Despite this, we are just beginning to understand how evolutionary pressures forced each of the domains to adapt to the hazards of an aerobic world. Through regulation of cellular fluxes of O$_2 \cdot^-$ and H$_2$O$_2$, organisms are able to control O$_2$ metabolism, thereby assuring the survival of the many diverse species that inhabit this planet.

In recent years, another free radical, NO$\cdot$, which has been found to exhibit a broad spectrum of physiologic and immunologic functions, has caught the fancy of the scientific community. In the following chapter, we will begin to search the many complex activities attributed to this very simple free radical.

References

Abdalla, D.S.P., Campa, A., and Monteiro, H.P. (1992). Low density lipoprotein oxidation by stimulated neutrophils and ferritin. *Atherosclerosis* **97**: 149–159.

Abo, A., Boyhan, A., West, I., Thrasher, A.J., and Segal, A.W. (1992). Re-constitution of neutrophil NADPH oxidase activity in the cell-free system by four components: p$^{67-phox}$, p$^{47-phox}$, p21rac1, and cytochrome b$_{-245}$. *J. Biol. Chem.* **267**: 16767–16770.

Abo, A., Pick, E., Hall, A., Totty, N., Teahan, C.G., and Segal, A.W.

(1991). Activation of the NADPH oxidase involves the small GTP-binding protein p21^{rac1}. *Nature* **353**: 668–670.

Afanas'ev, I.B. (1989). Reactivity of superoxide ion, in: *Superoxide Ion: Chemistry and Biological Implications*, Vol. I, pp. 41–61, CRC Press, Boca Raton, FL.

Aikens, J., and Dix, T.A. (1991). Perhydroxyl radical (HOO·) initiated lipid peroxidation. The role of fatty acid hydroperoxides. *J. Biol. Chem.* **266**: 15091–15098.

Aisen, P., and Leibman, A. (1972). Lactoferrin and transferrin: A comparative study. *Biochim. Biophys. Acta* **257**: 314–323.

Aisen, P., Aasa, R., Malmström, B.G., and Vänngård, T. (1967). Bicarbonate and the binding of iron to transferrin. *J. Biol. Chem.* **242**: 2484–2490.

Aisen, P., Leibman, A., and Zweier, J. (1978). Stoichiometric and site characteristics of the binding of iron to human transferrin. *J. Biol. Chem.* **253**: 1930–1937.

Albrich, J.M., Gilbaugh, J.H., III, Callahan, K.B., and Hurst, J.K. (1986). Effects of the putative neutrophil-generated toxin, hypochlorous acid, on membrane permeability and transport systems of *Escherichia coli*. *J. Clin. Invest.* **78**: 177–184.

Albrich, J.M., McCarthy, C.A., and Hurst, J.K. (1981). Biological reactivity of hypochlorous acid: Implications for microbicidal mechanisms of leukocyte myeloperoxidase. *Proc. Natl. Acad. Sci. USA* **78**: 210–214.

Alexander, M.S., Husney, R.M., and Sagone, A.L., Jr. (1986). Metabolism of benzoic acid by stimulated polymorphonuclear cells. *Biochem. Pharmacol.* **20**: 3649–3651.

Allen, D.R., Wallis, G.L., and McCay, P.B. (1994). Catechol adrenergic agents enhance hydroxyl radical generation in xanthine oxidase systems containing ferritin: Implications for ischemia/reperfusion. *Arch. Biochem. Biophys.* **315**: 235–243.

Allgood, G.S., and Perry, J.J. (1986). Characterization of a manganese- containing catalase from the obligate thermophile *Thermoleophilum album*. *J. Bacteriol.* **168**: 563–567.

Ambruso, D.R., and Johnston, R.B., Jr. (1981). Lactoferrin enhances hydroxyl radical production by human neutrophils, neutrophil particulate fractions and an enzymatic generating system. *J. Clin. Invest.* **67**: 352–360.

Ambruso, D.R., Bentwood, B., Henson, P.M., and Johnston, R.B., Jr., (1984). Oxidative metabolism of cord blood neutrophils: Relationship to content and degranulation of cytoplasmic granules. *Pediatr. Res.* **18**: 1148–1153.

Anderson, B.F., Baker, H.M., Dodson, E.J., Norris, G.E., Rumball, S.V., Waters, J.M., and Baker, E.N. (1987). Structure of human lactoferrin at 3.2-Å resolution. *Proc. Natl. Acad. Sci. USA* **84**: 1769–1773.

Anderson, B.F., Baker, H.M., Norris, G.E., Rumball, S.V., and Baker, E.N. (1990). Apolactoferrin structure demonstrates ligand-induced conformational changes in transferrins. *Nature* **344**: 784–787.

Arneson, R.M. (1970). Substrate-induced chemiluminescence of xanthine oxidase and aldehyde oxidase. *Arch. Biochem. Biophys.* **136**: 352–360.

Aruoma, O.I., and Halliwell, B. (1987). Superoxide-dependent and ascorbate-dependent formation of hydroxyl radicals from hydrogen peroxide in the presence of iron are lactoferrin and transferrin promoters of hydroxyl-radical generation? *Biochem. J.* **241**: 273–278.

Aruoma, O.I., Bomford, A., Polson, R.J., and Halliwell, B. (1988). Nontransferrin-bound iron in plasma from hemochromatosis patients: Effect of phlebotomy therapy. *Blood* **72**: 1416–1419.

Asada, K., and Kanematsu, S. (1976). Reactivity of thiols with superoxide radicals. *Agric. Biol. Chem.* **40**: 1891–1892.

Asada, K., Kanematsu, S., Okaka, S., and Hayakawa, T. (1980). Phylogenic distribution of three types of superoxide dismutase in organisms and in cell organelles, in: *Chemical and Biochemical Aspects of Superoxide and Superoxide Dismutases* (Bannister, J.V., and Hill, H.A.O., eds.), pp. 136–153, Elsevier/North-Holland, New York.

Asada, K., Yoshikawa, K., Takahashi, M.A., Maeda, Y., and Enmanji, K.

(1975). Superoxide dismutases from a blue-green algae, *Plectonema boryanum. J. Biol. Chem.* **250**: 2801–2807.

Asmus, K.-D. (1990). Sulfur-centered free radicals. *Methods Enzymol.* **186**: 168–180.

Augusto, O., Beilan, H.S., and Ortiz de Montellano, P.R. (1982). The catalytic mechanism of cytochrome P-450. Spin-trapping evidence for one-electron substrate oxidation. *J. Biol. Chem.* **257**: 11288–11295.

Augusto, O., Weingrill, C.L.V., Schreier, S., and Amemiya, H. (1986). Hydroxyl radical formation as a result of the interaction between primaquine and reduced pyridine nucleotides: Catalysis by hemoglobin and microsomes. *Arch. Biochem. Biophys.* **244**: 147–155.

Aust, S.D., and Svingen, B.A. (1982). The role of iron in enzymatic lipid peroxidation, in: *Free Radicals in Biology* (Pryor, W.A., ed.), Vol. V, pp. 1–28, Academic Press, New York.

Avissar, N., Kerl, E.A., Baker, S.S., and Cohen, H.J. (1994). Extracellular glutathione peroxidase mRNA and protein in human cell lines. *Arch. Biochem. Biophys.* **309**: 239–246.

Avissar, N., Whitin, J.C., Allen, P.Z., Wagner, D.D., Liegey, P., and Cohen, H.J. (1989). Plasma selenium-dependent glutathione peroxidase: Cell of origin and secretion. *J. Biol. Chem.* **264**: 15850–15855.

Babcock, G.T., and Wikstrom, M. (1992). Oxygen activation and the conservation of energy in cell respiration. *Nature* **356**: 301–309.

Babior, B.M. (1978). Oxygen-dependent microbial killing by phagocytes. *N. Engl. J. Med.* **298**: 659–678.

Babior, B.M., Kipnes, R.S., and Curnutte, J.T. (1973). Biological defense mechanisms: The production by leukocytes of superoxide, a potential bactericidal agent. *J. Clin. Invest.* **52**: 741–744.

Baehner, R.I., and Nathan, D.G. (1967). Leukocyte oxidase: Defective activity in chronic granulomatous disease. *Science* **155**: 835–836.

Balazs, R., and Brooksbank, B.W.L. (1985). Neurochemical approaches to the pathogenesis of Down's syndrome. *J. Ment. Defic. Res.* **29**: 1–14.

Baldridge, C.W., and Gerard, R.W. (1933). The extra respiration of phagocytosis. *Am. J. Physiol.* **103**: 235–236.

Baldwin, D.A. (1980). The kinetics of iron release from human transferrin by EDTA. Effects of salts and detergents. *Biochim. Biophys. Acta* **623**: 183–188.

Baldwin, D.A., Jenny, E.R., and Aisen, P. (1984). The effect of human serum transferrin and milk lactoferrin on hydroxyl radical formation from superoxide and hydrogen peroxide. *J. Biol. Chem.* **259**: 13391–13394.

Balla, G., Jacob, H.S., Balla, J., Rosenberg, M., Nath, K., Apple, F., Eaton, J.W., and Vercellotti, G.M. (1992). Ferritin: A cytoprotective antioxidant strategem of endothelium. *J. Biol. Chem.* **267**: 18148–18153.

Balla, G., Vercellotti, G.M., Eaton, J.W., and Jacob, H.S. (1990). Iron loading of endothelial cells augments oxidant damage. *J. Lab. Clin. Med.* **116**: 546–554.

Banci, L., Bertini, I., Cabelli, D., Hallewell, R.A., Luchinat, C., and Viezzoli, M.S. (1990). Investigation of copper-zinc superoxide dismutase Ser-137 and Ala-137 mutants. *Inorg. Chem.* **29**: 2398–2403.

Banci, L., Cabelli, D.E., Getzoff, E.D., Hallewell, R.A., and Viezzoli, M.S. (1993). An essential role for the conserved Glu-133 in the anion interaction with superoxide dismutase. *J. Inorg. Biochem.* **50**: 89–100.

Bannister, J.V., and Bannister, W.H. (1985). Production of oxygen-centered radicals by neutrophils and macrophages as studied by electron spin resonance (ESR). *Environ. Health Perspect.* **64**: 37–43.

Bannister, J.V., Bannister, W.H., and Rotilio, G. (1987). Aspects of the structure, function, and applications of superoxide dismutase. *CRC Crit. Rev. Biochem.* **22**: 111–180.

Bannister, J.V., Bellavite, P., Serra, M.C., Thornalley, P.J., and Rossi, F. (1982). An EPR study of the production of superoxide radicals by neutrophil NADPH oxidase. *FEBS Lett.* **145**: 323–326.

Barber, D.J.W., and Thomas, J.K. (1978). Reactions of radicals with lecithin bilayers. *Radiat. Res.* **74**: 51–65.

Bates, G.W., Billups, C., and Saltman, P. (1967). The kinetics and mechanism of iron (III) exchange between chelates and transferrin. I. The complexes of citrate and nitrilotriacetic acid. *J. Biol. Chem.* **242**: 2810–2815.

Bayraktutan, U., Draper, N., Kang, D., and Shah, A.M. (1998). Expression of functional neutrophil type NADPH oxidase in cultured rat coronary microvascular endothelial cells. *Cardiovasc. Res.* **38**: 256–262.

Bean, J.W. (1945). Effects of oxygen at increased pressure. *Physiol. Rev.* **25**: 1–147.

Beck, B.L., Tabatabai, L.B., and Mayfield, J.E. (1990). A protein isolated from *Brucella abortus* is a Cu-Zn superoxide dismutase. *Biochemistry* **29**: 372–376.

Beckman, K.B., and Ames, B.N. (1998). The free radical theory of aging matures. *Physiol. Rev.* **78**: 547–581.

Behar, D., Czapski, G., Rabani, J., Dorfman, L.M., and Schwarz, H.A. (1970). The acid dissociation constant and decay kinetics of the perhydroxyl radical. *J. Phys. Chem.* **74**: 3209–3212.

Behnke, A.R. (1940). High atmosphereic pressures: Physiological effects of increased and decreased pressure; application of these findings to clinical medicine. *Ann. Int. Med.* **13**: 2217–2228.

Behnke, A.R., Forbes, H.S., and Motley, E.P. (1936). Circulatory and visual effects of oxygen at 3 atmospheres pressure. *Am. J. Physiol.* **114**: 436–442.

Bellavite, P. (1988). The superoxide-forming enzymatic system of phagocytes. *Free Radical Biol. Med.* **4**: 225–261.

Benatti, U., Morelli, A., Guida, L., and DeFlora, A. (1983). The production of activated oxygen species by an interaction of methemoglobin with ascorbate. *Biochem. Biophys. Res. Commun.* **111**: 980–987.

Benedetti, A., Casini, A.F., Ferrali, M., and Comporti, M. (1977). Studies on relationships between carbon tetrachloride-induced alterations of liver microsomal lipids and impairment of glucose-6-phosphatase activity. *Exp. Mol. Pathol.* **27**: 309–323.

Benov, L.T., and Fridovich, I. (1994). *Escherichia coli* expresses a copper-and zinc-containing superoxide dismutase. *J. Biol. Chem.* **269**: 25310-25314.

Benov, L., and Fridovich, I. (1996). Functional significance of the Cu,Zn SOD in *Escherichia coli*. *Arch. Biochem. Biophys.* **327**: 249–253.

Benov, L., Chang, L.Y., Day, B., and Fridovich, I. (1995). Copper, zinc superoxide dismutase in *Escherichia coli*: Periplasmic localization. *Arch. Biochem. Biophys.* **319**: 508–511.

Bernofsky, C. (1991). Nucleotide chloramine and neutrophil- mediated cytotoxicity. *FASEB J.* **5**: 295–300.

Bert, P. (1943). *Barometric Pressure: Researches in Experimental Physiology* (Translated by M.A. and F.A. Hitchock), pp. 709–851, College Book Co; Columbus, OH.

Beyer, W.F., Jr., and Fridovich, I. (1985). Pseudocatalase from *Lactobacillus plantarum*: Evidence for a homopentameric structure containing two atoms of manganese per subunit. *Biochemistry* **24**: 6460–6467.

Beyer, W.F., Jr., and Fridovich, I. (1987). Effect of hydrogen peroxide on the iron-containing superoxide dismutase of *Escherichia coli*. *Biochemistry* **26**: 1251–1257.

Bielski, B.H.J. (1978). Reevaluation of the spectral and kinetic properties of the HO_2 and O_2^- free radicals. *Photochem. Photobiol.* **28**: 645–649.

Bielski, B.H. J., and Allen, A.O. (1977). Mechanism of the disproportionation of superoxide radicals. *J. Phys. Chem.* **81**: 1048–1050.

Bielski, B.H.J., and Cabelli, D.E. (1991). Highlights of current research involving superoxide and perhydroxyl radicals in aqueous solutions. *Int. J. Radiat. Biol.* **59**: 291–319.

Bielski, B.H.J., Cabelli, D.E., Arudi, R.L., and Ross, A.B. (1985). Reactivity of HO_2/O_2^- radicals in aqueous solution. *J. Phys. Chem. Ref. Data* **14**: 1041–1100.

Biemond, P., Swaak, A.J.G., van Eijk, H.G., and Koster, J.F. (1988). Superoxide dependent iron release from ferritin in inflammatory diseases. *Free Radical Biol. Med.* **4**: 185–198.

Bisby, R.H., and Parker, A.W. (1995). Reaction of ascorbate with the α-tocopheroxyl radical in micellar and bilayer systems. *Arch. Biochem. Biophys.* **317**: 170–178.

Björnstedt, M., Xue, J., Huang, W., Åkesson, B., and Holmgren (1994). The thioredoxin and glutaredoxin systems are efficient electron donors to human plasma glutathione peroxidase. *J. Biol. Chem.* **269**: 29382–29384.

Black, S. (1963). The biochemistry of sulfur-containing compounds. *Ann. Rev. Biochem.* **32**: 399–418.

Bluard-Deconinck, J., Williams, J., Evans, R.W., van Snick, J., Osinski, P.A., and Masson, P.L. (1978). Iron-binding fragments from the N-terminal and C-terminal regions of human lactoferrin. *Biochem. J.* **171**: 321–327.

Böger, R.H., Bode-Böger, S.M., Mügge, A., Kienke, S., Brandes, R., Dwenger, A., and Frölich, J.C. (1995). Supplementation of hypercholesterolaemic rabbits with L-arginine reduces the vascular release of superoxide anions and restores NO production. *Atherosclerosis* 117: 273–284.

Bokoch, G.M., Bohl, B.P., and Chuang, T.-H. (1994). Guanine nucleotide exchange regulates membrane translocation of Rac/Rho GTP-binding proteins. *J. Biol. Chem.* **269**: 31674–31679.

Bol, D.K., and Yasbin, R.E. (1990). Characterization of an inducible oxidative stress system in *Bacillus subtilis*. *J. Bacteriol.* **172**: 3503–3506.

Bolann, B.J., Henriksen, H., and Ulvik, R.J. (1992). Decay kinetics of $O_2\cdot^-$ studied by direct spectrophotometry. Interaction with catalytic and non-catalytic substances. *Biochim. Biophys. Acta* **1156**: 27–33.

Bolkenius, F.N., Grisar, J.M., and De Jong, W. (1991). A water-soluble quaternary ammonium analog of α-tocopherol, that scavenges lipoperoxyl, superoxyl and hydroxyl radicals. *Free Radical Res. Commun.* **14**: 363–372.

Bolli, R., Jeroudi, M.O., Patel, B.S., Aruoma, O.I., Haliwell, B., Lai, E.K., and McCay, P.B. (1989). Marked reduction of free radical generation and contractile dysfunction by antioxidant therapy begun at the time of reperfusion. Evidence that myocardial "stunning" is a manifestation of reperfusion injury. *Circ. Res.* **65**: 607–622.

Bolli, R., Patel, B.S., Jeroudi, M.O., Li, X.-Y., Triana, J.F., Lai, E.K., and McCay, P.B. (1990). Iron-mediated radical reactions upon reperfusion contribute to myocardial "stunning." *Am. J. Physiol.* **259**: H1901–H1911.

Borders, C.L., Jr., and Fridovich, I. (1985). A comparison of the effects of cyanide, hydrogen peroxide, and phenylglyoxal on eucaryotic and procaryotic Cu,Zn superoxide dismutases. *Arch. Biochem. Biophys.* **241**: 472–476.

Boveris, A., and Cadenas, E. (1982). Production of superoxide radicals and hydrogen peroxide in mitochondria, in: *Superoxide Dismutases* (Oberley, L.W., ed.), Vol. II, pp. 15–30, CRC Press, Boca Raton, FL.

Boveris, A., and Chance, B. (1973). The mitochondrial generation of hydrogen peroxide. *Biochem. J.* **134**: 707–716.

Boveris, A.N., Oshino, N., and Chance, B. (1972). The cellular production of hydrogen peroxide. *Biochem. J.* **128**: 617–630.

Bowling, A.C., Schultz, J.B., Brown, R.H., Jr., and Beal, M.F. (1993). Superoxide dismutase activity, oxidative damage, and mitochondrial energy metabolism in familial and sporadic amyotrophic lateral sclerosis. *J. Neurochem.* **61**: 2322–2325.

Bowry, V.W., and Ingold, K.U. (1995). Extraordinary kinetic behavior of the α-tocopheroxyl (vitamin E) radical. *J. Org. Chem.* **60**: 5456–5467.

Bray, R.C., Cockle, S.A., Fielden, E.M., Roberts, P.B., Rotilio, G., and Calabrese, L. (1974). Reduction and inactivation of superoxide dismutase by hydrogen peroxide. *Biochem. J.* **139**: 43–48.

Brines, R.D., and Brock, J.H. (1983). The effect of trypsin and chymotrypsin on the *in vitro* antimicrobial and iron-binding properties of lactoferrin in human milk and bovine colostrum: Unusual resistance of human apolactoferrin to proteolytic digestion. *Biochim. Biophys. Acta* **759**: 229–235.

Britigan, B.E., and Edeker, B.L. (1991). *Pseudomonas* and neutrophil products modify transferrin and lactoferrin to create conditions that favor hydroxyl radical formation. *J. Clin. Invest.* **88**: 1092–1102.

Britigan, B.E., Coffman, T.J., and Buettner, G.R. (1990). Spin trap evidence for the lack of significant hydroxyl radical production during the respiration burst of human phagocytes using a spin adduct resistant to superoxide mediated destruction. *J. Biol. Chem.* **265**: 2650–2656.

Britigan, B.E., Cohen, M.S., and Rosen, G.M. (1987). Detection of the production of oxygen-centered free radicals by human neutrophils using spin trapping techniques: A critical perspective. *J. Leukocyte Biol.* **41**: 349–362.

Britigan, B.E., Hassett, D.J., Rosen, G.M., Hamill, D.R., and Cohen, M.S. (1989). Neutrophil degranulation inhibits potential hydroxyl radical formation: Differential impact of myeloperoxidase and lactoferrin release on hydroxyl radical production by iron supplemented neutrophils assessed by spin trapping. *Biochem. J.* **264**: 447–455.

Britigan, B.E., Hayek, M.B., Doebbeling, B.N., and Fick, R.B., Jr. (1993). Transferrin and lactoferrin undergo proteolytic cleavage in the *Pseudomonas aeruginosa*-infected lungs of patients with cystic fibrosis. *Infect. Immun.* **61**: 5049–5055.

Britigan, B.E., Rasmussen, G.T., and Cox, C.D. (1994). The pseudomonas siderophore pyochelin enhances neutrophil-mediated endothelial cell injury. *Am. J. Physiol.* **266**: L192–L198.

Britigan, B.E., Ratcliffe, H.R., Buettner, G.R., and Rosen, G.M. (1996). Binding of myeloperoxidase to bacteria: Effect on hydroxyl radical formation and susceptibility to oxidant-mediated killing. *Biochim. Biophys. Acta* **1290**: 231–240.

Britigan, B.E., Roeder, T.L., Rasmussen, G.T., Shasby, D.M., McCormick, M.L., and Cox, C.D. (1992a). Interaction of the *Pseudomonas aeruginosa* secretory products pyocyanin and pyochelin generates hydroxyl radical and causes synergistic damage to endothelial cells: Implications for pseudomonas-associated tissue injury. *J. Clin. Invest.* **90**: 2187–2196.

Britigan, B.E., Roeder, T.L., and Shasby, D.M. (1992b). Insight into the nature and site of oxygen-centered free radical generation by endothelial cell monolayers using a novel spin trapping technique. *Blood* **79**: 699–707.

Britigan, B.E., Rosen, G.M., Chai, Y., and Cohen, M.S. (1986a). Do human neutrophils make hydroxyl radical? Detection of free radicals generated by human neutrophils activated with a soluble or particulate stimulus using electron paramagnetic resonance spectrometry. *J. Biol. Chem.* **261**: 4426–4431.

Britigan, B.E., Rosen, G.M., Thompson, B.Y., Chai, Y., and Cohen, M.S. (1986b). Stimulated neutrophils limit iron-catalyzed hydroxyl radical formation as detected by spin trapping techniques. *J. Biol. Chem.* **261**: 17026–17032.

Brittenham, G.M., Griffith, P.M., Nienhuis, A.W., McLaren, C.E., Young, N.S., Tucker, E.E., Allen, C.J., Farrell, D.E., and Harris, J.W. (1994). Efficacy of deferoxamine in preventing complications of iron overload in patients with thalassemia major. *N. Engl. J. Med.* **331**: 567–573.

Broderick, D.J., Deagen, J.T., and Whanger, P.D. (1987). Properties of glutathione peroxidase isolated from human plasma. *J. Inorg. Biochem.* **30**: 299–308.

Buettner, G.R. (1993). The pecking order of free radicals and antioxidants: Lipid peroxidation, α-tocopherol and ascorbate. *Arch. Biochem. Biophys.* **300**: 535–543.

Buettner, G.R., and Chamulitrat, W. (1990). The catalytic activity of iron in synovial fluid as monitored by the ascorbate free radical. *Free Radical Biol. Med.* **8**: 55–56.

Buettner, G.R., and Jurkiewicz, B.A. (1994). Ascorbate free radical as a marker for oxidative stress: An EPR study. *Free Radical Biol. Med.* **14**: 49–55.

Buettner, G.R., Doherty, T.P., and Patterson, L.K. (1983). The kinetics of the reaction of superoxide radical with Fe(III) complexes of EDTA, DETAPAC, and HEDTA. *FEBS Lett.* **158**: 143–146.

Bull, C., and Fee, J.A. (1985). Steady-state kinetic studies of superoxide dismutases: Properties of the iron containing protein from *Escherichia coli*. *J. Am. Chem. Soc.* **107**: 3295–3304.

Bullen, J.J., Rogers, H.J., and Griffith, E. (1978). Role of iron in bacterial infection. *Curr. Top. Microbiol. Immunol.* **80**: 1–35.

Burk, R.F., and Lane, J.M. (1979). Ethane production and liver necrosis in rats after administration of drugs and other chemicals. *Toxicol. Appl. Pharmacol.* **50**: 467–478.

Burk, R.F., Nishiki, K., Lawrence, R.A., and Chance, B. (1978). Peroxide removal by a selenium-dependent and selenium-independent glutathione peroxidase in hemoglobin-free perfused rat liver. *J. Biol. Chem.* **253**: 43–46.

Burton, G.W., and Ingold, K.U. (1981). Autoxidation of biological molecules. 1. The antioxidant activity of vitamin E and related chain-breaking phenolic antioxidants. *J. Am. Chem. Soc.* **103**: 6472–6477.

Burton, G.W., and Ingold, K.U. (1989). Vitamin E as an *in vitro* and *in vivo* antioxidant. *Ann. N.Y. Acad. Sci.* **570**: 7–22.

Burton, G.W., Hughes, L., and Ingold, K.U. (1983). Antioxidant activity of phenols related to vitamin E. Are there chain-breaking antioxidants better than α-tocopherol? *J. Am. Chem. Soc.* **105**: 5950–5951.

Busciglio, J., and Yankner, B.A. (1995). Apoptosis and increased generation of reactive oxygen species in Down's syndrome in neurons *in vitro*. *Nature* **378**: 776–779.

Butler, J., and Halliwell, B. (1982). Reaction of iron-EDTA chelates with the superoxide radical. *Arch. Biochem. Biophys.* **218**: 174–178.

Cabelli, D.E., Allen, D., Bielski, B.H.J., and Holcman, J. (1989). The interaction between Cu (I) superoxide dismutase and hydrogen peroxide. *J. Biol. Chem.* **264**: 9967–9971.

Cadenas, E., and Ernster, L. (1990). Quinoid compounds: High-performance liquid chromotography with electrochemical detection. *Methods Enzymol.* **186**: 180–196.

Cadenas, S., Rojas, C., Pérez-Campo, R., López-Torres, M., and Barja, G. (1995). Vitamin E protects guinea pig liver from lipid peroxidation without depressing levels of antioxidants. *Int. J. Biochem. Cell Biol.* **27**: 1175–1181.

Candeias, L.P., Patel, K.B., Stratford, M.R.L., and Wardman, P. (1993). Free hydroxyl radicals are formed on reaction between the neutrophil-derived species superoxide anion and hypochlorous acid. *FEBS Lett.* **333**: 151–153.

Capozzi, G., and Modena, G. (1974). Oxidation of thiols, in: *The Chemistry of Thiol Groups* (Patai, S., ed.), Part 2, pp. 785–839, John Wiley & Sons, London.

Carlioz, A., and Touati, D. (1986). Isolation of superoxide dismutase mutants in *Escherichia coli*: Is superoxide dismutase necessary for aerobic life? *EMBO J.* **5**: 623–630.

Carlsson, L.M., Jonsson, J., Edlund, T., and Marklund, S.L. (1995). Mice lacking extracellular superoxide dismutase are more sensitive to hyperoxia. *Proc. Natl. Acad. Sci. USA* **92**: 6264–6268.

Carlsson, L.M., Marklund, S.L., and Edlund, T. (1996). The rat extracellular superoxide dismutase dimer is converted to a tetramer by the exchange of a single amino acid. *Proc. Natl. Acad. Sci. USA* **93**: 5219–5222.

Carmagnol, F., Sinet, P.M., and Jerome, H. (1983). Selenium-dependent and non-selenium dependent glutathione peroxidases in human tissue extracts. *Biochim. Biophys. Acta* **759**: 49–57.

Carr, A.C., van den Berg, J.J., and Winterbourn, C.C. (1998). Differential reactivities of hypochlorous and hypobromous acids with purified *Escherichia coli* phospholipid: Formation of haloamines and halohydrins. *Biochim. Biophys. Acta* **1392**: 254–264.

Ceballos, I., Delabar, J.M., Nicole, A., Lynch, R.E., Hallewell, R.A., Kamoun, P. and Sinet, P.M. (1988). Expression of transfected human CuZn superoxide dismutase gene in mouse L cells and NS20Y neuroblastoma cells induces enhancement of glutathione peroxidase activity. *Biochim. Biophys. Acta* **949**: 58–64.

Chan, A.C. (1993). Partners in defense, vitamin E and vitamin C. *Can J. Physiol. Pharmacol.* **71**: 725–731.

Chance, B., Greenstein, D.S., and Roughton, F.J.W. (1952). The mechanism of catalase action. I. Steady-state analysis. *Arch. Biochem. Biophys.* **37**: 301–321.

Chance, B., Sies, H., and Boveris, A. (1979). Hydroperoxide metabolism in mammalian organs. *Physiol. Rev.* **59**: 527–605.

Chang, D.J., Ringold, G.M., and Heller, R.A. (1992). Cell killing and induction of manganous superoxide dismutase by tumor necrosis factor-α is mediated by lipoxygenase metabolites of arachidonic acid. *Biochem. Biophys. Res. Commun.* **188**: 538–546.

Chanock, S.J., El Benna, J., Smith, R.M., and Babior, B.M. (1994). The respiratory burst oxidase. *J. Biol. Chem.* **269**: 24519–24522.

Chevion, M. (1988). A site-specific mechanism for free radical induced biological damage: The essential role of redox-active transition metals. *Free Radical Biol. Med.* **5**: 27–37.

Choi, H.J., Kang, S.W., Yang, C.H., Rhee, S.G., and Ryu, S.E. (1988). Crystal structure of a novel human peroxidase enzyme at 2.0 Å resolution. *Nat. Struct. Biol.* **5**: 400–406.

Chung, T.D.Y., and Raymond, K.N. (1993). Lactoferrin: The role of conformational changes in its iron binding and release. *J. Am. Chem. Soc.* **115**: 6765–6768.

Church, S.L., Grant, J.W., Ridnour, L.A., Oberley, L.W., Swanson, P.E., Meltzer, P.S., and Trent, J.M. (1993). Increased manganese superoxide dismutase expression suppresses the malignant phenotype of human melanoma cells. *Proc. Natl. Acad. Sci USA* **90**: 3113–3117.

Claiborne, A., and Fridovich, I. (1979). Purification of the o-dianisidine peroxidase from *Escherichia coli* B. Physiochemical characterization and analysis of its dual catalatic and peroxidatic activities. *J. Biol. Chem.* **254**: 4245–4252.

Claiborne, A., Malinowski, D.P., and Fridovich, I. (1979). Purification and characterization of hydroperoxidase II of *Escherichia coli* B. *J. Biol. Chem.* **254**: 11664–11668.

Clark, R.A., Volpp, B.D., Leidal, K.G., and Nauseef, W.M. (1990). Two cytosolic components of the human neutrophil respiratory burst oxidase translocate to the plasma membrane during cell activation. *J. Clin. Invest.* **85**: 714–721.

Coddeville, B., Stratil, A., Wieruszeksi, J.-M., Strecker, G., Montreuil, J., and Spik, G. (1989). Primary structure of horse serotransferrin glycans. Demonstration that heterogeneity is related to the number of glycans and to the presence of N-acetylneuramininc acid and N-acetyl-4-O- acetylneuraminic acid. *Eur. J. Biochem.* **186**: 583–590.

Coddeville, B., Strecker, G., Wieruszeski, J.-M., Vliegenthart, J.F.G., Van Halbeek, H., Peter-Katalinic, J., Egge, H., and Spik, G. (1992). Heterogeneity of bovine lactotransferrin glycans. Characterization of α-D-Gal*p*-(1→3)-β-D-Gal- and α-NeuAc-(2→6)-β-D-Gal*p*-NAc-(1→4)-β-D-GlcNAc-substituted N-linked glycans. *Carbohydr. Res.* **236**: 145–164.

Coffman, T.J., Cox, C.D., Edeker, B.L., and Britigan, B.E. (1990). Possible role of bacterial siderophores in inflammation. Iron bound to the *Pseudomonas* siderophore pyochelin can function as a hydroxyl radical catalyst. *J. Clin. Invest.* **86**: 1030–1037.

Cohen, G., and Hochstein, P. (1963). Glutathione peroxidase: The primary agent for the elimination of hydrogen peroxide in erythrocytes. *Biochemistry* **2**: 1420–1428.

Cohen, M.S., Britigan, B.E., Hassett, D.J., and Rosen, G.M. (1988a). Do human neutrophils form hydroxyl radical? Evaluation of an unresolved controversy. *Free Radical Biol. Med.* **5**: 81–88.

Cohen, M.S., Britigan, B.E., Hassett, D.J., and Rosen, G.M. (1988b). Phagocytes, O_2 reduction, and hydroxyl radical. *Rev. Infect. Dis.* **10**: 1088–1096.

Cohen, M.S., Mao, J., Rasmussen, G.T., Serody, J.S., and Britigan, B.E. (1992). Interaction of lactoferrin and lipopolysaccharide (LPS): Effects on the antioxidant property of lactoferrin and the ability of LPS to prime human neutrophils for enhanced superoxide formation. *J. Infect. Dis.* **166**: 1375–1378.

Cohn, Z.A., and Morse, S.I. (1960). Functional and metabolic properties of polymorphonuclear leuocytes. II. The influence of a lipopolysaccharide endotoxin. *J. Exp. Med.* **111**: 689–704.

Conger, A.D., and Fairchild, L.M. (1952). Breakage of chromosomes by oxygen.

Proc. Natl. Acad. Sci. USA **38**: 289–299.

Cosgrove, J.P., Borish, E.T., Church, D.F., and Pryor, W.A. (1985). The metal-mediated formation of hydroxyl radical by aqueous extracts from cigarette tar. *Biochem. Biophys. Res. Commun.* **132**: 390–396.

Cox, C.D. (1980). Iron uptake with ferri-pyochelin and ferric citrate by *Pseudomonas aeruginosa*. *J. Bacteriol.* **142**: 581–587.

Cox, C.D., and Adams, P. (1985). Sidero-phore activity of pyoverdin for *Pseudomonas aeruginosa*. *Infect. Immun.* **48**: 130–138.

Cox, C.D., and Graham, R. (1979) Isolation of an iron-binding compound from *Pseudomonas aeruginosa*. *J. Bacteriol.* **137**: 357–364.

Cox, C.D., Rinehart, K.L., Jr., Moore, M.L., and Cook, C.J., Jr., (1981). Pyochelin: Novel structure of an iron-chelating growth promotor for *Pseudomonas aeruginosa*. *Proc. Natl. Acad. Sci. USA* **78**: 4256–4260.

Craven, P.A., DeRubertis, F.R., and Pratt, D.W. (1979). Electron spin resonance study of the role of NO· catalase in the activation of guanylate cyclase by NaN_3 and NH_2OH. Modulation of enzyme responses by heme proteins and their nitrosyl derivatives. *J. Biol. Chem.* **254**: 8213–8222.

Cross, C.E., Halliwell, B., and Borish, E.T. (1987). Oxygen radicals and human disease. *Ann. Intern. Med.* **107**: 526–545.

Csallany, A.S., and Draper, H.H. (1960). Determination of N,N′-diphenyl-p-phenylenediamine in animal tissues. *Proc. Soc. Exp. Biol. Med.* **104**: 739–742.

Czapski, G., and Bielski, B.H.J. (1993). Absorption spectra of the ·OH and $O·^-$ radicals in aqueous solutions. *Radiat. Phys. Chem.* **41**: 503–505.

Das, K.C., Lewis-Molock, Y., and White, C.W. (1997). Elevation of manganese superoxide dismutase gene expression by thioredoxin. *Am. J. Respir. Cell Molec. Biol.* **17**: 713–726.

Daugherty, A., Dunn, J.L., Rateri, D.L., and Heinecke, J.W. (1994). Myeloper-oxidase, a catalyst for lipoprotein oxidation, is expressed in human

atherosclerotic lesions. *J. Clin. Invest.* **94**: 437–444.

DeGraff, W.G., Krishna, M.C., Kaufman, D., and Mitchell, J.B. (1992). Nitrox-ide-mediated protection against X-ray- and neocarzinostatin-induced DNA damage. *Free Radical Biol. Med.* **13**: 479–487.

Deisseroth, A., and Dounce, A.L. (1970). Catalase: Physical and chemical properties, mechanism of catalysis, and physiological role. *Physiol. Rev.* **50**: 319–375.

De Jong, G., Van Dijk, J.P., and Van Eijk, H.G. (1990). The biology of transferrin. *Clin. Chim. Acta* **190**: 1–46.

De Keulenaer, G.W., Alexander, R.W., Ushio-Fukai, M., Ishizaka, N., and Griendling, K.K. (1998). Tumour necrosis factor α activates a p22[phox]-based NADH oxidase in vascular smooth muscle. *Biochem. J.* **329**: 653–657.

De Leo, F.R., Ulman, K.V., Davis, A.R., Jutila, K.L., and Quinn, M.T. (1996). Assembly of the human NADPH oxi-dase involves binding of p67[phox] and flavocytochrome b to a common functional domain in p47[phox]. *J. Biol. Chem.* **271**: 17013–17020.

de Mendez, I., Garrett, M.C., Adams, A.G., and Leto, T.L. (1994). Role of p67[phox] SH3 domains in assembly of the NADPH oxidase system. *J. Biol. Chem.* **269**: 16326–16332.

de Mendez, I., Homayounpour, N., and Leto, T.L. (1997). Specificity of p47[phox] SH3 domain interactions in NADPH oxidase assembly and activation. *Mol. Cell. Biol.* **17**: 2177–2185.

Deng, H.-X., Hentati, A., Tainer, J.A., Iqbal, Z., Cayabyab, A., Hung, W.-Y., Getzoff, E.D., Hu, P., Herzfeldt, B., Roos, R.P., Warner, C., Deng, G., Soriano, W., Smyth, C., Parge, H.E., Ahmed, A., Roses, A.D., Hallewell, R.A., Pericak-Vance, M.A., and Sid-dique, T. (1993). Amyotrophic lateral sclerosis and structural defects in Cu,Zn superoxide dismutase. *Science* **261**: 1047–1051.

Dennery, P.A., Spitz, D.R., Yang, G., Tatarov, A., Lee, C.S., Shegog, M.L., Poss, K.D. (1998). Oxygen toxicity and iron accumulation in the lungs of mice lacking heme oxygenase-2. *J. Clin. Invest.* **101**: 1001–1011.

Dikalov, S., Khramstov, V., and Simmer, G. (1996). Determination of rate constants of the reactions of thiols with superoxide radical by electron paramagnetic resonance: Critical remarks on spectrophotometric approaches. *Arch. Biochem. Biophys.* **326**: 207–218.

Diplock, A.T. (1985). Vitamin E, in: *Fat-Soluble Vitamins* (Diplock, A.T., ed.), pp. 154–224, Technomic Publishing Co., Lancaster, PA.

Doba, T., Burton, G.W., and Ingold, K.U. (1985). Antioxidant and co- antioxidant effect of vitamin C. The effect of vitamin C, either alone or in the presence of vitamin E or a water-soluble vitamin E analog, upon the peroxidation of aqueous multilamellar phospholipid liposomes. *Biochim. Biophys. Acta* **835**: 298–303.

Doba, T., Burton, G.W., Ingold, K.U., and Matsuo, M. (1984). α-Tocopheroxyl decay: Lack of effect of oxygen. *J. Chem. Soc., Chem. Commun.* 461–462.

Domigan, N.M., Charlton, T.S., Duncan, M.W., Winterbourn, C.C., and Kettle, A.J. (1995). Chlorination of tyrosyl residues in peptides by myeloperoxidase and human neutrophils. *J. Biol. Chem.* **270**: 16542–16548.

Dorseuil, O., Reibel, L., Bokoch, G.M., Camonis, J., and Gacon, G. (1996). The rac target NADPH oxidase p67^{phox} interacts preferentially with rac2 rather than rac1. *J. Biol. Chem.* **271**: 83–88.

Dounce, A.L. (1983). A proposed mechanism for the catalatic action of catalase. *J. Theor. Biol.* **105**: 553–567.

Eaton, S.S., Dubach, J., Eaton, G.R., Thurman, G., and Ambruso, D.R. (1990). Electron spin echo envelope modulation evidence for carbonate binding to iron(III) and copper(II) transferrin and lactoferrin. *J. Biol. Chem.* **265**: 7138–7141.

Eisenstein, R.S., Garcia-Mayol, D., Pettingell, W., and Munro, H.N. (1991). Regulation of ferritin and heme oxygenase synthesis in rat fibroblasts by different forms of iron. *Proc. Natl. Acad. Sci. USA* **88**: 688–692.

Eiserich, J.P., Cross, C.E., Jones, A.D., Halliwell, B., and van der Vliet, A. (1996). Formation of nitrating and chlorinating species by reaction of nitrite with hypochlorous acid. A novel mechanism for nitric oxide- mediated protein modification. *J. Biol. Chem.* **271**: 19199–19208.

Eiserich, J.P., Hristova, M., Cross, C.E., Jones, A.D., Freeman, B.A., Halliwell, B.A., and van der Vliet, A. (1998). Formation of nitric oxide-derived inflammatory oxidants by myeloperoxidase in neutrophils. *Nature* **391**: 393–397.

Eling, T.E., Curtis, J.F., Harman, L.S., and Mason, R.P. (1986). Oxidation of glutathione to its thiyl free radical metabolite by prostaglandin H synthase. A potential endogenous substrate for the hydroperoxidase. *J. Biol. Chem.* **261**: 5023–5028.

Ellerby, L.M., Cabelli, D.E., Graden, J.A., and Valentine, J.S. (1996). Copper- zinc superoxide dismutase: Why not pH-dependent? *J. Am. Chem. Soc.* **118**: 6556–6561.

Elroy-Stein, O., and Groner, Y. (1988). Impaired neurotransmitter uptake in PC12 cells overexpressing human Cu/Zn-superoxide dismutase—Implication for gene dosage effects in Down syndrome. *Cell* **52**: 259–267.

Elroy-Stein, O., Bernstein, Y., and Groner, Y. (1986). Overproduction of human Cu/Zn-superoxide dismutase in transfected cells: Extenuation of paraquat-mediated cytotoxicity and enhancement of lipid peroxidation. *EMBO J.* **5**: 615–622.

Emerit, J., and Chaudiere, J. (1989). Free radicals and lipid peroxidation in cell pathology, in: *CRC Handbook of Free Radicals and Antioxidants in Biomedicine* (Miquel, J., Quintanilha, A.T., and Weber, H., eds.), pp. 177–185, CRC Press, Boca Raton, FL.

Erickson, R.W., Malawista, S.E., Garrett, M.C., Van Blaricom, G., Leto, T.L., and Curnutte, J.T. (1992). Identification of a thermolabile component of the human neutrophil NADPH oxidase. A model for chronic granulomatous disease caused by deficiency of the p67-phox cytosolic component. *J. Clin. Invest.* **89**: 1587–1595.

Esworthy, R.S., Chu, F.-F., Geiger, P., Girotti, A.W., and Doroshow, J.H. (1993). Reactivity of plasma glutathione peroxidase with hydroperoxide substrates and glutathione. *Arch. Biochem. Biophys.* **307**: 29–34.

Evans, R.W., and Williams, J. (1978). Studies of the binding of different iron donors to human serum transferrin and isolation of iron-binding fragments from the N-and C-terminal regions of the protein. *Biochem. J.* **173**: 543–552.

Fairlamb, A.H., and Cerami, A. (1985). Identification of a novel, thiol- containing co-factor essential for glutathione reductase enzyme activity in trypanosomatids. *Mol. Biochem. Parasitol.* **14**: 187–198.

Fairlamb, A.H., and Cerami, A. (1992). Metabolism and functions of trypanothione in the kinetoplastida. *Annu. Rev. Microbiol.* **46**: 695–729.

Fairlamb, A.H., Blackburn, P., Ulrich, P., Chait, B.T., and Cerami, A. (1985). Trypanothione: A novel bis(glutathionyl)spermidine cofactor for glutathione reductase in Trypanosomatids. *Science* **227**: 1485–1487.

Farr, S.B., D'Ari, R., and Touati, D. (1986). Oxygen-dependent mutagenesis in *Escherichia coli* lacking superoxide dismutase. *Proc. Natl. Acad. Sci. USA* **83**: 8268–8272.

Fee, J.A., and Bull, C. (1986). Steady-state kinetic studies of superoxide dismutase. Saturative behavior of the copper- and zinc-containing protein. *J. Biol. Chem.* **261**: 13000–13005.

Fee, J.A., and Valentine, J.S. (1977). Chemical and physical properties of superoxide, in: *Superoxide and Superoxide Dismutases* (Michelson, A.M., McCord, J.M., and Fridovich, I., eds.), pp. 19–60, Academic Press New York.

Fenton, H.J.H. (1898). Note on the oxidation of certain acids in presence of iron. *J. Chem. Soc. Proc.* **14**: 119–120.

Fielden, E.M., Roberts, P.B., Bray, R.C., Lowe, D.J., Mautner, G.N., Rotilio, G., and Calabrese, L. (1974). The mechanism of action of superoxide dismutase from pulse radiolysis and electron paramagnetic resonance. Evidence that only half the active sites function in catalysis. *Biochem. J.* **139**: 49–60.

Finan, P., Shimizu, Y., Gout, I., Hsuan, J., Truong, O., Butcher, C., Bennett, P., Waterfield, M.D., and Kellie, S.

(1994). An SH3 domain and proline-rich sequence mediate an interaction between two components of the phagocyte NADPH oxidase complex. *J. Biol. Chem.* **269**: 13752–13755.

Finkelstein, E., Rosen, G.M., and Rauckman, E.J. (1980a). Spin trapping. Kinetics of the reaction of superoxide and hydroxyl radicals with nitrones. *J. Am. Chem. Soc.* **102**: 4994–4999.

Finkelstein, E., Rosen, G.M., and Rauckman, E.J. (1980b). Spin trapping of superoxide and hydroxyl radical: Practical aspects. *Arch. Biochem. Biophys.* **200**: 1–16.

Finkelstein, E., Rosen, G.M., and Rauckman, E.J. (1982). Production of hydroxyl radical by decomposition of superoxide spin trapped adducts. *Mol. Pharmacol.* **21**: 262–265.

Finkelstein, R.A., Sciortino, C.V., and McIntosh, M.A. (1983). Role of iron in microbe–host interactions. *Rev. Infect. Dis.* **5**: 5759–5777.

Fita, I., and Rossman, M.G. (1985). The active center of catalase. *J. Mol. Biol.* **185**: 21–37.

Flohé, L. (1982). Glutathione peroxidase brought into focus, in: *Free Radicals in Biology* (Pryor, W.A., ed.), Vol. V, pp. 223–254, Academic Press, New York.

Flohé, L., and Günzler, W.A. (1974). Glutathione peroxidase, in: *Glutathione* (Flohé, L., Benohr, H.C., Sies, H., Waller, H.D., and Wendel, A., eds.), pp. 133–145, Academic Press, New York.

Flohé, L., Günzler, W.A., Jung, G., Schaich, E., and Schneider, F. (1971). Glutathion-peroxidase, II. Substratspezifität und hemmbarkeit durch substratanaloge. *Hoppe-Seyler's Z. Physiol. Chem.* **352**: 159–169.

Flohé, L., Günzler, W.A., and Shock, H.H. (1973). Glutathione peroxidase: A selenoenzyme. *FEBS Lett.* **32**: 132–134.

Flohé, L., Loschen, G., Günzler, W.A., and Eichele, E. (1972). Glutathione peroxidase, V. The kinetic mechanism. *Hoppe-Seyler's Z. Physiol. Chem.* **353**: 987–999.

Foley, A.A., and Bates, G.W. (1988). The influence of inorganic anions on the formation and stability of Fe^{3+} transferrin-anion complexes. *Biochim. Biophys. Acta* **965**: 154–162.

Folz, R.J., Guan, J., Seldin, M.F., Oury, T.D., Enghild, J.J., and Crapo, J.D. (1997). Mouse extracellular superoxide dismutase: Primary structure, tissue-specific gene expression, chromosomal localization, and lung *in situ* hybridization. *Am. J. Respir. Cell Molec. Biol.* **17**: 393–403.

Fong, K.L., McCay, P.B., Poyer, J.L., Misra, H.P., and Keele, B.B. (1976). Evidence for superoxide-dependent reduction of Fe^{3+} and its role in enzyme-generated hydroxyl radical formation. *Chem.-Biol. Interact.* **15**: 77–89.

Forman, H.J., and Fridovich, I. (1973). Superoxide dismutase: A comparison of rate constants. *Arch. Biochem. Biophys.* **158**: 396–400.

Foroozan, R., Ruedi, J.M., and Babior, B.M. (1992). The reduction of cytochrome b_{558} and the activity of the respiratory burst oxidase from human neutrophils. *J. Biol. Chem.* **267**: 24400–24407.

Forrest, C.B., Forehand, J.R., Axtell, R.A., Roberts, R.L., and Johnston, R.B., Jr. (1988). Clinical features and current management of chronic granulomatous disease. *Hematol./Oncol. Clin. North Am.* **2**: 253–266.

Forstrom, J.W., Stults, F.H., and Tappel, A.L. (1979). Rat liver cytosolic glutathione peroxidase: Reactivity with linoleic acid hydroperoxide and cumene hydroperoxide. *Arch. Biochem. Biophys.* **193**: 51–55.

Forstrom, J.W., Zakowski, J.J., and Tappel, A.L. (1978). Identification of the catalytic site of rat liver glutathione peroxidase as selenocysteine. *Biochemistry* **17**: 2639–2644.

Frederiks, W.M., and Bosch, K.S. (1997). Localization of superoxide dismutase activity in rat tissues. *Free Radical Biol. Med.* **22**: 241–248.

Freeman, J.L., Abo, A., and Lambeth, J.D. (1996). Rac "insert region" is a novel effector region that is implicated in the activation of NADPH oxidase, but not PAK65. *J. Biol. Chem.* **271**: 19794–19801.

Fridovich, I. (1974). Evidence for the symbiotic origin of mitochondria. *Life Sci.* **14**: 819–826.

Fridovich, I. (1978). The biology of oxygen radicals. *Science* **201**: 875–880.

Fridovich, I. (1982). Oxygen toxicity in prokaryotes: The importance of superoxide dismutase, in: *Superoxide Dismutase* (Oberley, L.W., ed.), Vol. I, pp. 79–88, CRC Press, Boca Raton, FL.

Fridovich, I. (1995). Superoxide radical and superoxide dismutases. *Annu. Rev. Biochem.* **64**: 97–112.

Fridovich, I. (1997). Superoxide anion radical ($O_2 \cdot ^-$), superoxide dismutases, and related matters. *J. Biol. Chem.* **272**: 18515–18517.

Fridovich, I., and Handler, P. (1958). Xanthine oxidase. IV. Participation of iron in internal electron transport. *J. Biol. Chem.* **233**: 1581–1585.

Fridovich, I., and Handler, P. (1961). Detection of free radicals generated during enzymatic oxidations by the initiation of sulfite oxidation. *J. Biol. Chem.* **236**: 1836–1840.

Fujii, H., Johnson, M.K., Finnegan, M.G., Miki, T., Yoshida, L.S., and Kakinuma, K. (1995). Electron spin resonance studies on neutrophil cytochrome b_{558}. Evidence that low-spin heme iron is essential for $O_2 \cdot ^-$ generating activity. *J. Biol. Chem.* **270**: 12685–12689.

Fukui, T., Ishizaka, N., Rajagopalan, S., Laursen, J.B., Capers, Q., IV, Taylor, W.R., Harrison, D.G., de Leon, H., Wilcox, J.N., and Griendling, K.K. (1997). p22phox mRNA expression and NADPH oxidase activity are increased in the aortas from hypertensive rats. *Circ. Res.* **80**: 45–51.

Galley, H.F., Davies, M.J., and Webster, N.R. (1996). Ascorbyl radical formation in patients with sepsis: Effect of ascorbate loading. *Free Radical Biol. Med.* **20**: 139–143.

Gannon, D.E., Varani, J., Phan, S.H., Ward, J.H., Kaplan, J., Till, G.O., Simon, R.H., Ryan, U.S., and Ward, P.A. (1987). Source of iron in neutrophil-mediated killing of endothelial cells. *Lab. Invest.* **57**: 37–44.

Gebicki, J.M., and Bielski, B.H.J. (1981). Comparisions of the capacities of the perhydroxyl and the superoxide radicals to initiate chain oxidation of linoleic acid. *J. Am. Chem. Soc.* **103**: 7020–7022.

Gerschman, R., Gilbert, D.L., Nye, S.W., Dwyer, P., and Fenn, W.O. (1954).

Oxygen poisoning and x-irradiation: A mechanism in common. *Science* **119**: 623–626.

Gerstein, M., Anderson, B.F., Norris, G.E., Baker, E.N., Lesk, A.M., and Chothia, C. (1993). Domain closure in lactoferrin. Two hinges produce a see-saw motion between alternative close-packed interfaces. *J. Mol. Biol.* **234**: 357–372.

Getzoff, E.D., Cabelli, D.E., Fisher, C.L., Parge, H.E., Viezzoli, M.S., Banci, L., and Hallewell, R.A. (1992). Faster superoxide dismutase mutants designed by enhancing electrostatic guidance. *Nature* **358**: 347–351.

Getzoff, E.D., Tainer, J.A., Weiner, P.K., Kollman, P.A., Richardson, J.S., and Richardson, D.C. (1983). Electrostatic recognition between superoxide and copper, zinc superoxide dismutase. *Nature* **306**: 287–290.

Gilbert, D.L. (1981). Perspective on the history of oxygen and life, in: *Oxygen and Living Processes. An Interdisciplinary Approach* (Gilbert, D.L., ed.), pp. 1–43, Springer-Verlag, New York.

Girotti, A.W. (1985). Mechanisms of lipid peroxidation. *J. Free Radical Biol. Med.* **1**: 87–95.

Glascott, P.A., Jr., Gilfor, E., Serroni, A., and Farber, J.L. (1996). Independent antioxidant action of vitamins E and C in cultured rat hepatocytes intoxicated with allyl alcohol. *Biochem. Pharmacol.* **52**: 1245–1252.

Goldstein, S., Meyerstein, D., and Czapski, G. (1993). The Fenton reagents. *Free Radical Biol. Med.* **15**: 435–445.

Gottsch, J.D., Pou, S., Bynoe, L.A., and Rosen, G.M. (1990). Hematogenous photosensitization. A mechanism for the development of age-related macular degeneration. *Invest. Ophthalmol. Vis. Sci.* **31**: 1674–1682.

Grace, S.C. (1990). Phylogenetic distribution of superoxide dismutase supports an endosymbiotic origin for chloroplasts and mitochondria. *Life Sci.* **47**: 1875–1886.

Graf, E., Mahoney, J.R., Bryand, R.G., and Eaton, J.W. (1984). Iron-catalyzed hydroxyl radical formation. *J. Biol. Chem.* **259**: 3620–3624.

Gralla, E.B., and Valentine, J.S. (1991). Null mutants of *Saccharomyces cerevisiae* Cu,Zn superoxide dismutase: Characterization and spontaneous mutation rates. *J. Bacteriol.* **173**: 5918–5920.

Granger, D.N., Rutili, G., and McCord, J.M. (1981). Superoxide radicals in feline intestinal ischemia. *Gastroenterology* **81**: 22–29.

Green, M.R., Hill, H.A.O., Okolow-Zubkowska, M.J., and Segal, A.W. (1979). The production of hydroxyl and superoxide radicals by stimulated human neutrophils—measurements by EPR spectroscopy. *FEBS Lett.* **100**: 23–26.

Greenlee, L., Fridovich, I., and Handler, P. (1962). Chemiluminescence induced by operation of iron-flavoproteins. *Biochemistry* **1**: 779–783.

Gregory, E.M., and Fridovich, I. (1973). The induction of superoxide dismutase by molecular oxygen. *J. Bacteriol.* **114**: 543–548.

Gregory, E.M., and Fridovich, I. (1974). Visualization of catalase on acrylamide gels. *Anal. Biochem.* **58**: 57–62.

Grierson, L., Hildenbrandt, K., and Bothe, E. (1992). Intramolecular transformation reaction of the glutathione thiyl radical into a non-sulphur-centred radical: A pulse-radiolysis and EPR study. *Int. J. Radiat. Biol.* **62**: 265–277.

Griffith, T.M., Edwards, D.H., Lewis, M.J., Newby, A.C., and Henderson, A.H. (1984). The nature of endothelium-derived vascular relaxant factor *Nature* **308**: 645–647.

Grinstein, S., and Furuya, W. (1988). Assessment of Na^+–H^+ exchange activity in phagosomal membranes of human neutrophils. *Am. J. Physiol.* **254**: C272–C285.

Griswold, C.M., Matthews, A.L., Bewley, K.E., and Mahaffey, J.W. (1993). Molecular characterization and rescue of acatalasemic mutants of *Drosophilia melanogaster*. *Genetics* **134**: 781–788.

Grossmann, A., and Wendel, A. (1983). Non-reactivity of the selenoenzyme glutathione peroxidase with enzymatically hydroperoxidized phospholipids. *Eur. J. Biochem.* **135**: 549–552.

Grossmann, J.G., Neu, M., Evans, R.W., Lindley, P.F., Appel, H., and Hasnain, S.S. (1993). Metal-induced conformational changes in transferrins. *J. Mol. Biol.* **229**: 585–590.

Grossmann, J.G., Neu, M., Pantos, E., Schwab, F.J., Evans, R.W., Townes-Andrews, E., Lindley, P.F., Appel, H., Thies, W.-G., and Hasnain, S.S. (1992). X-ray solution scattering reveals conformational changes upon iron uptake in lactoferrin, serum and ovo-transferrins. *J. Mol. Biol.* **225**: 811–819.

Guengerich, F.P., and Macdonald, T.L. (1984). Chemical mechanisms of catalysis by cytochrome P-450: A unified view. *Acc. Chem. Res.* **17**: 9–16.

Guengerich, F.P, Vaz, A.D.N., Raner, G.N., Pernecky, S.J., and Coon, M.J. (1997). Evidence for a role of a perferryl-oxygen complex, FeO^{3+}, in the N-oxygenation of amines by cytochrome P-450 enzymes. *Mol. Pharmacol.* **51**: 147–151.

Günzler, W.A., Hartmut, V., Müller, I., and Flohé, L. (1972). Glutathion- peroxidase, VI. Die reaktion der glutathion-peroxidase mit verschiedenen hydroperoxiden. *Hoppe-Seyler's Z. Physiol. Chem.* **353**: 1001–1004.

Gurney, M.E., Pu, H., Chiu, A.Y., Dal Canto, M.C., Polchow, C.Y., Alexander, D.D., Caliendo, J., Hentati, A., Kwon, Y.W., Deng, H.-X., Chen, W., Zhai, P., Sufit, R.L., and Siddique, T. (1994). Motor neuron degeneration in mice that express a human Cu,Zn superoxide dismutase mutation. *Science* **264**: 1772–1775.

Gutteridge, J.M.C. (1986). Iron promoters of the Fenton reaction and lipid peroxidation can be released from haemoglobin by peroxides. *FEBS Lett.* **201**: 291–295.

Gutteridge, J.M.C. (1990). Superoxide-dependent formation of hydroxyl radicals from ferric-complexes and hydrogen peroxide: An evaluation of fourteen iron chelators. *Free Radical Res. Commun.* **9**: 119–125.

Gutteridge, J.M.C., Paterson, S.K., Segal, A.W., and Halliwell, B. (1981). Inhibition of lipid peroxidation by the iron-binding protein lactoferrin. *Biochem. J.* **199**: 259–261.

Gutteridge, J.M.C., Rowley, D.A., Griffiths, E., and Halliwell, B. (1985). Low- molecular-weight iron complexes and oxygen radical reactions in idiopathic haemochromatosis. *Clin. Sci.* **68**: 463–467.

Haber, F., and Weiss, J. (1934). The catalytic decomposition of hydrogen peroxide by iron salts. *Proc. R. Soc. Lond., Ser. A* **147**: 332–351.

Halliwell, B. (1989). Protection against tissue damage *in vivo* by desferrioxamine: What is its mechanism of action. *Free Radical Biol. Med.* **7**: 645–651.

Halliwell, B., and Gutteridge, J.M.C. (1981). Formation of a thiobarbituric acid-reactive substance from deoxyribose in the presence of iron salts. The role of superoxide and hydroxyl radicals. *FEBS Lett.* **128**: 347–351.

Halliwell, B., and Gutteridge, J.M.C. (1984). Oxygen toxicity, oxygen radicals, transition metals and disease. *Biochem. J.* **219**: 1–14.

Halliwell, B., and Gutteridge, J.M.C. (1985). The role of transition metals in superoxide-mediated toxicity, in: *Superoxide Dismutase* (Oberley, L.W., ed.), Vol. III, pp. 45–82, CRC Press, Boca Raton, FL.

Halliwell, B., and Gutteridge, J.M.C. (1986). Oxygen free radicals and iron in relation to biology and medicine: Some problems and concepts. *Arch. Biochem. Biophys.* **246**: 501–514.

Halliwell, B., and Gutteridge, J.M.C. (1990). Role of free radicals and catalytic metals in human disease: An overview. *Methods Enzymol.* **186**: 1–85.

Halliwell, B., Aruoma, O.I., Mufti, G., and Bomford, A. (1988a). Bleomycindetectable iron in serum from leukaemic patients before and after chemotherapy: Therapeutic implications for treatment with oxidant-generating drugs. *FEBS Lett.* **241**: 202–204.

Halliwell, B., Aruoma, O.I., Wasil, M., and Gutteridge, J.M.C. (1988b). The resistance of transferrin, lactoferrin and caeruloplasmin to oxidative damage. *Biochem. J.* **256**: 311.

Hamazaki, S., Okada, S., Li, J.-L., Toyokuni, S., and Midorikawa, O. (1989). Oxygen reduction and lipid peroxidation by iron chelates with special reference to ferric nitrilotriacetate. *Arch. Biochem. Biophys.* **272**: 10–17.

Harel, S., and Kanner, J. (1989). Haemoglobin and myoglobin as inhibitors of hydroxyl radical generation in a model system of "iron redox" cycle. *Free Radical Res. Commun.* **6**: 1–10.

Harel, S., Salan, M.A., and Kanner, J. (1988). Iron release from metmyoglobin, methaemoglobin and cytochrome c by a system generating hydrogen peroxide. *Free Radical Res. Commun.* **5**: 11–19.

Harman, D. (1956). Aging: A theory based on free radical and radiation chemistry. *J. Gerontol.* **11**: 298–300.

Harman, L.S., Carver, D.K., Schreiber, J., and Mason, R.P. (1986). One- and two-electron oxidation of reduced glutathione by peroxidase. *J. Biol. Chem.* **261**: 1642–1648.

Harris, C.A., Derbin, K.S., Hunte-McDonough, B., Krauss, M.R., Chen, K.T., Smith, D.N., and Epstein, L.B. (1991). Manganese superoxide dismutase is induced by IFN-γ in multiple cell types. Synergistic induction by IFN-γ and tumor necrosis factor or IL-1. *J. Immunol.* **147**: 149–154.

Harris, W.R. (1985). Thermodynamics of anion binding to human serum transferrin. *Biochemistry* **24**: 7412–7418.

Hassan, H.M. (1988). Biosynthesis and regulation of superoxide dismutases. *Free Radical Biol. Med.* **5**: 377–385.

Hassan, H.M., and Fridovich, I. (1977). Enzymatic defenses against toxicity of oxygen and of streptonigrin in *Escherichia coli* K12. *J. Bacteriol.* **129**: 1574–1583.

Hassan, H.M., and Fridovich, I. (1978). Regulation of the synthesis of catalase and peroxidase in *Escherichica coli*. *J. Biol. Chem.* **253**: 6445–6450.

Hassett, D.J., and Cohen, M.S. (1989). Bacterial adaptation to oxidative stress: Implications for pathogenesis and interaction with phagocytic cells. *FASEB J.* **3**: 2574–2582.

Hatchikian, E.C., and Henry, T.A. (1977). An iron-containing superoxide dismutase from the strict anaerobe *Desulfovibrio desulfuricans* (Norway 4). *Biochimie* **59**: 153–161.

Hawkins, H.L., and Davies, M.J. (1998). Hypochlorite-induced damage to proteins: Formation of nitrogen-centered radicals from lysine residues and their role in protein fragmentation. *Biochem. J.* **332**: 617–625.

Hayakawa, T., Suzuki, K., Suzuki, S., Andrews, P.C., and Babior, B.M. (1986). A possible role for protein phosphorylation in the activation of the respiratory burst in human neutrophils. Evidence from studies with cells from patients with chronic granulomatous disease. *J. Biol. Chem.* **261**: 9109–9115.

Hazen, S.L., and Heinecke, J.W. (1997). 3-Chlorotyrosine, a specific marker of myeloperoxidase-catalyzed oxidation, is markedly elevated in low density lipoprotein isolated from human atherosclerotic intima. *J. Clin. Invest.* **99**: 2075–2081.

Hazen, S.L., d'Avignon, A., Anderson, M.M., Hsu, F.F., and Heinecke, J.W. (1998a). Human neutrophils employ the myeloperoxidase-hydrogen peroxide-chloride system to oxidize α-amino acids to a family of reactive aldehydes: Mechanistic studies identifying labile intermediates along the reaction pathway. *J. Biol. Chem.* **273**: 4997–5005.

Hazen, S.L., Gaut, J.P., Hsu, F.F., Crowley, J.R., d'Avignon, A., and Heinecke, J.L. (1997). *p*-Hydroxyphenylacetaldehyde, the major product of L-tyrosine oxidation by the myeloperoxidase–H_2O_2–chloride system of phagocytes, covalently modifies ε-amino groups of protein lysine residues. *J. Biol. Chem.* **272**: 16990–16998.

Hazen, S.L., Hsu, F.F., d'Avignon, A., and Heinecke, S.L. (1998b). Human neutrophils employ myeloperoxidase to convert α-amino acids to a battery of reactive aldehydes: A pathway for aldehyde generation at sites of inflammation. *Biochemistry* **37**: 6864–6873.

Hazen, S.L., Hsu, F.F., and Heinecke, J.W. (1996). *p*-Hydroxyphenylacetaldehyde is the major product of L-tyrosine oxidation by activated phagocytes: A chloride-dependent mechanism for the conversion of free amino acids into reactive aldehydes by myeloperoxidase. *J. Biol. Chem.* **271**: 1861–1867.

Heffner, J.E., and Repine, J.E. (1989). Pulmonary strategies of antioxidant defense. *Am. Rev. Respir. Dis.* **140**: 531–554.

Heinecke, J.W., Li, W., Daehnke, H.L., III, and Goldstein, J.A. (1993a). Dityrosine, a specific marker of oxidation, is synthesized by the myeloperoxidase-hydrogen peroxide system of human neutrophils and macrophages. *J. Biol. Chem.* **268**: 4069–4077.

Heinecke, J.W., Li, W., Francis, G.A., and Goldstein, J.A. (1993b). Tyrosyl radical generated by myeloperoxidase catalyzes the oxidative cross- linking of proteins. *J. Clin. Invest.* **91**: 2866–2872.

Henry, L.E.A., and Hall, D.O. (1977). Superoxide dismutases in green algae: An evolutionary survey, in: *Photosynthetic Organelles: Structure and Function* (Miyachi, S., Katoh, S., Fujita, Y., and Shibata, K., eds.), pp. 377–382, Japanese Society of Plant Physiologists and Center for Academic Publications, Japan.

Hensley, K., Carney, J.M., Mattson, M.P., Aksenova, M., Harris, M., Wu, J.F., Floyd, R.A., and Butterfield, D.A. (1994). A model for β-amyloid aggregation and neutrotoxicity based on free radical generation by the peptide: Relevance to Alzheimer disease. *Proc. Natl. Acad. Sci. USA* **91**: 3270–3274.

Hershko, C., Link, G., Tzahor, M., Kaltwasser, J.P., Athias, P., Grynberg, A., and Pinson, A. (1993). Anthracycline toxicity is potentiated by iron and inhibited by deferoxamine: Studies in rat heart cells in culture. *J. Lab. Clin. Med.* **122**: 245–251.

Heyworth, P.G., and Badwey, J.A. (1990). Continuous phosphorylation of both the 47 and 49 kDa proteins occurs during superoxide production by neutrophils. *Biochim. Biophys. Acta* **1052**: 299–305.

Heyworth, P.G., Bohl, B.P., Bokoch, G.M., and Curnutte, J.T. (1994). Rac translocates independently of the neutrophil NADPH oxidase components p47^{phox} and p67^{phox}. Evidence for its interaction with flavocytochrome b$_{558}$. *J. Biol. Chem.* **269**: 30749–30752.

Heyworth, P.G., Curnutte, J.T., Nauseef, W.M., Volpp, B.D., Pearson, D.W., Rosen, H., and Clark, R.A. (1991). Neutrophil nicotinamide adenine dinucleotide phosphate oxidase assembly. Translocation of p47-*phox* and p67-*phox* requires interaction between p47-*phox* and cytochrome b$_{558}$. *J. Clin. Invest.* **87**: 352–356.

Hill, H.A.O. (1978). The superoxide ion and the toxicity of molecular oxygen, in: *New Trends in Bioinorganic Chemistry* (Williams, R.J.P., and Da Silva, J.R.F., eds.), pp. 173–208, Academic Press, New York.

Hill, H.A.O. (1979). The chemistry of dioxygen and its reduction products, in: *Oxygen Free Radicals and Tissue Damage* (Fitzsimons, D.W., ed.), Ciba Foundation Series 65 (new series), pp. 5–17, Excerpta Medica, Amsterdam.

Hinson, J.A., Monks, T.J., Hong, M., Highet, R.J., and Pohl, L.R. (1982). 3-(Glutathion-S-yl)acetaminophen: A biliary metabolite of acetaminophen. *Drug Metab. Dispos.* **10**: 47–50.

Hjalmarsson, S., Marklund, S.L., Engstrom, A., and Edlung, T. (1987). Isolation and sequence of complementary DNA encoding human extracellular superoxide dismutase. *Proc. Natl. Acad. Sci. USA* **84**: 6340–6344.

Hodgson, E.K., and Fridovich, I. (1975a). The interaction of bovine erythocyte superoxide dismutase with hydrogen peroxide: Inactivation of the enzyme. *Biochemistry* **14**: 5294–5298.

Hodgson, E.K., and Fridovich, I. (1975b). The interaction of bovine erythocyte superoxide dismutase with hydrogen peroxide: Chemiluminescence and peroxidation. *Biochemistry* **14**: 5299–5303.

Holmes, B., Page A.R., and Good, R.A. (1967). Studies of the metabolic activity of leukocytes from patients with a genetic abnormality of a phagocyte function. *J. Clin. Invest.* **46**: 1422–1432.

Horton, P.J., Borders, C.L., Jr., and Beyer, W.F., Jr. (1989). Inactivation of human arginine-143, lysine-143, and isoleucine-143 Cu,Zn superoxide dismutases by hydrogen peroxide: Multiple mechanisms for inactivation. *Arch. Biochem. Biophys.* **269**: 114–124.

Hurst, J.R., and Barrette, W.C., Jr. (1989). Leukocyte oxygen activation and microbial oxidative toxins. *CRC Crit. Rev. Biochem. Molec. Biol.* **24**: 271–328.

Hurst, J.K., Barrette, W.C., Jr., Michel, B.R., and Rosen, H. (1991). Hypochlorous acid and myeloperoxidase-catalyzed oxidation of iron-sulfur clusters in bacterial respiratory dehydrogenase. *Eur. J. Biochem.* **202**: 1275–1282.

Iacopetta, B.J., and Morgan, E.H. (1983). The kinetics of transferrin endocytosis and iron uptake from transferrin in rabbit reticulocytes. *J. Biol. Chem.* **258**: 9108–9115.

Ignarro, L.J., Harbison, R.G., Wood, K.S., and Kadowitz, P.J. (1986). Activation of purified soluble guanylate cyclase by endothelium-derived relaxing factor from intrapulmonary artery and vein: Stimulation by acetylcholine, bradykinin and arachidonic acid. *J. Pharmacol. Exp. Ther.* **237**: 893–900.

Imlay, J.A., and Fridovich, I. (1992). Suppression of oxidative envelope damage by pseudoreversion of a superoxide dismutase-deficient mutant of *Escherichia coli. J. Bacteriol.* **174**: 953–961.

Imlay, J.A., and Linn, S. (1988). DNA damage and oxygen radical toxicity. *Science* **240**: 1302–1309.

Imlay, J.A., Chin, S.M., and Linn, S. (1988). Toxic DNA damage by hydrogen peroxide through the Fenton reaction *in vivo* and *in vitro*. *Science* **240**: 640–642.

Inanami, O., Johnson, J.L., McAdara, J.K., Benna, J.E., Faust, L.P., Newburger, P.E., and Babior, B.M. (1998). Activation of the leukocyte NADPH oxidase by phorbol ester requires the phosphorylation of p47PHOX on serine 303 or 304. *J. Biol. Chem.* **273**: 9539–9543.

Isogai, Y., Iizuka, T., and Shiro, Y. (1995). The mechanism of electron donation to molecular oxygen by phagocytic cytochrome b$_{558}$. *J. Biol. Chem.* **270**: 7853–7857.

Iyer, G.Y.N., Islam, M.F., and Quastel, J.H. (1961). Biochemical aspects of phagocytosis. *Nature* **192**: 535–541.

Jacob, G.S., and Orme-Johnson, W.H. (1979). Catalase of *Neurospora crassa*. 1. Induction, purification and physical properties. *Biochemistry* **18**: 2967–2975.

Janero, D.R. (1990). Malondialdehyde and thiobarbituric acid-reactivity as diagnostic indices of lipid peroxidation and peroxidative tissue injury. *Free Radical Biol. Med.* **9**: 515–540.

Jin, D.-Y., Chae, H.Z., Rhee, S.G., and Jeang, K.-T. (1997). Regulatory role for a novel human thioredoxin peroxidase in NF-κB activation. *J. Biol. Chem.* **272**: 30952–30961.

Johnston, W.D., Jr., Mei, B., and Cohn, Z.A. (1977). The separation, long-term cultivation, and maturation of the human monocyte. *J. Exp. Med.* **146**: 1613–1626.

Jones, S.A., Hancock, J.T., Jones, O.T., Neubauer, A., and Topley, N. (1995). The expression of NADPH oxidase components in human glomerular mesangial cells: Detection of protein and mRNA for p47phox, p67phox, and p22phox. *J. Am. Soc. Nephrol.* **5**: 1483–1491.

Jones, S.A., O'Donnell, V.B., Wood, J.D., Broughton, J.P., Hughes, E.J., and Jones, O.T.G. (1996). Expression of phagocyte NADPH oxidase components in human endothelial cells. *Am. J. Physiol.* **271**: H1626–H1634.

Joshi, P., and Dennis, P.P. (1993). Structure, function and evolution of the family of superoxide dismutase proteins from halophilic archaebacteria. *J. Bacteriol.* **175**: 1572–1579.

Juckett, M., Zheng, Y., Yuan, H., Pastor, T., Antholine, W., Weber M., and Vercellotti, G. (1998). Heme and the endothelium. Effects of nitric oxide on catalytic iron and heme degradation by heme oxygenase. *J. Biol. Chem.* **273**: 23388–23397.

Kanematsu, S., and Asada, K. (1978a). Superoxide dismutase from an aerobic photosynthetic bacterium, *Chromatium vinosum. Arch. Biochem. Biophys.* **185**: 473–482.

Kanematsu, S., and Asada, K. (1978b). Crystalline ferric superoxide dismutase from an anaerobic green sulfur bacterium, *Chlorobium thiosulfatophilum. FEBS Lett.* **91**: 94–98.

Kang, S.W., Baines, I.C., and Rhee, S.G. (1998). Characterization of a mammalian peroxiredoxin that contains one conserved cysteine. *J. Biol. Chem.* **273**: 6303–6311.

Karlsson, K., and Marlund, S.L. (1987). Heparin-induced release of extracellular superoxide dismutase to human blood plasma. *Biochem. J.* **242**: 55–59.

Karlsson, K., and Marklund, S.L. (1988a). Extracellular superoxide dismutase in the vascular system of mammals. *Biochem. J.* **255**: 223–228.

Karlsson, K., and Marklund, S.L. (1988b). Plasma clearance of human extracellular-superoxide dismutase C in rabbits. *J. Clin. Invest.* **82**: 762–766.

Karlsson, K., and Marklund, S.L. (1989). Binding of human extracellular-superoxide dismutase C to cultured cell lines and to blood cells. *Lab. Invest.* **60**: 659–666.

Karlsson, K., Lindahl, U., and Marklund, S.L. (1988). Binding of human extracellular superoxide dismutase C to sulphated glucosaminoglycans. *Biochem. J.* **256**: 29–33.

Karnovsky, M.L., and Badwey, J.A. (1983). Determinants of the production of active oxygen species by granulocytes and macrophages. *J. Clin. Chem. Clin. Biochem.* **21**: 545–553.

Katoh, S., Toyama, J., Kodama, I., Akita, T., and Abe, T. (1992). Deferoxamine, an iron chelator, reduces myocardial injury and free radical generation in isolated neonatal rabbit hearts subjected to global ischaemia–reperfusion. *J. Mol. Cell. Cardiol.* **24**: 1267–1275.

Kelm, M., and Schrader, J. (1990). Control of coronary vascular tone by nitric oxide. *Circ. Res.* **66**: 1561–1575.

Kettle, A.J. (1996). Neutrophils convert tyrosyl residues in albumin to chlorotyrosine. *FEBS Lett.* **379**: 103–106.

Keyer, K., and Imlay, J.A. (1996). Superoxide accelerates DNA damage by elevating free-iron levels. *Proc. Natl. Acad. Sci. USA* **93**: 13635–13640.

Keyer, K., Gort, A.S., and Imlay, J.A. (1995). Superoxide and the production of oxidative DNA damage. *J. Bacteriol.* **177**: 6782–6790.

Keyse, S.M., and Tyrrell, R.M. (1989). Heme oxygenase is the major 32-kDa stress protein induced in human skin fibroblasts by UVA radiation, hydrogen peroxide, and sodium arsenite. *Proc. Natl. Acad. Sci. USA* **86**: 99–103.

Keyse, S.M., Applegate, L.A., Tromvoukis, Y., and Tyrrell, R.M. (1990). Oxidant stress leads to transcriptional activation of the human heme oxygenase gene in cultured skin fibroblasts. *Mol. Cell. Biol.* **10**: 4967–4969.

Khangulov, S.V., Barynin, V.V., and Antonyuk-Barynina, S.V. (1990). Manganese-containing catalase from *Thermus thermophilus* peroxide-induced redox transformation of manganese ions in presence of specific inhibitors of catalase activity. *Biochim. Biophys. Acta* **1020**: 25–33.

Kharitonov, V.G., Sundquist, A.R., and Sharma, V.S. (1994). Kinetics of nitric oxide autoxidation in aqueous solution. *J. Biol. Chem.* **269**: 5881–5883.

Kihara, T., Sakata, S., and Ikeda, M. (1995). Direct detection of ascorbyl radical in experimental brain injury: Microdialysis and an electron spin resonance spectroscopic study. *J. Neurochem.* **65**: 282–286.

Kim, I.H., Kim, K., and Rhee, S.G. (1989). Induction of an antioxidant protein of *Saccharomyces cerevisiae* by O_2, Fe^{3+}, or 2-mercaptoethanol. *Proc. Natl. Acad. Sci. USA* **86**: 6018–6022.

Klebanoff, S.J. (1993). Reactive nitrogen intermediates and antimicrobial activity: Role of nitrite. *Free Radical Biol. Med.* **14**: 351–360.

Klebanoff, S.J., and Hamon, C.B. (1972). Role of myeloperoxidase-mediated anti-microbial systems in intact leukocytes. *J. Reticuloendothel. Soc.* **12**: 170–196.

Klebanoff, S.J., and Rosen, H. (1978). Ethylene formation by polymorphonuclear leukocytes. Role of myeloperoxidase. *J. Exp. Med.* **148**: 490–506.

Kleinberg, M.E., Malech, H.L., and Rotrosen, D. (1990). The phagocyte 47-kilodalton cytosolic oxidase protein is an early reactant in activation of the respiratory burst. *J. Biol. Chem.* **265**: 15577–15583.

Klug, D., Fridovich, I., and Rabani, J. (1972). A direct demonstration of the catalytic action of superoxide through the use of pulse radiolysis. *J. Biol. Chem.* **247**: 4839–4842.

Klug-Roth, D., Fridovich, I., and Rabani, J. (1973). Pulse radiolytic investigation of superoxide catalyzed disproportionation. Mechanism for bovine superoxide dismutase. *J. Am. Chem. Soc.* **95**: 2786–2790.

Kojima, N., and Bates, G.W. (1981). The formation of Fe^{3+}-transferrin-CO_3^{2-} via the binding and oxidation of Fe^{2+}. *J. Biol. Chem.* **256**: 12034–12039.

Kono, Y., and Fridovich, I. (1983). Isolation and characterization of the pseudocatalase of *Lactobacillus plantarum*. *J. Biol. Chem.* **258**: 6015–6019.

Koppenol, W.H. (1981). The physiologic role of the charge distribution on superoxide dismutase, in: *Oxygen and Oxy-*

Radicals in Chemistry and Biology (Rodgers, M.A.J., and Powers, E.L., eds.), pp. 671–674, Academic Press, New York.

Koppenol, W.H. (1993a). A thermodynamic appraisal of the radical sink hypothesis. *Free Radical Biol. Med.* **14**: 91–94.

Koppenol, W.H. (1993b). The centennial of the Fenton reaction. *Free Radical Biol. Med.* **15**: 645–651.

Korthuis, R.J., Carden, D.L., and Granger, D.N. (1992). Cellular dysfunction induced by ischemia/reperfusion: Role of reactive oxygen metabolites and granulocytes, in: *Biological Consequences of Oxidative Stress* (Spatz, L., and Bloom, D., eds.), pp. 50–77, Oxford Press, London.

Koshkin, V., Lotan, O., and Pick, E. (1996). The cytosolic component p47^{phox} is not a *sine qua non* participant in the activation of NADPH oxidase but is required for optimal superoxide production. *J. Biol. Chem.* **271**: 30326–30329.

Kosower, N.S., and Kosower, E.M. (1976). The glutathione–glutathione disulfide system, in: *Free Radicals in Biology* (Pryor, W.A., ed.), Vol. II, pp. 55–84, Academic Press, New York.

Kosower, N.S., and Kosower, E.M. (1978). The glutathione status of cells. *Int. Rev. Cytol.* **54**: 109–160.

Kreck, M.L., Uhlinger, D.J., Tyagi, S.R., Inge, K.L., and Lambeth, J.D. (1994). Participation of the small molecular weight GTP-binding protein Rac1 in cell-free activation and assembly of the respiratory burst oxidase. *J. Biol. Chem.* **269**: 4161–4168.

Kroll, J.S., Langford, P.R., and Loynds, B.M. (1991). Copper-zinc superoxide dismutase of *Haemophilus influenzae* and *H. parainfluenzae*. *J. Bacteriol.* **173**: 7449–7457.

Kukielka, E., and Cederbaum, A.I. (1994). Ferritin stimulation of hydroxyl radical production by rat liver nuclei. *Arch. Biochem. Biophys.* **308**: 70–77.

Kurata, S., and Nakajima, H. (1990). Transcriptional activation of the heme oxygenase gene by TPA in mouse M1 cells during their differentiation to macrophage. *Exp. Cell Res.* **191**: 89–94.

Kuthan H., and Ullrich, V. (1982). Oxidase and oxygenase function of the micro-somal cytochrome P450 monooxygenase system. *Eur. J. Biochem.* **126**: 583–588.

Kuthan, H., Tsuji, H., Graf, H., Ullrich, V., Werringloer, J., and Estabrook R.W. (1978). Generation of superoxide anion as a source of hydrogen peroxide in a reconstituted monooxygenase system. *FEBS Lett.* **91**: 343–345.

Kvietys, P.R., Inaven, W., Bacon, B.R., and Grisham, M.B. (1989). Xanthine oxidase-induced injury to endothelium: role of intracellular iron and hydroxyl radical. *Am. J. Physiol.* **257**: H1640–H1646.

Lavelle, F., McAdam, M.E., Fielden, E.M., and Roberts, P.B. (1977). A pulse-radiolysis study of the catalytic mechanism of the iron-containing superoxide dismutase from *Photobacterium leiognathi. Biochem. J.* **161**: 3–11.

Lawrence, R.A., and Burk, R.F. (1976). Glutathione peroxidase activity in selenium-deficient rat liver. *Biochem. Biophys. Res. Commun.* **71**: 952–958.

Lawrence, R.A., and Burk, R.F. (1978). Species, tissue and subcellular distribution of non-Se-dependent glutathione peroxidase activity. *J. Nutr.* **108**: 211–215.

Le, C.T., Hollaar, L., Van der Valk, E.J.M., and Van der Laarse, A. (1994). Desferrioxamine protects myocytes against peroxide-induced myocyte damage without affecting glutathione redox cycle turnover. *J. Mol. Cell. Cardiol.* **26**: 877–887.

Leeuwenburgh, C., Rasmussen, J.E., Hsu, F.F., Mueller, D.M., Pennathur, S., and Heinecke, J.W. (1997). Mass spectrometric quantification of markers for protein oxidation by tyrosyl radical, copper, and hydroxyl radical in low density lipoprotein isolated from human atherosclerotic plaques. *J. Biol. Chem.* **272**: 3520–3526.

Leto, T.L., Adams, A.G., and de Mendez, I. (1994). Assembly of the phagocyte NADPH oxidase: Binding of Src homology 3 domains to proline-rich targets. *Proc. Natl. Acad. Sci. USA* **91**: 10650–10654.

Liebler, D.C. (1993). The role of metabolism in the antioxidant function of vitamin E. *Crit. Rev. Toxicol.* **23**: 147–169.

Liebler, D.C., Kaysen, K.L., and Kennedy, T.A. (1989). Redox cycles of vitamin E:

68   FREE RADICALS

Hydrolysis and ascorbic acid dependent reduction of 8a-(alkyldioxy)tocopherones. *Biochemistry* **28**: 9772–9777.

Lin, J.H.-C., Villalon, P., Martasek, P., and Abraham, N.G. (1990). Regulation of heme oxygenase gene expression by cobalt in rat liver and kidney. *Eur. J. Biochem.* **192**: 577–582.

Liochev, S.L., and Fridovich, I. (1994). The role of $O_2 \cdot^-$ in the production of HO·: *In vitro* and *in vivo*. *Free Radical Biol. Med.* **16**: 29–33.

Lloyd, R.V., Hanna, P.M., and Mason, R.P. (1997). The origin of the hydroxyl radical oxygen in the Fenton reaction. *Free Radical Biol. Med.* **22**: 885–888.

Locksley, R.M., Wilson, C.B., and Klebanoff, S.J. (1982). Role for endogenous and acquired peroxidase in the toxoplasmacidal activity of murine and human mononuclear phagocytes. *J. Clin. Invest.* **69**: 1099–1111.

Loewen, P. (1984). Isolation of catalase-deficient *Escherichia coli* mutants and genetic mapping of *katE*, a locus that affects catalase activity. *J. Bacteriol.* **157**: 622–626.

Loewen, P.C., and Switala, J. (1986). Purification and characterization of catalase HPII from *Escherichia coli* K12. *Biochem. Cell Biol.* **64**: 638–646.

Loewen, P.C., and Switala, J. (1988). Purification and characterization of spore-specific catalase-2 from *Bacillus subtilis*. *Biochem. Cell Biol.* **66**: 707–714.

Loewen, P., Switala, J., and Triggs-Raine, B.L. (1985). Catalase HPI and HPII in *Escherichia coli* are induced independently. *Arch. Biochem. Biophys.* **243**: 144–149.

Long, C.A., and Bielski, B.H.J. (1980). Rate of reaction of superoxide radical with chloride-containing species. *J. Phys. Chem.* **84**: 555–557.

Longo, V.D., Gralla, E.B., and Valetine, J.S. (1996). Superoxide dismutase activity is essential for stationary phase survival of *Saccharomyces cerevisiae*: Mitochondrial production of toxic oxygen species *in vivo*. *J. Biol. Chem.* **271**: 12275–12280.

Loven, D.P., and Oberley, L.W. (1982). Free radicals, insulin action and diabetes. in: *Superoxide Dismutases* (Oberley, L.W., ed.), Vol. III, pp. 151–189, CRC Press, Boca Raton, FL.

Lukacs, G.L., Rotstein, O.D., and Grinstein, S. (1990). Phagosomal acidification is mediated by a vacuolar-type H^+-ATPase in murine macrophages. *J. Biol. Chem.* **265**: 21009–21107.

Lukacs, G.L., Rotstein, O.D., and Grinstein, S. (1991). Determinants of the phagosomal pH in macrophages. *In situ* assessment of vacuolar H^+-ATPase activity, counterion conductance, and H^+ "leak." *J. Biol. Chem.* **266**: 24540–24548.

MacGillivray, R.T.A., Mendez, E., Shewale, J.G., Sinha, S.K., Lineback-Zins, J., and Brew, K. (1983). The primary structure of human serum transferrin. The structures of the seven cyanogen bromide fragments and the assembly of the complete structure. *J. Biol. Chem.* **258**: 3543–3553.

Machlin, L.J., and Bendich, A. (1987). Free radical tissue damage: Protective role of antioxidant nutrients. *FASEB J.* **1**: 441–445.

Maddipati, K.R., and Marnett, L.J. (1987). Characterization of the major hydroperoxide-reducing activity of human plasma. Purification and properties of a selenium-dependent glutathione peroxidase. *J. Biol. Chem.* **262**: 17398–17403.

Magliozzo, R.S., and Marcinkeviciene, J.A. (1997). The role of Mn(II)-peroxidase activity of mycobacterial catalase-peroxidase in activation of the antibiotic isoniazid. *J. Biol. Chem.* **272**: 8867–8870.

Malech, H.L., and Nauseef, W.M. (1997). Primary inherited defects in neutrophil function: Etiology and treatment. *Seminars Hematol.* **34**: 279–290.

Malinski, T., Bailey, F., Zhang, Z.G., and Chopp, M. (1993). Nitric oxide measured by a porphyrinic microsensor in rat brain after transient middle cerebral artery occlusion. *J. Cereb. Blood Flow Metab.* **13**: 355–358.

Mao, G.D., Thomas, P.D., and Poznansky, M.J. (1994). Oxidation of spin trap 5,5-dimethyl-1-pyrroline-1-oxide in an electron paramagnetic resonance study of the reaction of methemoglobin with hydrogen peroxide. *Free Radical Biol. Med.* **16**: 493–500.

Maples, K.R., Eyer, P., and Mason, R.P. (1990). Aniline-, phenylhydroxylamine-, nitrosobenzene-, and nitrobenzene-induced hemoglobin thiyl free radical formation *in vivo* and *in vitro*. *Mol. Pharmacol.* **37**: 311–318.

March, J. (1992). *Advanced Organic Chemistry, Reactions, Mechanisms and Structure*, 4th edition, p. 741, John Wiley & Sons, New York.

Marklund, S.L. (1982). Human copper-containing superoxide dismutase of high molecular weight. *Proc. Natl. Acad. Sci. USA* **79**: 7634–7638.

Marklund, S.L. (1984a). Properties of extracellular superoxide dismutase from human lung. *Biochem. J.* **220**: 269–272.

Marklund, S.L. (1984b). Extracellular superoxide dismutase and other superoxide dismutase isoenzymes in tissues from nine mammalian species. *Biochem. J.* **222**: 649–655.

Marklund, S.L. (1985). Product of extracellular-superoxide dismutase catalysis. *FEBS Lett.* **184**: 237–239.

Marklund, S.L. (1990). Expression of extracellular superoxide dismutase by human cell lines. *Biochem. J.* **266**: 213–219.

Marklund, S.L. (1992). Regulation by cytokines of extracellular superoxide dismutase and other superoxide dismutase isoenzymes in fibroblasts. *J. Biol. Chem.* **267**: 6696–6701.

Marquez, L.A., and Dunford, H.B. (1994). Chlorination of taurine by myeloperoxidase. Kinetic evidence for an enzyme-bound intermediate. *J. Biol. Chem.* **269**: 7950–7956.

Mason, R.P., and Rao, D.N.R. (1990). Thiyl free radical metabolites of thiol drugs, glutathione and proteins. *Methods Enzymol.* **186**: 319–329.

Masuda, A., Longo, D.L., Kobayashi, Y., Appella, E., Oppenheim, J.J., and Matsushima, K. (1988). Induction of mitochondrial manganese superoxide dismutase by interleukin 1. *FASEB J.* **2**: 3087–3091.

Mattson, M.P. (1994). Calcium and neuronal injury in Alzheimer's Disease. Contributions of β-amyloid precursor protein mismetabolism, free radicals, and metabolic compromise. *Ann. N.Y. Acad. Sci.* **747**: 50–76.

May, J.M., Qu, Z.-C., and Morrow, J.D. (1996). Interaction of ascorbate and α-tocopherol in resealed human erythrocyte ghosts. Transmembrane electron transfer and protection from lipid peroxidation. *J. Biol. Chem.* **271**: 10577–10582.

Mayeno, A.N., Curran, A.J., Roberts, R.L., and Foote, C.S. (1989). Eosinophils preferentially use bromide to generate halogenating agents. *J. Biol. Chem.* **264**: 5660–5668.

McAdam, M.E., Fox, R.A., Lavelle, F., and Fielden, E.M. (1977a). A pulse-radiolysis study of manganese-containing superoxide dismutase from *Bacillus stearothermophilus*. A kinetic model for the enzyme action. *Biochem. J.* **165**: 71–79.

McAdam, M.E., Lavelle, F., Fox, R.A., and Fielden, E.M. (1977b). A pulse-radiolysis study of the manganese-containing superoxide dismutase from *Bacillus stearothermophilus*. Further studies on the properties of the enzyme. *Biochem. J.* **165**: 81–87.

McBride, R.K., Stone K.K., and Marin, M.G. (1993). Oxidant injury alters barrier function of ferret tracheal epithelium. *Am. J. Physiol.* **264**: L165–L174.

McCord, J.M., and Fridovich, I. (1968). The reduction of cytochrome c by milk xanthine oxidase. *J. Biol. Chem.* **243**: 5753–5760.

McCord, J.M., and Fridovich, I. (1969). Superoxide dismutase—an enzymic function for erythrocuprein (hemocuprein). *J. Biol. Chem.* **244**: 6049–6055.

McCord, J.M., Keele, B.B., Jr., and Fridovich, I. (1971). An enzyme-based theory of obligate anaerobiosis: The physiological function of superoxide dismutase. *Proc. Natl. Acad. Sci. USA* **68**: 1024–1027.

McCormick, M.L., Buettner, G.R., and Britigan, B.E. (1998b). Endogenous superoxide dismutase levels regulate iron-dependent hydroxyl radical formation in *Escherichia coli* exposed to hydrogen peroxide. *J. Bacteriol.* **180**: 622–625.

McCormick, M.L., Gout, J.P., Buettner, G.R., Britigan, B.E., and Heinecke, J.W. (1998a). Electron paramagnetic resonance detection of free tyrosyl radical generated by myeloperoxidase, lactoperoxidase, and horseradish peroxidase. *J. Biol. Chem.* **273**: 32030–32037.

McCormick, M.L., Roeder, T.L., Railsback, M.A., and Britigan, B.E. (1994). Eosinophil peroxidase-dependent hydroxyl radical generation by human eosinophils. *J. Biol. Chem.* **269**: 27914–27919.

Mead, J.F. (1976). Free radical mechanisms of lipid damage and consequences for cellular membranes, in: *Free Radicals in Biology* (Pryor, W.A., ed.), Vol. I, pp. 51–68, Academic Press, New York.

Michelson, A.M., Puget, K., Durosay, P., and Bonneau, J.C. (1977). Clinical aspects of the dosage of erythrocuprein, in: *Superoxide and Superoxide Dismutases* (Michelson, A.M., McCord, J.M., and Fridovich, I., eds.), pp. 467–499, Academic Press, London.

Miller, R.A., and Britigan, B.E. (1995). Protease-cleaved iron-transferrin augments oxidant-mediated endothelial cell injury via hydroxyl radical formation. *J. Clin. Invest.* **95**: 2491–2500.

Miller, R.W., and Massey, V. (1965). Dihydroorotic dehydrogenase II. Oxidation and reduction of cytochrome c. *J. Biol. Chem.* **240**: 1466–1472.

Mills, G.C. (1957). Hemoglobin catabolism. I. Glutathione peroxidase, an erythrocyte enzyme which protects hemoglobin from oxidative damage. *J. Biol. Chem.* **229**: 189–197.

Mitani, K., Fujita, H., Kappas, A., and Sassa, S. (1992). Heme oxygenase is a positive acute-phase reactant in human Hep3B hepatoma cells. *Blood* **79**: 1255–1259.

Miyachi, Y., Yoshioka, A., Imamura, S., and Niwa, Y. (1987). Decreased hydroxyl radical generation from polymorphonuclear leucocytes in the presence of D-penicillamine and thiopronine. *J. Clin. Lab. Immunol.* **22**: 81–84.

Miyaski, K.T., Zambon, J.J., Jones, C.A., and Wilson, M.E. (1987). Role of high-avidity binding of human neutrophil myeloperoxidase in the killing of *Actinobaccillus actinomycetemcomitans Infect. Immun.* **55**: 1029–1036.

Mönig, J., Asmus, K.-D., Forni, L.G., and Willson, R.L. (1987). On the reaction of molecular oxygen with thiyl radicals: A re-examination. *Int. J. Radiat. Biol.* **52**: 589–602.

Monteiro, H.P., Vile, G.F., and Winterbourn, C.C. (1989). An iron chelator is not required for reductive iron release from ferritin by radical generating systems. *Free Radical Res. Commun.* **7**: 33–35.

Morel, I., Cillard, J., Lescoat, G., Sergent, O., Pasdeloup, N., Ocaktan, A.Z., Abdallah, M.A., Brissot, P., and Cillard, P. (1992). Antioxidant and free radical scavenging activities of the iron chelators pyoverdin and hydroxypyrid-4-ones in iron-loaded hepatocyte cultures: Comparison of their mechanism of protection with that of desferrioxamine. *Free Radical Biol. Med.* **13**: 499–508.

Morel, I., Sergent, O., Cogrel, P., Lescoat, G., Pasdeloup, N., Brissot, P., Cillard, P., and Cillard, J. (1995). EPR study of antioxidant activity of the iron chelators pyoverdin and hydroxypyrid-4-one in iron-loaded hepatocyte culture: Comparison with that of desferrioxamine. *Free Radical Biol. Med.* **18**: 303–310.

Morgan, M.S., Van Trieste, P.F., Garlick, S.M., Mahon, M.J., and Smith, A.L. (1988). Ultraviolet molar absorptivities of aqueous hydrogen peroxide and hydroperoxyl ion. *Analyt. Chim. Acta* **215**: 325–329.

Motohasi, N., and Mori, I. (1983). Superoxide dependent formation of hydroxyl radical catalyzed by transferrin. *FEBS Lett.* **157**: 197–199.

Munday, R., and Winterbourn, C.C. (1989). Reduced glutathione in combination with superoxide dismutase as an important biological antioxidant defence mechanism. *Biochem. Pharmacol.* **38**: 4349–4352.

Nakagawa, K., Kanno, H., and Miura, Y. (1997). Detection and analyses of ascorbyl radical in cerebrospinal fluid and serum of acute lymphoblastic leukemia. *Anal. Biochem.* **254**: 31–35.

Nakamura, H., Nakamura, K., and Yodoi, J. (1997). Redox regulation of cellular activation. *Ann. Rev. Immunol.* **15**: 351–369.

Nakamura, M. (1990). Lactoferrin-mediated formation of oxygen radicals by NADPH-cytochrome P-450 reductase system. *J. Biochem.* **107**: 395–399.

Nakamura, W., Hosoda, S., and Hayashi, K. (1974). Purification and properties of rat liver glutathione peroxidase. *Biochim. Biophys. Acta* **358**: 251–261.

Nakayama, K. (1992). Nucleotide sequence of *Streptococcus mutans* superoxide dismutase gene and isolation of insertion mutants. *J. Bacteriol.* **174**: 4928–4934.

Nakayama, K. (1994). Rapid viability loss on exposure to air in a superoxide dismutase-deficient mutant of *Porphysomonas gingivalis*. *J. Bacteriol.* **176**: 1939–1943.

Nanda, A., Romanek, R., Curnutte, J.T., and Grinstein, S. (1994). Assessment of the contribution of the cytochrome b moiety of the NADPH oxidase to the transmembrane H^+ conductance of leukocytes. *J. Biol. Chem.* **269**: 27280–27285.

Nath, K.A. (1994). The functional significance of induction of heme oxygenase by oxidant stress. *J. Lab. Clin. Med.* **123**: 461–463.

Natvig, D.O., Imlay, K., Touati, D., and Hallewell, R.A. (1987). Human copper-zinc superoxide dismutase complements superoxide dismutase-deficient *Escherichia coli* mutants. *J. Biol. Chem.* **262**: 14697–14701.

Neilands, J.B. (1981). Microbial iron compounds. *Ann. Rev. Biochem.* **50**: 715–731.

Neilands, J.B. (1993). Siderophores. *Arch. Biochem. Biophys.* **302**: 1–3.

Nguyen, S.A.K., Craig, A., and Raymond, K.N. (1993). Transferrin: The role of conformational changes in iron removal by chelators. *J. Am. Chem. Soc.* **115**: 6758–6764.

Niki, E. (1987). Interaction of ascorbate and α-tocopherol. *Ann. N.Y. Acad. Sci.* **498**: 186–198.

Nishikimi, M. (1975). Oxidation of ascorbic acid with superoxide anion generated by the xanthine–xanthine oxidase system. *Biochem. Biophys. Res. Commun.* **63**: 463–468.

Nunoi, H., Rotrosen, D., Gallin, J.L., and Malech, H.L. (1988). Two forms of autosomal chronic granulomatous disease lack distinct neutrophil cytosol factors. *Science* **242**: 1298–1301.

Oberley, L.W., and Buettner, G.R. (1979). Role of superoxide dismutase in cancer: A review. *Cancer Res.* **39**: 1141–1149.

Oberley, L.W., and Oberley, T.D. (1988). Role of antioxidant enzymes in cell immortalization and transformation. *Molec. Cell. Biochem.* **84**: 147–153.

O'Connell, M., Halliwell, B., Moorhouse, C.P., Aruoma, O.I., Baum, H., and Peters, T.J. (1986). Formation of hydroxyl radicals in the presence of ferritin and haemosiderin. Is haemosiderin formation a biological protective mechanism? *Biochem. J.* **234**: 727–731.

Okamura, N., Curnutte, J.T., Roberts, R.L., and Babior, B.M. (1988). Relationship of protein phosphorylation to the activation of the respiratory burst in human neutrophils. Defects in the phosphorylation of a group of closely related 48-kDa proteins in two forms of chronic granulomatous disease. *J. Biol. Chem.* **263**: 6777–6782.

Olakanmi, O., McGowan, S.E., Hayek, M.B., and Britigan, B.E. (1993). Iron sequestration by macrophages decreases the potential for extracellular hydroxyl radical formation. *J. Clin. Invest.* **91**: 889–899.

O'Neill, R., Davies, S., Fielden, E.M., Calabrese, L., Capo, C., Marmocchi, F., Natoli, G., and Rotilio, G. (1988). The effects of pH and various salts upon the activity of a series of superoxide dismutases. *Biochem. J.* **251**: 41–46.

Oshino, N., Chance, B., Sies, H., and Bücher, T. (1973). The role of H_2O_2 generation in perfused rat liver and the reaction of catalase compound I and hydrogen donors. *Arch. Biochem. Biophys.* **154**: 117–131.

Otto, B.R., Verweij-van Vught, A.M.J.J., and MacLaren, D.M. (1992). Transferrins and heme-compounds as iron sources for pathogenic bacteria. *Crit. Rev. Microbiol.* **18**: 217–233.

Oury, T.D., Chang, L.,-Y., Marklund, S.L., Day, B.J., and Crapo, J.D. (1994). Immunocytochemical localization of extracellular superoxide dismutase in human lung. *Lab. Invest.* **70**: 889–898.

Oury, T.D., Crapo, J.D., Valnickova, Z., and Enghild, J.J. (1996). Human extracellular superoxide dismutase is a tetramer composed of two disulphide-linked dimers: A simplified, high-yield purification of extracellular superoxide dismutase. *Biochem. J.* **317**: 51–57.

Packer, J.E., Slater, T.F., and Willson, R.L. (1979). Direct observation of a free radical interaction between vitamin E and vitamin C. *Nature* **278**: 737–738.

Palmer, C., Roberts, R.L., and Bero, C. (1994). Deferoxamine posttreatment reduces ischemic brain injury in neonatal rats. *Stroke* **25**: 1039–1045.

Palmer, R.M.J., Ferrige, A.G., and Moncada, S. (1987). Nitric oxide release accounts for the biological activity of endothelium-derived relaxing factor. *Nature* **327**: 524–526.

Park, J.-W., Hoyal, C.R., Benna, J.E., and Barbior, B.M. (1997). Kinase- dependent activation of the leukocyte NADPH oxidase in a cell-free system. Phosphorylation of membranes and p47^{PHOX} during oxidase activation. *J. Biol. Chem.* **272**: 11035–11043.

Parkos, C.A., Allen, R.A., Cochrane, C.G., and Jesaitis, A.J. (1987). Purified cytochrome b from human granulocyte plasma membrane is comprised of two polypeptides with relative molecular weights of 91,000 and 22,000. *J. Clin. Invest.* **80**: 732–742.

Paul, B. and Sbarra, A.J. (1968). The role of the phagocyte in host-parasite interactions. XIII. The direct quantitative estimation of H_2O_2 in phagocytic cells. *Biochim. Biophys. Acta* **156**: 168–178.

Paul, B.B., Jacobs, A.A., Strauss, R.R., and Sbarra, A.J. (1970). Role of the phagocyte in host–parasite infections. XXIV. Aldehyde generation by the myeloperoxidase–H_2O_2–chloride system: A possible *in vivo* mechanism of action. *Infect. Immun.* **2**: 414–418.

Pauling, L. (1960). *The Nature of the Chemical Bond*, pp. 47–57, Cornell University Press, Ithaca, NY.

Penner-Hahn, J.E. (1992). Structural properties of the manganese site in manganese catalases, in: *Manganese Redox Enzymes* (Pecoraro, E.L., ed.), pp. 29–45, VCH Publishers, New York.

Perly, B., Smith, I.C.P., Hughes, L., Burton, G.W., and Ingold, K.U. (1985). Estimation of the location of natural α-tocopherol in lipid bilayers by ^{13}C-NMR spectroscopy. *Biochim. Biophys. Acta* **819**: 131–135.

Pietri, S., Culcasi, M., Stella, L., and Cozzone, P.J. (1990). Ascorbyl radical as a reliable indicator of free-radical-mediated myocardial ischemic and post-ischemic injury. A real-time continuous-flow ESR study. *Eur. J. Biochem.* **193**: 845–854.

Pinto, R.E., and Bartley, W. (1969). The effect of age and sex on glutathione reductase and glutathione peroxidase activities and on aerobic glutathione oxidation in rat liver homogenates. *Biochem. J.* **112**: 109–115.

Porter, N.A. (1984). Chemistry of lipid peroxidation. *Methods Enzymol.* **104**: 273–282.

Poss, K.D., and Tonegawa, S. (1997a). Reduced stress defense in heme oxygenase 1-deficient cells. *Proc. Natl. Acad. Sci. USA* **94**: 10925–10930.

Poss, K.D., and Tonegawa, S. (1997b). Heme oxygenase 1 is required for mammalian iron reutilization. *Proc. Natl. Acad. Sci. USA* **94**: 10919–10924.

Pou, S., Cohen, M.S., Britigan, B.E., and Rosen, G.M. (1989). Spin trapping and human neutrophils: Limits of detection of hydroxyl radical. *J. Biol. Chem.* **264**: 12299–12302.

Pou, S., Pou, W.S., Bredt, D.S., Snyder, S.H., and Rosen, G.M. (1992). Generation of superoxide by purified nitric oxide synthase. *J. Biol. Chem.* **267**: 24173–24176.

Pou, S., Ramos, C.L., Gladwell, T., Renks, E., Centra, M., Young, D., Cohen, M.S., and Rosen, G.M. (1994). A kinetic approach to the selection of a sensitive spin trapping system for the detection of hydroxyl radical. *Anal. Biochem.* **217**: 76–83.

Prohaska, J.R. (1980). The glutathione peroxidase activity of glutathione-S-transferases. *Biochim. Biophys. Acta* **611**: 87–98.

Prohaska, J.R., and Ganther, H.E. (1977). Glutathione peroxidase activity of the glutathione-S-transferases purified from rat liver. *Biochem. Biophys. Res. Commun.* **76**: 437–445.

Pryor, W.A. (1986). Oxy-radicals and related species: Their formation, lifetimes, and reactions. *Ann. Rev. Physiol.* **48**: 657–667.

Pryor, W.A., and Tang, R.H. (1978). Ethylene formation from methional. *Biochem. Biophys. Res. Commun.* **81**: 498–503.

Pryor, W.A., Cornicelli, J.A., Devall, L.J., Tait, B., Trivedi, B.K., Witiak, D.T., and Wu, M. (1993). A rapid screening test to determine the antioxidant potencies of natural and synthetic antioxidants. *J. Org. Chem.* **58**: 3521–3532.

Puget, K., and Michelson, A.M. (1974). Isolation of a new copper-containing superoxide dismutase bacteriocuprein. *Biochem. Biophys. Res. Commun.* **58**: 830–838.

Puppo, A., and Halliwell, B. (1988). Formation of hydroxyl radicals from hydrogen peroxide in the presence of iron. Is haemoglobin a biological Fenton reagent? *Biochem. J.* **249**: 185–190.

Quinn, M.T. (1995). Low-molecular weight GTP-binding proteins and leukocyte signal transduction. *J. Leukocyte Biol.* **58**: 263–276.

Quintiliani, M., Badiello, R., Tamba, M., Esfandi, A., and Gorin, G. (1977). Radiolysis of glutathione in oxygen-containing solutions of pH 7. *Int. J. Radiat. Biol.* **32**: 195–202.

Rajagopalan, K.V., and Handler, P. (1964). Hepatic aldehyde oxidase. II. Differential inhibition of electron transfer to various electron acceptors. *J. Biol. Chem.* **239**: 2022–2026.

Rakita, R.M., and Rosen, H. (1991). Penicillin-binding protein inactivation by human neutrophil myeloperoxidase. *J. Clin. Invest.* **88**: 750–754.

Rakita, R.M., Michel, B.R., and Rosen, H. (1989). Myeloperoxidase mediated inhibition of microbial respiration: Damage to *Escherichia coli* ubiquinol oxidase. *Biochemistry* **28**: 3031–3036.

Rall, T.W., and Lehninger, A.L. (1952). Glutathione reductase of animal tissue. *J. Biol. Chem.* **194**: 119–130.

Ramos, C.L., Pou, S., Britigan, B.E., Cohen, M.S., and Rosen, G.M. (1992). Spin trapping evidence for myeloperoxidase-dependent hydroxyl radical formation by human neutrophils and monocytes. *J. Biol. Chem.* **267**: 8307–8312.

Ramsey, P.G., Martin, T., Chi, E., and Klebanoff, S.J. (1982). Arming of mononuclear phagocytes by eosinophil peroxidase bound to *Staphylococcus aureus. J. Immunol.* **128**: 415–420.

Rauckman, E.J., Rosen, G.M., and Kitchell, B.B. (1979). Superoxide radical as an intermediate in the oxidation of hydroxylamines by mixed function amine oxidase. *Mol. Pharmacol.* **15**: 131–137.

Reif, D.W. (1992). Ferritin as a source of iron for oxidative damage. *Free Radical Biol. Med.* **12**: 417–427.

Reif, D.W., Schubert, J., and Aust, S.D. (1988). Iron release from ferritin and lipid peroxidation by radiolytically generated reducing radicals. *Arch. Biochem. Biophys.* **264**: 283–243.

Ren, B., Huang, W., Åkesson, B., and Ladenstein, R. (1997). The crystal structure of seleno-glutathione peroxidase from human plasma at 2.9 Å resolution. *J. Mol. Biol.* **268**: 869–885.

Rienecke, L.A., Rau, J.M., and McCay, P.B. (1994). Characteristics of an oxidant formed during iron (II) autooxidation. *Free Radical Biol. Med.* **16**: 485–492.

Repine, J.E., Eaton, J.W., Anders, M.W., Hoidal, J.R., and Fox, R.B. (1979). Generation of hydroxyl radical by enzymes, chemicals and human phagocytes *in vitro.* Detection with the anti-inflammatory agent, dimethyl sulfoxide. *J. Clin. Invest.* **64**: 1642–1651.

Rocha, E.R., Selby, T., Coleman, J.P., and Smith, C.J. (1996). Oxidative stress response in an anaerobe, *Bacteroides fragilis*: A role for catalase in protection against hydrogen peroxide. *J. Bacteriol.* **178**: 6895–6903.

Roginsky, V.A., and Stegmann, H.B. (1994). Ascorbyl radical as natural indicator of oxidative stress: Quantitative regularities. *Free Radical Biol. Med.* **17**: 93–103.

Rose, R.D., and Bode, A.M. (1993). Biology of free radical scavengers: An evaluation of ascorbate. *FASEB J.* **7**: 1135–1142.

Rosen, D.R., Siddique, T., Patterson, D., Figlewicz, D.A., Sapp, P., Hentati, A., Donaldson, D., Goto, J., O'Regan, J.P., Deng, H.-X., Rahmani, Z., Krizus, A., McKenna-Yasek, D., Cayabyab, A., Gaston, S.M., Berger, R., Tanzi, R.E., Halperin, J.J., Herzfeldt, B., Van den Bergh, R., Hung, W.-Y., Bird, T., Deng, G., Mulder, D.W., Smyth, C., Laing, N.G., Soriano, E., Pericak-Vance, M.A., Haines, J., Rouleau, G.A., Gusella, J.S., Horvitz,

H.R., and Brown, R.H., Jr. (1993). Mutations in Cu/Zn superoxide dismutase gene are associated with familial amyotrophic lateral sclerosis. *Nature* **362**: 59–62.

Rosen, G.M., Britigan, B.E., Cohen, M.S., Ellington, S.P., and Barber, M.J. (1988). Detection of phagocyte-derived free radicals with spin trapping techniques: Effects of temperature and cellular metabolism. *Biochim. Biophys. Acta* **969**: 236–241.

Rosen, G.M., Pou, S., Ramos, C.L., Cohen, M.S., and Britigan, B.E. (1995). Free radicals and phagocytic cells. *FASEB J.* **9**: 200–209.

Rosen, G.M., Rauckman, E.J., Ellington, S.P., Dahlin, D.C., Christie, J.L., and Nelson, S.D. (1984). Reduction and glutathione conjugation reactions of N-acetyl-p-benzoquinone imine and two dimethylated analogues. *Mol. Pharmacol.* **25**: 151–157.

Rosen, H., and Klebanoff, S.J. (1979a). Bactericidal activity of a superoxide anion-generating system. A model for the polymorphonuclear leukocytes. *J. Exp. Med.* **149**: 27–39.

Rosen, H., and Klebanoff, S.J. (1979b). Hydroxyl radical generation by polymorphonuclear leukocytes measured by electron spin resonance spectroscopy. *J. Clin. Invest.* **64**: 1725–1729.

Rosen, H., Michel, B.R., vanDevanter, D.R., and Hughes, J.P. (1998). Differential effects of myeloperoxidase-derived oxidants on *Escherichia coli* DNA replication. *Infect. Immun.* **66**: 2655–2659.

Rosen, H., Orman, J., Ratika, R.M., Michel, B.R., and VanDevanter, D.R. (1990). Loss of DNA-membrane interactions and cessation of DNA synthesis in myeloperoxidase-treated *Escherichia coli*. *Proc. Natl. Acad. Sci. USA* **87**: 10048–10052.

Rosen, H., Rakita, R.M., Waltersdorph, A.M., and Klebanoff, S.J. (1987). Myeloperoxidase-mediated damage to the succinate oxidase system of *Escherichia coli*. Evidence for selective inactivation of the dehydrogenase component. *J. Biol. Chem.* **262**: 15004–15010.

Ross, D., Norbeck, K., and Moldéus, P. (1985). The generation and subsequent fate of glutathionyl radicals in biological systems. *J. Biol. Chem.* **260**: 15028–15032.

Rothman, R.J., Serroni, A., and Farber, J.L. (1992). Cellular pool of transient ferric iron, chelatable by deferoxamine and distinct from ferritin, that is involved in oxidative cell injury. *Mol. Pharmacol.* **42**: 703–710.

Rotilio, G., Bray, R.C., and Fielden, E.M. (1972). A pulse radiolysis study of superoxide dismutase. *Biochim. Biophys. Acta* **268**: 605–609.

Rotrosen, D., Kleinberg, M.E., Nunoi, H., Leto, T., Gallin, J.I., and Malech, H.L. (1990). Evidence for a functional cytoplasmic domain of phagocyte oxidase cytochrome b_{558}. *J. Biol. Chem.* **265**: 8745–8750.

Rotrosen, D., Yeung, C.L., Leto, T.L., Malech, H.L., and Kwong, C.H. (1992). Cytochrome b_{558}: The flavin-binding component of the phagocyte NADPH oxidase. *Science* **256**: 1459–1462.

Rotruck, J.T., Pope, A.L., Ganther, H.E., Swanson, A.B., Hafeman, D.G., and Hoekstra, W.G. (1973). Selenium: Biochemical role as a component of glutathione peroxidase. *Science* **179**: 588–590.

Roubal, W.T. (1971). Free radicals, malonaldehyde and protein damage in lipid protein systems. *Lipids* **6**: 62–64.

Roubal, W.T., and Tappel, A.L. (1966). Polymerization of proteins induced by free radical-lipid peroxidation. *Arch. Biochem. Biophys.* **113**: 150–155.

Rubanyi, G.M., Lorenz, R.R., and Vanhoutte, P.M. (1985). Bioassay of endothelium-derived relaxing factor(s): Inactivation by catecholamines. *Am. J. Physiol.* **249**: H95–H101.

Rush, J.D., Maskos, Z., and Koppenol, W.H. (1990). Distinction between hydroxyl radical and ferryl species. *Methods Enzymol.* **186**: 148–156.

Ryan, T.P., and Aust, S.D. (1992). The role of iron in oxygen-mediated toxicities. *Crit. Rev. Toxicol.* **22**: 119–141.

Sadrzadeh, S.M.H., Graf, E., Panter, S.S., Hallaway, P.E., and Eaton, J.W. (1984). Hemoglobin: A biologic Fenton reagent. *J. Biol. Chem.* **259**: 14354–14356.

Sagone, A.L., Jr., and Husney, R.M. (1987). Oxidation of salicylates by stimulated granulocytes. Evidence that these drugs act as free radical scavengers in biological systems. *J. Immunol.* **138**: 2177–2183.

Sagone, A.L., Jr., Decker, M.A., Wells, R.M., and Democko, C.A. (1980a). A new method for the detection of hydroxyl radical production by phagocytic cells. *Biochim. Biophys. Acta* **628**: 90–97.

Sagone, A.J., Jr., Wells, R.M. and Democko, C.A. (1980b). Evidence that ·OH production by human PMNs is related to prostaglandin metabolism. *Inflammation* **4**: 65–71.

Sakamoto, H., and Touati, D. (1984). Cloning of the iron superoxide dismutase gene (*sodB*) in *Escherichia coli* K12. *J. Bacteriol.* **159**: 418–420.

Sampson, J.B., Ye, Y-Z., Rosen, H., and Beckman, J.S. (1998). Myeloperoxidase and horseradish peroxidase catalyze tyrosine nitration in proteins from nitrite and hydrogen peroxide. *Arch. Biochem. Biophys.* **356**: 207–213.

Sandström, J., Karlsson, K., Edlund, T., and Marklund, S.L. (1993). Heparin-affinity patterns and composition of extracellular superoxide dismutase in human plasma and tissues. *Biochem. J.* **294**: 853–857.

Saran, M., and Bors, W. (1989). Oxygen radicals acting as chemical messengers: A hypothesis. *Free Radical Res. Commun.* **7**: 213–220.

Sathyamoorthy, M., de Mendez, I., Adams, A.G., and Leto, T.L. (1997). p40^{phox} Down-regulates NADPH oxidase activity through interaction with its SH3 domain. *J. Biol. Chem.* **272**: 9141–9146.

Saunders, E.L., Maines, M.D., Meredith, M.J., and Freeman, M.L. (1991). Enhancement of heme oxygenase-1 synthesis by glutathione depletion in Chinese hamster ovary cells. *Arch. Biochem. Biophys.* **288**: 368–373.

Sbarra, A.J., and Karnovsky, M.L. (1959). The biochemical basis of phagocytosis. I. Metabolic changes during the ingestion of particles by polymorphonuclear leukocytes. *J. Biol. Chem.* **234**: 1355–1362.

Schlabach, M.R., and Bates, G.W. (1975). The synergistic binding of anions and Fe^{3+} by transferrin: Implications for the interlocking sites hypothesis. *J. Biol. Chem.* **250**: 2182–2188.

Schonbaum, G.R., and Chance, B. (1976). Catalase, in: *The Enzymes* (Boyer, P.D., ed.), 3rd edition, Vol. 13, pp. 363–408, Academic Press, New York.

Schöneich, C. (1995). Kinetics of thiol reactions. *Methods Enzymol.* **251**: 45–55.

Schraufstatter, I.U., Browne, K., Harris, A., Hyslop, P.A., Jackson, J.H., Quehenberger, O., and Cochrane, C.G. (1990). Mechanism of hypochlorite injury to target cells. *J. Clin. Invest.* **85**: 554–562.

Schreiber, J., Foureman, G.L., Hughes, M.F., Mason, R.P., and Eling, T.E. (1989). Detection of glutathione thiyl free radical catalyzed by prostaglandin H synthase present in keratinocytes. Study of cooxidation in a cellular system. *J. Biol. Chem.* **264**: 7936–7943.

Schuckelt, R., Brigelius-Flohé, R., Maiorino, M., Roveri, A., Reumkens, J., Strassburger, W., Ursini, F., Wolf, B., and Flohé, L. (1991). Phospholipid hydroperoxide glutathione peroxidase is a seleno-enzyme distinct from the classical glutathione peroxidase as evident from cDNA and amino acid sequencing. *Free Radical Res. Commun.* **14**: 343–361.

Schwartz, R.M., and Dayhoff, M.O. (1978). Origins of prokaryotes, eukaryotes, mitochondria and chloroplasts. *Science* **199**: 395–403.

Seddon, J.M., Ajani, U.A., Sperduto, R.D., Hiller, R., Blair, N., Burton, T.C., Farber, M.D., Gragoudas, E.S., Haller, J., Miller, D.T., Yannuzzi, L.A., and Willett, W. (1994). Dietary carotenoids, vitamins A, C, and E, and advanced age-related macular degeneration. *JAMA* **272**: 1413–1420.

Segal, A.W. (1987). Absence of both cytochrome b$_{245}$ subunits from neutrophils in X-linked chronic granulomatous disease. *Nature* **326**: 88–91.

Segal, A.W. (1989). The electron transport chain of the microbicidal oxidase of phagocytic cells and its involvement in the molecular pathology of chronic granulomatous disease. *J. Clin. Invest.* **83**: 1785–1793.

Segal, A.W., Cross, A.R., Garcia, R.C., Borregaard, N., Valerius, N.H., Soothill, J.H., and Jones, O.T.G. (1983). Absence of cytochrome b$_{245}$ in chronic granulomatous disease. A multicenter European evaluation of its incidence and relevance. *N. Engl. J. Med.* **308**: 245–251.

Segal, A.W., Heyworth, P.G., Cockcroft, S., and Barrowman, M.M. (1985). Stimulated neutrophils from patients with autosomal recessive chronic granulomatous disease fail to phosphorylate a M_r-44,000 protein. *Nature* **316**: 547–549.

Selvaraj, R.J., Zgliczynski, J.M., Paul, B.B., and Sbarra, A.J. (1978). Enhanced killing of myeloperoxidase coated bacteria in the myeloperoxidase–H_2O_2–Cl^- system. *J. Infect. Dis.* **137**: 481–485.

Sevanian, A., Muakkassah-Kelly, S.F., and Montsetruque, S. (1983). The influence of phospholipase A_2 and glutathione peroxidase on the elimination of membrane lipid peroxides. *Arch. Biochem. Biophys.* **223**: 441–452.

Sharma, M.K., Buettner, G.R., Spencer, K.T., and Kerber, R.E. (1994). Ascorbyl radical as a real-time marker of free radical generation in briefly ischemic and reperfused hearts. An electron paramagnetic resonance study. *Circ. Res.* **74**: 650–658.

Shilling, C.W., and Adams, B.H. (1933). A study of the convulsive seizures caused by breathing oxygen at high pressures. *U.S. Naval Med. Bull.* **31**: 112–121.

Shimazaki, K., Kawaguchi, A., Sato, T., Ueda, Y., Tomimura, T., and Shimamura, S. (1993). Analysis of human and bovine milk lactoferrins by rotofor and chromatofocusing. *Int. J. Biochem.* **25**: 1653–1658.

Shinmoto, H., Dosako, S., and Nakajima, I. (1992). Anti-oxidant activity of bovine lactoferrin on iron/ascorbate induced lipid peroxidation. *Biosci. Biotechnol. Biochem.* **56**: 2079–2080.

Shongwe, M.S., Smith, C.A., Ainscough, E.W., Baker, H.M., Brodie, A.M., and Baker, E.N. (1992). Anion binding by human lactoferrin: Results from crystallographic and physicochemical studies. *Biochemistry* **31**: 4451–4458.

Sies, H., and Moss, K.M. (1978). A role of mitochondrial glutathione peroxidase in modulating miochondrial oxidations in liver. *Eur. J. Biochem.* **84**: 377–383.

Sies, H., and Summer, K.H. (1975). Hydroperoxide-metabolizing systems in rat liver. *Eur. J. Biochem.* **57**: 503–512.

Sies, H., Stahl, W., and Sundquist, A.R. (1992). Antioxidant functions of vitamins. Vitamins E and C, beta-carotene, and other carotenoids. *Ann. N.Y. Acad. Sci.* **669**: 7–20.

Sinet, P.M. (1982). Metabolism of oxygen derivatives in Down's syndrome. *Ann. N.Y. Acad. Sci.* **396**: 83–94.

Singh, R.J., Karoui, H., Gunther, M.R., Beckman, J.S., Mason, R.P., and Kalyanaraman, B. (1998). Reexamination of the mechanism of hydroxyl radical adducts formed from the reaction between familial amyotrophic lateral sclerosis-associated Cu,Zn superoxide dismutase mutants and H_2O_2. *Proc. Natl. Acad. Sci. USA* **95**: 6675–6680.

Singh, S., and Hider, R.C. (1988). Colorimetric detection of the hydroxyl radical: Comparison of the hydroxyl radical-generating ability of various iron complexes. *Anal. Biochem.* **171**: 47–54.

Smith, C.V., Jones, D.P., Guenthner, T.M., Lash, L.H., and Lauterburg, B.H. (1996). Compartmentation of glutathione: Implications for the study of toxicity and disease. *Toxicol. Appl. Pharmacol.* **140**: 1–12.

Smith, J.B., Cusumano, J.C., and Babbs, C.F. (1990). Quantitative effects of iron chelators on hydroxyl radical production by the superoxide-driven Fenton reaction. *Free Radical Res. Commun.* **8**: 101–106.

Smith, M.A., Sayre, L.M., Monnier, V.M., and Perry, G. (1995). Radical AGEing in Alzheimer's disease. *Trends Neurosci.* **18**: 172–176.

Smith, T.J., Haque, S., and Drummond, G.S. (1991). Induction of heme oxygenase mRNA by cobalt protoporphyrin in rat liver. *Biochim. Biophys. Acta* **1073**: 221–224.

Sohal, R.S., and Weindruch, R. (1996). Oxidative stress, caloric restriction and aging. *Science* **273**: 59–63.

Sohal, R.S., Agarwal, A., Agarwal, S., and Orr, W.C. (1995). Simultaneous overexpression of copper- and zinc-containing superoxide dismutase and catalase retards age-related oxidative damage and increases metabolic potential in *Drosophila melanogaster*. *J. Biol. Chem.* **270**: 15671–15674.

Spik, G., Coddeville, B., Legrand, D., Mazurier, J., Leger, D., Goavec, M., and Montreuil, J. (1985). A comparative study of the primary structure of glycans from various sero-, lacto- and ovotransferrins. Role of human lactotransferrin glycans, in: *Proteins of Iron Storage and Transport* (Spik, G., Montreuil, J., Crichton, R.R., and Mazurier, J., eds.). pp. 47–51, Elsevier Science Publishers B.V., Amsterdam.

Spik, G., Coddeville, B., and Montreuil, J. (1988). Comparative study of the primary structures of sero-, lacto- and ovotransferrin glycans from different species. *Biochimie* 70: 1459–1469.

Spik, G., Coddeville, B., Strecker, G., Montreuil, J., Regoeczi, E., Chindemi, P.A., and Rudolph, J.R. (1993). Carbohydrate microheterogeneity of rat serotransferrin—Determination of glycan primary structures and characterization of a new type of trisialylated diantennary glycan. *Eur. J. Biochem.* 195: 397–405.

Staal, G.E.J., Visser, J., and Veeger, C. (1969). Purification and properties of glutathione reductase of human erythrocytes. *Biochim. Biophys. Acta* 185: 39–48.

Stabel, T.J., Sha, Z., and Mayfield, J.E. (1994). Periplasmic location of *Brucella abortus* Cu/Zn superoxide dismutase. *Vet. Microbiol.* 38: 307–314.

Starke, P.E., and Farber, J.L. (1985). Ferric iron and superoxide ions are required for the killing of cultured hepatocytes by hydrogen peroxide: Evidence for the participation of hydroxyl radicals formed by an iron-catalyzed Haber-Weiss reaction. *J. Biol. Chem.* 260: 10099–10104.

Steinbeck, M.L., Hegg, G.G., and Karnovsky, M.J. (1991). Arachidonate activation of the neutrophil NADPH-oxidase. Synergistic effects of protein phosphatase inhibitors compared with protein kinase activators. *J. Biol. Chem.* 266: 16336–16342.

Steinberg, D., Parthasarathy, S., Carew, T.E., Khoo, J.C., and Witztum, J.L. (1989). Beyond cholesterol. Modifications of low-density lipoprotein that increase its atherogenicity. *N. Engl. J. Med.* 320: 915–924.

Steinman, H.M. (1978). The amino acid sequence of mangano superoxide dismutase from *Escherichia coli* B. *J. Biol. Chem.* 253: 8708–8720.

Steinman, H.M. (1982a). Superoxide dismutases: Protein chemistry and structure–function relationships, in: *Superoxide Dismutase* (Oberley, L.W., ed.), Vol. I, pp.11–68, CRC Press, Boca Raton, FL.

Steinman, H.M. (1982b). Copper-zinc superoxide dismutase from *Caulobacter crescentus* CB15. *J. Biol. Chem.* 257: 10283–10293.

Steinman, H.M. (1985). Bacteriocuprein superoxide dismutases in pseudomonads. *J. Bacteriol.* 162: 1255–1260.

Steinman, H.M. (1987). Bacteriocuprein superoxide dismutase of *Photobacterium leiognathi*. Isolation and sequence of the gene and evidence for a precursor form. *J. Biol. Chem.* 262: 1882–1887.

Steinman, H.M., and Ely, B. (1990). Copper-zinc superoxide dismutase of *Caulobacter crescentus*: Cloning, sequencing, and mapping of the gene and periplasmic location of the enzyme. *J. Bacteriol.* 172: 2901–2910.

Steinman, H.M., and Hill, R.L. (1973). Sequencer homologies among bacterial and mitochondrial superoxide dismutases. *Proc. Natl. Acad. Sci. USA* 70: 3725–3729.

Steinman, H.M., Weinstein, L., and Brenowitz, M. (1994). The manganese superoxide dismutase of *Escherichia coli* K-12 associates with DNA. *J. Biol. Chem.* 269: 28629–28634.

Stocker, R. (1990). Induction of haem oxygenase as a defence against oxidative stress. *Free Radical Res. Commun.* 9: 101–112.

Storz, G., Christman, M.F., Sies, H., and Ames, B.N. (1987). Spontaneous mutagenesis and oxidative damage to DNA in *Salmonella typhimurium Proc. Natl. Acad. Sci. USA* 84: 8917–8921.

Stoyanovsky, D.A., Goldman, R., Jonnalagadda, S.S., Day, B.W., Claycamp, H.G., and Kagan, V.E. (1996). Detection and characterization of the electron paramagnetic resonance-silent glutathionyl-5,5-dimethyl-1-pyrroline N-oxide adduct derived from redox cycling of phenoxyl radicals in model systems and HL-60 cells. *Arch. Biochem. Biophys.* 330: 3–11.

Strobel, H.W., Lu, A.Y.H., Heidema, J., and Coon, M.J. (1970). Phosphatidylcholine requirement in the enzymatic reduction of hemoprotein P-450 and in fatty acid, hydrocarbon, and drug hydroxylation. *J. Biol. Chem.* **245**: 4851–4854.

Sullivan, S.G., Baysal, E., and Stern, A. (1992). Inhibition of hemin-induced hemolysis by desferrioxamine: Binding of hemin to red cell membranes and the effects of alteration of membrane sulfhydryl groups. *Biochim. Biophys. Acta* **1104**: 38–44.

Sumimoto, H., Hata, K., Mizuki, K., Ito, T., Kage, Y., Sakaki, Y., Fukumaki, Y., Nakamura, M., and Takeshige, K. (1996). Assembly and activation of the phagocyte NADPH oxidase. Specific interaction of the N-terminal Src homology 3 domain of p47^{phox} with p22^{phox} is required for activation of the NADPH oxidase. *J. Biol. Chem.* **271**: 22152–22158.

Sutherland, M.W., and Gebicki, J.M. (1984). A reaction between the superoxide free radical and lipid hydroperoxide in sodium linoleate micelles. *Arch. Biochem. Biophys.* **214**: 1–11.

Sutton, H.C. (1985). Efficiency of chelated iron compounds as catalysts for the Haber-Weiss reaction. *J. Free Radical Biol. Med.* **1**: 195–202.

Sutton, H.C., and Winterbourn, C.C. (1989). On the participation of higher oxidation states of iron and copper in Fenton reactions. *Free Radical Biol. Med.* **6**: 53–60.

Suzaki, E., Kawai, E., Kodama, Y., Suzaki, T., and Masujima, T. (1994). Quantitative analysis of superoxide anion generation in living cells by using chemiluminescence video microscopy. *Biochim. Biophys. Acta* **1201**: 328–332.

Swartz, H.M., and Dodd, N.J.F. (1981). The role of ascorbic acid on radical reactions *in vivo*, in: *Oxygen and Oxy-Radicals in Chemistry and Biology* (Rogers, M.J.A., and Powers, E.L., eds.), pp. 161–168, Academic Press, New York.

Switala, J., Triggs-Raine, B.L., and Loewen, P.C. (1990). Homology among bacterial catalase genes. *Can. J. Microbiol.* **36**: 728–731.

Szebeni, J., and Toth, K. (1986). Lipid peroxidation in hemoglobin-containing liposomes. Effects of membrane phospholipid composition and cholesterol content. *Biochim. Biophys. Acta* **857**: 139–145.

Szebeni, J., Winterbourn, C.C., and Carrell, R.W. (1984). Oxidative interactions between haemoglobin and membrane lipid. *Biochem. J.* **220**: 685–692.

Taha, Z., Kiechle, F., and Malinski, T. (1992). Oxidation of nitric oxide by oxygen in biological systems monitored by porphyrinic sensor. *Biochem. Biophys. Res. Commun.* **188**: 734–739.

Takahashi, K., and Cohen, H.J. (1986). Selenium-dependent glutathione peroxidase protein and activity: Immunological investigations on cellular and plasma enzymes. *Blood* **68**: 640–645.

Takahashi, K., Akasaka, M., Yamamoto, Y., Kobayashi, C., Mizoguchi, J., and Koyama, J. (1990). Primary structure of human plasma glutathione peroxidase deduced from cDNA sequences. *J. Biochem.* **108**: 145–148.

Takahashi, K., Avissar, N., Whitin, J., and Cohen, H. (1987). Purification and characterization of human plasma glutathione peroxidase: A selenoglycoprotein distinct from the known cellular enzyme. *Arch. Biochem. Biophys.* **256**: 677–686.

Takahashi, M., and Asada, K. (1982). A flash-photometric method for determination of reactivity of superoxide: Application to superoxide dismutase assay. *J. Biochem.* **91**: 889–896.

Takemoto, T., Zhang, Q.-M., and Yonei, S. (1998). Different mechanisms of thioredoxin in its reduced and oxidized forms in defense against hydrogen peroxide in *Escherichia coli*. *Free Radical Biol. Med.* **24**: 556–562.

Tamba, M., Simone, G., and Quintiliani, M. (1986). Interactions of thiyl free radicals with oxygen: A pulse radiolysis study. *Int. J. Radiat. Biol.* **50**: 595–600.

Tappel, A.L., Brown, W.D., Zalkin, H., and Maier, V.P. (1961). Unsaturated lipid peroxidation catalyzed by hematin compounds and its inhibition by vitamin E. *J. Am. Oil Chem. Soc.* **38**: 5–9.

Tappel, M.E., Chaudiere, J., and Tappel, A.L. (1982). Glutathione peroxidase activities of animal tissues. *Comp. Biochem. Physiol.* **73B**: 945–949.

Tauber, A.I., and Babior, B.M. (1977). Evidence for hydroxyl radical production by human neutrophils. *J. Clin. Invest.* **60**: 374–379.

Tauber, A.I., Borregaaard, N., Simons, E., and Wright, J. (1983). Chronic granulomatous disease: A syndrome of phagocyte oxidase deficiencies. *Medicine* **62**: 286–308.

Tesfamariam, B. (1994). Free radicals in diabetic endothelial cell dysfunction. *Free Radical Biol. Med.* **16**: 383–391.

Test, S.T., Lampert, M.B., Ossanna, P.J., Thoene, J.G., and Weiss, S.J. (1984). Generation of nitrogen-chlorine oxidants by human phagocytes. *J. Clin. Invest.* **74**: 1341–1349.

Thomas, E.L., Grisham, M.B., Melton, D.F., and Jefferson, M.M. (1985). Evidence for a role of taurine in the *in vitro* oxidative toxicity of neutrophils toward erythrocytes. *J. Biol. Chem.* **260**: 3321–3329.

Thomas, J.P., Maiorino, M., Ursini, F., and Girotti, A.W. (1990). Protective action of phospholipid hydroperoxide glutathione peroxidase against membrane-damaging lipid peroxidation: *In situ* reduction of phospholipid and cholesterol hydroperoxides. *J. Biol. Chem.* **265**: 454–461.

Thomas, M.J., Mehl, K.S., and Pryor, W.A. (1982). The role of superoxide in xanthine oxidase-induced autoxidation of linoleic acid. *J. Biol. Chem.* **257**: 8343–8347.

Thomas, M.J., Sutherland, M.W., Arudi, R.L., and Bielski, B.H.L. (1984). Studies of the reactivity of HO_2/O_2^- with unsaturated hydroperoxides in ethanolic solvents. *Arch. Biochem. Biophys.* **233**: 772–775.

Thompson, A.B., Bohling, T., Heires, A., Linder, J., and Rennard, S.I. (1991). Lower respiratory tract iron burden is increased in association with cigarette smoking. *J. Lab. Clin. Med.* **117**: 493–499.

Tibell, L., Hjalmarsson, K., Edlund, T., Skogman, G., Engstrom, A., and Marklund, S.L. (1987). Expression of human extracellular superoxide dismutase in Chinese hamster ovary cells and characterization of the product. *Proc. Natl. Acad. Sci. USA* **84**: 6634–6638.

Touati, D. (1983). Cloning and mapping of the manganese superoxide dismutase gene (*sodA*) of *Escherichia coli*. *J. Bacteriol.* **154**: 1078–1087.

Touati, D. (1988). Molecular genetics of superoxide dismutases. *Free Radical Biol. Med.* **5**: 393–402.

Touati, D. (1992). Regulation and protective role of the microbial superoxide dismutases, in: *Molecular Biology of Free Radical Scavenging System* (Scandalios, J.G., ed.), pp. 231–261, Cold Spring Harbor Laboratory Press, New York.

Touati, D., Tardat, B., and Compan, I. (1991). Regulation of MnSOD in *E. coli* in response to environmental stimuli, in: *Oxidative Damage and Repair. Chemical, Biological and Medical Aspects* (Davies, K.J.A., ed.), pp. 13–18, Pergamon Press, Oxford, England.

Triggs-Raine, B.L., Doble, B.W., Mulvey, M.R., Sorby, P.A., and Loewen P.C. (1988). Nucleotide sequence of *katG*, encoding catalase HPI of *Escherichia coli*. *J. Bacteriol.* **170**: 4415–4419.

Tyrrell, R.M., and Basu-Modak, S. (1994). Transient enhancement of heme oxygenase 1 mRNA accumulation: A marker of oxidative stress to eukaryotic cells. *Methods Enzymol.* **234**: 224–235.

Uhlinger, D.J., Taylor, K.L., and Lambeth, J.D. (1994). p67-*phox* enhances the binding of p47-*phox* to the human neutrophil respiratory burst oxidase complex. *J. Biol. Chem.* **269**: 22095–22098.

Ursini, F., Maiorino, M., and Gregolin, C. (1985). The selenoenzyme phospholipid hydroperoxide glutathione peroxidase. *Biochim. Biophys. Acta* **839**: 62–70.

Ursini, F., Maiorino, M., Valente, M., Ferri, L., and Gregolin, C. (1982). Purification from pig liver of a protein which protects liposomes and biomembranes from peroxidative degradation and exhibits glutathione peroxidase activity on phosphatidylcholine hydroperoxides. *Biochim. Biophys. Acta* **710**: 197–211.

Ushio-Fukai, M., Zafari, A.M., Fukui, T., Ishizaka, N., and Griendling, K.K. (1996). p22^{phox} is a critical component of the superoxide-generating NADH/NADPH oxidase system and regulates angiotensin II-induced hypertrophy in vascular smooth muscle cells. *J. Biol. Chem.* **271**: 23317–23321

van Asbeck, B.S., Hillen, F.C., Boonen, H.C.M., de Jong, Y., Dormans, J.A.M.A., van der Wal, N.A.A., Marx, J.J.M., and Sangster, B. (1989). Continuous intravenous infusion of deferoxamine reduces mortality by paraquat in vitamin E-deficient rats. *Am. Rev. Respir. Dis.* **139**: 769–773.

Van Den Berg, J.J.M., Roelofsen, B., Op Den Kamp, J.A.F., and Van Deenen, L.L.M. (1989). Cooperative protection by vitamins E and C of human erythrocytes membranes against peroxidation. *Ann. N.Y. Acad. Sci.* **570**: 527–529.

van der Vliet, A., Eiserich, J.P., Halliwell, B., and Cross, C.E. (1997). Formation of reactive nitrogen species during peroxidase-catalyzed oxidation of nitrite: A potential additional mechanism of nitric oxide-dependent toxicity. *J. Biol. Chem.* **272**: 7617–7625.

Varani, J., Dame, M.K., Gibbs, D.F., Taylor, C.G., Weinberg, J.M., Shayevitz, J., and Ward, P.A. (1992). Human umbilical vein endothelial cell killing by activated neutrophils: Loss of sensitivity to injury is accompanied by decreased iron content during *in vitro* culture and is restored with exogenous iron. *Lab. Invest.* **66**: 708–714.

Varotsis, C.A., and Babcock, G.T. (1995). Photolytic activity of early intermediates in dioxygen activation and reduction by cytochrome oxidase. *J. Am. Chem. Soc.* **117**: 11260–11269.

Vile, G.F., Basu-Modak, S., Waltner, C., and Tyrrell, R.M. (1994). Heme oxygenase 1 mediates an adaptive response to oxidative stress in human skin fibroblasts. *Proc. Natl. Acad. Sci. USA* **91**: 2607–2610.

Volpp, B.D., Nauseef, W.M., and Clark, R.A. (1988) Two cytosolic neutrophil oxidase components absent in autosomal chronic granulomatous disease. *Science* **242**: 1295–1297.

von Ossowski, I. (1993). Characterization of *katE* and its products, catalase HPII, from *Escherichia coli* by sequence analysis and site-directed mutagenesis.

Ph.D. thesis, Department of Microbiology, University of Manitoba, Winnipeg, Canada.

von Ossowski, I., Hausner, G., and Loewen, P.C. (1993). Molecular evolutionary analysis based on the amino acid sequence of catalase. *J. Mol. Evol.* **37**: 71–76.

von Ossowski, I., Mulvey, M.R., Leco, P.A., Borys, A., and Loewen, P.C. (1991). Nucleotide sequence of *Escherichia coli katE*, which encodes catalase HPII. *J. Bacteriol.* **173**: 514–520.

Wagner, B.A., Buettner, G.R., and Burns, C.P. (1996). Vitamin E slows the rate of free radical-mediated lipid peroxidation in cells. *Arch. Biochem. Biophys.* **334**: 261–267.

Wahlländer, A., Soboll, S., and Sies, H. (1979). Hepatic mitochondrial and cytosolic glutathione content and the subcellular distribution of GSH-S-transferases. *FEBS Lett.* **97**: 138–140.

Waldo, G.S., and Penner-Hahn, J.E. (1995). Mechanism of manganese catalase peroxide disproportionation: Determination of manganese oxidation states during turnover. *Biochemistry* **34**: 1507–1512.

Waldo, G.S., Fronko, R.M., and Penner-Hahn, J.E. (1991). Inactivation and reactivity of manganese catalase: Oxidation-state asignments using X-ray absorption spectroscopy. *Biochemistry* **30**: 10486–10490.

Ward, P.A., and Mulligan, M.S. (1992). Leukocyte oxygen products and tissue damage, in: *Molecular Basis of Oxative Damage by Leukocytes* (Jesaitis, A.J., and Dratz, E.A., eds.), pp. 139–147, CRC Press, Boca Raton, FL.

Ward, P.A., Till, G.O., Kunkel, R., and Beauchamp, C. (1983). Evidence for the role of hydroxyl radical in complement and neutrophil-dependent tissue injury. *J. Clin. Invest.* **72**: 789–801.

Wardman, P., and von Sonntag, C. (1995). Kinetic factors that control the fate of thiyl radicals in cells. *Methods Enzymol.* **251**: 31–45.

Warner, B.B., Stuart, L., Gebb, S., and Wispé, J.R. (1996). Redox regulation of manganese superoxide dismutase. *Am. J. Physiol.* **271**: L150–L158.

Weiss, J. (1944). Radiochemistry of aqueous solutions. *Nature* **153**: 748–750.

Weiss, S.J. (1982). Neutrophil-mediated methemoglobin formation in the erythrocyte. The role of superoxide and hydrogen peroxide. *J. Exp. Med.* **147**: 316–323.

Weiss, S.J. (1989). Mechanisms of disease. *N. Engl. J. Med.* **320**: 365–376.

Weiss, S.J., Lampert, M.B., and Test, S.T. (1983). Long-lived oxidants generated by human neutrophils: Characterization and bioactivity. *Science* **222**: 625–628.

Weiss, S.J., Rustagi, P.K., and LeBuglio, A.F. (1978). Human granulocytes generation of hydroxyl radical. *J. Exp. Med.* **147**: 316–323.

Weiss, S.J., Test, S.T., Eckmann, C.M., Roos, D., and Regiani, S. (1986). Brominating oxidants generated by human eosinophils. *Science* **234**: 200–203.

Wendel, A. (1980). Glutathione peroxide, in: *Enzymatic Basis of Detoxification* (Jakoby, W.B., ed.), Vol. 1, pp. 333–354, Academic Press, New York.

Wendel, A., and Cikryt, P. (1980). The level and half-life of glutathione in human plasma. *FEBS Lett.* **120**: 209–211

Wendenbaum, S., Demange, P., Dell, A., Meyer, J.M., and Abdallah, M.A. (1983). The structure of pyoverdine PA, the siderophore of *Pseudomonas aeruginosa. Tetrahedron Lett.* **24**: 4877–4880.

White, R.E. (1991). The involvement of free radicals in the mechanisms of monooxygenases. *Pharmacol. Ther.* **49**: 21–42.

White, R.E. (1994). The importance of one-electron transfers in the mechanism of cytochrome P-450, in: *Cytochrome P450 Biochemistry, Biophysics and Molecular Biology* (Lechner, M.C., ed.), pp. 333–340, John Libbey Eurotext, Paris.

Wiedau-Pazos, M., Goto, J.J., Rabizadeh, S., Gralla, E.B., Roe, J.A., Lee, M.K., Valentine, J.S., and Bredesen, D.E. (1996). Altered reactivity of superoxide dismutase in familial amyotrophic lateral sclerosis. *Science* **271**: 516–518.

Williams, P.H., and Carbonetti, N.H. (1986). Iron, siderophores, and the pursuit of virulence: Independence of the aerobactin and enterochelin iron uptake systems in *Escherichia coli. Infect. Immun.* **51**: 942–947.

Williams, R.E., Zweier, J.L., and Flaherty, J.T. (1991). Treatment with deferoxamine during ischemia improves functional and metabolic recovery and reduces reperfusion-induced oxygen radical generation in rabbit hearts. *Circulation* **83**: 1006–1014.

Wills, E.D. (1971). Effects of lipid peroxidation on membrane bound enzymes of the endoplasmic reticulum. *Biochem. J.* **123**: 983–991.

Winkler, B.S., Orselli, S.M., and Rex, T.S. (1994). The redox couple between glutathione and ascorbic acid: A chemical and physiological perspective. *Free Radical Biol. Med.* **17**: 333–349.

Winterbourn, C.C. (1983). Lactoferrin-catalyzed hydroxyl radical production: Additional requirements for a chelating agent. *Biochem. J.* **210**: 15–19.

Winterbourn, C.C. (1993). Superoxide as an intracellular radical sink. *Free Radical Biol. Med.* **14**: 85–90.

Winterbourn, C.C., and Malloy, A.L. (1988). Susceptibilities of lactoferrin and transferrin to myeloperoxidase-dependent loss of iron-binding capacity. *Biochem. J.* **250**: 613–616.

Winterbourn, C.C., and Metodiewa, D. (1994). The reaction of superoxide with reduced glutathione. *Arch. Biochem. Biophys.* **314**: 284–290.

Winterbourn, C.C., and Metodiewa, D. (1995). Reaction of superoxide with glutathione and other thiols. *Methods Enzymol.* **251**: 81–86.

Winterbourn, C.C., Monteiro, H.P., and Galilee, C.F. (1990). Ferritin-dependent lipid peroxidation by stimulated neutrophils: Inhibition by myeloperoxidase-derived hypochlorous acid but not by endogenous lactoferrin. *Biochim. Biophys. Acta* **1055**: 179–185.

Winterbourn, C.C., Pichorner, H., and Kettle, A.J. (1997). Myeloperoxidase-dependent generation of a tyrosine peroxide by neutrophils. *Arch. Biochem. Biophys.* **338**: 15–21.

Winterbourn, C.C., Van den Berg, J.J.M., Roltman, E., and Kuypers, F.A. (1992). Chlorohydrin formation from unsaturated fatty acids reacted with hypochlorous acid. *Arch. Biochem. Biophys.* **296**: 547–555.

Winterbourn, C.C., Vile, G.F., and Monteiro, H.P. (1991). Ferritin, lipid peroxidation and redox-cycling xeno-biotics. *Free Radical Biol. Med.* **12–13**: 107–114.

Witting, L.A. (1980). Vitamin E and lipid antioxidants in free-radical-initiated reactions, in: *Free Radicals in Biology* (Pryor, W.A., ed.), Vol. IV, pp. 295–319, Academic Press, New York.

Wong, G.H.W., and Goeddel, D.V. (1988). Induction of manganous superoxide dismutase by tumor necrosis factor: Possible protective mechanism. *Science* **242**: 941–944.

Wong, P.C., Pardo, C.A., Borchelt, D.R., Lee, M.K., Copeland, N.G., Jenkins, N.A., Sisodia, S.S., Cleveland, D.W., and Price, D.L. (1995). An adverse property of a familial ALS-linked SOD1 mutation causes motor neuron disease characterized by vacuolar degeneration of mitochondria. *Neuron* **14**: 1105–1116.

Wright, C.D., and Nelson, R.D. (1998). Candidacidal activity of myeloperoxidase: Characterization of myeloperoxidase-yeast complex formation. *Biochem. Biophys. Res. Commun.* **154**: 809–817.

Yamamoto, Y., and Takahashi, K. (1993). Glutathione peroxidase isolated from plasma reduces phospholipid hydroperoxides. *Arch. Biochem. Biophys.* **305**: 541–545.

Yamazaki, I., and Piette, L.H. (1990). ESR spin-trapping studies on the reaction of Fe^{2+} ions with H_2O_2-reactive species in oxygen toxicity in biology. *J. Biol. Chem.* **265**: 13589–13594.

Yamazaki, I., Piette, L.H., and Grover, T.A. (1990). Kinetic studies on spin trapping of superoxide and hydroxyl radicals generated in NADPH–cytochrome P-450 reductase paraquat systems: Effect of iron chelates. *J. Biol. Chem.* **265**: 652–659.

Yim, M.B., Chock, P.B., and Stadtman, E.R. (1990). Copper,zinc superoxide dismutase catalyzes hydroxyl radical production from hydrogen peroxide. *Proc. Natl. Acad. Sci. USA* **87**: 5006–5010.

Yim, M.B., Chock, P.B., and Stadtman, E.R. (1993). Enzyme function of copper,zinc superoxide dismutase as a free radical generator. *J. Biol. Chem.* **268**: 4099–4105.

Ying, L., He, J., and Furmanski, P. (1994). Iron-induced conformational change in human lactoferrin: Demonstration by sodium dodecyl sulfate-polyacrylamide gel electrophoresis and analysis of effects of iron binding to the N and C lobes of the molecule. *Electrophoresis* **15**: 244–250.

Yoshida, T., Tanaka, M., Sotomatsu, A., and Hirai, S. (1995). Activated microglia causes superoxide-mediated release of iron from ferritin. *Neurosci. Lett.* **190**: 21–24.

Yoshpe-Purer, Y., Henis, Y., and Yashphe, J. (1977). Regulation of catalase level in *Escherichia coli* K12. *Can. J. Microbiol.* **23**: 84–91.

Yu, L., Quinn, M.T., Cross, A.R., and Dinauer, M.C. (1998). Gp91^{phox} is the heme binding subunit of the superoxide-generating NADPH oxidase. *Proc. Natl. Acad. Sci. USA* **95**: 7993–7998.

Zafari, A.M., Ushio-Fukai, M., Akers, M., Yon, Q., Shah, A., Harrison, D.G., Taylor, W.R., and Griendling, K.K. (1998). The role of NADH/NADPH oxidase-derived H_2O_2 in angiotensin II-induced vascular hypertrophy. *Hypertension* **32**: 488–495.

Zager, R.A., and Foerder, C.A. (1992). Effects of inorganic iron and myoglobin on *in vitro* proximal tubular lipid peroxidation and cytotoxicity. *J. Clin. Invest.* **89**: 989–995.

Zarley, J.H., Britigan, B.E., and Wilson, M.E. (1991). Hydrogen peroxide-mediated toxicity for *Leishmania donovani chagasi* promastigotes. Role of hydroxyl radical and protection by heat shock. *J. Clin. Invest.* **88**: 1511–1521.

Zhao, R., Lind, J., Merényi, G., and Eriksen, T.E. (1994). Kinetics of one-electron oxidation of thiols and hydrogen abstraction by thiyl radicals from α-amino C–H bonds. *J. Am. Chem. Soc.* **116**: 12010–12015.

3

Nitric Oxide

A Simple Molecule with Complex Biologic Functions

In view of the fact that the properties of EDRF
released from endothelial cells of the rabbit aorta by
acetylcholine are, in many respects, similar to those of
NO· generated in acidified nitrite solutions, it is proposed that EDRF is NO·.

Furchgott, 1988

While production of free radicals, including $O_2\cdot^-$ and HO·, has generally been
associated with cytotoxicity, the recognition that nitric oxide (NO·) controls a
myriad of important physiological activities, including the regulation of vascular
tone (Furchgott, 1988; Ignarro, et al., 1988; Ma, et al., 1993), ushered in a new era
in free radical research. Undeniably, NO· represents a novel class of transient
messengers, as this free radical augments cell–cell communications and governs
many intracellular events (Moncada and Higgs, 1991; Snyder, 1992; Förstermann,
et al., 1994, 1995; Huang, et al., 1993, 1994, 1995; Rubbo, et al., 1996; Lander,
1997). Besides these critical homeostatic functions, NO· appears to be an integral
part of host immune response, particularly effective against a number of intracellular pathogens (Nathan and Hibbs, 1991; De Groote and Fang, 1995;
MacMicking, et al., 1995; Wei, et al., 1995). These and similar findings have had
a profound influence on what would otherwise be a narrow view of free radicals in
biology.

As in many scientific endeavors, the discovery that endothelium-derived relaxing factor (EDRF) was NO· came from many seemingly unrelated observations,
whose connection only became clearer during the past decade. Publications of an
earlier era, typified by Beck, et al. (1961), primarily focused on the ability of nitro-
containing compounds to dilate blood vessels, thereby controlling vascular tone.
By the late 1970s, however, a number of reports surfaced that suggested that the
relaxation of the vascular smooth muscle by endothelium-dependent pathways and
the pharmacological activity of nitrogen oxide-containing compounds, such as
nitroglycerin, had a common element, linking what were previously thought to
be disparate regulatory mechanisms for vascular control (Figure 3.1) (Axelsson, *et*

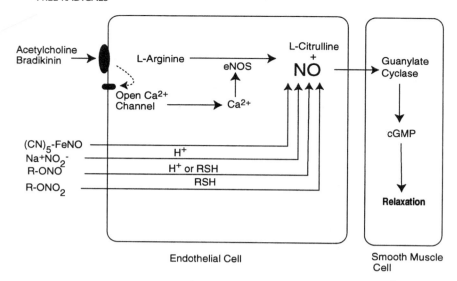

Figure 3.1 Mechanism of NO· production by endothelial cells. Receptor activation, by either acetylcholine or bradikinin, leads to intracellular Ca^{2+} binding to eNOS (NOS III)—essential for the metabolic oxidation of L-arginine to L-citrulline and NO·. Once formed, this free radical diffuses from this cell and into a smooth muscle cell, where NO· binds to guanylate cyclase. This reaction initiates a series of cellular events, including the formation of cGMP that leads to vascular smooth muscle relaxation. Examples of ·NO-releasing compounds include $(CN)_5FeNO$, nitroprusside; $NaNO_2$, sodium nitrite; R–ONO, organic nitrite esters; $R–ONO_2$, organic nitrate esters such as nitroglycerin. The reaction of nitrite or nitrate esters with thiols, RSH, can result in the formation of S-nitroso compounds that will release NO· upon intracellular hydrolysis.

al., 1979; Ignarro, et al., 1981, 1984; Holzmann, 1982; Diamond and Chu, 1983). It was soon realized, however, that the unifying thread was NO·, which, when bound to guanylate cyclase, mediated transient increases in intracellular levels of 3′,5′-cyclic guanosine monophosphate (cGMP) (Craven and DeRubertis, 1978, 1983; Ignarro, et al., 1982). This cascade, which was later shown to be a short-lived intracellular signal transduction pathway (Ignarro, 1991), is a critical element in the regulation of vascular tone. These findings, while verifying the important role that NO· plays in governing vascular response, supported the clinical use of nitrogen oxide-containing compounds, first demonstrated by Brunton (1867) over a hundred years ago.

Meanwhile, evidence from other avenues of biologic research suggested an important immunologic function for NO· (Figure 3.2). While studying the metabolism of nitrosamines, a class of carcinogens, it was found that germ-free rats excreted more nitrate than they ingested (Green, et al., 1981b). At about the same time, it was reported that nitrate levels in humans increased during infection (Green, et al., 1981a). The commonality of these observations, although not immediately apparent, became evident with the finding that murine macrophages were

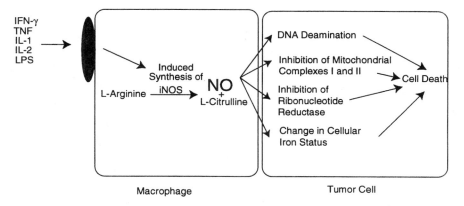

Figure 3.2 A number of cytokines can induce the synthesis of iNOS (NOS II) in macrophages and other cells. Nitric oxide is produced during the NOS II-catalyzed oxidation of L-arginine to L-citrulline. As in the case of the endothelial cell, NO· diffuses from this leukocyte to, for instance, a tumor cell. By reacting with critical cell targets, it is believed that this free radical plays an integral part in the antitumor activity of macrophages.

capable of producing an "activated" nitrogen oxide from L-arginine that coincidentally not only could catalyze the N-nitrosation of amines to nitrosamines (Stuehr and Marletta, 1985; Iyengar, *et al.*, 1987; Miwa, *et al.*, 1987), but also could mediate cytotoxicity (Hibbs, *et al.*, 1987a). One important outgrowth from these studies was that a family of *N*-guanidino-substituted L-arginine analogs blocked macrophage-mediated cell toxicity in a stereospecific manner (Hibbs, *et al.*, 1987a, 1987b). These specific antagonists have proven to be useful tools in defining various physiologic functions attributal to NO·. Finally, with the demonstration that NO· was the putative intermediate released during macrophage metabolism of L-arginine (Hibbs, *et al.*, 1988; Marletta, *et al.*, 1988; Stuehr and Nathan, 1989; Nathan, 1992), a new era in host immune response had begun, centered around the chemical reactions of this free radical.

The hypothesis that NO· can act as a second messenger in the brain has its foundation in studies conducted as early as 1977, when it was determined that this free radical activated soluble guanylate cyclase from crude homogenates of this tissue (Miki, *et al.*, 1977). The significance of this finding would, nevertheless, remain dormant for nearly a decade, even though it was known that L-arginine was the endogenous activator of this enzyme (Deguchi and Yoshioka, 1982). With the localization and the purification of a neuronal nitric oxide synthase (NOS I), capable of oxidizing L-arginine to L-citrulline and NO· (Bredt and Snyder, 1990; Bredt, *et al.*, 1990), the road linking NO· to different physiologic functions in the central nervous system (Demas, *et al.*, 1997), which had started nearly a quarter of a century ago, was now established. Nevertheless, it should be pointed out that the versatility of this free radical in controlling a myriad of brain functions will undoubtedly result in new and provocative discoveries (Snyder, 1992; Huang, *et al.*, 1993; Son, *et al.*, 1996) (Figure 3.3).

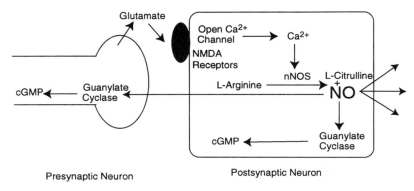

Figure 3.3 One current theory suggests that in the central nervous system, glutamate, through the NMDA receptors, stimulates the production of NO· by nNOS (NOS I). As depicted, NO· can act as either an anterograde or a retrograde transient cell messenger.

Chemistry of Nitric Oxide

Even though NO· is a free radical, it exhibits a chemical inertness toward dimerization that is uncharacteristic of most carbon-centered free radicals. This inherent electronic stability is the result of an overlap of the 2p orbitals on nitrogen and oxygen, leading to a filled π orbital. The unpaired electron is, therefore, accommodated in the π* antibonding orbital (Fontecave and Pierre, 1994). In valence bond terms, we show that NO· can be represented by the resonance structures:

$$:\overset{.}{N}\!\!=\!\!\overset{..}{O}: \quad\longleftrightarrow\quad :\overset{-}{N}\!\!=\!\!\overset{..}{\overset{+}{O}}:$$

The tendency of NO· not to dimerize, forming N_2O_2 in aqueous solutions at physiologic pH, has been attributed to the stabilization produced by the nearly complete resonance between the two structures shown above and to the consequential formation of a partial three-electron bond (Pauling, 1960). Despite this, NO· appears to form other intermediates in aqueous buffers. For instance, a recent kinetic study of N-nitrosation of morphine in oxygenated NO· solutions at physiologic pH is consistent with the generation of nitrous anhydride (N_2O_3) as the reactive species responsible for this reaction (Lewis, et al., 1995; Owens, et al., 1995).

In aerobic aqueous solutions, NO· combines with O_2 according to

$$4NO\cdot + O_2 + 2H_2 \rightarrow 4NO_2^- + 4H^+$$

in which the kinetics of the reaction dictate a second order in NO· and a first order in O_2 (Pogrebnaya, et al., 1975; Awad and Stanbury, 1993; Ford, et al., 1993;

Ignarro, *et al.*, 1993; Wink, *et al.*, 1993b; Kharitonov, *et al.*, 1994; Lewis and Deen, 1994; Pires, *et al.*, 1994; Goldstein and Czapski, 1995a, 1996; Lewis, *et al.*, 1995):

$$-d[NO\cdot]/dt = 4k[NO\cdot]^2[O_2]$$

The rate constant for the disappearance of NO· is, therefore, dependent on the concentration of the reactants. In the case where $[NO\cdot]_0 > [O_2]_0$ at 22–25°C, the rate constant has been found to vary from 2.05 to $2.9 \times 10^6 \, M^{-2} \, s^{-1}$ (Awad and Stanbury, 1993; Ford, *et al.*, 1993; Kharitonov, *et al.*, 1994; Lewis and Deen, 1994; Goldstein and Czapski, 1995a, 1996). When $[O_2]_0 > [NO\cdot]_0$, the rate constant appears to range from 1.8 to $11.5 \times 10^6 \, M^{-2} \, s^{-1}$ at 22–25°C (Pogrebnaya, *et al.*, 1975; Awad and Stanbury, 1993; Kharitonov, *et al.*, 1994; Pires, *et al.*, 1994; Goldstein and Czapski, 1995a). However, in hydrophobic environments, the rate of reaction of NO· with O_2 appears to be approximately 300 times faster than it is in aqueous solutions (Liu, *et al.*, 1998).

The slow third-order reaction of NO· in oxygenated buffers and the neutrality of this free radical (Stamler, *et al.*, 1992c) allow NO· to easily diffuse through membranes (Subczynski, *et al.*, 1996) to biologic targets of interest (Wood and Garthwaite, 1994). Coupled with this, the surprisingly long half-life of 3 min at the highest reported cellular level of 4 µM under autoxidation conditions (Tables 2.3 and 3.1) makes NO· an ideal transient second messenger molecule (Lancaster, 1994). Finally, given the greater reactivity of NO· toward O_2 in hydrophobic pockets, such as biologic membranes and other lipid sites within the cell, than in aqueous compartments, differences in local environment may play an important role in the disproportionate regulation of NO· bioactivition by contributing to its rapid removal (Liu, *et al.*, 1998).

Table 3.1 Tissue Levels of Nitric Oxide in Response to Stimulation

[NO·] (µM)	Tissue	Stimulus	References
0.35–5.0	Endothelium in aorta	Bradykinin	1,2
0.05–5.0	Cultured endothelial cells	Bradykinin	1, 3–5
0.13–0.85	Smooth muscle in aorta	Brandykinin	1, 2
0.5	Cultured smooth muscle	Interleukin-1β	4
1.0–5.0	Isolated platelets	Collagen	6
0.01–0.08	Brain slice	Electrical	7, 8
1.0–4.0	Brain (*in vivo*)	Ischemia	9, 10
0.4–1.0	Cultured astrocytes	Interferon-γ/LPS	4
0.124	Human veins (*in vivo*)	Bradykinin	11
0.38	Mesenteric arteries	A23187	12

Source: Brown, 1995a.

1 = Malinski and Taha, 1992; 2 = Malinski, *et al.*, 1993a; 3 = Tsukahara, *et al.*, 1993; 4 = Malinski, *et al.*, 1993b; 5 = Blatter, *et al*, 1995; 6 = Malinski, *et al.*, 1993c; 7 = Shibuki, 1990; 8 = Shibuki and Okada, 1991; 9 = Malinski, *et al.*, 1993d; 10 = Zhang, *et al.*, 1995b; 11 = Vallence, *et al.*, 1995; 12 = Tschudi, *et al.*, 1996.

As expected, NO· reacts with other free radicals at near-diffusion controlled rates. For instance, NO· combines with ethyl alcohol radicals, either the α- or the β-isomer, and organic peroxyl radicals with rate constants approaching 3×10^9 $M^{-1} s^{-1}$ and $1-3 \times 10^9 M^{-1} s^{-1}$, respectively (Padmaja and Huie, 1993; Czapski, *et al.*, 1994). The reaction of $O_2 \cdot^-$ with NO· at pH > 6 is similarly rapid, ranging from 3.8 to $6.7 \times 10^9 M^{-1} s^{-1}$ (Huie and Padmaja, 1993; Goldstein and Czapski, 1995b; Kobanyashi, *et al.*, 1995), and by one account as high as $19 \times 10^9 M^{-1} s^{-1}$ (Kissner, *et al.*, 1997).

Transition metal ions either incorporated into proteins or bound as an Fe–S cluster of metalloproteins are critical targets for NO· (Deisseroth and Dounce, 1970; Doyle and Hoekstra, 1981; Tsai, 1994; Bohle and Hung, 1995; Brown, 1995b; Gorbunov, *et al.*, 1995; Bouton, *et al.*, 1996; Hoshino, *et al.*, 1996; Jia, *et al.*, 1996; Morlino, *et al.*, 1996). Surprisingly, the association rate constant for this binding is small, ranging from 10^3 to $10^7 M^{-1} s^{-1}$ and depends, to a large extent, on the protein structure at the heme pocket (Sharma, *et al.*, 1987). Even under the most ideal of conditions, these rate constants are several orders of magnitude slower than is the reaction of NO· with other free radicals (Huie and Padmaja, 1993; Padmaja and Huie, 1993; Czapski, *et al.*, 1994; Goldstein and Czapski, 1995b). Despite this, the resulting competition between NO· and O_2 for binding sites on heme-containing proteins may have serious implications (Hori, *et al.*, 1992; Nakano, *et al.*, 1996). For illustrative purposes, consider cytochrome c oxidase, the terminal enzyme of the respiratory electron transport chain (Brown and Cooper, 1994; Cleeter, *et al.*, 1994; Schweizer and Richter, 1994; Takehara, *et al.*, 1995; Torres, *et al.*, 1996). In this case, as little as 60 nM of NO· can half-inhibit this enzyme at cellular O_2 levels approaching 30 μM (Brown, 1995a).

In theory, NO· can either accept one electron, forming NO^- or donate one electron, generating NO^+. Based on the redox potential of model compounds, it has been suggested that few available oxidizing agents found in biological systems are able to oxidize NO· to NO^+ (Fontecave and Pierre, 1994). In contrast, the redox potential of NO·/NO^- is about 0.25 V versus the normal hydrogen electrode, which would render the reduction of NO· to NO^- a facile reaction (Fontecave and Pierre, 1994). For instance, Fe^{2+} can serve as a reducing agent. Once formed, NO^- would seek biologic pathways to rapidly lose an electron, producing NO·.

The coordination chemistry of NO· to transition metal represents another important aspect of the chemistry of this free radical. In the presence of redox active metal ions, such as chelated-Fe^{3+}, NO· can form NO–metal complexes that can then undergo charge transfer, generating either NO^+ or NO^-. In fact, the charge transfer species NO^+ has been postulated to be the intermediate responsible for the nitrosation reaction of proteins induced by NO· (Fontecave and Pierre, 1994). Transition metal ions might, therefore, play a critical role in the redox chemistry of NO·. Despite this, it is important to emphasize that much remains to be determined as to the reactivity of NO· in biological systems, especially with regard to reactions that involve electron transfer and the coordination of this free radical to metal-containing proteins. With this in mind, it has been proposed that NO· is associated with neurodegeneration, whereas NO^+ is neuroprotective (Lei, *et al.*, 1992; Lipton, *et al.*, 1993, Lipton and Stamler, 1994).

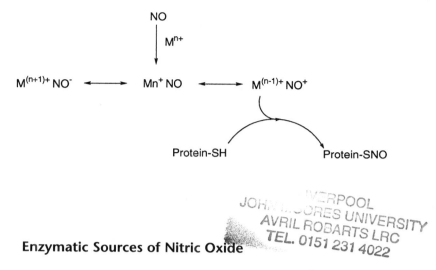

Enzymatic Sources of Nitric Oxide

There is little doubt that 1987 was another watershed year[1] in the exploration of free radicals in biology, as two independent reports[2] (Ignarro, et al., 1987; Palmer, et al., 1987) demonstrated that NO· could mimic the vasodilation attributal to endothelium-derived relaxation factor (Furchgott and Zawadzki, 1980). Unlike the indiscriminate generation of other free radicals that frequently arise, secondary to cell metabolism, production of NO·, like $O_2 \cdot^-$ in phagocytic cells, is regulated through receptor-mediated events (Moncada and Higgs, 1991, 1993). From these initial observations (Furchgott and Zawadzki, 1980), others discovered that NO· was derived from L-arginine (Palmer, et al., 1988a, 1988b) through the catalytic action of the monooxygenase, nitric oxide synthase (NOS) (Bredt and Snyder, 1990; Mayer, et al., 1990; Schmidt, et al., 1991; Stuehr, et al., 1991a; Yui, et al., 1991).

Currently, there appear to be two distinct classes of NOS—a constitutive and an inducible form of the enzyme (Table 3.2) (Janssens, et al., 1992; Marsden, et al., 1992; Geller, et al., 1993; Nakane, et al., 1993; Martin, et al., 1994; Förstermann, et al., 1995). The constitutive isozyme is associated with endothelial cells (NOS III) lining vascular beds (Hecker, et al., 1990; Pollock, et al., 1991; Lamas, et al., 1992; Sessa, et al., 1992) and myocytes (Balligand, et al., 1995), whereas another basal isoform (NOS I) is localized to the central and peripheral neurons (Dawson, et al., 1992; Sheng, et al., 1992), spinal cord (Dun, et al., 1992), sympathetic ganglia and adrenal glands (Dun, et al., 1993; Sheng, et al., 1993), skeletal muscles (Nakane, et al., 1993; Kobzik, et al., 1994), β cells of pancreatic islets (Schmidt, et al., 1992a), cells of the anterior pituitary gland (Duvilanski, et al., 1995), and epithelial cells from various organs (Schmidt, et al., 1992b; Asano, et al., 1994). In the case of skeletal muscle, it appears that both NOS I and NOS II play an important role in the adjustment in contractile response (Gath, et al., 1996). Here, NOS II appears to be constitutively expressed in certain muscle cells (Gath, et al., 1996). While the above list of NOS distribution is extensive, it is not meant, however, to be exhaustive. In fact, with time, NOS I and NOS III will undoubtedly be found to be ubiquitously distributed in many other tissues.

Table 3.2 Nomenclature for Isozymes of Nitric Oxide Synthase[a]

Numerical	Description	Definition
NOS I	nNOS	A neuronal constitutive enzyme whose activity is regulated by intracellular Ca^{2+}. nNOS is characterized as a low-output NO· enzyme.
NOS II	iNOS	A cytokine-inducible enzyme whose activity is independent of intracellular Ca^{2+}. iNOS, prototypically expressed by murine macrophases, is characterized as a high-output NO· enzyme.
NOS III	eNOS	An endothelial constitutive enzyme whose activity is regulated by intracellular Ca^{2+}. eNOS is characterized as a low-output NO· enzyme.

Source: Nathan and Xie, 1994b and Förstermann, *et al.*, 1995.

[a]Distinct isozymes have been isolated and classified as either constitutive or inducible NOS. The use of the terms constitutive or inducible NOS may be confusing, since biosyntheses of the different isoforms can be enhanced by various stimuli. To avoid this confusion, it has been proposed that all NOS isozymes be classified as either NOS I (nNOS) for neuronal, NOS III (eNOS) for endothelial, and NOS II (iNOS) for immunological (Nathan and Xie, 1994b, Förstermann, *et al.*, 1995; Rosen, *et al.*, 1995).

Another distinct isozyme of this enzyme, NOS II was initially discovered in macrophages isolated from several different rodent species (Iyengar, *et al.*, 1987; Granger, *et al.*, 1988; Marletta, *et al.*, 1988; Hevel, *et al.*, 1991; Stuehr, *et al.*, 1991a; Lyons, *et al.*, 1992; Xie, *et al.*, 1992; Deng, *et al.*, 1993).[3] Unlike the constitutive enzymes NOS I and NOS III, NOS II has been expressed in a number of primary and immortalized cells upon exposure to a variety of cytokines, such as lipopolysaccharide and interferon-γ (Nathan, 1992; Bandaletova, *et al.*, 1993; Charles, *et al.*, 1993; Geller, *et al.*, 1993; Kleeman, *et al.*, 1993; Kobzik, *et al.*, 1993; Asano, *et al.*, 1994; Balligand, *et al.*, 1994; Hattori, *et al.*, 1994; Martin, *et al.*, 1994; Melillo, *et al.*, 1994; Robbins, *et al.*, 1994; Rosenkranz-Weiss, *et al.*, 1994; Weisz, *et al.*, 1994; de-Mello, *et al.*, 1996; Xie and Gross, 1997). This isozyme has been, for instance, induced in human chondrocytes upon treatment with interleukin 1β (Charles, *et al.*, 1993), in human hepatocytes after treatment with tumor necrosis factor-α, interleukin 1, interferon-γ and lipopolysaccharide (Geller, *et al.*, 1993; Duval, *et al.*, 1996), in human alveolar macrophages of patients with idiopathic pulmonary fibrosis, who were subsequently exposed *in vitro* to *Mycobacterium bovis bacillus Calmette-Guérin* (Nozaki, *et al.*, 1997) and in human macrophages following *in vitro* ligation of the low affinity IgE receptor FcεRII/CD23 (Vouldoukis, *et al.*,1995). Similarly, NOS II has been observed in human lung macrophages from individuals with acute bronchopneumonia (Tracey, *et al.*, 1994), tuberculosis (Nicholson, *et al.*, 1996), and in brains of patients with multiple sclerosis (Bagasra, *et al.*, 1995). Because of this unique feature, this isozyme has been given the name inducible nitric oxide synthase.

Nitric oxide synthase I is primarily a soluble migrating enzyme with a molecular mass of 150–160 kDa (Bredt and Synder, 1990; Mayer, *et al.*, 1990; Schmidt, *et al.*, 1991, Bredt, 1996). Some particulate activity has, however, been shown, particularly in skeletal muscle, where it appears to be membrane bound (Hiki, *et al.*, 1992; Matsumoto, *et al.*, 1993; Nakane, *et al.*, 1993; Kobzik, *et al.*, 1994). Unlike cytochrome P-450, with which it shares many common features (McMillian, *et al.*, 1992; Stuehr and Ikeda-Saito, 1992; White and Marletta, 1992; Masters, *et al.*, 1996), NOS I is a single, bidomain protein of identical subunits (Sheta, *et al.*, 1994; McMillian and Masters, 1995; Riveros-Moreno, *et al.*, 1995; Klatt, *et al.*, 1996). It contains the flavins FAD and FMN (Bredt, *et al.*, 1991; Mayer, *et al.*, 1991; Lowe, *et al.*, 1996), an iron protoporphyrin IX (Klatt, *et al.*, 1992; McMillian, *et al.*, 1992; Stuehr and Ikeda-Saito, 1992; White and Marletta, 1992; McMillian and Masters, 1993), and binding sites for NADPH, L-arginine, and calmodulin, the latter associated with Ca^{2+} (Schmidt, *et al.*, 1992c; Abu-Soud and Stuehr, 1993; Abu-Soud, *et al.*, 1994; Su, *et al.*, 1995; Zhang, *et al.*, 1995a; Persechini, *et al.*, 1996). The pyridine nucleotide, NADPH, donates electrons through the flavins (Iyengar, *et al.*, 1987; Pou, *et al.*, 1992) to the terminal acceptor, the heme (Leone, *et al.*, 1991; Griffith and Stuehr, 1995). This event catalyzes the oxidation of L-arginine to L-citrulline and NO·. Interestingly, recent studies suggest that NO· can reversibly react with the heme, forming an inactive ferrous nitrosyl complex (Rengasamy and Johns, 1993; Griscavage, *et al.*, 1994; Abou-Soud, *et al.*, 1995). As such, this free radical controls its own synthesis. An additional cofactor, tetrahydro-L-biopterin (BH$_4$) has been reported to be bound to a sequential site on this monooxygenase; however, its participation in enzymic catalysis remains controversial[4] (Mayer, *et al.*, 1990, 1991, 1995; Giovanelli, *et al.*, 1991; Stuehr, *et al.*, 1991a; Schmidt, *et al.*, 1992c; McMillian and Masters, 1993; Pufahl and Marletta, 1993; Griscavage, *et al.*, 1994; Klatt, *et al.*, 1995, 1996; Nishimura, *et al.*, 1995; Gorren, *et al.*, 1996; Saura, *et al.*, 1996; Witteveen, *et al.*, 1996). One rather surprising finding is that the active dimeric conformation of NOS I remains, even in the absence of BH$_4$ and L-arginine (Klatt, *et al.*, 1995).

Protein–protein interactions regulate NOS I activity and target this enzyme to specific subcellular loci (Bredt, 1996; Venema, *et al.*, 1997b). For instance, the binding of Ca^{2+} to calmodulin activates the enzyme, allowing electron transfer between the oxidized and reduced state of the enzyme (Masters, *et al.*, 1996; Lee and Stull, 1998), whereas a newly discovered 10-kDa protein, known as PIN, has been found to inhibit formation of NO· (Jaffrey and Snyder, 1996). This isozyme of NOS is targeted to specific sites by interaction with postsynaptic density proteins PSD-95 and PSD-93 and to the sarcolemma of skeletal muscle cells by α1-syntrophin in the membrane cytoskeleton dystrophin complex (Brenman, *et al.*, 1995, 1996a, 1996b). Protein binding is at the PDZ domain of a 230-amino acid N-terminal extension (Bredt, 1996).

Nitric oxide synthase III shares cofactor requirements, primary amino acid sequence, and prosthetic groups with NOS I (Marsden, *et al.*, 1992; Chen, *et al.*, 1996; Rodríguez-Crespo, *et al.*, 1996). There are, nevertheless, a number of peculiarities unique to this isozyme (Hellermann and Solomonson, 1997; Venema, *et al.*, 1997a). It appears that NOS III is initially N-myristolyated (Pollock, *et al.*, 1992). This allows the enzyme to enter the Golgi, where it is then palmitoylated (Robinson, *et al.*, 1995; Sessa, *et al.*, 1995; García-Cardeña, *et al.*, 1996b). The trans-Golgi network relocates NOS III to the caveolae, plasmalemmal domains within the plasma membrane that regulate enzyme activity (Pollock, *et al.*, 1991;

Hecker, *et al.*, 1994; Shaul, *et al.*, 1996; Song, *et al.*, 1996; Ju, *et al.*, 1997; García-Cardeña, *et. al.*, 1997). This compartmentalization, which most likely results from hydrophobic interactions between the N-myristoylated palmitoylated protein and membrane phospholipids, permits the activated enzyme to secrete the labile NO· into the surrounding milieu (Busconi and Michel, 1993, 1994; Liu and Sessa, 1994; Michel, *et al.*, 1997). Thereafter, NOS III is phosphorylated, allowing its subcellular trafficking from the cellular membrane to the cytosol (Michel, *et al.*, 1993; Hecker, *et al.*, 1994; García-Cardeña, *et al.*, 1996a, 1996b; Prabhakar, *et al.*, 1998). This phosphorylation leads to a deactivation of the enzyme. This biochemical reaction may serve as a negative feedback mechanism, controlling the production of NO· (Michel, *et al.*, 1993). Several observations point to this thesis. First, upon stimulation with bradykinin, endothelial cell-release of NO· occurs within seconds, returning to baseline over the next 5 min (Malinski and Taha, 1992). Second, phosphorylation is at its highest point after 5 min, at a time when NO· secretion is not detectable (Michel, *et al.*, 1993).

Despite the constitutive nature of NOS I and NOS III, levels of enzyme expression can be modulated by chemical stimuli or changes in physiologic status. For instance, mRNA levels for NOS I and NOS III increase in response to shear-stress, estrogen, and hypoxia (Nishida, *et al.*, 1992; Kuchan and Frangos, 1994; Weiner, *et al.*, 1994; O'Neill, 1995; Arnet, *et al.*, 1996; Ayajiki, *et al.*, 1996; Corson, *et al.*, 1996; Fukaya and Ohhashi, 1996); whereas, upon prolonged exposure to tumor necrosis factor-α, a slight decrease in expression of mRNA NOS III has been noted (Lamas, *et al.*, 1992; Nishida, *et al.*, 1992; Yoshizumi, *et al.*, 1993).

Nitric oxide synthase II, a predominately soluble, dimeric protein comprising two identical subunits of molecular mass of 130 kDa, contains the same prosthetic groups associated with the other NOS family of enzymes (Kwon, *et al.*, 1990; Hevel, *et al.*, 1991; Stuehr, *et al.*, 1991a; Stuehr and Ikeda-Saito, 1992; Baek, *et al.*, 1993; Cho, *et al.*, 1995; Ghosh and Stuehr, 1995; Chabin, *et al.*, 1996; Sari, *et al.*, 1996; Siddhanta, *et al.*, 1996; Grant, *et al.*, 1997). In contrast to the constitutive isoforms NOS I and NOS III, NOS II secretes NO· with resting cellular levels of Ca^{2+} that avidly binds to calmodulin (Cho, *et al.*, 1992; Hevel and Marletta, 1992; Lyons, *et al.*, 1992; Xie, *et al.*, 1992; Sakai, *et al.*, 1993; Nathan and Xie, 1994a; Stevens-Truss and Marletta, 1995; Ruan, *et al.*, 1996); this occurs for a surprisingly prolonged period of time (Green, *et al.*, 1994; Vodovotz, *et al.*, 1994). As with the other NOS isozymes, BH$_4$ is an essential cofactor for enzyme activity, even though its catalytic function remains uncertain (Kwon, *et al.*, 1989; Tayeh and Marletta, 1989; Hevel and Marletta, 1992; Baek, *et al.*, 1993; Siddhanta, *et al.*, 1996).

Although cellular expression of NOS II is controlled at the transcriptional level (Xie, *et al.*, 1992; Geller, *et al.*, 1993; Nathan and Xie, 1994b), there is evidence that protranscriptional alterations in the enzyme are essential for activation (Baek, *et al.*, 1993). In particular, NOS II is synthesized as a monomer, then dimerizes to generate NO· (Albakri and Stuehr, 1996). The formation of the "active" enzyme is surprisingly slow, as after 8 hr of cytokine induction there is an equal mixture of NOS II monomer and dimer (Baek, *et al.*, 1993), with dimerization limited by the presence and reaction of NO· with heme (Albakri and Stuehr, 1996). This free radical can further limit its own production through binding to the sixth ligand position of the iron protoporphyrin IX (Wang, *et al.*, 1994; Hurshman and Marletta, 1995). In this manner, this free radical can act in a negative feedback pathway when saturating concentrations of L-arginine would, otherwise, produce toxic levels of NO·. In other situations, by promoting iron loss, NO· actually enhances mRNA and protein expression (Weiss, *et al.*, 1994). Despite these known mechanisms for the regulation of NO·, once the enzyme is expressed, continued secretion of this free radical goes unabated for extended periods.

Target Sites for Nitric Oxide

The biological targets of NO· have been only partially identified (Figure 3.4). The best characterized of these is guanylate cyclase, which is activated by NO· through the reversible attachment to the iron located at its functional site (Craven and DeRubertis, 1978; Ignarro, *et al.*, 1982; Henry, *et al.*, 1991; Ignarro, 1991; Lincoln and Cornwell, 1993; Tsai, 1994; Stone and Marletta, 1996). Likewise, NO· activates heme oxygenase, thereby offering a degree of cytoprotection from oxidant stress (Motterlini, *et al.*, 1996). For other heme-containing proteins, such as cytochrome P-450 (Khatsenko, *et al.*, 1993; Wink, *et al.*, 1993b; Carlson and Billings, 1996; Donato, *et al.*, 1997), lipoxygenase and cyclooxygenase (Kanner, *et al.*, 1992), and xanthine oxidase (Fukahori, *et al.*, 1994), NO· is inhibitory. Similar findings have been noted for glutathione peroxidase (Asahi, *et al.*, 1995). In this case, the inactivation of this enzyme through the formation of selenenyl sulfide by NO· results in enhanced levels of peroxides with the accompanying damage to cellular membranes (Asahi, *et al.*, 1997).

Not surprisingly, NO· has also been reported to bind to Fe–S centers such as those found in *Clostridium botulinum* (Reddy, *et al.*, 1983). Unlike the NO· associated with heme-containing proteins, the reaction of this free radical with Fe–S cluster proteins frequently inactivates the enzyme, as suggested for mitochondrial aconitase (Drapier and Hibbs, 1988), NADH–ubiquinone oxidoreductase, and NADH–succinate oxidoreductase (Drapier and Hibbs, 1986; Hibbs, *et al.*, 1988).

Of particular interest is the role that NO· plays in regulating iron metabolism through its reaction with iron-regulatory proteins, the apo forms of mammalian cytoplasmic aconitase (Hibbs, *et al.*, 1988; Weiss, *et al.*, 1994; Pantopoulos and Hentze, 1995; Pantopoulos, *et al.*, 1996). This family of proteins controls cellular transport of iron by binding to iron-responsive elements, structural motifs within the untranslated regions of mRNAs, which are involved in storage, uptake, and utilization of iron (Pantopoulos, *et al.*, 1994; Butler, *et al.*, 1995; Oria, *et al.*, 1995; Pantopoulos and Hentze, 1995; Juckett, *et al.*, 1996). The iron-regulatory proteins "sense" iron levels in the cell and bind to iron-responsive elements when cells are in an iron starvation state. Nitric oxide appears to play an important role in this process (Pantopoulos, *et al.*, 1994). Here, this free radical has been shown to inactivate aconitase enzyme activity (Gardner, *et al.*, 1997), with a simultaneous rise in the level of the iron-regulatory protein 1 binding to RNA (Drapier, *et al.*, 1993; Weiss, *et al.*, 1993; Kennedy, *et al.*, 1997). Other findings, however, cast doubt that NO·, per se, is responsible for the inactivation of aconitases. Rather, it has been suggested that peroxynitrite, (ONOO$^-$), derived from the reaction of

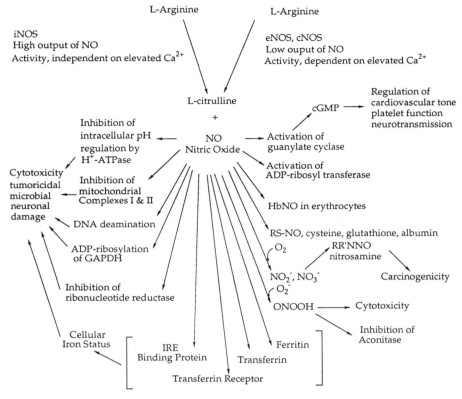

Figure 3.4 Potential cellular targets for NO·. (Adapted from Henry, *et al.*, 1993.)

NO· with $O_2\cdot^-$, inhibits these enzymes (Castro, *et al.*, 1994; Hausladen and Fridovich, 1994).

Several years ago, it was proposed that NO· acts as a cell-messenger molecule through activation of cGMP-independent pathways (Lander, *et al.*, 1993, 1995a). From these and other in-depth studies, it was suggested that this free radical, by interacting with the *ras* oncogene product p21^{ras}, regulates cell signaling (Lander, *et al.*, 1995a, 1995b, 1996). More recent inquiries have identified a cysteine residue, Cys^{118}, located within a highly conserved region of the *ras* superfamily and the GTP-binding protein motif, as the critical site of S-nitrosylation (Lander, *et al.*, 1997). The presence of a thiol at such a critical domain suggests the importance of redox-triggering pathways in cellular regulation of signaling. One engaging aspect of this theory is that a molecular switch may, indeed, be under the control of cellular oxidant levels. As such, GSH concentrations may indirectly regulate signal transduction pathways by acting as a rheostat for the S-nitrosylation of p21^{ras} (Lander, *et al.*, 1997). Future studies will undoubtedly test the validity of this hypothesis.

Physiologic Actions of Nitric Oxide

The ubiquitous nature of NO· attests to its multifaceted role in controlling a myriad of physiologic responses. Early studies verified its status in the relaxation of vascular smooth muscle (Moncada, *et al.*, 1991) and the inhibition of platelet aggregation and adhesion (Furlong, *et al.*, 1987; Radomski, *et al.*, 1987a; Sneddon and Vane, 1988; Groves, *et al.*, 1993), thought to occur through its effects on cGMP and synergy with prostacyclin (Radomski, *et al.*, 1987b). More recent inquiries demonstrate the ability of NO· to hinder neutrophil rolling, through decreased expression of P-selectin (Gauthier, *et al.*, 1994) and adherence to the endothelium by blocking the binding of the neutrophilic β_2 integrin Mac-1 (macrophage antigen-1, CD11b/CD18) to ICAM-1 (intercellular adhesion molecule-1), and VLA-4 (very late antigen, $\alpha_4\beta_1$ integrin) to VCAM-1 (vascular cell adhesion molecule-1), on endothelial cells (Kubes, *et al.*, 1991; Ma, *et al.*, 1993; Moilanen, *et al.*, 1993; Fujita, *et al.*, 1994; De Caterina, *et al.*, 1995; Peng, *et al.*, 1995). Finally, this free radical can modulate the proliferation of smooth and skeletal muscle cells that may significantly impact atherosclerosis and related diseases (Garg and Hassid, 1989; Cooke, *et al.*, 1992; Flavahan, 1992; Nakane, *et al.*, 1993; Asano, *et al.*, 1994; Kobzik, *et al.*, 1994; Davies, *et al.*, 1995), as well as intimal hyperplasia, an important etiologic factor in the development of restenosis (McNamara, *et al.*, 1993; Guo, *et al.*, 1994; Hamon, *et al.*, 1994; Tarry and Makhoul, 1994; Von der Leyen, *et al.*, 1995; Lee, *et al.*, 1996).

As a transient neurotransmitter, NO· is par excellence (Snyder, 1992, 1995). Its action in the central nervous system is multifaceted, reflecting a breadth of sources, multiple targets, and variable biologic outcomes (Garthwaite, *et al.*, 1988; Bredt, *et al.*, 1990; Bredt and Synder, 1992; Huang, *et al.*, 1993; Rosen, *et al.*, 1995; Arancio, *et al.*, 1996). In the peripheral nervous system, this free radical has been found to behave as a mediator in the nonadrenergic noncholinergic control of muscular relaxation (Gillespie, *et al.*, 1989; Gibson, *et al.*, 1990; Li and Rand, 1990; Boeckxstaens, *et al.*, 1991; Liu, *et al.*, 1991; Burnett, *et al.*, 1992; Sheng, *et al.*, 1992).

The well-documented microbicidal activity of nitrite and nitrate suggests an important role for NO· in the control of intracellular pathogens (De Groote and Fang, 1995), perhaps through the binding of NO· to cellular iron (Mabbott and Sternberg, 1995). The microbicidal properties of this free radical are supported, in part, by the finding that phagocytic-derived NO· (Kotake, *et al.*, 1995) exhibits antimicrobial activity against such microorganisms as *Candida albicans* (Cenci, *et al.*, 1993; Vazquez-Torres, *et al.*, 1995; Fierro, *et al.*, 1996), *Cryptococcus neoformans* (Granger, *et al.*, 1988, 1990; Alspaugh and Granger, 1991), *Entamoeba histolytica* (Lin and Chadee, 1992), *Leishmania donvani* (Green, *et al.*, 1991), *Leishmania major* (Green, *et al.*, 1990a, 1990b, 1994; Liew, *et al.*, 1990; Li, *et al.*, 1992; Cunha, *et al.*, 1993; Assreuy, *et al.*, 1994; Vouldoukis, *et al.*, 1995), *Mycobacterium avium* (Denis, 1991), *Mycobacterium leprae* (Adams, *et al.*, 1991), *Mycobacterium tuberculosis* (Flesch and Kaufmann, 1991), *Naegleria fowleri* (Fischer-Stenger and Marciano-Cabral, 1992), *Plasmodium falciparum* (Rockett, *et al.*, 1991), *Pneumocystis carinii* (Sherman, *et al.*, 1991), *Schistosoma mansoni* (James and Glaven, 1989; Oswald, *et al.*, 1994), *Toxoplasma gondii*

(Adams, *et al.*, 1990), *Trypanosoma cruzi* (Muñoz-Fernández, *et al.*, 1992; Metz, *et al.*, 1993; Vespa, *et al.*, 1994; Abrahamsohn and Coffman, 1995), *Trypanosoma brucei* (Sternberg, *et al.*, 1994), and *Trypanosoma musculi* (Vincendeau and Daulouède, 1991). In the case of *Candida albicans*, when murine macrophages are stimulated to secrete both $O_2 \cdot^-$ and NO·, thereby producing $ONOO^-$, considerable candidacidal activity has been imputed to this peroxide (Vazquez-Torres, *et al.*, 1996).

Cytokines, which induce NOS II in cells where it is not constitutively expressed, have been found to display antiviral activity (Duersken-Hughes, *et al.*, 1992; Croen, 1993; Karupiah, *et al.*, 1993; Harris *et al.*, 1995; Karupiah and Harris, 1995; MacLean, *et al.*, 1998; Pope, *et al.*, 1998). Based on this general observation, there has been considerable interest in whether NO· is the pharmacophore responsible for interferon-induced host immune response against viral invasion. Various study designs, including the transduction of the gene for NOS II into epithelial cells and B lymphocytes and a battery of inhibitory experiments, support an antiviral role for this free radical (Croen, 1993; Karupiah, *et al.*, 1993; Mannick, *et al.*, 1994; Harris, *et al.*, 1995; Karupiah and Harris, 1995). Production of NO· and subsequent products may, however, contribute to tissue injury resulting from some viral infections (Akaike, *et al.*, 1996; Adler, *et al.*, 1997).

In other settings, NO· has been considered to be an important cytoprotective agent by scavenging different free radicals (Rubanyi, *et al.*, 1991; Wink, *et al.*, 1993a). As such, NO· may become a valuable therapeutic agent in the modulation of vascular damage secondary to ischemia/reperfusion injury (Johnson, *et al.*, 1990; Siegfried, *et al.*, 1992a, 1992b; Cooke and Tsao, 1993; Isono, *et al.*, 1993; Lefer, *et al.*, 1993; Ma, *et al.*, 1993).

Toxicity Attributed to Nitric Oxide

Considering the breadth of physiologic actions to which NO· is intimately linked (Moncada and Higgs, 1991, 1993), one must be prudent and exercise caution when attempting to assign specific toxic events to this free radical (Brüne, *et al.*, 1995). For the most part, toxicity attributed to NO· arises as the result of a change in homeostasis. For instance, there is a rapid efflux of NO· associated with an ischemic event (Malinski, *et al.*, 1993d; Loscalzo and Welch, 1995; Zhang, *et al.*, 1995b; Zweier, *et al.*, 1995). During reperfusion, diminished availability of NO· allows neutrophils, drawn to the site of inflammation, to adhere to the endothelium, predisposing this tissue to multiple toxic insults associated with leukocyte-secreted oxidants and proteolytic enzymes (Lefer, *et al.*, 1993; Ma, *et al.*, 1993; Kurose, *et al.*, 1995; Maulik, *et al.*, 1995).

Despite this, there have been a number of reported toxicities directly ascribed to this free radical, such as the NO· deamination of deoxynucleosides, and deoxynucleotides, inhibition in DNA synthesis (Nguyen, *et al.*, 1992; Wink, *et al.*, 1992; Lepoivre, *et al.*, 1994; Spencer, *et al.*, 1996; Tamir, *et al.*, 1996), and S-nitrosylation of proteins (Stamler, *et al.*, 1992a). The *in vivo* significance of these reactions, however, remains to be determined. In the central nervous system, excessive activation of NOS via *N*-methyl-D-aspartate (NMDA) receptor-mediated Ca^{2+} influx has been proposed as a pathway for excitotoxicity (Dawson, *et al.*, 1991, 1993). Support for this theory comes, in part, from the observation that inhibited generation of NO· attenuated NMDA-mediated neurotoxicity (Izumi, *et al.*, 1992; Kollegger, *et al.*, 1993).

Nitric oxide may also induce cytotoxicity by binding to iron heme in a number of key enzymes, including cytochrome c oxidase in the respiratory chain (Bastian, *et al.*, 1994; Brown, 1995a; Lizasoain, *et al.*, 1996), cytochrome P-450 (Hori, *et al.*, 1992; Khatsenko, *et al.*, 1993; Wink, *et al.*, 1993a; Stadler, *et al.*, 1994; Kim, *et al.*, 1995; Carlson and Billings, 1996; Gergel, *et al.*, 1997), and a variety of other heme-containing enzymes (Lancaster and Hibbs, 1990; Kim, *et al.*, 1995). Similarly, by inhibiting the apoproteins of certain isozymes of cytochrome P-450, NO· selectively blocks the metabolism of specific substrates (Khatsenko and Kikkawa, 1997). Likewise, NO· allows the release of iron bound to transferrin by reacting with this enzyme (Lee, *et al.*, 1994; Richardson, *et al.*, 1995). Potential toxicities from the resultant iron–nitrosyl chelate remain uncertain. For instance, it has been suggested that this iron–nitrosyl chelate is unable to redox cycle, thereby attenuating the oxidant properties of this metal ion (Kanner, *et al.*, 1991). In other situations, the reversibility of this complex might, however, exasperate the formation of oxygen-centered free radicals at potentially sensitive cellular sites (Reif and Simmons, 1990).

Enzymology of Nitric Oxide Synthase

The similarity of the NOS family of enzymes (Bredt, *et al.*, 1991; McMillian, *et al.*, 1992; Clement, *et al.*, 1993; Marletta, 1994a; Masters, *et al.*, 1996) to the well-characterized drug-metabolizing system, cytochrome P-450 (Guengerich and Macdonald, 1984; Porter and Coon, 1991), has allowed insightful debates as to possible mechanisms for the oxidation of L-arginine to L-citrulline and NO·. From these discussions, several very reasonable models for the enzymology of NOS have been put forth (Feldman, *et al.*, 1993; Klatt, *et al.*, 1993; Marletta, 1993, 1994a; Korth, *et al.*, 1994; Rosen, *et al.*, 1995; Lancaster and Stuehr, 1996; Crane, *et al.*, 1997; Siddhanta, *et al.*, 1998).

Based on current observations, the constitutive enzymes are regulated by the binding of Ca^{2+} to calmodulin, incorporated into the tertiary structure of the enzyme (Knowles, *et al.*, 1989; Mayer, *et al.*, 1989; Bredt and Snyder, 1990, Förstermann, *et al.*, 1990; Nakane, *et al.*, 1991; Schmidt, *et al.*, 1991; Lowenstein and Snyder, 1992; Côté and Roberge, 1996; Venema, *et al.*, 1996; Crane, *et al.*, 1997; Gerber, *et al.*, 1997). This allows the formation of NO· from L-arginine (Abu-Soud, *et al.*, 1994). For murine macrophage NOS II, Ca^{2+} is so tightly bound to calmodulin that supplementation from intracellular stores appears to be unnecessary, even though there are examples of Ca^{2+} chelates inhibiting enzyme activity from other cellular sources of NOS II (Marletta, 1994b).

The initial step in the generation of NO· is the transport of electrons as a hydride from NADPH to the oxidized flavin, FAD. After abstraction of a proton from the surrounding milieu, the fully reduced flavin $FADH_2$ is ready to transfer an electron to FMN. This occurs through a reaction known as disproportionation. For illustrative purposes, consider the hydrogen atom transfer from 1,4-dihydroxybenzene to 1,4-benzoquinone, resulting in the formation of the corresponding semiquinone free radical (Figure 3.5). Likewise, the reaction of $FADH_2$ with FMN leads to FADH·/FMNH· (Figure 3.6) (White, 1991, 1994). As is the case for cytochrome P-450 (Guengerich and Macdonald, 1984), the electron donation from FMNH· to Fe^{3+} gives the reduced heme, Fe^{2+} and FADH·/FMN $\Leftrightarrow$ FAD/FMNH· (Galli, *et al.*, 1996) (Figures 3.7 and 3.8) Binding of O_2 in the sixth ligand position would give the hypothetical intermediate, $[Fe^{2+} \cdots O_2]$ (Groves and Watanabe, 1986; Abu-Soud, *et al.*, 1997). In the absence of L-arginine, O_2 accepts an electron from the heme, generating $O_2 \cdot^-$ (Pou, *et al.*, 1992; Wever, *et al.*, 1997; Vásquez-Vivar, *et al.*, 1998; Xia, *et al.*, 1998a, 1998b). When L-arginine is present, however, the binding of the guanidino nitrogen in an ordered position near the heme (Tierney, *et al.*, 1998) allows a second electron to be shunted, from either FAD/FMNH· or perhaps from tetrahydro-L-biopterin (Mayer, *et al.*, 1995; Abu-Soud, *et al.*, 1997). Transfer of this second electron yields a superoxide-ferrous complex, $[(Fe^{2+})(O_2^-)(Arg)]^+$, or its resonance equivalent, the hydroperoxyl-ferric species, $[(Fe^{3+})(HO_2^-)(Arg)]^+$ (Figure 3.7).

For cytochrome P-450, the electron-rich thiolate near the hydroperoxyl complex promotes the decomposition of this species through either a heterolytic or a homolytic reaction (White, 1991; Liu, *et al.*, 1995; Harris and Loew, 1996). In the case of the heterolytic pathway, the perferryl–oxygen complex $[Fe^{5+}=O]^{3+}$ would be the proximal oxidant (Groves, *et al.*, 1981; Dolphin, *et al.*, 1997; Guengerich, *et al.*, 1997; Harris, *et al.*, 1998), whereas for the homolytic route, this burden would fall on Fe^{4+}–OH. Although most evidence, based on product identification, supports the $[Fe^{5+}=O]^{3+}$ complex, sufficient data to rule out Fe^{+4}–OH as the catalytic species has not been forthcoming. For NOS, the thiolate appears to be a cysteine residue (Chen, *et al.*, 1994); however, the identity of the proximal oxidant and the mechanism of its formation remain to be resolved.

It is assumed that the initial step in the reaction of the $[Fe^{5+}=O]^{3+}$ complex of cytochrome P-450 with a hydrocarbon is an oxidation prior to the addition of HO· to the electrophilic center of the substrate (Augusto, *et al.*, 1982). Recently, an alternative mechanism has been proposed in which a concerted insertion of an

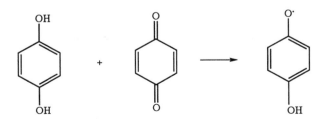

Figure 3.5 The reaction 1,4-dihydroxybenzene and 1,4-benzoquinone is representative of a disproportionation, resulting in the formation of the corresponding semiquinone free radical.

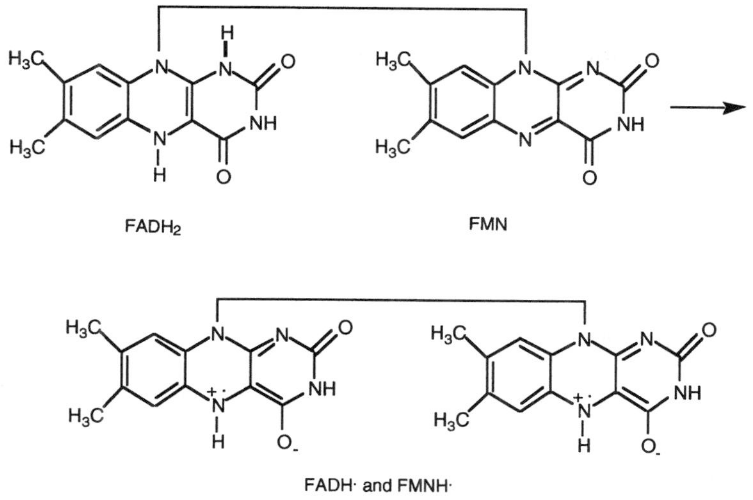

Figure 3.6 The disproportionation of flavins FADH$_2$–FMN of NOS. As illustrated in Figure 3.5, this reaction leads to the stable semiquinones, FADH· and FMNH·.

activated oxygen atom into the hydrocarbon is envisioned. This is followed by a Lewis acid-catalyzed ionic rearrangement of the corresponding alcohol (Newcomb and Chestney, 1994). Using a hypersensitive free radical probe to distinguish between these two possible pathways, kinetic studies point to an inconsistency with the current free radical model for cytochrome P-450-mediated carbon hydroxylation reactions (Newcomb, et al., 1995). Contrary to this, recent isotope effect studies on the dealkylation of N,N-dialkylanilines by mammalian and bacterial cytochrome P-450 support the free radical pathway (Karki, et al., 1995; Guengerich, et al., 1996, 1997). These conflicting findings, unfortunately, leave open the question as to the mechanism by which NOS catalyzes the production of NO·.

It has been theorized that the perferryl complex of NOS abstracts a hydrogen atom from the guanidinyl moiety of L-arginine, giving Fe^{4+}–OH (Klatt, et al., 1993). Within this complex, rapid transfer of HO· to one of the nitrogen atoms of the guanidinyl group would lead to N-hydroxyl-L-arginine (Figure 3.7) (Iyengar, et al., 1987; Marletta, et al., 1988; White and Marletta, 1992). Verification of this pathway is primarily based on the finding that NO· can be derived from N$^{\omega}$-hydroxyl-L-arginine (Stuehr, et al., 1991b; Pufahl, et al., 1992). Although N$^{\omega}$-hydroxyl-L-arginine has been observed under some experimental conditions, other data suggest this intermediate is not released by the enzyme, and further oxidation to L-citrulline and NO· is kinetically favored (Stuehr, et al., 1991b). Despite this, the hydroxylation step must be considered surprising when one realizes the infrequent cytochrome P-450-mediated oxidation of amines with

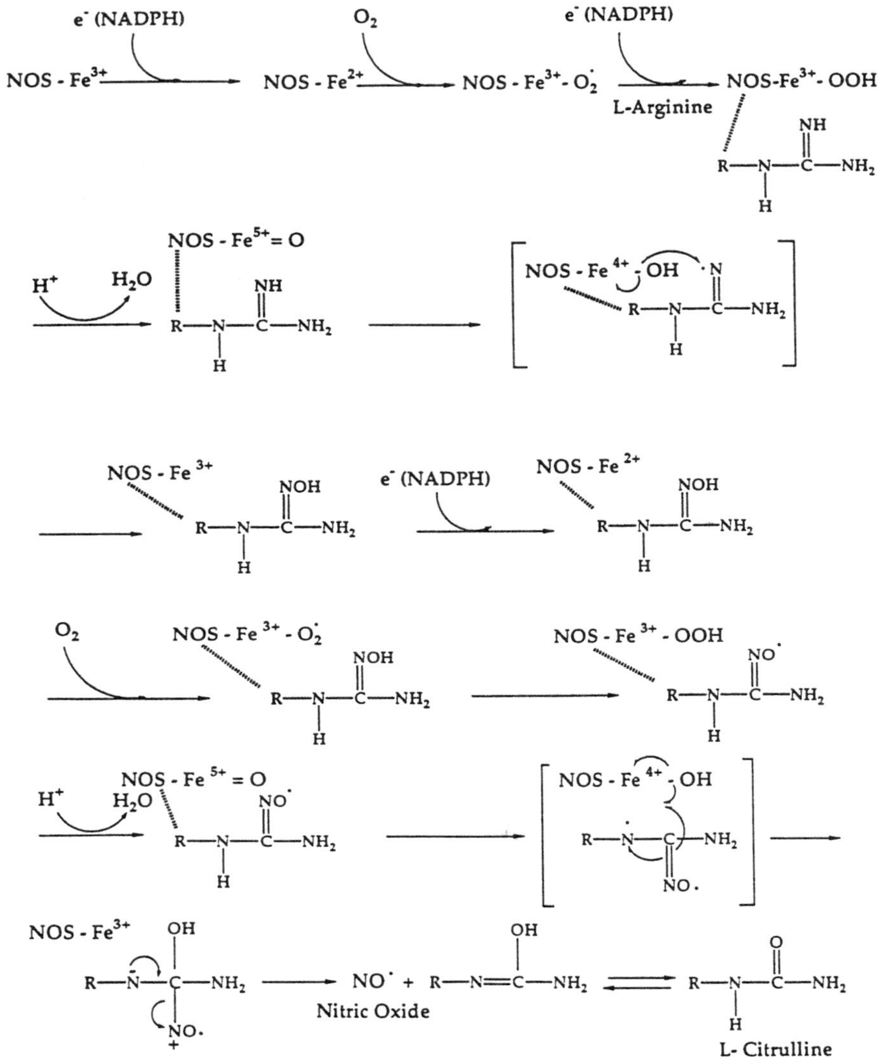

Figure 3.7 Proposed pathway for NO· generation catalyzed by NOS. This scheme shows the formation and involvement of $[Fe^{5+}=O]^{3+}$ in the oxidation of L-arginine to N-hydroxyl-L-arginine and N-hydroxyl-L-arginine to L-citrulline and NO·. The mechanism by which the electrons are transferred from NADPH to the enzyme through the flavins FAD–FMN is depicted. (Taken from Rosen, *et al.*, 1995).

pK_as $\gg$ 8–9 (Watkins and Gorrod, 1987; Ziegler, 1991; Seto and Guengerich, 1993). There are, of course, exceptions to this general rule (Bondon, *et al.*, 1989; Seto and Guengerich, 1993; Guengerich, *et al.*, 1997). Nevertheless, in this particular case, the guanidinyl moiety with a pK_a of 12.5 must promote this reaction by charge delocalization through the guanidinium group.

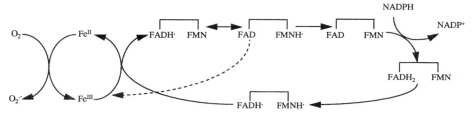

Figure 3.8   A proposed model that illustrates $O_2 \cdot^-$ generation from NOS I in the absence of
L-arginine. Initial electron flow is from NADPH to FAD, resulting in $FADH_2$. After disproportionation with FMN, two semiquinone free radicals, FADH· and FMNH· allow
electron transport to the heme iron, giving Fe^{2+}. In the absence of L-arginine, O_2 is the
preferred substrate, resulting in $O_2 \cdot^-$ formation. As there still remains one electron on FAD/
FMNH·, the iron heme is again reduced, with subsequent transfer to O_2. Thus, in the
absence of L-arginine, for every mole of NADPH oxidized, there are two moles of $O_2 \cdot^-$
generated.

Since L-citrulline is a metabolite of NOS-catalyzed oxidation, there has been
considerable interest in determining which of the guanidinyl nitrogen atoms is
hydroxylated by this enzyme. This is by no means a trivial task to resolve, since
each nitrogen atom is equivalent. Recently, however, a series of elegant studies
using differentially labeled $^{15}N^{\omega}$-hydroxyl-L-arginines have found the 2-position to
be the primary site of enzymic oxidation (Stuehr, et al., 1991b). Equally as fasci-
nating is the singularity of substrate specificity exhibited by this enzyme towards
the L-isomer of either arginine or N^{ω}-hydroxyl-L-arginine, since the D-isomers did
not result in generation of NO· (Stuehr, et al., 1991b). This observation points to a
subsequent regulatory mechanism governing the secretion of NO·.

Once N^{ω}-hydroxyl-L-arginine has been generated, there are a number of potential routes to NO· (Craven, et al., 1979; Fukuto, et al., 1993; Mansuy, et al., 1995;
Pufahl, et al., 1995). One pathway envisions a hydrolytic reaction, giving L-citrulline and N-hydroxylamine. The ensuing oxidation by catalase would produce NO·
(Craven, et al., 1979). There are several lines of evidence that suggest this sequential scheme is not operative. First, for example, N-hydroxylamine is not oxidized to
NO· by crude cytosol fractions of neuroblastoma cells that contain NOS in the
absence of catalase, even though this same preparation will produce NO· from L-
arginine (Pou, et al., 1991). Second, the oxygen atom introduced into L-citrulline
and NO· has its origin in O_2 and not H_2O (Kwon, et al., 1990; Leone, et al., 1991).

Alternative oxidative reactions envision either cytochrome P-450 (Schott, et al.,
1994; Mansuy, et al., 1995) or $O_2 \cdot^-$ (Sennequier, et al., 1995; Vetrovsky, et al.,
1996; Modolell, et al., 1997) as the catalysts in the generation of NO·. Likewise, an
oxazirine intermediate, produced during the metabolism of N^{ω}-hydroxyl-L-argi-
nine, has been suggested to rearrange, yielding L-citrulline and HNO·. Loss of an
electron gives NO· (Fukuto, et al., 1993).

Although each of these oxidative mechanisms deserves serious consideration, a
reaction pathway in which N^{ω}-hydroxyl-L-arginine is the source of the second
electron is an equally enticing hypothesis (Rauckman, et al., 1982; Korth, et al.,

1994). Stoichiometric studies support this thesis, as the oxidation of N^{ω}-hydroxyl-L-arginine by NOS requires only one electron from NADPH (Stuehr, *et al.*, 1991b). Since the actual oxidation is a two-electron process, what may be the source of the second electron? Some have suggested a role for tetrahydro-L-biopterin (Stuehr, *et al.*, 1991b; Mayer, *et al.*, 1995). Other studies, however, have proposed pathways in which a concerted intermolecular nucleophilic attack by a ferri-hydroperoxy intermediate, $[Fe^{3+}-OO^-]$, mediates the reaction (Figure 3.9) (Feldman, *et al.*, 1993; Klatt, *et al.*, 1993; Marletta, 1993; Korth, *et al.*, 1994; Mansuy, *et al.*, 1995; Pufahl, *et al.*, 1995). If one considers the substrate-participation route, once the perferryl complex of NOS, $[Fe^{5+}=O]^{3+}$, is formed, it would abstract a hydrogen atom from either one of the guanidino nitrogens of nitroxyl-L-arginine, giving $Fe^{4+}-OH$ and the corresponding diradical. Rapid addition of HO·, followed by intramolecular rearrangement, would lead to NO· and L-citrulline (Figure 3.7).

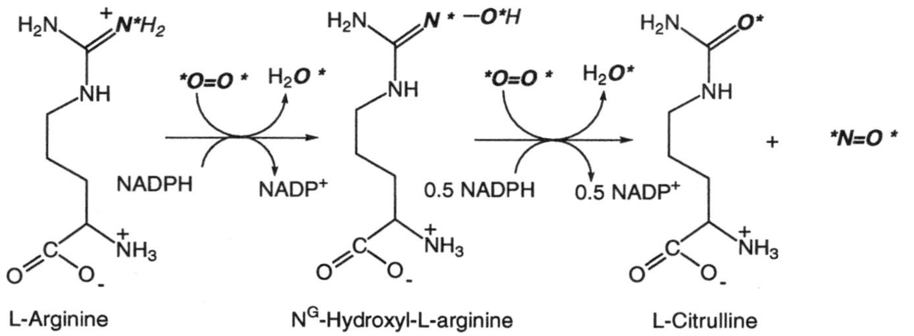

L-Arginine   N^G-Hydroxyl-L-arginine   L-Citrulline

Nitric oxide synthase-catalyzed oxidation of L-arginine is unusual, requiring only 1.5 mol of NADPH and 2 mol of O_2 to give NO· and L-citrulline (Stuehr, *et al.*, 1991b). The oxidation is, therefore, a five-electron process—three from NADPH and two from O_2. Since NADPH normally donates electrons to flavoproteins (like NOS) as a hydride, resulting in $FADH_2/FMN$, where does the remaining electron reside? There are several possibilities worth examining. Let us first consider the case where the unpaired electron is on the nucleotide, NADP·. Reaction of NADP· with O_2 would result in $O_2·^-$ production (Bielski and Chan, 1976; Anderson, 1980; Farrington, *et al.*, 1980). With this model, for every mole of NO· produced by the enzyme, an equal amount of $O_2·^-$ would be

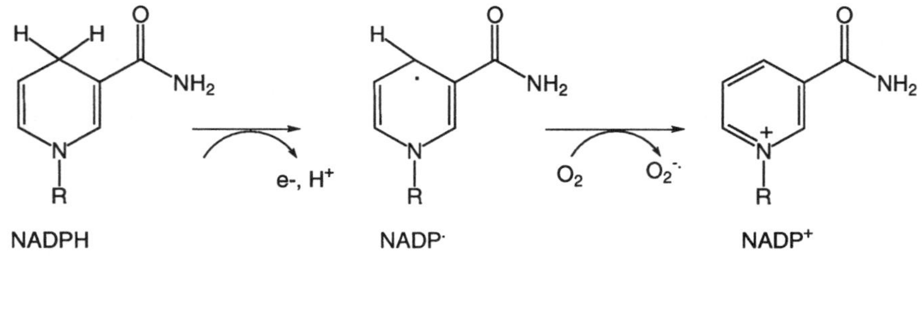

NADPH   NADP·   NADP$^+$

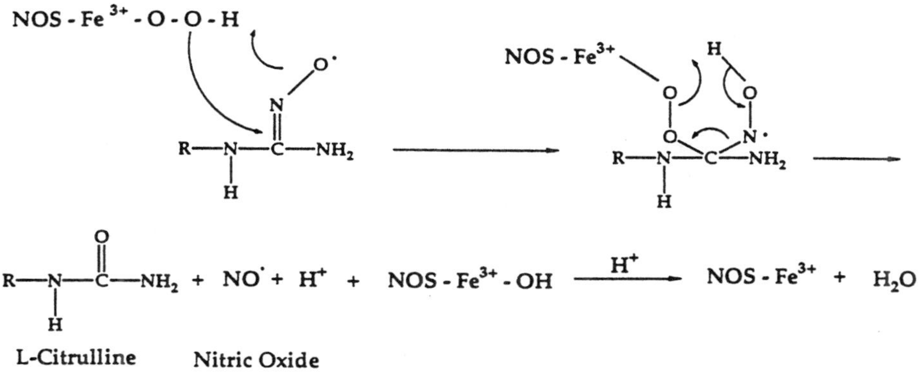

NOS - Fe^{3+} - O - O - H

$$R—N(H)—C(=N—...)—NH_2 \longrightarrow NOS-Fe^{3+} \cdots \longrightarrow$$

$$R—N(H)—C(=O)—NH_2 + NO^{\cdot} + H^+ + NOS-Fe^{3+}-OH \xrightarrow{H^+} NOS-Fe^{3+} + H_2O$$

L-Citrulline   **Nitric Oxide**

Figure 3.9 Mechanism of NO· generation catalyzed by NOS as proposed by Korth, *et al.* (1994). In contrast to Figure 3.7, this scheme presents a concerted pathway by which [Fe^{3+}–OO·] oxidizes *N*-hydroxyl-L-arginine to L-citrulline and NO·. (Taken from Rosen, *et al.*, 1995.)

obtained from the reaction of O_2 with the nucleotide free radical. This is a teleological unattractive theory, since, for example, at physiological pH, $O_2 \cdot^-$ reacts with NO· to form ONOO$^-$ at near-diffusion-controlled rates, ranging from 4.3 to $19 \times 10^9 \, M^{-1} s^{-1}$ (Huie and Padmaja, 1993; Goldstein and Czapski, 1995b; Kobanyashi, *et al.*, 1995; Kissner, *et al.*, 1997). Under these conditions, secretion of NO· from one cell to another could only transpire through elimination of $O_2 \cdot^-$. This would require SOD to be either associated with or proximal to NOS. Currently, there is no evidence to support this hypothesis. Alternatively, the electron could be stabilized on one of the flavins of NOS, giving FADH$_2$–FMNH· as the steady-state condition of this enzyme (Figure 3.8). With this model, 2 mol of NO· would be produced while NOS would consume 3 mol of NADPH. Precedent for such a hypothesis comes from studies of the enzymology of cytochrome P-450. In this case, it is envisioned that FADH$_2$–FMNH· from NADPH–cytochrome P-450 reductase is the steady-state form of this enzyme, allowing the transport of an electron to the iron protoporphyrin IX incorporated into the terminal oxidase, cytochrome P-450 (Iyanagi and Mason, 1973; Vermilion and Coon, 1978; Vermilion, *et al.*, 1981; Guengerich, 1983; White, 1994).

In the absence of L-arginine, electron movement through either NOS I, NOS II or NOS III leads to significant production of $O_2 \cdot^-$ (Pou, *et al.*, 1992, 1999; Wever, *et al.*, 1997; Vásquez-Vivar, *et al.*, 1998; Xia, *et al.*, 1998a, 1998b). With the addition of L-arginine, in a dose-dependent manner, inhibition in $O_2 \cdot^-$ generation was noted, with NOS II requiring considerably higher concentrations of substrate than was found for NOS I (Xia, *et al.*, 1998a; Pou, *et al.*, 1999). Surprisingly, L-arginine had no impact on NOS III formation of $O_2 \cdot^-$ (Vásquez-Vivar, *et al.*, 1998; Xia, *et al.*, 1998b). Based on these finding, there should be conditions where NO· and $O_2 \cdot^-$ are both generated, the rate of each free radical produced would be dependent on the concentration of L-arginine. For NOS II, electron transport from the flavin to the heme domains is apparently not very tightly coupled, since some leakage results in formation of $O_2 \cdot^-$ from the flavin site even in the presence of high levels of L-arginine (Xia, *et al.*, 1998a). In contrast, the heme domain of NOS I and NOS III (Vásquez-Vivar, *et al.*, 1998; Xia, *et al.*, 1998b; Pou, *et al.*, 1999) appears to be the origin of $O_2 \cdot^-$ formation (Figure 3.8), as is the case for cytochrome P-450 (White, 1994). Therefore, during NOS I oxidation of L-arginine to L-citrulline and NO·, direct competition with O_2 results in $O_2 \cdot^-$ formation. At saturating levels of O_2, the ratio of NO· and $O_2 \cdot^-$ is dependent upon the concentration of L-arginine. Thus, in the case of NOS I and NOS II, there might be cellular conditions that might promote $ONOO^-$ formation, albeit at a low flux. Validity for such a theory comes from experiments with cells, containing either NOS I or NOS II. Under defined conditions, $ONOO^-$ was detected; the source of $O_2 \cdot^-$, was, however, not defined (Xia, *et al.*, 1996; Xia and Zweier, 1997).

Glutathione—A Cell Transporter of Nitric Oxide

Considering that cellular levels of GSH are in the millimolar range (Kosower and Kosower, 1978), it has been postulated that this and other thiols may store, transport, and deliver NO· to physiologic sites of action (Ignarro, *et al.*, 1981; Stamler, *et al.*, 1992a, 1992b; Clancy, *et al.*, 1994; Boese, *et al.*, 1995, 1997; Kharitonov, *et al.*, 1995; Nikitovic and Holmgren, 1996; Pietraforte, *et al.*, 1996; Singh, *et al.*, 1996) and augument potential toxicities associated with high levels of this free radical (Wakulich and Tepperman, 1997):

$$2RSNO \rightarrow RSSR + 2NO \cdot$$

Yet, the fundamental question remains as to whether NO· per se, or the more long-lived *S*-nitrosothiols, such as *S*-nitrosoglutathione (GSNO) (Jia, *et al.*, 1996; Nikitovic and Holmgren, 1996), are the putative species responsible for many of the physiologic actions attributable to this free radical. Of particular concern in this inquiry is the nature of the species that combines with RSH. One theory is that N_2O_3 is the proximal reactant (Kharitonov, *et al.*, 1995):

$$2NO \cdot + O_2 \rightarrow 2 \cdot NO_2$$

$$\cdot NO_2 + NO \cdot \rightarrow N_2O_3$$

$$N_2O_3 + RSH \rightarrow HNO_2 + RSNO$$

Alternatively, $\cdot NO_2$ could combine with the thiol. In this case, thiyl free radical (RS·) would be the intervening agent in the formation of RSNO (Goldstein and Czapski, 1996):

$$\cdot NO_2 + RSH \rightarrow RS \cdot + HNO_2$$

$$RS \cdot + NO \cdot \rightarrow RSNO$$

However, one contemporary study has suggested, based on a series of inhibit experiments, that GS· was not produced in this reaction (Hogg, *et al.*, 1996). A more recent inquiry, though, was able to spin trap GS·, pointing to ·NO$_2$ as the more likely intermediate in the aerobic nitrosation of thiols, such as GSH (Pou and Rosen, 1998). Given the likelihood that GS· is a cellular by-product of NO· metabolism, what might be the fate of this free radical? One likely scenario is the formation of O$_2$·$^-$, arising as a consequence of the aerobic reaction of GS· with GS$^-$ (Winterbourn and Metodiewa, 1994):

$$GS \cdot + GS^- \rightarrow GSSG \cdot^-$$

$$GSSG \cdot^- + O_2 \rightarrow GSSG + O_2 \cdot^-$$

As GSH may be considered a scavenger for NO·, this thiol can be viewed as a cytoprotective agent, attenuating the lethal effects of NO· and products derived from this free radical (Wink, *et al.*, 1994; Bolaños, *et al.*, 1996). In this regard, it has been found that GSH-depleted cells are considerably more susceptible to the toxic effects of NO· than are cells whose physiologic levels of this thiol are maintained (Luperchio, *et al.*, 1996; Petit, *et al.*, 1996).

Peroxynitrite—A Wolf in Sheep's Clothing?

By the mid-1980s, at a time predating the identification of EDRF as NO·, it became clear that variable fluxes of O$_2$·$^-$ could modulate the biologic activity associated with EDRF (Gryglewski, *et al.*, 1986; Moncada, *et al.*, 1986; Rubanyi and Vanhoutte, 1986a, 1986b). Within the decade, however, the search for this attenuating property of O$_2$·$^-$ led many to explore the reaction between these two free radicals, resulting in formation of ONOO$^-$, which exhibited a half-life of approximately 1 s at 37°C and pH 7.4 (Beckman, *et al.*, 1990; Saran, *et al.*, 1990; Huie and Padmaja, 1993; Goldstein and Czapski, 1995b; Pryor and Squadrito, 1995). Subsequent chemical transformations and biologic consequences have likewise been defined (Liu, *et al.*, 1994; Pryor and Squadrito, 1995; Darley-Usmar, *et al.*, 1995; Fukuto and Ignarro, 1997).

Although the reaction of O$_2$·$^-$ with NO· was first reported in 1985 (Blough and Zafiriou, 1985), the fate of this peroxide has remained, surprisingly, in doubt. Some accounts, represented by Halfpenny and Robinson (1952), Beckman, *et al.* (1990), Hogg, *et al.* (1992), and Crow, *et al.*, (1994) suggested that decomposition of peroxynitrous acid led to the formation of HO· (Pryor, *et al.*, 1996):

$$NO \cdot + O_2 \cdot^- \rightarrow ONOO^-$$

$$ONOO^- + H^+ \rightleftarrows ONOOH$$

$$ONOOH \leftrightarrow [HO \cdots NO_2]_{cage}$$

$$[HO \cdots NO_2]_{cage} \rightarrow HO \cdot + \cdot NO_2$$

In this scheme, ONOOH, which has a pK_a of 6.8 at 37°C (Koppenol, *et al.*, 1992), undergoes homolytic cleavage to give HO· and NO_2· —a novel mechanism of HO· generation that is not dependent upon redox active transition metal ions (Coddington, *et al.* 1999). Contrary to this, others have argued, based on thermo-dynamic and kinetic considerations, that ONOOH decomposition does not lead to HO· (Koppenol, *et al.*, 1992; Shi, *et al.*, 1994; Bartlett, *et al.*, 1995; Lemercier, *et al.*, 1995; Goldstein, *et al.*, 1996) and that the observed HO·-like reactivity of ONOOH was derived from a vibrationally excited intermediate of the *trans*-peroxynitrous acid, ONOOH* (Koppenol, *et al.*, 1992; Bartlett, *et al.*, 1995; Pryor, *et al.*, 1996; Pfeiffer, *et al.*, 1997):

$$NO \cdot + O_2 \cdot^- \rightarrow ONOO^-$$

$$ONOO^- + H^+ \rightleftarrows ONOOH$$

$$ONOOH \rightleftarrows ONOOH^*$$

$$ONOOH^* \rightarrow H^+ + NO_3^-$$

A more recent theoretical study has suggested that the concerted pathway for rearrangement of ONOOH* involves a stepwise process that leads to hydrogen-bonded radical pairs, HO· and ·NO_2 (Houk, *et al.*, 1996). As these radical pairs are produced in a "cage," there is no barrier to recombination. Thus, one would surmise that the concentration of "free" HO· would be small. This is in line with experiments in which the amount of detectable HO· at physiologic pH was low— approximately 1–4% of the initial concentration of $ONOO^-$ (Yang, *et al.*, 1992; Augusto, *et al.*, 1994; van der Vliet, *et al.*, 1994; Pou, *et al.*, 1995; Kaur, *et al.*, 1997). More recent data suggest that this value may be as high as 10% (Richeson, *et al.*, 1998). These observations parallel the finding that the yield of HO· from ONOOH depends on the pH of the solution (Crow, *et al.*, 1994). As the pH increases from 6.8 to 8, the amount of HO· produced decreases precipitously. One must, therefore, question the presumption that ONOOH-generated HO· is a contributor to the toxicological profile of this peroxide (Henry, *et al.*, 1993).

In contrast, the reaction of $ONOO^-$ with carbon dioxide results in the forma-tion of the nitrocarbonate anion, whose oxidant properties are of considerable biologic importance (Lymar and Hurst, 1995; Denicola, *et al.*, 1996; Uppu, *et al.*, 1996; Goldstein and Czapski, 1999). Nitrocarbonate can oxidize substrates via one- and two-electron pathways, as well as nitrosylate a wide variety of compounds (Gow, *et al.*, 1996; Uppu, *et al.*, 1996; Berlett, *et al.*, 1998; DuVall, *et al.*, 1998):

$$ONOO^- + CO_2 \rightarrow ONOOCO_2^- \rightarrow O_2NOCO_2^-$$

Some studies suggest that ONOOH may directly elicit toxicity through the oxidation of thiols and thiol-containing proteins (Radi, *et al.*, 1991a; Gatti, *et al.*, 1994) and membrane lipids (Radi, *et al.*, 1991b). In the case of methionine, this peroxide has been shown to catalyze one- and two-electron reactions through a selective "hydroxyl radical-like oxidant" (Pryor, *et al.*, 1994). Similarly, ONOOH has been found to damage sodium ion channels (Bauer, *et al.*, 1992; Hu, *et al.*, 1994; DuVall, *et al.*, 1998), disrupt the calcium ion channels in myocardium (Elliott, 1996; Ishida, *et al.*, 1996), and inhibit glutamate transporters (Trotti, *et al.*, 1996), prostacyclin synthase (Zou and Ullrich, 1996), glutathione peroxidase (Padmaja, *et al.*, 1998), MnSOD (Yamakura, *et al.*, 1998), creatine kinase (Stachowiak, *et al.*, 1998), dehydratase clusters (Keyer and Imlay, 1997), and cytochrome oxidase (Sharpe and Cooper, 1998). Yet other investigations have reported this peroxide to be either less than or no more toxic than $O_2 \cdot^-$ (Oury, *et al.*, 1992; Lafon-Cazal, *et al.*, 1993; Lefer, *et al.*, 1997). Depending on experimental conditions, NO· can either stimulate or inhibit $O_2 \cdot^-$ and H_2O_2 dependent lipid peroxidation and accompanying toxicity (Berisha, *et al.*, 1994; Rubbo, *et al.*, 1994).

Another potential pathway by which ONOOH can initiate tissue injury is through its ability to nitrate phenols, including tyrosines of SOD and other proteins, critical to cell function. The reaction appears coincidentally to be catalyzed by SOD and Fe^{3+}–EDTA (Beckman, *et al.*, 1992; Ischiropoulos, *et al.*, 1992b). Of particular biologic interest are the X-ray diffraction studies, demonstrating the ability of $ONOO^-$ to nitrate the tyrosine-108 residue of CuZnSOD (Smith, *et al.*, 1992). This nitration has been postulated to occur through electron transfer from $ONOO^-$ to Cu^{2+}, followed by the formation of the strong oxidant nitronium ion, NO_2^+ (Ischiropoulos, *et al.*, 1992b):

$$ONOO^- + Cu^{2+}, ZnSOD \rightarrow ONOO–Cu^{1+}, ZnSOD$$

$$ONOO–Cu^{1+}, ZnSOD + H^+ \rightarrow NO_2^+ + HO–Cu^{1+}, ZnSOD$$

A critical factor to consider when evaluating the cytotoxic potential of $ONOO^-$ is the high concentration frequently required to evoke an untoward reaction (Radi, *et al.*, 1991a; Zhu, *et al.*, 1992; Gatti, *et al.*, 1994; Moro, *et al.*, 1994). For instance, a 50-µM dose of $ONOO^-$ was sufficient to inhibit prostacyclin synthase by 50% (Zou and Ullrich, 1996); 100 µM was required to similarly antagonize the glutamate transporter (Trotti, *et al.*, 1996) and 250 µM was essential to elicit microbicial activity towards *E. coli* (Zhu, *et al.*, 1992). Contrast this with a typical low flux rate of 0.1 µM min^{-1} per 10^6 cells (Ischiropoulos, *et al.*, 1992a) produced by stimulated phagocytic cells (Ding, *et al.*, 1988; Albina, *et al.*, 1989; McCall, *et al.*, 1989; Schmidt, *et al.*, 1989). At these steady-state concentrations, unless $ONOO^-$ is produced at a site sensitive to oxidative metabolism, such as DNA, the small amount of HO· and/or oxidative intermediate generated would be eliminated by endogenous free radical scavengers or enzymes before significant toxicity could arise (Fukuto and Ignarro, 1997). Therefore, formation of HO· from the decomposition of $ONOO^-$ might not be an important pathway in the toxicological profile of ONOO$^-$ (Zhu, *et al.*, 1992; Moro, *et al.*, 1994). This, coupled with the recent observation that glutathione peroxidase protects against oxidative damage, illustrates the cytoprotective capacity of cells (Sies, *et al.*, 1997). Despite this, ONOO$^-$ might be an important contributor to certain disease states, such as atherosclerosis (Darley-Usmar, *et al.*, 1992; Beckman, *et al.*, 1994; White, *et al.*, 1994). Future investigations will undoubtedly unravel the role of ONOO$^-$ in the control of human disease and physiology—whether it is a cytotoxic or a cytoprotective agent.

Cellular Defenses Against NO·

As described in Chapter 2, our understanding of the mechanisms by which cells protect themselves from O_2-centered radicals has evolved over the past four decades. We have only just begun such inquiries as to how cells limit their susceptibility to injury from NO· and products derived from this free radical. It appears that low-molecular-weight thiols, as they do for O_2-derived reactive species, contribute to cellular resistance to NO· and ONOO$^-$. In the case of *Salmonella typhimurium*, homocysteine has recently been implicated as a critical element in limiting ·NO-mediated toxicity and contributing to virulence in a mouse model of infection (De Groote, *et al.*, 1996). In the case of Salmonella there is evidence that the presence of a periplasmic SOD serves a protective role against ·NO-initiated cytotoxicity (De Groote, *et al.*, 1997). The mechanism presumably involves the removal of $O_2 \cdot^-$ by Cu,ZnSOD, thereby decreasing the potential for ONOO$^-$ formation upon exposure of the organism to NO·.

It has been recognized for a number of years that exposure of various bacterial species to NO· leads to the induction of a number of genes and gene products. For example, the soxRS regulon in *E. coli* is induced by exposure to not only $O_2 \cdot^-$ but also NO· and/or S-nitrosothiols (Nunoshiba, *et al.*, 1993, 1995; Hausladen, *et al.*, 1996). Products of these regulons enhance the resistance of the organism to $O_2 \cdot^-$ and/or NO·. In the case of *Mycobacterium tuberculosis*, three genes have been cloned that confer resistance to NO· (Ehrt, *et al.*, 1997). Some of the gene products responsible for these events are now becoming defined. In the case of *M. tuberculosis* and perhaps *Salmonella*, one of the gene products appears to be a subunit of alkyl hydroperoxide reductase, a member of the periroxin family (Chen, *et al.*, 1998). This enzyme had previously been shown to reduce organic hydroperoxides (Chae, *et al.*, 1994; Cha, *et al.*, 1995; Ferrante, *et al.*,1995) and it exhibits considerable sequence homology to another enzyme implicated in antioxidant defense, thioredoxin reductase (Tartaglia, *et al.*, 1990; Chae, *et al.*, 1994).

As for *E. coli*, a NO-dioxygenase responsible for the catabolism of NO· has emerged as the likely candidate to protect this bacterium from ·NO-mediated cytotoxicity (Gardner, *et al.*, 1998a). Recent evidence suggests that the function of this dioxgenase, which is mediated by bacterial flavohemoglobin, is to react with and consume NO· and S-nitrosothiols without the formation of ONOO$^-$ (Crawford and Goldberg, 1998; Gardner, *et al.*, 1998b; Membrillo-Hernández, *et al.*, 1999). Regulation of flavohemoglobin has previously been shown to be impacted by various forms of oxidative stress (Poole, *et al.*, 1996; Zhao, *et al.*,

1996; Membrillo-Hernández, *et al.*, 1997; Buisson and Labbe-Bois, 1998). Flavo-hemoglobins are found in many organisms of the archae, bacteria and eucarya domains (Zhu and Riggs, 1992), suggesting that the protective role of these enzymes against NO· could be widespread.

Concluding Thoughts

During the past decade, we have come to recognize the diversity of cells that generate NO·. We are now just beginning to appreciate the divergent and multi-faceted role that this free radical plays in controlling cellular function. Although there are a number of inhibitors of NOS, some of which appear to exhibit high affinity for this enzyme, profound discussions of the physiologic roles of NO· are ultimately linked to the detection of this free radical and the ability of specific NOS antagonists to attenuate NO· production. In subsequent chapters, we will begin to evaluate the variety of analytic methods most often used to identify many of the important biologically generated free radicals, including NO·.

References

Abrahamsohn, I.A., and Coffman, R.L. (1995). Cytokine and nitric oxide regulation of the immunosuppression in *Trypanosoma cruzi* infection. *J. Immunol.* **155**: 3955–3963.

Abu-Soud, H.M., and Stuehr, D.J. (1993). Nitric oxide synthases reveal a role for calmodulin in controlling electron transfer. *Proc. Natl. Acad. Sci. USA* **90**: 10769–10772.

Abu-Soud, H.M., Gachhui, R., Raushel, F.M., and Stuehr, D.J. (1997). The ferrous–dioxy complex of neuronal nitric oxide synthase. Divergent effects of L-arginine and tetrahydrobiopterin on its stability. *J. Biol. Chem.* **272**: 17349–17353.

Abu-Soud, H.M., Wang, J., Rousseau, D.L., Fukuto, J.M., Ignarro, L.J., and Stuehr, D.J. (1995). Neuronal nitric oxide synthase self-inactivates by forming a ferrous–nitrosyl complex during aerobic catalysis. *J. Biol. Chem.* **270**: 22997–23006.

Abu-Soud, H.M., Yoho, L.L., and Stuehr, D.J. (1994). Calmodulin controls neutronal nitric-oxide synthase by a dual mechanism. Activation of intra- and interdomain electron transfer. *J. Biol. Chem.* **269**: 32047–32050.

Adams, L.B., Franzblau, S.G., Vavrin, Z., Hibbs, J.B., Jr., and Krahenbuhl, J.L. (1991). L-Arginine-dependent macrophage effector functions inhibit metabolic activity of *Mycobacterium leprae*. *J. Immunol.* **147**: 1642–1646.

Adams, L.B., Hibbs, J.B., Jr., Taintor, R.R., and Krahenbuhl, J.L. (1990). Microbiostatic effect of murine-activated macrophages for *Toxoplasma gondii*: Role for synthesis of inorganic nitrogen oxides from L-arginine. *J. Immunol.* **144**: 2725–2729.

Adler, H., Beland, J.L., Del-Pan, N.C., Kobzik, L., Brewer, J.P., Martin, T.R., and Rimm, I.J. (1997). Suppression of Herpes simplex virus type 1 (HSV-1)-induced pneumonia in mice by inhibition of inducible nitric oxide synthase (iNOS, NOS2). *J. Exp. Med.* **185**: 1533–1540.

Akaike, T., Noguchi, Y., Ijiri, S., Setoguchi, K., Suga, M., Zheng, Y.M., Dietzschold, B., and Maeda, H. (1996). Pathogenesis of influenza virus- induced pneumonia: Involvement of both nitric oxide and oxygen radicals. *Proc. Natl. Acad. Sci. USA* **93**: 2448–2453.

Albakri, Q.A., and Stuehr, D.J. (1996). Intracellular assembly of inducible NO synthase is limited by nitric oxide-mediated changes in heme insertion and availability. *J. Biol. Chem.* **271**: 5414–5421.

Albina, J.E., Caldwell, M.D., Henry, W.L., Jr., and Mills, C.D. (1989). Regulation of macrophage function by L-arginine. *J. Exp. Med.* **169**: 1021–1029.

Alspaugh, J.A., and Granger, D.L. (1991). Inhibition of *Cryptococcus neoformans* replication by nitrogen oxides supports the role of these molecules as effectors of macrophage-mediated cytostasis. *Infect. Immun.* **59**: 2291–2296.

Anderson R.F. (1980). Energetics of the one-electron steps in the $NAD^+/$ NADH redox couple. *Biochim. Biophys. Acta* **590**: 277–280.

Arancio, O., Kiebler, M., Lee, C.J., Lev-Ram, V., Tsien, R.Y., Kandel, E.R., and Hawkins, R.D. (1996). Nitric oxide acts directly in the presynaptic neuron to produce long-term potentiation in cultured hippocampal neurons. *Cell* **87**: 1025–1035.

Arnet, U.A., McMillan, A., Dinerman, J.L., Ballermann, B., and Lowenstein, C.J. (1996). Regulation of endothelial nitric-oxide synthase during hypoxia. *J. Biol. Chem.* **271**: 15069–15073.

Asahi, M., Fujii, J., Suzuki, K., Seo, H.G., Kuzuya, T., Hori, M., Tada, M., Fujii, S., and Taniguchi, N. (1995). Inactivation of glutathione peroxidase by nitric oxide. Implication for cytotoxicity. *J. Biol. Chem.* **270**: 21035- 21039.

Asahi, M., Fujii, J., Takao, T., Kuzuya, T., Hori, M., Shimonishi, Y., and Taniguchi, N. (1997). The oxidation of seleocysteine is involved in the inactivation of glutathione peroxidase by nitric oxide. *J. Biol. Chem.* **272**: 19152–19157.

Asano, K., Chee, C.B.E., Gaston, B., Lilly, C.M., Gerard, C., Drazen, J.M., and Stamler, J.S. (1994). Constitutive and inducible nitric oxide synthase gene expression, regulation, and activity in human lung epithelial cells. *Proc. Natl. Acad. Sci. USA* **91**: 10089–10093.

Assreuy, J., Cunha, F.Q., Epperlein, M., Noronha-Dutra, A., O'Donnell, C.A., Liew, F.Y., and Moncada, S. (1994). Production of nitric oxide and superoxide by activated macrophages and killing of *Leishmania major*. *Eur. J. Immunol.* **24**: 672–676.

Augusto, O., Beilan, H.S., and Ortiz de Montellano, P.R. (1982). The catalytic mechanism of cytochrome P-450. Spin-trapping evidence for one- electron substrate oxidation. *J. Biol. Chem.* **257**: 11288–11295.

Augusto, O., Gatti, R.M., and Radi, R. (1994). Spin-trapping studies of peroxynitrite decomposition and of 3-morpholinosydnonime N-ethylcarbamide autooxidation: Direct evidence for metal-independent formation of free radical intermediates. *Arch. Biochem. Biophys.* **310**: 118–125.

Awad, H.H., and Stanbury, D.M. (1993). Autoxidation of NO in aqueous solution. *Int. J. Chem. Kinet.* **25**: 375–381.

Axelsson, K.L., Wikberg, J.E.S., and Andersson, R.G.G. (1979). Relationship between nitroglycerin, cyclic GMP and relaxation of vascular smooth muscle. *Life Sci.* **24**: 1779–1786.

Ayajiki, K., Kindermann, M., Hecker, M., Fleming, I., and Busse, R. (1996). Intracellular pH and tyrosine phosphorylation but not calcium determine shear stress-induced nitric oxide production in native endothelial cells. *Circ. Res.* **78**: 750–758.

Baek, K.J., Thiel, B.A., Lucas, A., and Stuehr, D.J. (1993). Macrophage nitric oxide synthase subunits. Purification, characterization, and role of prosthetic groups and substrate in regulating their association into a dimeric enzyme. *J. Biol. Chem.* **268**: 21120–21129.

Bagasra, O., Michaels, F.H., Zheng, Y.M., Bobroski, L.E., Spitsin, S.V., Fu, Z.F., Tawadros, R., and Koprowski, H. (1995). Activation of the inducible form of nitric oxide synthase in brains of patients with multiple sclerosis. *Proc. Natl. Acad. Sci. USA* **92**: 12041–12045.

Balligand, J.-L., Kobzik, L., Han, X., Kaye, D.M., Belhassen, L., O'Hara, D.S., Kelly, R.A., Smith, T.W., and Michel, T. (1995). Nitric oxide-dependent parasympathetic signaling is due to activation of constitutive endothelial (type III) nitric oxide synthase in cardiac myocytes. *J. Biol. Chem.* **270**: 14582–14586.

Balligand, J.-L., Ungureanu-Longrois, D., Simmons, W.W., Pimental, D., Malinski, T.A., Kapturczak, M., Taha, Z., Lowenstein, C.J., Davidoff, A.J., Kelly, R.A., Smith, T.W., and Michel, T. (1994). Cytokine-inducible nitric oxide synthase (iNOS) expression in cardiac myocytes. Characterization and regulation of iNOS expression and detection of iNOS activity in single cardiac myocytes *in vitro*. *J. Biol. Chem.* **269**: 27580–27588.

Bandaletova, T., Brouet, I., Bartsch, H., Sugimura, T., Esumi, H., and Ohshima, H. (1993). Immunohistochemical localization of an inducible form of nitric oxide synthase in various organs of rats treated with *Propionibacterium acnes* and lipopolysaccharide. *APMIS* **101**: 330–336.

Bartlett, D., Church, D.F., Bounds, P.L., and Koppenol, W.H. (1995). The kinetics of the oxidation of L-ascorbic acid by peroxynitrite. *Free Radical Biol. Med.* **18**: 85–92.

Bastian, N.R., Yim, C.-Y., Hibbs, J.B., Jr., and Samlowski, W.E. (1994). Induction of iron-derived EPR signals in murine cancers by nitric oxide. Evidence for multiple intracellular targets. *J. Biol. Chem.* **269**: 5127–5131.

Bauer, M.L., Beckman, J.S., Bridges, R.J., Fuller, C.M., and Matalon, S. (1992). Peroxynitrite inhibits sodium uptake in rat colonic membrane vesicles. *Biochim. Biophys. Acta* **1104**: 87–94.

Beck, W., Schrire, V., Vogelpoel, L., Nellen, M., and Swanepoel, A. (1961). Hemodynamic effects of amyl nitrite and phenylephrine on the normal human circulation and their relation to changes in cardiac murmurs. *Am. J. Cardiol.* **8**: 341–349.

Beckman, J.S., Beckman, T.W., Chen, J., Marshall, P.A., and Freeman B.A. (1990). Apparent hydroxyl radical production by peroxynitrite: Implications for endothelial injury from nitric oxide and superoxide. *Proc. Natl. Acad. Sci. USA* **87**: 1620–1624.

Beckman, J.S., Ischiropoulos, H., Zhu, L., van der Woerd, M., Smith, C., Chen, J.,

Harrison, J., Martin, J.C., and Tasi, M. (1992). Kinetics of superoxide dismutase- and iron-catalyzed nitration of phenolics by peroxynitrite. *Arch. Biochem. Biophys.* **298**: 438–445.

Beckman, J.S., Ye, Y.Z., Anderson, P.G., Chen, J., Accavitti, M.A., Tarpey, M.M., and White, C.R. (1994). Extensive nitration of protein tyrosines in human atherosclerosis detected by immunohistochemistry. *Biol. Chem. Hoppe-Seyler* **375**: 81–84.

Berisha, H.I., Pakbaz, H., Absood, A., and Said, S.I. (1994). Nitric oxide as a mediator of oxidant lung injury due to paraquat. *Proc. Natl. Acad. Sci. USA* **91**: 7445–7449.

Berlett, B.S., Levine, R.L., and Stadtman, E.R. (1998). Carbon dioxide stimulates peroxynitrite-mediated nitration of tyrosine residues and inhibits oxidation of methionine residues of glutamine synthetase: Both modifications mimic effects of adenylylation. *Proc. Natl. Acad. Sci. USA* **95**: 2784–2789.

Bielski B.H.J., and Chan, P.C. (1976). Reevaluation of the kinetics of lactate dehydrogenase-catalyzed chain oxidation of nicotinamide adenine dinucleotide by superoxide radicals in the presence of ethylenediaminetetraacetate. *J. Biol. Chem.* **251**: 3841–3844.

Blatter, L.A., Taha, Z., Mesaros, S., Shacklock, P.S., Wier, W.G., and Malinski, T. (1995). Simultaneous measurement of Ca^{2+} and nitric oxide in bradykinin-stimulated vascular endothelial cells. *Circ. Res.* **76**: 922–924.

Blough, N.V., and Zafiriou, O.C. (1985). Reaction of superoxide with nitric oxide to form peroxynitrite ion alkaline aqueous solution. *Inorg. Chem.* **24**: 3502–3504.

Boeckxstaens, G.E., Pelckmans, P.A., Bogers, J.J., Bult, H., De Man, J.G., Oosterbosch, L., Herman, A.G., and Van Maercke, Y.M. (1991). Release of nitric oxide upon stimulation of nonadrenergic noncholingeric nerves in the rat gastric fundus. *J. Pharmacol. Exp. Ther.* **256**: 441–447.

Boese, M., Keese, M.A., Becker, K., Busse, R., and Mülsch, A. (1997). Inhibition of glutathione reductase by dinitrosyl–iron–dithiolate complex. *J. Biol. Chem.* **272**: 21767–21773.

Boese, M., Mordvintcev, P.I., Vanin, A.F., Busse, R., and Mülsch, A. (1995). S-Nitrosation of serum albumin by dinitrosyl-iron complex. *J. Biol. Chem.* **270**: 29244–29249.

Bohle, D.S., and Hung, C.-H. (1995). Ligand-promoted rapid nitric oxide dissociation from ferrous porphyrin nitrosyls. *J. Am. Chem. Soc.* **117**: 9584–9585.

Bolaños, J.P., Heales, S.J.R., Peuchen, S., Barker, J.E., Land, J.M., and Clark, J.B. (1996). Nitric oxide-mediated mitochondrial damage: A potential neuroprotective role for glutathione. *Free Radical Biol. Med.* **21**: 997–1001.

Bondon, A., Macdonald, T.L., Harris, T.M., and Guengerich, F.P. (1989). Oxidation of cycloalkylamines by cytochrome P-450. Mechanism-based inactivation, adduct formation, ring expansion, and nitrone formation. *J. Biol. Chem.* **264**: 1988–1997.

Bouton, C., Raveau, M., and Drapier, J.-C. (1996). Modulation of iron regulatory protein functions. Further insights into the role of nitrogen- and oxygen-derived reactive species. *J. Biol. Chem.* **271**: 2300–2306.

Bredt, D.S. (1996). Targeting of nitric oxide to its targets. *Proc. Soc. Exp. Biol. Med.* **211**: 41–48.

Bredt, D.S., and Snyder, S.H. (1990). Isolation of nitric oxide synthetase, a calmodulin-requiring enzyme. *Proc. Natl. Acad. Sci. USA* **87**: 682–685.

Bredt, D.S., and Snyder, S.H. (1992). Nitric oxide, a novel neuronal messenger. *Neuron* **8**: 3–11.

Bredt, D.S., Hwang, P.M., Glatt, C.E., Lowenstein, C., Reed, R.R., and Snyder, S.H. (1991). Cloned and expressed nitric oxide synthase structurally resembles cytochrome P-450 reductase. *Nature* **351**: 714–718.

Bredt, D.S., Hwang, P.M., Snyder, S.H. (1990). Localization of nitric oxide synthase indicating a neural role for nitric oxide. *Nature* **347**: 768–770.

Brenman, J.E., Chao, D.S., Gee, S.H., McGee, A.W., Craven, S.E., Santillano, D.R., Wu, Z., Huang, F., Xia, H., Peters, M.F., Froehner, S.C., and Bredt, D.S. (1996a). Interaction of nitric oxide synthase with the postsynaptic density protein PSD-95 and α1-syntropin mediated by PDZ domains. *Cell* **84**: 757–767.

Brenman, J.E., Chao, D.S., Xia, H., Aldape, K., and Bredt, D.S. (1995). Nitric oxide synthase complexed with dystrophin and absent from skeletal muscle sarcolemma in Duchenne muscular dystrophy. *Cell* **82**: 743–752.

Brenman, J.E., Christopherson, K.S., Craven S.E., McGee, A.W., and Bredt, D.S. (1996b). Cloning and characterization of postsynaptic density 93, a nitric oxide synthase interacting protein. *J. Neurosci.* **16**: 7407–7415.

Brown, G.C. (1995a). Nitric oxide regulates mitochondrial respiration and cell functions by inhibiting cytochrome oxidase. *FEBS Lett.* **369**: 136–139.

Brown, G.C. (1995b). Reversible binding and inhibition of catalase by nitric oxide. *Eur. J. Biochem.* **232**: 188–191.

Brown, G.C., and Cooper, C.E. (1994). Nanomolar concentrations of nitric oxide reversibly inhibit synaptosomal respiration by competing with oxygen at cytochrome oxidase. *FEBS Lett.* **356**: 295–298.

Brüne, B., Meßmer, U.K., and Sandau, K. (1995). The role of nitric oxide in cell injury. *Toxicol. Lett.* **82/83**: 233–237.

Brunton, T.L. (1867). On the use of nitrite of amyl in angina pectoris. *Lancet* **2**: 97–98.

Buisson, N., and Labbe-Bois, R. (1998). Flavohemoglobin expression and function in *Saccharomyces cerevisiae*. No relationship with respiration and complex response to oxidative stress. *J. Biol. Chem.* **273**: 9527–9533.

Burnett, A.L., Lowenstein, C.J., Bredt, D.S., Chang, T.S.K., and Snyder, S.H. (1992). Nitric oxide: A physiologic mediator of penile erection. *Science* **257**: 401–403.

Busconi, L., and Michel, T. (1993). Endothelial nitric oxide synthase. N-terminal myristoylation determines subcellular localization. *J. Biol. Chem.* **268**: 8410–8413.

Busconi, L., and Michel, T. (1994). Endothelial nitric oxide synthase membrane targeting. Evidence against involvement of a specific myristate receptor. *J. Biol. Chem.* **269**: 25016–25020.

Butler, A.R., Flitney, F.W., and Williams, D.L.H. (1995). NO, nitrosonium ion, nitroxide ions, nitrosothiols and iron-nitrosyls in biology: A chemist's perspective. *Trends Pharmacol. Sci.* **16**: 18–22.

Calaycay, J.R., Kelly, T.M., MacNaul, K.L., McCauley, E.D., Qi, H., Grant, S.K., Griffin, P.R., Klatt, T., Raju, S.M., Nussler, A.K., Shah, S., Weidner, J.R., Williams, H.R., Wolfe, G.C., Geller, D.A., Billiar, T.R., MacCoss, M., Mumford, R.A., Tocci, M.J., Schmidt, J.A., Wong, K.K., and Hutchinson, N.I. (1996). Expression and immunoaffinity purification of human inducible nitric-oxide synthase. Inhibition studies with 2-amino-5,6-dihydro-4H-1,3-thiazine. *J. Biol. Chem.* **271**: 28212–28219.

Carlson, T.J., and Billings, R.E. (1996). Role of nitric oxide in the cytokine-mediated regulation of cytochrome P-450. *Mol. Pharmacol.* **49**: 796–801.

Castro, L., Rodriguez, M., and Radi, R. (1994). Aconitase is readily inactivated by peroxynitrite, but not by its precursor, nitric oxide. *J. Biol. Chem.* **269**: 29409–29415.

Cenci, E., Romani, L., Mencacci, A., Spaccapelo, R., Schiaffella, E., Puccetti, P., and Bistoni, F. (1993). Interleukin-4 and interleukin-10 inhibit nitric oxide-dependent macrophage killing of *Candida albicans*. *Eur. J. Immunol.* **23**: 1034–1038.

Cha, M.K., Kim, H.-K., and Kim, I.H. (1995). Thioredoxin-linked "thiol peroxidase" from periplasmic space of *Escherichia coli*. *J. Biol. Chem.* **270**: 28635–28641.

Chabin, R.M., McCauley, E., Calaycay, J.R., Kelly, T.M., MacNaul, K.L., Wolfe, G.C., Hutchinson, N.I., Madhusudanaraju, S., Schmidt, J.A., Kozarich, J.W., and Wong, K.K. (1996). Active-site structure analysis of recombinant human inducible nitric oxide synthase using imidazole. *Biochemistry* **35**: 9567–9575.

Chae, H.Z., Robison, K., Poole, L.B., Church, G., Storz, G., and Rhee, S.G. (1994). Cloning and sequencing of thiol-specific antioxidant from mammalian brain: Alkyl hydroperoxide reductase and thiol-specific antioxidant

define a large family of antioxidant enzymes. *Proc. Natl. Acad. Sci. USA* **91**: 7017–7021.

Charles, I.G., Palmer, R.M.J., Hickery, M.S., Bayliss, M.T., Chubb, A.P., Hall, V.S., Moss, D.W., and Moncada S. (1993). Cloning, characterization, and expression of a cDNA encoding an inducible nitric oxide synthase from the human chrondrocyte. *Proc. Natl. Acad. Sci. USA* **90**: 11419–11423.

Chen, L., Xie, Q.W., and Nathan, C. (1998). Alkyl hydroperoxide reductase subunit C (AhpC) protects bacterial and human cells against reactive nitrogen intermediates. *Molec. Cell* **1**: 795–805.

Chen, P.-F., Tsai, A.-L., Berka, V., and Wu, K.K. (1996). Endothelial nitric- oxide synthase. Evidence for bidoman structure and successful reconstitution of catalytic activity from two separate domains generated by a baculovirus expression system. *J. Biol. Chem.* **271**: 14631–14635.

Chen, P.-F., Tsai, A.-L., and Wu, K.K. (1994). Cysteine 184 of endothelial nitric oxide synthase is involved in heme coordination and catalytic activity. *J. Biol. Chem.* **269**: 25062–25066.

Cho, H.J., Martin, E., Xie, Q.-W., Sassa, S., and Nathan, C. (1995). Inducible nitric oxide synthase: Identification of amino acid residues essential for dimerization and binding of tetrahydrobiopterin. *Proc. Natl. Acad. Sci. USA* **92**: 11514–11518.

Cho, H.J., Xie, Q.-W., Calaycay, J., Mumford, R.A., Swiderek, K.M., Lee, T.D., and Nathan, C. (1992). Calmodulin is a subunit of nitric oxide synthase from macrophages. *J. Exp. Med.* **176**: 599–604.

Clancy, R.M., Levartovsky, D., Leszczynska-Piziak, J., Yegudin, J., and Abramson, S.B. (1994). Nitric oxide reacts with intracellular glutathione and activates the hexose monophosphate shunt in human neutrophils: Evidence for S-nitrosoglutathione as a bioactive intermediate. *Proc. Natl. Acad. Sci. USA* **91**: 3680–3684.

Cleeter, M.W.J., Cooper, J.M., Darley-Usmar, V.M., Moncada, S., and Schapira, A.H.V. (1994). Reversible inhibition of cytochrome c oxidase, the terminal enzyme of the mitochondrial respiratory chain, by nitric oxide. Implications for neurodegenerative disease. *FEBS Lett.* **345**: 50–54.

Clement, B., Schultze-Mosgau, M.-H., and Wohlers, H. (1993). Cytochrome P-450 dependent N-hydroxylation of a guanidine (debrisoquine), microsomal catalyzed reduction and further oxidation of the N-hydroxy-guanidine metabolite to the urea derivative. Similarity with the oxidation of arginine to citrulline and nitric oxide. *Biochem. Pharmacol.* **46**: 2249–2267.

Coddington, J.W., Hurst, J.K., and Lymar, S.V. (1999). Hydroxyl radical formation during peroxynitrous acid decomposition. *J. Am. Chem. Soc.* **121**: 2438–2443.

Cooke, J.P., and Tsao, P.S. (1993). Cytoprotective effects of nitric oxide. *Circulation* **88**: 2451–2454.

Cooke, J.P., Singer, A.H., Taso, P., Zera, P., Rowan, R.A., and Billingham, M.E. (1992). Antiatherogenic effects of L-arginine in the hypercholesterolemic rabbit. *J. Clin. Invest.* **90**: 1168–1172.

Corson, M.A., James, N.L., Latta, S.E., Nerem, R.M., Berk, B.C., and Harrison, D.G. (1996). Phosphorylation of endothelial nitric oxide synthase in response to fluid shear stress. *Circ. Res.* **79**: 984–991.

Côté, J.-F., and Roberge, A.G. (1996). Nitric oxide synthase in cat brain: Cofactors-enzyme-substrate interaction. *Free Radical Biol. Med.* **21**: 109–115.

Crane, B.R., Arvai, A.S., Gachhui, R., Wu, C., Ghosh, D.K., Getzoff, E.D., Stuehr, D.J., and Tainer, J.A. (1997). The structure of nitric oxide synthase oxygenase domain and inhibitor complexes. *Science* **278**: 425–431.

Craven, P.A., and DeRubertis, F.R. (1978). Restoration of the responsiveness of purified guanylate cyclase to nitroso-guanidine, nitric oxide, and related activators by heme and hemeproteins. Evidence for involvement of the paramagnetic nitrosyl–heme complex in enzyme activation. *J. Biol. Chem.* **253**: 8433–8443.

Craven, P.A., and DeRubertis, F.R. (1983). Requirement for heme in the activation of purified guanylate cyclase by nitric oxide. *Biochim. Biophys. Acta* **745**: 310–321.

Craven, P.A., DeRubertis, F.R., and Pratt, D.W. (1979). Electron spin resonance study of the role of NO· catalase in the activation of guanylate cyclase by NaN$_3$ and NH$_2$OH. Modulation of enzyme responses by heme proteins and their nitrosyl derivatives. *J. Biol. Chem.* **254**: 8213–8222.

Crawford, M.J., and Goldberg, D.E. (1998). Role for the *Salmonella* flavohemoglobin in protection from nitric oxide. *J. Biol. Chem.* **273**: 12543–12547.

Croen, K.D. (1993). Evidence for an antiviral effect of nitric oxide. Inhibition of herpes simplex virus type 1 replication. *J. Clin. Invest.* **91**: 2446–2452.

Crow, J.P., Spruell, C., Chen, J., Gunn, C., Ischiropoulos, H., Tsai, M., Smith, C.D., Radi, R., Koppenol, W.H., and Beckman, J.S. (1994). On the pH-dependent yield of hydroxyl radical products from peroxynitrite. *Free Radical Biol. Med.* **16**: 331–338.

Cunha, F.Q., Assreuy, J., Xu, D., Charles, I., Liew, F.Y., and Moncada, S. (1993). Repeated induction of nitric oxide synthase and leishmanicidal activity in murine macrophages. *Eur. J. Immunol.* **23**: 1385–1388.

Czapski, G., Holcum, J., and Bielski, B.H.J. (1994). Reactivity of nitric oxide with simple short-lived radicals in aqueous solutions. *J. Am. Chem. Soc.* **116**: 11465–11469.

Darley-Usmar, V., Hogg, N., O'Leary, V., Wilson, M.T., and Moncada, S. (1992). The simultaneous generation of superoxide and nitric oxide can initate lipid peroxidation in human low density lipoprotein. *Free Radical Res. Commun.* **17**: 9–20.

Darley-Usmar, V., Wiseman, H., and Halliwell, B. (1995). Nitric oxide and oxygen radicals: A question of balance. *FEBS Lett.* **369**: 131–135.

Davies, M.G., Dalen, H., Kim, J.H., Barber, L., Svendsen, E., and Hagen, P.-O. (1995). Control of accelerated vein graft atheroma with the nitric oxide precursor: L-Arginine. *J. Surg. Res.* **59**: 35–42.

Dawson, T.M., Dawson, V.L., and Snyder, S.H. (1992). A novel neuronal messenger molecule in brains: The free radical, nitric oxide. *Ann. Neurol.* **32**: 297–311.

Dawson, V.L., Dawson, T.M., Bartley, D.A., Uhl, G.R., and Snyder, S.H. (1993). Mechanism of nitric oxide mediated neurotoxicity in primary brain cultures. *J. Neurosci.* **13**: 2651–2661.

Dawson, V.L., Dawson, T.M., London, E.D., Bredt, D.S., and Snyder, S. (1991). Nitric oxide mediates glutamate neurotoxicity in primary cortical cultures. *Proc. Natl. Acad. Sci. USA* **88**: 6368–6371.

De Caterina, R., Libby, P., Peng, H.-B., Thannickal, V.J., Rajavashisth, T.B., Gimbrone, M.A., Jr., Shin, W.S., and Liao, J.K. (1995). Nitric oxide decreases cytokine-induced endothelial activation. Nitric oxide selectively reduces endothelial expression of adhesion molecules and proinflammatory cytokines. *J. Clin. Invest.* **96**: 60–68.

De Groote, M.A., and Fang, F.C. (1995). NO inhibitions: Antimicrobial properties of nitric oxide. *Clin. Infect. Dis.* **21**(Suppl. 2): S162–S165.

De Groote, M.A., Ochsner, U.A., Shiloh, M.U., Nathan, C., McCord, J.M., Dinauer, M.C., Libby, S.J., Vazques-Torres, A., Xu, Y., and Fang, F.C. (1997). Periplasmic superoxide-dismutase protects *Salmonella* from products of phagocyte NADPH oxidase and nitric oxide synthase. *Proc. Natl. Acad. Sci. USA* **94**: 13997–14001.

De Groote, M.A., Testerman, T. Xu, Y., Stauffer, G., and Fang, F.C. (1996). Homocysteine antagonism of nitric oxide-related cytostasis in *Salmonella typhimurium*. *Science* **272**: 414–417.

Deguchi, T., and Yoshioka, M. (1982) L-Arginine identified as an endogenous activator for soluble guanylate cyclase from neuroblastoma cells. *J. Biol. Chem.* **257**: 10147–10151.

Deisseroth, A., and Dounce, A.L. (1970). Catalase: Physical and chemical properties, mechanism of catalysis, and physiological role. *Physiol. Rev.* **50**: 319–375.

Demas, G.E., Eliasson, M.J., Dawson, T.M., Dawson, V.L., Kriegsfeld, L.J., Nelson, R.J., and Snyder, S.H. (1997). Inhibition of neuronal nitric oxide

synthase increases aggressive behavior in mice. *Mol. Med.* **3**: 610–616.

de-Mello, M.A.R., Flodström, M., and Eizirik, D.L. (1996). Ebselen and cyto-kine-induced nitric oxide synthase expression in insulin- producing cells. *Biochem. Pharmacol.* **52**: 1703–1709.

Deng, W., Thiel, B., Tannenbaum, C.S., Hamilton, T., and Stuehr, D.J. (1993). Synergistic cooperation between T cell lymphokines for induction of the nitric oxide synthase gene in murine peritoneal macrophages. *J. Immunol.* **151**: 322–329.

Denicola, A., Freeman, B.A., Trujillo, M., and Radi, R. (1996). Peroxynitrite reaction with carbon dioxide/bicarbonate: Kinetics and influence on peroxynitrite-mediated oxidation. *Arch. Biochem. Biophys.* **333**: 49–58.

Denis, M. (1991). Tumor necrosis factor and granulocyte macrophage- colony stimulating factor stimulate human macrophages to restrict growth of virulent *Mycobacterium avium* and to kill avirulent *M. avium*: Killing effector mechanism depends on the generation of reactive nitrogen intermediates. *J. Leukocyte Biol.* **49**: 380–387.

Diamond, J., and Chu, E.B. (1983). Possible role for cyclic GMP in endothelium-dependent relaxation of rabbit aorta by acetylcholine. Comparison with nitroglycerin. *Res. Commun. Chem. Pathol. Pharmacol.* **41**: 369–381.

Ding, A.H., Nathan, C.F., and Stuehr, D.J. (1988). Release of reactive nitrogen intermediates and reactive oxygen intermediates from mouse peritoneal macrophages. Comparison of activating cytokines and evidence for independent production. *J. Immunol.* **141**: 2407–2412.

Dolphin, D., Traylor, T.G., and Xie, L.Y. (1997). Polyhaloporphyrins: Unusual ligands for metals and metal-catalyzed oxidations. *Acc. Chem. Res.* **30**: 251–259.

Donato, M.T., Guillén, M.I., Jover, R., Castell, J.V., and Gómez-Lechón, M.J. (1997). Nitric oxide-mediated inhibition of cytochrome P450 by interferon-γ in human hepatocytes. *J. Pharmacol. Exp. Ther.* **281**: 484–490.

Doyle, M.P., and Hoesktra, J.W. (1981). Oxidation of nitrogen oxides by bound dioxygen hemoproteins. *J. Inorg. Biochem.* **14**: 351–358.

Drapier, J.C., and Hibbs, J.B., Jr. (1986). Murine cytotoxic activated macrophages inhibit aconitase in tumor cells. Inhibition involves the iron-sulfur prosthetic group and is reversible. *J. Clin. Invest.* **78**: 790–797.

Drapier, J.C., and Hibbs, J.B., Jr. (1988). Differentiation of murine macrophages to express nonspecific cytotoxicity for tumor cells results in L-arginine-dependent inhibition of mitochondrial iron-sulfur enzymes in the macrophage effector cells. *J. Immunol.* **140**: 2829–2838.

Drapier, J.C., Hirling, H., Wietzerbin, J., Kaldy, P., and Kühn, L.C. (1993). Biosynthesis of nitric oxide activates iron regulatory factor in macrophages. *EMBO J.* **12**: 3643–3649.

Duersken-Hughes, P.J., Day, D.B., Laster, S.M., Zachariades, N.A., Aquino, L., and Gooding, L.R. (1992). Both tumor necrosis factor and nitric oxide participate in lysis of simian virus 40-transformed cells by activated macrophages. *J. Immunol.* **149**: 2114–2122.

Dun, N.J., Dun, S.L., Förstermann, U., and Tseng, L.F. (1992). Nitric oxide synthase immunoreactivity in rat spinal cord. *Neurosci. Lett.* **147**: 217–220.

Dun, N.J., Dun, S.L., Wu, S.Y., and Förstermann, U. (1993). Nitric oxide synthase immunoreactivity in rat superior cervical ganglia and adrenal glands. *Neurosci. Lett.* **158**: 51–54.

Duval, D.L., Miller, D.R., Collier, J., and Billings, R.E. (1996). Characterization of hepatic nitric oxide synthase: Identification as the cytokine-inducible form primarily regulated by oxidants. *Mol. Pharmacol.* **50**: 277–284.

DuVall, M.D., Zhu, S., Fuller, C.M., and Matalon, S. (1998). Peroxynitrite inhibits amiloride-sensitive Na^+ currents in *Xenopus* oocytes expressing $\alpha\beta\gamma$-rENaC. *Am. J. Physiol.* **274**: C1417–C1423.

Duvilanski, B.H., Zambruno, C., Seilicovich, A., Pisera, D., Lasaga, M., Diaz, M. del C., Belova, N., Rettori, V., and McCann, S.M. (1995). Role of nitric oxide in control of prolactin release by the adenohypophysis. *Proc. Natl. Acad. Sci. USA* **92**: 170–174.

Ehrt, S., Shiloh, M.U., Ruan, J., Choi, M., Gunzburg, S., Nathan, C., Xie, Q., and Riley, L.W. (1997). A novel antioxidant gene from *Mycobacterium tuberculosis*. *J. Exp. Med.* **186**: 1885–1896.

Elliott, S.J. (1996). Peroxynitrite modulates receptor-activated Ca^{2+} signaling in vascular endothelial cells. *Am. J. Physiol.* **270**: L954-L961.

Farrington, JA., Land, E.J., and Swallow, J. (1980). The one-electron reduction potentials of NAD. *Biochim. Biophys. Acta* **590**: 273–276.

Feldman, P.L., Griffith, O.W., and Stuehr, D.J. (1993). The surprising life of nitric oxide. *Chem. Eng. News,* **71**(51): 26–38.

Ferrante, A.A., Augliera, J., Lewis, K., and Klibanov, A.M. (1995). Cloning of an organic solvent-resistance gene in *Escherichia coli*: The unexpected role of alkylhydroperoxide reductase. *Proc. Natl. Acad. Sci. USA* **92**: 7617–7621.

Fierro, I.M., Barja-Fidalgo, C., Cunha, F.Q., and Ferreira, S.H. (1996). The involvement of nitric oxide in the anti-*Candida albicans* activity of rat neutrophils. *Immunology* **89**: 295–300.

Fischer-Stenger, K., and Marciano-Cabral, F. (1992). The arginine- dependent cytolytic mechanism plays a role in destruction of *Naegleria fowleri* amoebae by activated macrophages. *Infect. Immun.* **60**: 5126–5131.

Flavahan, N.A. (1992). Atherosclerosis or lipoprotein-induced endothelial dysfunction. Potential mechanisms underlying reduction in EDRF/nitric oxide activity. *Circulation* **85**: 1927–1938.

Flesch, I.E.A., and Kaufmann, S.H.E. (1991). Mechanisms involved in mycobacterial growth inhibition by gamma interferon-activated bone marrow macrophages: Role of reactive nitrogen intermediates. *Infect. Immun.* **59**: 3213–3218.

Fontecave M., and Pierre, J.-L. (1994). The basic chemistry of nitric oxide and its possible biologic reactions. *Bull. Soc. Chim. Fr.* **131**: 620–631.

Ford, P.C., Wink, D.A., and Stanbury, D.M. (1993). Autoxidation kinetics of aqueous nitric oxide. *FEBS Lett.* **326**: 1–3.

Förstermann, U., Closs, E.I., Pollock, J.S., Nakane, M., Schwarz, P., Gath, I., and Kleinert, H. (1994). Nitric oxide synthase isozymes. Characterization, purification, molecular cloning and functions. *Hypertension* **23**: 1121–1131.

Förstermann, U., Gath, I., Schwarz, P., Closs, E.I., and Kleinert, H. (1995). Isoforms of nitric oxide synthase. Properties, cellular distribution and expressional control. *Biochem. Pharmacol.* **50**: 1321–1332.

Förstermann, U., Gorsky, L.D., Pollock, J.S., Ishii, K., Schmidt, H.H.H.W., Heller, M., and Murad, F. (1990). Hormone-induced biosynthesis of endothelium-derived relaxation factor/nitric oxide-like material in N1E-115 neuroblastoma cells requires calcium and calmodulin. *Mol. Pharmacol.* **38**: 7–13.

Fujita, H., Morita, I., and Murota, S.-I. (1994). A possible mechanism for vascular endothelial cell injury elicited by activated leukocytes: A significant involvement of adhesion, CD11/CD18, and ICAM-1. *Arch. Biochem. Biophys.* **309**: 62–69.

Fukahori, M., Ichimori, K., Ishida, H., Nakagawa, H., and Okino, H. (1994). Nitric oxide reversibly suppresses xanthine oxidase activity. *Free Radical Res.* **21**: 203–212.

Fukaya, Y., and Ohhashi, T. (1996). Acetylcholine- and flow-induced production and release of nitric oxide in arterial and venous endothelial cells. *Am. J. Physiol.* **270**: H99-H106.

Fukuto, J.M., and Ignarro, L.J. (1997). *In vivo* aspects of nitric oxide (NO) chemistry: Does peroxynitrite (⁻OONO) play a major role in cytotoxicity? *Acc. Chem. Res.* **30**: 149–152.

Fukuto, J.M., Stuehr, D.J., Feldman, P.L., Bova, M.P., and Wong, P. (1993). Peracid oxidation of an N-hydroxyguanidine compound: A chemical model for the oxidation of N^{ω}-hydroxyl-L-arginine by nitric oxide synthase. *J. Med. Chem.* **36**: 2666–2670.

Furchgott, R.F. (1988). Studies on relaxation of rabbit aorta by sodium nitrite: The basis for the proposal that the acid-activatable inhibitory factor from bovine retractor penis is inorganic nitrite and the endothelium-derived relaxing factor is nitric oxide, in: *Vasodilatation: Vascular Smooth Muscle, Peptides, Autonomic Nerves and Endothelium* (Vanhoutte, P.M., ed.), pp. 401–414, Raven Press, New York.

Furchgott, R.F., and Zawadzki, J.V. (1980). The obligatory role of endothelial cells on the relaxation of arterial smooth muscle by acetylcholine. *Nature* **288**: 373–376.

Furlong, B., Henderson, A.H., Lewis, M.J., and Smith, J.A. (1987). Endothelium-derived relaxing factor inhibits *in vitro* platelet aggregation. *Br. J. Pharmacol.* **90**: 687–692.

Galli, C., MacArthur, R., Abu-Soud, H.M., Clark, P., Stuehr, D.J., and Brudvig, G.W. (1996). EPR spectroscopic characterization of neuronal NO synthase. *Biochemistry* **35**: 2804–2810.

García-Cardeña, G., Fan, R., Stern, D.F., Liu, J., and Sessa, W.C. (1996a). Endothelial nitric oxide synthase is regulated by tyrosine phosphorylation and interactions with caveolin-1. *J. Biol. Chem.* **271**: 27237–27240.

García-Cardeña, G., Martasek, P., Masters, B.S.M., Skidd, P.M., Couet, J., Li, S., Lisanti, M.P., and Sessa, W.C. (1997). Dissecting the interaction between nitric oxide synthase (NOS) and caveolin. Functional signficance of the NOS caveolin binding domain *in vivo*. *J. Biol. Chem.* **272**: 25437–25440.

García-Cardeña, G., Oh, P., Liu, J., Schnitzer, J.E., and Sessa, W.C. (1996b). Targeting of nitric oxide synthase to endothelial cell caveolae via palmitoylation: Implications for nitric oxide signaling. *Proc. Natl. Acad. Sci. USA* **93**: 6448–6453.

Gardner, P.R., Costantino, G., and Salzman, A.L. (1998a). Constitutive and adaptive detoxification of nitric oxide in *Escherichia coli*: Role of nitric-oxide dioxygenase in the protection of aconitase. *J. Biol. Chem.* **273**: 26528–26533.

Gardner, P.R., Costantino, G., Szabó, C., and Salzman, A.L. (1997). Nitric oxide sensitivity of the aconitases. *J. Biol. Chem.* **272**: 25071–25076.

Gardner, P.R., Gardner, A.M., Martin, L.A., and Salzman, A.L. (1998b). Nitric oxide dioxygenase: An enzymic function for flavohemoglobin. *Proc. Natl. Acad. Sci. USA* **95**: 10378–10383.

Garg, U.C., and Hassid, A. (1989). Nitric oxide-generating vasodilators and

8-bromocyclic guanosine monophosphate inhibit mitogenesis and proliferation of cultured rat vascular smooth muscle cells. *J. Clin. Invest.* **83**: 1774–1777.

Garthwaite, J., Charles, S.L., and Chess-Williams, R. (1988). Endothelium-derived relaxing factor release on activation of NMDA receptors suggests role as intercellular messenger in the brain. *Nature* **336**: 385–388.

Gath, I., Closs, E.I., Gödtel-Armbrust, U., Schmitt, S., Nakane, M., Wessler, I., and Förstermann, U. (1996). Inducible NO synthase II and neuronal NO synthase I are constitutively expressed in different structures of guinea pig skeletal muscle: Implications for contractile function. *FASEB J.* **10**: 1614–1620.

Gatti, R.M., Radi, R., and Augusto, O. (1994). Peroxynitrite-mediated oxidation of albumin to the protein-thiyl free radical. *FEBS Lett.* **348**: 287–290.

Gauthier, R.T.W., Davenpeck, K.L., and Lefer, A.M. (1994). Nitric oxide attenuates leukocyte–endothelial interactions via P-selectin in splanchnic ischemia-reperfusion. *Am. J. Physiol.* **267**: G562-G568.

Geller, D.A., Lowenstein, C.J., Shapiro, R.A., Nussler, A.K., Di Silvio, M., Wang S.C., Nakayama, D.K., Simmons, R.L., Snyder, S.H., and Billiar, T.R. (1993). Molecular cloning and expression of inducible nitric oxide synthase from human hepatocytes. *Proc. Natl. Acad. Sci. USA* **90**: 3491–3495.

Gerber, N.C., Rodriquez-Crespo, I., Nishida, C.R., and Ortiz de Montellano, P.R. (1997). Active site topologies and cofactor-mediated conformational changes of nitric-oxide synthases. *J. Biol. Chem.* **272**: 6285–6290.

Gergel, D., Misík, V., Riesz, P., and Cederbaum, A.I. (1997). Inhibition of rat and human cytochrome P4502E1 catalytic activity and reactive oxygen radical formation by nitric oxide. *Arch. Biochem. Biophys.* **337**: 239–250.

Ghosh, D.K., and Stuehr, D.J. (1995). Macrophage NO synthase: Characterization of isolated oxygenase and reductase domains reveal a head-to-head subunit interaction. *Biochemistry* **34**: 801–807.

Gibson, A., Mirzazadeh, S., Hobbs, A.J., and Moore, P.K. (1990). L-N^G-Monomethyl arginine and L-N^G-nitro arginine inhibit non-adrenergic noncholingeric relaxation of the mouse anococcygeus muscle. *Br. J. Pharmacol.* **99**: 602–606.

Gillespie, J.S., Liu, X., and Martin, W. (1989). The effects of L-arginine and N^G-monomethyl L-arginine on the response of the rat anococcygeus muscle to NANC nerve stimulation. *Brit. J. Pharmacol.* **98**: 1080–1082.

Giovanelli, J., Campos, K.L., and Kaufman, S. (1991). Tetrahydrobiopterin, a cofactor for rat cerebellar nitric oxide synthase, does not function as a reactant in the oxygenation of arginine. *Proc. Natl. Acad. Sci. USA* **88**: 7091–7095.

Goldstein, S., and Czapski, G. (1995a). Kinetics of nitric oxide autoxidation in aqueous solution in the absence and presence of various reductants. The nature of the oxidizing intermediates. *J. Am. Chem. Soc.* **117**: 12078–12084.

Goldstein, S., and Czapski, G. (1995b). The reaction of NO· with $O_2\cdot^-$ and $HO_2\cdot$: A pulse radiolysis study. *Free Radical Biol. Med.* **19**: 505–510.

Goldstein, S., and Czapski, G. (1996). Mechanism of the nitrosation of thiols and amines by oxygenated ·NO solutions: The nature of the nitrosating intermediates. *J. Am. Chem. Soc.* **118**: 3419–3425.

Goldstein, S., and Czapski, G. (1999). Viscosity effects on the reaction of peroxynitrite with CO_2: Evidence for radical formation in a solvent cage. *J. Am. Chem. Soc.* **121**: 2444–2447.

Goldstein, S., Squadrito, G.L., Pryor, W.A., and Czapski, G. (1996). Direct and indirect oxidations by peroxynitrite, neither involving the hydroxyl radical. *Free Radical Biol. Med.* **21**: 965–974.

Gorbunov, N.V., Osipov, A.N., Day, B.W., Zayas-Rivera, B., Kagan, V.E., and Elsayed, N.M. (1995). Reduction of ferrylmyoglobin and ferrylhemoglobin by nitric oxide: A protective mechanism against ferryl hemoprotein-induced oxidations. *Biochemistry* **34**: 6689–6699.

Gorren, A.C.F., List, B.M., Schrammel, A., Pitters, E., Hemmens, B., Werner, E.R., Schmidt, K., and Mayer, B. (1996). Tetrahydrobiopterin-free neuronal nitric oxide synthase: Evidence for two identical highly anticooperative pteridine binding sites. *Biochemistry* **35**: 16735–16745.

Gow, A., Duran, D., Thom, S.R., and Ischiropoulos, H. (1996). Carbon dioxide enhancement of peroxynitrite-mediated protein tyrosine nitration. *Arch. Biochem. Biophys.* **333**: 42–48.

Granger, D.L., Hibbs, J.B., Jr., Perfect, J.R., and Durack, D.T. (1988). Specific amino acid (L-arginine) requirement for microbiostatic activity of murine macrophages. *J. Clin. Invest.* **81**: 1129–1136.

Granger, D.L., Hibbs, J.B., Jr., Perfect, J.R., and Durack, D.T. (1990). Metabolic fate of L-arginine in relation to microbiostatic capability of murine macrophages. *J. Clin. Invest.* **85**: 264–273.

Grant, S.K., Green, B.G., Wang, R., Pacholok, S.G., and Kozarich, J.W. (1997). Characterization of inducible nitric-oxide synthase by cytochrome P-450 substrates and inhibitors. Inhibition by chlorzoxazone. *J. Biol. Chem.* **272**: 977–983.

Green, L.C., Ruiz De Luzuriaga, K., Warner, D.A., Rand, W., Istfan, N., Young, V.R., and Tannenbaum, S.R. (1981a). Nitrate biosynthesis in man. *Proc. Natl. Acad. Sci. USA* **78**: 7764–7768.

Green, L.C., Tannenbaum, S.R., and Goldman, R. (1981b). Nitrate synthesis in the germ free and conventional rat. *Science* **212**: 56–58.

Green, S.J., Mellouk, S., Hoffman, S.L., Meltzer, M.S., and Nacy, C.A. (1990a). Cellular mechanisms of nonspecific immunity to intracellular infection: Cytokine-induced synthesis of toxic nitrogen oxides from L-arginine by macrophages and hepatocytes. *Immunol. Lett.* **25**: 15–20.

Green, S.J., Melzer, M.S., Hibbs, J.B., Jr., and Nacy, C.A. (1990b). Activated macrophages destroy intracellular *Leishmania major* amastigotes by an L-arginine-dependent killing mechanism. *J. Immunol.* **144**: 278–283.

Green, S.J., Nacy, C.A., and Melzer, M.S. (1991). Cytokine-induced synthesis of nitrogen oxides in macrophages: A protective host response to *Leishmania* and other intracellular pathogens. *J. Leukocyte Biol.* **50**: 93–103.

Green, S.J., Scheller, L.F., Marletta, M.A., Seguin, M.C., Koltz, F.W., Slayter, M., Nelson, B.J., and Nacy, C.A. (1994). Nitric oxide: Cytokine-regulation of nitric oxide in host resistance to intracellular pathogens. *Immunol. Lett.* **43**: 87–94.

Griffith, O.W., and Stuehr, D.J. (1995). Nitric oxide synthase: Properties and catalytic mechanism. *Annu. Rev. Physiol.* **57**: 707–736.

Griscavage, J.M., Fukuto, J.M., Komori, Y., and Ignarro, L.J. (1994). Nitric oxide inhibits neuronal nitric oxide synthase by interacting with the heme prosthetic group. Role of tetrahydrobiopterin in modulating the inhibitory action of nitric oxide. *J. Biol. Chem.* **269**: 21644–21649.

Groves, J.T., and Watanabe, Y. (1986). Oxygen activation by metalloporphyrins related to peroxidase and cytochrome P-450. Direct observation of the O–O bond cleavage step. *J. Am. Chem. Soc.* **108**: 7834–7836.

Groves, J.T., Haushalter, R.C., Nakamura, M., Nemo, T.E., and Evans, B.J. (1981). High-valent iron–porphyrin complexes related to peroxidase and cytochrome P-450. *J. Am. Chem. Soc.* **103**: 2884–2886.

Groves, P.H., Lewis, M.J., Cheadle, H.A., and Penny, W.J. (1993). SIN-1 reduces platelet adhesion and platelet thrombus formation in a porcine model of balloon angioplasty. *Circulation* **87**: 590–597.

Gryglewski, R.J., Palmer, R.M.J., and Moncada, S. (1986). Superoxide anion is involved in the breakdown of endothelial-derived relaxing factor. *Nature* **320**: 454–456.

Guengerich, F.P. (1983). Oxidation-reduction properties of rat liver cytochrome P-450 and NADPH-cytochrome P-450 reductase related to catalysis in reconstituted systems. *Biochemistry* **22**: 2811–2820.

Guengerich, F.P., and Macdonald, T.L. (1984). Chemical mechanisms of catalysis by cytochrome P-450: A unified view. *Acc. Chem. Res.* **17**: 9–16.

Guengerich, F.P., Vaz, A.D.N., Raner, G.N., Pernecky, S.J., and Coon, M.J. (1997). Evidence for a role of a perferryl–oxygen complex, FeO^{3+}, in the N-oxygenation of amines by cytochrome P-450 enzymes. *Mol. Pharmacol.* **51**: 147–151.

Guengerich, F.P., Yun, C.H., and Macdonald, T.L. (1996). Evidence for a 1-electron oxidation mechanism in N-dealkylation of N,N-dialkyl-anilines by cytochrome P-450 2B1. Kinetic hydrogen isotope effects, linear free energy relationships, comparisons with horseradish peroxidase, and studies with oxygen surrogates. *J. Biol. Chem.* **271**: 27321–27329.

Guo, J.P., Milhoan, K.A., Tuan, R.S., and Lefer, A.M. (1994). Beneficial effect of SPM-5185, a cysteine-containing nitric oxide donor, in rat carotid artery intimal injury. *Circ. Res.* **75**: 77–84.

Halfpenny, E., and Robinson, P.L. (1952). Pernitrous acid. The reaction between hydrogen peroxide and nitrous acid, and the properties of an intermediate product. *J. Chem. Soc.* 928–938.

Hamon, M., Vallet, B., Bauters, C., Wernert, N., McFadden, E.P., Lablanche, J.-M., Dupuis, B., and Bertrand, M.E. (1994). Long-term oral administration of L-arginine reduces intimal thickening and enhances neoendothelium-dependent acetylcholine-induced relaxation after arterial injury. *Circulation* **90**: 1357–1362.

Harris, D.L., and Loew, G.H. (1996). Investigation of the proton-assisted pathway to formation of the catalytically active, ferryl species of P450s by molecular dynamics studies of P450eryF. *J. Am. Chem. Soc.* **118**: 6377–6387.

Harris, D., Loew, G., and Waskell, L. (1998). Structure and spectra of ferrous dioxygen and reduced ferrous dioxygen model cytochrome P450. *J. Am. Chem. Soc.* **120**: 4308–4318.

Harris, N., Buller, R.M.L., and Karupiah, G. (1995). Interferon-gamma induced, nitric oxide mediated inhibition of vaccinia viral replication. *J. Virology.* **69**: 910–915.

Hattori, Y., Campbell, E.B., and Gross, S.S. (1994). Argininosuccinate synthetase mRNA and activity are induced by immunostimulants in vascular smooth muscle. Role in the regeneration of arginine for nitric oxide synthesis. *J. Biol. Chem.* **269**: 9405–9408.

Hausladen, A., and Fridovich, I. (1994). Superoxide and peroxynitrite inactivated aconitases, but nitric oxide does not. *J. Biol. Chem.* **269**: 29405–29408.

Hausladen, A., Privalle, C.T., Keng, T., DeAngelo, J., and Stamler, J.S. (1996). Nitrosative stress: Activation of the transcription factor OxyR. *Cell* **86**: 719–729.

Hecker, M., Mülsch, A., Bassenge, E., Förstermann, U., and Busse, R. (1994). Subcellular localization and characterization of nitric oxide synthase(s) in endothelial cells: Physiological implications. *Biochem. J.* **299**: 247–252.

Hecker, M., Sessa, W.C., Harris, H.J., Änggård, E.E., and Vane, J.R. (1990). The metabolism of L-arginine and its significance for the biosynthesis of endothelium-derived relaxing factor: Cultured endothelial cells recycle L-citrulline to L-arginine. *Proc. Natl. Acad. Sci. USA* **87**: 8612–8616.

Hellermann, G.R., and Solomonson, L.P. (1997). Calmodulin promotes dimerization of the oxygenase domain of human endothelial nitric- oxide synthase. *J. Biol. Chem.* **272**: 12030–12034.

Henry, Y., Ducrocq, C., Drapier, J.-C., Servent, D., Pellant, C., and Guissani, A. (1991). Nitric oxide, a biological effector. Electron paramagnetic resonance detection of nitrosyl–iron–protein complexes in whole cells. *Eur. Biophys. J.* **20**: 1–15.

Henry, Y., Lepoivre, M., Drapier, J.-C., Ducrocq, C., Boucher, J.-L., and Guissani, A. (1993). EPR characterization of molecular targets for NO in mammalian cells and organelles. *FASEB J.* **7**: 1124–1134.

Hevel, J.M., and Marletta, M.A. (1992). Macrophage nitric oxide synthase: Relationship between enzyme-bound tetrahydrobiopterin and synthase activity. *Biochemistry* **31**: 7160–7165.

Hevel, J.M., White, K.A., and Marletta, M.A. (1991). Purification of the inducible murine macrophage nitric oxide synthase. Identification as a flavoprotein. *J. Biol. Chem.* **266**: 22789–22791.

Hibbs, J.B., Jr., Taintor, R.R., and Vavrin, Z. (1987a) Macrophage cytotoxicity: Role of L-Arginine deiminase and imino nitrogen oxidation to nitrite. *Science* **235**: 473–476.

Hibbs, J.B., Jr., Taintor, R.R., Vavrin, Z., and Rachlin, E.M. (1988). Nitric oxide: A cytotoxic activated macrophage effector molecule. *Biochem. Biophys. Res. Commun.* **157**: 87–94.,

Hibbs, J.B., Jr., Vavrin, Z., and Taintor, R.R. (1987b). L-arginine is required for expression of the activated macrophage effector mechanism causing selective metabolic inhibition in target cells. *J. Immunol.* **138**: 550–565.

Hiki, K., Hattori, R., Kawai, C., and Yui, Y. (1992). Purification of insoluble nitric oxide synthase from rat cerebellum. *J. Biochem.* **111**: 556–558.

Hogg, N., Darley-Usmar, V.M., Wilson, M.T., and Moncada, S. (1992). Production of hydroxyl radicals from the simultaneous generation of superoxide and nitric oxide. *Biochem. J.* **281**: 419–424.

Hogg, N., Singh, R.J., and Kalyanaraman, B. (1996). The role of glutathione in the transport and catabolism of nitric oxide. *FEBS Lett.* **382**: 223–228.

Holzmann, S. (1982). Endothelium-induced relaxation by acetylcholine associated with larger rises in cyclic GMP in coronary arterial strips. *J. Cyclic Nucleotide Res.* **8**: 409–419.

Hori, H., Masuya, F., Tsubaki, M., Yoshikawa, S., and Ichikawa, Y. (1992). Electron and stereochemical characterizations of intermediates in the photolysis of ferric cytochrome P450ssc nitrosyl complexes. Effects of cholesterol and its analogues on ligand binding. *J. Biol. Chem.* **267**: 18377–18381.

Hoshino, M., Maeda, M., Konishi, R., Seki, H., and Ford, P.C. (1996). Studies on the reaction mechanism for reductive nitrosylation of ferrihemoproteins in buffer solutions. *J. Am. Chem. Soc.* **118**: 5702–5707.

Houk, K.N., Condroski, K.R., and Pryor, W.A. (1996). Radical and concerted mechanisms in oxidations of amines, sulfides, and alkenes by peroxynitrite, peroxynitrous acids and the peroxynitrite-CO_2 adduct: Density functional theory transition structures and energetics. *J. Am. Chem. Soc.* **118**: 13002–13006.

Hu, P., Ischiropoulos, H., Beckman, J.S., and Matalon, S. (1994). Peroxynitrite inhibition of oxygen consumption and sodium transport in alveolar type II cells. *Am. J. Physiol.* **266**: L628–L634.

Huang, P.L., Dawson, T.M., Bredt, D.S., Snyder, S.H., and Fishman, M.C. (1993). Targeted disruption of the neuronal nitric oxide synthase gene. *Cell* **75**: 1273–1286.

Huang, P.L., Huang, Z., Mashimo, H., Bloch, K.D., Moskowitz, M.A., Bevan, J.A., and Fishman, M.C. (1995). Hypertension in mice lacking the gene for endothelial nitric oxide synthase. *Nature* **377**: 239–242.

Huang, Z., Huang, P.L., Panahian, N., Dalkara, T., Fishman, M.C., and Moskowitz, M.A. (1994). Effects of cerebral ischemia in mice deficient in neuron nitric oxide synthase. *Science* **265**: 1883–1885.

Huie, R.E., and Padmaja, S. (1993). The reaction of NO with superoxide. *Free Radical Res. Commun.* **18**: 195–199.

Hurshman, A.R., and Marletta, M.A. (1995). Nitric oxide complexes of inducible nitric oxide synthase: Spectral characterization and effect on catalytic activity. *Biochemistry* **34**: 5627–5634.

Ignarro, L.J. (1991). Signal transduction mechanisms involving nitric oxide. *Biochem. Pharmacol.* **41**: 485–490.

Ignarro, L.J., Buga, G.M., Wood, K.S., Byrns, R.E., and Chaudhuri, G. (1987). Endothelium-derived relaxing factor produced and released from artery and vein is nitric oxide. *Proc. Natl. Acad. Sci. USA* **84**: 9265–9269.

Ignarro, L.J., Burke, T.M., Wood, K.S., Wolin, M.S., and Kadowitz, P.J. (1984). Association between cyclic GMP accumulation and acetylcholine-elicited relaxation of bovine intrapulmonary artery. *J. Pharmacol. Exp. Ther.* **228**: 682–690.

Ignarro, L.J., Byrns, R.E., and Wood, K.S. (1988). Biochemical and pharmacological properties of endothelium-derived relaxing factor and its similiarity to nitric oxide radical, in: *Vasodilatation: Vascular Smooth Muscle, Peptides,*

Autonomic Nerves and Endothelium (Vanhoutte, P.M., ed.), pp. 427–435, Raven Press, New York.

Ignarro, L.J., Degnan, J.N., Baricos, W.H., Kadowitz, P.J., and Wolin, M.S. (1982). Activation of purified guanylate cyclase by nitric oxide requires heme: Comparison of heme-deficient, heme-reconstituted and heme- containing forms of soluble enzyme from bovine lung. *Biochim. Biophys. Acta* **718**: 49–59.

Ignarro, L.J., Fukuto, J.M., Griscavage, J.M., Rogers, N.E., and Byrns, R.E. (1993). Oxidation of nitric oxide in aqueous solution to nitrite but not nitrate: Comparison with enzymatically formed nitric oxide from L-arginine. *Proc. Natl. Acad. Sci. USA* **90**: 8103–8107.

Ignarro, L.J., Lippton, H., Edwards, J.C., Baricos, W.H., Hyman, A.L., Kadowitz, P.J., and Gruetter, C.A. (1981). Mechanism of vascular smooth muscle relaxation by organic nitrates, nitrites, nitroprusside and nitric oxide: Evidence for the involvement of S-nitrosothiols as active intermediates. *J. Pharmacol. Exp. Ther.* **218**: 739–749.

Ischiropoulos, H., Zhu, L., and Beckman, J.S. (1992a). Peroxynitrite formation from macrophage-derived nitric oxide. *Arch. Biochem. Biophys.* **298**: 446–451.

Ischiropoulos, H., Zhu, L., Chen, J., Tsai, M., Martin, J.C., Smith, C.D., and Beckman, J.S. (1992b). Peroxynitrite-mediated tyrosine nitration catalyzed by superoxide dismutase. *Arch. Biochem. Biophys.* **298**: 431–437.

Ishida, H., Ichimori, K., Hirota, Y., Fukahori, M., and Nakazawa, H. (1996). Peroxynitrite-induced cardiac myocyte injury. *Free Radical Biol. Med.* **20**: 343–350.

Isono, T., Sato, N., Koibuchi, Y., Sakai, S., Yamamoto, T., Ozaki, R., Mori, J., Kohsaka, M., and Ohtsuka, M. (1993). Effect of FK409, a novel nitric oxide donor, on acute experimental myocardial ischemia. *Jpn. J. Pharmacol.* **62**: 315–324.

Iyanagi, T., and Mason, H.S. (1973). Some properties of hepatic reduced nicotinamide adenine dinucleotide phosphate-cytochrome c reductase. *Biochemistry* **12**: 2297–2308.

Iyengar, R., Stuehr, D.J., and Marletta, M.A. (1987). Macrophage synthesis of nitrite, nitrate and N-nitrosoamines: Precursors and role of the respiratory burst. *Proc. Natl. Acad. Sci. USA* **84**: 6369–6373.

Izumi, Y., Benz, A.M., Clifford, D.B., and Zorumski, C.F. (1992). Nitric oxide inhibitors attenuate N-methyl-D-aspartate excitotoxicity in rat hippocampal slices. *Neurosci. Lett.* **135**: 227–230.

Jaffrey, S.R., and Snyder, S.H. (1996). PIN: An associated protein inhibitor of neuronal nitric oxide synthase. *Science* **274**: 774–777.

James, S.L., and Glaven, J. (1989). Macrophage cytotoxicity against schistosomula of *Schistosoma mansoni* involves arginine-dependent production of reactive nitrogen intermediates. *J. Immunol.* **143**: 4208–4212.

Janssens, S.P., Shimouchi, A., Quertermous, T., Bloch, D.B., and Bloch, K.D. (1992). Cloning and expression of a cDNA encoding human endothelial-derived relaxing factor/nitric oxide synthase. *J. Biol. Chem.* **267**: 14519–14522.

Jia, L., Bonaventura, C., Bonaventura, J., and Stamler, J.S. (1996). S-Nitrosohaemoglobin: A dynamic activity of blood involved in vascular control. *Nature* **380**: 221–226.

Johnson, G., III, Tsao, P.S., Mulloy, D., and Lefer, A.M. (1990). Cardioprotective effects of acidified sodium nitrite in myocardial ischemia with reperfusion. *J. Pharmacol. Exp. Ther.* **252**: 35–41.

Ju, H., Zou, R., Venema, V.J., and Venema, R.C. (1997). Direct interaction of endothelial nitric-oxide synthase and caveolin-1 inhibits synthase activity. *J. Biol. Chem.* **272**: 18522–18525.

Juckett, M.B., Weber, M., Balla, J., Jacob, H.S., and Vercellotti, G.M. (1996). Nitric oxide donors modulate ferritin and protect endothelium from oxidative injury. *Free Radical Biol. Med.* **20**: 63–73.

Kanner, J., Harel, S., and Granit, R. (1991). Nitric oxide as an antioxidant. *Arch. Biochem Biophys.* **289**: 130–136.

Kanner, J., Harel, S., and Granit, R. (1992). Nitric oxide, an inhibitor of lipid oxidation by lipoxygenase, cyclooxygenase and hemoglobin. *Lipids* **27**: 46–49.

Karki, S.B., Dinnocenzo, J.P., Jones, J.P., and Korzekwa, K.R. (1995). Mechanism of oxidative amine dealkylation of substituted N,N-dimethylanilines by cytochrome P-450: Application of isotope effect profiles. *J. Am. Chem. Soc.* **117**: 3657–3664.

Karupiah, G., and Harris, N. (1995). Inhibition of viral replication by nitric oxide and its reversal by ferrous sulfate and tricarboxylic acid cycle metabolites. *J. Exp. Med.* **181**: 2171–2179.

Karupiah, G., Xie, Q.-W., Buller, M.L., Nathan, C., Duarte, C., and MacMicking, J.D. (1993). Inhibition of viral replication by interferon-γ-induced nitric oxide synthase. *Science* **261**: 1445–1448.

Kaur, H., Whiteman, M., and Halliwell, B. (1997). Peroxynitrite-dependent aromatic hydroxylation and nitration of salicylate and phenylalanine. Is hydroxyl radical involved? *Free Radical Res.* **26**: 71–82.

Kennedy, M.C., Antholine, W.E., and Beinert, H. (1997). An EPR investigation of the products of the reaction of cytosolic and mitochondrial aconitases with nitric oxide. *J. Biol. Chem.* **272**: 20340–20347.

Keyer, K., and Imlay, J.A. (1997). Inactivation of dehydratase [4Fe–4S] clusters and disruption of iron homeostasis upon cell exposure to peroxynitrite. *J. Biol. Chem.* **272**: 27652–27659.

Kharitonov, V.G., Sundquist, A.R., and Sharma, V.S. (1994). Kinetics of nitric oxide autoxidation in aqueous solution. *J. Biol. Chem.* **269**: 5881–5883.

Kharitonov, V.G., Sundquist, A.R., and Sharma, V.S. (1995). Kinetics of nitrosation of thiols by nitric oxide in the presence of oxygen. *J. Biol. Chem.* **270**: 28158–28164.

Khatsenko, O., and Kikkawa, Y. (1997). Nitric oxide differentially affects constitutive cytochrome P-450 isoforms in rat liver. *J. Pharmacol. Exp. Ther.* **280**: 1463–1470.

Khatsenko, O.G., Gross, S.S., Rifkind, A.B., and Vane J.R. (1993). Nitric oxide is a mediator of the decrease in cytochrome P-450-dependent metabolism caused by immunostimulants.

Proc. Natl. Acad. Sci. USA **90**: 11147–11151.

Kim, Y.-M., Bergonia, H.A., Müller, C., Pitt, B.R., Watkins, W.D., and Lancaster, J.R., Jr. (1995). Loss and degradation of enzyme-bound heme induced by cellular nitric oxide synthesis. *J. Biol. Chem.* **270**: 5710–5713.

Kissner, R., Nauser, T., Bugnon, P., Lye, P.G., and Koppenol, W.H. (1997). Formation and properties of peroxynitrite as studied by laser flash photolysis, high-pressure stopped-flow technique, and pulse radiolysis. *Chem. Res. Toxicol.* **10**: 1285–1292.

Klatt, P., Pfeiffer, S., List, B.M., Lehner, D., Glatter, O., Bächinger, H.P., Werner, E.R., Schmidt, K., and Mayer, B. (1996). Characterization of heme-deficient neuronal nitric-oxide synthase reveals a role for heme in subunit dimerization and binding of the amino acid substrate and tetrahydrobiopterin. *J. Biol. Chem.* **271**: 7336–7342.

Klatt, P., Schmidt, K., Lehner, D., Glatter, O., Bächinger, H.P., and Mayer, B. (1995). Structural analysis of porcine brain nitric oxide synthase reveals a role for tetrahydrobiopterin and L-arginine in the formation of SDS-resistant dimer. *EMBO J.* **14**: 3687–3695.

Klatt, P., Schmidt, K., and Mayer, B. (1992). Brain nitric oxide synthase is a haemoprotein. *Biochem. J.* **288**: 15–17.

Klatt, P., Schmidt, K., Uray, G., and Mayer, B. (1993). Multiple catalytic functions of brain nitric oxide synthase. Biochemical characterization, cofactor-requirement, and the role of N^ω-hydroxy-L-arginine as an intermediate. *J. Biol. Chem.* **268**: 14781–14787.

Kleeman, R., Rothe, H., Kolb-Bachofen, V., Xie, Q.-W., Nathan, C., Martin, S., and Kolb, H. (1993). Transcription and translation of inducible nitric oxide synthase in the pancreas of prediabetic BB rats. *FEBS Lett.* **328**: 9–12.

Knowles, R.G., Palacios, M., Palmer, R.M.J., and Moncada, S. (1989). Formation of nitric oxide from L-arginine in the central nervous system: A transduction mechanism for stimulation of the soluble guanylate cyclase. *Proc. Natl. Acad. Sci. USA* **86**: 5159–5162.

Kobayashi, K., Miki, M., and Tagawa, S. (1995). Pulse-radiolysis study of the reaction of nitric oxide with super-oxide. *J. Chem. Soc., Dalton Trans.* 2885–2889.

Kobzik, L., Bredt, D.S., Lowenstein, C.J., Drazen, J., Gaston, B., Sugarbaker, D., and Stamler, J.S. (1993). Nitric oxide synthase in human and rat lung: Immunocytochemical and histochemical localization. *Am. J. Respir. Cell Mol. Biol.* **9**: 371–377.

Kobzik, L., Reid, M.B., Bredt, D.S., and Stamler, J.S. (1994). Nitric oxide in skeletal muscle. *Nature* **372**: 546–548.

Kollegger, H., McBean, G.J., and Tipton, K.F. (1993). Reduction of striatal N-methyl-D-aspartate toxicity by inhibition of nitric oxide synthase. *Biochem. Pharmacol.* **45**: 260–264.

Koppenol, W.H., Moreno, J.J., Pryor, W.A., Ischiropoulos, H., and Beckman, J.S. (1992). Peroxynitrite, a cloaked oxidant formed by nitric oxide and superoxide. *Chem. Res. Toxicol.* **5**: 834–842.

Korth, H.-G., Sustmann, R., Thater, C., Butler, A.R., and Ingold, K.U. (1994). On the mechanism of the nitric oxide synthase-catalyzed conversion of N^{ω}-hydroxy-L-arginine to citrulline and nitric oxide. *J. Biol. Chem.* **269**: 17776–17779.

Kosower, N.S., and Kosower, E.M. (1978). The glutathione status of cells. *Int. Rev. Cytol.* **54**: 109–160.

Kotake, Y., Tanigawa, T., Tanigawa, M., and Ueno, I. (1995). Spin trapping isotopically-labeled nitric oxide produced from [^{15}N]L-arginine and [^{17}O]dioxygen by activated macrophages using a water soluble Fe^{++}-dithiocarbamate spin trap. *Free Radical Res.* **23**: 287–295.

Kubes, P., Suzuki, M., and Granger, D.N. (1991). Nitric oxide: An endogenous modulator of leukocyte adhesion. *Proc. Natl. Acad. Sci. USA* **88**: 4651–4655.

Kuchan, M.J., and Frangos, J.A. (1994). Role of calcium and calmodulin in flow-induced nitric oxide production in endothelial cells. *Am. J. Physiol.* **266**: C628–C636.

Kurose, I., Wolf, R., Grisham, M.B., Aw, T.Y., Specian, R.D., and Granger, D.N. (1995). Microvascular responses to inhibition of nitric oxide production.

Role of active oxidants. *Circ. Res.* **76**: 30–39.

Kwon, N.S., Nathan, C.F., Gilker, C., Griffeth, O.W., Matthews, D.E., and Steuhr D.J. (1990). L-Citrulline production from L-arginine by macrophage nitric oxide synthase: The ureido oxygen derives from dioxygen. *J. Biol. Chem.* **265**: 13442–13445.

Kwon, N.S., Nathan, C.F., and Stuehr, D.J. (1989). Reduced biopterin as a cofactor in the generation of nitrogen oxides by murine macrophages. *J. Biol. Chem.* **264**: 20496–20501.

Lafon-Cazal, M., Culcasi, M., Gaven, F., Pietri, S., and Bockaert, J. (1993). Nitric oxide, superoxide and peroxynitrite: Putative mediators of NMDA-induced cell death in cerebellar granule cells. *Neuropharmacology* **32**: 1259–1266.

Lamas, S., Marsden, P.A., Li, G.K., Tempst, P., and Michel, T. (1992). Endothelial nitric oxide synthase: Molecular cloning and characterization of a distinct constitutive enzyme isoform. *Proc. Natl. Acad. Sci. USA* **89**: 6348–6352.

Lancaster, J.R., Jr. (1994). Simulation of the diffusion and reaction of endogenously produced nitric oxide. *Proc. Natl. Acad. Sci. USA* **91**: 8137–8141.

Lancaster, J.R., Jr., and Hibbs, J.B., Jr. (1990). EPR demonstration of iron-nitrosyl complex formation by cytotoxic activated macrophages. *Proc. Natl. Acad. Sci. USA* **87**: 1223–1227.

Lancaster, J.R., Jr., and Stuehr, D.J. (1996). The intracellular reactions of nitric oxide in the immune system and its enzymatic synthesis, in: *Nitric Oxide. Principles and Actions* (Lancaster, J., Jr., ed.), pp. 139–175, Academic Press, San Diego, CA.

Lander, H.M. (1997). An essential role for free radicals and derived species in signal transduction. *FASEB J.* **11**: 118–124.

Lander, H.M., Hajjar, D.P., Hempstead, B.L., Mirza, U.A., Chait, B.T., Campbell, S., and Quilliam, L.A. (1997). A molecular redox switch on p21^{ras}. Structural basis for the nitric oxide-p21^{ras} interaction. *J. Biol. Chem.* **272**: 4323–4326.

Lander, H.M., Millbank, A.J., Tauras, J.M., Hajjar, D.P., Hempstead, B.L.,

Schwartz, G.D., Kraemer, R.T., Mirza, U.A., Chalt, B.T., Burk, S.C., and Quilliam, L.A. (1996). Redox regulation of cell signalling. *Nature* **381**: 380–381.

Lander, H.M., Ogiste, J.S., Pearce, S.F.A., Levi, R., and Novogrodsky, A. (1995a). Nitric oxide-stimulated guanine nucleotide exchange on p21^{ras}. *J. Biol. Chem.* **270**: 7017–7020.

Lander, H.M., Ogiste, J.S., Teng, K.K., and Novogrodsky, A. (1995b). p21^{ras} As a common signaling target of reactive free radicals and cellular redox stress. *J. Biol. Chem.* **270**: 21195–21198.

Lander, H.M., Sehajpal, P.K., and Novogrodsky, A. (1993). Nitric oxide signaling: A possible role for G proteins. *J. Immunol.* **151**: 7182–7187.

Laubach, V.E., Garvey, E.P., and Sherman, P.A. (1996). High-level expression of human inducible nitric oxide synthase in Chinese hamster ovary cells and characterization of the purified enzyme. *Biochem. Biophys. Res. Commun.* **218**: 802–807.

Lee, J.S., Adrie, C., Jacob, H.J., Roberts, J.D., Jr., Zapol, W.M., and Bloch, K.D. (1996). Chronic inhalation of nitric oxide inhibits neointimal formation after balloon-induced arterial injury. *Circ. Res.* **78**: 337–342.

Lee, M., Arosio, P., Cozzi, A., and Chasteen, D. (1994). Identification of the EPR-active iron–nitrosyl complexes in mammalian ferritins. *Biochemistry* **33**: 3679–3687.

Lee, S.-J., and Stull, J.T. (1998). Calmodulin-dependent regulation of inducible and neuronal nitric-oxide synthase. *J. Biol. Chem.* **273**: 27430–27437.

Lefer, D.J., Nakanishi, K., Johnston, W.E., and Vinten-Johansen, J. (1993). Antineutrophil and myocardial protecting actions of a novel nitric oxide donor after acute myocardial ischemia and reperfusion in dogs. *Circulation* **88**: 2337–2350.

Lefer, D.J., Scalia, R., Campbell, B., Nossuli, T., Hayward, R., Salamon, S., Grayson, J., and Lefer, A.M. (1997). Peroxynitrite inhibits leukocyte–endothelial cell interactions and protects against ischemia-reperfusion injury in rats. *J. Clin. Invest.* **99**: 684–691.

Lei, S.Z., Pan, Z.-H., Aggarwal, S.K., Chen, H.-S.V., Hartman, J., Sucher, N.J., and Lipton S.A. (1992). Effect of nitric-oxide production on the redox modulatory site of the NMDA receptor–channel complex. *Neuron* **8**: 1087–1099.

Lemercier, J.-N., Squadrito, G.L., and Pryor, W.A. (1995). Spin trap studies on the decomposition of peroxynitrite. *Arch. Biochem. Biophys.* **321**: 31–39.

Leone, A.M., Palmer, R.M.J., Knowles, R.G., Francis, P.L., Ashton, D.S., and Moncada, S. (1991). Constitutive and inducible nitric oxide synthases incorporate molecular oxygen into both nitric oxide and citrulline. *J. Biol. Chem.* **266**: 23790–23795.

Lepoivre, M., Flaman, J.-M., Bobé, P., Lemaire, G., and Henry, Y. (1994). Quenching of the tyrosyl free radical of ribonucleotide reductase by nitric oxide. Relationship to cytostasis induced in tumor cells by cytotoxic macrophages. *J. Biol. Chem.* **269**: 21891–21897.

Lewis, R.S., and Deen, W.M. (1994). Kinetics of the reaction of nitric oxide with oxygen in aqueous solutions. *Chem. Res. Toxicol.* **7**: 568–574.

Lewis, R.S., Tannenbaum, S.R., and Deen, W.M. (1995). Kinetics of N-nitrosation in oxygenated nitric oxide solutions at physiological pH: Role of nitrous anhydride and effects of phosphate and chloride. *J. Am. Chem. Soc.* **117**: 3933–3939.

Li, C.G., and Rand, M.J. (1990). Nitric oxide and vasoactive intestinal polypeptide mediate non-adrenergic noncholinergic inhibitory transmission to smooth muscle of the rat gastric fundus. *Eur. J. Pharmacol.* **191**: 303–309.

Li, Y., Severn, A., Rogers, M.V., Palmer, R.M.J., Mondada, S., and Liew, F.Y. (1992). Catalase inhibits nitric oxide synthesis and the killing of intracellular *Leishmania major* in murine macrophages. *Eur. J. Immunol.* **22**: 441–446.

Liew, F.Y., Millott, S., Parkinson, C., Palmer, R.M.J., and Moncada, S. (1990). Macrophage killing of *Leishmania* parasite *in vivo* is mediated by nitric oxide from L-arginine. *J. Immunol.* **144**: 4794–4797.

Lin, J.Y., and Chadee, K. (1992). Macrophage cytotoxicity against *Entamoeba histolytica* trophozoites is mediated by nitric oxide from L-arginine. *J. Immunol.* **148**: 3999–4005.

Lincoln, T.M., and Cornwell, T.L. (1993). Intracellular cyclic GMP receptor proteins. *FASEB J.* **7**: 328–338.

Lipton, S.A., and Stamler, J.S. (1994). Actions of redox-related congers of nitric oxide at the NMDA receptor. *Neuropharmacology* **33**: 1229–1233.

Lipton, S.A., Choi, Y.-B., Pan, Z.-H., Lei, S.Z., Chen, H.-S. V., Sucher, N.J., Loscalzo, J., Singel, D.J., and Stamler, J.S. (1993). A redox-based mechanism for the neuroprotective and neurodestructive effects of nitric oxide and related nitroso-compounds. *Nature* **369**: 626–632.

Liu, A., Beckman, J.S., and Ku, D.D. (1994). Peroxynitrite, a product of superoxide and nitric oxide, produces coronary vasorelaxation in dogs. *J. Pharmacol. Exp. Ther.* **268**: 1114–1121.

Liu, H.I., Sono, M., Kadkhodayan, S., Hager, L.P., Hedman B., Hodgson, K.O., and Dawson, J.H. (1995). X-ray absorption near edge studies of cytochrome P-450-CAM, chloroperoxidase, and myoglobin. Direct evidence for the electron releasing character of a cysteine thiolate proximal ligand. *J. Biol. Chem.* **270**: 10544–10550.

Liu, J., and Sessa, W.C. (1994). Identification of covalently bound amino-terminal myristic acid in endothelial nitric oxide synthase. *J. Biol. Chem.* **269**: 11691–11694.

Liu, X., Gillespie, J.S., Gibson, I.F., and Martin, W. (1991). Effects of N^G-substituted analogues of L-arginine on NANC relaxation of the anococcygeus and retractor penis muscles and the bovine penile artery. *Brit. J. Pharmacol.* **104**: 53–58.

Liu, X., Muller, M.J.S., Joshi, M.S., Thomas, D.D., and Lancaster, J.L., Jr. (1998). Accelerated reaction of nitric oxide with O_2 within the hydrophobic interior of biological membranes. *Proc. Natl. Acad. Sci. USA* **95**: 2175–2179.

Lizasoain, I., Moro, M.A., Knowles, R.G., Darley-Usmar, V., and Moncada, S. (1996). Nitric oxide and peroxynitrite exert distinct effects on mitochondrial respiration which are differently blocked by glutathione or glucose. *Biochem. J.* **314**: 877–880.

Loscalzo, J., and Welch, G. (1995). Nitric oxide and its role in the cardiovascular system. *Prog. Cardiovasc. Dis.* **38**: 87–104.

Lowe, P.N., Smith, D., Stammers, D.K., Riveros-Moreno, V., Moncada, S., Charles, I., and Boyhan, A. (1996). Identification of the domains of neuronal nitric oxide synthase by limited proteolysis. *Biochem. J.* **314**: 55–62.

Lowenstein, C.J., and Snyder, S.H. (1992). Nitric oxide, a novel biologic messenger. *Cell* **70**: 705–707.

Luperchio, S., Tamir, S., and Tannenbaum, S.R. (1996). NO-induced oxidative stress and glutathione metabolism in rodent and human cells. *Free Radical Biol. Med.* **21**: 513–519.

Lymar, S.V., and Hurst, J.K. (1995). Rapid reaction between peroxynitrite ion and carbon dioxide: Implications for biological activity. *J. Am. Chem. Soc.* **117**: 8867–8868.

Lyons, C.R., Orloff, G.J., and Cunningham, J.M. (1992). Molecular cloning and functional expression of an inducible nitric oxide synthase from a murine macrophage cell line. *J. Biol. Chem.* **267**: 6370–6374.

Ma, X.L., Weyrich, A.S., Lefer, D.J., and Lefer, A.M. (1993). Diminished basal nitric oxide release after myocardial ischemia and reperfusion promotes neutrophil adherence to coronary endothelium. *Circ. Res.* **72**: 403–412.

Mabbott, N., and Sternberg, J. (1995). Bone marrow nitric oxide production and development of anemia in *Trypanosoma brucei*-infected mice. *Infect. Immun.* **63**: 1563–1566.

MacLean, A, Wei, X.Q., Huang, F.P., Al-Alem, U.A., Chan, W.L., and Liew, F.Y. (1998). Mice lacking inducible nitric-oxide synthase are more susceptible to herpes simplex virus infection despite enhanced Th1 cell responses. *J. Gen. Virology* **79**: 825–830.

MacMicking, J.D., Nathan, C., Hom, G., Chartrain, N., Fletcher, D.S., Trumbauerr, M., Stevens, K., Xie, Q.-W., Sokol, K., Hutchinson, N., Chen, H., and Mudgett, J.S. (1995). Altered responses to bacterial infection and endotoxic shock in mice lacking inducible nitric oxide synthase. *Cell* **81**: 641–650.

Malinski, T., and Taha, Z. (1992). Nitric oxide release from a single cell measured *in situ* by a porphyrinic-based microsensor. *Nature* **358**: 676–678.

Malinski, T., Bailey, F., Zhang, Z.G., and Chopp, M. (1993a). Nitric oxide measured by a porphyrinic microsensor in rat brain after transient middle cerebral artery occlusion. *J. Cereb. Blood Flow Metab.* **13**: 355–358.

Malinski, T., Kapturczak, M., Dayharsh, J., and Bohr, D. (1993b). Nitric oxide synthase activity in genetic hypertension. *Biochem. Biophys. Res. Commun.* **194**: 654–658.

Malinski, T., Radomski, M.W., Taha, Z., and Moncada, S. (1993c). Direct electrochemical measurement of nitric oxide released from human platelets. *Biochem. Biophys. Res. Commun.* **194**: 960–965.

Malinski, T., Taha, Z., Grunfeld, S., Patton, S., Kapturczak, M., and Tomboulian, P. (1993d). Diffusion of nitric oxide in the aorta wall monitored *in situ* by prophyrinic microsensors. *Biochem. Biophys. Res. commun.* **193**: 1076–1082.

Mannick, J.B., Asano, K., Izumi, K., Kieff, E., and Stamler, J.S. (1994). Nitric oxide produced by human B lymphocytes inhibits apoptosis and Epstein-Barr virus reactivation. *Cell* **79**: 1137–1146.

Mansuy, D., Boucher, J.L., and Clement, B. (1995). On the mechanism of nitric oxide formation upon oxidative cleavage of C=N(OH) bonds by NO-synthases and cytochromes P450. *Biochimie* **77**: 661–667.

Marletta, M.A. (1993). Nitric oxide synthase structure and mechanism. *J. Biol. Chem.* **268**: 12231–12234.

Marletta, M.A. (1994a). Nitric oxide synthase: Aspects concerning structure and catalysis. *Cell* **78**: 927–930.

Marletta, M.A. (1994b). Approaches towards selective inhibition of nitric oxide synthase. *J. Med. Chem.* **37**: 1899–1907.

Marletta, M.A., Yoon, P.S., Iyengar, R., Leaf, C.D., and Wishnok, J.S. (1988). Macrophage oxidation of L-arginine to nitrite and nitrate: Nitric oxide is an intermediate. *Biochemistry* **27**: 8706–8711.

Marsden, P.A., Schappert, K.T., Chen, H.S., Flowers, M., Sundell, C.L., Wilson, J.N, Lamas, S., and Michel, T. (1992). Molecular cloning and characterization of human endothelial nitric oxide synthase. *FEBS Lett.* **307**: 287–293.

Martin, E., Nathan, C., and Xie, Q.-W. (1994). Role of interferon regulatory factor 1 in induction of nitric oxide synthase. *J. Exp. Med.* **180**: 977–984.

Masters, B.S.S., McMillan, K., Sheta, E.A., Nishimura, J.S., Roman, L.J., and Martasek, P. (1996). Neuronal nitric oxide synthase, a modular enzyme formed by convergent evolution: Structure studies of a cysteine thiolate-ligand heme protein that hydroxylates L-arginine to produce NO· as a cellular signal. *FASEB J.* **10**: 552–558.

Matsumoto, T., Pollock, J.S., Nakane, M., and Förstermann, U. (1993). Developmental changes of cytosolic and particulate nitric oxide synthase in rat brain. *Dev. Brain Res.* **73**: 199–203.

Maulik, N., Engelman, D.T., Watanabe, M., Engelman, R.M., Maulik, G., Cordis, G.A., and Das, D.K. (1995). Nitric oxide signaling in ischemic heart. *Cardiovasc. Res.* **30**: 593–601.

Mayer, B., John, M., and Böhme, E. (1990). Purification of a Ca^{2+}/calmodulin-dependent nitric oxide synthase from porcine cerebellum. Cofactor-role of tetrahydrobiopterin. *FEBS Lett.* **277**: 215–219.

Mayer, B., John, M., Heinzel, B., Werner, E., Wachter, H., Schultz, G., and Böhme, E. (1991). Brain nitric oxide synthase is a biopterin- and flavin-containing multi-functional oxido-reductase. *FEBS Lett.* **288**: 187–191.

Mayer, B., Klatt, P., Werner, E.R., and Schmidt, K. (1995). Kinetics and mechanism of tetrahydrobiopterin-induced oxidation of nitric oxide. *J. Biol. Chem.* **270**: 655–659.

Mayer, B., Schmidt, K., Humbert, P., and Böhme, E. (1989). Biosynthesis of endothelium-derived relaxing factor: A cytosolic enzyme in porcine aortic endothelial cells Ca^{2+}-dependently converts L-arginine into an activator of soluble guanylyl cyclase. *Biochem. Biophys. Res. Commun.* **164**: 678–685.

McCall, T.B., Boughton-Smith, N.K., Palmer, R.M.J., Whittle, B.J.R., and Moncada, S. (1989). Synthesis of nitric oxide from L-arginine by neutrophils. Release and interaction with superoxide anion. *Biochem. J.* **261**: 293–296.

McMillian, K. and Masters, B.S.S. (1993). Optical difference spectrophotometry as a probe of rat brain nitric oxide synthase heme-substrate interaction. *Biochemistry* **32**: 9875–9880.

McMillian, K., and Masters, B.S.S. (1995). Prokaryotic expression of the heme- and flavin-binding domains of rat neuronal nitric oxide synthase as distinct polypeptides: Identification of the heme-binding proximal thiolate ligand as cysteine-415. *Biochemistry* **34**: 3686–3693.

McMillian, K., Bredt, D.S., Hirsch, D.J., Snyder, S.H., Clark, J.E., and Masters, B.S.S. (1992). Cloned, expressed rat cerebellar nitric oxide synthase contains stoichiometric amounts of heme, which binds carbon monoxide. *Proc. Natl. Acad. Sci. USA* **89**: 11141–11145.

McNamara, D.B., Bedi, B., Aurora, H., Tena, L., Ignarro, L.J., Kadowitz, P.J., and Akers, D.L. (1993). L-Arginine inhibits balloon catheter-induced intimal hyperplasia. *Biochem. Biophys. Res. Commun.* **193**: 291–296.

Melillo, G., Cox, G.W., Biragyn, A., Sheffler, L.A., and Varesio, L. (1994). Regulation of nitric oxide synthase mRNA expression by interferon-γ and picolinic acid. *J. Biol. Chem.* **269**: 8128–8133.

Membrillo-Hernández, J., Coopamah, M.D., Anjum, M.F., Stevanin, T.M., Kelly, A., Hughes, M.N., and Poole, R.K. (1999). The flavohemoglobin of *Escherichia coli* confers resistance to a nitrosating agent, a "nitric oxide releaser," and paraquat and is essential for transcriptional responses to oxidative stress. *J. Biol. Chem.* **274**: 748–754.

Membrillo-Hernández, J., Kim, S.O., Cook, G.M., and Poole, R.K. (1997). Paraquat regulation of *hmp* (flavohemoglobin) gene expression in *Escherichia coli* K-12 is SoxRS independent but

modulated by σ^s. *J. Bacteriol.* **179**: 3164–3170.

Metz, G., Carlier, Y., and Vray, B. (1993). *Trypanosoma cruzi* upregulates nitric oxide release by IFN-γ-preactivated macrophages, limiting cell infection independently of the respiratory burst. *Parasite Immunol.* **15**: 693–699.

Michel, J.B., Feron, O., Sacks, D., and Michel, T. (1997). Reciprocal regulation of endothelial nitric-oxide synthase by Ca^{2+}-calmodulin and caveolin. *J. Biol. Chem.* **272**: 15583–15586.

Michel, T., Ki, G.K., and Busconi, L. (1993). Phosphorylation and subcellular translocation of endothelial nitric oxide synthase. *Proc. Natl. Acad. Sci. USA* **90**: 6252–6256.

Miki, N., Kawabe, Y., and Kuriyama, K. (1977) Activation of cerebral guanylate cyclase by nitric oxide. *Biochem. Biophys. Res. Commun.* **75**: 851–856.

Miwa, M., Stuehr, D.J., Marletta, M.A., Wishnok, J.S., and Tannenbaum, S.R. (1987) Nitrosation of amines by stimulated macrophages. *Carcinogenesis* **8**: 955–958.

Modolell, M., Eichmann, K., and Soler, G. (1997). Oxidation of N^G-hydroxyl-L-arginine to nitric oxide mediated by respiratory burst: An alternative pathway to NO synthesis. *FEBS Lett.* **401**: 123–126.

Moilanen, E., Vuorinen, P., Kankaanranta, H., Metsä-Ketelä, T., and Vapaatalo, H. (1993). Inhibition by nitric oxide-donors of human polymorphonuclear leucocyte function. *Br. J. Pharmacol.* **109**: 852–858.

Moncada, S., and Higgs, E.A. (1991). Endogenous nitric oxide: Physiology, pathology and clinical relevance. *Eur. J. Clin. Invest.* **21**: 361–374.

Moncada, S., and Higgs, A. (1993). The L-arginine–nitric oxide pathway. *N. Engl. J. Med.* **329**: 2002–2012.

Moncada, S., Higgs, E.A., Hodson, H.F., Knowles, R.G., Lopez-Jaramillo, P., McCall, T., Palmer, R.M.J., Radomski, M.W., Rees, D.D., and Schulz, R. (1991). The L-arginine:Nitric oxide pathway. *J. Cardiovasc. Pharmacol.* **17**(suppl 3): S1–S9.

Moncada, S., Palmer, R.M.J., and Gryglewski, R.J. (1986). Mechanism of action of some inhibitors of endothelium-derived relaxing factor. *Proc. Natl. Acad. Sci. USA* **83**: 9164–9168.

Morlino, E.A., and Rodgers, M.A.J. (1996). Nitric oxide deligation from nitrosyl complexes of two transition metal prophyrins: A photokinetic investigation. *J. Am. Chem. Soc.* **118**: 11798–11804.

Moro, M.A., Darley-Usmar, V.M., Goodwin, D.A., Read, N.G., Zamora-Pino, R., Feelisch, M., Radomski, M.W., and Moncada, S. (1994). Paradoxical fate and biological action of peroxynitrite on human platelets. *Proc. Natl. Acad. Sci. USA* **91**: 6702–6706.

Motterlini, R., Foresti, R., Intaglietta, M., and Winslow, R.M. (1996). NO-mediated activation of heme oxygenase: Endogenous cytoprotection against oxidative stress to endothelium. *Am. J. Physiol.* **270**: H107-H114.

Muñoz-Fernández, M.A., Fernández, M.A., and Fresno, M. (1992). Activation of human macrophages for the killing of intracellular *Trypanosoma cruzi* by TNF-α and IFN-γ through a nitric oxide-dependent mechanism. *Immunol. Lett.* **33**: 35–40.

Nakane, M., Mitchell, J., Förstermann, U., and Murad, F. (1991). Phosphorylation by calcium calmodulin-dependent protein kinase II and protein kinase C modulates the activity of nitric oxide synthase. *Biochem. Biophys. Res. Commun.* **180**: 1396–1402.

Nakane, M., Schmidt, H.H.H.W., Pollock, J.S., Förstermann, U., and Murad, F. (1993). Cloned human brain nitric oxide synthase is highly expressed in skeletal muscle. *FEBS Lett.* **316**: 175–180.

Nakano, R., Sato, H., Watanabe, A., Ito, O., and Shimizu, T. (1996). Conserved Glu[318] at the cytochrome P450 1A2 distal site is crucial in the nitric oxide complex stability. *J. Biol. Chem.* **271**: 8570–8574.

Nathan, C. (1992). Nitric oxide as a secretory product of mammalian cells. *FASEB J.* **6**: 3051–3064.

Nathan, C., and Hibbs, J.B., Jr. (1991) Role of nitric oxide synthesis in macrophage antimicrobial activity. *Curr. Opin. Immunol.* **3**: 65–70.

Nathan, C., and Xie, Q.-W. (1994a). Regulation of biosynthesis of nitric oxide. *J. Biol. Chem.* **269**: 13725–13728.

Nathan, C., and Xie, Q.-W. (1994b). Nitric oxide synthases: Roles, tolls and controls. *Cell* **78**: 915–918.

Newcomb, M., and Chestney, D.L. (1994). A hypersensitive mechanistic probe for distinguishing between radical and carbocation intermediates. *J. Am. Chem. Soc.* **116**: 9753–9754.

Newcomb, M., Le Tadic, M.-H., Putt, D.A., and Hollenberg, P.F. (1995). An incredibly fast apparent oxygen rebound rate constant for hydrocarbon hydroxylation by cytochrome P-450 enzymes. *J. Am. Chem. Soc.* **117**: 3312–3313.

Nguyen, T., Brunson, D., Crespi, C.L., Penman, B.W., Wishnok, J.S., and Tannenbaum, S.R. (1992). DNA damage and mutation in human cells exposed to nitric oxide *in vitro*. *Proc. Natl. Acad. Sci. USA* **89**: 3030–3034.

Nicholson, S., Bonecini-Almeida, M.D.G., Silva, J.R.L.E., Nathan, C., Xie, Q.W., Mumford, R., Weidner, J.R., Calaycay, J., Geng, J., Boechat, N., Linhares, C., Rom, W., and Ho, J.L. (1996). Inducible nitric oxide synthase in pulmonary alveolar macrophages from patients with tuberculosis. *J. Exp. Med.* **183**: 2293–2302.

Nikitovic, D., and Holmgren, A. (1996). S-Nitrosoglutathione is cleaved by the thioredoxin system with liberation of glutathione and redox regulating nitric oxide. *J. Biol. Chem.* **271**: 19180–19185.

Nishida, K., Harrison, D.G., Navas, J.P., Fisher, A.A., Dockery, S.P., Uematsu, M., Nerem, R.M., Alexander, R.W., and Murphy, T.J. (1992). Molecular cloning and characterization of the constitutive bovine aortic endothelial cell nitric oxide synthase. *J. Clin. Invest.* **90**: 2092–2096.

Nishimura, J., Martasek, P., McMillan, K., Salerno, J.C., Liu, Q., Gross, S.S., and Masters, B.S.S. (1995). Modular structure of neuronal nitric oxide synthase: Localization of the arginine binding site and modulation by pterin. *Biochem. Biophys. Res. Commun.* **210**: 288–294.

Nozaki, Y., Hasegawa, Y., Ichiyama, S., Nakashima, I., and Shimokata, K. (1997). Mechanism of nitric oxide-dependent killing of *Mycobacterium bovis* BCG in human alveolar macrophages. *Infect. Immun.* **65**: 3644–3647.

Nunoshiba, T., DeRojas-Walker, T., Tannenbaum, S.R., and Demple, B. (1995). Roles of nitric oxide in inducible resistance of *Escherichia coli* to activated murine macrophages. *Infect. Immun.* **63**: 794–798.

Nunoshiba, T., DeRojas-Walker, T., Wishnok, J.S., Tannenbaum, S.R., and Demple, B. (1993). Activation by nitric oxide of an oxidative-stress response that defends *Escherichia coli* against activated macrophages. *Proc. Natl. Acad. Sci. USA* **90**: 9993–9997.

O'Neill, W.C. (1995). Flow-mediated NO release from endothelial cells is independent of K^+ channel activation or intracellular Ca^{2+}. *Am. J. Physiol.* **269**: C863–C869.

Oria, R., Sánchez, L., Houston, T., Hentze, M.W., Liew, F.Y., and Brock, J.H. (1995). Effect of nitric oxide on expression of transferrin receptor and ferritin and on cellular iron metabolism in K562 human erythroleukemia cells. *Blood* **85**: 2962–2966.

Oswald, I.P., Eltoum, I., Wynn, T.A., Schwartz, B., Caspar, P., Paulin, D., Sher, A., and James, S.L. (1994). Endothelial cells are activated by cytokine treatment to kill an intravascular parasite, *Schistosoma mansoni*, through the production of nitric oxide. *Proc. Natl. Acad. Sci. USA* **91**: 999–1003.

Oury, T.D., Ho, Y.-E., Piantadosi, C.A., and Crapo, J.D. (1992). Extracellular superoxide dismutase, nitric oxide and central nervous system O_2 toxicity. *Proc. Natl. Acad. Sci. USA* **89**: 9715–9719.

Owens, M.W., Milligan, S.A., and Grishman, M.B. (1995). Nitric oxide-dependent N-nitrosating activity of rat pleural mesothelial cells. *Free Radical Res.* **23**: 371–378.

Padmaja, S., and Huie, R.E. (1993). The reaction of nitric oxide with organic peroxyl radicals. *Biochem. Biophys. Res. Commun.* **195**: 539–544.

Padmaja, S., Squadrito, G.L., and Pryor, W.A. (1998). Inactivation of glutathione peroxidase by peroxynitrite. *Arch. Biochem. Biophys.* **349**: 1–6.

Palmer, R.J.M., Ashton, D.S., and Moncada, A. (1988a) Vascular endothelial cells synthesize nitric oxide from L-arginine. *Nature* **333**: 664–666.

Palmer, R.M.J., Ferrige, A.G., and Moncada, S. (1987). Nitric oxide release accounts for the biological activity of endothelium-derived relaxing factor. *Nature* **327**: 524–526.

Palmer, R.M.J., Rees, D.D., Ashton, D.S., and Moncada, S. (1988b). L-Arginine is the physiological precursor for the formation of nitric oxide in endothelium-derived relaxation. *Biochem. Biophys. Res. Commun.* **153**: 1251–1256.

Pantopoulos, K., and Hentze, M.W. (1995). Nitric oxide signaling to iron-regulatory protein: Direct control of ferritin mRNA translation and transferrin receptor mRNA stability in transfected fibroblasts. *Proc. Natl. Acad. Sci. USA* **92**: 1267–1271.

Pantopoulos, K., Weiss, G., and Hentze, M.W. (1994). Nitric oxide and the post-transcriptional control of cellular iron traffic. *Trends Cell Biol.* **4**: 82–86.

Pantopoulos, K., Weiss, G., and Hentze, M.W. (1996). Nitric oxide and oxidative stress (H_2O_2) control mammalian iron metabolism by different pathways. *Mol. Cell. Biol.* **16**: 3781–3788.

Pauling, L. (1960). *The Nature of the Chemical Bond*, pp. 343–351, Cornell University Press, Ithaca, NY.

Peng, H.-B., Libby, P., and Liao, J.K. (1995). Induction and stabilization of IκBα by nitric oxide mediates inhibition of NF-κB. *J. Biol. Chem.* **270**: 14214–14219.

Persechini, A., White, H.D., and Gansz, K.J. (1996). Different mechanisms for Ca^{2+} dissociation from complexes of calmodulin with nitric oxide synthase or myosin light chain kinase. *J. Biol. Chem.* **271**: 62–67.

Petit, J.-L., Nicaise, M., Lepoivre, M., Guissani, A., and Lemaire, G. (1996). Protection by glutathione against the antiproliferative effects of nitric oxide. Dependence on kinetics of NO release. *Biochem. Pharmacol.* **52**: 205–212.

Pfeiffer, S., Gorren, A.C.F., Schmidt, K., Werner, E.R., Hansert, B., Bohle, D.S. and Mayer, B. (1997). Metabolic fate of peroxynitrite in aqueous solution. Reaction with nitric oxide and pH-dependent decomposition to nitrite and oxygen in a 2:1 stoichiometry. *J. Biol. Chem.* **272**: 3465–3470.

Pietraforte, D., Mallozzi, C., Di Stasi, A.M.M., Scorza, G., and Minetti, M. (1996). Protein thiyl free radicals produced by nitric oxide and nitric oxide donors. *Res. Chem. Intermed.* **22**: 481–489.

Pires, M., Rossi, M.J., and Ross, D.S. (1994). Kinetic and mechanistic aspects of the NO oxidation by O_2 in aqueous phase. *Int. J. Chem. Kinet.* **26**: 1207–1227.

Pogrebnaya, V.L., Usov, A.P., Baranov, A.V., Nesterenko, A.I., and Bez'yazy-chyi, P.I. (1975). Oxidation of nitric oxide by oxygen in the liquid phase. *J. Appl. Chem. USSR (Engl. Transl.)* **48**: 1004–1007.

Pollock, J.S., Förstermann, U., Mitchell, J.A., Warner, T.D., Schmidt, H.H.H.W., Nakane, M., and Murad, F. (1991). Purification and characterization of particulate endothelium-derived relaxing factor synthase from cultured and native bovine aortic endothelial cells. *Proc. Natl. Acad. Sci. USA* **88**: 10480–10484.

Pollock, J.S., Klinghofer, V., Förstermann, U., and Murad, F. (1992). Endothelial nitric oxide synthase is myristylated. *FEBS Lett.* **309**: 402–404.

Poole, R.K., Anjum, M.F., Membrillo-Her-nández, J., Kim, S.O., Hughes, M.N., and Stewart, V. (1996). Nitric oxide, nitrite, and Fnr regulation of *hmp* (fla-vohemoglobin) gene expression in *Escherichia coli* K-12. *J. Bacteriol.* **178**: 5487–5492.

Pope, M., Marsden, P.A. Cole, E., Sloan, S., Fung, L.S., Ning, Q., Ding, J.W., Lei-bowitz, J.L., Phillips, M.J., and Levy, G.A. (1998). Resistance to murine hepatitis virus strain 3 is dependent on production of nitric oxide. *J. Virology* **72**: 7084–7090.

Porter, T.D., and Coon, M.J. (1991). Cyto-chrome P-450. Multiplicity of isoforms, substrates, and catalytic and regulatory mechanisms. *J. Biol. Chem.* **266**: 13469–13472.

Pou, S., and Rosen, G.M. (1998). Genera-tion of thiyl radical by nitric oxide: A spin trapping study. *J. Chem. Soc., Per-kin Trans.* 2, 1507–1512.

Pou, S., Keaton, L., Suricharmorn, W., and Rosen, G.M. (1999). Mechanism of superoxide generation by neuronal nitric oxide synthase. *J. Biol. Chem.* **274**: 9573–9580

Pou, S., Nguyen, S.Y., Gladwell, T., and Rosen, G.M. (1995). Does peroxyni-trite generate hydroxyl radical? *Bio-chim. Biophys. Acta* **1244**: 62–68.

Pou, S., Pou, W.S., Bredt, D.S., Snyder, S.H., and Rosen, G.M. (1992). Gener-ation of superoxide by purified brain nitric oxide synthase. *J. Biol. Chem.* **267**: 24173–24176.

Pou, S., Pou, W.S., Rosen, G.M., and El-Fakahany, E.E. (1991). N-Hydroxyl-amine is not an intermediate in the con-version of L-arginine to an activator of soluble guanylate cyclase in neuroblas-toma N1E-115 cells. *Biochem. J.* **273**: 547–552.

Prabhakar, P., Thatte, H.S., Goetz, R.M., Cho, M.R., Golan, D.E., and Michel, T. (1998). Receptor-regulated translo-cation of endothelial nitric-oxide synthase. *J Biol. Chem.* **273**: 27383–27388.

Pryor, W.A., and Squadrito, G.L. (1995). The chemistry of peroxynitrite: A pro-duct from the reaction of nitric oxide with superoxide. *Am. J. Physiol.* **268**: L699-L722.

Pryor, W.A., Jin, X., and Squadrito, G.L. (1994). One and two electron oxida-tions of methionine by peroxynitrite. *Proc. Natl. Acad. Sci. USA* **91**: 11173–11177.

Pryor, W.A., Jin, X., and Squadrito, G.L. (1996). Insensitivity of the rate of decomposition of peroxynitrite to changes in viscosity: Evidence against free radical formation. *J. Am. Chem. Soc.* **118**: 3125–3128.

Pufahl, R.A., and Marletta, M.A. (1993). Oxidation of N^G-hydroxyl-L-arginine by nitric oxide synthase: Evidence for the involvement of the heme in cataly-sis. *Biochem. Biophys. Res. Commun.* **193**: 963–970.

Pufahl, R.A., Nanjappan, P.G., Woodard, R.W., and Marletta, M.A. (1992). Mechanistic probes of N-hydroxyl-ation of L-arginine by the inducible ni-tric oxide synthase from murine macro-phages. *Biochemistry* **31**: 6822–6828.

Pufahl, R.A, Wishnok, J.S., and Marletta, M.A. (1995). Hydrogen peroxide-supported oxidation of N^G-hydroxy-L-arginine by nitric oxide synthase. *Biochemistry* **34**: 1930–1941.

Radi, R., Beckman, J.S., Bush, K.M., and Freeman, B.A. (1991a). Peroxynitrite oxidation of sulfhydryls. The cytotoxic potential of superoxide and nitric oxide. *J. Biol. Chem.* **266**: 4244–4250.

Radi, R., Beckman, J.S., Bush, K.M., and Freeman, B.A. (1991b). Peroxidation-induced membrane lipid peroxidation: The cytotoxic potential of superoxide and nitric oxide. *Arch. Biochem. Biophys.* **288**: 481–487.

Radomski, M.W., Palmer, R.M.J., and Moncada, S. (1987a). Endogenous nitric oxide inhibits human platelet adhesion to vascular endothelium. *Lancet* **ii**, 1057–1058.

Radomski, M.W., Palmer, R.M.J., and Moncada, S. (1987b). Comparative pharmacology of endothelium-derived relaxing factor, nitric oxide and prostacyclin in platelets. *Br. J. Pharmacol.* **92**: 181–187.

Rauckman, E.J., Rosen, G.M., and Cavagnaro, J. (1982). Norcocaine nitroxide. A potential hepatotoxic metabolite of cocaine. *Mol. Pharmacol.* **21**: 458–463.

Reddy, D., Lancaster, J.R., Jr., and Cornforth, D.P. (1983). Nitrite inhibition of *Clostridium botulinum*: Electron spin resonance detection of iron–nitric oxide complexes. *Science* **221**: 769–770.

Reif, D.W., and Simmons, R.D. (1990). Nitric oxide mediates iron release from ferritin. *Arch. Biochem. Biophys.* **283**: 537–541.

Rengasamy, A., and Johns, R.A. (1993). Regulation of nitric oxide synthase by nitric oxide. *Mol. Pharmacol.* **44**: 124–128.

Richardson, D.R., Neumannova, V., and Ponka, P. (1995). Nitrogen monoxide decreases iron uptake from transferrin but does not mobilise iron from pre-labelled neoplastic cells. *Biochim. Biophys. Acta* **1266**: 250–260.

Richeson, C.E., Mulder, P., Bowry, V.W., and Ingold, K.U. (1998). The complex chemistry of peroxynitrite: New insights. *J. Am. Chem. Soc.* **120**: 7211–7219.

Riveros-Moreno, V., Heffernan, B., Torres, B., Chubb, A., Charles, I., and Moncada, S. (1995). Purification to homogeneity and characterisation of rat brain recombinant nitric oxide synthase. *Eur. J. Biochem.* **230**: 52–57.

Robbins, R.A., Springall, D.R., Warren, J.B., Kwon, O.J., Buttery, L., Wilson, A.J., Adcock, I.M., Riveros-Moreno, V., Moncada, S., Polak, J., and Barnes, P.I. (1994). Inducible nitric oxide synthase is increased in murine lung epithelial cells by cytokine stimulation. *Biochem. Biophys. Res. Commun.* **198**: 835–843.

Robinson, L.J., Busconi, L., and Michel, T. (1995). Agonist-modulated palmitoylation of endothelial nitric oxide synthase. *J. Biol. Chem.* **270**: 995–998.

Rockett, K.A., Awburn, M.M., Cowden, W.B., and Clark, I.A. (1991). Killing of *Plasmodium falciparum in vitro* by nitric oxide derivatives. *Infect. Immun.* **59**: 3280–3283.

Rodríguez-Crespo, I., Gerber, N.C., and Ortiz de Montellano, P.R. (1996). Endothelial nitric-oxide synthase. Expression in *Escherichia coli*, spectroscopic characterization, and role of tetrahydrobiopterin in dimer formation. *J. Biol. Chem.* **271**: 11462–11467.

Rosen, G.M., Eccles, C.U., and Pou, S. (1995). Nitric oxide in the brain: Looking at two sides of the same coin. *Neurologist* **1**: 311–325.

Rosenkranz-Weiss, P., Sessa, W.C., Milstein, S., Kaufman, S., Watson, C.A., and Pober, J.S. (1994). Regulation of nitric oxide synthesis by proinflammatory cytokines in human umbilical vein endothelial cells. *J. Clin. Invest.* **93**: 2236–2243.

Ruan, J., Xie, Q.-W., Hutchinson, N., Cho, H., Wolfe, G.C., and Nathan, C. (1996). Inducible nitric oxide synthase requires both the canonical calmodulin-binding domain and additional sequences in order to bind calmodulin and produce nitric oxide in the absence of free Ca^{2+}. *J. Biol. Chem.* **271**: 22679–22686.

Rubanyi, G.M., and Vanhoutte, P.M. (1986a). Oxygen-derived free radicals, endothelium, and responsiveness of vascular smooth muscle. *Am. J. Physiol.* **250**: H815-H822.

Rubanyi, G.M., and Vanhoutte, P.M. (1986b). Superoxide anions and hyperoxia inactivate endothelial derived relaxing factor. *Am. J. Physiol.* **250**: H822-H827.

Rubanyi, G.M., Ho, E.H., Cantor, E.H., Lumma, W.C., and Parker-Botelho, L.H. (1991). Cytoprotective function of nitric oxide: Inactivation of superoxide radicals produced by human leukocytes. *Biochem. Biophys. Res. Commun.* **181**: 1392-1397.

Rubbo, H., Darley-Usmar, V., and Freeman, B.A. (1996). Nitric oxide regulation of tissue free radical injury. *Chem. Res. Toxicol.* **9**: 809-820.

Rubbo, H., Radi, R., Trujilo, M., Telleri, R., Kalyanaraman, B., Barnes, S., Kirk, M., and Freeman, B.A. (1994). Nitric oxide regulation of superoxide and peroxynitrite-dependent lipid peroxidation. Formation of novel nitrogen-containing oxidized lipid derivatives. *J. Biol. Chem.* **269**: 26066-26075.

Sakai, N., Kaufman, S., and Milstein, S. (1993). Tetrahydrobiopterin is required for cytokine-induced nitric oxide production in a murine macrophage cell line (RAW 264). *Mol. Pharmacol.* **43**: 6-10.

Saran, M., Michel, C., and W. Bors, W. (1990). Reaction of nitric oxide with superoxide. Implications for the action of endothelial-derived relaxing factor (EDRF). *Free Radical Res. Commun.* **10**: 221-226.

Sari, M.-A., Booker, S., Jaouen, M., Vadon, S., Boucher, J.-L., Pompon, D., and Mansuy, D. (1996). Expression in yeast and purification of functional macrophage nitric oxide synthase. Evidence for cysteine-194 as iron proximal ligand. *Biochemistry* **35**: 7204-7213.

Saura, M., Pérez-Sala, D., Cañada, F.J., and Lamas, S. (1996). Role of tetrahydrobiopterin availability in the regulation of nitric-oxide synthase expression in human mesangial cells. *J. Biol. Chem.* **271**: 14290-14295.

Schmidt, H.H.H.W., Gagne, G.D., Nakane, M., Pollock, J.S., Miller, M.F., and Murad, F. (1992b). Mapping of neural nitric oxide synthase in the rat suggests frequent co-localization with NADPH diaphorase but not with soluble guanylyl cyclase, and novel paraneural functions for nitrinergic signal trans-

duction. *J. Histochem. Cytochem.* **40**: 1439-1456.

Schmidt, H.H.H.W., Pollock, J.S., Nakane, M., Gorsky, L.D., Förstermann, U., and Murad F. (1991). Purification of a soluble isoform of guanylyl cyclase-activating-factor synthase. *Proc. Natl. Acad. Sci. USA* **88**: 365-369.

Schmidt, H.H.H.W, Seifert, R., and Böhme, E. (1989). Formation and release of nitric oxide from human neutrophils and HL-60 cells induced by a chemotactic peptide, platelet activating factor and leukotriene B$_4$. *FEBS Lett.* **244**: 357-360.

Schmidt, H.H.H.W., Smith, R.M., Nakane, M., and Murad, F. (1992c). Ca^{2+}/calmodulin-dependent NO synthase type I: A bipteroflavoprotein with Ca^{2+}/calmodulin-independent diaphorase and reductase activities. *Biochemistry* **31**: 3242-3249.

Schmidt, H.H.H.W., Warner, T.D., Ishii, K., Sheng, H., and Murad, F. (1992a). Insulin secretion from pancreatic B cells caused by L-arginine-derived nitrogen oxides. *Science* **255**: 721-723.

Schott, C.A., Goven, C.M., Vetrovsky, P., Berton, C.C., and Stoclet, J.C. (1994). Exogenous N^G-hydroxyl-L-arginine causes nitrite production in vascular smooth muscle cells in the absence of nitric oxide synthase activity. *FEBS Lett.* **341**: 203-207.

Schweizer, M., and Richter, C. (1994). Nitric oxide potently and reversibly deenergizes mitochondria at low oxygen tension. *Biochem. Biophys. Res. Commun.* **204**: 169-175.

Sennequier, N., Boucher, J.-L., Battioni, P., and Mansuy, D. (1995). Superoxide anion efficiently performs the oxidative cleavage of C=NOH bonds of amidoximes and N-hydroxyguanidines with formation of nitrogen oxides. *Tetrahedron Lett.* **36**: 6059-6062.

Sessa, W.C., García-Cardeña, G., Liu, J., Keh, A., Pollock, J.S., Bradley, J., Thiru, S., Braverman, I.M., and Desai, K.M. (1995). The Golgi association of endothelial nitric oxide synthase is necessary for the efficient synthesis of nitric oxide. *J. Biol. Chem.* **270**: 17641-17644.

Sessa, W.C., Harrison, J.K., Barber, C.M., Zeng, D., Durieux, M.E., D'Angelo,

D.D., Lynch, K.R., and Peach, M.J. (1992). Molecular cloning and expression of a cDNA encoding endothelial cell nitric oxide. *J. Biol. Chem.* **267**: 15274–15276.

Seto, Y., and Guengerich, F.P. (1993). Partitioning between N-dealkylation and N-oxygenation in the oxidation of N,N-dialkylarylamines catalyzed by cytochrome P450 2B1. *J. Biol. Chem.* **268**: 9986–9997.

Sharma, V.S., Traylor, T.G., Gardiner R., and Mizukami, H. (1987). Reaction of nitric oxide with heme proteins and model compounds of hemoglobin. *Biochemistry* **26**; 3837–3843.

Sharp, M.A., and Cooper, C.E. (1998). Interaction of peroxynitrite with mitochrondrial cytochrome oxidase: Catalytic production of nitric oxide and irreverisble inhibition of enzyme activity. *J. Biol. Chem.* **47**: 30961–30972.

Shaul, P.W., Smart, E.J., Robinson, L.J., German, Z., Yuhanna, I.S., Ying, Y., Anderson, R.G.W., and Michel, T. (1996). Acylation targets endothelial nitric-oxide synthase to plasmalemmal caveolae. *J. Biol. Chem.* **271**: 6518–6522.

Sheng, H., Gagne, G.D., Matsumoto, T., Miller, M.F., Förstermann, U., and Murad, F. (1993). Nitric oxide synthase in bovine superior cervical ganglion. *J. Neurochem.* **61**: 1120–1126.

Sheng, H., Schmidt, H.H.H.W., Nakane, M., Mitchell, J.A., Pollock, J.S., Förstermann, U., and Murad, F. (1992). Characterization and localization of nitric oxide synthase in non-adrenergic non-cholinergic nerves from bovine retractor penis muscle. *Br. J. Pharmacol.* **106**: 768–773.

Sherman, M.P., Loro, M.L., Wong, V.Z., and Taskin, D.P. (1991). Cytokine- and *Pneumocystis carinii*-induced L-arginine oxidation by murine and human pulmonary alveolar macrophages. *J. Protozool.* **38**: 234S-236S.

Sheta, E.A., McMillan, K., and Masters, B.S.S. (1994). Evidence for a bidomain structure of constitutive cerebellar nitric oxide synthase. *J. Biol. Chem.* **269**: 15147–15153.

Shi, X., Lenhart, A., and Mao, Y. (1994). ESR spin trapping investigation on peroxynitrite decomposition: No evidence for hydroxyl radical production. *Biochem. Biophys. Res. Commun.* **203**: 1515–1521.

Shibuki, K. (1990). An electrochemical microprobe for detecting nitric oxide release in brain tissue. *Neurosci. Res.* **9**: 69–76.

Shibuki, K., and Okada, D. (1991). Endogenous nitric oxide release required for long-term synpatic depression in the cerebellum. *Nature* **349**: 326–328.

Siddhanta, U. Presta, A., Fan, B., Wolan, D., Rousseau, D.L., and Stuehr, D.J. (1998). Domain swapping in inducible nitric-oxide synthase. Electron transfer occurs between flavin and heme groups located on adjacent subunits in the dimer. *J. Biol. Chem.* **273**: 18950–18958.

Siddhanta, U., Wu, C., Abu-Soud, H.M., Zhang, J., Ghosh, D.K., and Stuehr, D.J. (1996). Heme iron reduction and catalysis by a nitric oxide synthase heterodimer containing one reductase and two oxygenase domains. *J. Biol. Chem.* **271**: 7309–7312.

Siegfried, M.R., Carey, C., Ma, X.-L., and Lefer, A.M. (1992a). Beneficial effects of SPM-5185, a cysteine-containing NO donor in myocardial ischemia–reperfusion. *Am. J. Physiol.* **263**: H771–H777.

Siegfried, M.R., Erhardt, J., Rider, T., Ma, X.-L., and Lefer, A.M. (1992b). Cardioprotection and attenuation of endothelial dysfunction by organic nitric oxide donors in myocardial ischemia-reperfusion. *J. Pharmacol. Exp. Ther.* **260**: 668–675.

Sies, H., Sharov, V.C., Koltz, L.-O., and Briviba, K. (1997). Glutathione peroxidase protects against peroxynitrite-mediated oxdiations. A new function for selenoproteins as peroxynitrite reductase. *J. Biol. Chem.* **272**: 27812–27817.

Singh, R.J., Hogg, N., Joseph, J., and Kalyanaraman, B. (1996). Mechanism of nitric oxide release from S-nitrosothiols. *J. Biol. Chem.* **271**: 18596–18603.

Smith, C.D., Carson, M., van der Woerd, M., Chen, J., Ischiropoulos, H., and Beckman, J.S. (1992). Crystal structure of peroxynitrite-modified bovine

Cu,Zn superoxide dismutase. *Arch. Biochem. Biophys.* **299**: 350–355.

Sneddon, J.M., and Vane, J.R. (1988). Endothelium-derived relaxing factor reduces platelet adhesion to bovine endothelial cells. *Proc. Natl. Acad. Sci. USA* **85**: 2800–2804.

Snyder, S.H. (1992). Nitric oxide: First in a new class of neurotransmitters? *Science* **257**: 494–496.

Snyder, S.H. (1995). No endothelial NO· *Nature* **377**: 196–197.

Son, H., Hawkins, R.D., Martin, K., Kiebler, M., Huang, P.L., Fishman, M.C., and Kandel, E.R. (1996). Long-term potentiation is reduced in mice that are doubly mutant in endothelial and neuronal nitric oxide synthase. *Cell* **87**: 1015–1023.

Song, K.S., Li, S., Okamoto, T., Quilliam, L.A., Sargiacomo, M., and Lisanti, M.P. (1996). Co-purification and direct interaction of Ras with caveolin, an integral membrane protein of caveolae microdomains. Detergent-free purification of caveolae membranes. *J. Biol. Chem.* **271**: 9690–9697.

Spencer, J.P.E., Wong, J., Jenner, A., Aruoma, O.I., Cross, C.E., and Halliwell, B. (1996). Base modification and strand breakage in isolated calf thymus DNA and in DNA from human skin epidermal keratinocytes exposed to peroxynitrite or 3-morpholinosydnonime. *Chem. Res. Toxicol.* **9**: 1152–1158.

Stachowiak, O., Dolder, M., Wallimann, T., and Richter, C. (1998). Mitochondrial creatine kinase is a prime target of peroxynitrite-induced modification and inactivation. *J. Biol. Chem.* **273**: 16694–16699.

Stadler, J., Trockfeld, J., Schmalix, W.A., Brill, T., Siewert, J.R., Griem, H., and Doehmer, J. (1994). Inhibition of cytochromes P4501A by nitric oxide. *Proc. Natl. Acad. Sci. USA* **91**: 3559–3564.

Stamler, J.S., Jaraki, O., Osborne, J., Simon, D.I., Keaney, J., Vita, J., Singel, D., Valeri, C.R., and Loscalzo, J. (1992a). Nitric oxide circulates in mammalian plasma primarily as an S-nitroso adduct of serum albumin. *Proc. Natl. Acad. Sci. USA* **89**: 7674–7677.

Stamler, J.S., Simon, D.I., Osborne, J.A., Mullins, M.E., Jaraki, O., Michel, T., Singel, D.J., and Loscalzo, J. (1992b). S-Nitrosylation of proteins with nitric oxide: Synthesis and characterization of biologically active compounds. *Proc. Natl. Acad. Sci. USA* **89**: 444–448.

Stamler, J.S., Singel, D.J., and Loscalzo, J. (1992c). Biochemistry of nitric oxide and its redox-activated forms. *Science* **258**: 1898–1902.

Sternberg, J., Mabbott, N., Sutherland, I., and Liew, F.Y. (1994). Inhibition of nitric oxide synthesis leads to reduced parasitemia in murine *Trypanosoma brucei* infection. *Infect. Immun.* **62**: 2135–2137.

Stevens-Truss, R., and Marletta, M.A. (1995). Interaction of calmodulin with the inducible murine macrophage nitric oxide synthase. *Biochemistry* **34**: 15638–15645.

Stone, J.R., and Marletta, M.A. (1996). Spectral and kinetic studies on the activation of soluble guanylate cyclase by nitric oxide. *Biochemistry* **35**: 1093–1099.

Stuehr, D.J., and Ikeda-Saito, M. (1992). Spectral characterization of brain and macrophage nitric oxide synthase. Cytochrome P-450-like hemeproteins that contain a flavin semiquione radical. *J. Biol. Chem.* **267**: 20547–20550.

Stuehr, D.J., and Marletta, M.A. (1985). Mammalian nitrate biosynthesis: Mouse macrophages produce nitrite and nitrate in response to *Escherichia coli* lipopolysaccharide. *Proc. Natl. Acad. Sci. USA* **82**: 7738–7742.

Stuehr, D.J., and Nathan, C.F. (1989). Nitric oxide: A macrophage product responsible for cytostasis and respiratory inhibition in tumor target cells. *J. Exp. Med.* **169**: 1543–1545.

Stuehr, D.J., Cho, H.J., Kwon, N.S., Weise, M.F., and Nathan, C.F. (1991a). Purification and characterization of the cytokine-induced macrophage nitric oxide synthase: An FAD- and FMN-containing flavoprotein. *Proc. Natl. Acad. Sci. USA* **88**: 7773–7777.

Stuehr, D.J., Kwon, N.S., Nathan, C.F., Griffith, O.W., Feldman, P.L., and Wiseman, J. (1991b). N^{ω}-Hydroxy-L-arginine is an intermediate in the biosynthesis of nitric oxide from L-arginine. *J. Biol. Chem.* **266**: 6259–6263.

Su, Z., Blazing, M.A., Fan, D., and George, S.E. (1995). The calmodulin–nitric oxide synthase interaction. Critical role of the calmodulin latch domain in enzyme activation. *J. Biol. Chem.* **270**: 29117–29122.

Subczynski, W., Lomnicka, M., and Hyde, J.S. (1996). Permeability of nitric oxide through lipid bilayer membranes. *Free Radical Res.* **24**: 343–349.

Takehara, Y., Kanno, T., Yoshioka, T., Inoue, M., and Utsumi, K. (1995). Oxygen-dependent regulation of mitochondrial energy metabolism by nitric oxide. *Arch. Biochem. Biophys.* **323**: 27–32.

Tamir, S., Burney, S., and Tannenbaum, S.R. (1996). DNA damage by nitric oxide. *Chem. Res. Toxicol.* **9**: 821–827.

Tarry, W.C., and Makhoul, R.C. (1994). L-Arginine improves endothelium-dependent vasorelaxation and reduces intimal hyperplasia after balloon angioplasty. *Arterioscler. Thromb.* **14**: 938–943.

Tartaglia, L.A., Storz, G., Brodsky, M.H., Lai, A., and Ames, B.N. (1990). Alkyl hydroperoxide reductase from *Salmonella typhimurium*. Sequence and homology to thioredoxin reductase and other flavoprotein disulfide oxidoreductases. *J. Biol. Chem.* **265**: 10535–10540.

Tayeh, M.A., and Marlette, M.A. (1989). Macrophage oxidation of L-arginine to nitric oxide, nitrite, and nitrate. Tetrahydrobiopterin is required as a cofactor. *J. Biol. Chem.* **264**: 19654–19658.

Tierney, D.L., Martasek, P., Doan, P.E., Masters, B.S.S., and Hoffman, B.M. (1998). Location of guanidino nitrogen of L-arginine substrate bound to neuronal nitric oxide synthase (nNOS): Determination by Q-band pulsed ENDOR spectroscopy. *J. Am. Chem. Soc.* **120**: 2983–2984.

Torres, J., Darley-Usmar, V., and Wilson, M.T. (1996). Inhibition of cytochrome c oxidase in turnover by nitric oxide: Mechanism and implications for control of respiration. *Biochem. J.* **312**: 169–173.

Tracey, W.R., Xue, C., Klinghofer, V., Barlow, J., Pollock, J.S., Förstermann, U., and Johns, R.A. (1994). Immunochemical detection of inducible NO synthase in human lung. *Am. J. Physiol.* **266**: L722–L727.

Trotti, D., Rossi, D., Gjesdal, O., Levy, L.M., Racagni, G., Danbolt, N.C., and Volterra, A. (1996). Peroxynitrite inhibits glutamate transporter subtypes. *J. Biol. Chem.* **271**: 5976–5979.

Tsai, A.-L. (1994). How does NO activate hemeproteins? *FEBS Lett.* **341**: 141–145.

Tschudi, M.R., Mesaros, S., Lüscher, T.F., and Malinski, T. (1996). Direct *in situ* measurement of nitric oxide in mesenteric resistance arteries. Increased decomposition by superoxide in hypertension. *Hypertension* **27**: 32–35.

Tsukahara, H., Gordienko, D.V., and Goligorsky, M.S. (1993). Continuous monitoring of nitric oxide release from human umbilical vein endothelial cells. *Biochem. Biophys. Res. Commun.* **193**: 722–729.

Tzeng, E., Billiar, T.R., Robbins, P.D., Loftus, M., and Stuehr, D.J. (1995). Expression of human inducible nitric oxide synthase in a tetrahydrobiopterin (H_4B)-deficient cell line: H_4B promotes assembly of enzyme subunits into an active dimer. *Proc. Natl. Acad. Sci. USA*, **92**: 11771–11775.

Uppu, R.M., Squadrito, G.L., and Pryor, W.A. (1996). Acceleration of peroxynitrite oxidations by carbon dioxide. *Arch. Biochem. Biophys.* **327**: 335–343.

Vallance, P., Patton, S., Bhagat, K., MacAllister, R., Radomski, M., Moncada, S., and Malinski, T. (1995). Direct measurement of nitric oxide in human beings. *Lancet* **345**: 153–154.

van der Vliet, A., O'Neill, C.H., Halliwell, B., Cross, C.E., and Kaur, H. (1994). Aromatic hydroxylation and nitration of phenylalanine and tyrosine by peroxynitrite. *FEBS Lett.* **339**: 89–92.

Vásquez-Vivar, J., Kalyanaraman, B., Martásek, P., Hogg, N., Masters, B.S.S., Karoui, H., Tordo, P., and Pritchard, K.A., Jr. (1998). Superoxide generation by endothelial nitric oxide synthase: The influence of cofactors. *Proc. Natl. Acad. Sci. USA* **95**: 9220–9225.

Vazquez-Torres, A., Jones-Carson, J., and Balish, E. (1996). Peroxynitrite contributes to the candidacidal activity of nitric oxide-producing macrophages. *Infect. Immun.* **64**: 3127–3133.

Vazquez-Torres, A., Jones-Carson, J., Warner, T., and Balish, E. (1995). Nitric oxide enhances resistance of SCID mice to mucosal candidiasis. *J. Infect. Dis.* **172**: 192–198.

Venema, R.C., Ju, H., Zou, R., Ryan, J.W., and Venema, V.J. (1997a). Subunit interactions of endothelial nitric-oxide synthase. Comparisons to the neuronal and inducible nitric-oxide synthase isoforms. *J. Biol. Chem.* **272**: 1276–1282.

Venema, R.C., Sayegh, H.S., Kent, J.D., and Harrison, D.G. (1996). Identification, characterization, and comparison of the calmodulin-binding domains of the endothelial and inducible nitric oxide synthases. *J. Biol. Chem.* **271**: 6435–6440.

Venema, V.J., Ju, H., Zou, R., and Venema, R.C. (1997b). Interaction of neuronal nitric-oxide synthase with caveolin-3 in skeletal muscle. Identification of a novel caveolin scaffolding/inhibitory domain. *J. Biol. Chem.* **272**: 28187–28190.

Vermilion, J.L., and Coon, M.J. (1978). Identification of the high and low potential flavins of liver microsomal NADPH-cytochrome P-450 reductase. *J. Biol. Chem.* **253**: 8812–8819.

Vermilion, J.L., Ballou, D.P., Massey, V., and Coon, M.J. (1981). Separate roles for FMN and FAD in catalysis by liver microsomal NADPH-cytochrome P-450 reductase. *J. Biol. Chem.* **256**: 266–277.

Vespa, G.R., Cunha, F.Q., and Silva, J.S. (1994). Nitric oxide is involved in control of *Trypanosoma cruzi*-induced parasitemia and directly kills the parasite *in vitro*. *Infect. Immun.* **62**: 5177–5182.

Vetrovsky, P., Stoclet, J.-C., and Entlicher, G. (1996). Possible mechanism of nitric oxide production from N^G-hydroxyl-L-arginine or hydroxylamine by superoxide ion. *Int. J. Biochem. Cell Biol.* **28**: 1311–1318.

Vincendeau, P., and Daulouède, S. (1991). Macrophage cytostatic effect on *Trypanosoma musculi* involves an L-arginine-dependent mechanism. *J. Immunol.* **146**: 4338–4343.

Vodovotz, Y., Kwon, N.S., Popischil, M., Manning, J., Paik, J., and Nathan C. (1994). Inactivation of nitric oxide synthase after prolonged incubation of mouse macrophages with IFN-γ and bacterial lipopolysaccharide. *J. Immunol.* **152**: 4110–4118.

Von der Leyen, H.E., Gibbons, G.H., Morishita, R., Lewis, N.P., Zhang, L., Nakajima, M., Kaneda, Y., Cooke, J.P., and Dzau, V.J. (1995). Gene therapy inhibiting neointimal vascular lesion: *In vivo* transfer of endothelial cell nitric oxide synthase gene. *Proc. Natl. Acad. Sci. USA* **92**: 1137–1141.

Vouldoukis, I., Riveros-Moreno, V., Dugas, B., Ouaaz, F., Bécherel, P., Debré, P., Moncada, S., and Mossalayi, M.D. (1995). The killing of *Leishmania major* by human macrophages is mediated by nitric oxide induced after ligation of the FcεRII/CD23 surface antigen. *Proc. Natl. Acad. Sci. USA* **92**: 7804–7808.

Wakulich, C.A., and Tepperman, B.L. (1997). Role of glutathione in nitric oxide mediated injury to rat gastric mucosal cells. *Eur. J. Pharmacol.* **319**: 333–341.

Wang, J., Rousseau, D.L., Abu-Soud, H.M., and Stuehr, D.J. (1994). Heme coordination of NO in NO synthase. *Proc. Natl. Acad. Sci. USA* **91**: 10512–10516.

Watkins, P.J., and Gorrod, J.W. (1987). Studies on the *in vitro* biological oxidation of trimethoprim. *Eur. J. Drug Metab. Pharmacokinet.* **12**: 245–251.

Wei, X.-Q., Charles, I.G., Smith, A., Ure, J., Feng, G.-J., Huang, F.-P., Xu, D., Muller, W., Moncada, S., and Liew, F.Y. (1995). Altered immune response in mice lacking inducible nitric oxide synthase. *Nature* **375**: 408–411.

Weiner, C.P., Lizasoain, I., Baylis, S.A., Knowles, R.G., Charles, I.G., and Moncada, S. (1994). Induction of calcium-dependent nitric oxide synthases by sex hormones. *Proc. Natl. Acad. Sci. USA* **91**: 5212–5216.

Weiss, G., Goossen, B., Doppler, W., Fuchs, D., Pantopoulos, K., Werner-Felmayer, G., Wachter, H., and Hentze, M.W. (1993). Translational regulation via iron-responsive elements by the nitric oxide/NO-synthase pathway. *EMBO J.* **12**: 3651–3657.

Weiss, G., Werner-Felmayer, G., Werner, E.R., Grünewald, K., Wachter, H., and Hentze, M.W. (1994). Iron regulates nitric oxide synthase activity by controlling nuclear transcription. *J. Exp. Med.* **180**: 969–976.

Weisz, A., Oguchi, S., Cicatiello, L., and Esumi, H. (1994). Dual mechanism for the control of inducible-type NO synthase gene expression in macrophages during activation by interferon-γ and bacterial lipopolysaccharide. Transcriptional and posttranscriptional regulation. *J. Biol. Chem.* **269**: 8324–8333.

Wever, R.M.F., van Dam, T., van Rijn, H.J.M., De Groot, F., and Rabelink, T.J. (1997). Tetrahydrobiopterin regulates superoxide and nitric oxide generation by recombinant endothelial nitric oxide synthase. *Biochem. Biophys. Res. Commun.* **237**: 340–344.

White, C.R., Brock, T.A., Chang, L.-Y., Crapo, J., Briscoe, P., Ku, D., Bradley, W.A., Gianturco, S.H., Gore, J., Freeman, B.A., and Tarpey, M.M. (1994). Superoxide and peroxynitrite in atherosclerosis. *Proc. Natl. Acad. Sci. USA* **91**: 1044–1048.

White, K.A., and Marletta, M.A. (1992). Nitric oxide synthase is a cytochrome P-450 type hemoprotein. *Biochemistry* **31**: 6627–6631.

White, R.E. (1991). The involvement of free radicals in the mechanism of monooxygenases. *Pharmacol. Ther.* **49**: 21–42.

White, R.E. (1994). The importance of one-electron transfers in the mechanism of cytochrome P-450, in: *Cytochrome P450 Biochemistry, Biophysics and Molecular Biology* (Lechner, M.C., ed.), pp. 333–340, John Libbey Eurotext, Paris.

Wink, D.A., Hanbauer, I., Krishna, M.C., DeGraff, W., Gamson, J., and Mitchell, J.B. (1993a). Nitric oxide protects against cellular damage and cytotoxicity from reactive oxygen species. *Proc. Natl. Acad. Sci. USA* **90**: 9813–9817.

Wink, D.A., Kasprzak, K.S., Maragos, C.M., Elespuru, R.K., Misra, M., Dunams, T.M., Cebula, T.A., Koch, W.H., Andrews, A.W., Allen, J.S., and Keefer, L.K. (1992). DNA deaminating ability and genotoxicity of nitric oxide and its progenitors. *Science* **254**: 1001–1003.

Wink, D.A., Nims, R.W., Darbyshire, J.F., Christodoulou, D., Hanbauer, I., Cox, G.W., Laval, F., Laval, J., Cook, J.A., Krishna, M.C., DeGraff, W.G., and Mitchell, J.B. (1994). Reaction kinetics for nitrosation of cysteine and glutathione in aerobic nitric oxide solutions at neutral pH. Insights into the fate and physiological effects of intermediates generated in the NO/O$_2$ reaction. *Chem. Res. Toxicol.* **7**: 519–525.

Wink, D.A., Osawa, Y., Darbyshire, J.F., Jones, C.R., Eshenaur, S.C., and Nims, R.W. (1993b). Inhibition of cytochromes P450 by nitric oxide and a nitric oxide-releasing agent. *Arch. Biochem. Biophys.* **300**: 115–123.

Winterbourn, C.C., and Metodiewa, D. (1994). The reaction of superoxide with reduced glutathione. *Arch. Biochem. Biophys.* **314**: 284–290.

Witteveen, C.F.B., Giovanelli, J., and Kaufman, S. (1996). Reduction of quinonoid dihydrobiopterin to tetrahydrobiopterin by nitric oxide synthase. *J. Biol. Chem.* **271**: 4143–4147.

Wood, J., and Garthwaite, J. (1994). Models of the diffusional spread of nitric oxide: Implications for neural nitric oxide signalling and its pharmacological properties. *Neuropharmacology* **33**: 1235–1244.

Xia, Y., and Zweier, J.L. (1997). Superoxide and peroxynitrite generation from inducible nitric oxide synthase in macrophages. *Proc. Natl. Acad. Sci. USA* **94**: 6954–6958.

Xia, Y., Dawson, V.L., Dawson, T.M., Snyder, S.H., and Zweier, J.L. (1996). Nitric oxide synthase generates superoxide and nitric oxide in arginine-depleted cells leading to peroxynitrite-mediated cellular injury. *Proc. Natl. Acad. Sci. USA* **93**: 6770–6774.

Xia, Y., Roman, L.J., Masters, B.S.S., and Zweier, J.L. (1998a). Inducible nitric-oxide synthase generates superoxide from the reductase domain. *J. Biol. Chem.* **273**: 22635–22639.

Xia, Y., Tsai, A.-L., Berka, V., and Zweier, J.L. (1998b). Superoxide generation from endothelial nitric-oxide synthase. A Ca^{2+}/calmodulin- dependent and tetrahydrobiopterin regulatory process. *J. Biol. Chem.* **273**: 25804–25808.

Xie, L., and Gross, S.S. (1997). Argininosuccinate synthetase overexpression in vascular smooth muscle cells potentiates immunostimulant-induced NO production. *J. Biol. Chem.* **272**: 16624–16630.

Xie, Q.-W., Cho, H.J., Calaycay, J., Mumford, R.A., Swiderek, K.M., Lee, T.D., Ding, A., Troso, T., and Nathan, C. (1992). Cloning and characterization of inducible nitric oxide synthase from mouse macrophages. *Science* **256**: 225–228.

Yamakura, F., Taka, H., Fujimura, T., and Murayama, K. (1998). Inactivation of human manganese-superoxide dismutase by peroxynitrite is caused by exclusive nitration of tyrosine 34 to 3-nitrotyrosine. *J. Biol. Chem.* **273**: 14085–14089.

Yang, G., Candy, T.E.G., Boaro, M., Wilkin, H.E., Jones, P., Nazhat, N.B., Saadalla-Nazhat, R.A., and Blake, D.R. (1992). Free radical yields from the homolysis of peroxynitrous acid. *Free Radical Biol. Med.* **12**: 327–330.

Yoshizumi, M., Perrella, M.A., Burnett, J.C., Jr., and Lee, M.-E. (1993). Tumor necrosis factor downregulates an endothelial nitric oxide synthase mRNA by shortening its half-life. *Circ. Res.* **73**: 205–209.

Yui, Y., Hattori, R., Kosuga, K., Eizawa, H., Hiki, K., Ohkawa, S., Ohnish, K., Terao, S., and Kawai, C. (1991). Calmodulin-independent nitric oxide synthase from rat polymorphonuclear neutrophils. *J. Biol. Chem.* **266**: 3369–3371.

Zhang, M., Yuan, T., Aramin, J.M[...] Vogel, H.J. (1995a). Interaction o[...] modulin with its binding domain of [...] cerebellar nitric oxide synthase. A mul[...] tinuclear NMR study. *J. Biol. Chem.* **270**: 20901–20907.

Zhang, Z.G., Chopp, M., Bailey, F., and Malinski, T. (1995b). Nitric oxide changes in the rat brain after transient middle cerebral artery occlusion. *J. Neurolog. Sci.* **128**: 22–27.

Zhao, X.J., Raitt, D.V., Burke, P., Clewell, A.S., Kwast, K.E., and Poyton, R.O. (1996). Function and expression of flavohemoglobin in *Saccharomyces cerevisiae*. Evidence for a role in the oxidative stress response. *J. Biol. Chem.* **271**: 25131–25138.

Zhu, H., and Riggs, A.F. (1992). Yeast flavohemoglobin is an ancient protein related to globins and a reductase family. *Proc. Natl. Acad. Sci. USA* **89**: 5015–5019.

Zhu, L., Gunn, C., and Beckman, J.S. (1992). Bactericidal activity of peroxynitrite. *Arch. Biochem. Biophys.* **298**: 452–457.

Ziegler, D.M. (1991). Unique properties of the enzymes of detoxification. *Drug Metab. Dispos.* **19**: 847–852.

Zou, M.-H., and Ullrich, V. (1996). Peroxynitrite formed by simultaneous generation of nitric oxide and superoxide selectively inhibits bovine aortic prostacyclin synthase. *FEBS Lett.* **382**: 101–104.

Zweier, J.L., Wang, P., and Kuppusamy, P. (1995). Direct measurement of nitric oxide generation in the ischemic heart using electron paramagnetic resonance spectroscopy. *J. Biol. Chem.* **270**: 304–307.

...nods of Free Radical
...tection

The reduction of cytochrome c was shown to be mediated by univalently reduced oxygen (the superoxide radical), which is produced by xanthine oxidase and liberated into free solution.

McCord and Fridovich, 1969

Central to the study of free radical events at a cellular level is the ability to detect and characterize these reactive species. Of those instruments that make identification possible, electron paramagnetic resonance (EPR) spectroscopy, which draws upon the magnetic properties of unpaired electrons, offers the highest degree of selectivity. Nuclear magnetic resonance (NMR) spectroscopy, which is dependent upon the diamagnetic nature of odd-numbered nuclei, such as ^1H or ^{13}C, is a viable alternative. Since free radicals are paramagnetic, these compounds, when added to a water solution, alter the relaxivity of the surrounding environment (Koenig, 1986). This change in relaxation times, both T_1 (spin–lattice or longitudinal) and T_2, (spin–spin or transverse), has allowed the successful development of contrast-enhancing agents for magnetic resonance imaging (MRI) (Lauterbur, 1973; Weinmann, et al., 1984). For instance, this imaging modality has been successfully used to localize free radicals, as their spin trapped adducts, in vivo (Janzen, et al., 1990) and it has surfaced as a reliable method to estimate the concentration of paramagnetic species, such as nitroxides, in homogeneous solutions (Keana, et al., 1987), as well as in vivo (Rosen, et al., 1987; Pou, et al., 1995b). Finally, nitroxides with MRI or electron paramagnetic resonance imaging (EPRI) have been used to obtain reasonable estimates of O_2 concentrations in tissues (Halpern, et al., 1994, 1996; Gallez, et al., 1996).

With current EPR spectrometers, the lowest theoretical limit of detection in an aqueous solution is about $10 nM$. However, considerably greater concentrations, in the range of $0.1–10 \mu M$, are required to record and resolve detailed spectra with reasonable signal-to-noise ratios (Borg, 1976). Within these parameters, only stable free radicals that accumulate to measurable levels can be identified by

EPR spectroscopy at temperatures acceptable to the preponderance of life forms. Examples of stable free radicals of biologic interest include those derived from vitamin C (Yamazaki and Piette, 1961), vitamin E (Das, *et al.*, 1970), flavins (Ehrenberg, *et al.*, 1967), including quinones, and a broad spectrum of other xenobiotics (Mason, 1982).

For shorter-lived free radicals, such as $O_2 \cdot^-$, slowing the rate of free radical elimination by lowering the temperature of the reaction mixture, through rapid submersion into liquid nitrogen, enhances the opportunity to identify specific paramagnetic species. This experimental approach has successfully been used to demonstrate the formation of $O_2 \cdot^-$ during the metabolism of xanthine by xanthine oxidase (Knowles, *et al.*, 1969). Confirmation of the assigned EPR spectrum to that of $O_2 \cdot^-$ soon followed (Nilsson, *et al.*, 1969). Although simple from a study design perspective, rapid quenching of samples does, nevertheless, present significant limitation in interpreting experimental data. Since the free radical is no longer in a fluid environment, its anticipated anisotropy is lost, frequently obscuring the identity of this reactive species. With the nearly 10 ms required to freeze the reaction, unanticipated artifacts may enter the picture (Bolton, *et al.*, 1972). Even under carefully controlled conditions (Zweier, *et al.*, 1987), sample preparation can sometimes result in spurious EPR signals (Baker, *et al.*, 1988; Nakazawa, *et al.*, 1988; van der Kraaij, *et al.*, 1989). Continuous-flow EPR spectroscopy in combination with signal averaging is, in some cases, a more viable alternative, although the requirement for large quantities of reagents may, in other situations, be self-limiting (Yamazaki, *et al.*, 1960; Bolton, *et al.*, 1972).

Alternative methods take advantage of the reactivity of the initiating free radical. Typical of these techniques are those in which a probe, by reacting with a specific free radical, generates a stable product that can be identified through one of a variety of analytic methods. These reactions frequently involve electron transfer, leading to a change in oxidation state of a metal ion. Similarly, addition of a free radical to either a metal ion complex or an electrophilic center results in the formation of a product with characteristics that can be linked to the initiating free radical. Since the stable product can accumulate, these indirect methods can often detect free radicals at low flux rates for prolonged periods of time (Bynoe, *et al.*, 1992).

Even though the focus of this book is the application of spin trapping to biologically generated free radicals, we would be neglectful not to describe, though briefly, alternative methods for the identification of these reactive intermediates. However, for more extensive presentations on various assays for free radical detection, the interested reader should consult the many reviews that have appeared in the literature over the past several decades. A few of the more popular accounts were written by McCord, *et al.* (1977), Bors, *et al.* (1982), Afanas'ev (1989), Weber (1990), and Archer (1993).

Superoxide

Since $O_2 \cdot^-$ can behave as either a one-electron reductant or a one-electron oxidant, there are, not surprisingly, a myriad of indirect methods to detect this free radical—the most popular of which is the reduction of ferricytochrome c (Table 4.1).

Table 4.1 Methods of Quantitating Superoxide Production[a]

Indicator	Product	Method of Detection	Rate Constant ($M^{-1} s^{-1}$)	Reference(s)
Reduction				
Ferricytochrome c	Ferrocytochrome c	Optical, $E_{550} = 21 \, mM^{-1} \, cm^{-1}$	6×10^5	1–8
Tetranitromethane	Nitroform anion	Optical, $E_{350} = 15 \, mM^{-1} \, cm^{-1}$	1.9×10^9	9, 10
Nitroblue tetrazolium	Formazon	Optical, $E_{560} = 28.6 \, mM^{-1} \, cm^{-1}$	5.8×10^4	8, 11–13
OXANO/Thiols	OXANOH	EPR spectroscopy	—	14
Nitroxide-fluorophore	Hydroxylamines	EPR spectroscopy	—	15
Thiols		Fluorescence, $E_{ex} = 393 \, nm$, $E_{em} = 482 \, nm$	—	15
Oxidation				
Epinephrine	Adrenochrome	Optical, $E_{310} = 22 \, mM^{-1} \, cm^{-1}$	5.6×10^4	16, 17
Tiron	Semiquinone	Optical, $E_{300} = 2.3 \, mM^{-1} \, cm^{-1}$	1.7×10^7	18
OXANOH	OXANO	EPR spectroscopy	6.7×10^3	19
Others				
Luminol	Aminophthalate	Chemiluminescence, $E_\lambda = 427 \, nm$	—	20, 21
Lucigenin	Biacridane	Chemiluminescence, $E_\lambda = 470 \, nm$	—	20, 21
Luciferin analog		Chemiluminescence, $E_\lambda = 460 \, nm$		22
Luciferin analog		Chemiluminescence, $E_\lambda = 465 \, nm$		23, 24

Data adapted from Ellington, *et al.*, 1988.

[a] Rate constants for the reaction of $O_2 \cdot^-$ with indicators are at pH $\approx$ 7.8.

1 = Massey, 1959; 2 = Van Gelder and Slater, 1962; 3 = Land and Swallow, 1971; 4 = Butler, *et al.*, 1975; 5 = Simic, *et al.*, 1975; 6 = Koppenol, *et al.*, 1976; 7 = Seki, *et al.*, 1976, at pH 7.3 = $5 \times 10^5 \, M^{-1} \, s^{-1}$; 8 = McCord, *et al.*, 1997; 9 = Rabani *et al.*, 1965; 10 = Forman and Fridovich, 1973; 11 = Miller, 1970; 12 = Halliwell, 1976; 13 = Bielski, *et al.*, 1980; 14 = Finkelstein, *et al.*, 1984; 15 = Pou, *et al.*, 1995a; at pH 7.4; 16 = Bors, *et al.*, 1978a; 17 = Bors, *et al.*, 1978b; 18 = Bors, *et al.*, 1979; 19 = Rosen, *et al.*, 1982; 20 = Seitz and Neary, 1976; 21 = Campbell, *et al.*, 1985; 22 = Nakano, *et al.*, 1986; 23 = Nishida, *et al.*, 1989, 24 = Nakano, *et al.*, 1995.

This is primarily due to the ease of experimental design, inexpensive instrumenta-tion, and the unique spectral absorption maximum of the product, ferrocyto-chrome c (Massey, 1959; Kuthan, *et al.*, 1982). Unfortunately, this reaction is not unique to $O_2 \cdot^-$ in that other reductants, including ascorbate and thiols (McCord, *et al.*, 1977), as well as enzymes such as cytochrome P-450 reductase (Azzi, *et al.*, 1975; Kuthan, *et al.*, 1978; Finkelstein, *et al.*, 1981), will mediate this one-electron reduction, thereby mimicking the presence of this free radical. Because of this limitation, inclusion of SOD in the experimental design is essential to overcome the otherwise lack of specificity. In some cases, however, when even the SOD control leaves the result in doubt, alternative methods of $O_2 \cdot^-$ detection should be sought. This point is best illustrated by the finding that spin trapping/ EPR spectroscopy documented the formation of $O_2 \cdot^-$ by NOS I (Pou, *et al.*, 1992), whereas this free radical was unobserved when ferricytochrome c was the indicator molecule employed (Mayer, *et al.*, 1991; Heinzel, *et al.*, 1992; Klatt, *et al.*, 1992; Pou, *et al.*, 1992).

In some situations where biological preparations may be contaminated with cytochrome c oxidase, a lack of ferricytochrome c reduction can frequently lead to erroneous conclusions (McCord, *et al.*, 1977; Kuthan, *et al.*, 1978). Finally, ferrocytochrome c can be oxidized by H_2O_2 and peroxynitrite. The reaction with H_2O_2 is sufficiently slow as to not significantly underestimate initial rates of $O_2 \cdot^-$ production (Bynoe, *et al.*, 1992). In contrast, its interaction with peroxynitrite is rapid, with a second order rate constant approaching $2.3 \times 10^5 \, M^{-1} \, s^{-1}$ (Thomson, *et al.*, 1995). This may, in some experimental designs, preclude the use of the cytochrome c assay for detecting $O_2 \cdot^-$.

Two other indicators that respond to the reductive properties of $O_2 \cdot^-$, namely tetranitromethane and nitroblue tetrazolium, have received considerably less attention than ferricytochrome c (Table 4.1). In the case of tetranitromethane, this is primarily the result of its unpleasant odor and lack of specificity towards $O_2 \cdot^-$ (Michelson, 1977), even though the reaction of the probe with this free radical is nearly at diffusion-controlled rates (Rabani, *et al.*, 1965). Although nitroblue tetrazolium was one of the earliest reporters of $O_2 \cdot^-$ (Rajagopalan and Handler, 1964; Halliwell, 1976), the lack of specificity towards this free radical (Auclair, *et al.*, 1978) has, for the most part, limited its use to clinical screening for chronic granulomatous disease (Babior and Woodman, 1990).

The conversion of epinephrine to adrenochrome is the best known of the cate-chol oxidative assays for $O_2 \cdot^-$ (Table 4.1). In this assay, epinephrine is oxidized by this free radical to a semiquinone free radical, which can then reduce O_2 to produce more $O_2 \cdot^-$ and a quinone (McCord and Fridovich, 1969; Cohen and Heikkila, 1977). This quinone is further oxidized, resulting in the eventual formation of the colored adrenochrome. In addition, autoxidation of epinephrine, especially in the presence of trace metal ions, like Fe^{3+}, can lead to copious amounts of artificially produced $O_2 \cdot^-$ (Misra and Fridovich, 1972a, 1972b). Because $O_2 \cdot^-$ acts as both a free radical chain initiator and a propagator during adrenochrome formation, especially at higher pHs, this oxidative reaction frequently overestimates the flux of $O_2 \cdot^-$.

Epinephrine oxidation to adrenochrome represents one of the most complicated and artifact-prone assays for $O_2 \cdot^-$; on the other hand, Tiron, 1,2-dihydroxybenzene-3,5-disulfonate typifies the simplest of the catechol-type probes for this free radical (Table 4.1) (Miller and Rapp, 1973). Although the semiquinone free radical, arising from the oxidation of Tiron by $O_2 \cdot^-$, is stable sufficiently long to be observable by EPR spectroscopy at ambient temperatures, this reaction is not unique to $O_2 \cdot^-$, but can also be generated by HO· (Bors, et al., 1979). Because of the uncertain nature of the initiating free radical, the Tiron method has lost most of its initial glamor.

Chemiluminescence is the emission of photons from a vibrationally excited species as it relaxes to its ground state (Campbell, et al., 1985). Like other methods, the product, arising from the reaction of a reactive species with an indicator molecule, is the emitter of this energy (Table 4.1). Not surprisingly, the sensitivity afforded by single-photon counting has made chemiluminescence an attractive assay for detecting free radicals in a number of *in vitro* and *in situ* models (Cadenas, et al., 1984; Campell, et al., 1985; Murphy and Sies, 1990; Henry, et al., 1991; Kumar, et al., 1991; Filatov, et al., 1995). Thus, initial studies, correlating $O_2 \cdot^-$ production from model systems with light emission, were most encouraging (Greenlee, et al., 1962; Vorhaben and Steele, 1967; Arneson, 1970). Adaptation of this technique to cell suspensions (Allen, et al., 1974) and isolated tissues (Boveris, et al., 1980) seemed to support the early findings. Further excitement arose with the development of specific enhancers of low-level chemiluminescence, such as luminol (Seitz and Neary, 1976; Weimann, et al., 1984; Schulte-Herbrüggen and Sies, 1989; Sharov, et al., 1989), lucigenin (Seitz and Neary, 1976; Weimann, et al., 1984), or luciferin analogs (Michelson and Isambert, 1973; Gotoh and Niki, 1994). Although increased specificity has allowed some insightful inquiries as to free radical production from stimulated neutrophils (Edwards, 1987; Gyllenhammar, 1987; Suzaki, et al., 1994) and *in situ* organ preparations (Takahashi, et al., 1990), questionable specificity still remains a troublesome issue (Baxendale, 1973; Henry and Michelson, 1977; Smith, et al., 1991) and restricts their application to certain experimental conditions.

Attempts at assigning phagocyte-derived $O_2 \cdot^-$ to light emission typifies the limitations of chemiluminescence. Early studies attributed enhanced light emission to bacterial phagocytosis and oxidative events, including formation of singlet O_2 (Allen, et al., 1972; Grebner, et al., 1977; Nelson, et al., 1977; Hatch, et al., 1978). However, more extensive inquiries suggested that chemiluminescence was derived from a more complex series of biologic events, only a portion of which was initiated by free radicals (Cormier and Prichard, 1968; Webb, et al., 1974; Allen, 1975a, 1975b; Cheson, et al., 1976; Tsan, 1977; Harvath, et al., 1978; DeChatelet, et al., 1982; Dahlgren and Stendahl, 1983). In fact, the nature of the reactive oxygen species responsible for light emission has been a topic of much debate, in which $O_2 \cdot^-$, H_2O_2, and HOCl have been suggested as possible sources (Webb, et al., 1974; Cheson, et al., 1976; Rosen and Klebanoff, 1976; DeChatelet, et al., 1982; Vilim and Wilhelm, 1989).

An additional concern has recently surfaced with the use of lucigenin-associated chemiluminescence as a means of identifying $O_2 \cdot^-$ in biologic systems. Lucigenin must first undergo reduction to a cationic free radical. Upon subsequent exposure of this radical to $O_2 \cdot^-$, the lucigenin free radical undergoes an additional one-electron reduction to dioexathane. Decay of this latter species generates detectable light. It has been reported that lucigenin, per se, can actually stimulate $O_2\cdot^-$ formation as, in the presence of O_2, the initial cationic free radical can undergo autooxidation, resulting in reduction of O_2 to $O_2\cdot^-$. Thus, it has been argued that lucigenin cannot be used as a reliable indicator of $O_2\cdot^-$ in biological systems (Vásquez-Vivar, et al., 1997; Liochev and Fridovich, 1997, 1998; Heiser, et al., 1998). However, it has been suggested that since the magnitude of $O_2\cdot^-$ generated is quite small, concentrations of lucigenin above those usually employed in experimental systems would be required to generate the magnitude of $O_2\cdot^-$ needed to confound data interpretation (Li, et al., 1998). These uncertainties have left many in search of more reliable alternatives.

Hydrogen Peroxide

Although not a free radical, H_2O_2 is, nonetheless, an oxidant, which plays an important role in many biologic events. As discussed earlier, H_2O_2 is formed through the spontaneous and SOD-catalyzed disproportionation of $O_2\cdot^-$. Thus, in biologic reactions in which H_2O_2 is produced exclusively through this pathway, measurement of the rate of $O_2\cdot^-$ production, using one of the methods described above, provides an accurate estimation of H_2O_2 concentration:

$$O_2\cdot^- + O_2\cdot^- + 2H^+ \rightarrow H_2O_2 + O_2$$

Alternatively, measurement of H_2O_2 flux can, in some models, give reliable estimations of $O_2\cdot^-$ production. Detection of H_2O_2 by stimulated phagocytic cells has, for instance, been used as a marker of $O_2\cdot^-$ generation (Root and Metcalf, 1977). In some cases, when there are competing reactions, $O_2\cdot^-$ production may not correlate well with the rate of H_2O_2 formation, or H_2O_2 formation may occur independently of $O_2\cdot^-$ generation (Fridovich, 1970; Masters and Holmes, 1977; Chance, et al., 1979; Del Rio, et al., 1992; Liochev and Fridovich, 1995). Consequently, a variety of methods have been developed to detect this peroxide. Several of the more popular methodologies have been those in which a peroxidase, such as horseradish peroxidase (HRP), has been included in the reaction mixture along with a substrate capable of being oxidized by the product of the reaction of H_2O_2 with the peroxidase (Table 4.2). For instance, oxidation of phenolsulfonephthalein (phenol red) by HRP and H_2O_2 leads to the formation of a chromophore that absorbs at 610 nm. Although this assay is procedurally simple, thereby providing a straightforward analytic measure of this peroxide, the lack of sensitivity is troublesome (Pick and Keisari, 1980; Pick and Mizell, 1981; Pick, 1986).

When detection limits are a critical factor in a experimental design, a fluorometric adaptation of the HRP/H_2O_2 procedure, in which scopoletin is substituted for phenol red, is an attractive alternative to spectrophotometric assays (Andrea, 1955; Perschke and Broda, 1961; Root, et al., 1975; Root and Metcalf, 1977). However, its use can be problematic as the correlation between fluorescence and concentration of H_2O_2 is not consistent over a broad range. Optimally, SOD needs to be included in the reaction mixture because $O_2\cdot^-$, under some conditions, may

Table 4.2 Methods of Quantitating Hydrogen Peroxide Production

Indicator	Product	Method of Detection	Reference(s)
Phenol red	HRP oxidation	Optical, E_{610}	1–3
Scopoletin	HRP oxidation	Fluorescence, $E_{ex} = 350\,nm$	4–7
		$E_{em} = 460\,nm$	
Homovanillic acid	HRP oxidation	Fluorescence, $E_{ex} = 312\,nm$	8–11
		$E_{em} = 420\,nm$	
$2',7'$-Dichlorofluorescin	$2',7'$-Dichlorofluorescein	Fluorescence, $E_{ex} = 501\,nm$ $E_{em} = 521\,nm$	12–17
Luminol	Aminophthlate	Chemiluminescence, $E_\lambda = 427\,nm$	18, 19
Catalase	O_2	O_2 electrode	20
H_2O_2 electrode	—	Polarimetric	21, 22
Aminotriazole	Inactivation of catalase	Polarimetric	23–26
Diaminebenzidine	Oxidized diamine–benzidine polymer	Conventional or electron microscopy	27–32

1 = Pick and Keisari, 1980; 2 = Pick and Mizell, 1981; 3 = Pick, 1986; 4 = Andrea, 1955; 5 = Labato and Briggs, 1987; 6 = Root, et al., 1975; 7 = Root and Metcalf, 1977; 8 = Guilbault, et al., 1968; 9 = Ruch, et al., 1983; 10 = Baggiolini, et al., 1986; 11 = Kettle, et al., 1994; 12 = Keston and Brandt, 1965; 13 = Cathcart, et al., 1983; 14 = Halliwell and Gutteridge, 1981; 15 = Bass, et al., 1983; 16 = Bass, et al., 1986; 17 = Ubezio and Civoli, 1994; 18 = Seitz and Neary, 1976; 19 = Campbell, et al., 1985; 20 = Britigan, et al., 1991; 21 = Test and Weiss, 1984; 22 = Test and Weiss, 1986; 23 = Margoliash, et al., 1960; 24 = Gee, et al., 1970; 25 = Kinnula, et al., 1991; 26 = Kinnula, et al., 1992; 27 = Briggs, et al., 1975; 28 = Labato and Briggs, 1985; 29 = Halbhuber, et al., 1988; 30 = Hoffstein, et al., 1988; 31 = Babbs, 1994; 32 = Dannenberg, et al., 1994.

interfere with detection of H_2O_2 (Kettle, *et al.*, 1994). Other substrates, such as homovanillic acid (Guilbault, *et al.*, 1968; Ruch, *et al.*, 1983; Baggiolini, *et al.*, 1986) and p-hydroxyphenylacetic acid (Panus, *et al.*, 1993), have been substituted for scopoletin.

The H_2O_2-mediated oxidation of $2',7'$-dichlorofluorescein, which is catalyzed by a number of peroxidases, leads to the fluorescent $2',7'$-dichlorofluorescein (Keston and Brandt, 1965; Cathcart, *et al.*, 1983). This method has been adapted to detect intracellular generation of H_2O_2 by phagocytes and endothelial cells in combination with flow cytometry (Halliwell and Gutteridge, 1981; Bass, *et al.*, 1983, 1986; Rothe and Valet, 1990; Carter, *et al.*, 1994). Although technical problems, such as the source of the oxidant (Zhu, *et al.*, 1994), have limited the broad application to other cellular systems, recent studies, in which H_2O_2 was detected by human adrenocarcinoma cells exposed to doxorubicin, are most encouraging (Ubezio and Civoli, 1994). Nevertheless, it is important to recognize that the exact mechanism and species responsible for $2',7'$-dichlorofluorescein oxidation in biological systems is poorly understood (Marchiesi, *et al.*, 1999). Hydrogen peroxide does not directly oxidize $2',7'$-dichlorofluorescein, and thus the magnitude of fluorescence cannot be automatically equated with the amount of H_2O_2 formed. Further, fluorescence, in this setting, can be influenced by a number of factors other than H_2O_2. These would include: activity and access to intracellular hydroperoxides, quenching of fluorescence by other biomolecules, and variation in cellular uptake and stability of the probe.

The reaction of catalase with H_2O_2 generates H_2O and O_2 (Schonbaum and Chance, 1976). Thus, the determination of catalase-enhanced O_2 generation provides a direct estimation of H_2O_2 production. Using a Clark-type electrode capable of measuring dissolved O_2, assays for H_2O_2 have been developed based on the increased formation of O_2 in the presence of catalase (Britigan, et al., 1991). Similar polarometric methods have been used to identify this peroxide from phagocytic cells (Test and Weiss, 1984, 1986). However, poor sensitivity and the frequent need to change the membrane on the electrode are limitations that need to be addressed. Finally, in cell preparations, which contain peroxidases and catalase, this method is frequently inadequate to give accurate estimations of intracellular production of this peroxide. Despite this, there have been efforts to obtain a more realistic quantification of cellular levels of H_2O_2 through the inhibition of catalase by aminotriazole (Margoliash, et al., 1960; Gee, et al., 1970; Kinnula, et al., 1991, 1992). However, caution should be exercised when calculating H_2O_2 fluxes in the presence of aminotriazole, as this agent is ineffective in antagonizing glutathione peroxidase. Data from those studies would lead to an underestimation of the cellular flux of H_2O_2.

Although histochemical staining does not provide quantitative measures of H_2O_2 production, this method is, nevertheless, an invaluable tool in determining qualitative estimates of tissue peroxide concentrations (Briggs, et al., 1975; Labato and Briggs, 1985; Halbhuber, et al., 1988; Hoffstein, et al., 1988; Babbs, 1994; Dannenberg, et al., 1994). Typical of this assay is the oxidation of diaminobenzidine by H_2O_2. The resulting polymers of diaminobenzidine are insoluble in aqueous and organic solvents, do not migrate from their cellular site of formation, and are detectable by light or electron microscopy. Recent modifications may result in greater specificity towards H_2O_2 (Babbs, 1994).

Hydroxyl Radical

Identification of HO· in the biological milieu is filled with a myriad of difficulties, simply because this free radical is so reactive. Hydroxyl radical reacts with nearly everything it collides with at rates approaching diffusion controlled (von Sonntag and Schuchmann, 1994). In complex biological systems, assigning specific oxidative products to reactions that involve this free radical is speculative. Even with inclusion of scavengers for HO·, a high degree of uncertainty remains a serious issue, which is not easily resolved even with the best of assay procedures.

Despite these real concerns, there have been significant advances in assay development, which can lead to reasonable estimates of HO· production *in vitro* and *in vivo* (Bors, et al., 1982; Karam, et al., 1991; Shigenaga and Ames, 1991; Aruoma, 1994; Kaur and Halliwell, 1994a; Simic, 1994). The most frequently employed methods fall into three groups (Table 4.3). The first measures gaseous hydrocarbons (usually from sulfur-containing compounds) using gas chromatography (GC), frequently in combination with mass spectrometry (MS). Typical of these techniques is the estimation of ethylene production from either methional (Beauchamp and Fridovich, 1970; Tauber and Babior, 1977; Cohen and Cederbaum, 1979; Hoidal, et al., 1979; Winterbourn, 1981, 1983;

Table 4.3 Methods of Quantitating Hydroxyl Radical

Indicator	Product	Method of Detection	Reference(s)
Methional, KMB	Ethylene	GC/MS	1–17
DMSO	Formaldehyde/methane	GC/MS	18–22
DMSO	Methanesulfinic acid	Optical, E_{425}	23–27
DMSO	Methanesulfinic acid	HPLC	28
Salicylate	Hydroxylated products	Optical, fluorescence	29–40
Benzoate		GC/MS	
Phenylalanine nucleosides		HPLC/MS	
2-Deoxyribose	Malondialdehyde	Optical, fluorescence	41–50
[^{14}C]Benzoate	$^{14}CO_2$	Radiometry	51–53

1 = Yang, 1967; 2 = Beauchamp and Fridovich, 1970; 3 = Bors; et al., 1976; 4 = Tauber and Babior, 1977; 5 = Klebanoff and Rosen, 1978; 6 = Pryor and Tang, 1978; 7 = Weiss, et al., 1978; 8 = Cohen and Cederbaum, 1979; 9 = Hoidal, et al., 1979; 10 = Ambruso and Johnston, 1981; 11 = Winterbourn, 1981; 12 = Karnovsky and Badwey, 1983; 13 = Janco and English, 1983; 14 = Winterbourn, 1983; 15 = Speer, et al., 1985; 16 = Winston, et al., 1986; 17 = Pryor, et al., 1994; 18 = Repine, et al., 1979; 19 = Cohen and Cederbaum, 1980; 20 = Klein, et al., 1980; 21 = Klein, et al., 1981; 22 = Shimazaki, et al., 1991; 23 = Babbs and Gale, 1987; 24 = Babbs and Griffin, 1989; 25 = Babbs and Steiner, 1990; 26 = Steiner and Babbs, 1990; 27 = Pou, et al., 1995c; 28 = Scaduto, 1995; 29 = Halliwell, 1978; 30 = Richmond, et al., 1981; 31 = Floyd, et al., 1984; 32 = Alexander, et al., 1986; 33 = Sagone and Husney, 1987; 34 = Maskos, et al., 1990; 35 = Kaur, et al., 1988; 36 = Halliwell, et al., 1991; 37 = Sun, et al., 1993; 38 = Kaur and Halliwell, 1994a; 39 = Shigenaga, et al., 1994; 40 = van der Vliet, et al., 1994; 41 = Halliwell and Gutteridge, 1981; 42 = Gutteridge, 1984; 43 = Winterbourn and Sutton, 1986; 44 = Winterbourn, 1986; 45 = Winterbourn, 1987; 46 = Greenwald, et al., 1989; 47 = Coffman, et al., 1990; 48 = Tadolini and Cabrini, 1990; 49 = Winterbourn, 1991; 50 = Aruoma, 1994; 51 = Cohen and Cederbaum, 1980; 52 = Sagone, et al., 1980a; 53 = Green, et al., 1985.

Karnovsky and Badwey, 1983) or the related thioether, 2-keto-4-thiomethyl-butryic acid (KMB) (Bors, et al., 1976; Rosen and Klebanoff, 1978; Weiss, et al., 1978; Cohen and Cederbaum, 1979; Hoidal, et al., 1979; Ambruso and Johnston, 1981; Janco and English, 1983; Karnovsky and Badwey, 1983; Winterbourn, 1983; Speer, et al., 1985; Winston, et al., 1986). Despite the ability to detect HO· in vivo by monitoring ethylene production in human exhalants, the lack of specificity is troublesome. For instance, peroxidase-derived oxidants, peroxynitrite, and other free radicals have been shown to generate ethylene from methional or KMB (Yang, 1967; Klebanoff and Rosen, 1978; Pryor and Tang, 1978; Pryor, et al., 1994).

Methane and formaldehyde are among the reaction products that result from the reaction of dimethylsulfoxide (DMSO) with HO·. The ease of detection using standard GC/MS procedures (Repine, et al., 1979; Cohen and Cederbaum, 1980; Klein, et al., 1980, 1981) must be balanced against the uncertain specificity of the reaction (Shimazaki, et al., 1991). The measurement of methanesulfinic acid, derived from the reaction of HO· with DMSO, is a reasonable spectrophotometric alternative for quantifying HO· production (Babbs and Gale, 1987; Babbs and Griffin, 1989; Babbs and Steiner, 1990; Steiner and Babbs, 1990). This analytic procedure has successfully been used to detect small concentrations of HO· produced during the decomposition of peroxynitrite (Pou, et al., 1995c). A recent report describes the development of high-performance liquid chromatography (HPLC) methods for determination of methanesulfinate (Scaduto, 1995). With this technique, enhanced sensitivity ensures the popularity of this method for the identification of this free radical.

The second general technique for identification of HO· involves the measurement of hydroxylated nucleosides, such as deoxyguanosine (Floyd, *et al.*, 1986; Shigenaga and Ames, 1991; Shigenaga, *et al.*, 1994; Yin, *et al.*, 1995; Herbert, *et al.*, 1996), and aromatic hydrocarbons, including salicylate (Halliwell, 1978; Richmond, *et al.*, 1981; Floyd, *et al.*, 1984; Sagone and Husney, 1987; Maskos, *et al.*, 1990; Udassin, *et al.*, 1991; Kim and Wells, 1996), benzoic acid (Alexander, *et al.*, 1986), and phenylalanine (Kaur, *et al.*, 1988; Sun, *et al.*, 1993; Kaur and Halliwell, 1994a, 1994b). Although the detection of hydroxylated hydrocarbons can be accomplished using a variety of instruments, such as spectrophotometry, spectrofluorometry, GC/MS, and HPLC/MS, careful analyses of the products are essential for optimal specificity (Halliwell, *et al.*, 1991; Montgomery, *et al.*, 1995).

The third general approach centers on the identification of carbonyl compounds derived from the reaction of HO· with 2-deoxyribose. One of the products of this reaction is malondialdehyde (MDA)—the same aldehyde formed during the peroxidation of many lipids (Janero, 1990). Quantification of MDA relies on the spectrophotometric or spectrofluorometric identification of a thiobarbituric acid (TBA) reaction product (Halliwell and Gutteridge, 1981; Gutteridge, 1984; Aruoma, 1994). Even though this is a simple assay for the detection of HO· (Winterbourn, 1986; Greenwald, *et al.*, 1989; Coffman, *et al.*, 1990; Britigan and Edeker, 1991), other strong oxidants, including a number of iron chelates appear, to behave toward 2-deoxyribose in a fashion similar to this free radical (Winterbourn and Sutton, 1986; Winterbourn, 1987; 1991). Thus, the selectivity of this assay toward HO· is in doubt, especially under conditions in which lipid peroxidation or changes in pH may influence the outcome of the findings (Tadolini and Cabrini, 1990; Winterbourn, 1991). Beyond issues of specificity, localization of 2-deoxyribose at potential intracellular sites of HO· production remains uncertain.

Hydroxyl radical decarboxylation of [^{14}C]benzoic acid results in the release of $^{14}CO_2$, which can be quantified through radiometric assays (Cohen and Cederbaum, 1980; Sagone, *et al.*, 1980a, 1980b; Winston and Cederbaum, 1982). As with alternative assays for HO·, other oxidants can confound data interpretation (Green, *et al.*, 1985).

Lipid Peroxidation

Formation of lipid peroxides and hydroperoxides is intimately associated with the direct or indirect effects of free radical production in biological systems. Plasma membranes serve as the principal site for the generation of these compounds. Over the years, an enormous number of methodologies have been developed to quantitate lipid peroxidation, whether such processes occur *in vitro* or *in vivo*. A detailed description of these assays is beyond the focus of this book. Nevertheless, a familiarity with alternative procedures is an important first step in understanding the benefits and limitations of applying spin trapping to the identification of lipid free radicals.

A variety of aldehydes are produced during the peroxidation of lipids, including those associated with biological membranes. Among these is MDA, whose generation has formed the basis for one of the oldest and most popular assays for membrane peroxidation (Esterbauer and Cheeseman, 1990; Jentzsch, et al., 1996). When heated in the presence of TBA, the presence of MDA results in the formation of a chromophore, which exhibits a λ maximum at 532 nm (Bernheim, et al., 1948; Yu and Sinnhuber, 1957; Warren, 1959; Dennis and Shibamoto, 1989; Kosugi and Kikugawa, 1989; Draper and Hadley, 1990; Janero, 1990). Despite the uncertain source of this absorbance (Draper and Hadley, 1990; Janero, 1990), this assay continues to receive considerable attention. It should be pointed out that other compounds besides MDA can lead to reaction products with similar absorption properties (Kosugi and Kikugawa, 1989; Draper and Hadley, 1990; Janero, 1990). Therefore, specificity and, in some cases, sensitivity (Thomas, et al., 1989) remain serious drawbacks, even though GC/MS (Van Kuijk, et al., 1990a; Bachi, et al., 1996), HPLC (Esterbauer, et al., 1984; Chirico, 1994; Miyazawa, et al., 1994; Sattler, et al., 1994; Yamamoto, 1994), fluorometry (Akasaka, et al., 1993), and electrochemical detection (Korytowski, et al., 1993) have greatly aided the identification of lipid peroxides, as well as products derived from these oxidized lipids, such as MDA. In the biological milieu, however, lipid extraction, a time-consuming and often error-prone procedure, is a prerequisite to analyses. With such limitations, dye-sensitive chemiluminescence has again surfaced as an attractive alternative to these well-defined methods, despite reservations as to the source of the photoemission (Hiramitsu, et al., 1994; Zamburlini, et al., 1995a, 1995b).

Aldehydes other than MDA are produced during lipid peroxidation (Esterbauer and Zollner, 1989; Esterbauer and Cheeseman, 1990; Van Kuijk, et al., 1990a, 1990b) and their detection serves as the basis for alternative assays. A typical method involves the trans-esterification of fatty acid esters to products analyzable by GC/MS. Similarly, 4-hydroxynonenal, among the 4-hydroxyalkenals formed during cleavage of lipid peroxides, can be directly detected by either HPLC or GC/MS (Esterbauer and Zollner, 1989; Esterbauer and Cheeseman, 1990; Van Kuijk, et al., 1990b). The reactivity of aldehydes towards 2,4-dinitrophenylhydrazine, yielding the corresponding hydrazone, has been the basis of a simple and reliable method to identify the parent carbonyl, as this reaction creates stable derivatives that can easily be identified through typical analytical methods, such as TLC or HPLC (Esterbauer and Cheeseman, 1990).

The metal-catalyzed decomposition of lipid peroxides results in the formation of pentane and ethane (Roberts, et al., 1983; Müller and Sies, 1984; Dillard and Tappel, 1989; Pitkanen, et al., 1989; Burns and Wagner, 1991). Although these gases are minor by-products of lipid peroxidation, the high sensitivity of GC/MS can frequently offset low yields of these hydrocarbons (Smith, 1991). As pentane and ethane can be detected from exhaled breath, on-line measurement of these gases can be used to estimate the degree of lipid peroxidation occurring in vivo (Roberts, et al., 1983; Dillard and Tappel, 1989; Jeejeebhoy, 1991; Smith, 1991).

Nitric Oxide

The demonstration that NO· exhibited many of the physiologic properties attributed to EDRF was based on the correlation between the bioassay—relaxation of precontracted aortic strips—and several analytic measurements of this free radical (Ignarro, *et al.*, 1987; Palmer, *et al.*, 1987). The relaxation of vascular smooth muscle is dependent upon the activation of the heme-containing cytosolic enzyme, guanylate cyclase, by NO· (Craven and DeRubertis, 1978, 1983; Ignarro, *et al.*, 1984, 1986). By adding ^{3}H-guanosine to a crude preparation of guanylate cyclase, estimation of cyclic ^{3}H-GMP formation is a sensitive and accurate indication of NO· generation (Table 4.4) (Pou, *et al.*, 1990, 1991). This assay is, however, not unique to NO· (Rapoport, 1986; Pou, *et al.*, 1991). Therefore, verification that this free radical was responsible for ^{3}H-GMP formation requires inhibition by a number of NOS antagonists.

Bioassays for NO·, whether derived from relaxation of precontracted aortic rings (Furchgott and Zawadzki, 1980; Palmer, *et al.*, 1987; Sessa, *et al.*, 1990) or inhibition in platelet aggregation (Sneddon and Vane, 1988), remain one of the premier methods of detecting this free radical. Although sensitive by most standards, these bioassays fall short of providing reliable quantitative estimates of NO· production.

There are a number of analytic assays[1] for NO· (Archer, 1993; Hevel and Marletta, 1994; Kostka, 1995). Of those methods, the simplest technologically is the diazotization procedure, in which levels of inorganic nitrite are estimated after reaction with sulfanilic acid and *N*-(1-naphthyl)ethylenediamine (Bell, *et al.*, 1963, Green, *et al.*, 1982; Tracey, *et al.*, 1990; Szabó, *et al.*, 1996). The ease of this measurement makes this a popular assay, even though the specificity of the reaction is questionable and detection limits are in the micromolar range (Tracey, *et al.*, 1990; Snell, *et al.*, 1996). Enhanced sensitivity can be gained by estimating nitrite levels through its reaction with 2,3-diaminonaphthalene to form the fluorescent

Table 4.4 Methods of Quantitating Nitric Oxide

Indicator	Product	Method of Detection	Reference(s)
^{32}GTP	Cyclic ^{32}P-GMP	Radiometry	1–5
^{3}H-Guanosine	Cyclic ^{3}H-GMP	Radioimmunoassay	6, 7
L-[^{3}H]Arginine	L-[^{3}H]Citrulline	Radiometry	8, 9
N-(1-Naphthyl)-ethylenediamine	Diazotization	Optical, E_{548}	10, 11
Oxyhemoglobin	Methemoglobin	Optical, E_{401}	12, 13
Ozone	·NO$_2$*	Chemiluminescence	14–16
Luminol	Aminophthalate	Chemiluminescence, $E_\lambda = 427$ nm	17

1 = Craven and DeRubertis, 1978; 2 = Ignarro, *et al.*, 1987; 3 = Shikano, *et al.*, 1988; 4 = Mayer, *et al.*, 1990; 5 = Pou, *et al.*, et al., 1991; 6 = Pou, *et al.*, 1994; 7 = Schmidt, *et al.*, 1991; 8 = Bredt and Snyder, 1990; 9 = Hecker, *et al.*, 1990; 10 = Green, *et al.*, 1982; 11 = Tracey, *et al.*, 1990; 12 = Murphy and Noack, 1994; 13 = Hevel and Marletta, 1994; 14 = Zafiriou and McFarland, 1980; 15 = Pai, *et al.*, 1987; 16 = Menon, *et al.*, 1991; 17 = Kikuchi, *et al.*, 1993.

compound, 1-(H)-naphthotriazole (Misko, *et al.*, 1993). An alternative HPLC approach has received some attention, even though limits are only in the millimolar range (Kok, *et al.*, 1983). A reliable spectrophotometric method depends upon the oxidation of reduced oxyhemoglobin [Hb(Fe^{2+})O$_2$] by this free radical to methemoglobin Hb(Fe^{3+}) (Doyle and Hoekstra, 1981; Ignarro, *et al.*, 1987; Kelm, *et al.*, 1988; Murphy and Noack, 1994). As this reaction is both rapid and sensitive, with nanomolar thresholds, real-time measurements of biologically generated NO· are practical:

$$NO \cdot + Hb(Fe^{2+})O_2 \rightarrow Hb(Fe^{3+}) + NO_3^-$$

Chemiluminescence is the most sensitive of the commercially available assays for NO·. The determination of this free radical is dependent upon its reaction with ozone (O$_3$) (Zafiriou and McFarland, 1980; Pai, *et al.*, 1987; Menon, *et al.*, 1991). With this method, detection limits for NO· are in the 20–50 picomol range (Palmer, *et al.*, 1987; Marletta, *et al.*, 1988; Archer and Cowan, 1991; Brien, *et al.*, 1991; Gustafsson, *et al.*, 1991; Kikuchi, *et al.*, 1993):

$$NO \cdot + O_3 \rightarrow \cdot NO_2{}^* + O_2$$

$$\cdot NO_2^* \rightarrow \cdot NO_2 + h\nu$$

Mass spectroscopy offers the opportunity to measure NO· in exhalants by trapping the free radical as nitrosothioproline, through the reaction of thioproline with NO· (Gustafsson, *et al.*, 1991). Although the sensitivity of this assay parallels that found with chemiluminescence, the cost and maintenance of a mass spectrometer are limiting factors in the widespread adpation of this technology.

Using electrochemical detection coupled with a porphyrinic-coated electrode microsensor, it has become possible to identify NO· in either single cells or tissue preparations (Malinski and Taha, 1992; Taha, *et al.*, 1992; Malinski, *et al.*, 1993; Tschudi, *et al.*, 1996). Typical of these experiments are those in which the microsensor is placed on the surface of an endothelial cell. From these studies, it appears that endothelial cells produce $\approx 1 \times 10^{-20}$ mol of NO· per cell, about a factor of 2–4 lower than earlier estimates (Marletta, 1989; Moncada, *et al.*, 1991). Other electrochemical sensors are available from a wide variety of commercial sources. Use of such instruments typically gives NO· fluxes in the picomolar range (Shibuki, 1990; Wink, *et al.*, 1995; Christodoulou, *et al.*, 1996; Gutierrez, *et al.*, 1996).

Peroxynitrite

As noted in other sections, it seems probable that conditions exist *in vivo* in which formation of NO· would occur with the simultaneous generation of O$_2$·$^-$, either by the same or neighboring cells. In addition, several recent studies have suggested that during NOS oxidation of L-arginine to NO· and L-citrulline, O$_2$·$^-$ may be a subsequent by-product of electron transport through the enzyme. Thus, at a fixed concentration of O_2, the ratio of each free radical is dependent upon the concentration of L-arginine (Pou, *et al.*, 1992, 1999; Schmidt, *et al.*, 1996; Vásquez-Vivar, *et al.*, 1998; Xia, *et al.*, 1998a, 1998b). Given the high probability that NO· and O_2·⁻ would be generated in close proximity to each other and that ONOO⁻ would be the proximal product derived from the reaction of NO· and O_2·⁻ (Saran, *et al.*, 1990; Huie and Padmaja, 1993), considerable effort has been placed on the development of techniques to assess the formation of this hydroperoxide in cellular systems[2].

Tyrosine and similarly related phenols are among the preferred substrates for ONOO⁻. However, the resulting dityrosine is not unique to this reaction. For instance, H_2O_2 and peroxidases, such as horseradish peroxidase and myeloperoxidase, can lead to dityrosine formation (Aeschbach, *et al.*, 1976; Heinecke, *et al.*, 1993; Winterbourn, *et al.*, 1997), whereas 3-nitrotyrosine is a more specific indicator of ONOO⁻. However, as noted in chapter 2, it is now apparent that peroxidase-catalyzed oxidation of nitrite can, likewise, result in 3-nitrotyrosine (van der Vliet, *et al.*, 1997; Eiserich, *et al.*, 1998).

Based on this chemistry, HPLC analysis in combination with spectrophotometry or spectrofluorometry allows the identification of nitrated and hydroxylated products of tyrosine (Beckman, *et al.*, 1992; Ischiropoulos, *et al.*, 1992; Crow and Ischiropoulos, 1996), as well as phenylalanine (van der Vliet, *et al.*, 1994, 1996; Ramezanian, *et al.*, 1996) and similarly related phenols (Sampson, *et al.*, 1996). Likewise, a simple fluorometric method that follows the oxidation of dihydrorhodamine has received considerable attention (Kooy, *et al.*, 1994). Detection limits for ONOO⁻ can be markedly enhanced through the use of chemiluminescence (Radi, *et al.*, 1993). However, luminol-dependent luminescence can occur from the formation of other oxidants, such as those derived from the reaction of H_2O_2 and myeloperoxidase (DeChatelet, *et al.*, 1982).

The recent development of a monoclonal antibody that recognizes 3-nitrotyrosine residues in proteins (Beckman, *et al.*, 1994; Ye, *et al.*, 1996) has broadened the ability to probe for the formation of ONOO⁻ in complex biological systems. Such antibodies can be used for both immunoblot and immunohistochemical analyses of cellular or tissue production of ONOO⁻. However, once again recent evidence suggests that such products can occur independent of ONOO⁻ (van der Vliet, *et al.*, 1997; Eiserich, *et al.*, 1998), particularly at *in vivo* sites where myeloperoxidase may be present. This may further limit the usefulness of this approach. Despite this, a variety of techniques for the detection of reaction products derived from ONOO⁻ have been developed. As discussed herein, the "optimal" detection system inevitably varies with the experimental design and the question to be addressed.

Concluding Thoughts

Investigating the biologic reactions of a specific free radical are ultimately linked to the detection and characterization of this reactive species. Although many of the methods discussed offer a high degree of sensitivity, the origin of the measurement is frequently uncertain. Furthermore, the inadequacy of these techniques to detect free radicals in sophisticated biologic systems is a perplexing problem, limiting future advancement. In the end, the search for a better technology has led many scientists to spin trapping/EPR spectroscopy. In forthcoming chapters, we will begin to describe how this method holds great promise for the *in vivo in situ* identification of free radicals in real time and at their site of evolution.

References

Aeschbach, R., Amado, R., and Neukom, H. (1976). Formation of dityrosine cross-links in proteins by oxidation of tyrosine residues. *Biochim. Biophys. Acta* **439**: 292–301.

Afanas'ev, I.B. (1989). Detection of superoxide ion, in: *Superoxide Ion: Chemistry and Biological Implications*, Vol. I, pp. 13–16, CRC Press, Boca Raton, FL.

Akasaka, K., Ohrui, H., Meguro, H., and Tamura, M. (1993). Determination of triacylglycerol and cholesterol ester hydroperoxides in human plasma by high-performance liquid chromatography with fluorometric postcolumn detection. *J. Chromatogr.* **617**: 205–211.

Alexander, M.S., Husney, R.M., and Sagone, A.L., Jr. (1986). Metabolism of benzoic acid by stimulated polymorphonuclear cells. *Biochem. Pharmacol.* **20**: 3649–3651.

Allen, R.C. (1975a). Halide dependence of the myeloperoxidase- mediated antimicrobial system of the polymorphonuclear leukocyte in the phenomenon of electronic excitation. *Biochem. Biophys. Res. Commun.* **63**: 675–683.

Allen, R.C. (1975b). The role of pH in the chemiluminescent response of the myeloperoxidase-halide-HOOH antimicrobial system. *Biochem. Biophys. Res. Commun.* **63**: 684–691.

Allen, R.C., Stjernholm, R.L., and Steele, R.H. (1972). Evidence for the generation of an electronic excitation states(s) in human polymorphonuclear leukocytes and its participation in bactericidal activity. *Biochem. Biophys. Res. Commun.* **47**: 679–684.

Allen, R.C., Yevich, S.J., Orth, R.W., and Steele, R.H. (1974). The superoxide anion and singlet molecular oxygen: Their role in the microbicidal activity of the polymorphonuclear leukocytes.

Biochem. Biophys. Res. Commun. **60**: 909–917.

Ambruso, D.R., and Johnston, R.B., Jr. (1981). Lactoferrin enhances hydroxyl radical production by human neutrophils, neutrophil particulate fractions and an enzymatic generating system. *J. Clin. Invest.* **67**: 352–360.

Andrea, W.A. (1955). A sensitive method for the estimation of hydrogen peroxide in biological materials. *Nature* **175**: 859–860.

Archer, A. (1993). Measurement of nitric oxide in biological models. *FASEB J.* **7**: 349–360.

Archer, A., and Cowan, N.J. (1991). Acetylcholine causes endothelium dependent vasodilation but does not stimulate nitric oxide production by rat pulmonary arteries or elevate endothelial cytosol calcium concentration. *Circ. Res.* **68**: 1569–1581.

Arneson, R.M. (1970). Substrate-induced chemiluminescence of xanthine oxidase and aldehyde oxidase. *Arch. Biochem. Biophys.* **136**: 522–360.

Aruoma, O. (1994). Deoxyribose assay for detecting hydroxyl radicals. *Methods Enzymol.* **233**: 57–66.

Auclair, C., Torres, M., and Hakim, J. (1978). Superoxide anion involvement in NBT reduction catalyzed by NADPH-cytochrome P-450 reductase: A pitfall. *FEBS Lett.* **89**: 26–28.

Azzi, A., Montecucco, C., and Richter, C. (1975). The use of acetylated ferricytochrome c for the detection of superoxide radicals produced by biological membranes. *Biochem. Biophys. Res. Commun.* **65**: 597–603.

Babbs, C.F. (1994). Histochemical methods for localization of endothelial superoxide and hydrogen peroxide generation in perfused organs. *Methods Enzymol.* **233**: 619–630.

Babbs, C.F., and Gale, M.J. (1987). Colorimetric assay for methanesulfinic acid in biological samples. *Anal. Biochem.* **163**: 67–73.

Babbs, C.F., and Griffin, D.G. (1989). Scatchard analysis of methane sulfinic acid production from dimethyl sulfoxide: A method to quantify hydroxyl radical formation in physiologic systems. *Free Radical Biol. Med.* **6**: 493–503.

Babbs, C.F., and Steiner, M.G. (1990). Detection and quantitation of hydroxyl radical using dimethyl sulfoxide as molecular probe. *Methods Enzymol.* **186**: 137–147.

Babior, B.M., and Woodman, R.C. (1990). Chronic granulomatous disease. *Semin. Hematol.* **27**: 247–259.

Bachi, A., Zuccato, E., Baraldi, M., Fanelli, R., and Chiabrando, C. (1996). Measurement of urinary 8-EPI-prostaglandin $F_{2\alpha}$, a novel index of lipid peroxidation *in vivo*, by immunoaffinity extraction/gas chromatography–mass spectrometry. Basal levels in smokers and nonsmokers. *Free Radical Biol. Med.* **20**: 619–624.

Baggiolini, M., Ruch, W., and Cooper, P.H. (1986). Measurement of hydrogen peroxide production by phagocytes using homovanillic acid and horseradish peroxidase. *Methods Enzymol.* **132**: 395–400.

Baker, J.E., Felix, C.C., Olinger, G.N., and Kalyanaraman, B. (1988). Myocardial ischemia and reperfusion: Direct evidence for free radical generation by electron spin resonance spectroscopy. *Proc. Natl. Acad. Sci. USA* **85**: 2786–2789.

Bass, D.A., Olbrantz, P., Szejda, P., Seeds, M.S., and McCall, C.E. (1986). Subpopulations of neutrophils with increased oxidative product formation in blood of patients with infection. *J. Immunol.* **136**: 860–866.

Bass, D.A., Parce, J.W., DeChatelet, L.R., Szejda, P., Seeds, M.S., and Thomas, M. (1983). Flow cytometric studies of oxidative product information by neutrophils: A graded response to membrane stimulation. *J. Immunol.* **130**: 1910–1917.

Baxendale, J.H. (1973). Pulse radiolysis study of the chemiluminescence from luminol (5-amino-2,3-dihydrophthalazine-1,4-dione). *J. Chem. Soc., Faraday I* **69**: 1665–1677.

Beauchamp, C., and Fridovich, I. (1970). A mechanism for the production of ethylene from methional. The generation of the hydroxyl radical by xanthine oxidase. *J. Biol. Chem.* **245**: 4641–4646.

Beckman, J.S., Ischiropoulos, H., Zhu, L., van der Woerd, M., Smith, C., Chen, J., Harrison, J., Martin, J.C., and Tsai, M. (1992). Kinetics of superoxide dismutase- and iron-catalyzed nitration of phenolics by peroxynitrite. *Arch. Biochem. Biophys.* **298**: 438–445.

Beckman, J.S., Ye, Y.Z., Anderson, P.G., Chen, J., Accavitti, M.A., Tarpey, M.M., and White C.R. (1994). Extensive nitration of protein tyrosines in human atherosclerosis detected by immunohistochemistry. *Biol. Chem. Hoppe-Seyler* **375**: 81–88.

Bell, F.K., O'Neill, J.J., and Burgison, R.M. (1963). Determination of the oil/water distribution coefficients of glyceryl trinitrate and two similar nitrate esters. *J. Pharm. Sci.* **52**: 637–639.

Bernheim, F., Bernheim, M.L.C., and Wilbur, K.M. (1948). The reaction between thiobarbituric acid and the oxidation products of certain lipids. *J. Biol. Chem.* **174**: 257–264.

Bielski, B.H.J., Shiue, G.G., and Bajuk, S. (1980). Reduction of nitro blue tetrazolium by CO_2^- and $O_2 \cdot^-$ radicals. *J. Phys. Chem.* **84**: 830–833.

Bolton, J.R., Borg, D.C., and Swartz, H.M. (1972). Experimental aspects of biological electron spin resonance studies, in: *Biological Applications of Electron Spin Resonance* (Swartz, H.M., Bolton, J.R., and Borg, D.C., eds.), pp. 64–118, Wiley-Interscience, New York.

Borg, D.C. (1976). Application of electron spin resonance in biology, in: *Free Radicals in Biology* (Pryor, W.A., ed.), Vol. I, pp. 69–147, Academic Press, New York.

Bors, W., Lengfelder, E., Saran, M., Fuchs, C., and Michel, C. (1976). Reactions of oxygen radical species with methional: A pulse radiolysis study. *Biochem. Biophys. Res. Commun.* **70**: 81–87.

Bors, W., Michel, C., Saran, M., and Lengfelder, E. (1978a). Kinetic investigations of the autoxidation of adrenalin. *Z. Naturforsch.* **33c**: 891–896.

Bors, W., Saran, M., Lengelder, E., Michel, C., Fuchs, C., and Frenzel, C. (1978b). Detection of oxygen radicals in biological reactions. *Photochem. Photobiol.* **28**: 629–638.

Bors, W., Saran, M., and Michel, C. (1979). Pulse-radiolytic investigations of catechols and catecholamines. II. Reactions of Tiron with oxygen radical species. *Biochim. Biophys. Acta* **582**: 537–542.

Bors, W., Saran, M., and Michel, C. (1982). Assays of oxygen radicals: Methods and mechanisms, in: *Superoxide Dismutase* (Oberley, L.W., ed.), Vol. II, pp. 31–62, CRC Press, Boca Raton, FL.

Boveris, A., Cadenas, E., Reiter, R., Filipkowski, M., Nakase, Y., and Chance, B. (1980). Organ chemiluminescence: Noninvasive assay for oxidative radical reactions. *Proc. Natl. Acad. Sci. USA* **77**: 347–351.

Bredt, D.S., and Snyder, S.H. (1990). Isolation of nitric oxide synthetase, a calmodulin-requiring enzyme. *Proc. Natl. Acad. Sci. USA* **87**: 682–685.

Brien, J., McLaughlin, B., Nakatsu, K., and Marks, G.S. (1991). Quantitation of nitric oxide formation from nitrovasodilator drugs by chemiluminescence analysis of headspace gas. *J. Pharmacol. Methods*, **25**: 19–27.

Briggs, R.T., Karnovsky, M.L., and Karnovsky, M.J. (1975). Cytochemical demonstration of hydrogen peroxide in polymorphonuclear leukocyte phagosomes. *J. Cell Biol.* **64**: 254–260.

Britigan, B.E., and Edeker, B.L. (1991). *Pseudomonas* and neutrophil products modify transferrin and lactoferrin to create conditions that favor hydroxyl radical formation. *J. Clin. Invest.* **88**: 1092–1102.

Britigan, B.E., Roeder, T.L., and Buettner, G.R. (1991). Spin traps inhibit formation of hydrogen peroxide via the dismutation of superoxide: Implications for spin trapping the hydroxyl free radical. *Biochim. Biophys. Acta* **1075**: 213–222.

Burns, C.P., and Wagner, B.A. (1991). Heightened susceptibility of fish oil polyunsaturate-enriched neoplastic cells to ethane generation during lipid peroxidation. *J. Lipid Res.* **32**: 79–87.

Butler, J., Jayson, G.G., and Swallow, A.J. (1975). The reaction between the superoxide anion radial and cytochrome c. *Biochim. Biophys. Acta* **408**: 215–222.

Bynoe, L.A., Gottsch, J.D., Pou, S., and Rosen, G.M. (1992). Light-dependent generation of superoxide from human erythrocytes. *Photochem. Photobiol.* **56**: 353–356.

Cadenas, E., Boveris, A., and Chance, B. (1984). Low-level chemiluminescence of biological systems, in: *Free Radicals in Biology* (Pryor, W.A., ed.), Vol. VI, pp. 211–242, Academic Press, New York.

Campbell, A.K., Hallett, M.B., and Weeks, I. (1985). Chemiluminescence as an analytical tool in cell biology and medicine. *Methods Biochem. Anal.* **31**: 317–416.

Carter, W.O., Narayan, P.K., and Robinson, J.P. (1994). Intracellular hydrogen peroxide and superoxide anion detection in endothelial cells. *J. Leukocyte Biol.* **55**: 253–258.

Cathcart, R., Schwiers, E., and Ames, B.N. (1983). Detection of picomole levels of hydroperoxides using a fluorescent dichlorofluorescein assay. *Anal. Biochem.* **134**: 111–117.

Chance, B., Sies, H., and Boveris, A. (1979). Hydroperoxide metabolism in mammalian organs. *Physiol. Rev.* **59**: 527–605.

Cheson, B.D., Christensen, R.L., Sperling, R., Kohler, B.E., and Babior, B.M. (1976). The origin of the chemiluminescence of phagocytosing granulocytes. *J. Clin. Invest.* **58**: 789–796.

Chirico, S. (1994). High-performance liquid chromatography-based thiobarbituric acid test. *Methods Enzymol.* **233**: 314–318.

Christodoulou, D., Kudo, S., Cook, J.A., Krishna, M.C., Miles, A., Grisham, M.B., Murugesan, R., Ford, P.C., and Wink, D.A. (1996). Electrochemical methods for detection of nitric oxide. *Methods Enzymol.* **268**: 69–83.

Coffman, T.J., Cox, C.D., Edeker, B.L., and Britigan, B.E. (1990). Possible role of bacterial siderophores in inflammation. Iron bound to the *Pseudomonas* siderophore pyochelin can function as a hydroxyl radical catalyst. *J. Clin. Invest.* **86**: 1030–1037.

Cohen, G., and Cederbaum, A.I. (1979). Chemical evidence for production of hydroxyl radicals during microsomal electron transfer. *Science* **204**: 66–68.

Cohen, G., and Cederbaum, A.I. (1980). Microsomal metabolism of hydroxyl radical scavenging agents: Relationship to the microsomal oxidation of alcohols. *Arch. Biochem. Biophys.* **199**: 438–447.

Cohen, G., and Heikkila, R.E. (1977). *In vivo* scavenging of superoxide radicals by catecholamines, in: *Superoxide and Superoxide Dismutases* (Michelson, A.M., McCord, J.M., and Fridovich, I., eds.), pp. 351–365, Academic Press, New York.

Cormier, M.J., and P.M. Prichard, P.M. (1968). An investigation of the mechanism of the luminescent peroxidation of luminol by stopped flow techniques. *J. Biol. Chem.* **243**: 4706–4714.

Craven, P.A., and DeRubertis, F.R. (1978). Restoration of the responsiveness of purified guanylate cyclase to nitrosoguanidine, nitric oxide, and related activators by heme and hemeproteins. Evidence for involvement of the paramagnetic nitrosyl-heme complex in enzyme activation. *J. Biol. Chem.* **253**: 8433–8443.

Craven, P.A., and DeRubertis, F.R. (1983). Requirement for heme in the activation of purified guanylate cyclase by nitric oxide. *Biochim. Biophys. Acta* **745**: 310–321.

Crow, J.P., and Ischiropoulos, H. (1996). Detection and quantitation of nitrotyrosine residues in proteins: *In vivo* marker of peroxynitrite. *Methods Enzymol.* **269**: 185–194.

Dahlgren, C., and Stendahl, O. (1983). Role of myeloperoxidase in luminol-dependent chemiluminescence of polymorphonuclear leukocytes. *Infect. Immun.* **39**: 736–741.

Dannenberg, A.M., Jr., Schofield, B.H., Rao, J.B., Dinh, T.T., Lee, K., Boulay, M., Abe, Y., Tsuruta, J., and Steinbeck, M.J. (1994). Histochemical demonstration of hydrogen peroxide production by leukocytes in fixed-frozen tissue sections of inflammatory lesions. *J. Leukocyte Biol.* **56**: 436–443.

Das, M.R., Connor, H.D., Leniart, D.S., and Freed, J.H. (1970). An electron nuclear double resonance and electron spin resonance study of semiquinones related to vitamins K and E. *J. Am. Chem. Soc.* **92**: 2258-2268.

DeChatelet, L.R., Long, G.D., Shirley, P.S., Bass, D.A., Thomas, M.J., Henderson, F.W., and Cohen, M.S. (1982). Mechanism of the luminol-dependent chemiluminescence of human neutrophils. *J. Immunol.* **129**: 1589–1593.

Del Rio, L.A., Sandalio, L.S., Palma, J.M., Bueno, P., and Corpas, F.J. (1992). Metabolism of oxygen radicals in peroxisomes and cellular implications. *Free Radical Biol. Med.* **13**: 557–580.

Dennis, K.L., and Shibamoto, T. (1989). Gas chromatographic determination of malonaldehyde formed by lipid peroxidation. *Free Radical Biol. Med.* **7**: 187–192.

Dillard, C.J., and Tappel, A.L. (1989). Lipid peroxidation products in biologic tissues. *Free Radical Biol. Med.* **7**: 193–196.

Doyle, M.P., and Hoekstra, J.W. (1981). Oxidation of nitrogen oxides by bound dioxygen in hemoproteins. *J. Inorg. Biochem.* **14**: 351–358.

Draper, H.H., and Hadley, M. (1990). Malondialdehyde determination as index of lipid peroxidation. *Methods Enzymol.* **186**: 421–431.

Edwards, S.W. (1987). Luminol- and lucigenin-dependent chemiluminescence of neutrophils: Role of degranulation. *J. Clin. Lab. Immunol.* **22**: 35–39.

Ehrenberg, A., Müller, F., and Hemmerich, P. (1967). Basicity, visible spectra, and electron spin resonance of flavosemiquinone anions. *Eur. J. Biochem.* **2**: 286–293.

Eiserich, J.P., Hristova, M., Cross, C.E., Jones, A.D., Freeman, B.A., Halliwell, B.A., and van der Vliet, A. (1998). Formation of nitric oxide-derived inflammatory oxidants by myeloperoxidase in neutrophils. *Nature* **391**: 393–397.

Ellington, S.P., Strauss, K.E., and Rosen, G.M. (1988). Spin trapping of free radicals in whole cells, isolated organs and *in vivo*, in: *Cellular Antioxidant Defense Mechanisms* (Chow, C.K., ed.), Vol., I, pp. 41–57, CRC Press, Boca Raton, FL.

Esterbauer, H., and Cheeseman, K.H. (1990). Determination of aldehydic lipid peroxidation products: Malonaldehyde and 4-hydroxynonenal. *Methods Enzymol.* **186**: 407–421.

Esterbauer, H., and Zollner, H. (1989). Methods for determination of aldehydic lipid peroxidation products. *Free Radical Biol. Med.* **7**: 179–203.

Esterbauer, H., Lang, J., Zadravec, S., and Slater, T.F. (1984). Detection of malonaldehyde by high-performance liquid chromatography. *Methods Enzymol.* **105**: 319–328.

Filatov, M.V., Varfolomeeva, E.Y., and Ivanov, E.I. (1995). Flow cytofluorometric detection of inflammatory processes by measuring respiratory burst reaction of peripheral blood neutrophils. *Biochem. Mol. Med.* **55**: 116–121.

Finkelstein, E., Rosen, G.M., Patton, S.E., Cohen, M.S., and Rauckman, E.J. (1981). Effect of modification of cytochrome c on its reactions with superoxide and NADPH: cytochrome P-450 reductase. *Biochem. Biophys. Res. Commun.* **102**: 1008–1015.

Finkelstein, E., Rosen, G.M., and Rauckman, E.J. (1984). Superoxide-dependent reduction of nitroxides by thiols. *Biochim. Biophys. Acta* **802**: 90–98.

Floyd, R.A., Watson, J.J., and Wong, P.K. (1984). Sensitive assay of hydroxyl free radical formation utilizing high pressure liquid chromatography with electrochemical detection of phenol and salicylate hydroxylation products. *J. Biochem. Biophys. Methods* **10**: 221–235.

Floyd, R.A., Watson, J.J., Wong, P.K., Altimiller, D.H., and Richard, R.C. (1986). Hydroxyl free radical adducts of deoxyguanosine: Sensitive detection and mechanisms of formation. *Free Radical Res. Commun.* **1**: 163–172.

Forman, H.J., and Fridovich, I. (1973). Superoxide dismutase: A comparison of rate constants. *Arch. Biochem. Biophys.* **158**: 396–400.

Fridovich, I. (1970). Quantitative apects of the production of superoxide anion radical by milk xanthine oxidase. *J. Biol. Chem.* **245**: 4053–4057.

Furchgott, R.F., and Zawadzki, J.V. (1980). The obligatory role of endothelial cells on the relaxation of arterial smooth muscle by acetylcholine. *Nature* **288**: 373–376.

Gallez, B., Bacic, G., Goda, F., Jiang, J., O'Hara, J.A., Dunn, J.F., and Swartz, H.M. (1996). Use of nitroxides for assessing perfusion, oxygenation, and viability of tissues: *In vivo* EPR and MRI studies. *Magn. Reson. Med.* **35**: 97–106.

Gee, J.B.L., Vassallo, C.L., Bell, P., Kaskin, J., Basford, R.E., and Field, J.B. (1970). Catalase-dependent peroxidative metabolism in the alveolar macrophage during phagocytosis. *J. Clin. Invest.* **49**: 1280–1287.

Gotoh, N., and Niki, E. (1994). Measurement of superoxide reaction by chemiluminescence. *Methods Enzymol.* **233**: 154–160.

Grebner, J.V., Mills, E.L., Gray, B.H., and Quie, P.G. (1977). Comparison of phagocytic and chemiluminescence response of human polymorphonuclear neutrophils. *J. Lab. Clin. Med.* **89**: 153–159.

Green, L.C., Wagner, D.A., Glogowski, J., Skipper, P.L., Wishnok, J.S., and Tannenbaum, S.R. (1982). Analysis of nitrate, nitrite and [^{15}N]nitrate in biologic fluids. *Anal. Biochem.* **126**: 131–138.

Green, T.R., Fellman, J.H., and Eicher, A.L. (1985). Myeloperoxidase oxidation of sulfur-centered and benzoic acid hydroxyl radical scavengers. *FEBS Lett.* **192**: 33–36.

Greenlee, L., Fridovich, I., and Handler, P. (1962). Chemiluminescence induced by operation of iron-flavoproteins. *Biochemistry* **1**: 779–783.

Greenwald, R.A., Rush, S.W., Mark, S.A., and Weitz, Z. (1989). Conversion of superoxide generated by polymorphonuclear leukocytes to hydroxyl radical: A direct spectrophotometric detection system based on degradation of deoxyribose. *Free Radical Biol. Med.* **6**: 385–392.

Guilbault, G.C., Brignac, P.J., and Juneau, M. (1968). New substrates for the fluorometric determination of oxidative enzymes. *Anal. Chem.* **40**: 1256–1263.

Gustafsson, L.E., Leone, A.M., Persson, M.G., Wiklund, N.P., and Moncada, S. (1991). Endogenous nitric oxide is present in the exhaled air of rabbits, guinea pigs and humans. *Biochem. Biophys. Res. Commun.* **181**: 852–857.

Gutierrez, H.H., Nieves, B., Chumley, P., Rivera, A., and Freeman, B.A. (1996). Nitric oxide regulation of superoxide-dependent lung injury: Oxidant-protective actions of endogenously produced and exogenously administered nitric oxide. *Free Radical Biol. Med.* **21**: 43–52.

Gutteridge, J.M.C. (1984). Reactivity of hydroxyl and hydroxyl-like radicals discriminated by release of thiobarbituric acid-reactive material from deoxy sugars, nucleosides and benzoate. *Biochem. J.* **224**: 761–767.

Gyllenhammar, H. (1987). Lucigenin chemiluminescence in the assessment of neutrophil superoxide production. *J. Immunol. Methods* **97**: 209–213.

Halbhuber, K.-J., Gossrau, R., Möller, U., Hulstaert, C.E., Zimmermann, N., and Feuerstein, H. (1988). The cerium perhydroxide–diaminobenzidine (C3–H_2O_2–DAB) procedure: New methods for light microscopic phosphatase histochemistry and immunohistochemistry. *Histochemistry* **90**: 289–297.

Halliwell, B. (1976). An attempt to demonstrate a reaction between superoxide and hydrogen peroxide. *FEBS Lett.* **72**: 8–10.

Halliwell, B. (1978). Superoxide-dependent formation of hydroxyl radicals in the presence of iron chelates. Is it a mechanism of hydroxyl radical formation in biochemical systems? *FEBS Lett.* **92**: 321–326.

Halliwell, B., and Gutteridge, J.M.C. (1981). Formation of a thiobarbituric acid-reactive substance from deoxyribose in the presence of iron salts. The role of superoxide and hydroxyl radicals. *FEBS Lett.* **128**: 347–351.

Halliwell, B., Kaur, H., and Ingelman-Sundberg, M. (1991). Hydroxylation of salicylate as an assay for hydroxyl radicals: A cautionary note. *Free Radical Biol. Med.* **10**: 439–441.

Halpern, H.J., Yu, C., Peric, M., Bath, E., Bowman, M.K., Grdina, D.J., and Teicher, B.A. (1994). Oxymetry deep in tissues with low-frequency electron parmagnetic resonance. *Proc. Natl. Acad. Sci. USA* **91**: 13047–13051.

Halpern, H.J., Yu, C., Peric, M., Bath, E.D., Karczmar, G.S., River, J.N., Grdina, D.J., and Teicher, B.A. (1996). Measurement of differences in pO_2 in response to perfluorocarbon/carbogen in FSa and NFSa murine fibrosarcomas with low-frequency EPR oximetry. *Radiat Res.* **145**: 610–618.

Harvath, L., Amirault, H.J., and Andersen, B.R. (1978). Chemiluminescence of human and canine polymorphonuclear leukocytes in the absence of phagocytosis. *J. Clin. Invest.* **61**: 1145–1154.

Hatch, G.E., Gardner, D.E., and Menzel, D.B. (1978). Chemiluminescence of phagocytic cells caused by N-formylmethionyl peptides. *J. Exp. Med.* **147**: 182–195.

Hecker, M., Mitchell, J.A., Harris, H.J., Katsura, M., Thiemermann, C., and Vane, J.R. (1990). Endothelial cells metabolize N^G-monomethyl-L-arginine to L-citrulline and subsequently to L-arginine. *Biochem. Biophys. Res. Commun.* **167**: 1037–1043.

Heinecke, J.W., Li, W., Daehnke, H.L., III, and Goldstein, J.A. (1993). Dityrosine, a specific marker of oxidation, is synthesized by the myeloperoxidase–hydrogen peroxide system of human neutrophils and macrophages. *J. Biol. Chem.* **268**: 4069–4077.

Heinzel, B., John, M., Klatt, P., Böhme E., and Mayer, B. (1992). Ca^{2+}/calmodulin-dependent formation of hydrogen peroxide by brain nitric oxide synthase. *Biochem. J.* **281**: 627–630.

Heiser, I., Muhr, A., and Elstner, E.F. (1998). The production of OH-radical-type oxidant by lucigenin. *Z. Naturforsch. [C]* **53**: 9–14.

Henry, J.P., and Michelson, A.M. (1977). Superoxide and chemiluminescence, in: *Superoxide and Superoxide Dismutases* (Michelson, A.M., McCord, J.M., and Fridovich, I., eds.), pp. 284–290, Academic Press, New York.

Henry, T.D., Archer, S.L., Nelson, D., Weir, E.K., and From, A.H.L. (1991). Enhanced chemiluminescence as a measure of oxygen-derived free radical generation during ischemia and reperfusion. *Circ. Res.* **67**: 1453–1461.

Herbert, K.E., Evans, M.D., Finnegan, M.T.V., Farooq, S., Mistry, N., Podmore, I.D., Farmer, P., and Lunec, J. (1996). A novel HPLC procedure for the analysis of 8-oxoguanine in DNA. *Free Radical Biol. Med.* **20**: 467–473.

Hevel, J.M. and Marletta, M.A. (1994). Nitric-oxide synthase assays. *Methods Enzymol.* **233**: 250–258.

Hiramitsu, T., Arimoto, T., Ito, T., and Nakano, M. (1994). A new method for detecting lipid peroxidation by using dye sensitized chemiluminescence. *Adv. Exp. Med. Biol.* **366**: 401–402.

Hoffstein, S.T., Gennaro, D.E., and Meunier, P.C. (1988). Cytochemical demonstration of constitutive H_2O_2 production by macrophages in synovial tissue from rats with adjuvant arthritis. *Am. J. Pathol.* **130**: 120–125.

Hoidal, J.R., Beall, G.D., and Repine, J.E. (1979). Production of hydroxyl radical by human alveolar macrophages. *Infect. Immun.* **26**: 1088–1092.

Huie, R.E., and Padmaja, S. (1993). The reaction of NO with superoxide. *Free Radical Res. Commun.* **18**: 195–199.

Ignarro, L.J., Adams, J.B., Horwitz, P.M., and Wood, K.S. (1986). Activation of soluble guanylate cyclase by NO-hemoproteins involves NO-heme exchange. Comparison of heme-containing and hemedeficient enzyme forms. *J. Biol. Chem.* **261**: 4997–5002.

Ignarro, L.J., Buga, G.M., Wood, K.S., Byrns, R.E., and Chaudhuri, G. (1987). Endothelium-derived relaxing factor produced and released from artery and vein is nitric oxide. *Proc. Natl. Acad. Sci. USA* **84**: 9265–9269.

Ignarro, L.J., Burke, T.M., Wood, K.S., Wolin, M.S., and Kadowitz, P.J. (1984). Association between cyclic GMP accumulation and acetylcholine-elicited relaxation of bovine intrapulmonary artery. *J. Pharmacol. Exp. Ther.* **228**: 682–690.

Ischiropoulos, H., Zhu, L., Chen., J., Tsai, M., Martin, J.C., Smith, C.D., and Beckman, J.S. (1992). Peroxynitrite-mediated tyrosine nitration catalyzed by superoxide dismutase. *Arch. Biochem. Biophys.* **298**: 431–437.

Janco, R.L., and English, D. (1983). Regulation of monocyte oxidative metabolism: Chemotactic factor enhancement of superoxide release, hydroxyl radical generation, and chemiluminescence. *J. Lab. Clin. Med.* **102**: 890–898.

Janero, D.R. (1990). Malondialdehyde and thiobarbituric acid-reactivity as diagnostic indices of lipid peroxidation and peroxidative tissue injury. *Free Radical Biol. Med.* **9**: 515–540.

Janzen, E.G., Towner, R.A., and Yamashiro, S. (1990). The effect of phenyl *tert*-butyl nitrone (PBN) on CCl$_4$-induced rat liver injury detected by proton magnetic resonance imaging (MRI) *in vivo* and electron microscopy (EM). *Free Radical Res. Commun.* **9**: 325–335.

Jeejeebhoy, K.N. (1991). *In vivo* breath alkane as an index of lipid peroxidation. *Free Radical Biol. Med.* **10**: 191–193.

Jentzsch, A.M., Bachmann, H., Fürst, P., and Biesalski, H.K. (1996). Improved analysis of malondialdehyde in human body fluids. *Free Radical Biol. Med.* **20**: 251–256.

Karam, L.R., Bergtold, D.S., and Simic, M.G. (1991). Biomarkers of OH radicals *in vivo*. *Free Radical Res. Commun.* **12–13**: 11–16.

Karnovsky, M.L., and Badwey, J.A. (1983). Determinants of the production of active oxygen species by granulocytes and macrophages. *J. Clin. Chem. Clin. Biochem.* **21**: 545–553.

Kaur, H., and Halliwell, B. (1994a). Aromatic hydroxylation of phenylalanine as an assay for hydroxyl radicals. Measurement of hydroxyl radical formation from ozone and in blood from premature babies using improved HPLC methodology. *Anal. Biochem.* **220**: 11–15.

Kaur, H., and Halliwell, B. (1994b). Detection of hydroxyl radicals by aromatic hydroxylation. *Methods Enzymol.* **233**: 67–82.

Kaur, H., Fagerheim, Z., Grootveld, M., Puppo, A., and Halliwell, B. (1988). Aromatic hydroxylation of phenylalanine as an assay for hydroxyl radicals: Application to activated neutrophils and heme protein leghemoglobin. *Anal. Biochem.* **172**: 360–367.

Keana, J.F.W., Pou, S., and Rosen, G.M. (1987). Nitroxides as potential contrast enhancing agents for MRI application: Influence of structure on the rate of reduction by rat hepatocytes, whole liver homogenate, subcellular fractions, and ascorbate. *Magn. Reson. Med.* **5**: 525–536.

Kelm, M., Feelisch, M., Spahr, R., Piper, H.-M., Noack, E., and Schrader, J. (1988). Quantitative and kinetic characterization of nitric oxide and EDRF release from cultured endothelial cells. *Biochem. Biophys. Res. Commun.* **154**: 237–244.

Keston, A.S., and Brandt, R. (1965). The fluoromeric analysis of ultramicro quantities of hydrogen peroxide. *Anal. Biochem.* **11**: 1–5.

Kettle, A.J., Carr, A.C., and Winterbourn, C.C. (1994). Assays using horseradish peroxidase and phenolic substrates require superoxide dismutase for accurate determination of hydrogen peroxide production by neutrophils. *Free Radical Biol. Med.* **17**: 161–164.

Kikuchi, K., Nagano, T., Hayakawa, H., Hirata, Y., and Hirobe, M. (1993). Real time measurement of nitric oxide produced *ex vivo* by luminol-H_2O_2 chemiluminescence method. *J. Biol. Chem.* **268**: 23106–23110.

Kim, P.M., and Wells, P.G. (1996). Phenytoin-initiated hydroxyl radical formation: Characterization by enhanced salicylate hydroxylation. *Mol. Pharmacol.* **49**: 172–181.

Kinnula, V.L., Chang, L., Everitt, J.I., and Crapo, J.D. (1992). Oxidants and antioxidants in alveolar epithelial type II cells: *In situ*, freshly isolated, and cultured cells. *Am. J. Physiol.* **262**: L69–L77.

Kinnula, V.L., Everitt, J.I., Whorton, A.R., and Crapo, J.D. (1991). Hydrogen peroxide production by alveolar type II cells, alveolar macrophages, and endothelial cells. *Am. J. Physiol.* **261**: L84–L91.

Klatt, P., Henizel, B., John, M., Kastner, M., Böhme, E., and Mayer, B. (1992). Ca^{2+}/calmodulin-dependent cytochrome c reductase activity of brain nitric oxide synthase. *J. Biol. Chem.* **267**: 11374–11378.

Klebanoff, S.J., and Rosen, H. (1978). Ethylene formation by polymorphonuclear leukocytes. Role of myeloperoxidase. *J. Exp. Med.* **149**: 490–506.

Klein, S.M., Cohen, G., and Cederbaum, A.I. (1980). The interaction of hydroxyl radicals with dimethylsulfoxide produces formaldehyde. *FEBS Lett.* **116**: 220–222.

Klein, S.M., Cohen, G., and Cederbaum, A.I. (1981). Production of formaldehyde during the metabolism of dimethyl sulfoxide by hydroxyl radical generating systems. *Biochemistry* **20**: 6006–6012.

Knowles, P.F., Gibson, J.F., Pick, F.M., and Bray, R.C. (1969). Electron spin resonance evidence for enzymic reduction of oxygen to a free radical, the superoxide ion. *Biochem. J.* **111**: 53–58.

Koenig, S.H. (1986). Relaxometry of contrast agents, in: *Medical Magnetic Resonance Imaging and Spectroscopy. A Primer* (Budinger, T.F., and Margulis, A.R., eds.), pp. 99–108, The Society of Magnetic Resonance in Medicine, Berkeley, CA.

Kok, S.H., Buckle, K.A., and Wootton, M. (1983). Determination of nitrate and nitrite in water using high-performance liquid chromatography. *J. Chromatogr.* **260**: 189–192.

Kooy, N.W., Royall, J.A., Ischiropoulos, H., and Beckman, J.S. (1994). Peroxynitrite-mediated oxidation of dihydrorhodamine 123. *Free Radical Biol. Med.* **16**: 149–156.

Koppenol, W.H., Van Buuren, K.J.H., Butler, J., and Braams, R. (1976). The kinetics of the reduction of cytochrome c by the superoxide anion radical. *Biochim. Biophys. Acta* **449**: 157–168.

Korytowski, W., Bachowski, G.J., and Girotti, A.W. (1993). Analysis of cholesterol and phospholipid hydroperoxides by high-performance liquid chromatography with mercury drop electrochemical detection. *Anal. Biochem.* **213**: 111–119.

Kostka, P. (1995). Free radicals (nitric oxide). *Anal. Chem.* **67**: 411R–416R.

Kosugi, H., and Kikugawa, K. (1989). Potential thiobarbituric acid-reactive substances in peroxidized lipids. *Free Radical Biol. Med.* **7**: 205–207.

Kumar, C., Okuda, M., Ikai, I., and Chance, B. (1991). Organ chemiluminescence: Prospects and problems, in: *Oxidative Damage and Repair. Chemical, Biological and Medical Aspects* (Davis, K.J.A., ed.), pp. 434–437, Pergamon Press, Oxford, England.

Kuthan, H., Tsuji, H., Graf, H., Ullrich, V., Werringloer, J., and Estabrook R.W. (1978). Generation of superoxide anion as a source of hydrogen peroxide in a reconstituted monooxygenase system. *FEBS Lett.* **91**: 343–345.

Kuthan, H., Ullrich, V., and Estabrook, R. (1982). A quantitative test for superoxide radicals produced in biological systems. *Biochem. J.* **203**: 551–558.

Labato, M.A., and Briggs, R.T. (1985). Cytochemical localization of hydrogen peroxide generating sites in the rat thyroid gland. *Tissue Cell* **17**: 889–900.

Land, E.J., and Swallow, A.J. (1971). One-electron reactions in biochemical systems as studied by pulse radiolysis. *Arch. Biochem. Biophys.* **145**: 365–372.

Lauterbur, P.C. (1973). Image formation by induced local interactions: Examples employing nuclear magnetic resonance. *Nature* **242**: 190–191.

Li, Y., Zhu, H., Kuppusamy, P., Roubaud, V., Zweier, J.L., and Trush, M.A. (1998). Validation of lucigenin (bis-N-methylacridinium) as a chemilumigenic probe for detecting superoxide anion radical production by enzymatic and cellular systems. *J. Biol. Chem.* **273**: 2015–2023.

Liochev, S.I., and Fridovich, I. (1995). Superoxide from glucose oxidase or from nitroblue tetrazolium. *Arch. Biochem. Biophys.* **318**: 408–410.

Liochev, S.I., and Fridovich, I. (1997). Lucigenin (bis-N-methylacridinium) as a mediator of superoxide anion production. *Arch. Biochem. Biophys.* **337**: 115–120.

Liochev, S.I., and Fridovich, I. (1998). Lucigenin as mediator of superoxide production: Revisited. *Free Radical Biol. Med.* **25**: 926–928.

Malinski, T., and Taha, Z. (1992). Nitric oxide release from a single cell measured *in situ* by a porphyrinic-based microsensor. *Nature* **358**: 676–678.

Malinski, T., Bailey, F., Zhang, Z.G., and Chopp, M. (1993). Nitric oxide measured by a porphyrinic microsensor in rat brain after transient middle cerebral artery occlusion. *J. Cereb. Blood Flow Metab.* **13**: 355–358.

Marchiesi, E., Rota, C., Fann, Y.C., Chignell, C.F., and Mason, R.P. (1999). Photoreduction of the fluorescent dye 2'-7'-dichlorofluorescein: A spin trapping and direct electron spin resonance study with implications for oxidative stress measurements. *Free Radical Biol. Med.* **26**: 148–161.

Margoliash, E., Novogrodsky, A., and Scheijter, A. (1960). Irreversible reaction of 3-amino-1,2,4-triazole and related inhibitors with the protein catalase. *Biochem. J.* **74**: 339–348.

Marletta, M. (1989). Nitric oxide: Biosynthesis and biological significance. *Trends Biochem. Sci.* **14**: 488–492.

Marletta, M.A., Yoon, P.S., Iyengar, R., Leaf, C.D., and Wishnok, J.S. (1988). Macrophage oxidation of L-arginine to nitrite and nitrate: Nitric oxide is an intermediate. *Biochemistry* **27**: 8706–8711.

Maskos, Z., Rush, J.D., and Koppenol, W.H. (1990). The hydroxylation of the salicylate anion by a Fenton reaction and Γ-radiolysis: A consideration of the respective mechanisms. *Free Radical Biol. Med.* **8**: 153–162.

Mason, R.P. (1982). Free-radical intermediates in the metabolism of toxic chemicals, in: *Free Radicals in Biology* (Pryor, W.A., ed.), Vol. V, pp. 161–222, Academic Press, New York.

Massey, V. (1959). The microestimation of succinate and the extinction coefficient of cytochrome c. *Biochim. Biophys. Acta* **34**: 255–256.

Masters, C., and Holmes, R. (1977). Peroxisomes: New aspects of cell physiology and biochemistry. *Physiol. Rev.* **57**: 816–882.

Mayer, B., John, M., and Böhme, E. (1990). Purification of a Ca^{2+}/calmodulin-dependent nitric oxide synthase from porcine cerebellum. *FEBS Lett.* **277**: 215–219.

Mayer, B., John, M., Henizel, B., Werner, E.R., Wachter, H., Schultz, G., and Böhme, E. (1991). Brain nitric oxide synthase is a biopterin- and flavin-containing multi-functional oxido-reductase. *FEBS Lett.* **288**: 187–191.

McCord, J.M., and Fridovich, I. (1969). Superoxide dismutase. An enzymic function for erythrocuprein (Hemocuprein). *J. Biol. Chem.* **244**: 6049–6055.

McCord, J.M., Crapo, J.D., and Fridovich, I. (1977). Superoxide dismutase assays:

A review of methodology, in: *Superoxide and Superoxide Dismutases* (Michelson, A.M., McCord, J.M., and Fridovich, I., eds.), pp. 11–17, Academic Press, New York.

Menon, N., Patricza, J., Binder, T., and Bing, R. (1991). Reduction of biological effluents in purge and trap micro reaction vessels and detection of endothelium-derived nitric oxide (Endo) by chemiluminescence. *J. Mol. Cell. Cardiol.* **23**: 389–393.

Michelson, A.M. (1977). Production of superoxide by metal ions, in: *Superoxide and Superoxide Dismutases* (Michelson, A.M., McCord, J.M., and Fridovich, I., eds.), pp. 78–105, Academic Press, New York.

Michelson, A.M., and Isambert, M.F. (1973). Studies in bioluminescence. XI. Further studies on the *Pholas dactylus* system. Mechanism of luciferase. *Biochemie* **55**: 619–634.

Miller, R.W. (1970). Reactions of superoxide anion, catechols and cytochrome c. *Can. J. Biochem.* **48**: 935–939.

Miller, R.W., and Rapp, U. (1973). The oxidation of catechols by reduced flavins and dehydrogenases. An electron spin resonance study of the kinetics and initial products of oxidation. *J. Biol. Chem.* **248**: 6084–6090.

Misko, T.P., Schilling, R.J., Salvemini, D., Moore, W.M., and Currie, M.G. (1993). A fluorometric assay for the measurement of nitrite in biological samples. *Anal. Biochem.* **214**: 11–16.

Misra, H.P., and Fridovich, I. (1972a). The univalent reduction of oxygen by reduced flavins and quinones. *J. Biol. Chem.* **247**: 188–192.

Misra, H.P., and Fridovich, I. (1972b). The role of superoxide anion in the autoxidation of epinephrine and a simple assay for superoxide dismutase. *J. Biol. Chem.* **247**: 3170–3175.

Miyazawa, T., Fujimoto, K., Suzuki, T., and Yasuda, K. (1994). Determination of phospholipid hydroperoxides using luminol chemiluminescence–high-performance liquid chromatography. *Methods Enzymol.* **233**: 324–332.

Moncada, A., Palmer, R.M.J., and Higgs, E.A. (1991). Nitric oxide: Physiology, pathophysiology, and pharmacology. *Pharmacol. Rev.* **43**: 109–142.

Montgomery, J., Ste-Marie, L., Boismenu, D., and Vachon, L. (1995). Hydroxylation of aromatic compounds as indices of hydroxyl radical production: A cautionary note revisited. *Free Radical Biol. Med.* **19**: 927–933.

Müller, A., and Sies, H. (1984). Assay of ethane and pentane from isolated organs and cells. *Methods Enzymol.* **105**: 311–319.

Murphy, M.E., and Noack, E. (1994). Nitric oxide assay using hemoglobin method. *Methods Enzymol.* **233**: 240–250.

Murphy, M.E., and Sies, H. (1990). Visible-range low-level chemiluminescence in biological systems. *Methods Enzymol.* **186**: 595–610.

Nakano, M., Kikuyama, M., Hasegawa, T., Ito, T., Sakurai, K., Hiraishi, K., Hashimura, E., and Adachi, M. (1995). The first observation of $O_2 \cdot^-$ generation at real time *in vivo* from non-Kupffer sinusoidal cells in perfused rat liver during acute ethanol intoxication. *FEBS Lett.* **372**: 140–143.

Nakano, M., Sugioka, K., Ushijima, Y., and Goto, T. (1986). Chemiluminescence probe with *Cypridina* luciferin analog, 2-methyl-6-phenyl-3,7-dihydroimidazo[1,2-a]pyrazin-3-one, for estimating the ability of human granulocytes to generate $O_2 \cdot^-$. *Anal. Biochem.* **159**: 363–369.

Nakazawa, H., Ichimori, K., Shinozaki, Y., Okino, H., and Hori, S. (1988). Is superoxide demonstration by electron-spin resonance spectroscopy really superoxide? *Am. J. Physiol.* **255**: H213–H215.

Nelson, R.D., Herron, M.J., Schmidtke, J.R., and Simmons, R.L. (1977). Chemiluminescence response of human leukocytes: Influence of medium components on light production. *Infect. Immun.* **17**: 513–520.

Nilsson, R., Pick, F.M., Bray, R.D., and Fielden, M. (1969). ESR evidence for $O_2 \cdot^-$ as a long-lived transient irradiated oxygenated alkaline aqueous solution. *Acta Chem. Scand.* **23**: 2554–2556.

Nishida, A., Kimura, H., Nakano, M., and Goto, T. (1989). A sensitive and specific chemiluminescence method for estimating the ability of human granulocytes and monocytes to generate $O_2 \cdot^-$. *Clin. Chim. Acta* **179**: 177–182.

Pai, T.G., Payne, W.J., and LeGall, J. (1987). Use of a chemiluminescence detector for quantitation of nitric oxide produced in assays of denitrifying enzymes. *Anal. Biochem.* **166**: 150–157.

Palmer, R.M.J., Ferrige, A.G., and Moncada, S. (1987). Nitric oxide release accounts for the biological activity of endothelium-derived relaxing factor. *Nature* **327**: 524–526.

Panus, P.C., Radi, R., Chumley, P.H., Lillard, R.H., and Freeman, B.A. (1993). Detection of H_2O_2 release from vascular endothelial cells. *Free Radical Biol. Med.* **14**: 217–223.

Perschke, N., and Broda, E. (1961). Determination of very small amounts of hydrogen peroxide. *Nature* **190**: 257–258.

Pick, E. (1986). Microassays for superoxide and hydrogen peroxide production and nitroblue tetrazolium reduction using an enzyme immunoassay microplate reader. *Methods Enzymol.* **132**: 407–421.

Pick, E., and Keisari, Y. (1980). A simple colorimetric method for the measurement of hydrogen peroxide produced by cells in culture. *J. Immunol. Methods* **38**: 161–170.

Pick, E., and Mizell, D. (1981). Rapid microassays for the measurement of superoxide and hydrogen peroxide production by macrophages in culture using an automatic enzyme immunoassay reader. *J. Immunol. Methods* **46**: 211–226.

Pitkanen, O.M., Hallman, M., and Andersson, S.M. (1989). Determination of ethane and pentane in free oxygen radical-induced lipid peroxidation. *Lipids* **24**: 157–159.

Pou, S., Anderson, D.E., Surichamorn, W., Keaton, L.L., and Tod, M.L. (1994). Biological studies of a nitroso compound that releases nitric oxide upon illumination. *Mol. Pharmacol.* **46**: 709–715.

Pou, S., Bhan, A., Bhadti, V.S., Wu, S.Y., Hosmane, R.S., and Rosen, G.M. (1995a). The use of fluorophore-containing spin traps as potential probes to localize free radicals in cells with fluorescent imaging methods. *FASEB J.* **9**: 1085–1090.

Pou, S., Davis, P.L., Wolf, G.L., and Rosen, G.M. (1995b). Use of nitroxides as NMR contrast enhancing agents for joints. *Free Radical Res.* **23**: 353–364.

Pou, S., Keaton, L., Suricharmorn, W., and Rosen, G.M. (1999). Mechanism of superoxide generation by neuronal nitric oxide synthase. *J. Biol. Chem.* **274**: 9573–9580.

Pou, S., Nguyen, S.Y., Gladwell, T., and Rosen, G.M. (1995c). Does peroxynitrite generate hydroxyl radical? *Biochim. Biophys. Acta* **1244**: 62–68.

Pou, S., Pou, W.S., Bredt, D.S., Synder, S.H., and Rosen, G.M. (1992). Generation of superoxide by purified brain nitric oxide synthase. *J. Biol. Chem.* **267**: 24173–24176.

Pou, S., Pou, W.S., Rosen, G.M., and El-Fakahany, E.E. (1991). N-Hydroxylamine is not an intermediate in the conversion of L-arginine to an activator of soluble guanylate cyclase in neuroblastoma N1E-115 cells. *Biochem. J.* **273**: 547–552.

Pou, W.S., Pou, S., Rosen, G.M., and El-Fakahany, E.E. (1990). EDRF release is a common pathway in the activation of guanylate cyclase by receptor agonists and calcium ionophores. *Eur. J. Pharmacol.* **182**: 393–394.

Pryor, W.A., and Tang, R.H. (1978). Ethylene formation from methional. *Biochem. Biophys. Res. Commun.* **81**: 498–503.

Pryor, W.A., Jin, X., and Squadrito, G.L. (1994). One- and two-electron oxidations of methionine by peroxynitrite. *Proc. Natl. Acad. Sci. USA* **91**: 11173–11177.

Rabani, J., Mulac, W.A., and Matheson, M.S. (1965). The pulse radiolysis of aqueous tetranitromethane. I. Rate constants and the extinction coefficient of e^-_{aq}. II. Oxygenated solutions. *J. Phys. Chem.* **69**: 53–70.

Radi, R., Cosgrove, T.P., Beckman, J.S., and Freeman, B.A. (1993). Peroxynitrite-induced luminol chemiluminescence. *Biochem. J.* **290**: 51–57.

Rajagopalan, K.V., and Handler, P. (1964). Hepatic aldehyde oxidase. II. Differential inhibition of electron transfer to various electron acceptors. *J. Biol. Chem.* **239**: 2022–2026.

Ramezanian, M.S., Padmaja, S., and Koppenol, W.H. (1996). Nitration and hydroxylation of phenolic compounds by peroxynitrite. *Methods Enzymol.* **269**: 195–210.

Rapoport, R. (1986). Cyclic guanosine monophosphate inhibition of contraction may be mediated through inhibition of phosphatidylinositol hydrolysis in rat aorta. *Circ. Res.* **58**: 407–410.

Repine, J.E., Eaton, J.W., Anders, M.W., Hoidal, J.R., and Fox, R.B. (1979). Generation of hydroxyl radical by enzymes, chemicals, and human phagocytes *in vitro*. Detection with the anti-inflammatory agent, dimethyl sulfoxide. *J. Clin. Invest.* **64**: 1642–1651.

Richmond, R., Halliwell, B., Chauhan, J., and Darbre, A. (1981). Superoxide-dependent formation of hydroxyl radicals: Detection of hydroxyl radicals by the hydroxylation of aromatic compounds. *Anal. Biochem.* **118**: 328–335.

Roberts, R.J., Rendak, I., and Bucher, J.R. (1983). Lipid peroxidation in the newborn rat: Influence of fasting and hyperoxia on ethane and pentane in expired air. *Dev. Pharmacol. Ther.* **6**: 170–178.

Root, R.K., and Metcalf, J.A. (1977). H_2O_2 release from human granulocytes during phagocytosis: Relationship to superoxide anion formation and cellular catabolism of H_2O_2: Studies with normal and cytochalasin B-treated cells. *J. Clin. Invest.* **60**: 1266–1279.

Root, R.K., Metcalf, J., Oshino, N., and Chance, B. (1975). H_2O_2 release from human granulocytes during phagocytosis. I. Documentation, quantitation, and some regulating factors. *J. Clin. Invest.* **55**: 945–955.

Rosen, G.M., Finkelstein, E., and Rauckman, E.J. (1982). A method for the detection of superoxide in biological systems. *Arch. Biochem. Biophys.* **215**: 367–378.

Rosen, G.M., Griffeth, L.K., Brown, M.A., and Drayer, B.P. (1987). Intrathecal administration of nitroxides as potential contrast agents for MR imaging. *Radiology* **163**: 239–243.

Rosen, H., and Klebanoff, S.J. (1976). Chemiluminescence and superoxide production by myeloperoxidase-deficient leukocytes. *J. Clin. Invest.* **58**: 50–60.

Rosen, H., and Klebanoff, S.J. (1978). Bactericidal activity of a superoxide generating system: A model for the polymorphonuclear leukocyte. *J. Exp. Med.* **149**: 27–39.

Rothe, G., and Valet, G. (1990). Flow cytometric analysis of respiratory burst activity in phagocytes with hydroethidine and 2′,7′-dichlorofluorescin. *J. Leukocyte Biol.* **47**: 440–448.

Ruch, W., Cooper, P.H., and Baggiolini, M. (1983). Assay of H_2O_2 production by macrophages and neutrophils with homovanillic acid and horseradish peroxidase. *J. Immunol. Methods* **63**: 347–357.

Sagone, A.L., Jr., and Husney, R.M. (1987). Oxidation of salicylates by stimulated granulocytes. Evidence that these drugs act as free radical scavengers in biological systems. *J. Immunol.* **138**: 2177–2183.

Sagone, A.L., Jr., Decker, M.A., Wells, R.M., and Democko, C. (1980a). A new method for the detection of hydroxyl radical production by phagocytic cells. *Biochim. Biophys. Acta* **628**: 90–97.

Sagone, A.L., Jr., Wells, R.M., and Democko, C. (1980b). Evidence that ·OH production by human PMNs is related to prostaglandin metabolism. *Inflammation* **4**: 65–71.

Sampson, J.B., Rosen, H., and Beckman, J.S. (1996). Peroxynitrite-dependent tyrosine nitration catalyzed by superoxide dismutase, myeloperoxidase and horseradish peroxidase. *Methods Enzymol.* **269**: 210–218.

Saran, M., Michel, C., and Bors, W. (1990). Reaction of nitric oxide with superoxide. Implications for the action of endothelial-derived relaxing factor (EDRF). *Free Radical Res. Commun.* **10**: 221–226.

Sattler, W., Mohr, D., and Stocker, R. (1994). Rapid isolation of lipoproteins and assessment of their peroxidation by high-performance liquid chromatography postcolumn chemiluminescence. *Methods Enzymol.* **233**: 469–489.

Scaduto, R.C., Jr. (1995). Oxidation of DMSO and methanesulfinic acid by the hydroxyl radical. *Free Radical Biol. Med.* **18**: 271–277.

Schmidt, H.H.H.W, Hofmann, H., Schindler, U., Shutenko, Z.S., Cunningham, D.D., and Feelisch, M. (1996). No ·NO from NO synthase. *Proc. Natl. Acad. Sci. USA* **93**: 14492–14497.

Schmidt, H.H.H.W., Pollock, J.S., Nakane, M., Gorsky, L.D., Förstermann, U., and Murad F. (1991). Purification of a soluble isoform of guanylyl cyclase-activating-factor synthase. *Proc. Natl. Acad. Sci. USA* **88**: 365–369.

Schonbaum, G.R., and Chance, B. (1976). Catalase, in: *The Enzymes* (Boyer, P.D., ed.), Vol. 13, 3rd edition, pp. 363–408, Academic Press, New York.

Schulte-Herbrüggen, T., and Sies, H. (1989). The peroxidase/oxidase activity of soybean lipoxygenase-II. Triplet carbonyls and red photoemission during polyunsaturated fatty acid and glutathione oxidation. *Photochem. Photobiol.* **49**: 705–710.

Seitz, W.R., and Neary, M.P. (1976). Recent advances in bioluminescence and chemiluminescence assay. *Methods Biochem. Anal.* **23**: 161–188.

Seki, H., Ilan, Y.A., Ilan, Y., and Stein, G. (1976). Reactions of the ferri-ferrocytochrome c system with superoxide/oxygen and CO_2^-/CO_2 studied by fast pulse radiolysis. *Biochim. Biophys. Acta* **440**: 573–586.

Sessa, W.C., Hecker, M., Mitchell, J.A., and Vane, J.R. (1990). The metabolism of L-arginine and its significance for the biosynthesis of endothelium-derived relaxing factor: L-Glutamine inhibits the generation of L-arginine by cultured endothelial cells. *Proc. Natl. Acad. Sci. USA* **87**: 8607–8611.

Sharov, V.S., Kazamanov, V.A., and Vladimirov, Y.A. (1989). Selective sensitization of chemiluminescence resulted from lipid and oxygen radical reactions. *Free Radical Biol. Med.* **7**: 237–242.

Shibuki, K. (1990). An electrochemical microprobe for detecting nitric oxide release in brain tissue. *Neurosci. Res.* **9**: 69–76.

Shigenaga, M.K., and Ames, B.N. (1991). Assays for 8-hydroxy-2′-deoxyguanosine: A biomarker of *in vivo* oxidative DNA damage. *Free Radical Biol. Med.* **10**: 211–216.

Shigenaga, M.K., Aboujaoude, E.N., Chen, A., and Ames, B.N. (1994). Assays of oxidative DNA damage biomarkers 8-oxo-2′-deoxyguanosine and 8-oxoguanine in nuclear DNA and biological fluids by high-performance liquid chromatography with electrochemical detection. *Methods Enzymol.* **234**: 16–33.

Shikano, K., Long, C.L., Ohlstein, E.H., and Berkowitz, B.A. (1988). Comparative pharmacology of endothelium-derived relaxing factor and nitric oxide. *J. Pharmacol. Exp. Ther.* **247**: 873–881.

Shimazaki, K., Kawano, N., and Yoo, Y.C. (1991). Comparison of bovine, sheep and goat milk lactoferrins in their electrophoretic behavior, conformation, immunochemical properties and lectin reactivity. *Comp. Biochem. Physiol.* **98B**: 417–422.

Simic, M.G. (1994). DNA markers of oxidative process *in vivo*: Relevance to carcinogenesis and anticarcinogenesis. *Cancer Res. Suppl.* **54**: 1918s–1923s.

Simic, M.G., Taub, I.A., Tocci, J., and Hurwitz, P.A. (1975). Free radical reduction of ferricytochrome c. *Biochem. Biophys. Res. Commun.* **62**: 161–167.

Smith, C.V. (1991). Correlation and apparent contradictions in assessment of oxidant stress status *in vivo*. *Free Radical Biol. Med.* **10**: 217–224.

Smith, J.A., Baker, M.S., and Weidemann, M.J. (1991). Continuous monitoring of antioxidant capacity in red blood cells by luminol-enhanced chemiluminescence during oxidative stress, in: *Oxidative Damage and Repair. Chemical, Biological and Medical Aspects* (Davis, K.J.A., ed.), pp. 582–586, Pergamon Press, Oxford, England.

Sneddon, J.M., and Vane, J.R. (1988). Endothelium-derived relaxing factor reduces platelet adhesion to bovine endothelial cells. *Proc. Natl. Acad. Sci. USA* **85**: 2800–2804.

Snell, J.C., Colton, C.A., Chernyshev, O.N., and Gilbert, D.L. (1996). Location-dependent artifact for NO measurement using multiwell plates. *Free Radical Biol. Med.* **20**: 361–363.

Speer, C.P., Ambruso, D.R., Grimsley, J., and Johnston, R.B., Jr. (1985). Oxidative metabolism in cord blood monocytes and monocyte-derived macrophages. *Infect. Immun.* **50**: 919–921.

Steiner, M.G., and Babbs, C.F. (1990). Quantitation of the hydroxyl radical by reaction with dimethyl sulfoxide. *Arch. Biochem. Biophys.* **278**: 478–481.

Sun, J.-Z., Kaur, H., Halliwell, B., Li, X.-Y., and Bolli, R. (1993). Use of aromatic hydroxylation of phenylalanine to measure production of hydroxyl radicals after myocardial ischemia *in vivo*: Direct evidence for a pathogenetic role of the hydroxyl radical in myocardial stunning. *Circ. Res.* **73**: 534–549.

Suzaki, E., Kawai, E., Kodama, Y., Suzaki, T., and Masujima, T. (1994). Quantitative analysis of superoxide anion generation in living cells by using chemiluminescence video microsocopy. *Biochim. Biophys. Acta* **1201**: 328–332.

Szabó, C., Day, B.J., and Salzman, A.L. (1996). Evaluation of the relative contribution of nitric oxide and peroxynitrite to the suppression of mitochondrial respiration in immunostimulated macrophages using manganese mesoporphyrin superoxide dismutase mimetic and peroxynitrite scavenger. *FEBS Lett.* **381**: 82–86.

Tadolini, B., and Cabrini, L. (1990). The influence of pH on ·OH scavenger inhibition of damage to deoxyribose by Fenton reaction. *Mol. Cell. Biochem.* **94**: 97–104.

Taha, Z., Kiechle, F., and Malinski, T. (1992). Oxidation of nitric oxide by oxygen in biological systems monitored by porphyrinic sensor. *Biochem. Biophys. Res. Commun.* **188**: 734–739.

Takahashi, A., Nakano, M., Mashiko, S., and Inaba, H. (1990). The first observation of $O_2.^-$ generation in *in situ* lungs of rats treated with drugs to induce experimental acute respiratory distress syndrome. *FEBS Lett.* **261**: 369–372.

Tauber, A.I., and Babior, B.M. (1977). Evidence for hydroxyl radical production by human neutrophils. *J. Clin. Invest.* **60**: 374–379.

Test, S.T., and Weiss, S.J. (1984). Quantitative and temporal characterization of the extracellular H_2O_2 pool generated by human neutrophils. *J. Biol. Chem.* **259**: 399–405.

Test, S.T., and Weiss, S.J. (1986). Assay of the extracellular hydrogen peroxide pool generated by phagocytes. *Methods Enzymol.* **132**: 401–406.

Thomas, S.M., Jessup, W., Gebicki, J.M., and Dean, R.T. (1989). A continuous-flow automated assay for iodometric estimation of hydroperoxides. *Anal. Biochem.* **176**: 353–359.

Thomson, L., Trujillo, M., Telleri, R., and Radi, R. (1995). Kinetics of cytochrome c^{2+} oxidation by peroxynitrite: Implications for superoxide measurements in nitric oxide-producing biological systems. *Arch. Biochem. Biophys.* **319**: 491–497.

Tracey, W.R., Linden, J., Peach, M.J., and Johns, R.A. (1990). Comparison of spectrophotometric and biologic assays for nitric oxide (NO) and endothelium-derived relaxing factor (EDRF): Nonspecificity of the diazotization reaction for NO and failure to detect EDRF. *J. Pharmacol. Exp. Ther.* **252**: 922–928.

Tsan, M.-F. (1977). Stimulation of the hexose monophosphate shunt independent of hydrogen peroxide and superoxide production in rabbit alveolar macrophages during phagocytosis. *Blood* **50**: 935–945.

Tschudi, M.R., Mesaros, S., Lüscher, T.F., and Malinski, T. (1996). Direct *in situ* measurement of nitric oxide in mesenteric resistance arteries. Increased decomposition by superoxide in hypertension. *Hypertension* **27**: 32–35.

Ubezio, P., and Civoli, F. (1994). Flow cytometric detection of hydrogen peroxide production induced by doxorubicin in cancer cells. *Free Radical Biol. Med.* **16**: 509–516.

Udassin, R., Ariel, I., Haskel, Y., Kitrossky, N., and Chevion, M. (1991). Salicylate as an *in vivo* free radical trap: Studies on ischemic insult to the rat intestine. *Free Radical Biol. Med.* **10**: 1–6.

van der Kraaij, A.M.M., Koster, J.F., and Hagen, W.R. (1989). Reappraisal of the e.p.r. signals in (post)-ischaemic cardiac tissue. *Biochem. J.* **264**: 687–694.

van der Vliet, A., Eiserich, J.P., Halliwell, B., and Cross, C.E. (1997). Formation of reactive nitrogen species during peroxidase-catalyzed oxidation of nitrite.

A potential additional mechanism of nitric oxide-dependent toxicity. *J. Biol. Chem.* **272**: 7617–7625.

van der Vliet, A., Eiserich, J.P., Kaur, H., Cross, C.E., and Halliwell, B. (1996). Nitrotyrosine as biomarker for reactive nitrogen species. *Methods Enzymol.* **269**: 175–184.

van der Vliet, A., O'Neill, C.H., Halliwell, B., Cross, C.E., and Kaur, H. (1994). Aromatic hydroxylation and nitration of phenylalanine and tyrosine by peroxynitrite. *FEBS Lett.* **339**: 89–92.

Van Gelder, B.F., and Slater, E.C. (1962). The extinction coefficient of cytochrome c. *Biochim. Biophys. Acta* **58**: 593–595.

Van Kuijk, F.J.G.M., Thomas, D.W., Stephens, R.J., and Dratz, E.A. (1990a). Gas chromatrography–mass spectrometry assays for lipid peroxides. *Methods Enzymol.* **186**: 388–398.

Van Kuijk, F.J.G.M., Thomas, D.W., Stephens, R.J., and Dratz, E.A. (1990b). Gas chromatography–mass spectrometry of 4-hydroxynonenal in tissues. *Methods Enzymol.* **186**: 399–406.

Vásquez-Vivar, J., Hogg, N., Pritchard, K.A., Jr., Martásek, P., Kalanaraman, B. (1997). Superoxide anion formation from lucigenin: An electron spin resonance spin-trapping study. *FEBS Lett.* **403**: 127–130.

Vásquez-Vivar, J., Kalyanaraman, B., Martásek, P., Hogg, N., Masters, B.S.S., Karoui, H., Tordo, P., and Pritchard, K.A., Jr. (1998). Superoxide generation by endothelial nitric oxide synthase: The influence of cofactors. *Proc. Natl. Acad. Sci. USA* **95**: 9220–9225.

Vilim, V., and Wilhelm, J. (1989). What do we measure by a luminol-dependent chemiluminescence of phagocytes? *Free Radical Biol. Med.* **6**: 623–629.

von Sonntag, C., and Schuchmann, H.-P. (1994). Suppression of hydroxyl radical reactions in biological systems: Considerations based on competition kinetics. *Methods Enzymol.* **233**: 47–56.

Vorhaben, J.E., and Steele, R.H. (1967). Studies on the generation of electronic excitation states in a riboflavin–hydrogen peroxide–copper ascorbate redox system leading to chemiluminescence and/or aromatic hydroxylation. *Biochemistry* **6**: 1404–1412.

Warren, L. (1959). Thiobarbituric acid assay of sialic acids. *J. Biol. Chem.* **234**: 1971–1975.

Webb, L.S., Keeler, B.B., Jr., and Johnston, R.B., Jr. (1974). Inhibition of phagocytosis-associated chemiluminescence by superoxide dismutase. *Infect. Immun.* **9**: 1051–1056.

Weber, G.F. (1990). The measurement of oxygen-derived free radicals and related substances in medicine. *J. Clin. Chem. Clin. Biochem.* **28**: 569–603.

Weimann, A., Hildebrandt, A.G., and Kahl, R. (1984). Different efficiency of various synthetic antioxidants towards NADPH induced chemiluminescence in rat liver microsomes. *Biochem. Biophys. Res. Commun.* **125**: 1033–1038.

Weinmann, H.-J., Brasch, R.C., Press, W.-R., and Wesbey, G.E. (1984). Characteristics of gadolinium-DTPA complex: A potential NMR contrast agent. *AJR* **142**: 619–624.

Weiss, S.J., Rustagi, P.K., and LeBuglio, A.F. (1978). Human granulocyte generation of hydroxyl radical. *J. Exp. Med.* **147**: 316–323.

Wink, D.A., Christodoulou, D., Ho, M., Krishna, M.C., Cook, J.A., Haut, H., Randolph, J.K., Sullivan, M., Coia, G., Murray, R., and Meyer, T. (1995). A discussion of electrochemical techniques for the detection of nitric oxide. *Methods: Companion Methods Enzymol.* **7**: 71–77.

Winston, G.W., and Cederbaum, A.I. (1982). Oxidative decarboxylation of benzoate to carbon dioxide by rat liver microsomes: A probe for oxygen radical production during microsomal electron transfer. *Biochemistry* **21**: 4265–4270.

Winston, G.W., Eibschutz, O.M., Strekas, T., and Cederbaum, A.I. (1986). Complex-formation and reduction of ferric iron by 2-oxo-4-thiomethylbutyric acid, and the production of hydroxyl radicals. *Biochem. J.* **235**: 521–529.

Winterbourn, C.C. (1981). Hydroxyl radical production in body fluids. Roles of metal ions, ascorbate and superoxide. *Biochem. J.* **198**: 125–131.

Winterbourn, C.C. (1983). Lactoferrin-catalyzed hydroxyl radical production: Additional requirements for a chelating agent. *Biochem. J.* **210**: 15–19.

Winterbourn, C.C. (1986). Myeloperoxidase as an effective inhibitor of hydroxyl radical production: Implications for the oxidative reactions of neutrophils. *J. Clin. Invest.* **78**: 545–550.

Winterbourn, C.C. (1987). The ability of scavengers to distinguish ·OH production in the iron catalyzed Haber-Weiss reaction. Comparison of four assays for ·OH. *Free Radical Biol. Med.* **3**: 33–39.

Winterbourn, C.C. (1991). Factors that influence the deoxyribose oxidation assay for Fenton reaction products. *Free Radical Biol. Med.* **11**: 353–360.

Winterbourn, C.C., and Sutton, H.C. (1986). Iron and xanthine oxidase catalyze formation of an oxidant species distinguishable from ·OH: Comparison with the Haber-Weiss reaction. *Arch. Biochem. Biophys.* **244**: 27–34.

Winterbourn, C.C., Pichorner, H., and Kettle, A.J. (1997). Myeloperoxidase-dependent generation of a tyrosine peroxide by neutrophils. *Arch. Biochem. Biophys.* **338**: 15–21.

Xia, Y., Dawson, V.L., Dawson, T.M., Snyder, S.H., and Zweier, J.L. (1996). Nitric oxide synthase generates superoxide and nitric oxide in arginine-depleted cells leading to peroxynitrite-mediated cellular injury. *Proc. Natl. Acad. Sci. USA* **93**: 6770–6774.

Xia, Y., Roman, L.J., Masters, B.S.S., and Zweier, J.L. (1998a). Inducible nitric-oxide synthase generates superoxide from the reductase domain. *J. Biol. Chem.* **272**: 22635–22639.

Xia, Y., Tsai, A.-L., Berka, V., and Zweier, J.L. (1998b). Superoxide generation from endothelial nitric-oxide synthase. A Ca^{2+}/calmodulin-dependent and tetrahydrobiopterin regulatory process. *J. Biol. Chem.* **273**: 25804–25808.

Yamamoto, Y. (1994). Chemiluminescence-based high-performance liquid chromatography assay of lipid hydroperoxides. *Methods Enzymol.* **233**: 319–324.

Yamazaki, I., and Piette, L.H. (1961). Mechanism of free radical formation and disappearance during the ascorbic acid oxidase and peroxidase reactions. *Biochim. Biophys. Acta* **50**: 62–69.

Yamazaki, I., Mason, H.S., and Piette, L. (1960). Identification, by electron paramagnetic resonance spectroscopy, of free radicals generated from substrates by peroxidase. *J. Biol. Chem.* **235**: 2444–2449.

Yang, S.F. (1967). Biosynthesis of ethylene. Ethylene formation from methional by horseradish peroxidase. *Arch. Biochem. Biophys.* **122**: 481–487.

Ye, Y.Z., Strong, M., Huang, Z.Q., and Beckman, J.S. (1996). Antibodies that recognize nitrotyrosine. *Methods Enzymol.* **269**: 201–209.

Yin, B., Whyatt, R.W., Perera, F.P., Randall, M.C., Cooper, T.B., and Santella, R.M. (1995). Determination of 8-hydroxydeoxyguanosine by an immunoaffinity chromatography–monoclonal antibody-based ELISA. *Free Radical Biol. Med.* **18**: 1023–1032.

Yu, T.C., and Sinnhuber, R.O. (1957). 2-Thiobarbituric acid method for the measurement of rancidity in fishery products. *Food Technol.* **11**: 104–108.

Zafiriou, O.C., and McFarland, M. (1980). Determination of trace levels of nitric oxide in aqueous solution. *Anal. Chem.* **52**: 1662–1667.

Zamburlini, A., Maiorino, M., Barbera, P., Pastorino, A.M., Roveri, A., Cominacini, L., and Ursini, F. (1995a). Measurement of lipid hydroperoxides in plasma lipoproteins by a new highly sensitive "single photon counting" luminometer. *Biochim. Biophys. Acta* **1256**: 233–240.

Zamburlini, A., Maiorino, M., Barbera, P., Roveri, A., and Ursini, F. (1995b). Direct measurement by single photon counting of lipid hydroperoxides in human plasma and lipoproteins. *Anal. Biochem.* **232**: 107–113.

Zhu, H., Bannerberg, G.L., Moldéus, P., and Shertzer, H.G. (1994). Oxidation pathways for the intracellular probe 2′,7′-dichlorofluorescin. *Arch. Toxicol.* **68**: 582–587.

Zweier, J.L., Flaherty, J.T., and Weisfeldt, M.L. (1987). Direct measurement of free radical generation following reperfusion of ischemic myocardium. *Proc. Natl. Acad. Sci. USA*, **84**: 1404–1407.

5

Spin Trapping Free Radicals

Historical Perspective

We wish to point out the potential value of the reaction of reactive free radicals with C-nitroso compounds as a probe for exploring the mechanisms of free-radical reactions in solution.

Chalfont, Perkins, and Horsfield, 1968

In the late 1960s, when our knowledge of the biological significance of free radicals, such as $O_2\cdot^-$, was in its infancy, chemists began to develop techniques to study "the mechanisms of free radical reactions" (Chalfont, *et al.*, 1968). One such approach focused on a rather simple reaction—the addition of free radicals to either nitrosoalkanes or nitrones (Mackor, *et al.*, 1966, 1967, 1968; Iwamura and Inamoto, 1967a, 1967b, 1970; Chalfont, *et al.*, 1968; Janzen and Blackburn, 1968; Lagercrantz and Forshult, 1968). The resulting nitroxides were remarkably stable due to resonance delocalization of the unpaired electron (Griller and Ingold, 1976) along the bond connecting the nitrogen and oxygen atoms (Rozantsev, 1970a):

From these early experiments, an innovative technology was born. At that time, few scientists, if any, realized the long-term significance of their observations. Nevertheless, within a few years this analytical tool began to open new vistas into the nature of cellular-derived free radicals.

No publication of that period was more influential in the development of methodologies to characterize free radicals than a brief communication written by Janzen and Blackburn (1968). In a little under two pages, the authors elegantly characterized the problem of detecting short-lived free radicals and presented a novel solution—the spin trapping technique.

Simply stated, spin trapping is a method by which exceedingly short-lived free radicals can be identified through their reaction with nitrones or nitrosoalkanes. The product of this reaction, a nitroxide—more generically known in the literature of this discipline as a spin trapped adduct—exhibits a lifetime that is considerably longer than the parent free radical:

| Nitrone | Free radical | Nitroxide, spin trapped adduct |

| Nitrosoalkane | Free radical | Nitroxide, spin trapped adduct |

The paramagnetic property of the nitroxide became an asset, since electron paramagnetic resonance (EPR) spectroscopy has proven to be a reliable and a sensitive instrumentation to detect these stable free radicals. Coincidentally, nitroxides absorb radiation over a wide range in the visible region of the electromagnetic spectrum, with a $\lambda_{max} \approx 410$ nm. Although spectrophotometry offers an inexpensive alternative to EPR spectroscopy, the poor sensitivity and the inability of this spectroscopic method to discriminate between various spin trapped adducts greatly limits its application.

Spin Traps

Nitrones are the most frequently used spin traps, as they are ideally suited for identifying the biologically important free radicals, $O_2\cdot^-$, HO·, and a number of secondary free radicals, derived primarily from the reaction of HO· with cellular and other targets. Although two structurally diverse groups of nitrones have received the most attention, principally the aromatic conjugated alkane-N-oxides, including α-phenyl-N-tert-butylnitrone (PBN) (Emmons, 1957), and the Δ^1-pyrroline-N-oxides, such as 5,5-dimethyl-1-pyrroline-N-oxide (DMPO) (Bonnett, et al., 1959), one familiar with the chemistry of other heterocycles could easily envision alternative ringed systems as spin traps. For instance, indol-2-one-N-oxides, benzazepine-N-oxides, isoindole-N-oxides, imidazole-N-oxides and, indol-2-one-N-oxides have been synthesized (Bond and Hooper, 1974; Carr, et al., 1994; Bernotas, et al., 1996a,1996b; Dikalov, et al., 1996), and appear to hold great promise as new classes of spin traps (Thomas, *et al.*, 1994, 1996; Dikalov, *et al.*, 1996; Nepveu, *et al.*, 1998).

In contrast to the enormous number of reports documenting the ability of nitrones to spin trap oxygen-centered and carbon-centered free radicals, the spin trapping literature devoted to nitrosoalkanes is substantially smaller, even though synthetic routes for these spin traps are considerably less demanding than are preparative schemes for nitrones (Kaur and Perkins, 1982). Furthermore, since the free radical adds directly to the nitrogen, additional hyperfine splittings result from its close proximity to the nitroxide. Despite these obvious advantages, the lack of excitement for nitrosoalkanes is based on two observations. First, these compounds are readily subject to an array of chemical reactions, giving nitroxides, that are frequently unrelated to the reaction under study (de Boer, 1982). This problem is best illustrated by the controversy over the reported spin trapping of $O_2 \cdot^-$ by sodium 3,5-dibromo-4-nitrosobenzenesulfonate (DBNBS) (Ozawa and Hanaki, 1986). Subsequent studies demonstrated that the source of the observed EPR spectrum was derived from the spin trapping of a sulfur trioxide anion free radical from DMSO, which was used as a solvent in the reaction mixture, and not from $O_2 \cdot^-$ (Stolze and Mason, 1987). Other more recent investigations have suggested alternative pathways for DBNBS-derived nitroxides (Mani and Crouch, 1989; Samuni, *et al.*, 1989; Nazhat, *et al.*, 1990). Second, the exceptionally poor stability of spin trapped adducts deduced from oxygen-centered free radicals greatly limits the use of these spin traps to specific free radicals or low temperatures (Wargon and Williams, 1972).

Nitroxides, although not considered to be spin traps in the classic sense of the term, have found an invaluable niche as reporters of free radical events (Hoffman, *et al.*, 1964; Brownlie and Ingold, 1967; Rozantsev, 1970b; Beckwith, *et al.*, 1986). Typical of these early studies was the formation of *O*-alkylhydroxylamines, generated during photolysis of chloroplasts in the presence of di-tert-butylnitroxide (Hoffman, *et al.*, 1964; Corker, *et al.*, 1966). Kinetic experiments have determined that the reaction of carbon-centered, but not oxygen-centered, free radicals with dialkyl nitroxides in aqueous solutions is rapid, with rate constants approaching $2-8 \times 10^8 \text{ M}^{-1} \text{ s}^{-1}$ and even faster in organic solvents (Ingold, 1984; Beckwith, *et al.*, 1988; Chateauneuf, *et al.*, 1988). Identification of the initial unstable free radical, however, is dependent upon other analytic methods, such as HPLC/mass spectrometry.

$$\begin{array}{ccc} R_1 & & R_1 \\ \diagdown & & \diagdown \\ N\text{---}O\cdot \ + \ R\cdot & \longrightarrow & N\text{---}O\text{---}R \\ \diagup & & \diagup \\ R_2 & & R_2 \end{array}$$

Although spin labels have been exceedingly valuable in studying the kinetics of free radical events in homogenous solutions (Ingold, 1973, 1984; Bowry and Ingold, 1992), their application to biologic systems has been more limited, owing to the ease of nitroxide reduction (Stier and Reitz, 1971; Goldberg, *et al.*, 1977;

Rosen and Rauckman, 1977; Rauckman, *et al.*, 1984; Swartz, *et al.*, 1995). In some situations, this reaction can frequently overwhelm the scavenging properties of these stable free radicals. Despite this, nitroxides are finding their place as *in vivo* radioprotectors in radiation oncology (Hahn, *et al.*, 1994) and as minimally invasive measures of whole-animal dynamics and metabolism (Utsumi, *et al.*, 1990; Takeshita, *et al.*, 1991; Miura, *et al.*, 1992).

Kinetics of Spin Trapping Free Radicals

By the mid-1970s, it became clear that nitrones could spin trap oxygen-centered free radicals like $O_2 \cdot^-$ and $HO \cdot$, yet the efficiency of these reactions was, at that time, poorly understood. It was assumed that the reaction of nitrones with free radicals was limited by the rate of free radical formation (Ingold, 1973). Thus, a few years later, when the rate constant for the reaction of DMPO with $HO \cdot$ was actually determined, the calculated value, $k \approx 2{-}3 \times 10^9 \, M^{-1} \, s^{-1}$, agreed well with the prevailing view (Finkelstein, *et al.*, 1980a; Marriott, *et al.*, 1980). In contrast to this, it was startling to discover the exceedingly slow reaction of $O_2 \cdot^-$ toward DMPO, with a rate constant at physiologic pH of $1.2{-}12 \, M^{-1} \, s^{-1}$, depending on experimental conditions (Finkelstein, *et al.*, 1979, 1980a; Yamazaki, *et al.*, 1990; Zang and Misra, 1992). Even at the pK_a of $HO_2 \cdot$, the rate constant was found to approach only $6.6 \times 10^3 \, M^{-1} \, s^{-1}$ (Finkelstein, *et al.*, 1980a). This finding is troublesome, considering the disproportionation rate constant of $O_2 \cdot^-$ at pH 7.4 is $\approx 3 \times 10^5 \, M^{-1} \, s^{-1}$ (see Table 2.2) and its reaction with SOD is even faster with a rate constant of $\approx 2 \times 10^9 \, M^{-1} \, s^{-1}$ (Getzoff, *et al.*, 1992). Attempts to enhance the reactivity of other spin traps toward $O_2 \cdot^-$ have not led to more fruitful results (Turner and Rosen, 1986; Fréjaville, *et al.*, 1994, Janzen, *et al.*, 1994, 1995; Janzen and Zhang, 1995).

Detection of Spin Trapped Adducts with Electron Paramagnetic Resonance Spectroscopy

Much of the power of the spin trapping technique is derived from the use of EPR spectroscopy and the ability of this spectroscopy to detect free radicals in turbid solutions of cell suspensions and, most recently, *in vivo* (Ferrari, *et al.*, 1990, 1994; Utsumi, *et al.*, 1990; Mäder, *et al.*, 1992, 1993, 1995; Komarov, *et al.*, 1993; Fujii, *et al.*, 1994; Hartell, *et al.*, 1994; Lai and Komarov, 1994; Halpern, *et al.*, 1995; Hiramatsu, *et al.*, 1995; Komarov and Lai, 1995; Jiang, *et al.*, 1995, 1996; Yoshimura, *et al.*, 1996; Liu, *et al.*, 1999). As early as 1936, changes in the absorption of high-frequency electromagnetic radiation by a paramagnetic species were observed when a magnetic field was applied to the sample (Gorter, 1936a, 1936b; Gorter and de L. Kronig, 1936). These observations established the first fundamental aspect of magnetic resonance—the development, in the paramagnetic sample, of a difference in the energy of its populations of unpaired electron magnetic moments, which are dependent upon their quantized orientation state relative to an externally imposed magnetic field (Figure 5.1). Part A of this illustration depicts the electron state population for a zero magnetic field. When a magnetic field is

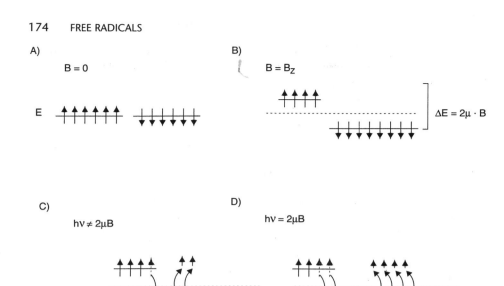

Figure 5.1 Representation of the energies and populations of magnetic substates (also referred to as orientational states in the text) under varied conditions. The representation depicts a simple two-energy level or two-orientational system. E refers to the energy of the magnetic substates; ΔE is the difference in energy between magnetic substates, ν is the frequency of the electromagnetic radiation to which the spin sample population is exposed; h is Planck's constant. The numbers of arrows/oriented lines refers to the substate population. (A) External magnetic field, B, is zero — no electromagnetic radiation. (B) External magnetic field is present. The population of the higher energy level is smaller than that of the lower energy level, as required by the Boltzmann distribution of energy-state populations at equilibrium. The direction of the magnetic field is down to generate the energy levels shown for the substate orientation shown. (C) External magnetic field is present, as well as electromagnetic radiation. The frequency of the electromagnetic radiation is not equal to the energy-level splitting; it is, therefore, not resonant. As shown, a small number of transitions are stimulated by the electromagnetic radiation. The number of transitions is proportional to the population of the magnetic substate. Thus, there are more transitions from lower energy state to higher energy state. The result is a small absorption of energy from the electromagnetic radiation. (D) External magnetic field is present, as well as electromagnetic radiation. The frequency of the electromagnetic radiation is such that the energy of an electromagnetic quantum matches the difference between the energy levels of the magnetic substate. The frequency of the magnetic radiation satisfies the resonance condition. There are many more transitions stimulated at the resonance, resulting in a larger absorption of energy from the electromagnetic radiation.

imposed on a paramagnetic compound, an energy-level splitting, ΔE, is created that is equal to the product of the magnetic field, B, and the component of the magnetic moment, μ in the direction of the magnetic field (Figure 5.1B). At thermal equilibrium, the number of electron magnetic moments in each orientation will differ as a function of the energy state. The higher energy state will have a smaller population than the lower energy state. Transitions of the electron state from one energy level, with its associated orientation, to other states are stimulated by exposing the electron magnetic moment state populations to electromagnetic radiation. For a simple two-state system, the transitions from the lower to the higher energy state will outnumber transitions from the higher to the lower energy state. This is so because there are more electron magnetic moments in the lower energy orientation state. The transition from the lower to the higher energy state will require the absorption of energy from electromagnetic radiation to conserve energy. A transition from the higher to the lower energy state will, on the other hand, contribute energy to the electromagnetic radiation. A larger number of events will involve absorption, because the population of the lower energy state is larger than that of the higher energy state. The net effect is an absorption of radiation much like a resistor absorbs energy in an electric circuit.

The failure to observe resonance in populations of nuclear spins derived from their long relaxation times (Gorter, 1936b), while the inability to observe resonance in electronic spin was related to the low operating frequency relative to the magnetic fields imposed (Gorter, 1936a, 1936c). Drawing upon these earlier accomplishments, Zavoiskii (1945a, 1945b) and Cummerow and Halliday (1946) discovered the resonant paramagnetic absorption property, which allowed the detection of the absorbed radiation. This was based on the fact that the rate at which transitions between orientation states occurs is dependent upon the frequency of the electromagnetic resonance, v, and the energy difference, ΔE, of those states created by the magnetic field. When the energy in a quantum wave packet of radiation, hv, matches ΔE, the denominators in both the semiclassical and quantum descriptions of the probability of transition become small. When the transition probability and the consequent rate of transition increase enormously, a large resonant absorption occurs. This phenomenon is depicted in Figure 5.1C and D. In a balanced electromagnetic circuit or electromagnetic bridge, analogous to the Wheatstone bridge for measuring resistance, the absorption of energy by a paramagnetic sample, which is seen as increased resistance, can be detected as the magnetic field is varied. Resonant absorption exists, therefore, as the match condition is fulfilled, giving a distinct absorption line.

The operating frequency of the electromagnetic radiation for current standard EPR spectrometers is approximately 10 gigahertz (GHz). This allows penetration (the depth of which the amplitude of the electromagnetic wave diminishes to $1/e$ or 37% of its surface value) of ≈ 1 mm into aqueous medium (Van Vleck, 1951; Johnson and Guy, 1972; Bottomley and Andrew, 1978; Röschmann, 1987). It is for this reason that the standard EPR aqueous quartz flat cell is designed with an internal thickness of 0.3 mm. Larger depths can be measured in samples with less electromagnetic loss.

Magnetic fields are imposed on the unpaired electron magnetic moment by many magnetic field sources. Other paramagnetic materials nearby will, therefore, create random and time-varying dipolar magnetic fields at the unpaired electron magnetic moment (Abragam, 1961). This will cause resonance at magnetic fields other than that simply calculated from the match condition based only on the imposed magnetic field. This will have the effect of broadening or smearing the absorption line. Spin exchange, a process deriving from the fundamental indistinguishability of electrons located near each other, is another important source of broadening (Anderson, 1954; Kivelson, 1957; Kaplan, 1958; Abragam, 1961; Freed, 1966; Eastman, *et al.*, 1969). Here, the observed electron is transiently exposed to a different magnetic environment through its exchange with other electrons in paramagnetic molecules with which its host molecule collides. Local concentrations of iron or oxygen have, therefore, substantial broadening effects.

Relationship of an Electron Paramagnetic Resonance Spectrum to a Specific Free Radical

For the detection of the spin trapped adduct by EPR spectroscopy, the most important nearby source of magnetic field is derived from the nuclei with magnetic moments in atoms in which the unpaired electron resides. The magnetic field of the nucleus will add to or subtract from the externally imposed magnetic field by a fixed amount, depending on its quantized orientation. This phenomenon results in splitting the absorption line. The magnitude of a splitting depends on the size of the nuclear magnetic moment and the fraction of the electron spin density distribution at the particular nucleus. Multiple factors contribute to the spin density distribution, not all of which are fully understood (Janzen, *et al.*, 1973). The varied magnitude of splittings from nuclei whose magnetic moments affect the electron results in a multiline EPR spectrum. These line splittings are often an important signature, allowing the characterization of a paramagnetic molecule (Figure 5.2) (Janzen and Liu, 1973).

From a specific splitting pattern, hyperfine splitting constants are calculated, whose values are used to discriminate between different spin trapped adducts (Buettner, 1987; Li and Chignell, 1991). This is one of the great strengths of this technology. Even small differences between free radicals can frequently result in significant spectral changes (Figure 5.2). The ability of DMPO to distinguish between $O_2 \cdot^-$ and HO· (Finkelstein, *et al.*, 1979), for example, has had a profound influence on our knowledge of the biologic generation of these free radicals (Britigan, *et al.*, 1986). Inclusion of either SOD or catalase into the reaction has been used to correlate the specific spectrum with either free radical (Finkelstein, *et al.*, 1979). However, such fingerprint patterns are not universally discernible with other nitrones. Typically, it is difficult to distinguish between carbon-centered and oxygen-centered free radicals spin trapped by either PBN or 4-POBN (Connor, *et al.*, 1986), owing to the inability of these atoms to influence the structure of the nitroxide. Therefore, independent verification is of particular importance when the source of the spin trapped adduct is in doubt. Confirmation may include the synthesis of the nitroxide, derived from the reaction of a free radical with the spin trap (Janzen and Blackburn, 1968; Rosen and Rauckman, 1980; Keana, 1984), isotope labeling and substitution (Poyer, *et al.*, 1980; Mottley, *et al.*, 1986; Haire, *et al.*, 1988; Motten, *et al.*, 1988; Pou, *et al.*, 1990; Barasch, *et al.*, 1994; Janzen, *et al.*, 1994), GC/MS (Janzen, *et al.*, 1990b), HPLC/MS (Janzen, *et al.*, 1990a; Iwahashi, *et al.*, 1991a, 1991b; Parker, *et al.*, 1991), EPR/HPLC (Kominami, *et al.*, 1976; Chen, *et al.*, 1994), a highly sensitive colorimetric method (Becker, 1996), or addition of a scavenger unique for the specific free radical under investigation (Britigan, *et al.*, 1987).

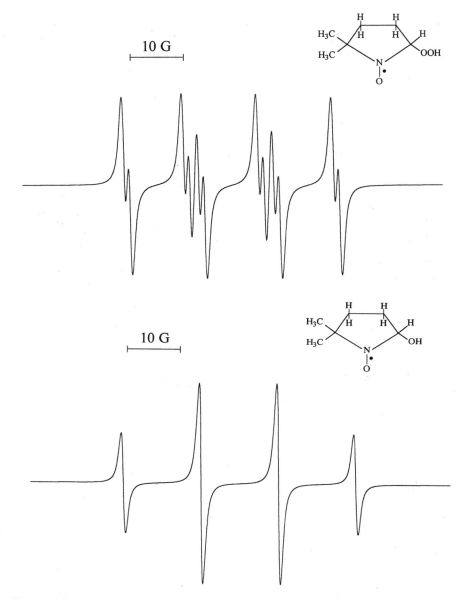

Figure 5.2 Aqueous solution simulation of the $O_2 \cdot^-$ and HO· spin trapped adducts of DMPO (5,5-dimethyl-1-pyrroline-*N*-oxide). (A) DMPO–OOH (2,2-dimethyl-5-hydro-peroxy-1-pyrrolinyloxyl). The hyperfine splitting constants at 9.5 GHz are $A_N = 14.3\,G$, $A_H{}^\beta = 11.4\,G$, and $A_H{}^\gamma = 1.3\,G$. (B) DMPO–OH (2,2-dimethyl-5-hydroxy-1-pyrrolinyloxyl). The hyperfine splitting constants at 9.5 GHz are $A_N = 14.9\,G$, $A_H{}^\beta = 14.9\,G$, and $A_H{}^\gamma = $ not detected with a standard EPR spectrometer in aqueous solutions. These nitroxides differ only in the number of oxygen atoms at position 5. Nonetheless, this results in significant changes in the EPR line splittings induced by the nitrogen nucleus, the hydrogen nucleus attached at position 5, and the nucleus of one of the hydrogen atoms bound at position 4. This can be seen in the substantial difference in the spectral signatures shown.

Stability of the Spin Trap and the Spin Trapped Adduct

Although linking the hyperfine splitting constants of a spin trapped adduct to a specific free radical has frequently been used to identify the parent species (Rosen and Klebanoff, 1979), enthusiasm for this practice must often be tempered by the fact that other, non-free radical reactions can inadvertently result in spurious nitroxide signals (Finkelstein, *et al.*, 1980b). For example, the ease of DMPO oxidation by adventitious iron in solutions (Makino, *et al.*, 1990) or associated with enzyme preparations raises an immediate concern as to the validity of experimental findings (Britigan, *et al.*, 1990). Furthermore, rapid conversion of DMPO–OOH to DMPO–OH by either chemical (Finkelstein, *et al.*, 1982) or enzymatic reactions (Rosen and Freeman, 1984) may sometimes lead to erroneous conclusions.

The principle underlying spin trapping is that the spin trapped adduct exhibits a lifetime sufficiently long to allow for its spectroscopic identification. In the case where experiments involve the biologic generation of free radicals, detection of the nitroxide is, therefore, restricted to a very narrow temperature range. This prerequisite often dictates the choice of spin trap. For instance, nitrosoalkanes are capable of spin trapping both carbon-centered and oxygen-centered free radicals. The spin trapped adduct derived from the oxygen-centered free radical is, however, unstable at ambient temperatures (Wargon and Williams, 1972). Likewise, the half-life of PBN–OH or 4-POBN–OH is less than a minute at room temperature, whereas the half-life of DMPO–OH under similar conditions is considerably longer (Finkelstein, *et al.*, 1980b; Kotake and Janzen, 1991).

Frequently, rapid elimination of the nitroxide, through intramolecular rearrangements (Rosen and Rauckman, 1984) or reductions, either chemical (Keana, *et al.*, 1987) or enzymatic (Rauckman, *et al.*, 1984), can lead to an underestimation of the breadth as well as the rate of free radical production. For example, the poor stability of DMPO-spin trapped adducts in the presence of stimulated neutrophils (Samuni, *et al.*, 1988; Pou, *et al.*, 1989) prevented the discovery of the unique myeloperoxidase-dependent pathway for HO· generation until a new spin trapping system could be developed (Ramos, *et al.*, 1992).

Localization of Spin Traps

An assumption common to the literature of this discipline is that spin traps diffuse into cells. Surprisingly, there is a dearth of evidence supporting this contention. Several independent studies suggest the intracellular localization of DMPO (Morgan, *et al.*, 1985, Samuni, *et al.*, 1986). Similar investigations with PBN and 4-POBN are less definitive (McCay, *et al.*, 1980; Morgan, *et al.*, 1985; Chen, *et al.*, 1990; Bynoe, *et al.*, 1991; Anderson, *et al.*, 1993), even though PBN has been used to spin trap trichloromethyl radical during the *in vivo* metabolism of carbon tetrachloride (McCay, *et al.*, 1980). Unfortunately, when similar experiments were undertaken with nitrosoalkanes, cytotoxicity from these compounds greatly limited distributional studies (Morgan, *et al.*, 1985; Streeter, *et al.*, 1990). Nevertheless, with the advent of isotope labeling (Janzen, *et al.*, 1986; Pou, *et al.*, 1990) in combination with chromium oxalate (Samuni, *et al.*, 1986; Kaneko, *et al.*, 1995), it is now possible to more accurately estimate the apportionment of spin trapped adducts between intra- and extracellular compartments.

Concluding Thoughts

As this century draws to a close, spin trapping/EPR spectroscopy will undoubtedly continue to play an important role in defining the physiologic function of free radicals in biological systems. The potential of combining confocal microscopy with spin trapping (Blough and Salmeen, 1990; Pou, *et al.*, 1995) opens exciting opportunities to localize and image free radicals within cells. Similarly, with the development of lower frequency EPR spectrometers (Halpern, *et al.*, 1989), the ability to *in vivo in situ* identify specific free radicals in real time is within our grasp. For these goals to be realized, however, new spin traps must be developed. Like most scientific advances, future synthetic routes will rely, to a large extent, on past experiences. In the next chapter, we will delve into the synthesis of a wide variety of nitrones and nitrosoalkanes, probes that are so essential to the future of free radical exploration in isolated cells and in animal models.

References

Abragam, A. (1961). *The Principles of Nuclear Magnetisms* (Marchall, W.C., and Wilkinson, D.H., eds.), The International Series of Monographs on Physics, Oxford University Press, New York.

Anderson, D.E., Yuan, X.-J., Tseng, C.-M., Rubin, L.J., Rosen, G.M., and Tod, M.L. (1993). Nitrone spin-traps block calcium channels and induce pulmonary artery relaxation independent of free radicals. *Biochem. Biophys. Res. Commun.* **193**: 878–885.

Anderson, P.W. (1954). A mathematical model for the narrowing of spectral lines by exchange or motion. *J. Phys. Soc. Japan* **9**: 316–339.

Barasch, D., Krishna, M.C., Russo, A., Katzhendler, J., and Samuni, A. (1994). Novel DMPO-derived ^{13}C-labeled spin traps yield identifiable stable nitroxides. *J. Am. Chem. Soc.* **116**: 7319–7324.

Becker, D.A. (1996). Highly sensitive colorimetric detection and facile isolation of diamagnetic free radical adducts of novel chromotropic nitrone spin trapping agents readily derived from guaiazulene. *J. Am. Chem. Soc.* **118**: 905–906.

Beckwith, A.L.J., Bowry, V.W., and Moad, G. (1988). Kinetics of the coupling reactions of the nitroxyl radical 1,1,3,3-tetramethylisoindoline-2-oxyl with carbon-centered radicals. *J. Org. Chem.* **53**: 1632–1641.

Beckwith, A.L.J., Bowry, V.W., O'Leary, M., Moad, G., Rizzardo, E., and Solomon, D.H. (1986). Kinetic data for coupling of primary alkyl radicals with a stable nitroxide. *J. Chem. Soc., Chem. Commun.* 1003–1004.

Bernotas, R.C., Adams, G., and Carr, A.A. (1996a). Synthesis of benzazepine-based nitrones as radical traps. *Tetrahedron* **52**: 6519–6526.

Bernotas, R.C., Thomas, C.E., Carr, A.A., Nieduzak, T.R., Adams, G., Ohlweiler, D.F., and Hay, D.A. (1996b). Synthesis and radical scavenging activity of 3,3-

dialkyl-3,4-dihydro-isoquinoline-2-oxides. *Bioorg. Med. Chem. Lett.* **6**: 1105–1110.

Blough, N.V., and Salmeen, I.T. (1990). Fluorescence detection of free radicals by nitroxide scavenging. *Free Radical Res. Commun.* **10**: 119–121.

Bond, C.C., and Hooper, M. (1974). Isatogen: 9. The synthesis of isatogens by the oxidation of 2-substituted 1-hydroxy-indoles. *Synthesis* 443.

Bonnett, R., Brown, R.F.C., Clark, V.M., Sutherland, I.O., and Todd, A. (1959). Experiments towards the synthesis of corrins. Part II. The preparation and reactions of Δ^1-pyrroline 1-oxides. *J. Chem. Soc.* 2094–2102.

Bottomley, P.A., and Andrew, E.R. (1978). RF magnetic field penetration, phase shift and power dissipation in biological tissues: Implications for NMR imaging. *Phys. Med. Biol.* **23**: 630–643.

Bowry, V.W., and Ingold, K.U. (1992). Kinetics of nitroxide radical trapping. 2. Structural effects. *J. Am. Chem. Soc.* **114**: 4992–4996.

Britigan, B.E., Cohen, M.S., and Rosen, G.M. (1987). Detection of the production of oxygen-centered free radicals by human neutrophils using spin trapping techniques: A critical perspective. *J. Leukocyte Biol.* **41**: 349–362.

Britigan, B.E., Rosen, G.M., Thompson, B.Y., Chai, Y., and Cohen, M.S. (1986). Stimulated human neutrophils limit iron-catalyzed hydroxyl radical formation as detected by spin trapping techniques. *J. Biol. Chem.* **261**: 17026–17032.

Britigan, B.E., Pou, S., Rosen, G.M., Lilleg, D.M., and Buettner, G.R. (1990). Hydroxyl radical is not a product of the reaction of xanthine oxidase and xanthine. The confounding problem of adventitious iron bound to xanthine oxidase. *J. Biol. Chem.* **265**: 17533–17538.

Brownlie, I.T., and Ingold, K.U. (1967). The inhibited autoxidation of styrene. Part VII. Inhibition by nitroxides and hydroxylamines. *Can. J. Chem.* **45**: 2427–2432.

Buettner, G.R. (1987). Spin trapping: ESR parameters of spin adducts. *Free Radical Biol. Med.* **3**: 259–303.

Bynoe, L.A., Pou, S., Gottsch, J.D., and Rosen, G.M. (1991). Light-dependent spin trapping of hydroxyl radical from human erythrocytes. *Biochem. Biophys. Res. Commun.* **179**: 1305–1310.

Carr, A.A., Thomas, G.E., Bernotas, R.C., and Ku, G. (1994). Cyclic nitrones, pharmaceutical composition thereof and their use in treating shock U.S. Patent Number 5,292,746.

Chalfont, G.R., Perkins, M.J., and Horsfield, A. (1968). A probe for homolytic reactions in solution. II. The polymerization of styrene. *J. Am. Chem. Soc.* **90**: 7141–7142.

Chateauneuf, J., Lusztyk, J., and Ingold, K.U. (1988). Absolute rate constants for the reactions of some carbon-centered radicals with 2,2,6,6-tetramethyl-piperidine-N-oxyl. *J. Org. Chem.* **53**: 1629–1632.

Chen, G., Bray, T.M., Janzen, E.G., and McCay, P.B. (1990). Excretion, metabolism and tissue distribution of a spin trapping agent, α-phenyl-N-*tert*-butyl-nitrone (PBN) in rats. *Free Radical Res. Commun.* **9**: 317–323.

Chen, G., Janzen, E.D., and Bray, T.M. (1994). Identification of 3-MI-derived N-centered radicals obtained from incubation of 3-MI with microsomal-NADPH system by EPR–HPLC spin trapping. *Free Radical Biol. Med.* **17**: 19–25.

Connor, H.D., Fischer, V., and Mason, R.P. (1986). A search for oxygen- centered free radicals in the lipoxygenase/linoleic acid system. *Biochem. Biophys. Res. Commun.* **141**: 614–621.

Corker, G.A., Klein, M.P., and Calvin, M. (1966). Chemical trapping of a primary quantum conversion product of photosynthesis. *Proc. Natl. Acad. Sci. USA* **56**: 1365–1369.

Cummerow, R.L., and Halliday, D. (1946). Paramagntic losses in two manganous salts. *Phys. Rev.* **70**: 433.

de Boer, Th.J. (1982). Spin-trapping in early and some recent nitroso chemistry. *Can. J. Chem.* **60**: 1602–1609.

Dikalov, S., Kirilyuk, I., and Grigor'ev, I. (1996). Spin trapping of O-, C-, and S-centered radicals and peroxynitrite by 2H-imidazole-1-oxides. *Biochem. Biophys. Res. Commun.* **218**: 616–622.

Eastman, M.P., Kooser, R.G., Das, M.R., and Freed, J.H. (1969). Studies of Heisenberg spin exchange in ESR spectra. I. Linewidth and saturation effects. *J. Chem. Phys.* **51**: 2690–2709.

Emmons, W.D. (1957). The preparation and properties of oxazirines. *J. Am. Chem. Soc.* **79**: 5739–5754.

Ferrari, M., Colacicchi, S., Gualtieri, G., Santini, M.T., and Sotgiu, A. (1990). Whole mouse nitroxide free radical pharmacokinetics by low frequency electron paramagnetic resonance. *Biochem. Biophys. Res. Commun.* **166**: 168–173.

Ferrari, M., Quaresima, V., Ursini, C.L., Alecci, M., and Sotgiu, A. (1994). *In vivo* electron paramagnetic resonance spectroscopy-imaging in experimental oncology: The hope and the reality. *Int. J. Radiat. Oncol. Biol. Phys.* **29**: 421–425.

Finkelstein, E., Rosen, G.M., and Rauckman, E.J. (1980a). Spin trapping. Kinetics of the reaction of superoxide and hydroxyl radicals with nitrones. *J. Am. Chem. Soc.* **102**: 4994–4999.

Finkelstein, E., Rosen, G.M., and Rauckman, E.J. (1980b). Spin trapping of superoxide and hydroxyl radical: Practical aspects. *Arch. Biochem. Biophys.* **200**: 1–16.

Finkelstein, E., Rosen, G.M., and Rauckman, E.J. (1982). Production of hydroxyl radical by decomposition of superoxide spin-trapped adducts. *Mol. Pharmacol.* **21**: 262–265.

Finkelstein, E., Rosen, G.M., Rauckman, E.J., and Paxton J. (1979). Spin trapping of superoxide. *Mol. Pharmacol.* **16**: 676–685.

Freed, J.H. (1966). Theory of saturation and double resonance effects in electron spin resonance spectra. II. Exchange vs. dipolar mechanisms. *J. Phys. Chem.* **71**: 38–51.

Fréjaville, C., Karoui, H., Tuccio, B., Le Moigne, F., Culcasi, M., Pietri, S., Lauricella, R., and Tordo, P. (1994). 5-Diethoxyphosphoryl-5-methyl-1-pyrroline N-oxide (DEPMPO): A new phosphorylated nitrone for efficient *in vitro* and *in vivo* spin trapping of oxygen-centred radicals. *J. Chem. Soc., Chem. Commun.* 1793–1794.

Fujii, H., Zhao, B., Koscielniak, J., and Berliner, L.J. (1994). *In vivo* EPR studies of the metabolic fate of nitrosobenzene in the mouse. *Magn. Reson. Med.* **31**: 77–80.

Getzoff, E.D., Cabelli, D.E., Fisher, C.L., Parge, H.E., Viezzoli, M.S., Banci, L., and Hallewell, R.A. (1992). Faster superoxide dismutase mutants designed by enhancing electrostatic guidance. *Nature* **358**: 347–351.

Goldberg, J.S., Rauckman, E.J., and Rosen, G.M. (1977). Bioreduction of nitroxides by *Staphylococcus aureus*. *Biochem. Biophys. Res. Commun.* **79**: 198–202.

Gorter, C.J. (1936a). Paramagnetic relaxation. *Physica* **3**: 503–514.

Gorter, C.J. (1936b). Negative result of an attempt to detect nuclear magnetic spins. *Physica* **3**: 995–998.

Gorter, C.J. (1936c). Paramagnetic relaxation in a transversal magnetic field. *Physica* **3**: 1006–1008.

Gorter, C.J., and de L. Kronig, R. (1936). The theory of absorption and dispersion in paramagnetic and dielectric media. *Physica* **3**: 1009–1020.

Griller, D., and Ingold, K.U. (1976). Persistent carbon-centered radicals. *Acc. Chem. Res.* **9**: 13–19.

Hahn, S.M., Krishna, C.M., Samuni, A., DeGraff, W., Cuscela, D.O., Johnstone, P., and Mitchell, J.B. (1994). Potential use of nitroxides in radiation oncology. *Cancer Res.* (Suppl.) **54**: 2006s–2010s.

Haire, D.L., Oehler, U.M., Krygsman, P.H., and Janzen, E.G. (1988). Correlation of radical structure with EPR spin adduct parameters: Utility of the ^{1}H, ^{13}C, and ^{14}N hyperfine splitting constants of aminoxyl adducts of PBN-*nitronyl*-^{13}C for three parameter scatter plots. *J. Org. Chem.* **53**: 4535–4542.

Halpern, H.J., Spencer, D.P., vanPolen, J., Bowman, M.K., Nelson, A.C., Dowey, E.M., and Teicher, B.A. (1989). Imaging radio frequency electron-spin-resonance spectrometer with high resolution and sensitivity for *in vivo* measurements. *Rev. Sci. Instrum.* **60**: 1040–1050.

Halpern, H.J., Yu, C., Barth, E., Peric, M., and Rosen, G.M. (1995). *In situ* detection, by spin trapping of hydroxyl radical markers produced from ionizing radiation in the tumor of a living mouse. *Proc. Natl. Acad. Sci. USA* **92**: 796–800.

Hartell, M.G., Borzone, G., Clanton, T.L., and Berliner, L.J. (1994). Detection of free radicals in blood by electron spin resonance in a model of respiratory failure in the rat. *Free Radical Biol. Med.* **17**: 467–472.

Hiramatsu, M., Oikawa, K., Noda, H., Mori, A., Ogata, T., and Kamada, H. (1995). Free radical imaging by electron spin resonance computed tomography in rat brain. *Brain Res.* **697**: 44–47.

Hoffmann, A.K., Feldman, A.M., Gelblum, E., and Hodgson, W.G. (1964). Mechanism of the formation of di-*t*-butylnitroxide from *t*-nitrobutane and sodium metal. *J. Am. Chem. Soc.* **86**: 639–646.

Ingold, K.U. (1973). Rate constants for free radical reactions in solution, in: *Free Radicals* (Kochi, J.K., ed.), Vol. I., pp. 37–112, Wiley-Interscience, New York.

Ingold, K.U. (1984). Reactions with a different radical, in: *Landolt- Börnstein. New Series. Radical Reaction Rates in Liquids* (Fischer, H., ed.), Vol. 13, Part c, pp. 181–200, Springer-Verlag, Berlin.

Iwahashi, H., Albro, P.W., McGown, S.R., Tomer, K.B., and Mason, R.P. (1991a). Isolation and identification of α-(4-pyridyl-1-oxide)-N-*tert*-butylnitrone radical adducts formed by the decomposition of the hydroperoxides of linoleic acid, linolenic acid and arachidonic acid by soybean lipoxygenase. *Arch. Biochem. Biophys.* **285**: 172–180.

Iwahashi, H., Parker, C.E., Mason, R.P., and Tomer, K.B. (1991b). Radical adducts of nitrosobenzene and 2-methyl-2-nitrosopropane with 12,13-epoxylinoleic acid radical, 12,13-epoxylinolenic acid radical and 14,15-epoxyarachidonic acid radical. Identification by h.p.l.c.–e.p.r. and liquid chromatography–thermospray–m.s. *Biochem. J.* **276**: 447–453.

Iwamura, M., and Inamoto, N. (1967a). Novel radical 1,3-addition to nitrones. *Bull. Chem. Soc. Japan* **40**: 702.

Iwamura, M., and Inamoto, N. (1967b). Novel formation of nitroxide radicals by radical addition to nitrones. *Bull. Chem. Soc. Japan* **40**: 703.

Iwamura, M., and Inamoto, N. (1970). Reactions of nitrones with free radicals. II. Formation of nitroxides. *Bull. Chem. Soc. Japan* **43**: 860–863.

Janzen, E.G., and Blackburn, B.J. (1968). Detection and identification of short-lived free radicals by an electron spin resonance trapping technique. *J. Am. Chem. Soc.* **90**: 5909–5910.

Janzen, E.G., and Liu, J.I.-P. (1973). Radical addition reactions of 5,5-dimethyl-1-pyrroline-1-oxide. ESR spin trapping with a cyclic nitrone. *J. Magn. Reson.* **9**: 510–512.

Janzen, E.G., and Zhang, Y.-K. (1995). Identification of reactive free radicals with a new ^{31}P-labeled DMPO spin trap. *J. Org. Chem.* **60**: 5441–5445.

Janzen, E.G., Evans, C.A., and Liu, J.I.-P. (1973). Factors influencing hyperfine splitting in the ESR spectra of five-membered ring nitroxides. *J. Magn. Reson.* **9**: 513–516.

Janzen, E.G., Krygsman, P.H., Lindsay, D.A., and Haire, D.L. (1990a). Detection of alkyl, alkoxyl, and alkyperoxyl radicals from the thermolysis of azobis-(isobutyronitrile) by ESR/spin trapping. Evidence for double spin adducts from liquid-phase chromatography and mass spectroscopy. *J. Am. Chem. Soc.* **112**: 8279–8284.

Janzen, E.G., Oehler, U.M., Haire, D.L., and Kotake, Y. (1986). ENDOR spectra of aminoxyls. Conformational study of alklyl and aryl spin adducts of deuterated α-phenyl-N-*tert*-butylnitrone (PBN) based on proton and β^{31}C hyperfine splittings. *J. Am. Chem. Soc.* **108**: 6858–6863.

Janzen, E.G., Towner, R.A., Krygsman, P.H., Haire, D.L., and Poyer, J.L. (1990b). Structure identification of free radicals by ESR and GC/MS of PBN spin adducts from the *in vitro* and *in vivo* rat liver metabolism of halothane. *Free Radical Res. Commun.* **9**: 343–351.

Janzen, E.G., Zhang, Y.-K., and Arimura, M. (1995). Synthesis and spin-trapping chemistry of 5,5-dimethyl-2-(trifluoromethyl)-1-pyrroline N-oxide. *J. Org. Chem.* **60**: 5434–5440.

Janzen, E.G., Zhang, Y.-K., and Haire, D.L. (1994). Synthesis of a novel nitrone, 2-phenyl-5,5-dimethyl-1-pyrroline N-oxide (nitronyl-^{13}C), for enhanced radical addend recognition and spin adduct persistence. *J. Am. Chem. Soc.* **116**: 3738–3743.

Jiang, J.J., Liu, K.J., Jordan, S.J., Swartz, H.M., and Mason, R.P. (1996). Detection of free radical metabolite formation using *in vivo* EPR spectroscopy: Evidence of rat hemoglobin thiyl formation following administration of phenylhydrazine. *Arch. Biochem. Biophys.* **330**: 266–270.

Jiang, J.J., Liu, K.J., Shi, X., and Swartz, H.M. (1995). Detection of short-lived free radicals by low-frequency electron paramagnetic resonance spin trapping in whole living animals. *Arch. Biochem. Biophys.* **319**: 570–573.

Johnson, C.C., and Guy, A.W. (1972). Nonionizing electromagnetic wave effects in biological materials and systems. *Proc. IEEE* **60**: 692–718.

Kaneko, M., Kodama, M., and Inoue, F. (1995). The localization of DMPO spin adducts of ·OH in endothelial cells exposed to hydrogen peroxide. *Toxicol. Lett.* **81**: 73–78

Kaplan, J.I. (1958). Exchange broadening in nuclear magnetic resonance. *J. Chem. Phys.* **28**: 278–281.

Kaur, H., and Perkins, M.J. (1982). From early spin-trapping experiments to acyl nitroxide chemistry: Synthesis of and spin trapping with a water-soluble nitrosopyrazole derivative. *Can. J. Chem.* **60**: 1587–1593.

Keana, J.F.W. (1984). Synthesis and chemistry of nitroxide spin labels, in: *Spin Labeling in Pharmacology* (Holtzman, J., ed.), pp. 1–85, Academic Press, New York.

Keana, J.F.W., Pou, S., and Rosen, G.M. (1987). Nitroxides as potential contrast enhancing agents for MRI applications: Influence of structure on the rate of reduction by rat hepatocytes, subcellular fractions and ascorbate. *Magn. Reson. Med.* **5**: 525–536.

Kivelson, D. (1957). Theory of the effects of exchange on the nuclear fine structure in the paramagnetic resonance spectra of liquids. *J. Chem. Phys.* **27**: 1087–1098.

Komarov, A.M., and Lai, C.-S. (1995). Detection of nitric oxide production in mice by spin-trapping electron paramagnetic resonance spectroscopy. *Biochim. Biophys. Acta* **1272**: 29–36.

Komarov, A., Mattson, D., Jones, M.M., Singh, P.K., and Lai, C.-S. (1993). *In vivo* spin trapping of nitric oxide in mice. *Biochem. Biophys. Res. Commun.* **195**: 1191–1198.

Kominami, S., Rokushika, S., and Hatano, H. (1976). Studies of short-lived radicals in the γ-irradiated aqueous solution of uridine-5′-monophosphate by the spin-trapping method and liquid chromatography. *Int. J. Radiat. Biol.* **30**: 525–534.

Kotake, Y., and Janzen, E.G. (1991). Decay and fate of the hydroxyl radical adduct of α-phenyl-N-*tert*-butylnitrone in aqueous media. *J. Am. Chem. Soc.* **113**: 9503–9506.

Lagercrantz, C., and Forshult, S. (1968). Trapping of free radicals formed by γ-irradiation of organic compounds. *Nature* **218**: 1247–1248.

Lai, C.-S., and Komarov, A.M. (1994). Spin trapping of nitric oxide produced *in vivo* in septic-shock in mice. *FEBS Lett.* **345**: 120–124.

Li, A.S.W., and Chignell, C.F. (1991). The NoH value in EPR spin trapping: A new parameter for the identification of 5,5-dimethyl-1-pyrroline-N-oxide spin adducts. *J. Biochem. Biophys. Methods* **22**: 83–87.

Liu, K.J., Miyake, M., Panz, T., and Swartz, H. (1999). Evaluation of DEPMPO as a spin trapping agent in biological systems. *Free Radical Biol. Med.* **26**: 714–721.

Mackor, A., Wajer, Th.A.J.W., and de Boer, Th.J. (1966). C-Nitroso compounds. Part I. The formation of nitroxides by photolysis of nitroso compounds as studied by electron spin resonance. *Tetrahedron Lett.* 2115–2123.

Mackor, A., Wajer, Th.A.J.W., and de Boer, Th.J. (1967). C-Nitroso compounds. Part III. Alkoxy-alkyl-nitroxides as intermediates in the reaction of alkoxyl-radicals with nitroso compounds. *Tetrahedron Lett.* 385–390.

Mackor, A., Wajer, Th.A.J.W., and de Boer, Th.J. (1968). C-Nitroso compounds. VI. Acyl-alkyl-nitroxides from acyl radicals and nitroso compounds as studied by ESR. *Tetrahedron* **24**: 1623–1631.

Mäder, K., Bacic, G., and Swartz, H.M. (1995). *In vivo* detection of anthralin-derived free radicals in the skin of hairless mice by low-frequency electron paramagnetic resonance spectroscopy. *J. Invest. Dermatol.* **104**: 514–517.

Mäder, K., Stösser, R., and Borchert, H.-H. (1992). Transcutaneous absorption of nitroxide radicals detected *in vivo* by means of X-band ESR. *Pharmazie* **47**: 946–947.

Mäder, K., Stösser, R., and Borchert, H.-H. (1993). Detection of free radicals in living mice after inhalation of DTBN by X-band ESR. *Free Radical Biol. Med.* **14**: 339–342.

Makino, K., Hagiwara, T., Hagi, A., Nishi, M., and Murakami, A. (1990). Cautionary note for DMPO spin trapping in the presence of iron ion. *Biochem. Biophys. Res. Commun.* **172**: 1073–1080.

Mani, V., and Crouch, R.K. (1989). Spin trapping of the superoxide anion: Complications in the use of the water-soluble nitroso-aromatic reagent DBNBS. *J. Biochem. Biophys. Methods* **18**: 91–96.

Marriott, P.R., Perkins, M.J., and Griller, D. (1980). Spin trapping for hydroxyl in water: A kinetic evaluation of two popular traps. *Can. J. Chem.* **58**: 803–807.

McCay, P.B., Noguchi, T., Fong, K.-L., Lai, E.K., and Poyer, J.L. (1980). Production of radicals from enzyme systems and the use of spin traps, in: *Free Radicals in Biology* (Pryor, W.A., ed), Vol. IV, pp. 155–186, Academic Press, New York.

Miura, Y., Utsumi, H., and Hamada, A. (1992). Effects of inspired oxygen concentration on *in vivo* redox reaction of nitroxide radicals in whole mice. *Biochem. Biophys. Res. Commun.* **182**: 1108–1114.

Morgan, D.D., Mendenhall, C.L., Bobst, A.M., and Rouster, S.D. (1985). Incorporation of the spin trap DMPO into cultured fetal mouse liver cells. *Photochem. Photobiol.* **42**: 93–94.

Motten, A.G., Levy, L.A., and London, R.E. (1988). Carbon-13 labeling of nitrone spin traps for enhanced adduct identification. *J. Magn. Reson.* **80**: 112–115.

Mottley, C., Connor, H.D., and Mason R.P. (1986). [^{17}O]oxygen hyperfine structure for the hydroxyl and superoxide radical adducts of the spin traps, DMPO, PBN and 4-POBN. *Biochem. Biophys. Res. Commun.* **141**: 622–628.

Nazhat, N.B., Yang, G., Allen, R.E., Blake, D.R., and Jones, P. (1990). Does 3,5-dibromo-4-nitrosobenzene sulphonate spin trap superoxide radicals? *Biochem. Biophys. Res. Commun.* **166**: 807–812.

Nepveu, F., Souchard, J.-P., Rolland, Y., Dorey, G., and Spedding, M. (1998). 2-2′-Pyridylisatogen, a selective allosteric modulator of P2 receptor, is a spin trapping agent. *Biochem. Biophys. Res. Commun.* **242**: 272–276.

Ozawa, T., and Hanaki, A. (1986). Spin-trapping of superoxide ion by a water-soluble, nitroso-aromatic spin-trap. *Biochem. Biophys. Res. Commun.* **136**: 657–664.

Parker, C.E., Iwahashi, H., and Tomer, K.B. (1991). Spin-trapped radicals: Determination by LC-TSP-MS and LC-ESI-MS. *J. Am. Soc. Mass Spectrom.* **2**: 413–418.

Pou, S., Bhan, A., Bhadti, V.S., Wu, S.Y., Hosmane, R.S., and Rosen, G.M. (1995). The use of fluorophore-containing spin traps as potential probes to localize free radicals in cells with fluorescent imaging methods. *FASEB J.* **9**: 1085–1090.

Pou, S., Cohen, M.S., Britigan, B.E., and Rosen, G.M. (1989). Spin trapping and human neutrophils: Limits of detection of hydroxyl radical. *J. Biol. Chem.* **264**: 12299–12302.

Pou, S., Rosen, G.M., Wu, Y., and Keana, J.F.W. (1990). Synthesis of deuterium- and ^{15}N-containing pyrroline-1-oxides: A spin trapping study. *J. Org. Chem.* **55**: 4438–4443.

Poyer, J.L., McCay, P.B., Lai, E.K., Janzen, E.G., and Davis, E.R. (1980). Confirmation of assignment of the trichloromethyl radical spin adduct detected by spin trapping during[13]C-carbon tetra-chloride metabolism *in vitro* and *in vivo*. *Biochem. Biophys. Res. Commun.* **94**: 1154–1160.

Ramos, C.L., Pou, S., Britigan, B.E., Cohen, M.S., and Rosen, G.M. (1992). Spin trapping evidence for myeloperoxidase-dependent hydroxyl radical formation by human neutrophils and monocytes. *J. Biol. Chem.* **267**: 8307–8312.

Rauckman, E.J., Rosen, G.M., and Griffeth, L.K. (1984). Enzymatic reactions of spin labels, in: *Spin Labeling in Pharmacology* (Holtzman, J., ed.), pp. 175–190, Academic Press, New York.

Röschmann, P. (1987). Radiofrequency penetration and absorption in the human body: Implications to high-field whole-body nuclear magnetic resonance imaging. *Med. Phys.* **14**: 922–931.

Rosen, G.M., and Freeman, B.A. (1984). Detection of superoxide by endothelial cells. *Proc. Natl. Acad. Sci. USA* **81**: 7269–7273.

Rosen, G.M., and Rauckman, E.J. (1977). Formation and reduction of a nitroxide radical by liver microsomes. *Biochem. Pharmacol.* **26**: 675–678.

Rosen, G.M., and Rauckman, E.J. (1980). Spin trapping of the primary radical involved in the activation of the carcinogen N-hydroxy-2 acetylamino-fluorene by cumen hydroperoxide–hematin. *Mol. Pharmacol.* **17**: 233–238.

Rosen, G.M., and Rauckman, E.J. (1984). Spin trapping of superoxide and hydroxyl radicals, in: *Methods of Enzymology* (Packer, L., ed.), Vol. 105, pp. 198–209, Academic Press, New York.

Rosen, H., and Klebanoff, S.J. (1979). Hydroxyl radical generation by polymorphonuclear leukocytes measured by electron spin resonance spectroscopy. *J. Clin. Invest.* **64**: 1725–1729.

Rozantsev, E.G. (1970a). *Free Nitroxyl Radicals*, pp. 1–10, Plenum Press, New York.

Rozantsev, E.G. (1970b). *Free Nitroxyl Radicals*, pp. 95–97, Plenum Press, New York.

Samuni, A., Black, C.D.V., Krishna, C.M., Malech, H.L., Bernstein, E.F., and Russo, A. (1988). Hydroxyl radical production by stimulated neutrophils reappraised. *J. Biol. Chem.* **263**: 13797–13801.

Samuni, A., Carmichael, A.J., Russo, A., Mitchell, J.B., and Reisz, P. (1986). On the spin trapping and ESR detection of oxygen-derived free radicals generated inside cells. *Proc. Natl. Acad. Sci. USA* **83**: 7593–7597.

Samuni, A., Samuni, A., and Swartz, H.M. (1989). Evaluation of dibromonitroso-benzene sulfonate as a spin trap in biological systems. *Free Radical Biol. Med.* **7**: 37–43.

Stier, A., and Reitz, I. (1971). Radical production in amine oxidation by liver microsomes. *Xenobiotica* **1**: 499–500.

Stolze, K., and Mason, R.P. (1987). Spin trapping artifacts in DMSO. *Biochem. Biophys. Res. Commun.* **143**: 941–946.

Streeter, A.J., Nims, R.W., Sheffels, P.R., Heur, Y.H., Yang C.S., Mico, B.A., Gombar, C.T., and Keefer, L.K. (1990). Metabolic denitrosation of N-nitrosodimethylamine *in vivo* in the rat. *Cancer Res.* **50**: 1144–1150.

Swartz, H.M., Sentjuric, M., and Kocherginsky, N. (1995). Metabolism of nitroxides and their products in cells, in: *Nitroxide Spin Labels. Reactions in Biology and Chemistry* (Kocherginsky, N., and Swartz, H.M., eds.), pp. 113–147, CRC Press, Boca Raton, FL.

Takeshita, K., Utsumi, H., and Hamada, A. (1991). ESR measurement of radical clearance in lung of whole mice. *Biochem. Biophys. Res. Commun.* **177**: 874–880.

Thomas, C.E., Ohlweiler, D.F., Carr, A.A., Nieduzak, T.R., Hay, D.A., Adams, G., Vaz, R., and Bernotas, R.C. (1996). Characterization of the radical trapping activity of a novel series of cyclic nitrone spin traps. *J. Biol. Chem.* **271**: 3097–3104.

Thomas, C.E., Ohlweiler, D.F., and Kalyanaraman, B. (1994). Multiple mechanisms for inhibition of low density lipoprotein oxidation by novel cyclic nitrone spin traps. *J. Biol. Chem.* **269**: 28055–28061.

Turner, M.J., III, and Rosen, G.M. (1986). Spin trapping of superoxide and hydroxyl radicals with substituted pyrroline 1-oxides. *J. Med. Chem.* **29**: 2439–2444.

Utsumi, H., Muto, E., Masuda, S., and Hamada, A. (1990). *In vivo* ESR measurement of free radicals in whole mice. *Biochem. Biophys. Res. Commun.* **172**: 1342–1348.

Van Vleck, J.H. (1951). The relationship between absorption and dispersion, in: *RLS: Propagation of Short Radio Waves* (Kerr, D.E., ed.), pp. 641–692, McGraw-Hill, New York.

Wargon, J.A., and Williams F. (1972). Electron spin resonance studies of radical trapping in the radiolysis of organic liquids. I. Evidence for the primary formation of the methoxy radical in methanol. *J. Am. Chem. Soc.* **94**: 7917–7918.

Yamazaki, I., Piette, L.H., and Grover, T.A. (1990). Kinetic studies on spin trapping of superoxide and hydroxyl radicals generated in NADPH–cytochrome P-450 reductase–paraquat sys-tems. Effect of iron chelates. *J. Biol. Chem.* **265**: 652–659.

Yoshimura, T., Yokoyama, H., Fujii, S., Takayama, F., Oikawa, K., and Kamada, H. (1996). *In vivo* EPR detection and imaging of endogenous nitric oxide in lipopolysaccharide-treated mice. *Nat. Biotechnol.* **14**: 992–994.

Zang, L.-Y., and Misra, H.P. (1992). EPR kinetic studies of superoxide radicals generated during the autoxidation of 1-methyl-4-phenyl-2,3-dihydropyridin-ium, a bioactivated intermediate of Parkinsonian-inducing neurotoxin 1-methyl-4-phenyl-1,2,3,6-tetrahydro-pyridine. *J. Biol. Chem.* **267**: 23601–23608.

Zavoiskii, E. (1945a). Paramagnetic relaxation of liquid solutions for perpendicular fields. *J. Phys.* **9**: 211–216.

Zavoiskii, E. (1945b). Spin-magnetic resonance in paramagnetic substances. *J. Phys.* **9**: 245.

6

The Synthesis of Spin Traps

In connection with a study of the spin trapping chemistry of water-soluble spin traps the pyridine N-oxide analogues of PBN have been synthesized.

Janzen, Wang, and Shetty, 1978

In many examples of scientific endeavors, useful discoveries are often not directly associated with the original intent of the research. In the case of cyclic nitrones, the initial design of these compounds was to address fundamental questions unrelated to spin trapping free radicals. Early studies[1] investigating the planar quadridentate ligands of vitamin B_{12}, for example, led to the development of a family of Δ^1-pyrroline 1-oxides, including 5,5-dimethyl-1-pyrroline-N-oxide (**5**), as models in the exploration of the physical structure of this unique ringed system (Bonnett, *et al.*, 1959b). More than a decade would elapse, however, until this Δ^1-pyrroline 1-oxide would be found to react with free radicals (Iwamura and Inamoto, 1967, 1970a, 1970b; Janzen and Liu, 1973), and several additional years before nitrone (**5**) would surface as the spin trap of choice for $O_2 \cdot^-$ and $HO \cdot$ (Habour, *et al.*, 1974; Harbour and Bolton, 1975; Buettner and Oberley, 1978; Buettner, *et al.*, 1978; Sealy, *et al.*, 1978; Finkelstein, *et al.*, 1979).

At the dawn of spin trapping, much effort was put forth to synthesize nitrones to which a free radical could readily undergo a 1,3-addition, thereby yielding a long-lived nitroxide (Iwamura and Inamoto, 1970a; Janzen, 1971). The early success, particularly in the biologic sciences, led to the widespread commercialization of these compounds, and, not surprisingly, to retarding efforts to synthesize a broad spectrum of these pyrroline N-oxides. With recent accomplishments realized from spin trapping free radicals in animal models at their site of formation and in real time (Halpern, *et al.*, 1995; Jiang, *et al.*, 1995, 1996; Liu, *et al.*, 1999), there is now an impetus to prepare new families of spin traps, aimed at addressing questions fundamental to the biologic role of free radicals. Since the design of these compounds will be, to a large extent, built upon nearly 40 years of synthetic experiences (Hamer and Macaluso, 1964), a compilation of preparative schemata for the most frequently used as well as novel nitrones and nitrosoalkanes is presented in this chapter in the hope of sparking an interest in the development of the next generation of spin traps.

Δ^1 PYRROLINE-N-OXIDES

5,5-Dimethyl-1-Pyrroline-N-Oxide

The preparation of 5,5-dimethyl-1-pyrroline-N-oxide (**5**) is not overly challenging, although there are steps that require care and attention (Bonnett, *et al.*, 1959a; Rosen and Rauckman, 1984; Janzen, *et al.*, 1989). The initial reaction is a simple Michael addition (March, 1992) in which acrolein is carefully added to the anion of 2-nitropropane (**2**), giving 4-methyl-4-nitro-1-pentanal (**3**) (Shechter, *et al.*, 1952). In the original experimental scheme, this anion was generated by slowly adding 2-nitropropane to sodium methoxide at low temperatures (Shechter, *et al.*, 1952). Alternative bases, such as benzyltrimethylammonium hydroxide (Janzen, *et al.*, 1989), have achieved similar results. To this mixture is added acrolein, freshly distilled over drierite to remove water, and stabilizer, at a rate to maintain the reaction temperature below 0°C. After stirring for several min, acidification by either bubbling hydrogen chloride (Shechter, *et al.*, 1952) or adding hydrochloric acid (Janzen, *et al.*, 1989) stops the reaction. Substituting acetic acid at this step has been found to increase the recoverable yield of the desired aldehyde, 4-methyl-4-nitro-1-pentanal (**3**).

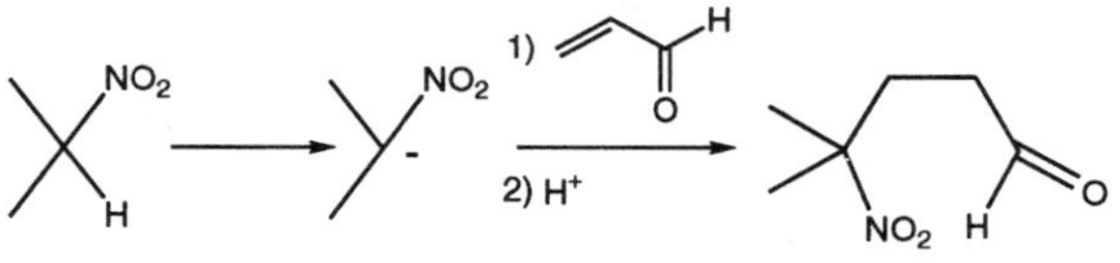

Isolation of 4-methyl-4-nitropentanal (**3**) requires considerable attention to experimental details. Evaporation of the reaction mixture at moderate temperatures and under reduced pressure, using a rotary evaporator, allows the solvent and excess 2-nitropropane to be removed, thereby minimizing decomposition. At this point, the remaining residue should be a clear liquid, with a slight yellow tinge. If, however, what remains after evaporation is "motor oil brown," this thick fluid should be discarded. Isolation of the desired aldehyde, through high-vacuum distillation of the clear, slightly yellow liquid, should be followed carefully. In particular, the distillation flask should not be overheated, as an explosion, due to polymerization of unreacted acrolein, is a real and serious danger. Typical yields of pure aldehyde (**3**) have been reported to range from 40 to 49% (Shechter, *et al.*, 1952; Rosen and Rauckman, 1984; Janzen, *et al.*, 1989).

The reduction of γ-nitro-carbonyl compounds by zinc dust has been found to be of general utility in the synthesis of Δ^1-pyrroline 1-oxides (Bonnett, *et al.*, 1959a).

In the case of aldehyde (**3**), it is advisable to initially protect the aldehyde, as 2-(3-methyl-3-nitrobutyl)-1,3-dioxolane (**4**), prior to reduction, thereby avoiding the formation of pyrrolidines (Bonnett, *et al.*, 1959a). Preparation of the acetal follows standard procedures in which ethylene glycol is added to a refluxing benzene solution containing aldehyde (**3**) in the presence of the catalyst, *p*-toluenesulfonic acid:

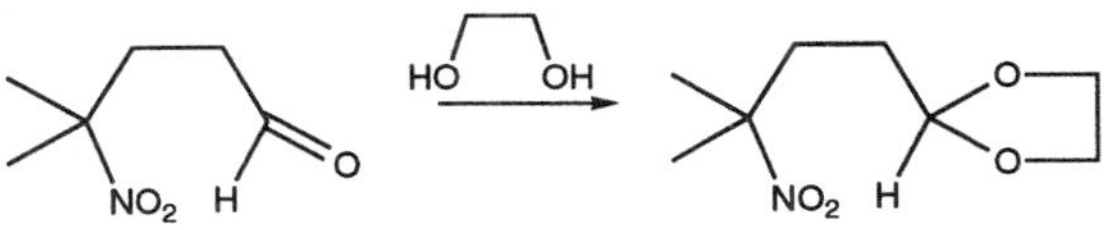

When including a Dean-Stark trap to collect the water, the desired acetal (**4**), can be obtained in excellent yields. Although there is some evidence that further purification of this crude 1,3-dioxolane (**4**) is unnecessary prior to reductive cyclization to 5,5-dimethyl-1-pyrroline-*N*-oxide (**5**), experience from several laboratories indicates that distillation of this dioxolane will make it considerably easier to purify the desired Δ^1-pyrroline 1-oxide in the following step. Subsequent reduction of an aqueous solution of the 1,3-dioxolane (**4**) with zinc dust and ammonium chloride, maintaining the reaction at 10–15°C, gives the corresponding hydroxylamine, which, upon acid hydrolysis to remove the protective group, leads to the desired spin trap, nitrone (**5**):

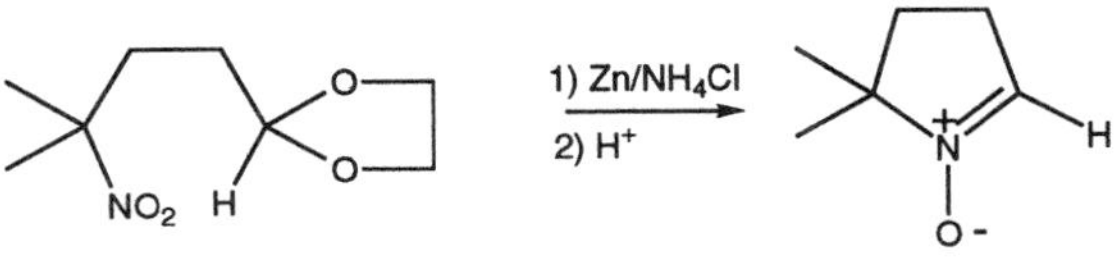

As will be discussed in chapter 9, nitrone (**5**) is susceptible to a variety of chemical reactions, some of which can result in the formation of aminoxyls, that are unrelated to the spin trapping of free radicals (Finkelstein, *et al.*, 1980; Makino, *et al.*, 1990). During the synthesis of 5,5-dimethyl-1-pyrroline-*N*-oxide (**5**), it is not unusual to observe an EPR spectrum from an aqueous solution of this nitrone (e.g., a typical concentration of 100 mM in 50 mM phosphate buffer) (Finkelstein, *et al.*, 1980; Janzen, *et al.*, 1989; Makino, *et al.*, 1990). The spectrum is a composite, in various amounts, of three different nitroxides. One, based on the three-line EPR spectrum with a hyperfine splitting constant of $A_N = 14.7$ G, is undoubtedly a 2,2,5,5-tetrasubstituted nitroxide. Another, presenting a six-line EPR spectrum with hyperfine splitting constants of $A_N = 15.3$ G and $A_H = 22$ G, is characteristic of a carbon-centered spin trapped adduct of nitrone (**5**). The third nitroxide, exhibiting a four-line EPR spectrum with a hyperfine splitting constant of $A_N = A_H = 14.9$ G, has been assigned to 2,2-dimethyl-5-hydroxy-1-pyrrolidinyloxyl.

Simple vacuum distillation of nitrone (**5**) does not always remove these unwanted nitroxide impurities. When this situation arises, purification can frequently be achieved through one of two approaches. Although filtration of an aqueous solution of this Δ^1-pyrroline 1-oxide (5) through activated charcoal has been successful at removing most of the observed paramagnetic species (Buettner and Oberley, 1978), care must be taken to minimize charcoal dissolution. Often this purification process introduces, in some experimental designs, other spurious EPR signals.

Column chromatography followed by bulb-to-bulb distillation has proven to be a viable alternative. The initial step includes the fractional elution of the nitrone through silica gel with a dichloromethane/methanol solvent mixture, increasing the polarity once the nitroxide has been removed. This purification step can be followed by thin-layer chromatography (TLC). At this point, TLC analysis of nitrone (5) will most often not disclose any impurities in the sample. Frequently, an aqueous solution (100 mM) of nitrone (5), even in the presence of DTPA (diethylenetriaminepentaacetic acid, 1 mM), will reveal a small EPR nitroxide signal. When this occurs, additional purification to the requirements essential for spin trapping can be accomplished by high-vacuum bulb-to-bulb Kugelrohr distillation at room temperature. At this point, nitrone (5) is invariably void of paramagnetic impurities and can be stored under either argon or nitrogen at $-78°C$.

Other synthetic schemata can be envisioned for the preparation of nitrone (5). One attractive approach calls for the addition of methylmagnesium bromide to 2-methyl-1-pyrroline-N-oxide (6) (Bradman and Conley, 1973). Copper (II)-catalyzed aerial oxidation of the corresponding hydroxylamine (7) results in formation of nitrone (5). Purification of nitrone (5) follows the two-step procedure described above.

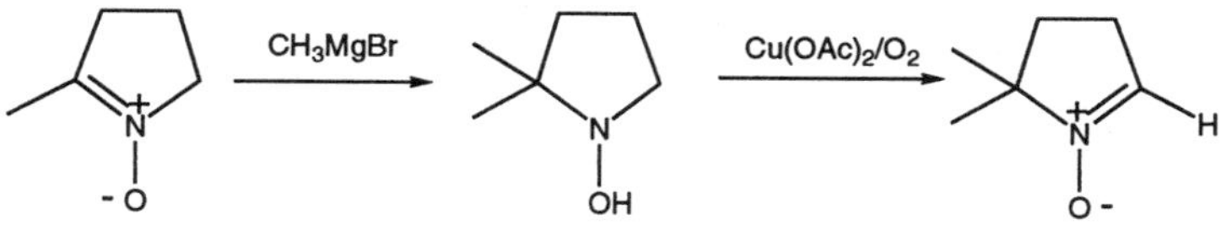

There have been several attempts to shorten the synthesis of 5,5-dimethyl-1-pyrroline-N-oxide (5). One approach calls for zinc reduction of aldehyde (3) in the presence of acetic acid (Haire, et al., 1986). Although these authors report an outstanding purified yield of 60%, other laboratories have not achieved this success. The variability in the percentage of the isolatable nitrone is due to the great difficulty in removing the by-products arising during zinc reduction.

Reactions Involving 5,5-Dimethyl-1-Pyrroline-N-Oxide

Even though the reactive nature of Δ^1-pyrroline 1-oxides toward nucleophilic attack has limited the use of this class of spin traps to specific free radicals under well-defined experimental conditions, in other settings, the propensity of this ring system to readily undergo addition reactions makes this family of compounds an ideal model to synthesize a wide variety of substituted nitrones. Typical of these reactions is the addition of methylmagnesium bromide to nitrone (**5**). Upon copper-catalyzed aerial oxidation of the intermediate hydroxylamine (**8**), 2,5,5-trimethyl-1-pyrroline-*N*-oxide (**9**) can be obtained in excellent yields (Bonnett, *et al.*, 1959a).

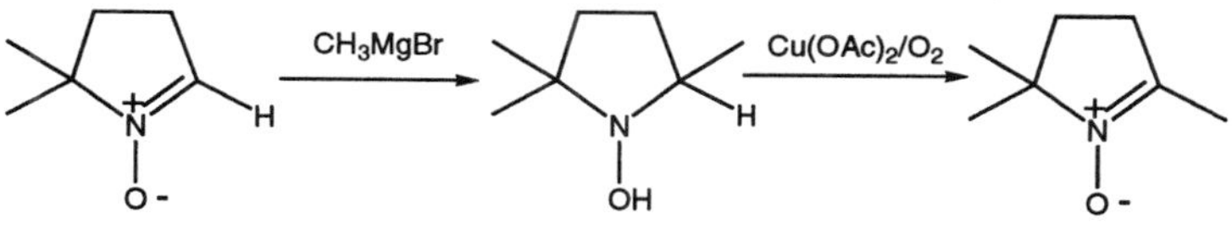

Another nucleophilic reaction of 5,5-dimethyl-1-pyrroline-*N*-oxide (**5**) includes the addition of hydrogen cyanide across this nitrone, which, after appropriate oxidation, affords 2-cyano-5,5-dimethyl-1-pyrroline-*N*-oxide (**10**) (Bonnett, *et al.*, 1959a). Hydrolysis of nitrone (**10**) leads to 2-carboxy-5,5-dimethyl-1-pyrroline-*N*-oxide (**11**) (Bonnett, *et al.*, 1959a):

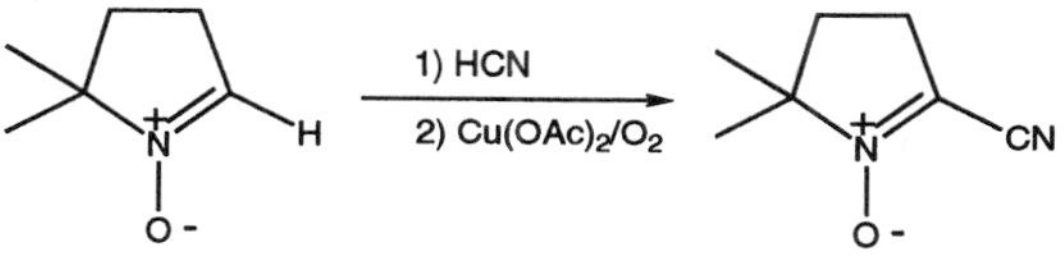

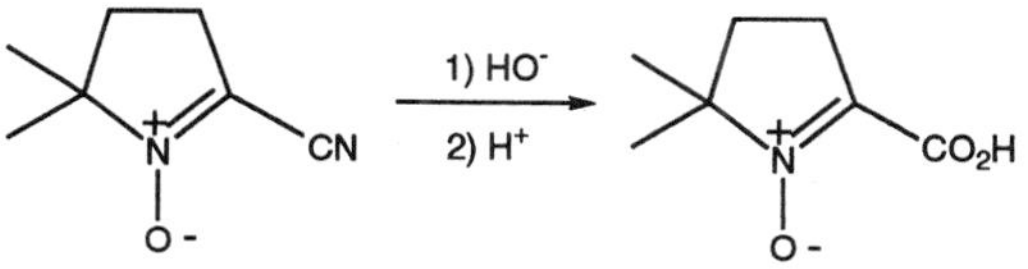

Alternatively, nitrone (**9**) can be synthesized in a two-step procedure. First, freshly distilled methyl vinyl ketone (**12**) is added to the anion of 2-nitropropane (**2**) at low temperatures to give 5-methyl-5-nitro-2-hexanone (**13**) (Shechter, *et al.*, 1952). Reductive cyclization with zinc dust and ammonium chloride results in reasonable yields of 2,5,5-trimethyl-1-pyrroline-*N*-oxide (**9**):

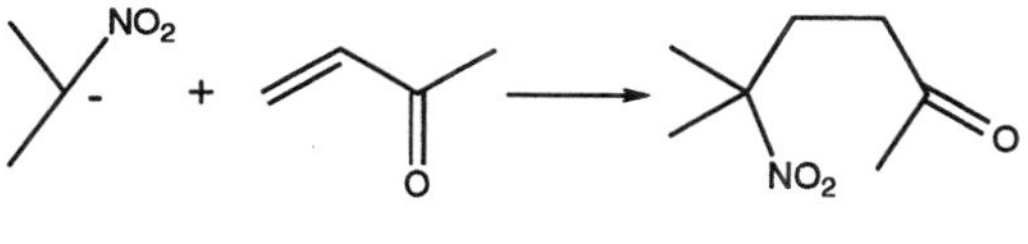

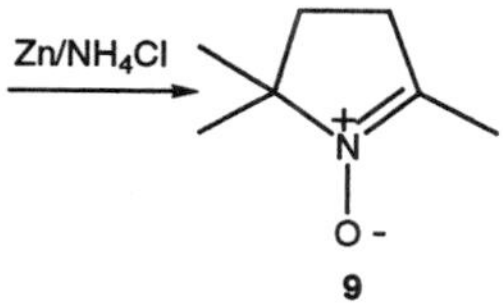

The reaction of 2,5,5-trimethyl-1-pyrroline-N-oxide (**9**) with various free radicals, such as $O_2\cdot^-$ and $HO\cdot$, results in a family of nitroxides with marked increased stability as compared to spin trapped adducts derived from 5,5-dimethyl-1-pyrroline-N-oxide (**5**). However, EPR spectra from these 2-substituted-2,5,5-trimethyl-1-pyrrolidinyloxyls lack the ability to discriminate between various spin trapped free radicals—all exhibiting similar hyperfine splitting constants, $A_N \approx 15\,G$.

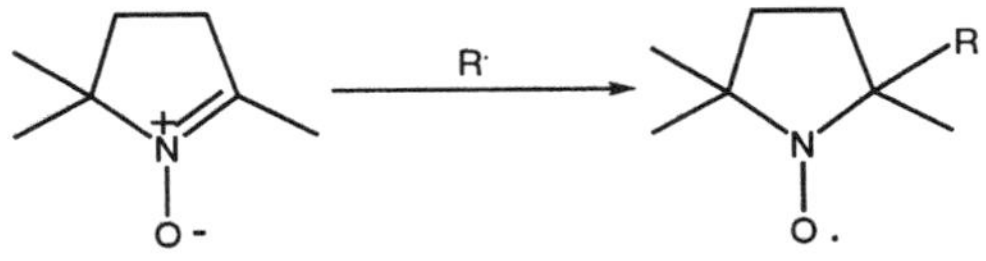

5,5-Dialkyl-, 4,5,5-, or 3,5,5-Trialkyl-1-Pyrroline-N-Oxides

Even though 5,5-dimethyl-1-pyrroline-N-oxide (**5**) is considered to be the prototype spin trap for $O_2\cdot^-$ and $HO\cdot$, alkyl groups at either the 4- or 5-position offer the opportunity to vary the lipophilicity of Δ^1-pyrroline 1-oxides without significantly altering the selectivity of these cyclic spin traps toward these free radicals. There are, surprisingly, only a few publications devoted to such efforts (Bonnett, *et al.*, 1959a; Haire and Janzen, 1982; Turner and Rosen, 1986; Zhang, *et al.*, 1991). The approach in each of these publications is similar to the general synthetic scheme described previously: prepare nitroalkanes, which in the presence of base will readily undergo Michael addition to acrolein. This step is followed by reductive cyclization of γ-nitro-acetals with zinc dust to give appropriate substituted-Δ^1-pyrroline N-oxides. For discussion purposes, several examples are presented (Bonnett, *et al.*, 1959a; Haire and Janzen, 1982; Turner and Rosen, 1986; Zhang, *et al.*, 1991):

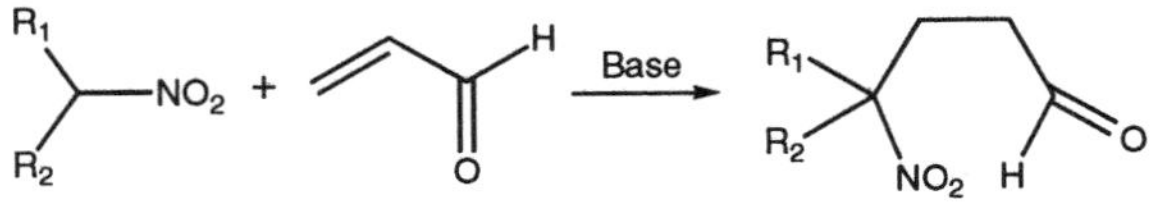

15 = $R_1 = CH_3\text{-}CH_2\text{-}CH_2\text{-}$, $R_2 = CH_3\text{-}$ **17** = $R_1 = CH_3\text{-}CH_2\text{-}CH_2\text{-}$, $R_2 = CH_3\text{-}$
16 = $R_1 = R_2 = CH_3\text{-}CH_2\text{-}CH_2\text{-}$ **18** = $R_1 = R_2 = CH_3\text{-}CH_2\text{-}CH_2\text{-}$

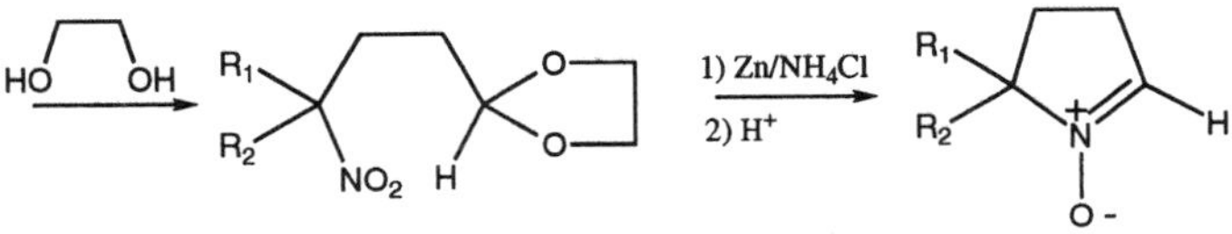

19 = $R_1 = CH_3\text{-}CH_2\text{-}CH_2\text{-}$, $R_2 = CH_3\text{-}$ **21** = $R_1 = CH_3\text{-}CH_2\text{-}CH_2\text{-}$, $R_2 = CH_3\text{-}$
20 = $R_1 = R_2 = CH_3\text{-}CH_2\text{-}CH_2\text{-}$ **22** = $R_1 = R_2 = CH_3\text{-}CH_2\text{-}CH_2\text{-}$

Lipophilic nitrones can likewise be prepared by the reaction of various Grignard reagents with nitrone (**6**) (Bonnett, *et al.*, 1959a; Zhang and Xu, 1989). After aerobic oxidation with copper (II) acetate, the desired nitrone is obtained in reasonable yields (**23, 24**) (Zhang and Xu, 1989):

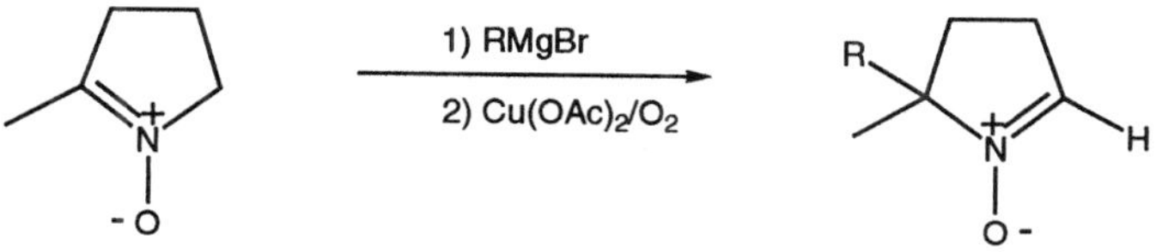

23 = R = CH$_3$-(CH$_2$)$_n$-   (n = 2,3,4,5,7,8,9,11,13,15)
24 = R = C$_6$H$_5$-

As with nitrone (**5**), addition of hydrogen cyanide, followed by oxidation and hydrolysis, affords a family of pH-dependent charged nitrones (**26**):

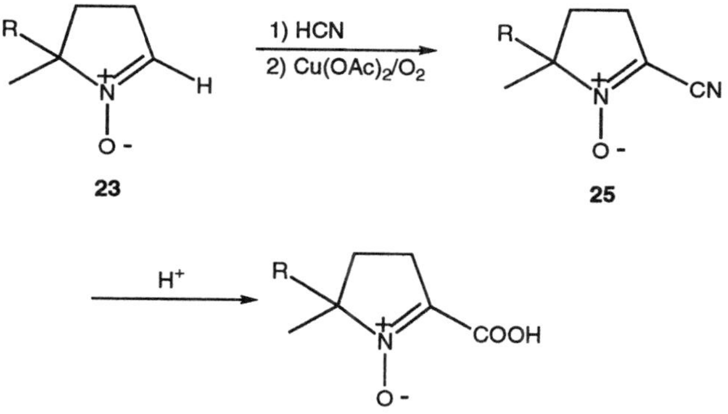

Placing substitutents at the 4-position requires, in the simplest cases, the base-catalyzed reaction of 2-nitropropane (**1**) with either crotonaldehyde (**27**) or cinnamaldehyde (**28**). At this point, protection of the corresponding γ-nitroaldehyde (**29, 30**) with ethylene glycol, followed by zinc reductive cyclization would result in the desired 4-alkyl or 4-aryl-Δ^1-pyrroline-*N*-oxide (**31, 32**) (Bonnett, *et al.*, 1959a; Black and Boscacci, 1976). Surprisingly, reasonable yields of either 4,5,5-trimethyl-1-pyrroline-*N*-oxide (**31**) or 4-phenyl-5,5-dimethyl-1-pyrroline-*N*-oxide (**32**) have been obtained by direct reduction of the appropriate γ-nitroaldehyde (**29, 30**) with zinc dust, rather than, as in the case for nitrone (**5**), the appropriate 1,3-dioxolane is preferentially reduced to the resulting Δ^1-pyrroline *N*-oxide (**31, 32**) (Bonnett, *et al.*, 1959a; Bapat and Black, 1968; Haire, *et al.*, 1988a; Zhang, *et al.*, 1991):

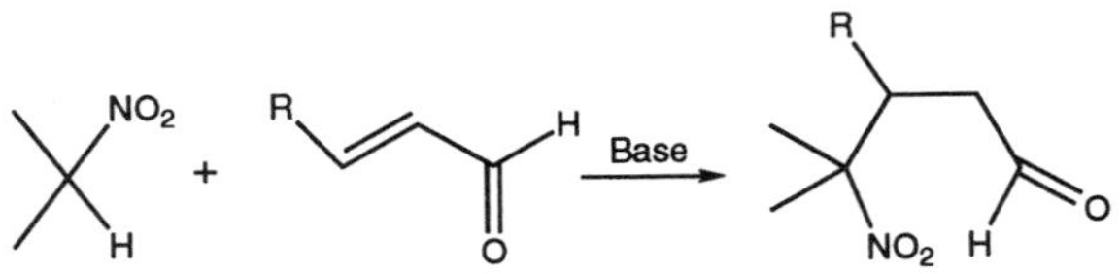

27 = R = CH₃-
28 = R = C₆H₅-

29 = R = CH₃-
30 = R = C₆H₅-

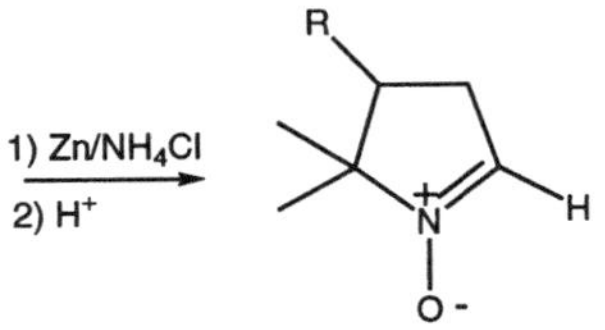

31 = R = CH₃-
32 = R = C₆H₅-

Pathways that involve the synthesis of 3-subtituted-Δ^1-pyrroline N-oxide draw upon the reactivity of 2-nitropropane (1) toward, for example, methacrolein (33). In this instance, catalytic reduction of the corresponding γ-nitroaldehyde (34) leads to 3,5,5-trimethyl-1-pyrroline-N-oxide (35) (Zhang, *et al.*, 1991):

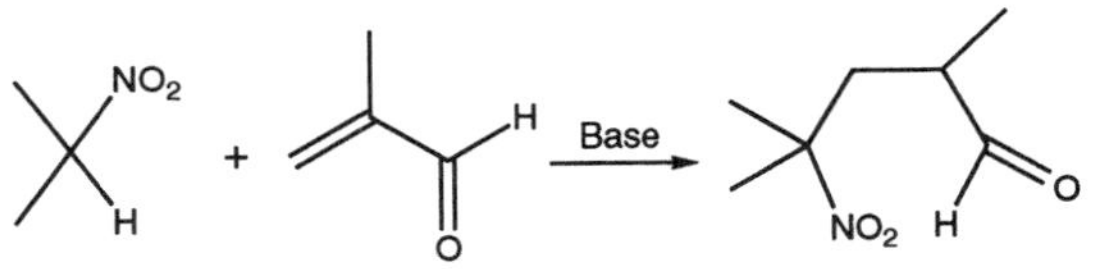

Future advances in the development of spin traps for $O_2 \cdot^-$ and HO· will undoubtedly require the functionalization of Δ^1-pyrroline N-oxides. A design strategy might be, for instance, the incorporation of a nitrone into a molecule with characteristics that would allow transport to defined loci within cells. This new spin trap would, therefore, make it feasible to detect specific free radicals at their site of formation. Various functional groups, such as esters, acids, alcohols, or aldehydes, are ideally suited as linking agents. Although several different preparative schemes are outlined below, one can, nevertheless, envision alternative synthetic designs.

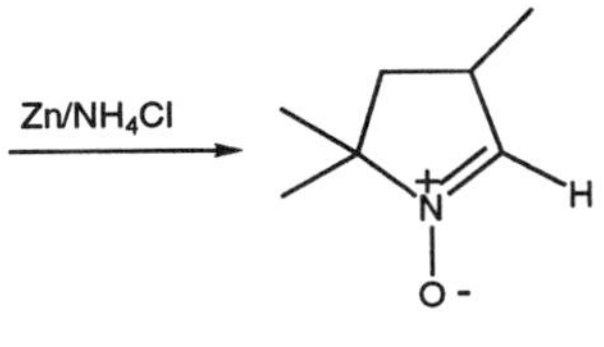

A novel approach to 5,5-dimethyl-4-hydroxymethyl-1-pyrroline-N-oxide (39) was developed based on the Michael addition of 2-nitropropane (1) to 2(5H)-furanone (36) in the presence of a base. The resulting 4-(1-methyl-1-nitroethyl)-tetrahydrofuran-2-one (37) upon reduction with diisobutylaluminum hydride afforded 2-hydroxy-4-(1-methyl-1-nitroethyl)tetrahydrofuran (**38**). Reductive cyclization with zinc dust led to the desired nitrone (**39**) in reasonable yields (Konaka, *et al.*, 1995):

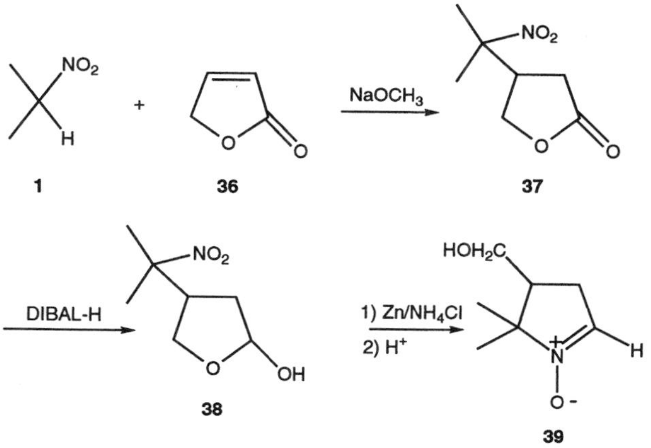

Flexibility in the positioning of carboxyl groups can be achieved using, for instance, ethyl α-nitropropionate (Kornblum, *et al.*, 1957) in the Michael addition to acrolein (**3**). After protecting the resulting aldehyde (**41**) with ethylene glycol, the 1,3-dioxolane (**42**) is reduced with zinc dust, resulting in, after hydrolysis of the corresponding ester, 5-carboxy-5-methyl-1-pyrroline-*N*-oxide (**44**) (Bonnett, *et al.*, 1959a):

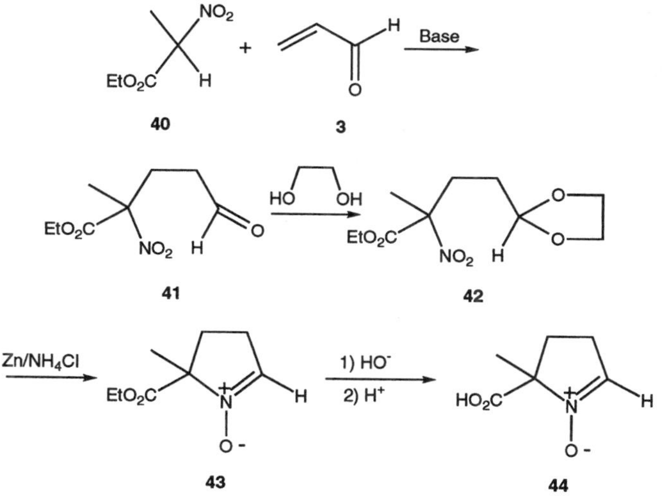

Enamines, with their well-known reactivity toward Michael addition, offer the opportunity to prepare long-chain nitrones with functional groups that can be tied to macromolecules. In so doing, spin traps can then be compartmentalized within cells. The preparation of 5,5-dimethyl-3-(2-ethoxycarbonylethyl)-1-pyrroline-*N*-

oxide (**49**) is illustrative of such a synthetic scheme. The difficult step in this sequence is the reaction of the enamine (**46**) (Stork, *et al.*, 1963) prepared by condensing 4-methyl-4-nitro-1-pentanal (**45**) and piperidine, with the electrophilic olefin, ethyl acrylate (**47**), to give ethyl 4-(2-methyl-2-nitropropyl)-5-oxopentano-ate (**48**). Reduction with zinc dust results in 5,5-dimethyl-3-(2-ethoxycarbony-lethyl)-1-pyrroline-*N*-oxide (**49**) (Zhang, *et al.*, 1990):

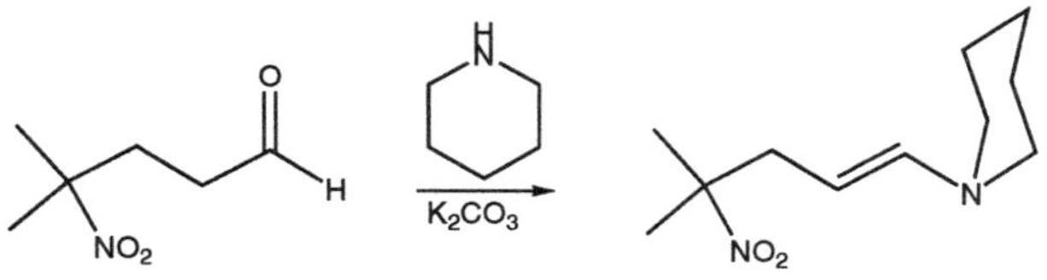

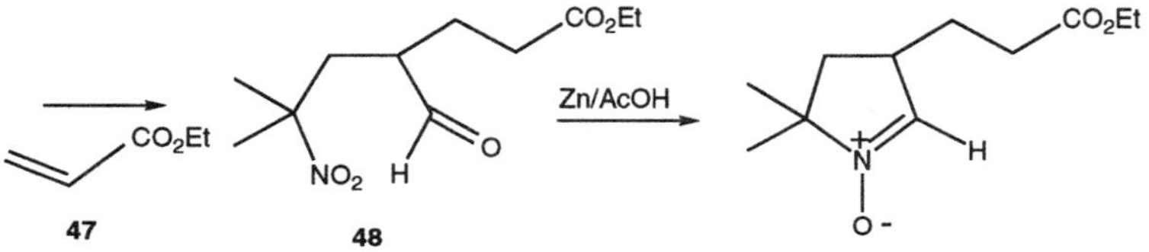

3,3-Dialkyl-5,5-Dimethyl-1-Pyrroline-*N*-Oxides

One of the advantages of using 5,5-dimethyl-1-pyrroline-*N*-oxide (**5**), to distinguish between $O_2 \cdot^-$ and HO· is the unique EPR spectrum assignable to the corresponding spin trapped adducts of these free radicals. In some situations, however, the conversion of nitroxide (**50**) to nitroxide (**51**) (Finkelstein, *et al.*, 1982; Rosen and Freeman, 1984) is so rapid that, in some experimental systems, it is not possible to verify the production of $O_2 \cdot^-$ with current spin trapping methods.

The synthesis of Δ^1-pyrroline 1-oxides with selectivity toward HO· seemed to be a reasonable option for this serious limitation. One synthetic avenue worth exploring was based on the finding that the second-order rate constant for spin trapping $O_2 \cdot^-$ is surprisingly small (Finkelstein, *et al.* 1980). Therefore, it was thought that by increasing the steric hindrance at position 3 of the pyrroline ring, by preparing 3,3,5,5-tetramethyl-1-pyrroline-*N*-oxide (**56**) and 3,3-diethyl-5,5-dimethyl-1-

pyrroline-N-oxide (**64**), it might be possible to prepare nitrones that will not spin trap $O_2 \cdot^-$ (Rosen and Turner, 1988).

Although encouraging results were obtained with nitrone (**64**), the use of this spin trap was found to be limited to specific experimental conditions, as it was easily oxidized in aerobic buffers contaminated with redox active metal ions. Furthermore, dangers encountered during its preparation ended its brief usage for all practical purposes (Rosen and Turner, 1988). In the intervening years, newer analogs, such as 4-[(2-ethoxycarbonyl)ethyl]-3,3,5,5-tetramethyl-1-pyrroline-N-oxide (**76**), have offered an alternative to nitrone (**64**), as this nitrone has similar spin trapping properties without the explosive potential associated with the synthesis of nitrone (**64**) (Arya, *et al.*, 1992).

The synthetic sequences described in the following sections, although designed specifically for these compounds, are general routes, allowing easy adaption to other Δ^1-pyrroline 1-oxides. For 3,5,5-tetramethyl-1-pyrroline-N-oxide (**56**), the key step is the Michael condensation of nitromethane (**52**) with 4-methyl-3-penten-2-one (**53**) in the presence of a base to afford 4,4-dimethyl-5-nitro-2-pentanone (**54**). Reductive cyclization with zinc dust and ammonium chloride results in the formation of 2,4,4-trimethyl-1-pyrroline-N-oxide (**55**). Addition of methylmagnesium bromide gives, after copper acetate-catalyzed aerial oxidation of the corresponding hydroxylamine, 3,3,5,5-tetramethyl-1-pyrroline-N-oxide (**56**) (Bonnett, *et al.*, 1959a):

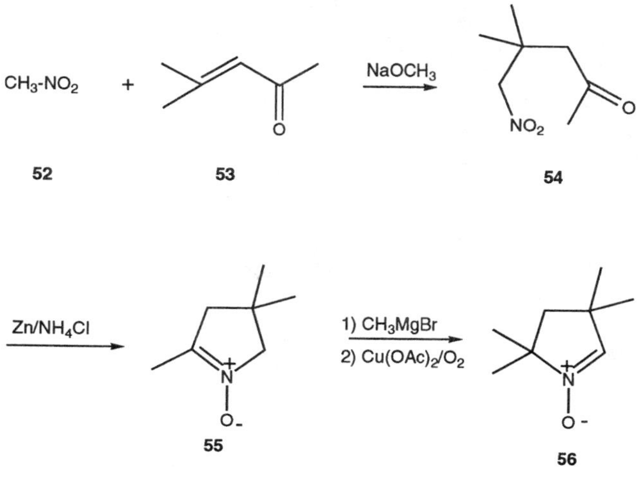

The initial step in the synthesis of 3,3-diethyl-5,5-dimethyl-1-pyrroline-N-oxide (**64**) (Rosen and Turner, 1988) is a Reformansky reaction in which ethyl 3-ethyl-3-hydroxypentanoate (**59**) is prepared by adding magnesium powder to a benzene solution containing ethyl bromoacetate (**57**) and 3-pentanone (**58**). Great care should be exercised when undertaking this reaction as, with time, the mixture becomes exothermic. At this point, if the reaction is not submerged into an ice bath, it will become more violent, eventually discharging the lachrymator, ethyl bromoacetate, from the reaction flask. Hydrolysis of ethyl 3-ethyl-3-hydroxypentanoate (**59**) followed by dehydration with acetic anhydride gives 3-ethyl-2-penten-oic acid (**60**). Methylation with methyl lithium in ether (Dalton and Stokes, 1975) affords 4-ethyl-3-hexene-2-one (**61**). Michael addition with nitromethane and potassium fluoride (Belsky, 1977) in a phase-transfer reaction yields the desired 4,4-diethyl-5-nitro-2-pentanone (**62**). Reductive cyclization with zinc and ammonium chloride affords nitrone (**63**). Reaction with methylmagnesium bromide, followed by copper acetate-catalyzed aerial oxidation of the corresponding hydroxylamine, results in 3,3-diethyl-5,5-dimethyl-1-pyrroline-*N*-oxide (**64**):

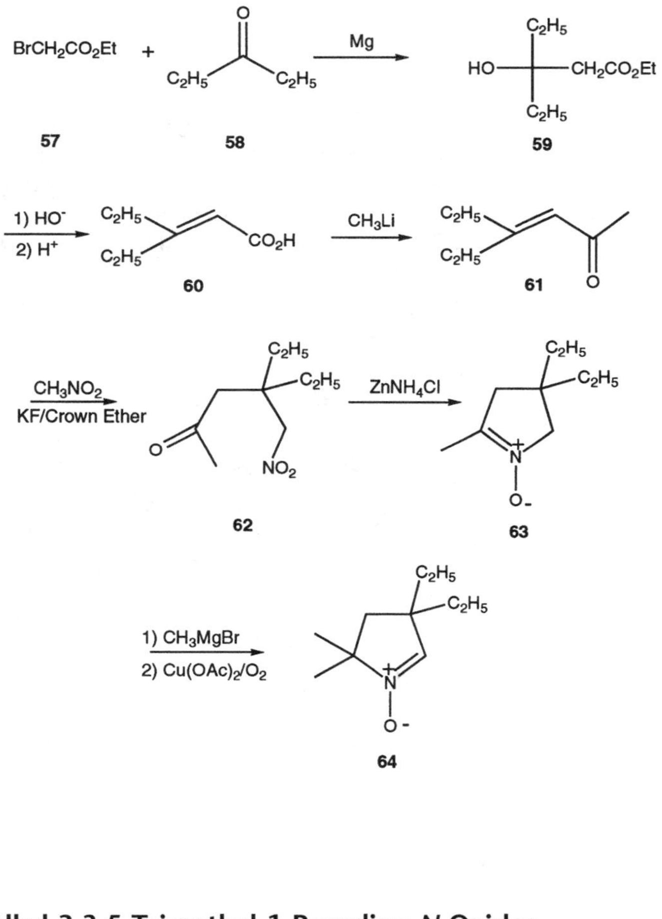

5-Alkyl-3,3,5-Trimethyl-1-Pyrroline-*N*-Oxides

One of the long-term goals of spin trapping is to localize spin traps at cellular sites, thought to be the source of free radicals. Considering that Δ^1-pyrroline-*N*-oxides are inherently hydrophilic, the challenge is to synthesize nitrones with lipophilic characteristics, thereby enhancing cellular uptake of these spin traps. One approach is to add long-chained aliphatic Grignard reagents to 2,4,4-trimethyl-1-pyrroline-*N*-oxide (**55**) (Bonnett, *et al.*, 1959a; Haire and Janzen, 1982; Turner and Rosen, 1986). Typical of this reaction sequence is the preparation of 5-hexadecyl-3,3,5-trimethyl-1-pyrroline-*N*-oxide (**65**) (Barker, *et al.*, 1985):

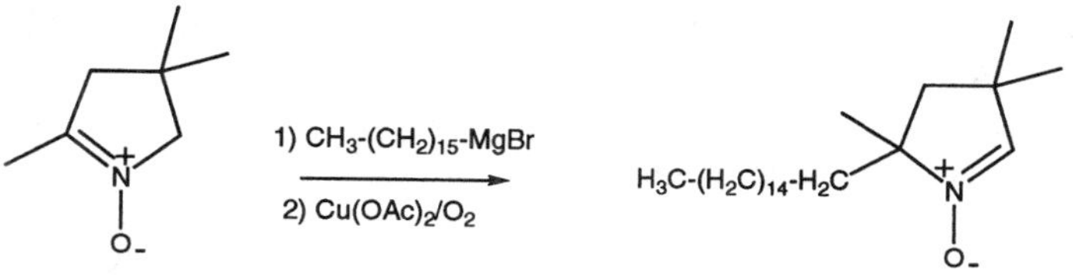

4-Carbethoxy-3,3,5,5-Tetramethyl-1-Pyrroline-*N*-Oxide

The key step in the synthesis of 4-carbethoxy-3,3,5,5-tetramethyl-1-pyrroline-*N*-oxide (**69**) is a Wittig reaction—the 1,3-cycloaddition of the lithio anion, derived from the reaction of butyl lithium with diethyl (*N,N*-dimethyliminomethyl)phosphonate (**66**) (Ratcliffe and Christensen, 1973; Dehnel and Lavielle, 1980), to ethyl 3,3-dimethylacrylate (**67**), giving the Δ^1-pyrroline, 4-carbethoxy-3,3,5,5-tetramethyl-1-pyrroline (**68**) (Dehnel, *et al.*, 1988). Reduction with sodium borohydride followed by sodium tungstate-catalyzed H_2O_2-mediated oxidation yields 4-carbethoxy-3,3,5,5-tetramethyl-1-pyrroline-*N*-oxide (**69**) (Dehnel, *et al.*, 1988; Murahashi *et al.*, 1990). Hydrolysis using standard procedures affords 4-carboxy-3,3,5,5-tetramethyl-1-pyrroline-*N*-oxide (**70**):

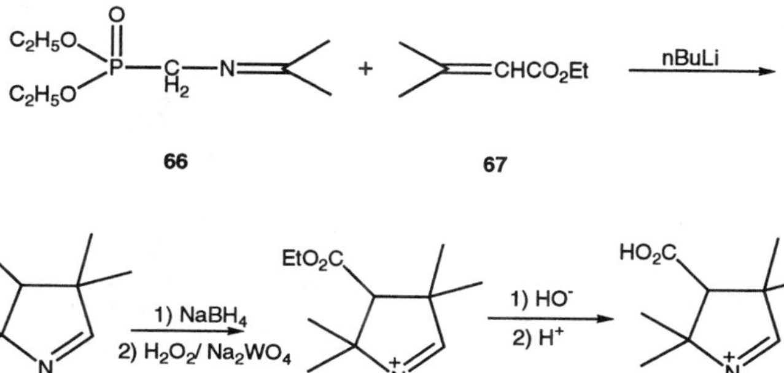

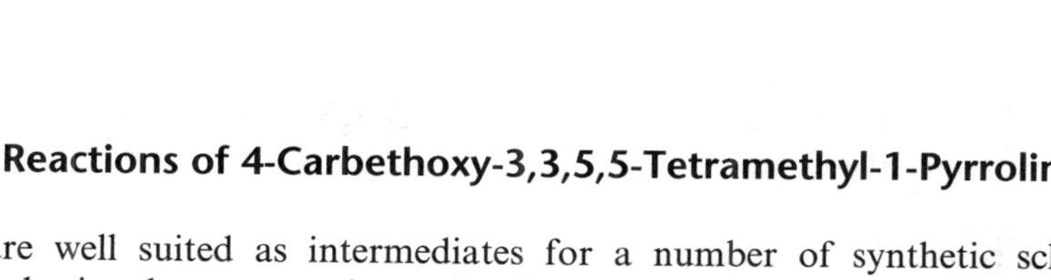

Reactions of 4-Carbethoxy-3,3,5,5-Tetramethyl-1-Pyrroline

Esters are well suited as intermediates for a number of synthetic schemata, particularly in the preparation of other reactive intermediates. Typical of these reactions is the preparation of 4-[2-(ethoxycarbonyl) ethyl]-3,3,5,5-

tetramethyl-1-pyrroline-*N*-oxide (**76**) (Arya, *et al.*, 1992). The initial step is a lithium aluminum hydride reduction of 4-carbethoxy-3,3,5,5-tetramethyl-1-pyrroline (**68**) to the corresponding alcohol, 3-(hydroxymethyl)-2,2,4,4-tetramethyl-1-pyrrolidine (**71**). Selective protection of this pyrrolidine with benzyl chloroformate followed by oxidation of the alcohol (**72**) with pyridinium chlorochromate gives the aldehyde (**73**) (Arya, *et al.*, 1992). A Wittig reaction of aldehyde (**73**) with (carbethoxymethylene)triphenylphosphorane in benzene at $\approx 60°C$ affords, after catalytic hydrogenation with Pd/C, 3-[2-(ethoxycarbonyl)ethyl]-2,2,4,4-tetramethyl-1-pyrrolidine (**75**). Oxidation with sodium tungstate/H_2O_2 leads to 4-[2-(ethoxycarbonyl)ethyl]-3,3,5,5-tetramethyl-1-pyrroline-*N*-oxide (**76**) (Arya, *et al.*, 1992):

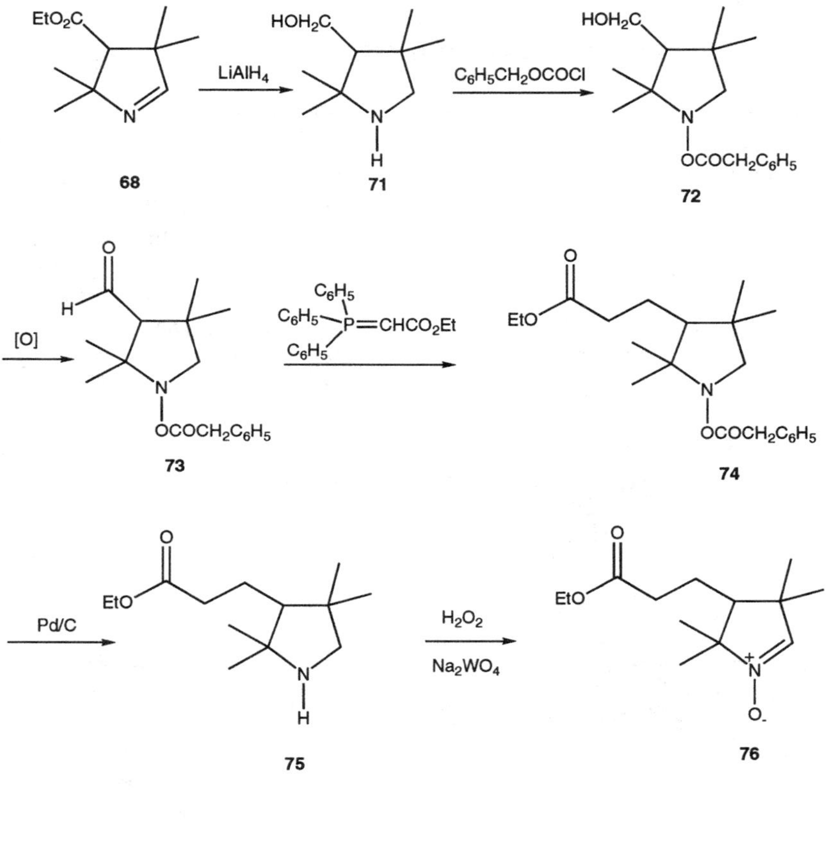

4-(*N*-Maleimidobutyryloxymethyl)-3,3,5,5-Tetramethyl-1-Pyrroline-*N*-Oxide

Maleimide-containing nitroxides, to which functional groups can be covalently attached, have been successfully used to study the fluidity of membranes by incorporating the spin labeled maleimide into thiol proteins (Butterfield, *et al.*, 1976;

Barber, *et al.*, 1983) or thiocholesterol (Huang, *et al.*, 1970). By labeling the plasma membrane of intact cells with a maleimide-containing nitroxide, it has recently become possible to investigate the effects of free radicals on the integrity of the cell (Rosen, *et al.*, 1983, 1999; Freeman, *et al.*, 1986). The synthesis of maleimide-linked Δ^1-pyrroline-*N*-oxides is, therefore, a logical evolution of the earlier work (Butterfield, *et al.*, 1976; Freeman, *et al.*, 1986) and allows for possible detection of free radicals at specific cellular sites. In the preparation of these spin traps, *N*-(benzyloxycarbonyl)-3-(hydroxymethyl)-2,2,4,4-tetramethylpyrrolidine (**72**) is an ideal intermediate. Thus, reaction of structure (**72**) with *N*-maleimidiobutryic acid (**77**) gives *N*-(benzyloxycarbonyl)-3-(*N*-maleimidobutyryloxymethyl)-2,2,4,4-tetramethylpyrrolidine (**78**). After removal of the protective group with hydrobromide in acetic acid, pyrroldine (**79**) is obtained. Oxidation with sodium tungstate/ H_2O_2 results in the formation of 4-(*N*-maleimidobutyryloxymethyl)-3,3,5,5-tetramethyl-1-pyrroline-*N*-oxide (**80**) (Arya, 1996):

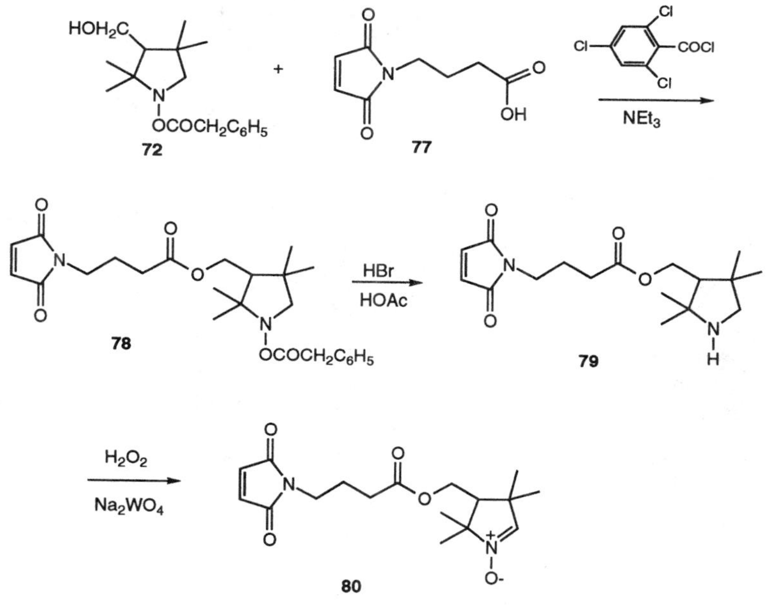

Other Cyclic Nitrones

As our knowledge of the involvement of free radicals in a myriad of human diseases continues to grow, there is an ever-increasing need to develop spin traps to identify specific free radicals at sites of pathophysiologic injury (Carney, *et al.*, 1991; Novelli, 1992; Pogrebniak, *et al.*, 1992). The ease by which nitrones can be incorporated into pharmacological agents with high affinities for specific conditions broadens the versatility of this class of spin traps. Recently, for instance, isoquinoline-*N*-oxides and benzoazepine-*N*-oxides have received considerable attention as potential therapeutic agents for septic shock and bacteremia, suggesting a previously unknown role for free radicals in these diseases (French, *et al.*, 1994).

The initial step in the preparation of 3,4-dihydro-3,3-dimethylisoquinoline-*N*-oxide (**85**) (Carr, *et al.*, 1994; Bernotas, *et al.*, 1996a) is the synthesis of *N*-(1,1-dimethyl-2-phenethyl)formamide (**82**) from β,β-dimethylphenethyl alcohol (**81**) (Shetty, 1961; Ritter and Kalish, 1964; Lavin, *et al.*, 1981). Ring closure using phosphorous pentoxide gives 3,4-dihydro-3,3-dimethylisoquinoline (**83**). Peracid oxidation to the corresponding oxazirino[3,2a]isoquinoline (**84**) followed by intramolecular rearrangement results in the formation of the desired isoquinoline-*N*-oxide (**85**):

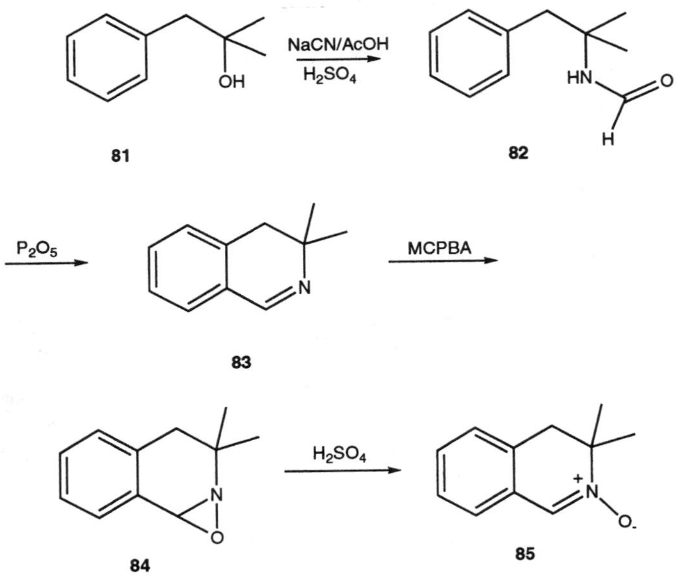

By varying the substituents on the β,β-dimethylphenethyl alcohols (**86**), isoquinoline-*N*-oxides with different degrees of lipophilicity can readily be synthesized (Bernotas, *et al.*, 1996a, 1996b; Dage, *et al.*, 1997). In this, however, cyclization with oxalyl chloride affords the desired isoquinoline (**87**). Upon either hydrolysis with sulfuric acid or heating structure (**87**) neat, the resulting substituted 3,3-dimethyl-3,4-dihydroisoquinoline (**88**) can be reduced by NaBH$_4$ to the substituted 1,2,3,4-tetrahydro-3,3-dimethylisoquinoline (**89**). Upon treatment with H$_2$O$_2$/Na$_2$WO$_4$, substituted 3,3-dimethyl-3,4-dihydroisoquinoline-*N*-oxides (**90**) can be obtained in reasonable yields (Bernotas, *et al.*, 1996a, 1996b). Alternatively, the direct oxidation of the isoquinoline (**88**) with H$_2$O$_2$/Na$_2$WO$_4$ affords the same spin trap:

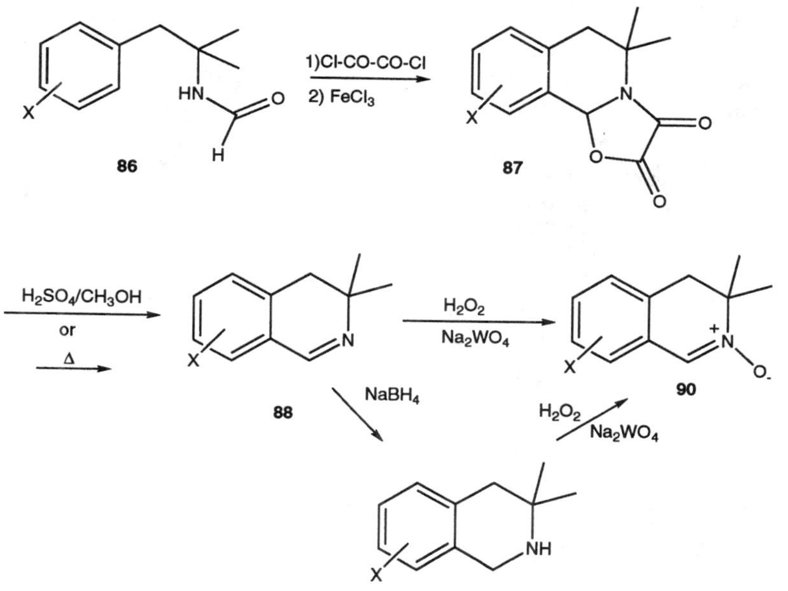

Using a similar synthetic scheme with minor modifications, 3,3-dimethyl-4,5-dihydro-3H-2-benzazepine-N-oxide (**95**) can likewise be prepared in reasonable yields (Carr, *et al.*, 1994; Bernotas, *et al.*, 1996c):

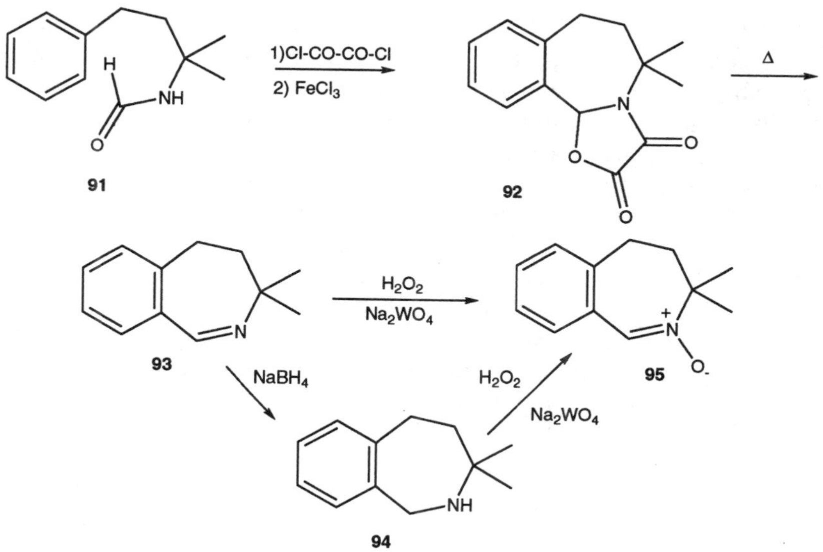

The recent disclosure that 2H-imidazole-N-oxides can react with a variety of free radicals of biologic interest (Dikalov, *et al.*, 1992, 1996) has ushered in a new class of spin traps. This finding is of particular importance, considering the specificity described for these compounds (Dikalov, *et al.*, 1992) and the ease of their preparation. For instance, the condensation of substituted α-hydroximino ketones (**96**) with various ketones (**97**) in the presence of ammonium acetate results in the formation of a family of 2*H*-imidazole-*N*-oxides (**98**) (Kirilyuk, *et al.*, 1991). By varying the lipophilicity of each reactant, 2*H*-imidazole-*N*-oxides can be synthesized with functional groups that would allow for the identification of specific free radicals in cellular and extracellular loci (Kirilyuk, *et al.*, 1991; Dikalov, *et al.*, 1996):

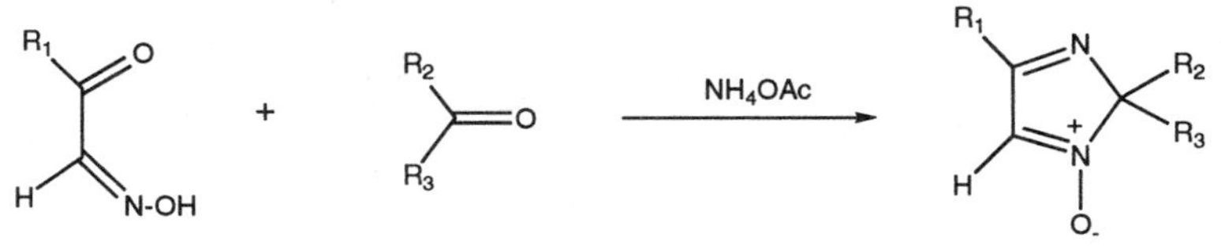

Isotope-Labeled Pyrroline-*N*-Oxides

Based on early *in situ* pharmacokinetic studies with a low-frequency EPR spectrometer (Rosen, *et al.*, 1988), enhanced sensitivity of the spin trapping method was found to be essential prior to attempts to spin trap free radicals *in vivo*. One approach, based on previous studies in which an ^{15}N–^{2}H-containing nitroxide was shown to exhibit a 10-fold increase in sensitivity as compared with its ^{14}N–^{1}H-analog (Beth, *et al.*, 1981), is to substitute ^{2}H and ^{15}N in place of ^{1}H and ^{14}N in nitrone (**5**) (Pou, *et al.*, 1990). Various ring and substitutent aliphatic hydrogens contribute to the unresolved broadening of the EPR spectral lines. Deuterium has a magnetic moment approximately 1/7 that of hydrogen, but higher multiplicity, giving an overall linewidth reduction by a factor of 4. This directly translates into an increased EPR spectral peak height by a factor of 4 for a given concentration of substrate. In turn, this will give a proportionately higher sensitivity toward $O_2\cdot^-$ and HO· (Halpern, *et al.*, 1993).

The acidic nature of the hydrogen at the 2-position of 5,5-dimethyl-1-pyrroline-*N*-oxide (**5**) makes it possible to synthesize substituted Δ^1-pyrroline 1-oxides through base treatment (Elsworth and Lamchen, 1970; Haire, *et al.*, 1988a). Of interest is the finding that gentle warming of this nitrone with NaOD and D_2O leads to the exchange of hydrogens at positions 2 and 3 through tautomerization, thereby resulting in the formation of 5,5-dimethyl[2,3,3-$^2H_3$]-1-pyrroline-*N*-oxide (**99**) (Pou, *et al.*, 1990):

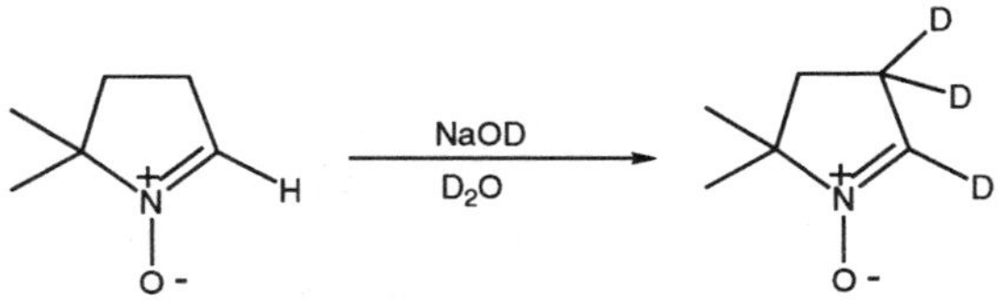

Unlike the preparation of nitrone (**99**), the perdeuteration of nitrone (**5**) requires de novo synthesis. An acid-catalyzed proton exchange of 5-nitropentan-2-one (**100**) seems to be a likely scheme to 5-nitro(1,1,1,3,3-$^2H_5$]pentan-2-one (**101**). As with earlier synthetic sequences, reductive cyclization of (**101**) with zinc dust results in reasonable yields of 1-([$^2H_3$]methyl)[2,2-$^2H_3$]-1-pyrroline-*N*-oxide (**102**). The addition of methyl-d_3-magnesium bromide affords 5,5-di([$^2H_3$]methyl) [4,4-$^2H_2$]-1-pyrroline-*N*-oxide (**103**). The final step of base-catalyzed deuterium exchange leads to the perdeuterated nitrone, 5,5-di([$^2H_3$]methyl)[2,3,3,4,4-$^2H_5$]-1-pyrroline-*N*-oxide (**104**) (Pou, *et al.*, 1990):

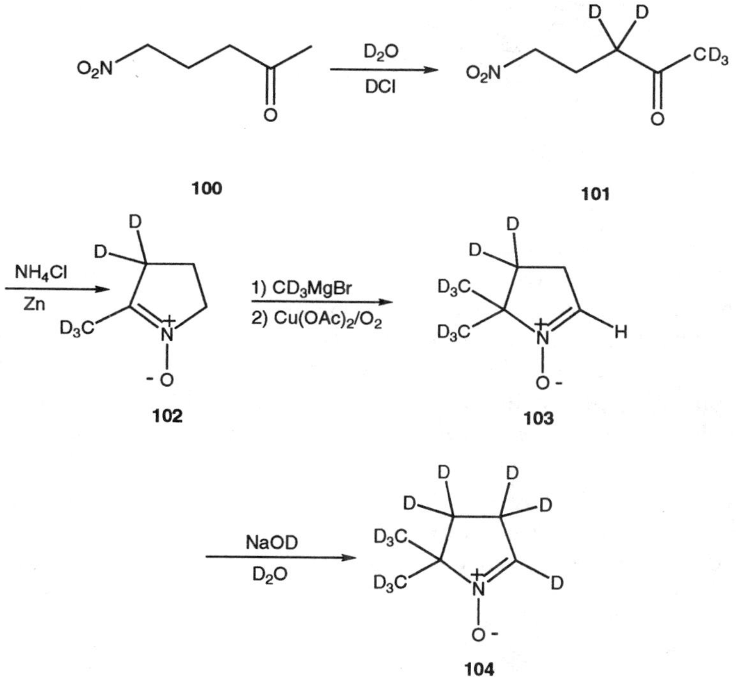

New experimental procedures were developed for the efficient synthesis of ^{15}N-deuterium-labeled analogs of nitrone (**5**). The initial step calls for the formation of the ketoxime (**106**). As anion (**107**) is required for the Michael reaction with acrolein, a modification of the method of Barnes and Patterson (1976) was adopted, yielding 4-([$^2H_3$]methyl)-4-[^{15}N]nitro)[5,5,5-$^2H_3$]pentanal (**108**). After protecting aldehyde (**108**) with ethylene glycol, 5,5-di([$^2H_3$]methyl)[^{15}N]-1-pyrroline-*N*-oxide (**110**) is obtained upon zinc reduction. Deuterium exchange at positions 2 and 3 is achieved by mild treatment with NaOD in D_2O, affording 5,5-di([$^2H_3$]methyl)[^{15}N,2,3,3-$^2H_3$]-1-pyrroline-*N*-oxide (**111**) (Pou, *et al.*, 1990):

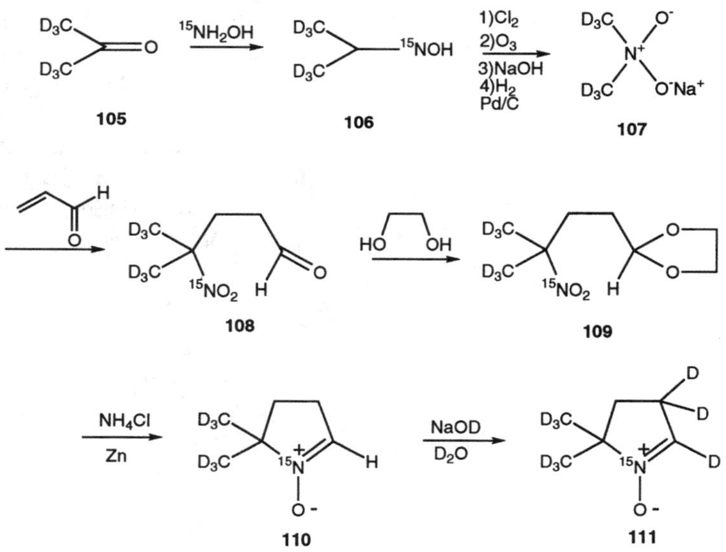

One of the perplexing problems associated with the spin trapping of HO· and O$_2$·$^-$ by 5,5-dimethyl-1-pyrroline-N-oxide (5) is the mediocre stability of the corresponding spin trapped adducts (see Tables 8.5 and 8.6). Despite efforts to overcome this limitation, until the recent syntheses of 2-substituted-5,5-dimethyl-1-pyrroline-N-oxides enriched with ^{13}C at either the α- or β-carbon, there were few solutions to the poor endurance of α-hydrogen-containing nitroxides, an indispensable element essential to the discriminating features of nitrone (5). The simplest of these Δ^1-pyrroline 1-oxides is 5,5-dimethyl-2-([^{13}C]-methyl)-1-pyrroline-N-oxide (112), obtained by the addition of [^{13}C]-methyl magnesium bromide to nitrone (5). After copper (II)-catalyzed aerial oxidation of the corresponding hydroxylamine, nitrone (112) can be isolated using standard methods (Barasch, *et al.*, 1994):

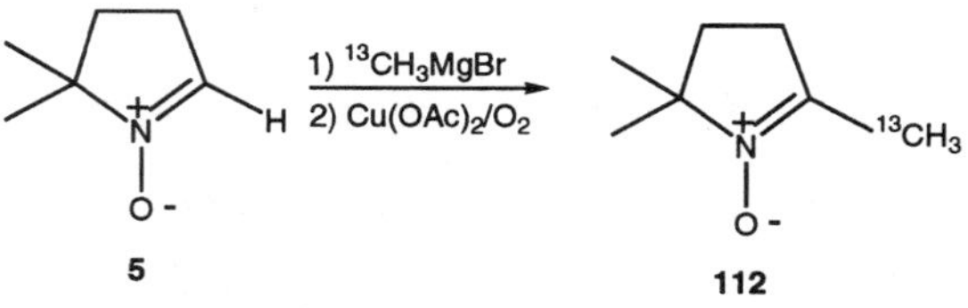

The initial step in the preparation of 2,5,5-trimethyl-1-pyrroline-N-oxide enriched with ^{13}C at the α-carbon requires the synthesis of 4-(diethylamino)-[2-^{13}C]-2-butanone (114) from [2-^{13}C]acetone (113). After a Michael addition with 2-nitropropane, the isolated 5-methyl-5-nitro-[2-^{13}C]-2-hexanone (115) is reduced with zinc dust, affording 2,5,5-trimethyl-[2-^{13}C]-1-pyrroline-N-oxide (116) (Barasch, *et al.*, 1994):

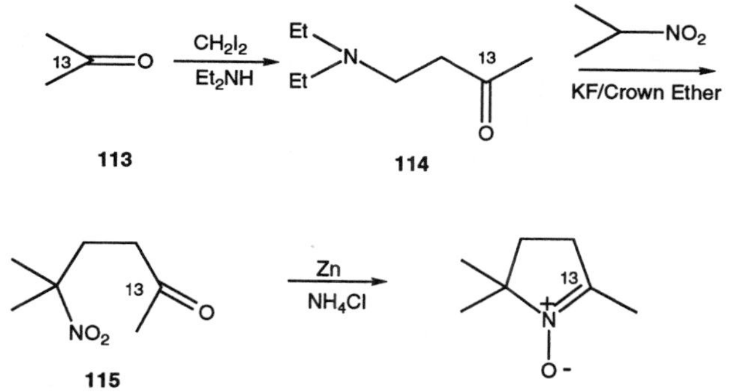

113   **114**

115   **116**

The synthesis of 5,5-dimethyl-2-phenyl-[2-^{13}C]-1-pyrroline-N-oxide (**121**), although similar to other Δ^1-pyrroline 1-oxides (Black and Boscacci, 1976), requires the preparation of phenylvinyl-[2-^{13}C]-ketone (**119**), obtained from the reaction of benzoyl-[2-^{13}C]-chloride (**117**) with vinyltributyltin (**118**) (Labadie, *et al.*, 1983). Upon condensation with 2-nitropropane in the presence of a base, the resulting nitroketone (**120**) is reduced with zinc dust, affording 5,5-dimethyl-2-phenyl-[2-^{13}C]-1-pyrroline-N-oxide (**121**) (Janzen, *et al.*, 1994):

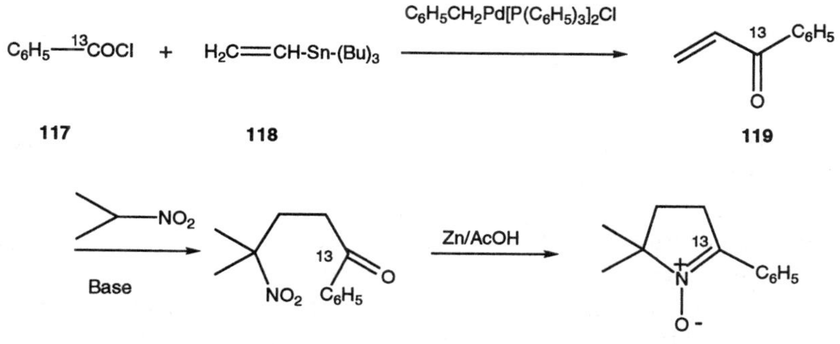

117   **118**   **119**

120   **121**

More recent synthetic efforts have taken a slightly different approach. Here, the additional hyperfine splittings associated with γ-fluorides, as observed from the trifluoromethyl spin trapped adduct of PBN (Janzen, *et al.*, 1973), have been incorporated into a Δ^1-pyrroline-N-oxide (Janzen, *et al.*, 1995). Specifically, 5,5-dimethyl-2-(trifluoromethyl)-1-pyrroline-N-oxide (**127**) has been prepared, based on the belief that additional stability and enhanced lipophilicity, as compared with nitrone (**5**), will allow identification of free radicals at intracellular sites of formation (Janzen, *et al.*, 1995). The initial step in the synthesis of nitrone (**127**) requires the sequential reduction of ethyl 4,4,4-trifluoroacetoacetone (**122**) by BH$_4$ and LiAlH$_4$ (Tordeux and Wakselman, 1982), giving 2,4-dihydroxy-1,1,1-trifluorobutane (**123**). Protecting the 1-hydroxy through tosylation (Tordeux and Wakselman 1982) results in 2-hydroxy-4-(tosyloxy)-1,1,1-trifluorobutane (**124**), which, after oxidation with the Dess-Martin reagent (Dess and Martin, 1983), affords 4-(tosyloxy)-1,1,1-trifluoro-2-butanone (**125**) (Linderman and Graves, 1989; Janzen, *et al.*, 1995). Condensing 4-(tosyloxy)-1,1,1-trifluoro-2-butanone (**125**) with the anion of 2-methyl-2-nitropropane (**2**) results in 5-methyl-5-nitro-1,1,1-trifluoro-2-hexanone (**126**). Reductive cyclization of (**126**) with zinc dust gives 5,5-dimethyl-2-(trifluoromethyl)-1-pyrroline-*N*-oxide (**127**) (Janzen, *et al.*, 1995):

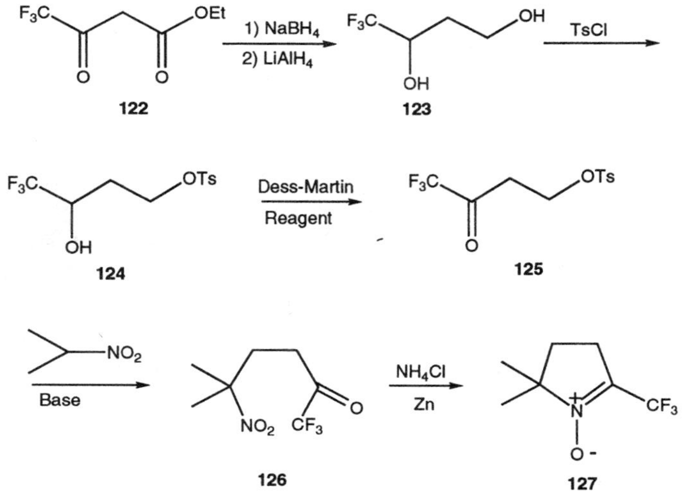

Finally, the introduction of a phosphorous group at position 2 of nitrone (**5**) offers several advantages over nitrone (**112**) (Janzen and Zhang, 1995). First, phosphorous hyperfine splittings are large. Therefore, the EPR spectrum from the resulting spin trapped adduct might be more sensitive to structural changes in the pyrrolidinoxyl ring than other 2-substituted-Δ^1-pyrroline-*N*-oxides. Second, as phosphorus has the same nuclear spin of 1/2 as does hydrogen, EPR spectra derived from 2-(diethylphosphono)-5,5-dimethyl-1-pyrroline-*N*-oxide (**128**) might be similar to those obtained with nitrone (**5**). This, of course, is an enormous advantage, as a first step in the analytical process of free radical identification is evaluating the spectral splitting pattern of the corresponding spin trapped adduct. The preparation of nitrone (**128**) relies on the addition of the lithio salt of diethyl phosphite to nitrone (**5**). After copper (II) acetate-catalyzed aerobic oxidation of the resulting hydroxylamine, 2-(diethylphosphono)-5,5-dimethyl-1-pyrroline-*N*-oxide (**128**) can be obtained in reasonable yields (Janzen and Zhang, 1995):

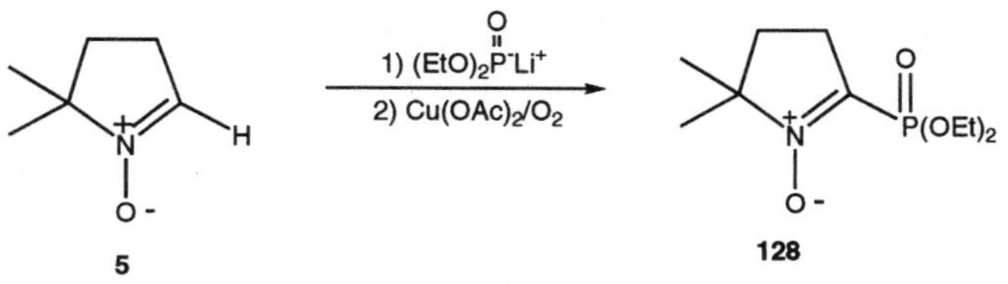

5-(Diethoxyphosphoryl)-5-Methyl-1-Pyrroline-*N*-Oxide

Phosphorus-containing nitrones provide an opportunity to incorporate a spin trap at sites where cellular metabolism may generate free radicals. As such, the phosphonate becomes a carrier for the spin trap—the reporter of free radical events (Tuccio, *et al.*, 1996). The preparation of this family of Δ^1-pyrroline-*N*-oxides can be obtained from a two-step procedure by initially bubbling ammonia into an ethanolic solution of 5-chloropentan-2-one (**129**) and diethyl phosphite. The resulting diethyl (2-methyl-2-pyrrolidinyl)phosphonate (**130**) is oxidized by *m*-chloroperbenzoic acid to 5-(diethoxyphosphoryl)-5-methyl-1-pyrroline-*N*-oxide (**131**):

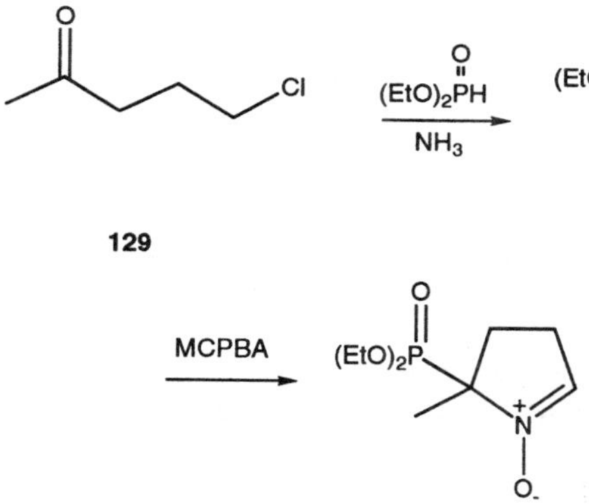

Using an experimental design similar to that reported for the formation of nitrone (**99**) (Pou, *et al.*, 1990), reaction of nitrone (**131**) with NaOD/D$_2$O affords a mixture of 5-(diethoxyphosphoryl)-5-methyl[2,3-^{2}H$_2$]-1-pyrroline-*N*-oxide (**132**) and 5-(diethoxyphosphoryl)-5-methyl[2,3,3,-^{2}H$_3$]-1-pyrroline-*N*-oxide (**133**) (Fréjaville, *et al.*, 1995).

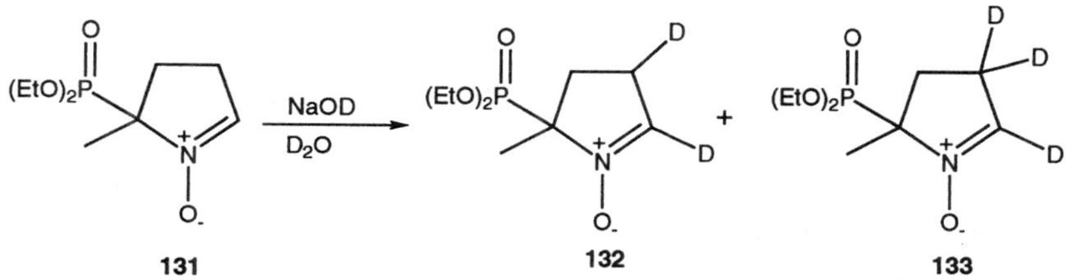

α-Aromatic-*tert*-Butylnitrones

As with the Δ^1-pyrroline-*N*-oxide family of cyclic nitrones, α-aryl-*tert*-butyl-nitrones have their origin at a time predating the development of the spin trapping technique (Emmons, 1957). The availability of a wide variety of aromatic aldehydes and the ease of preparation makes this class of nitrones ideal for exploring xenobiotic-generated free radicals. Although there are several synthetic schemata for the synthesis of α-aryl-*tert*-butylnitrones, each, nevertheless, is straightforward, and, as in the case of *N-tert*-butyl-α-phenylnitrone (**136**), requires the gentle heating of benzaldehyde (**134**) with *tert*-butylhydroxylamine (**135**) (Emmons, 1957). Frequently, a dehydrating agent, such as molecular sieves, can be added to the reaction mixture, shifting the equilibrium in the direction of the nitrone.

Alternative routes to α-aryl-*tert*-butylnitrones involve the condensation of *tert*-butylamine with aldehydes. Oxidation of the corresponding imines with peracetic acid gives reasonable yields of 2-*tert*-butyl-3-aryloxaziranes. Thermal conversion of these oxaziranes in refluxing acetonitrile results in high yields of the desired spin trap (Emmons, 1957). Typical of this reaction sequence is the preparation of *N*-benzylidene-*tert*-butylamine (**137**) from the reaction of benzaldehyde (**134**) with *tert*-butylamine. After peracid oxidation to 2-*tert*-butyl-3-phenyloxazirane (**138**), warming results in rearrangement to give *N-tert*-butyl-α-phenylnitrone (**136**). The necessity for 90% H_2O_2 and the danger accompanying the use of such a potentially explosive reagent makes this preparative route to α-aryl-*tert*-butylnitrones undesirable, considering the ease of other synthetic pathways.

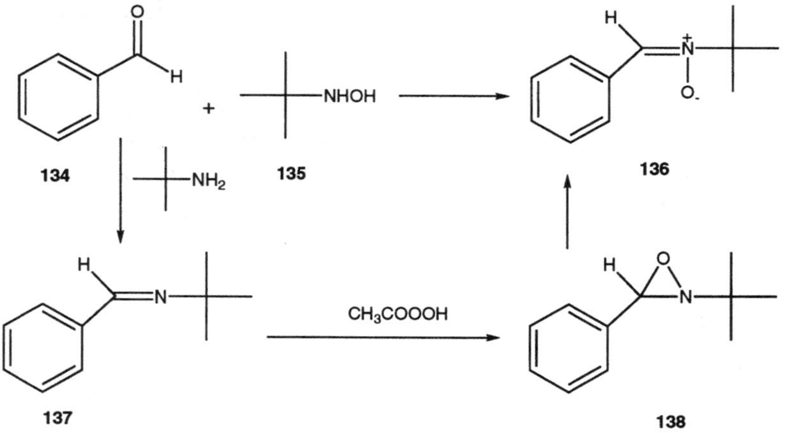

Likewise, reacting *N*-(*tert*-butyl)hydroxylamine (**135**) with 4-pyridinecarboxy-aldehyde *N*-oxide (**139**) affords α-(4-pyridyl 1-oxide)-*N-tert*-butylnitrone (**140**). Not surprisingly, α-aryl-*tert*-butylnitrones are susceptible to hydrolysis, the rate of which is pH dependent. For instance, the half-life of nitrone (**140**) at pH 2 is

≈ 14 min, whereas this nitrone is stable for more than 32 hr at neutral pH (Janzen, *et al.*, 1978).

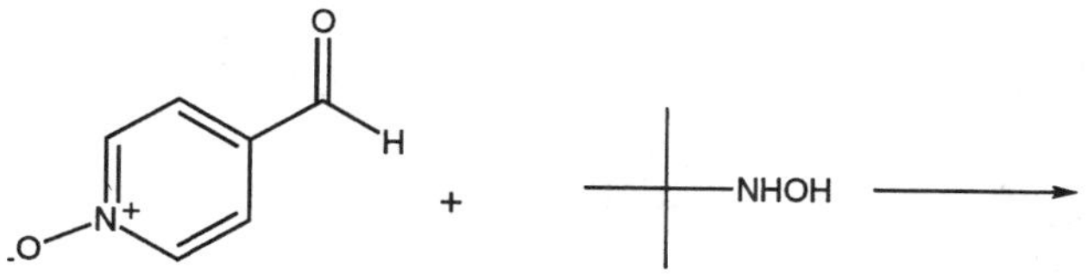

139   +   **135**

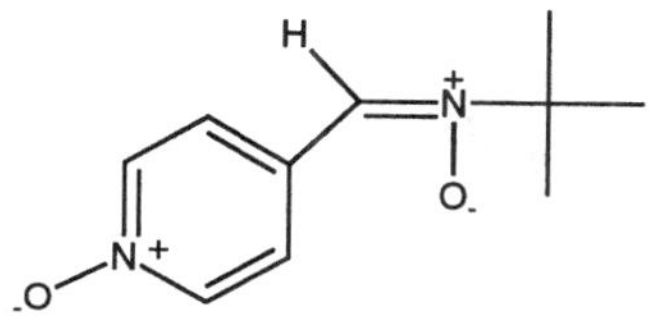

140

Despite the fact that α-aryl-*tert*-butylnitrones are zwitterions with polar characteristics, the lipophilic characteristics of these spin traps can vary greatly. For instance, α-(4-pyridyl 1-oxide)-N-*tert*-butylnitrone (**140**) is a thousand times more hydrophilic than is nitrone (**136**) (Pou, *et al.*, 1994a). Enhanced hydrophobicity can easily be achieved through specific substitution on the aromatic ring (Janzen and Coulter, 1984). In this case, acyclic nitrones, such as N-*tert*-butyl-α-(*p*-dode-cyloxy)phenylnitrone (**144**) (Huie and Cherry, 1985) or N-*tert*-butyl-α-(*p-tert*-butyl)phenylnitrone (**145**) (Janzen and Coulter, 1984) can be synthesized by incorporating, for instance, dodecyloxyl (**141**) or a *tert*-butyl (**142**) group as a hydrophilic tail on benzaldehyde (Janzen and Zawalski, 1978). Condensation of either of the aldehydes (**141**) or (**142**) with *in situ*-produced N-(*tert*-butyl)hydrox-ylamine (**135**), arising from the zinc reduction of 2-methyl-2-nitropropane (**143**), results in the desired spin trap (Huie and Cherry, 1985). Using similar synthetic routes, hydrophilic spin traps, including 4-methylpyridinium-*tert*-butylnitrone methyl sulfate (**149**) and sodium N-*tert*-butyl-α-(*o*-sulfonato)phenylnitrone (**151**) can be easily prepared (Janzen and Shetty, 1979; Janzen, *et al.*, 1979):

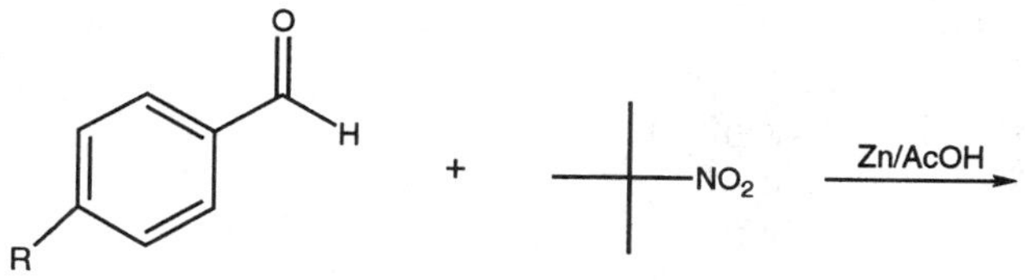

141 = R = CH₃-(CH₂)₁₁-
142 = R = (CH₃)₃C-   **143**

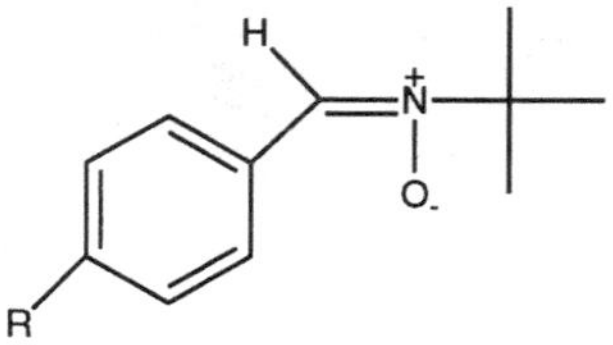

144 = R = CH₃-(CH₂)₁₁-
145 = R = (CH₃)₃C-

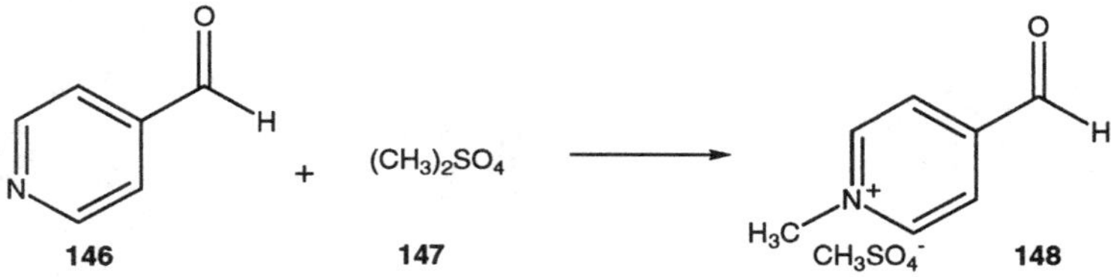

146   **147**   **148**

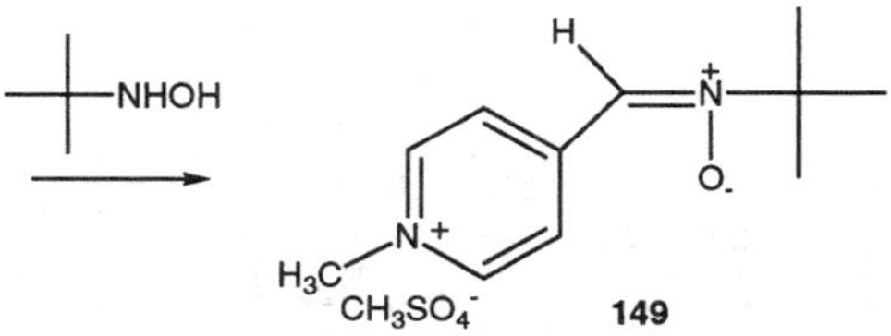

149

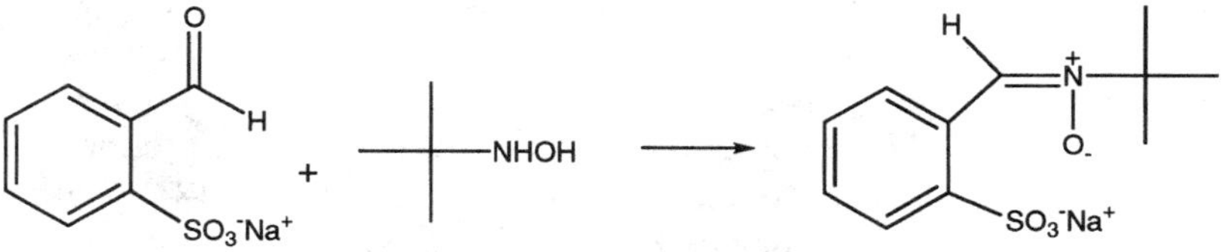

150   **135**   **151**

An alternative approach to enhancing the aqueous solubility of *N-tert*-butyl-α-phenylnitrone (**136**) is to replace a methyl group of the *N-tert*-butyl portion of this nitrone with β-hydroxyethyl. Typically, the synthesis of α-phenyl-*N*-[(1-hydroxy-2-methyl)-2-propyl]nitrone (**155**) can be realized by the reduction of 2-methyl-2-nitropropanol (**152**) in the presence of benzaldehyde (**134**) (Janzen and Zawalski, 1978). With a little ingenuity, three β-hydroxyethyl groups can be incorporated into the N-terminal of the nitrone, resulting in a spin trap with further enhanced hydrophilic properties. As in the case of α-phenyl-N-[(1-hydroxy-2-methyl)-2-propyl]nitrone (**156**), the preparation of nitrone (**157**) requires the *in situ* zinc reduction of the corresponding nitroalkane (**154**) with benzaldehyde (**134**) (Janzen and Zawalski, 1978):

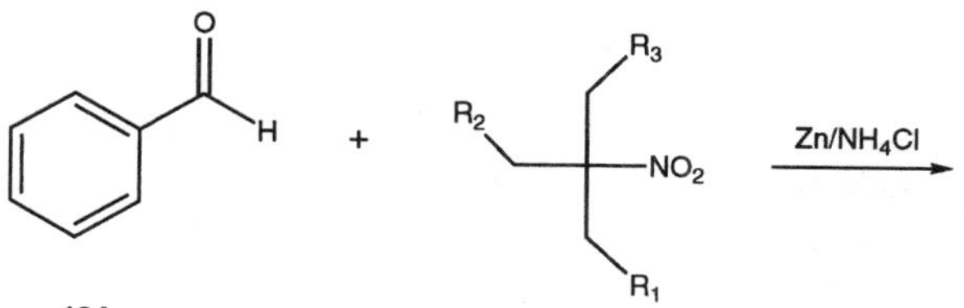

134

152 = R = OH, R_1 = R_2 = H
153 = R = R_1 = OH, R_2 = H
154 = R = R_1 = R_2 = OH

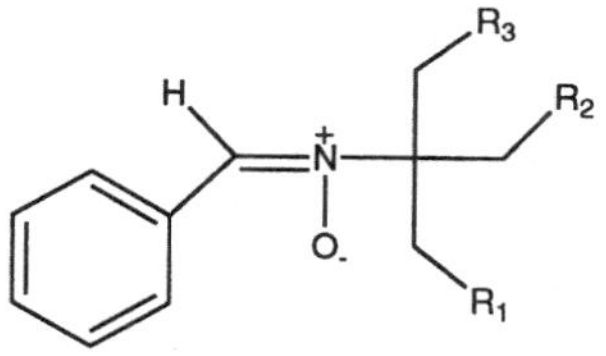

155 = R = OH, R_1 = R_2 = H
156 = R = R_1 = OH, R_2 = H
157 = R = R_1 = R_2 = OH

Through the use of linking agents, nitrones (**155–157**) can be covalently bound to surfaces, allowing the detection of free radicals at specific sites of formation (Bancroft, *et al.*, 1980). The preparation of benzoate esters of alcohols (**152–154**) enhances the yields of the desired acyclic nitrones. Thus, upon reduction of the nitrobenzoyl esters of (**158–160**) in the presence of benzaldehyde (**134**), the corresponding nitronyl alcohols (**161–163**) were obtained in reasonable yields (Janzen and Zawalski, 1978):

1) Benzoyl Chloride (BzCl)
2) Zn
3) Benzaldehyde

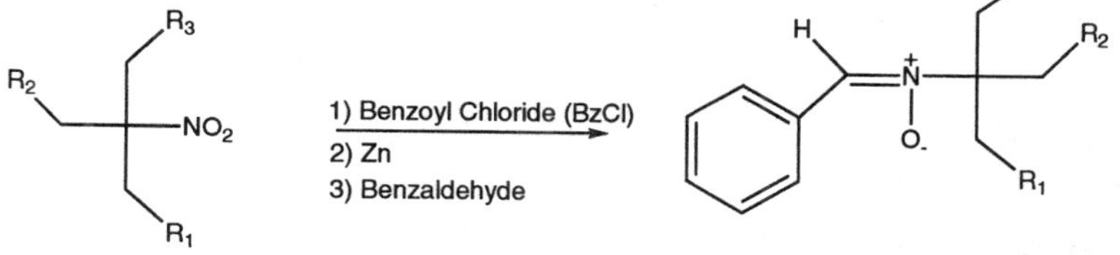

158 = R = OH, R_1 = R_2 = H
159 = R = R_1 = OH, R_2 = H
160 = R = R_1 = R_2 = OH

161 = R = OBz, R_1 = R_2 = H
162 = R = R_1 = OBz, R_2 = H
163 = R = R_1 = R_2 = OBz

Other variations on this general theme include the synthesis of α-phenyl-*N*-[(1-hydroxy-2-ethyl)-2-propyl]nitrone (**167**). In this case, after preparation of 3-nitro-3-methylbutanol (**165**), zinc reduction leads to the corresponding *N*-hydroxyl-amine (**166**). Condensation with benzaldehyde (**134**) affords the desired nitrone (**167**) (Janzen and Zawalski, 1978):

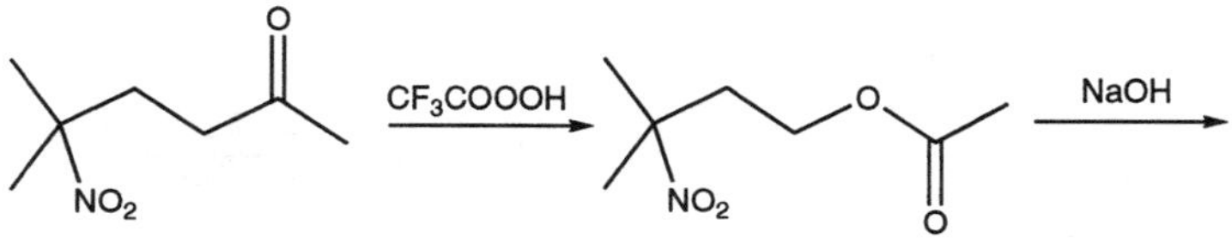

13 → **164**

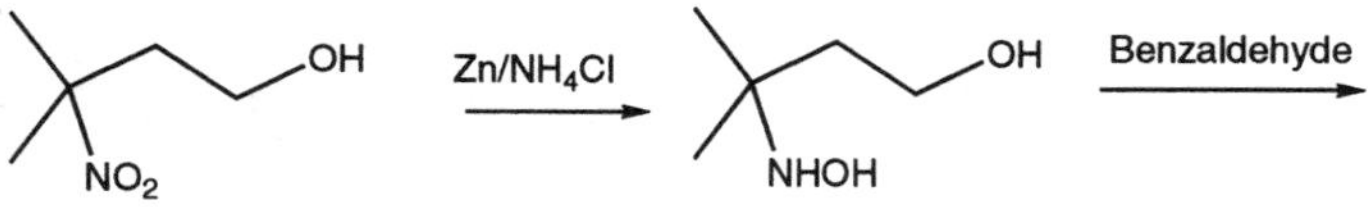

165 → **166**

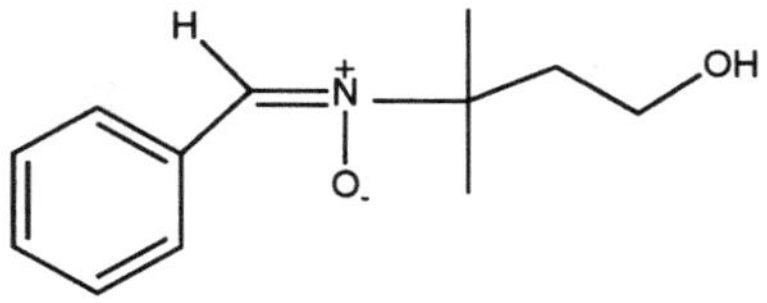

167

The selective reductive properties of sodium cyanoborohydride allow the direct formation of 4-methyl-4-nitropentanol (**168**) from 4-methyl-4-nitropentanal (**3**). As in the preparation of nitrone (**167**), initial zinc reduction of (**168**) followed by reaction with benzaldehyde (**134**) leads to nitrone (**170**) (Janzen and Zawalski, 1978):

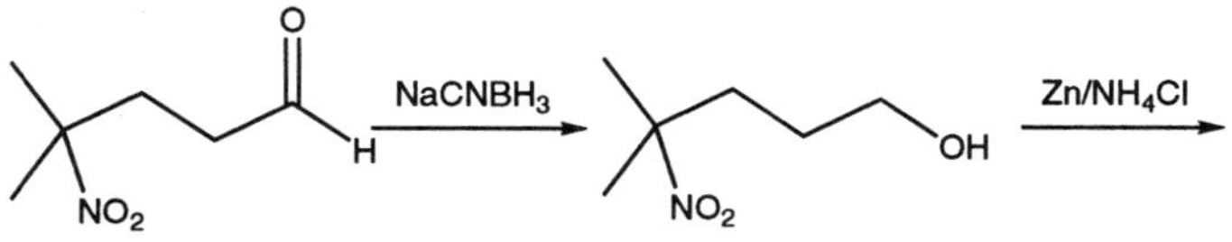

3 → **168**

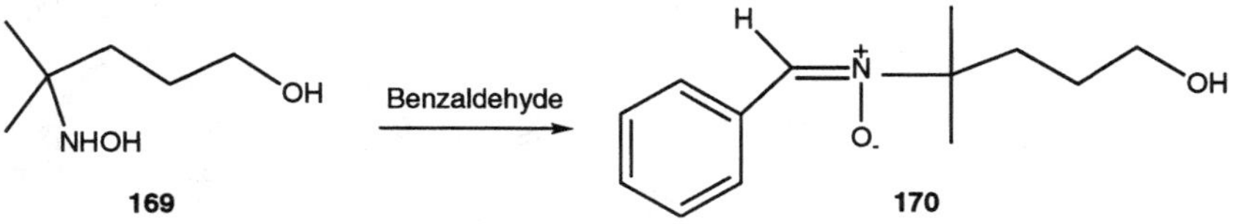

169 → **170**

Bromide is an example of another leaving group whose ease of removal makes this family of compounds ideal intermediates in the preparation of functionalized acyclic nitrones. Thus, condensation of benzaldehyde (**171**) with *N*-(*tert*-butyl)-hydroxylamine (**135**) yields the nitrone (**172**) (Keana and Pou, 1989):

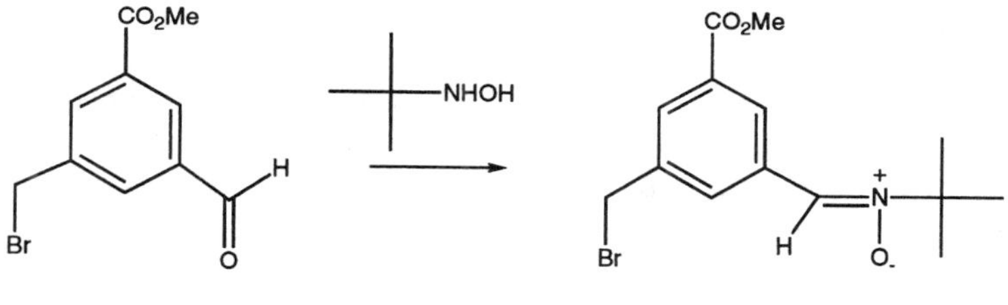

171   **172**

Although the discussion thus far has centered on the functionalization of a nitrone for its potential linkage to specific cellular sites, similar synthetic pathways have led to the formation of a fluorophore-containing spin trap (Pou, *et al.*, 1995). The rationale for the preparation of such a nitrone is based on the inability of EPR spectroscopy to image free radicals at a cellular level. Whereas confocal microscopy in combination with fluorescent probes containing different quantum yields can image cellular sites of free radical events. The design of the first family of these new spin traps, *N-tert*-butyl-α-4-[5-((2-carboxy)phenyl)-5-hydroxy-4-oxo-3-phenyl)-2-pyrrolin-1-yl]phenylnitrone (**176**), was based on the finding that nitroxides, when attached to fluorescent compounds, quench their quantum yields without altering the absorption and emission energies of these compounds (Blough and Simpson, 1987; Green, *et al.*, 1990). Reaction with a free radical enhances the fluorescence by eliminating the nitroxide (Kieber and Blough, 1990; Pou, *et al.*, 1993). The key step in the formation of the fluorophore-containing spin trap (**176**) is the reaction of *N-tert*-butyl-α-(*p*-amino)phenylnitrone (**175**) with fluorescamine in aqueous buffer at pH 9 (Pou, *et al.*, 1995):

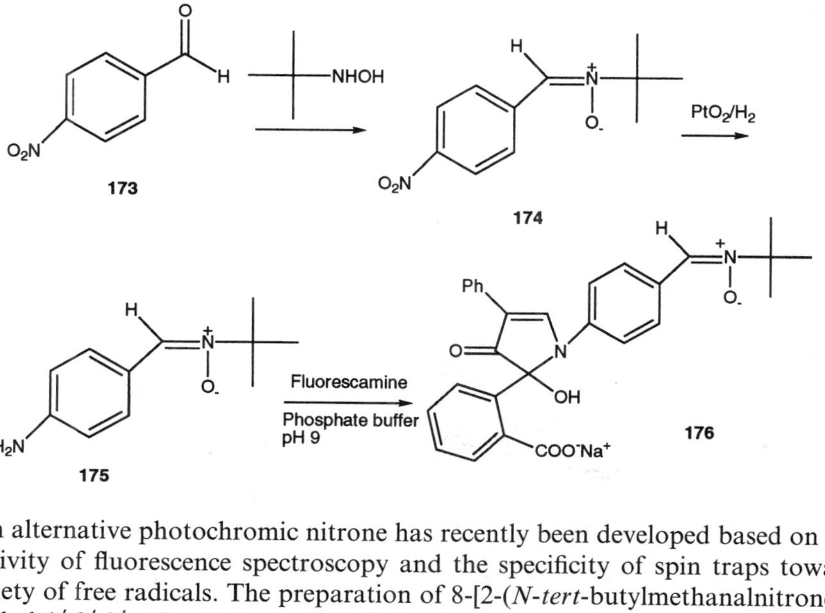

173   **174**

175

An alternative photochromic nitrone has recently been developed based on the sensitivity of fluorescence spectroscopy and the specificity of spin traps toward a variety of free radicals. The preparation of 8-[2-(*N-tert*-butylmethanalnitrone)]-6-methyl-1′,3′,3′,-trimethylspiro[2*H*-1   benzopyran   2,2′-indoline]   (**180**)   was surprisingly straightforward, requiring a two-step sequence in which perchlorate 1,2,3,3-tetramethylindoleninium (**177**) and 2,6-diformyl-6-methylphenol (**178**) were condensed under reflux to afford 8-formyl-6-methyl-1′,3′,3′-trimethylspiro[2*H*-1 benzopyran 2,2′-indoline] (**179**). After addition of N-(*tert*-butyl)hydroxylamine (**135**) to the aldehyde (**179**), the desired nitrone (**180**) was isolated from the reaction mixture (Luccioni-Houzé, *et al.*, 1996):

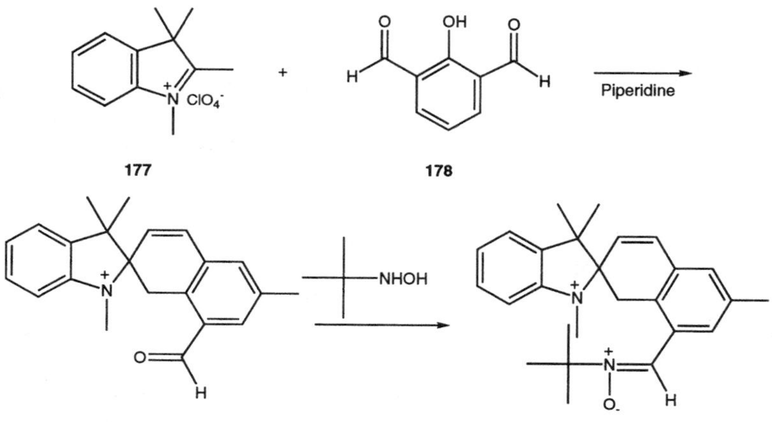

177 **178** **179** **180**

Drawing upon the previous success of fluorophore-containing spin traps (Pou, *et al.*, 1995), a chromotropic nitrone (**184**) was synthesized. Upon reaction with a free radical, a highly sensitive colorimetric product is obtained, which can be isolated using standard procedures (Becker, 1996):

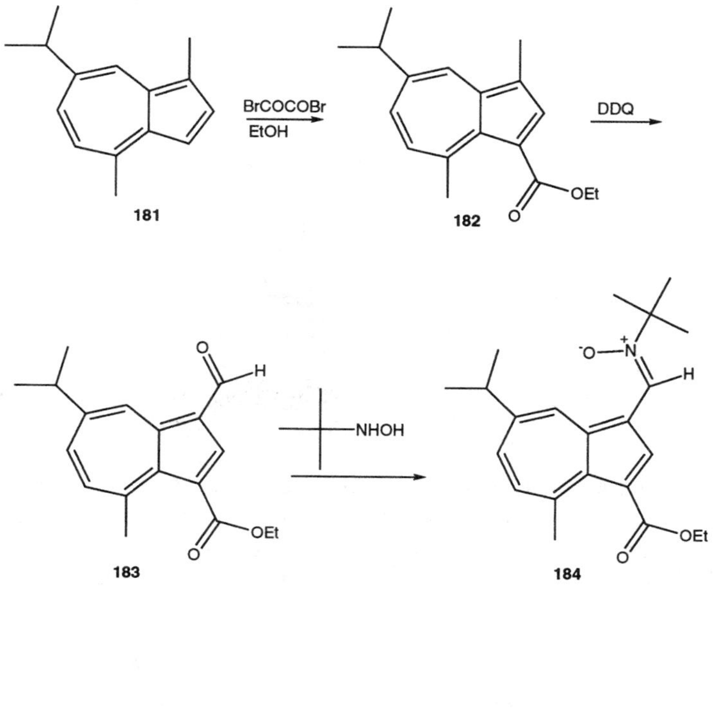

181 **182** **183** **184**

β-Phosphorylated-α-Aromatic-*tert*-Butylnitrones

One of the continued limitations of acyclic nitrones as spin traps for $O_2 \cdot^-$ and HO·
is the poor stability of the corresponding nitroxides (see Tables 8.6 and 8.7;
Finkelstein, *et al.*, 1979, 1980; Kotake and Janzen, 1991; Janzen, *et al.*, 1992).
Based on the recent finding that the half-life of the $O_2 \cdot^-$ spin trapped adduct of
5-diethoxyphosphoryl-5-methyl-1-pyrroline N-oxide (**131**) is significantly longer
than that observed for the corresponding nitroxide derived from the reaction of
nitrone (**5**) with this free radical (see Table 8.7; and Fréjaville, *et al.*, 1994), phosphorylated analogs of N-*tert*-butyl-α-phenylnitrone (**136**) and α-(4-pyridyl 1-
oxide)-N-*tert*-butylnitrone (**140**) have been prepared (Zeghdaoui, *et al.*, 1995).
Whether incorporation of a phosphate group into acyclic nitrones will significantly
enhance the biologic half-lives of the resulting spin trapped adducts of $O_2 \cdot^-$
remains uncertain (see Table 8.7; Roubaud, *et al.*, 1996).

The key step in the synthesis of N-benzylidene-1-diethoxyphosphoryl-1-methyl-
ethylamine N-oxide (**188**) and 1-diethoxyphosphoryl-1-methyl-N-[(oxidopyridin-
1-ium-4-yl)methylidene]ethylamine N-oxide (**189**) is the oxidation of 1-amino-
ethyl-1-methyl phosphonate (**186**) to 1-methyl-1-nitroethyl phosphonate (**187**).
Upon a one-pot reductive condensation with an appropriate aldehyde, the resulting nitrones (**188**) and (**189**) are obtained in acceptable yields (Zeghdaoui, *et al.*,
1995):

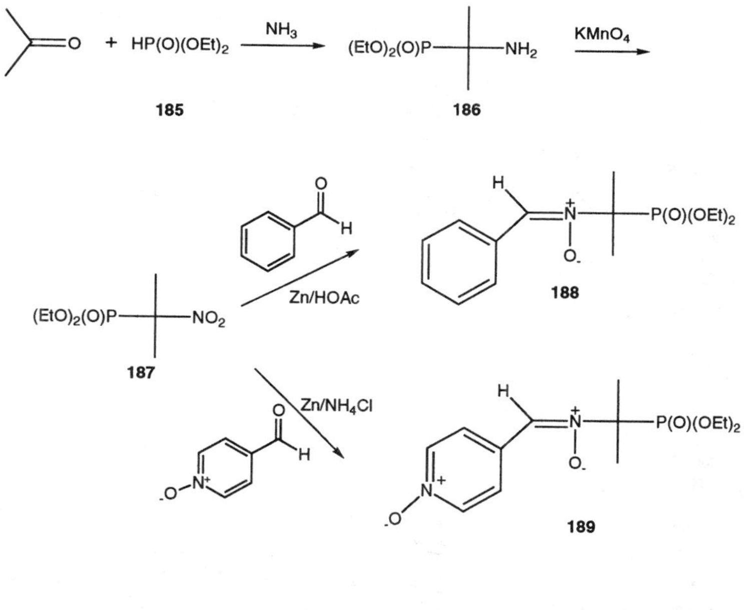

Isotope-Labeled α-Aromatic-*tert*-Butylnitrones

The rationale for the preparation of isotope-labeled acyclic nitrones is to
enhance the EPR signal intensity of a defined spin trapped adduct by decreasing

the hyperfine splittings and narrowing the spectral linewidths of the corresponding nitroxide (Pou, *et al.*, 1990; Zhang, *et al.*, 1991; Arya, *et al.*, 1992, Halpern, *et al.*, 1993; Haire and Janzen, 1994). The simplest of these synthetic sequences results in the formation of *N-tert*-butyl-α-phenyl[2,3,4,5,6-^{2}H$_5$]nitrone (**191**) from the reductive condensation of 2-methyl-2-nitropropane (**143**) with benz-[2,3,4,5,6-^{2}H$_5$]aldehyde (**190**) (Haire and Janzen, 1994). Similarly, *N-tert*-butyl[^{2}H$_9$]-α-phenylnitrone (**194**) and *N-tert*-butyl[^{2}H$_9$]-α-phenyl-[2,3,4,5,6-^{2}H$_5$]-nitrone (**195**) can be obtained from the appropriate deuterium-labeled reactants (Haire and Janzen, 1994):

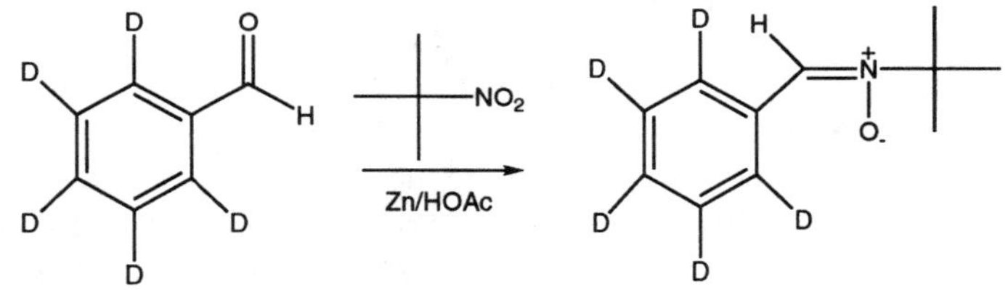

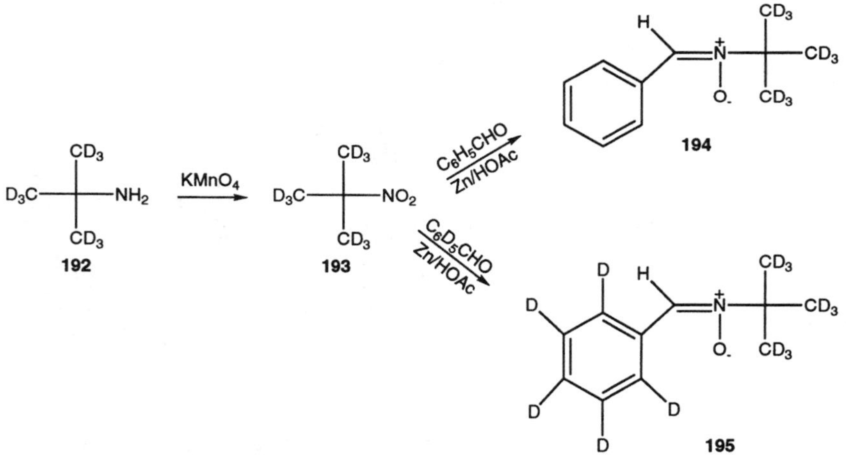

Substituting ^{15}N in place of ^{14}N in the spin trap (Pou, *et al.*, 1990; Zhang, 1996) has been found to increase the sensitivity of the EPR spectrum by a factor of 1.5 (Beth, *et al.*, 1981; Pou, *et al.*, 1990; Zhang, 1996). As in the case of ^{15}N-containing spin traps (**110**) and (**111**), the preparation of the ^{15}N-containing acyclic nitrone, *N-tert*-butyl-α-phenyl[^{15}N]nitrone (**201**) required the development of a new synthetic route to this spin trap. The difficult step in the preparation of nitrone (**201**) was the formation of 2-[^{15}N]amino-2-methylpropane (**199**) from [^{15}N]ammonium chloride (**196**). After oxidation of (**199**) to [^{15}N]nitro-2-methylpropane, reductive condensation with benzaldehyde affords *N-tert*-butyl-α-phenyl[^{15}N]nitrone (**201**) in low but acceptable yields (Zhang, 1996):

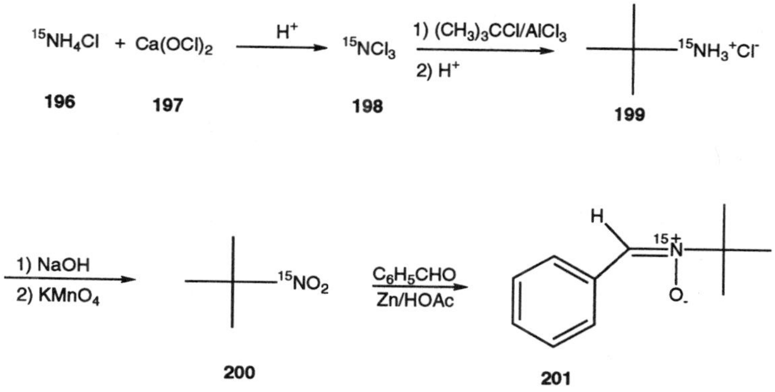

^{15}NH$_4$Cl
+ Ca(OCl)$_2$
H$^+$
^{15}NCl$_3$
1) (CH$_3$)$_3$CCl/AlCl$_3$
2) H$^+$
^{15}NH$_3^+$Cl$^-$
196
197
198
199
1) NaOH
2) KMnO$_4$
^{15}NO$_2$
C$_6$H$_5$CHO
Zn/HOAc
H
15
N
O
200
201

A continued limitation of spin trapping is the characterization of the free radical attached to the nitrone or nitrosoalkane. Even though analytic procedures have been developed, incorporation of a ^{13}C at the site to which the free radical adds might prove to be valuable in the further identification of the parent free radical, complementing other methods of analyses. The simplest acyclic nitrone is *N-tert*-butyl-α-phenyl[^{13}C]nitrone (**203**), prepared by subjecting benz[^{13}C]aldehyde (**202**) to 2-methyl-2-nitropropane (**143**) in the presence of zinc dust (Haire, *et al.*, 1988b):

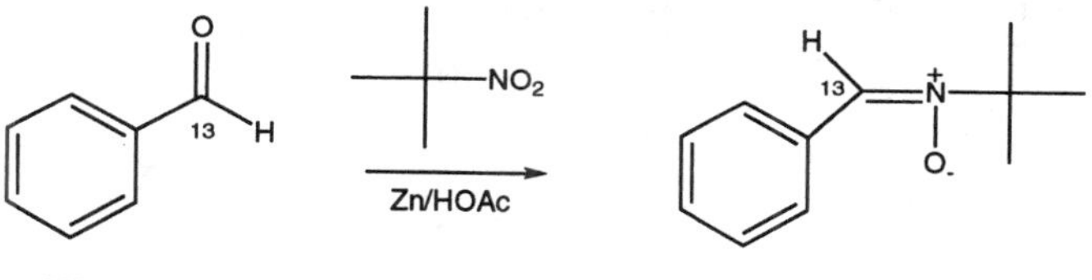

O
13
H
NO$_2$
Zn/HOAc
H
13
N
O

202 **203**

α-Alkoxyl-α-Aryl-*tert*-Butylnitrones

As reporters of $O_2\cdot^-$ and HO·, α-aryl-*tert*-butylnitrones are very limited, due, in part, to the rapidity by which the corresponding spin trapped adducts rearrange to nonparamagnetic species. Typical half-lives of these nitroxides are less than 1 min (Finkelstein, *et al.*, 1979, 1980). Although substitution at the α-position affords an enhanced degree of stability to the spin trapped adduct, the new nitroxide is not without concessions. For instance, the hyperfine splitting constant, A_H, so crucial in characterizing the initial unstable free radical, is lost with the presence of an α-alkyl or α-alkoxyl group. Despite this, α-alkoxyl-α-aryl-*tert*-butylnitrones will undoubtedly find a unique niche. One of the first reported syntheses of acyclic α-alkoxynitrones relied on the condensation of benzamide acetal (**204**) with *N*-methylhydroxylamine hydrochloride (**205**), giving a poor yield of an isomeric mixture of α-ethoxy-α-phenyl-*N*-methylnitrone (**206, 207**) (Ashburn and Coates, 1985):

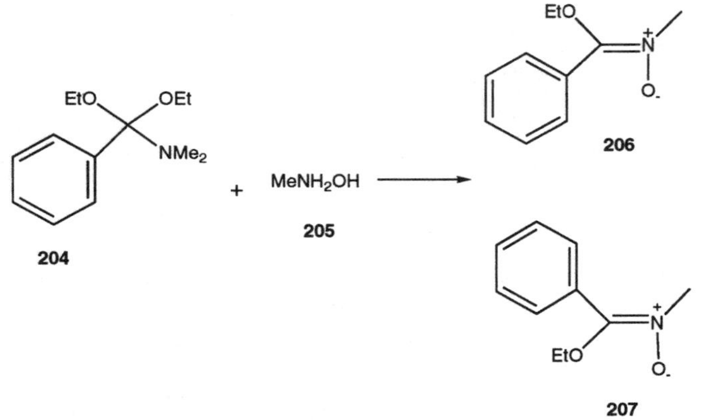

A more successful synthetic sequence calls for the alkylation of a hydroxamic acid (209) with methyl trifluoromethanesulfonate. After deprotonation, the desired acyclic α-alkoxynitrone (211) can be isolated in reasonable yields (Acken, *et al.*, 1989; Warshaw, *et al.*, 1989):

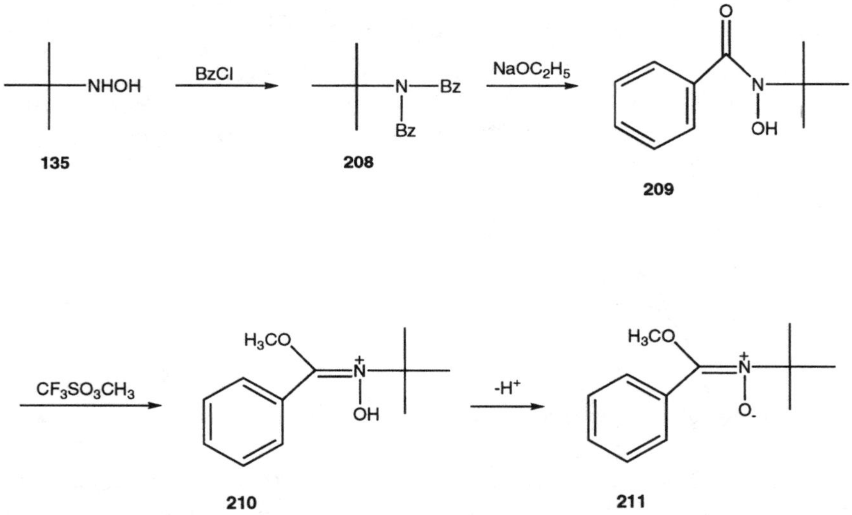

Nitrosoalkanes

The breadth of this class of spin traps is only as limiting as the imagination of the investigator. Such differing nitrosoalkanes as 2-methyl-2-nitrosopropane (212) (Calder, *et al.*, 1972), 2-(hydroxymethyl)-2-nitrosopropane (213) (De Groot, *et al.*, 1973), (sodium 3,5-dibromo-4-nitrosobenzenesulfonate (214) (Kaur, *et al.*, 1981), 2,3,5,6-tetramethylnitrosobenzene (215) (Terabe, *et al.*, 1973), 1-methyl-4-nitroso-3,5-diphenylpyrazole (216), sodium 3-(4-nitroso-3,5-diphenyl-1-pyrazolyl)propane-1-sulfonate (217) (Boyd and Norris, 1974; Kaur and Perkins, 1982), and 9-nitrosotriptycene (218) (Kaur, *et al.*, 1982) offer a diversity of structures with varying lipophilic characteristics:

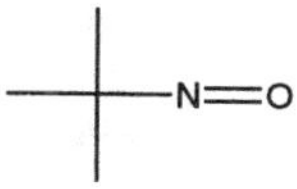

212

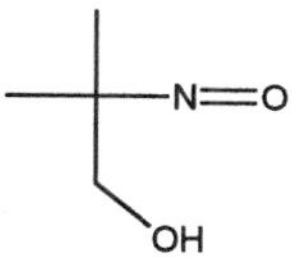

213

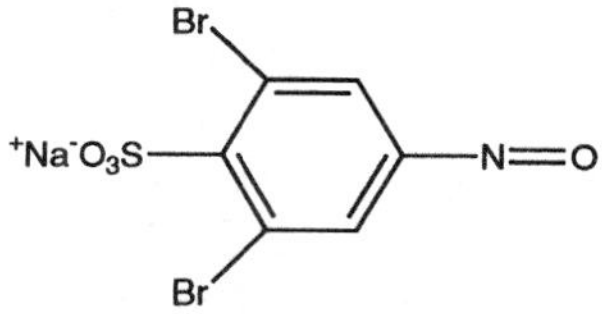

214

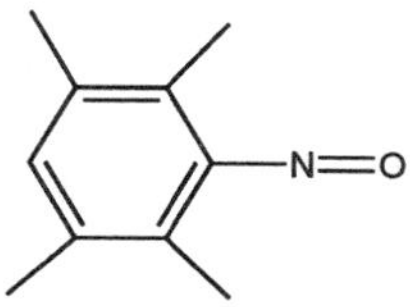

215

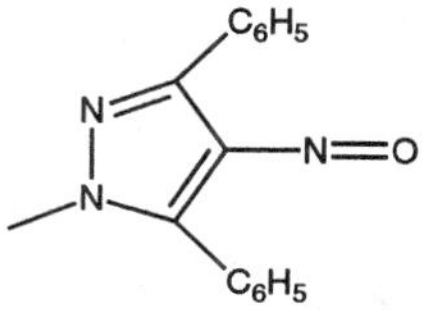

216

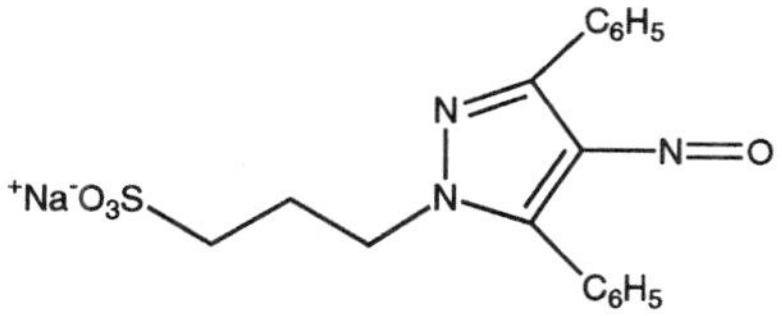

217

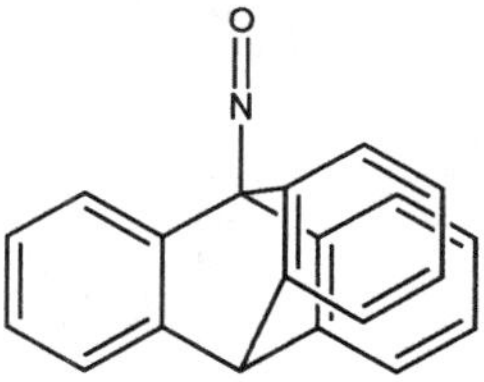

218

The initial step in the preparation of many of these compounds is the reduction of an appropriate nitroalkane to the corresponding *N*-hydroxylamine using reducing agents, such as aluminum amalgam (Calder, *et al.*, 1972), zinc dust (De Groot, *et al.*, 1973), or sodium borohydride (Kalyanaraman, *et al.*, 1979). Formation of the nitrosoalkane can be achieved by treatment with oxidants, such as silver carbonate (De Groot, *et al.*, 1973). For some nitrosobenzenes, alternative synthetic pathways, including nitrosation of aromatic rings (Terabe, *et al.*, 1973) and peroxidation of aromatic amines (Okazaki, *et al.*, 1969; Terabe and Konaka, 1969, 1971; Kaur, *et al.*, 1981), are viable routes to reasonable yields of these spin traps.

Although nitrosoalkanes are one of the easiest classes of spin traps to synthesize, there are nevertheless significant problems that limit their use in biologic systems. First, purification of the crude nitrosoalkane to the degree required for spin trapping is not a trivial task and requires continued care to eliminate nitroxide impurities. Second, questionable solubility in aqueous solutions, especially with substituted aromatic nitrosoalkanes, restricts the application of these spin traps to specific experimental designs. Finally, even under the best of conditions, the propensity of alkyl nitroso compounds, such as 2-methyl-2-nitrosopropane (212), to undergo light-induced homolytic cleavage, producing spurious nitroxide signals, may mask the ability to identify specific free radicals.

The initial step in the preparation of 2-methyl-2-nitrosopropane (212) is the aluminum–mercury reduction of 2-methyl-2-nitropropane (143) to N-(tert-butyl)-hydroxylamine (135). Considerable caution should be exercised when undertaking this reduction, as the reaction is slow to start and will take off without warning, overflowing the two condensors placed in series. Oxidation of N-(tert-butyl)hydroxylamine (135) with sodium hypobromite gives an outstanding yield of nitrosoalkane (212). When kept in the dark, the dimer of 2-methyl-2-nitrosopropane (219) is a colorless solid, stable for long periods of time at 0°C (Calder, et al., 1972). Upon dissolution in either organic or aqueous solvents, the liquid becomes blue, due to a mixture of monomer (212) and dimer (219). At 10°C, there is an equal mixture of each; thereafter, as the temperature increases, the monomer (212) begins to dominate, reaching levels of nearly 80% at 35°C (Stowell, 1971):

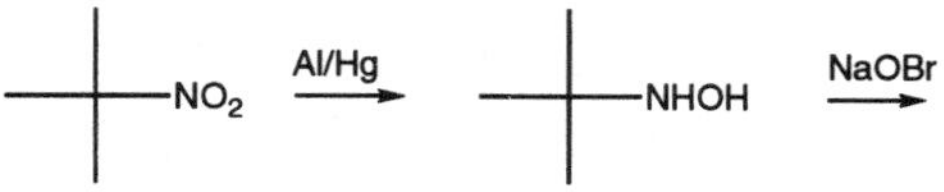

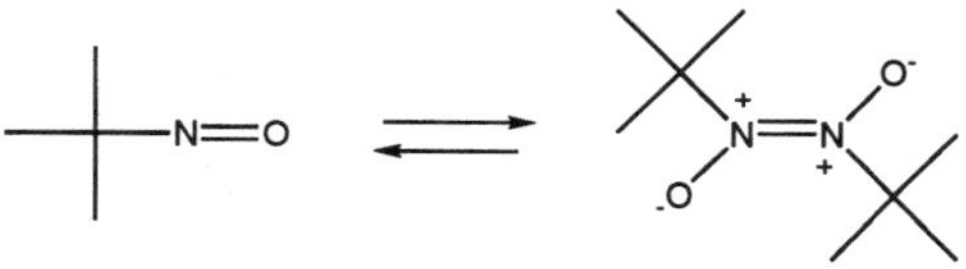

For 2-(hydroxymethyl)-2-nitrosopropane (213), the difficult step is the silver carbonate oxidation of hydroxylamine (220). When the reaction is conducted with efforts to minimize exposure to light, highly pure nitrosoalkane (213) can be expected:

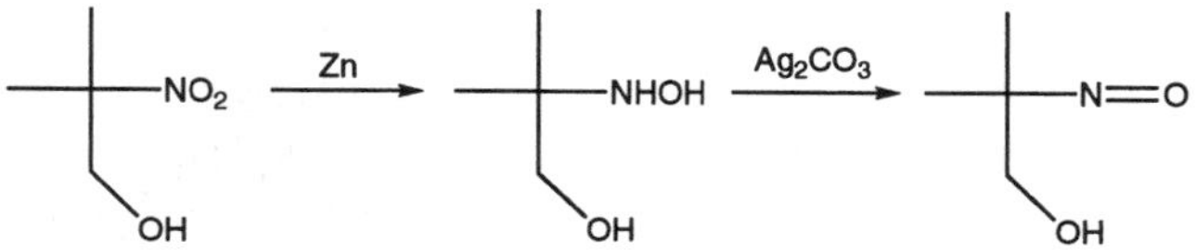

The synthesis of the substituted nitrosobenzene (214) from the corresponding amine (222) is straightforward, requiring the investigator to simply mix reagents together and wait several days. Purification can be achieved through standard recrystallization procedures (Kaur, et al., 1981):

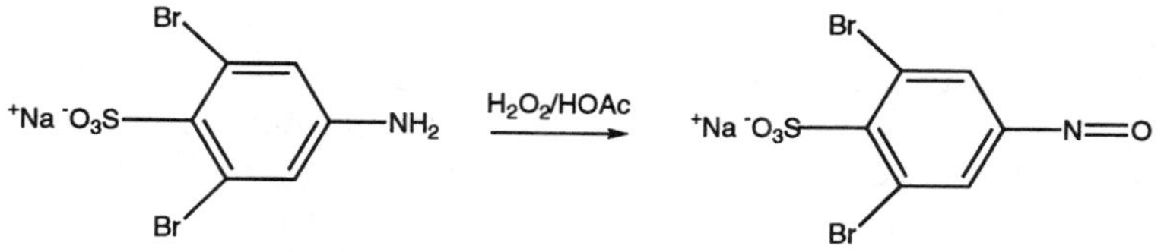

222   **214**

For 1-methyl-4-nitroso-3,5-diphenylpyrazole (**216**) and sodium 1-propylsulfate-4-nitroso-3,5-diphenylpyrazole (**217**), condensation of isonitrosodibenzoyl-methane (**223**) with the appropriately substituted hydrazines affords the desired pyrazoles (**216, 217**) in reasonable yields (Boyd and Norris, 1974; Kaur and Perkins, 1982):

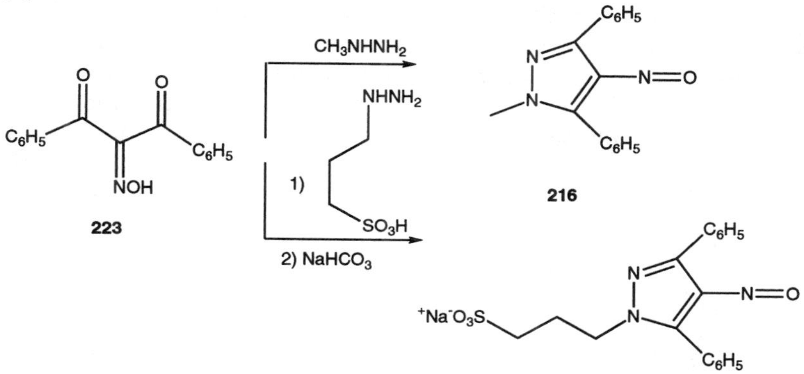

As with other classes of spin traps, isotope labeling enhances the sensitivity of these probes—especially important in biologic systems where the steady-state flux of free radicals is low. Typical of these preparative schemes is the synthesis of perdeuterated 2-methyl-2-nitrosopropane (**225**) by treatment of perdeuterated 2-methyl-2-aminopropane (**224**) with m-chloroperbenzoic acid (Holman and Perkins, 1971):

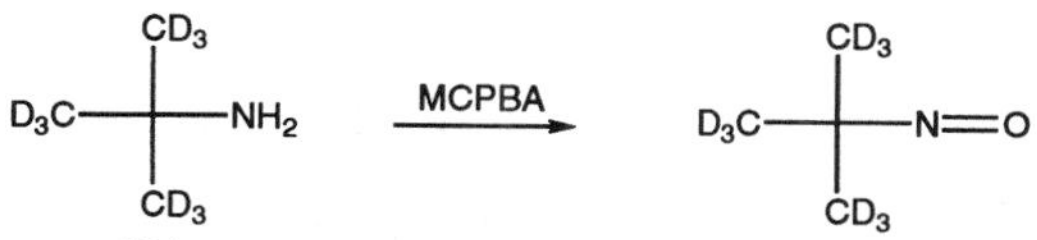

224   **225**

Spin Traps for Nitric Oxide

Nitric oxide is a free radical with a remarkably long lifetime in the biological milieu (see Table 2.3). Although this diminished and selective reactivity toward many biologic target suggests a high degree of specificity, an essential characteristic for a second messenger, this trait presents considerable problems when attempting to identify this free radical in living organisms. Typical techniques involve the reaction of NO· with iron-containing proteins, such as oxyhemoglobin (Ignarro, *et al.*, 1987; Archer, 1993) or iron chelates, including iron (II) diethylenedithiocarbamate ($Fe^{2+}(DETC)_2$) (Mordvintcev, *et al.*, 1991). Although this method and similar approaches have successfully detected this free radical in a number of experimental paradigms (Archer, 1993), it is, nevertheless, limited in its ability to localize NO· at specific tissue compartments. In contrast, nitrones and nitrosoalkanes, which can be directed to either intracellular or extracellular sites, are unable to spin trap NO· (Pou, *et al.*, 1994b).

Even though NO· can react with triplet biradicals, forming cyclic nitroxides (Korth, *et al.*, 1992), the short lifetimes of these biradicals preclude their use as effective spin trapping agents, except in specific chemical systems (Scaiano, 1982). On the other hand, a cis-conjugated diene, acting as a "biradical equivalent" has been shown to meet the energetics required to spin trap NO· (Korth, *et al.*, 1992, 1994). The heart of this "biradical equivalent" is the *in situ* generation of 7,7,8,8-tetramethyl-*o*-quinodimethane (**228**), produced during the photolysis of 1,1,3,3-tetramethyl-2-indanone (**227**) (Korth, *et al.*, 1992). Reaction with NO· forms the expected 1,1,3,3-tetramethylisoindolin-2-oxyl (**229**) (Korth, *et al.*, 1992):

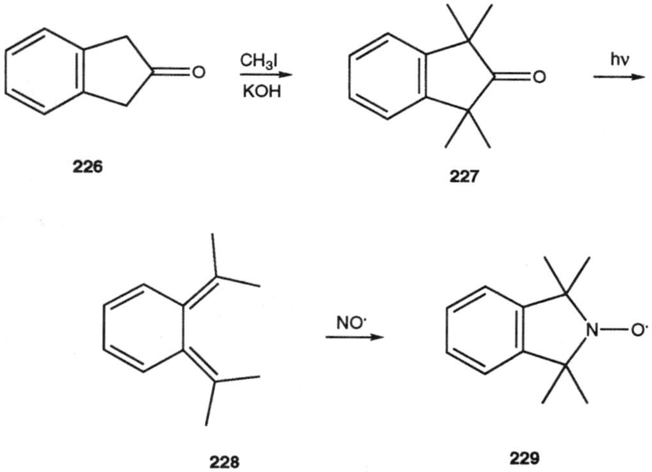

More recently, the water-soluble 5-carboxy-1,1,3,3-tetramethyl-2-indanone (**231**) has been prepared, which, upon photolysis at 254 nm, proceeds to give the desired *o*-quinodimethane (**232**). In the presence of cellularly produced NO·, an EPR spectrum corresponding to 5-carboxy-1,1,3,3-tetramethylisoindolin-2-oxyl (**233**) is recorded, confirming the utility of this method to identify NO· in the biological milieu (Korth, *et al.*, 1994). The key step is the reaction of the *in situ*-generated Grignard of 5-bromo-1,1,3,3-tetramethyl-2-indanone (**230**) with carbon dioxide. This reaction affords reasonable yields of 5-carboxy-1,1,3,3-tetramethyl-2-indanone (**231**) following standard isolation procedures (Korth, *et al.*, 1994):

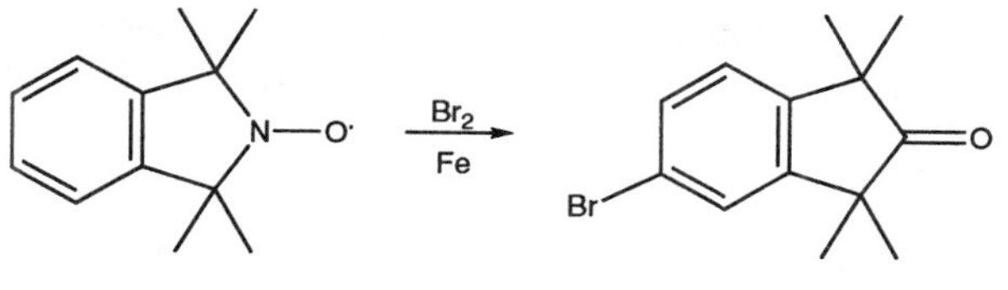

Although nitroxides can react with free radicals at rates approaching 1×10^9 $M^{-1} s^{-1}$ (Ingold, 1984), identification of the initial unstable free radical is frequently dependent upon analytic procedures other than EPR spectroscopy. Recent studies, however, have suggested that nitronyl nitroxides, such as 2-phenyl-4,4,5,5-tetramethylimidazolidine 3-oxide 1-oxyl (**234**), can react with NO·, resulting in an imino nitroxide, 2-phenyl-4,4,5,5-tetramethylimidazolidine 1-oxyl (**235**), with a hyperfine splitting pattern characteristic of this nitroxide (Akaike, *et al.*, 1993; Joseph, *et al.*, 1993; Konorev, *et al.*, 1995; Akaike and Maeda, 1996):

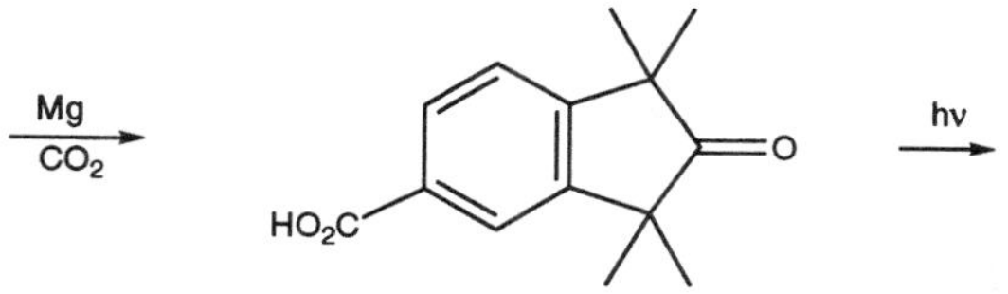

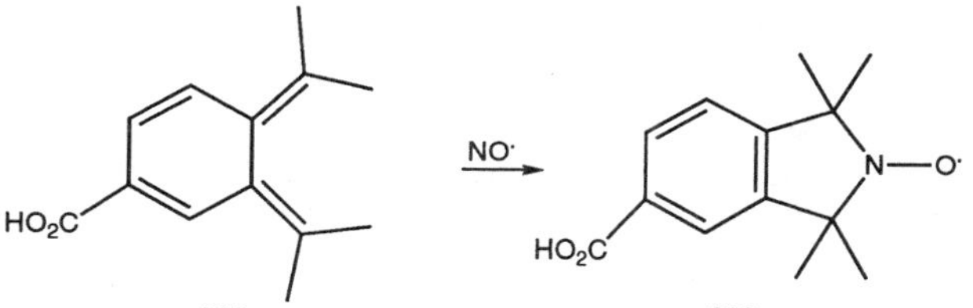

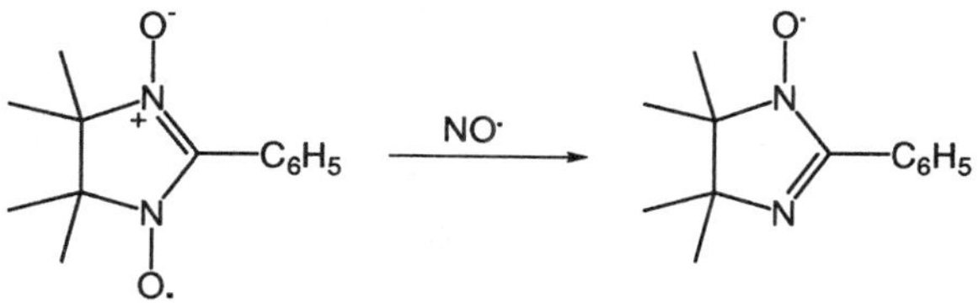

The initial step in the synthesis of nitronyl nitroxides is a simple condensation of 2,3-dimethyl-2,3-bis(hydroxylamino)butane (**236**) with an aldehyde, such as benzaldehyde (**134**), to give 1,3-dihydroxy-2-phenyl-4,4,5,5-tetramethylimidazolidine (**237**). Oxidation with lead dioxide affords the corresponding nitronyl nitroxide, 2-phenyl-4,4,5,5-tetramethylimidazolidine 3-oxide 1-oxyl (**234**) (Osiecki and Ullman, 1968; Ullman, *et al.*, 1972). By varying the nature of the aldehyde, nitronyl nitroxides with a wide range of lipophilic properties can be synthesized (Ullman, *et al.*, 1972; Akaike, *et al.*, 1993; Joseph, *et al.*, 1993; Konorev, *et al.*, 1995; Akaike and Maeda, 1996).

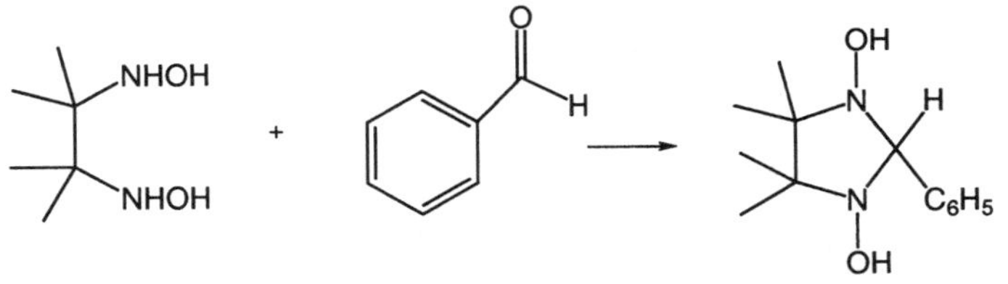

236   **134**   **237**

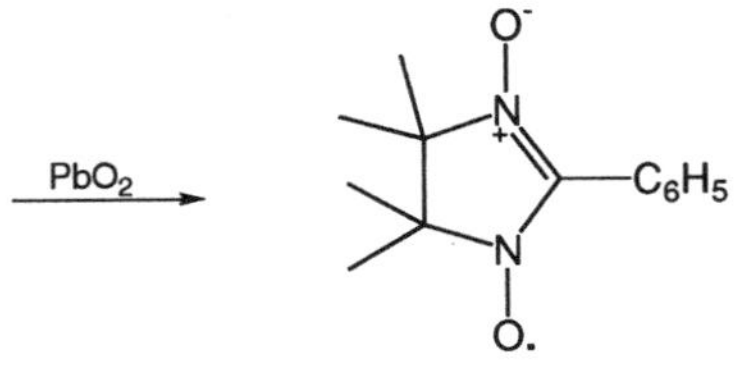

234

Concluding Thoughts

The future of spin trapping will undoubtedly depend on the development of spin traps whose characteristics allow detection of free radicals at unique sites. Microscopically, syntheses of spin traps attached to fluorescent chromophores will allow confocal microscopy to be linked to EPR spectroscopy (Pou, *et al.*, 1995). The resulting new technology will open new horizons in the study of free radical processes at a cellular level. On a macroscopic scale, recent developments in lower frequency EPR spectroscopy will greatly advance our understanding of free radical events in living animals. Yet, progress will unquestionably be bound to the ability to prepare spin traps that are site directed, thereby enhancing our knowledge of the physiologic role of free radicals in biological systems (Lai, and Komarov, 1994; Halpern, *et al.*, 1995; Yoshimura, *et al.*, 1996).

PREPARATIVE SCHEMATA

4-Methyl-4-nitropentanal (**3**)   A solution of sodium methoxide in methanol was prepared by carefully adding sodium metal (3.7 g, 160 mmol) to anhydrous methanol at 0°C. After the metal had dissolved in the solvent, 2-nitropropane (**1**) (106.8 g, 1.2 mol) in 50 mL of absolute methanol was added at such a rate that the temperature of the reaction mixture did not exceed 5–10°C. Once the addition of 2-nitropropane was completed, the temperature of the reaction was lowered to −20°C. Then, a mixture of acrolein (9 g, 10.7 mL, 160 mmol), freshly distilled over drierite, and 2-nitropropane (26.7 g, 300 mmol) was added dropwise over a 3-4 hr period, maintaining the temperature at ≈ −20°C. Once completed, the reaction was stirred for 2 hr more at room temperature, followed by the addition of glacial acetic acid (10 mL). At this point, the reaction mixture was decreased in volume by rotary evaporation at temperatures below 35°C. The remaining viscous oil was dissolved in ether (300 mL), then washed extensively with saturated solutions of $NaHCO_3$ and NaCl. The remaining ether solution was dried over anhydrous $MgSO_4$ and evaporated to dryness to give a liquid. Fractional distillation gave the aldehyde (3), 9.5 g (41%), bp, 98–100°C at 0.9 mm Hg (Shechter, et al., 1952). IR (Neat film) 1755 (C=O), 1540 (N–O), 1475 (N–O) cm^{-1}.

2-(3-Methyl-3-nitrobutyl)-1,3-dioxolane (4)   To a solution of 4-methyl-4-nitropentanal (3) (9.5 g, 69 mmol) dissolved in 50 mL of dry benzene (from azeotropic distillation) was added ethylene glycol (4.7 g, 76 mmol) and a catalytic amount of *p*-toluenesulfonic acid hydrate (0.2 g, 1 mmol). The reaction was heated under reflux until the calculated amount of water (in this case, 1.1 mL) was retained in a Dean-Stark trap. The benzene solution was cooled to 10°C, treated with a saturated aqueous $NaHCO_3$ solution, dried over anhydrous Na_2SO_4, and filtered. The remaining benzene solution was rotary evaporated to dryness to give, after distillation, 9.2 g (75%) of the 1,3-dioxolane (4), bp 102–105°C at 0.5 mm Hg (Bonnett, et al., 1959b). IR (Neat film)1540 (N–O), 1475 (N–O) cm^{-1}.

5,5-Dimethyl-1-pyrroline-N-*oxide* (5)   To a rapidly stirred solution of 2-(3-methyl-3-nitrobutyl)-1,3-dioxolane (4) (9.2 g, 57 mmol) and ammonium chloride (2.76 g, 52 mmol) in 56 mL of water at 10°C was added zinc dust (12.9 g, 197 mmol) in small portions so that the temperature did not rise above 15°C. After addition of the zinc was completed, the reaction was stirred for an additional 15–20 min, filtered, and the filter cake was washed with hot water (at approximately 70°C). The filtrate and washings were acidified with concentrated hydrochloric acid and left to stand overnight. The next day, this solution was heated to 75°C for 1 hr, before evaporation to one-third the original volume. The remaining liquid was made alkaline with NaOH (10%), rotary evaporated to near dryness, and extracted with chloroform. The chloroform solution was dried over anhydrous Na_2SO_4, filtered, and evaporated *in vacuo* to dryness to give, after distillation, 3.54 g (54%) of 5,5-dimethyl-1-pyrroline-N-oxide (5), bp 66–67°C at 0.6 mm Hg, which soldified in the collecting flask (Bonnett, et al., 1959a; Rosen and Rauckman, 1980, 1984). IR (Neat film) 1573 (C=N–O) cm^{-1}. NMR (CDCl$_3$) δ: 1.43 (s, 6H, CH$_3$), 2.14 (t, 2H, CH$_2$), 2.62 (t, 2H, CH$_2$), 6.80 (s, 1H, HC=N–O).

Most often, additional purification was required, since concentrated aqueous solutions of this nitrone (100 mM) contained spurious EPR spectra corresponding to a conglomerate of poorly defined nitroxides. The nitrone (5) (1 g), dissolved in dichloromethane, was added to a chromatographic column containing silica gel (mesh 230–400). Once nitroxides were removed, elution of the column with dichloromethane/methanol (95:5) afforded, after evaporation of the solvent, nitrone (5) as a colorless oil. Kugelrohr distillation, 20°C at 0.05 mm Hg, gave

900 mg of nitrone (**5**). When a sample (100 mM) of the spin trap was placed in sodium phosphate buffer (50 mM, pH 7.4) in the presence of DTPA (0.1 mM), the EPR scan was void of a signal. The remaining nitrone (**5**) was placed into small vials of approximately 50 mg and stored under argon (or nitrogen) at −78°C.

2-Methyl-1-pyrroline-N-oxide (**6**)   To a solution of 5-chloro-2-pentanone (freshly distilled, 12.10 g, 100 mmol) in ethanol (95%, 80 mL) was added a solution of hydroxylamine hydrochloride (7.7 g, 110 mmol) in water (40 mL). A solution of anhydrous K_2CO_3 (7.6 g, 55 mmol) dissolved in water (40 mL) was slowly added to the above reaction mixture. After addition was completed, this reaction was refluxed for 45 min, during which time the reaction changed from a colorless liquid to a deep brown oil. At this point, the reaction was cooled in an ice bath and basified with anhydrous K_2CO_3 to pH 9. To prevent decomposition of the nitrone, evaporation of the liquid was conducted at 25°C under high vacuum. By removing as much of the liquid as possible, an oil/salt mixture was left, to which methylene chloride was added. The methylene chloride solution was dried over anhydrous Na_2SO_4, filtered, and rotary evaporated to dryness to give, after distillation, 7.4 g (75%) of nitrone (**6**), bp 70–72°C at 0.4 mm Hg. IR (Neat film) 1610 (C=N–O) cm^{-1}. NMR ($CDCl_3$) δ: 4.02 (t, 2, J = 8.0 Hz, 5-CH_2), 2.92 (t, 2, J = 7.1 Hz, 3-CH_2), 2.03 (s, 3, 2-CH_3) in the midst of a complex multiplet, 2.4–1.88 (4-CH_2) (Brandman and Conley, 1973).

5,5-Dimethyl-1-pyrroline-N-oxide (**5**)   An ethereal solution of 2-methyl-1-pyrroline-*N*-oxide (**6**) (5 g, 50 mmol in 20 mL dry ether) was added at 0°C to an ethereal (40 mL) solution of methylmagnesium bromide (18 mL of a 3-M stock solution in ether, 55 mmol) cooled to 0°C. After the addition was completed, the reaction mixture was gently refluxed for 1 hr, then it was cooled in an ice bath to which a saturated NH_4Cl solution (5 mL) was carefully added with stirring. The ethereal solution was dried over anhydrous Na_2SO_4, filtered, and rotary evaporated to dryness to give an oil, 1-hydroxy-2,2-dimethylpyrrolidine (**7**). IR (Neat film) 3400–3300 (broad, N–OH) cm^{-1}.

Aerial oxidation of this crude oil to nitrone (**5**) was accomplished by bubbling O_2 (or air) into a solution of (**7**) in 60% ethanol (50 mL) containing copper acetate (1 g, 5.4 mmol) and ammonium hydroxide (30%, 5 mL) until the blue color persisted (Bonnett, *et al.*, 1959a). Although the original preparative sequence suggests that, with time, the blue color will return, it has been found that supplementation with addition of up to 1 g of copper acetate speeds the reaction without the loss of isolatable product. Once the reaction was completed, it was concentrated to 15–20 mL by rotary evaporation. The remaining solution was extracted with chloroform, dried over anhydrous Na_2SO_4, filtered, and rotary evaporated to dryness to give a blue liquid, which was purified by means of a two-step process. Initially, this residue is chromatographed using silica gel (mesh 230–400) methylene chloride and then increasing polarity of the solvent mixture to methylene chloride/methanol (95:5) after yellow impurities had been removed. Upon evaporation of this solvent, the remaining colorless oil was Kugelrohr distilled (20°C and 0.05 mm Hg) to give 2.2 g (40%) of nitrone (**5**).

2,5,5-Trimethyl-1-Pyrroline-N-*Oxide* (9)   A solution of 5,5-methyl-1-pyrroline-
N-oxide (5) (1 g, 8.8 mmol in 20 mL dry ether) was added at 0°C to a stirred
ethereal solution (40 mL) of methylmagnesium bromide (3.2 mL of a 3-M stock
solution in ether, 10 mmol). After addition was completed, the reaction mixture
was gently warmed for 1 hr. The reaction was then cooled in an ice bath, and a
saturated NH_4Cl solution (2.5 mL) was carefully added with stirring. The ethereal
solution was dried over anhydrous Na_2SO_4, filtered, and rotary evaporated to
dryness to give an oil, 1-hydroxy-2,5,5-trimethylpyrrolidine (8). IR (Neat film)
3400–3300 (broad, N–OH) cm^{-1}.

Continual aeration of this crude oil in ethanol (60%, 50 mL) with copper acetate
(0.5 g, 2.7 mmol) and ammonium hydroxide (30%, 3 mL) until a blue color persisted, gave, after standard isolation procedures, crude nitrone (9). In this case, the
addition of up to 0.5 g of copper acetate enhanced the reaction rate without any
apparent loss of product. Purification by column chromatography with silica gel
(mesh 230–400), methylene chloride/methanol afforded an oil, which upon
Kugelrohr distillation (20°C at 0.05 mm Hg) yielded 400 mg (35%) of nitrone (9)
(Bonnett, *et al.*, 1959a). IR (Neat film) 1570 (C=N–O) cm^{-1}.

2-Cyano-5,5-dimethyl-1-pyrroline-N-*oxide* (10)   To a stirred solution of 5,5-
methyl-1-pyrroline-*N*-oxide (5) (5 g, 44 mmol) and potassium cyanide (3.9 g,
60 mmol) in water (20 mL) at 0°C was added hydrochloric acid (2 M, 25 mL) at
such a rate that the temperature was not elevated above 5°C. Soon after the
addition was commenced, a white solid formed. After maintaining the reaction at
this temperature for 1.5 hr, the pH of the mixture was carefully elevated to 11 with
dilute sodium hydroxide solution (10%) while maintaining the temperature of the
mixture at 5°C. After continuously extracting the reaction mixture with ether, this
solution was dried over anhydrous Na_2SO_4, filtered, and rotary evaporated to
dryness to give an oil, which solidified upon the addition of light petroleum ether
(bp 40–60°C). Recrystallization from a mixture of ether–light petroleum ether gave
white crystals, mp 90–92°C of 2-cyano-1-hydroxy-5,5-dimethylpyrrolidine (4.8 g,
78%) (Bonnett, *et al.*, 1959a). IR (KBr pellet) 2240 (C≡N) cm^{-1}.

To a solution of 2-cyano-1-hydroxy-5,5-dimethylpyrrolidine (8.1 g, 58.3 mmol)
in 60% ethanol (100 mL) containing ammonium hydroxide (30%, 5 mL) was
added copper acetate (1.5 g, 8.3 mmol). The reaction was continually bubbled
with O_2 until a blue color returned. Concentration of this solution to 15–20 mL
by rotary evaporation, left, after extraction with chloroform, a blue oil. Vacuum
distillation (bp 110°C at 0.5 mm Hg) gave 5.8 g (73%) of desired nitrone (10)
(Bonnett, *et al.,* 1959a). IR (Neat film) 2220 (C≡N), 1540 (C=N–O) cm^{-1}.

2-Carboxy-5,5-dimethyl-1-pyrroline-N-*oxide* (11)   2-Cyano-5,5-dimethyl-1-pyr-
roline-*N*-oxide (10) (5 g, 36 mmol) was added to an aqueous sodium hydroxide
solution (10%, 40 mL). After refluxing for several hr, the solution was acidified to pH
1 with hydrochloric acid, extracted with chloroform, dried over anhydrous $MgSO_4$,
and evaporated to dryness to give 3.1 g (55%) of nitrone (11), as a white solid, a
portion of which was recrystallized from ether–light petroleum ether, mp 86°C
(Bonnett, *et al.*, 1959a).

5-Methyl-5-nitro-2-hexanone (**13**)   To a solution of sodium methoxide in absolute methanol (sodium metal—2.76 g, 120 mmol in 100 mL of anhydrous methanol) at 0°C was added 2-nitropropane (**1**) (10.7 g, 120 mmol) in 50 mL of anhydrous methanol, such that the temperature of the reaction mixture did not exceed 5°C. Once the addition of 2-nitroproprane was concluded, the temperature of the reaction was lowered to −20°C. Then, methyl vinyl ketone (**12**) (8.4 g, 9.72 mL, 120 mmol) freshly distilled, was added dropwise over a 1-hr period, maintaining the temperature at ≈ −20°C. Once completed, the reaction was stirred for 2 hr more at room temperature, followed by the addition of glacial acetic acid (10 mL). At this point, the reaction mixture was decreased in volume by rotary evaporation at temperatures below 35°C. The remaining viscous oil was dissolved in ether (300 mL) extensively washed with saturated solutions of NaHCO₃ and then NaCl, dried over anhydrous MgSO₄, and evaporated to dryness to give a liquid. Fractional distillation afforded the desired nitroketone (**13**), 7.5 g (60%), bp, 123–125°C at 10 mm Hg. (Shechter, *et al.*, 1952). IR (Neat film) 1760 (C=O), 1540 (N–O), 1475 (N–O) cm^{-1}.

2,5,5-Trimethyl-1-pyrroline-N-*oxide* (**9**)   To a rapidly stirred solution of 5-methyl-5-nitro-2-hexanone (**13**) (7.5 g, 48 mmol) and ammonium chloride (2.33 g, 44 mmol) in 50 mL of water at 10°C was added zinc dust (10.7 g, 165 mmol) in small portions so that the temperature did not rise above 15°C. The reaction was stirred for an additional 20 min, at which point it was filtered and the filter cake was washed with hot water (at approximately 70°C). The remaining liquid was made alkaline with NaOH (10%), rotary evaporated to near dryness and extracted with chloroform. The chloroform solution was dried over anhydrous Na₂SO₄, filtered, and rotary evaporated to dryness to give, after purification by column chromatography with silica gel (mesh 230–400) and methylene chloride/methanol followed by Kugelrohr distillation (20°C at 0.05 mm Hg) 3.3 g (54%) of nitrone (**9**) (Bonnett, *et al.*, 1959a; Lunt, 1964). IR (Neat film) 1585 (C=N–O) cm^{-1}.

4-Methyl-4-nitro-1-heptanal (**17**)   2-Nitropentane (**15**), prepared as described in Haire and Janzen (1982) (10 g, 85.5 mmol) and *N*-benzyltrimethylammonium hydroxide (40% in methanol, 0.57 mL, 1.37 mmol) were dissolved in dry tetrahydrofuran (THF, 75 mL) at temperatures ≈ 0°C. The dropwise addition of acrolein (4.21 g, 5 mL, 75.08 mmol), freshly distilled over drierite, in THF (25 mL) at ≈ 0°C, caused the colorless reaction to become brown-green. After stirring for an additional 4 hr at this temperature, hydrochloric acid (10%, 4.72 mL) was added to quench the reaction. After addition of methylene chloride, the reaction was washed with saturated NaCl, dried over anhydrous Na₂SO₄, filtered, and rotary evaporated to dryness to gave a viscous brown-green oil. Fractional distillation gave aldehyde (**17**) as a faint yellow oil, 7.6 g (58%), bp, 57°C at 0.01 mm Hg (Haire and Janzen, 1982). NMR (CDCl₃) δ: 1.54 (s, 3H, NO₂–C–CH₃), 9.96 (s, 1H, HC=O).

As in the case of 4-methyl-4-nitro-1-pentanal (**3**), great care should be taken when distilling crude 4-methyl-4-nitro-1-heptanal (**17**). If the distillate becomes brown-black, it is important to immediately remove the heat by placing the flask in an ice bath. Otherwise, thermal instability, due to the polymerization of unreacted acrolein, can result in an explosion.

2-(3-Methyl-3-nitrohexane)-1,3-dioxolane (**19**)   A solution containing 4-methyl-4-nitro-1-heptanal (**17**) (7.6 g, 43.88 mmol), ethylene glycol (3.23 g, 52 mmol), and a catalytic amount of *p*-toluenesulfonic acid hydrate (0.15 g, 767 μmol) in dry benzene (125 mL) was heated under reflux until the calculated amount of water (0.8 mL) was retained in a Dean-Stark trap. The reaction was cooled in an ice bath, treated with aqueous $NaHCO_3$ solution, dried over anhydrous Na_2SO_4, and filtered. The remaining organic solution was rotary evaporated to dryness to afford the desired 1,3-dioxolane (**19**) (6.86 g, 73%), bp 90°C at 0.1 mm Hg (Haire and Janzen, 1982). NMR ($CDCl_3$) δ: 1.57 (3H, NO_2–C–CH_3), 3.90 (m, 4H, OCH_2CH_2O), 4.85 (t, 1H, $(CH_2O)_2$C–H).

5-Methyl-5-propyl-1-pyrroline-N-oxide (**21**)   To a rapidly stirred solution containing 2-(3-methyl-3-nitrohexane)-1,3-dioxolane (**19**) (7.10 g, 32.67 mmol) and ammonium chloride (1.86 g, 34.77 mmol) in THF–H_2O (12:24 mL) was added zinc dust (90%, 8.66 g, 119.1 mmol) in small portions so that the reaction temperature was < 5°C. After addition of the zinc was concluded, the reaction was continually stirred at 25°C. The reduction typically required about 4 hr to complete, as suggested by following the disappearance of the 1,3-dioxolane by NMR spectroscopy. At this point, the reaction was filtered, and the filter cake was washed with hot water. The remaining solution was acidified with concentrated hydrochloric acid (37%) and heated to 70°C for 2 hr, to hydrolyze the acetal. The mixture was cooled and neutralized with $NaHCO_3$. After stirring for 1 hr, the reaction was filtered and extracted with pentane. This solution was dried over anhydrous Na_2SO_4, filtered, and rotary evaporated to dryness to give an oil, bp 56°C at 0.1 mm Hg. Further purification was achieved by column chromatography with silica gel (mesh 230–400) using increasing polar solvents from pentane, benzene, and then methylene chloride. This method resulted in the isolation of nitrone (**21**) (3 g, 65%) as a colorless hydroscopic oil (Haire and Janzen, 1982). IR (Neat film) 1560 (C=N–O), 1242 (N–O) cm^{-1}. NMR ($CDCl_3$) δ: 1.39 (s, 3H, O–N–C–CH_3), 2.15 (m, 2H, CH_2), 2.55 (m, 2H, CH_2), 6.82 (t, 1H, HC=N–O). The NMR location for the remaining 7 Hs was not presented in the paper cited here.

4-Nitro-4-propyl-1-heptanal (**18**)   A solution of 4-nitroheptane (**16**), prepared as described in Turner and Rosen (1986) (12 g, 83 mmol) and *N*-benzyltrimethyl-ammonium hydroxide (40% in methanol, 0.56 mL, 1.37 mmol) in dry tetrahydrofuran (THF, 50 mL) was cooled to ≈ 0°C. Acrolein (2.36 g, 2.8 mL, 42 mmol), freshly distilled over drierite, in THF (25 mL) was then added dropwise over a period of 20 min at such a rate as to maintain the reaction temperature at ≈ 0°C. After addition was completed, the reaction was stirred in an ice bath for 5 hr subsequently. The reaction was quenched with hydrochloric acid (3 mL, 1 M), diluted with methylene chloride (200 mL), washed with saturated NaCl solution, dried over anhydrous Na_2SO_4, filtered, and rotary evaporated to dryness to yield a light yellow oil. Vacuum distillation gave the desired aldehyde (**18**) (4.07 g, 49%), bp 100–102°C at 0.25 mm Hg (Turner and Rosen, 1986). IR (Neat film) 1723

(C=O), 1535 (N–O) cm^{-1}; NMR (CDCl$_3$) δ: 1.10–1.70 (m, 10H), 2.12 (m, 4H), 2.52 (m, 4H), 9.97 (s, 1H, HC=O).

As described for other synthetic schemata involving acrolein, vacuum distillation of 4-nitro-4-propyl-1-heptanal (**18**) should proceed with great care to avoid an explosion.

2-(3-Propyl-3-nitrohexane)-1,3-dioxolane (**20**)   A solution containing 4-nitro-4-propyl-1-heptanal (**18**) (3.5 g, 17 mmol), ethylene glycol (1.19 g, 19 mmol), and a catalytic amount of *p*-toluenesulfonic acid hydrate (50 mg, 253 μmol) in dry benzene (50 mL) was heated under reflux until the calculated amount of water (0.3 mL) was retained in a Dean-Stark trap. The benzene solution was then cooled in an ice bath, treated with aqueous NaHCO$_3$ solution, dried over anhydrous Na$_2$SO$_4$, and filtered. The remaining organic solution was rotary evaporated to dryness to give the 1,3-dioxolane (**20**) (3.21 g, 75%), bp 108–110°C at 0.1 mm Hg (Turner and Rosen, 1986). IR (Neat film) 1530 (N–O), 1350 (N–O) cm^{-1}, NMR (CDCl$_3$) δ: 0.85–1.55 (complex m, 12H), 1.85 (m, 6H), 3.85 (m, 4H, OCH$_2$CH$_2$O), 4.78 (t, 1H, CHO$_2$).

5,5-Dipropyl-1-pyrroline-N-oxide (**22**)   A solution of 2-(3-propyl-3-nitrohexane)-1,3-dioxolane (**20**) (2.6 g, 11 mol) and ammonium chloride (0.63 g, 11.8 mmol) in 60% ethanol (150 mL) was treated with zinc dust (3 g, 46.2 mmol), while the temperature of the reactions was maintained at < 15°C. After addition was completed, the reaction was stirred for 12 hr at room temperature, filtered, and the filter cake was washed with warm water. The combined filtrate and washings were acidified with concentrated hydrochloric acid (37%, 20 mL), and warmed to 70°C for 1 hr. Upon cooling the reaction in an ice bath, it was basified with potassium carbonate. The remaining mixture was concentrated by rotary evaporation ($T < 40°C$) to about 50 mL and extracted with methylene chloride. The combined extractions were dried over anhydrous Na$_2$SO$_4$, and filtered. The solvent was removed by rotary evaporation ($T < 35°C$) to give an oil. Purification was a two-step procedure. Initially, the oil was chromatographed using silica gel (mesh 230–400). Eluting the column with an increasing polar solvent mixture from methylene chloride to methylene chloride/methanol (95:5) gave an oil. This was followed by Kugelrohr distillation (30–50°C at 0.1 mm Hg), which afforded the desired nitrone (**22**) (1 g, 55%) as a colorless oil (Turner and Rosen, 1986). IR (Neat film) 1580 (C=N–O), 1205 (N–O) cm^{-1}, NMR (CCl$_4$) δ: 2.07 (d of t, 2H, CH$_2$CH$_2$C=N), 2.50 (m, 2H, CH$_2$C=N), 6.68 (m, 1H, HC=N–O).

5-n-Tetradecyl-5-methyl-1-pyrroline-N-oxide (**23**)   To a solution of *n*-tetradecylmagnesium chloride (120 mL, 120 mmol) in dry tetrahdyrofuran (50 mL) at 0°C was added 2-methyl-1-pyrroline-*N*-oxide (**6**) (2 g, 20 mmol) in tetrahydrofuran (50 mL) at a rate such that the temperature of the reaction did not rise above 2°C. After addition was completed, the reaction mixture was stirred at room temperature for 1 hr, cooled to 0°C, and a saturated NH$_4$Cl solution (5 mL) was carefully added with stirring. The tetrahydrofuran solution was dried over anhydrous Na$_2$SO$_4$, filtered, and rotary evaporated to dryness to give a yellow oil, 1-hydroxy-2-methyl-2-tetradecylpyrrolidine. This oil was treated with copper acetate (250 mg) in methanol (15 mL) containing NH$_4$OH (30%, 0.5 mL) and bubbled with air until the solution remained blue. Upon rotary evaporation, the remaining residue was taken up in chloroform and passed through a silica gel (mesh 230–400) column eluted with methylene chloride, tetrahydrofuran, and ethanol, successively, affording nitrone (**23**) as a solid (4.3 g, 73%), mp 34–35°C (Zhang and Xu, 1989). IR (Thin film) 1579 (C=N–O) cm^{-1}; NMR (CDCl$_3$) δ: 0.88 (t, 3H), 1.26 (s, 24H), 1.6–1.9 (m, 2H), 1.95–2.3 (m, 2H), 2.4–2.7 (m, 2H), 6.82 (t, J = 2.2–2.6 Hz, 1H).

5-Methyl-5-phenyl-1-pyrroline-N-*oxide* (**24**)   To a cooled (−20°C) solution of phenylmagnesium bromide (25 mL, 75.0 mmol) in dry ether (50 mL) was added 2-methyl-1-pyrroline-N-oxide (**6**) (4 g, 40 mmol) in ether (50 mL) at 0°C. After addition was completed, the reaction mixture was stirred at 0°C for 1 hr, at which point wet ether and a saturated NH$_4$Cl solution (5 mL) were carefully added with stirring. The ether solution was then dried over anhydrous Na$_2$SO$_4$, filtered, and rotary evaporated to dryness to give a yellow oil, 1-hydroxy-2-methyl-2-phenylpyrrolidine (4.1 g, 57%). This oil was treated with copper acetate (520 mg) in methanol (15 mL) containing NH$_4$OH (30%, 0.75 mL) and bubbled with air until the solution remained blue. Upon rotary evaporation, the remaining residue was taken up in chloroform and passed through a silica gel (mesh 230–400) column eluted with methylene chloride/methanol, yielding 5-methyl-5-phenyl-1-pyrroline-N-oxide (**24**) as a waxy oil. Upon sublimation at 40°C (0.02 mm Hg), nitrone (**24**) was isolated as a white solid (2 g, 28%), mp 68°C (Sankuratri, *et al.*, 1997). IR (Thin film) 1573 (C=N–O) cm^{-1}, NMR (CDCl$_3$) δ: 2.36 and 2.50 (m, 2H, H-4), 2.61 (m, 2H, H-3), 7.16 (t, J = 2.4 Hz, 1H, H-2), 7.30 (m, 5H, Ar).

3,4-Dimethyl-4-nitropentanal (**29**)   A solution of sodium methoxide in methanol was prepared by carefully adding sodium metal (1 g, 44 mmol) to anhydrous methanol at 0°C. After the metal has dissolved in the solvent, 2-nitropropane (25 g, 281 mmol) in 40 mL of anhydrous methanol was added at such a rate that the temperature of the reaction mixture did not exceed 5–10°C. To this solution was added freshly distilled crotonaldehyde (**27**) (17.5 g, 250 mmol), at such a rate that the temperature slowly rose to 60°C. After stirring for 3 hr at this temperature, the reaction was allowed to stand overnight at room temperature, at which point acetic acid (3 mL) was added. The solution was rotary evaporated to dryness, leaving an oil that was taken up into water. This solution was extracted with ether. The organic layer was washed with a saturated solution of NaHCO$_3$. After drying over anhydrous Na$_2$SO$_4$, the ether extraction was filtered and evaporated *in vacuo* giving, after distillation, aldehyde (**29**) (20 g, 51%), bp 80°C at 0.5 mm Hg (Bonnett, *et al.*, 1959a). NMR (CCl$_4$) δ: 0.92 (d, 3H, 3-CH$_3$), 1.50 (s, 6H, 2× CH$_3$), 2.20–2.83 (m, 3H, CH, CH$_2$), 9.63 (s, 1H, CHO) (Zhang, *et al.*, 1991).

4,5,5-Trimethyl-1-pyrroline-N-*oxide* (**31**)   To a stirred solution of 3,4-dimethyl-4-nitropentanal (**29**) (60 g, 377 mmol) and ammonium chloride (15 g, 283 mmol) in water (400 mL) at 15°C was added zinc powder (74 g, 1.14 mol) over a 3-hr period of time. After filtering the mixture and washing the filter cake with warm water (60°C), the solution was rotary evaporated (T < 40°C) to dryness. The residue was taken up into chloroform, dried over anhydrous Na_2SO_4, and filtered. The remaining chloroform solution was evaporated *in vacuo* to give the desired nitrone (**31**) (24.3 g, 51%), bp 85°C at 1 mm Hg (Bonnett, *et al.,* 1959a). IR (Neat film) 1572 (C=N–O) cm^{-1}. NMR (CCl$_4$) δ: 1.07 (d, 3H, 4-CH$_3$), 1.13 (s, 3H, 5-CH$_3$), 1.29 (s, 3H, 5-CH$_3$), 2.00–2.77 (m, 3H, CH, CH$_2$), 6.72 (s, 1H, CH=N) (Zhang, *et al.,* 1991).

2,4-Dimethyl-4-nitropentanal (**34**)   A solution of sodium methoxide in methanol was prepared by carefully adding sodium metal (1 g, 44 mmol) to anhydrous methanol at 0°C. After the metal had dissolved in the solvent, 2-nitropropane (25 g, 281 mmol) in 40 mL of anhydrous methanol was added at such a rate that the temperature of the reaction mixture did not exceed 5–10°C. To this solution was added freshly distilled methacrolein (**33**) (17.5 g, 20.7 mL, 250 mmol), allowing the temperature to rise to 10°C. After stirring for 1 hr at this temperature, the reaction was quenched by the addition of acetic acid (3 mL). The solution was rotary evaporated to dryness, leaving an oil that was taken up into water. This solution was extracted with ether. The organic layer was washed with a saturated solution of NaHCO$_3$. After drying over anhydrous Na$_2$SO$_4$, the ether extraction was filtered and evaporated *in vacuo* giving, after distillation, aldehyde (**34**) (7.2 g, 18%), bp 60–66°C at 0.5 mm Hg (Warner and Moe, 1952), bp 123–124°C at 18 mm Hg (Zhang, *et al.,* 1991). NMR (CCl$_4$) δ: 1.12 (d, 3H, CH$_3$), 1.5 (s, 3H, CH$_3$), 1.58 (s, 3H, CH$_3$), 1.73–2.7 (m, 3H, CH, CH$_2$), 9.5 (s, 1H, HC=O).

3,5,5-Trimethyl-1-pyrroline-N-*oxide* (**35**)   To a stirred solution of 2,4-dimethyl-4-nitropentanal (**34**) (6 g, 37.7 mmol) and ammonium chloride (1.5 g, 28.3 mmol) in water (40 mL) was added zinc powder (7.4 g, 114 mmol) at 15°C for over a 3-hr period of time. After filtering the mixture and washing the filter cake with warm water (60°C), the solution was rotary evaporated ($T < 40°C$) to dryness. The residue was taken up into chloroform, dried over anhydrous Na$_2$SO$_4$, filtered, and evaporated *in vacuo* to afford the desired nitrone (**35**) (3.1 g, 64%), bp 94–104°C at 3 mm Hg (Zhang, *et al.,* 1991). IR (Neat film) 1575 (C=N–O) cm^{-1}. NMR (CCl$_4$) δ: 1.18 (d, 3H, CH$_3$ at 3-position), 1.30 (s, 3H, CH$_3$ at 5-position), 1.35 (s, 3H, CH$_3$ at 5-position), 2.13–2.47 (q, 2H, CH$_2$), 2.60-3.10 (m, 1H, CH), 6.44 (s, 1H, HC=N–O).

4-(1-Methyl-1-nitroethyl)tetrahydrofuran-2-one (**37**)   A mixture of 2(5*H*)-furanone (**36**) (5 g, 60 mmol) and 2-nitropropane (**1**) (10 g, 110 mmol) was added dropwise to a methanolic solution of 2-nitropropane (**1**) (40.5 g, 450 mmol) containing sodium methoxide (3.3 g, 60 mmol) maintained at −20°C. After the addition was completed, the reaction was refluxed for 3 hr, cooled, and acidified with acetic acid. Upon removing the solvent under vacuum, the residue was taken up in water and extracted with chloroform. After drying the solution over anhydrous Na$_2$SO$_4$, it was filtered and evaporated *in vacuo* to give 4-(1-methyl-1-nitroethyl)tetrahydrofuran-2-one (**37**) (6.7 g, 64%) as colorless needles from hexane–ether, mp 67–69°C (Konaka, *et al.*, 1995).

2-Hydroxy-4-(1-methyl-1-nitroethyl)tetrahydrofuran (**38**)   A toluene solution of diisobutylaluminum hydride (1-M stock, 23 mL, 23 mmol) was added dropwise to a mixture of 4-(1-methyl-1-nitroethyl)tetrahydrofuran-2-one (**37**) (2.6 g, 15 mmol) in toluene (50 mL) under N_2 at −70°C. After the addition was completed, the reaction was stirred for 30 min at −70°C and, thereafter, it was poured over ice-water, acidified with acetic acid, and the toluene solution was separated. The remaining aqueous layer was extracted with chloroform. The combined organic solutions were dried over anhydrous Na_2SO_4, filtered, and evaporated *in vacuo*, affording 2-hydroxy-4-(1-methyl-1-nitroethyl)tetrahydrofuran (**38**) (1.1 g, 42%) as colorless solid from hexane–methylene chloride, mp 88–89°C (Konaka, *et al.*, 1995).

5,5-Dimethyl-4-hydroxymethyl-1-pyrroline-N-oxide (**39**)   Activated zinc dust (1.1 g, 170 mmol) was gradually added to a solution of methanol/water (5:10 mL) containing 2-hydroxy-4-(1-methyl-1-nitroethyl)tetrahydrofuran (**38**) (800 mg, 4.6 mmol) and ammonium chloride (250 mg, 4.6 mmol), and cooled in an ice bath. Once the addition of zinc was completed, the reaction was stirred for 30 min, filtered, and washed with a warm methanol–water mixture. The filtrate was acidified with dilute HCl and stored in a refrigerator overnight. After warming to 70°C for 40 min, the solution was cooled in an ice bath, and made alkaline with Na_2CO_3. After concentrating the remaining solution *in vacuo*, ethyl acetate was added to precipitate the product. Upon recrystallization from ethyl acetate, 5,5-dimethyl-4-hydroxymethyl-1-pyrroline-N-oxide (**39**) (360 mg, 55%) was obtained as colorless crystals, mp 87°C (Konaka, *et al.*, 1995).

Ethyl 2-methyl-2-nitro-5-oxopentanoate (**41**)   To a solution of sodium ethoxide in absolute ethanol (prepared by the addition of sodium metal, 0.3 g, 13 mmol to 100 mL of anhydrous ethanol) at 0°C was added ethyl 2-nitropropionate (**40**) (36 g, 245 mmol, synthesized as described in the literature [Kornblum, *et al.*, 1957]). After the addition was completed, acrolein (**2**) (11.75 g, 14 mL, 210 mmol), freshly distilled over drierite, in absolute ethanol (25 mL) was added at such a rate that the temperature of the reaction did not rise above 10°C. Once completed, the reaction was warmed to 40–50°C for 3 hr, then the mixture was cooled to 5°C and quenched by the addition of acetic acid (1 mL). After rotary evaporating the solution to dryness, the remaining liquid was taken up in benzene, dried over anhydrous $MgSO_4$, filtered, and evaporated *in vacuo*. Vacuum distillation gave ethyl 2-methyl-2-nitro-5-oxopentanoate (**41**) (31.5 g, 65%), bp 107°C at 0.5 mm Hg (Bonnett, *et al.*, 1959a). As in other reactions involving acrolein, distillation of aldehydo-ester should be undertaken with great care and caution, as overheating the distillation flask can result in an explosion.

2-(3-Ethoxycarbonyl-3-nitrobutyl)-1,3-dioxolane (**42**)   A solution containing ethyl 2-methyl-2-nitro-5-oxopentanoate (**41**) (30 g, 140 mmol), ethylene glycol (9.55 g, 154 mmol), and a catalytic amount of *p*-toluenesulfonic acid hydrate (500 mg, 2.6 mmol) in dry benzene (150 mL) were heated under reflux until the calculated amount of water (2.52 mL) was retained in a Dean-Stark trap. The benzene solution was cooled, treated with aqueous $NaHCO_3$ solution, dried over anhydrous Na_2SO_4, and filtered. The remaining organic solution was rotary evaporated to dryness to afford 2-(3-ethoxycarbonyl-3-nitrobutyl)-1,3-dioxolane (42) (27 g, 75%), bp 121–125°C at 0.5 mm Hg (Bonnett, *et al.*, 1959a).

5-Ethoxycarbonyl-5-methyl-1-pyrroline-N-*oxide* (43)   To a rapidly stirring mixture of 2-(3-ethoxycarbonyl-3-nitrobutyl)-1,3-dioxolane (42) (28.5 g, 111 mmol) and ammonium chloride (5.5 g, 104 mmol) in 50% ethanol (200 mL) was added zinc powder (41.5 g, 638 mmol), at such a rate that the temperature of the reaction was $\approx$ 30°C. After the addition of the zinc, the reaction was stirred at room temperature for 3 hr. Upon filtering the mixture, the remaining solution was evaporated to about 50 mL, extracted with chloroform, dried over anhydrous Na_2SO_4, filtered, and evaporated *in vacuo*, yielding the corresponding hydroxylamine as an oil. This residue was dissolved in dilute hydrochloric acid (0.3 N, 400 mL) and allowed to stand overnight at room temperature. After careful neutralization with ammonium hydroxide, the solution was rotary evaporated ($T < 40°C$) to dryness. The residue oil was taken up in chloroform, dried over anhydrous Na_2SO_4, and filtered. Upon evaporation to dryness, the remaining liquid was vaccum distilled, giving 5-ethoxycarbonyl-5-methyl-1-pyrroline-*N*-oxide (43) (11 g, 58%), bp 120°C at 0.3 mm Hg (Bonnett, *et al.*, 1959a). IR (Neat film) 1740 (C=O), 1585 (C=N–O) cm^{-1}.

5-Carboxy-5-methyl-1-pyrroline-N-*oxide* (44)   A dilute alkaline (2.5% NaOH, 20 mL) solution of 5-ethoxycarbonyl-5-methyl-1-pyrroline-*N*-oxide (43) (5 g) was refluxed for 1 hr. Upon cooling in an ice bath, the solution was passed through a Dowex 50 (H^+ form) column, collecting fractions, as determined by TLC (silica gel, methylene chloride/methanol, 9:1), until the presence of a product was no longer visible. The combined fractions were rotary evaporated ($T < 40°C$), giving the desired nitrone (44) (1.37 g, 77%) as a white solid, mp 135–136°C (with decomposition) from chloroform (Bonnett, *et al.*, 1959a).

Ethyl 4-(2-methyl-2-nitropropyl)-5-oxopentanoate (48)   To a mixture of piperidine (29.3 g, 345 mmol) and anhydrous K_2CO_3 (10 g, 72 mmol), cooled in an ice bath, was added 4-methyl-4-nitropentanal (45) (20 g, 138 mmol, prepared as described by Shechter, *et al.*, 1952) dropwise over 25 min, such that the temperature of the reaction did not rise above 2°C. After stirring overnight at ambient temperature, ether (300 mL) was added and the reaction was filtered. The remaining solution was rotary evaporated to dryness, affording the enamine (46). Without further purification, this oil (46) was taken up in dry acetonitrile (100 mL) at 2°C, to which ethyl acrylate (47) (18 g, 180 mmol) in acetonitrile (60 mL) was added dropwise over 1 hr, maintaining this low temperature. After addition was completed, the mixture was stirred for 5 hr at room temperature. Then, the reaction was refluxed for 40 hr, cooled in an ice bath, and diluted with acetic acid (10 mL in 60 mL of H_2O). Thereafter, this solution was refluxed for 8 hr. After cooling, NaCl (15 g) was added and the organic layer was dried over anhydrous Na_2SO_4, filtered, and evaporated to dryness. The residue oil was dissolved in benzene (250 mL) and passed through a short silica gel (mesh 230–400) column to remove black impurities. The remaining solution was evaporated to dryness and distilled to give 4.75 g (14%) of ethyl 4-(2-methyl-2-nitropropyl)-5-oxopentanoate

(48), bp 162–165°C at 0.5 mm Hg (Zhang, *et al.*, 1990). IR (Neat film) 1717 (C=O), 1527 (C=N–O) cm^{-1}. NMR (CDCl$_3$) δ: 1.24 (t, 3H, CH$_3$), 1.50 (s, 3H, CH$_3$), 1.60 (s, 3H, CH$_3$), 1.82–2.02 (m, 2H, CH$_2$), 2.21–2.68 (m, 5H, CH$_2$, CHCO, CH$_2$CO$_2$), 4.06 (q, 2H, CH$_2$O), 9.57 (s, 1H, HC=O).

5,5-Dimethyl-3-(2-ethoxycarbonylethyl)-1-pyrroline-N-*oxide* (49).   Zinc dust (2.76 g, 38 mmol) was added to a solution of ethyl 4-(2-methyl-2-nitropropyl)-5-oxopentanoate (48) (4.65 g, 19 mmol) in 95% ethanol (80 mL) at 3°C. Under brisk stirring, glacial acetic acid (5.06 g, 84 mmol) in 95% ethanol (20 mL) was added dropwise for 20 min, while the temperature of the reaction was maintained at < 8°C. The reaction was stirred for an additional 2 hr at this temperature and then stored in a refrigerator at 4°C for 20 hr. Upon filtering, the zinc salts were washed with 95% ethanol (20 mL). The remaining solution was then evaporated to dryness. The residue was taken up in chloroform, stored in the refrigerator at 4°C overnight, filtered, and evaporated to reduce the liquid. Passage through a silica gel (mesh 230–400) column gave 2.71 g (67%) of the desired nitrone (49) as a yellowish oil, which was further purified by distillation, bp 164–166°C at 1 mm Hg (Zhang, *et al.*, 1990). IR (Neat film) 1717 (C=O), 1567 (C=N–O) cm^{-1}. NMR (CCl$_4$) δ: 1.24 (t, 3H, CH$_3$), 1.28 (s, 3H, ring 5-CH$_3$), 1.35 (s, 3H, 5-CH$_3$), 1.60–1.93 (m, 2H, CH$_2$), 2.17 (d, 2H, ring CH$_2$), 2.31 (t, 2H, CH$_2$CO$_2$), 2.97 (m, 1H, ring 3-CH), 4.00 (q, 2H, CH$_2$O), 6.59 (d, 1H, HC=N–O).

4,4-Dimethyl-5-nitro-2-pentanone (54)   A mixture of 4-methyl-3-penten-2-one (53) (98 g, 1 mol), nitromethane (122 g, 5 mol) and diethylamine (7.3 g, 0.1 mol) was refluxed for 2–3 days. Excess reagents were removed by rotary evaporation (*T* < 40°C). The remaining liquid was taken up into chloroform and washed with dilute hydrochloric acid (10%). After drying the solution over anhydrous Na$_2$SO$_4$, it was filtered and rotary evaporated to dryness to give 4,4-dimethyl-5-nitro-2-pentanone (54) (92 g, 58%) after fractionally distilling the liquid three times, bp 118–120°C at 20 mm Hg (Kloetzel, 1947). IR (Neat film) 1775 (C=O), 1540 (N–O), 1475 (N–O) cm^{-1}.

2,4,4-Trimethyl-1-pyrroline-N-*oxide* (55)   To a stirred solution containing 4,4-dimethyl-5-nitro-2-pentanone (54) (20 g, 126 mmol), ammonium chloride (5.45 g, 103 mmol) in water (170 mL) at 0°C, was added zinc dust (27.3 gm, 420 mmol) over a 2-hr period. After addition was completed, the reaction mixture was vigorously stirred for 2 hr, and thereafter filtered and the filter cake washed with warm water. The filtrate and washings were combined, and evaporated to near dryness with a rotary evaporator, affording an oil, which was taken up into chloroform. The chloroform solution was dried over anhydrous Na$_2$SO$_4$, filtered, and rotary evaporated to dryness to give the desired (55) (11 g, 70%) after vacuum distillation, bp 76–78°C at 0.4 mm Hg (Bonnett, *et al.*, 1959a). IR (Neat film) 1613 (C=N–O) cm^{-1}.

3,3,5,5-Tetramethyl-1-pyrroline-N-*oxide* (56)   An ether solution of 2,4,4-trimethyl-1-pyrroline-*N*-oxide (55) (10 g, 78.8 mmol in 50 mL dry ether) was added to an ethereal (50 mL) solution of methylmagnesium bromide (29 mL of a 3-M stock solution in ether, 86.6 mmol) and cooled to 0°C. After addition was completed, the reaction mixture was gently refluxed for 1 hr, then it was cooled in an ice bath to which a saturated NH_4Cl solution (5 mL) was carefully added with stirring. The ether solution was dried over anhydrous Na_2SO_4, filtered, and rotary evaporated to dryness to give the oil, 1-hydroxy-2,2,4,4-tetramethylpyrrolidine. IR (Neat film) 3400–3300 (broad, N–OH) cm^{-1}.

Aerial oxidation of the crude hydroxylamine was undertaken by dissolving 1-hydroxy-2,2,4,4-tetramethylpyrrolidine in 60% ethanol (50 mL) to which copper acetate (3 g, 16.6 mmol) and ammonium hydroxide (30%, 10 mL) were added. This solution was bubbled with O_2 (or with air) until the blue color persisted (Bonnett, et al., 1959a). Although the original preparative sequence suggests that, with time, the blue color will return, it has been found that supplementation with addition of up to 200 mg of copper acetate enhances the rate of the reaction without the loss of isolatable product. Concentration of this solution to 15–20 mL by rotary evaporation, left, after extraction with chloroform, an oil, which upon column chromatography with silica gel (mesh 230–400) and methylene chloride/methanol yielded 6.6 g (60%) of 3,3,5,5-tetramethyl-1-pyrroline-N-oxide (56), bp 54–58°C at 0.5 mm Hg (Bonnett, et al., 1959a). This compound was further purified by Kugelrohr distillation (20°C at 0.05 mm Hg) giving pure nitrone (56), which was void of nitroxide impurities. IR (Neat film) 1585 (C=N–O) cm^{-1}.

3-Ethyl-2-pentenoic acid (60)   To a 2-L, three-necked round-bottom flask fitted with two condensors in series was added ethyl bromoacetate (57) (33.4 g, 21.4 mL, 200 mmol), 3-pentanone (58) (20 g, 230 mmol), and magnesium powder (6 g, 250 mmol) in dry benzene (100 mL). The reaction was stirred at room temperature in a well-vented fume hood in which an ice bath was placed near the reaction flask. As this mixture was stirring, the temperature began to rise. At this point, it is advisable to cool the reaction flask, as, without warning, the mixture rapidly moves up the condensors, pouring reactants and benzene all over the hood. Once the reaction had subsided, however, the mixture was warmed to 70°C for 1 hr, followed by decomposition with cold sulfuric acid. After the addition of more benzene (200 mL), the mixture was washed with water, then with dilute sodium hydroxide (10%), until the pH of the aqueous solution became neutral. The remaining benzene solution was then dried over anhydrous $MgSO_4$ and rotary evaporated to dryness to give crude ethyl 3-ethyl-3-hydroxypentanoate (59) (21 gm, 60%). IR (Neat film) 3500–3300 (O–H), 1730 (C=O) cm^{-1}.

Hydrolysis of crude ethyl 3-ethyl-3-hydroxypentanoate (59) (21 g) was accomplished by refluxing this ester, dissolved in methanol (100 mL), with a solution of potassium hydroxide (10 g in 20 mL of water) overnight. After rotary evaporation ($T < 35°C$), the remaining aqueous solution was extracted with ether to remove unreacted starting material. After acidification with dilute hydrochloric acid (10%), the mixture was extracted with ether, dried over anhydrous $MgSO_4$, and evaporated to dryness *in vacuo* to give crude 3-ethyl-3-hydroxypentanoic acid (11 g, 63%) (Kon and Linstead, 1925a; Rosen and Turner, 1988). IR (Neat film) 3500–3300 (O–H), 1720 (C=O) cm^{-1}.

Crude 3-ethyl-3-hydroxypentanoic acid (10 g, 68.5 mmol) was gently refluxed overnight in acetic anhydride (20.96 g, 205.5 mmol). This mixture was then steam distilled, collecting approximately 1 L of distillate. This distillate was saturated with NaCl and extracted continually with ether. The ether solution was dried over anhydrous $MgSO_4$, filtered, and evaporated to dryness *in vacuo* to afford an oil, which was fractionally distilled to yield 3-ethyl-3-pentenoic acid (**60**) (4.4 g, 50%), as a colorless liquid, bp 120–123°C at 21 mm Hg (Kon and Linstead, 1925b; Rosen and Turner, 1988). IR (Neat film) 3500–3100 (O–H), 1690 (C=O), 1650 (C=C) cm^{-1}.

4-Ethyl-3-hexene-2-one (**61**)   To a solution of 3-ethyl-3-pentenoic acid (**60**) (5 g, 39.1 mmol) in dry ether (200 mL) at 0°C was added methyllithium in ether (1.4 M, 56 mL, 78 mmol) dropwise over a 2-hr period. After the addition was completed, the reaction was stirred for 2 hr subsequently at 0°C, and then at room temperature overnight. At this point, the reaction was poured over a crushed ice/10% HCl solution. After the ether layer had separated, the aqueous mixture was extracted with ether. The combined ether solutions were dried over anhydrous $MgSO_4$, filtered, and evaporated to dryness, *in vacuo*. The remaining oil was fractionally distilled to give 4-ethyl-3-hexene-2-one (**61**) (2 g, 41%), bp 65–68°C at 20 mm Hg (Kon and Linstead, 1925a). IR (Neat film) 1690 (C=O), 1620 (C=C) cm^{-1}.

4,4-Diethyl-5-nitro-2-pentanone (**62**)   A solution of 4-ethyl-3-hexene-2-one (**61**) (31.6 g, 250 mmol), nitromethane (135 mL, 2.5 mol), anhydrous potassium fluoride (2.91 g, 50 mmol) and 18-crown-6-ether (3.3 g, 12.5 mmol) in dry acetonitrile was refluxed for 1 week under a nitrogen atmosphere. At this point, the reaction mixture was cooled, and the solvent and excess nitromethane were removed by rotary evaporation. The remaining residue was taken up in methylene chloride (200 mL), washed with dilute HCl (1 M), and dried over anhydrous $MgSO_4$. Removal of the solvent *in vacuo* affords an oil, which was fractionally distilled to give 4,4-diethyl-5-nitro-2-pentanone (**62**) (13 g, 28%) as a pale liquid, bp 89–91°C at 0.025 mm Hg (Rosen and Turner, 1988). IR (Neat film) 1710 (C=O), 1547 (N–O), 1380 (N–O) cm^{-1}, NMR ($CDCl_3$) δ: 0.82 (t, 6H, CH_2CH_3), 1.45 (m, 4H, CH_2CH_3), 2.14 (s, 3H, CH_3C=O), 2.52 (s, 2H, CH_2C=O), 4.61 (s, 2H, CH_2NO_2).

*4,4-Diethyl-2-methylpyrroline-*N-*oxide* (**63**)   Zinc dust (17.7 g, 272 mmol) was added in small portions to a solution of 4,4-diethyl-5-nitro-2-pentanone (**62**) (12.5 g, 66.7 mmol) and ammonium chloride (3.8 g, 71.7 mmol) in 60% ethanol (200 mL) at 10°C. The resulting suspension was stirred vigorously for 5 hr at 10–15°C. After filtering the mixture, the filter cake was washed with warm water, the combined filtrate and washings were concentrated to 20 mL *in vacuo*, and the remaining liquid was extracted with methylene chloride. The organic solution was dried over anhydrous Na_2SO_4, filtered, and rotary evaporated to dryness to afford the desired nitrone (**63**) (4.68 g, 45%) as a colorless liquid, bp 128–131°C at 0.5 mm Hg (Rosen and Turner, 1988). IR (Neat film) 1630 (C=N–O), 1225 (N–O) cm^{-1}, NMR ($CDCl_3$) δ: 0.85 (m, 6H, $2CH_2CH_3$), 1.47 (br m, 6H, $2C\underline{H}_2CH_3$ + $C\underline{H}_2C(CH_3)$=N), 2 (s, 3H, CH_3C=N), 3.73 (m, 2H, CH_2NO).

*3,3-Diethyl-5,5-dimethyl-1-pyrroline-*N-*oxide* (**64**)   A dry ethereal solution of 4,4-diethyl-2-methylpyrroline-*N*-oxide (**63**) (3 g, 19.3 mmol) was added dropwise to a cold solution ($T \approx 0°C$) of methyl magnesium bromide (3.2 M, 12 mL, 38.6 mmol) in ether, under a nitrogen atmosphere. After the Grignard reagent was added, the reaction was refluxed for 1 hr, then cooled in an ice bath. After slow addition of a saturated ammonium chloride (10 mL), the aqueous layer was extracted with ether and the combined ether solutions were dried over anhydrous MgSO$_4$, filtered, and evaporated *in vacuo*. The resulting hydroxylamine was dissolved in methanol (50 mL) to which copper acetate (50 mg, 0.28 mmol) and ammonium hydroxide (30%, 1 mL) were added. This solution was bubbled with O$_2$ (or with air) until the blue color persisted (Bonnett, *et al.*, 1959a). Frequently, additional copper acetate (50–100 mg) was found to speed the reaction to completion. When the solution remained blue, the mixture was evaporated to near dryness, taken up in chloroform, dried over anhydrous Na$_2$SO$_4$, filtered, and rotary evaporated to yield an oil. This oil was purified by silica gel (mesh 230–400) chromatography, eluting the column with methanol/methylene chloride (5:95) to remove the nitrone as a colorless oil. Kugelrohr distillation (45–48°C at 0.1 mm Hg), gave 3,3-diethyl-5,5-dimethyl-1-pyrroline-*N*-oxide (**64**) (1.91 g, 58%) (Rosen and Turner, 1988). IR (Neat film) 1580 (C=N–O), 1245 (N–O) cm^{-1}, NMR (CDCl$_3$) δ: 0.84 (t, 6H, 2CH$_2$C$\underline{H}_3$), 1.48 (s, q, 1OH), 1.90 (s, 2H), 6.63 (s, 1H, CH= N–O).

5-Hexadecyl-3,3,5-trimethyl-1-pyrroline-N-oxide (**65**)   Under a nitrogen atmosphere, an ether solution of 2,4,4-trimethyl-1-pyrroline-*N*-oxide (**55**) (1 g, 7.9 mmol in 50 mL dry ether) was added to an ethereal (50 mL) solution of hexadecylmagnesium iodide, prepared by adding magnesium metal to 1-iodohexadecane (3.52 g, 10 mmol). At this point, the reaction was refluxed for several hours under nitrogen and then cooled in an ice bath. The reaction was quenched by treatment with saturated ammonium chloride (5 mL) and the ether solution was separated. The remaining aqueous layer was extracted with ether and the combined ether solutions were dried over anhydrous MgSO$_4$, filtered, and evaporated *in vacuo*. The remaining oil, 5-hexadecyl-1-hydroxy-3,3,5-trimethyl-pyrrolidine was dissolved in methanol (10 mL) to which copper acetate (50 mg, 0.28 mmol) and ammonium hydroxide (30%, 1 mL) were added. This solution was bubbled with O$_2$ (or with air) until the blue color persisted (Bonnett, *et al.*, 1959a). When the solution remained blue, the mixture was evaporated to near dryness, taken up in ether, dried over anhydrous Na$_2$SO$_4$, filtered, and rotary evaporated to dryness, affording an oil. Crystallization from petroleum ether (30–60°C) gave 5-hexadecyl-3,3,5-trimethyl-1-pyrroline-*N*-oxide (**65**) as a white solid, mp 48–54°C (Cherry,[2] 1985; Barker, *et al.*, 1985). IR (KBr pellet) 1585 (C=N–O) cm^{-1}. NMR (CDCl$_3$) δ: 1.1–1.8 (multiple CH$_2$s with 3CH$_3$), 1.8 and 2.1 (d, 2H), 6.7 (s, 1H, CH= N–O).

4-Carbethoxy-3,3,5,5-tetramethyl-1-pyrroline (**68**)   Diethyl (aminomethyl)phosphonate (5 g, 31.3 mmol) (Ratcliffe and Christensen, 1973; Dehnel and Lavielle, 1980) and dry acetone (20 mL) were stirred at room temperature for 24–48 hr in the presence of molecular sieves (3A) and/or anhydrous MgSO$_4$. The progress of the reaction was monitored by removing a sample of the mixture. After evaporation of the acetone, the disappearance of the methylene group of the starting material at

3.26 δ in the NMR spectrum was followed. When the reaction had completed, the mixture was filtered, and evaporated *in vacuo*, to remove excess acetone. To the remaining enamine (66), tetrahydrofuran (10 mL, distilled from lithium aluminum hydride) was added and evaporated *in vacuo*. This process was repeated several times to ensure that the remaining enamine (66) was void of acetone.

Butyllithium (2.5 M, 19 mL, 48 mmol) was added to tetrahydrofuran (20 mL, distilled from lithium aluminum hydride) at 0°C under argon. At this point, the temperature was lowered to −78°C, followed by the slow addition of diethyl (*N,N*-dimethyliminomethyl)phosphonate (66) in tetrahydrofuran (15 mL). After stirring for 20 min at −78°C, a solution of ethyl 3,3-dimethylacrylate (67) (4.8 g, 4.83 mL, 37.5 mmol) in tetrahydrofuran (10 mL) was added at such a rate that the temperature of the reaction did not exceed −70°C. The reaction was maintained at −78°C for an additional 20 min, then elevated to ambient temperature and stirred for 3 hr. At this point, the reaction was cooled in an ice bath, and quenched by the addition of a sodium phosphate solution (pH 7.4, 20 mL). This aqueous solution was extracted with methylene chloride, dried over anhydrous $MgSO_4$, filtered, and rotary evaporated to dryness, giving 4-carbethoxy-3,3,5,5-tetramethyl-1-pyrroline (68) (2.2 g, 35%) as a colorless liquid, bp 44–47°C at 0.1 mm Hg (Dehnel, *et al.*, 1988). IR (Neat film) 1750 (C=O), 1630 (C=N) cm^{-1}. NMR ($CDCl_3$) δ: 1.2–1.4 (m, 15H, CH_3), 2.5 (s, 1H, $HCCO_2$), 4.15 (q, 2H, OCH_2), 7.1 (s, 1H, HC=N).

4-Carbethoxy-3,3,5,5-tetramethyl-1-pyrroline-N-oxide (69)   To a solution of 4-carbethoxy-3,3,5,5-tetramethyl-1-pyrroline (68) (1.4 g, 7.1 mmol) in absolute ethanol (15 mL) was added sodium borohydride (530 mg, 14.2 mmol). This reaction was stirred overnight at room temperature, at which point the mixture was evaporated *in vacuo*. The remaining residue was taken up into water (10 mL) and extracted with chloroform. The organic solution was dried over anhydrous Na_2SO_4. After filtration, the remaining solution was rotary evaporated, giving 4-carbethoxy-3,3,5,5-tetramethylpyrrolidine as an oil (1.25 g, 89%). The absence of an imine peak in the IR spectrum confirms the success of the reduction (Dehnel, *et al.*, 1988). IR (Neat liquid) 1750 (C=O) cm^{-1}.

This crude pyrrolidine was oxidized without further purification using the method of Murahashi, *et al.*, (1990). To a mixture of crude 4-carbethoxy-3,3,5,5-tetramethylpyrrolidine (1.25 g, 6.3 mmol) in methanol (10 mL) at 0°C under argon was added sodium tungstate dihydrate (83 mg, 250 μmol) and H_2O_2 (30%, 1.6 mL, 13.86 mmol). The reaction was continually stirred anaerobically at room temperature for 3 hr. Methanol was then removed *in vacuo*. To the remaining residue, saturated NaCl was added. After extraction with chloroform, the organic solution was dried over anhydrous Na_2SO_4, filtered, and rotary evaporated to dryness, giving 4-carbethoxy-3,3,5,5-tetramethyl-1-pyrroline-*N*-oxide (69) as an oil. IR (Neat liquid) 1750 (C=O), 1590 (C=N–O) cm^{-1}. Crystallization with hot pentane yielded the desired product (900 mg, 69%) as a white solid, mp 44–45°C (Dehnel, *et al.*, 1988). NMR ($CDCl_3$) δ: 1.1–1.6 (m, 15H, CH_3), 2.9 (s, 1H, $HCCO_2$), 4.2 (q, 2H, OCH_2), 6.61 (s, 1H, HC=N–O).

4-Carboxy-3,3,5,5-tetramethyl-1-pyrroline-N-oxide (70)   A sample of 4-carbethoxyl-3,3,5,5-tetramethyl-1-pyrroline-*N*-oxide (69) (1 g, 4.7 mmol) dissolved in ethanol (95%, 10 mL) was refluxed with dilute sodium hydroxide (1 M, 5 mL) for several hours. Upon cooling in an ice bath, the reaction was concentrated *in vacuo* and passed through a Dowex 50 (H^+ form) column until the presence of the product was no longer visible by TLC. Fractions were collected and rotary evaporated, giving 4-carboxyl-3,3,5,5-tetramethyl-1-pyrroline-*N*-oxide (**70**) (720 g, 83%) as a white solid from chloroform/hexane, mp > 240°C with decomposition (Arya,[3] 1995; Rosen,[3] 1995). NMR (D_2O) δ: 1.2–1.6 (m, 12H, CH_3), 2.9 (s, 1H, $HCCO_2$), 7.1 (s, 1H, HC=N–O).

3-(Hydroxymethyl)-2,2,4,4-tetramethyl-1-pyrrolidine (**71**)   To a mixture of lithium aluminum hydride (0.8 g, 20 mmol) in dry ether (10 mL) at 0°C was added an ethereal (100 mL) solution of 4-carbethoxy-3,3,5,5-tetramethyl-1-pyrroline (**68**) (4.95 g, 25.12 mmol) over 3 hr. At this point, the reaction was refluxed for 1 hr, then cooled in an ice bath and quenched by the careful addition of a NaOH/EtOH solution (10 mL), prepared by the mixing of 10% NaOH with 95% ethanol in a 1:1 ratio. The organic layer was dried over anhydrous Na_2SO_4, filtered, and rotary evaporated to dryness, giving 3-(hydroxymethyl)-2,2,4,4-tetramethyl-1-pyrrolidine (**71**) (3.6 g, 90%) as a white solid, mp 91–92°C from 95% ethanol (Arya, *et al.,* 1992). NMR ($CDCl_3$) δ: 0.85, 1.00, 1.01, 1.48 (4s, 12H, $4CH_3$), 1.45 (t, $J = 7.4$ Hz, 1H at H-3 position), 2.55 (AB, $J = 1.5$ Hz, 2H at H-5 position), 3.51, 3.54 (AB, $J = 5.6$ Hz, 2H, CH_2OH).

N-*(Benzyloxycarbonyl)-3-(hydroxymethyl)-2,2,4,4-tetramethyl-1-pyrrolidine* (**72**)   To a mixture of 3-(hydroxymethyl)-2,2,4,4-tetramethyl-1-pyrrolidine (**71**) (4.7 g, 30 mmol) and potassium carbonate (4.8 g, 35 mmol) in acetonitrile (100 mL) was added a solution of benzyl chloroformate (4.7 mL, 33 mmol) in acetonitrile (10 mL). After stirring at −25°C for 3 hr, the reaction is warmed to 20°C, quenched with sodium phosphate buffer (20 mL, 50 mM, pH 7.4), and extracted with methylene chloride. The combined organic extractions were dried over anhydrous Na_2SO_4, filtered, and evaporated *in vacuo*, giving a residual oil. After chromatography with silica gel (mesh 230–400) and ethyl acetate–hexane (1:3), *N*-(benzyloxycarbonyl)-3-(hydroxymethyl)-2,2,4,4-tetramethyl-1-pyrrolidine (**72**) was isolated as a white solid (8.1 g, 92%), mp 62–63°C (from ethyl acetate–hexane, 1:1) (Arya, *et al.*, 1992). NMR ($CDCl_3$) δ: 1.03, 1.15, 1.52, 1.57 (4s, 12H, $4CH_3$), 1.96 (t, $J = 7.3$ Hz, 1H at H-3 position), 2.62 (bs, 1H, OH), 3.16, 3.46 (AB, $J = 10$ Hz, 2H, H-5 position), 3.81 (AB, $J = 7.1$ Hz, 2H, C$\underline{H}_2$OH), 5.15 (s, 2H, OC$\underline{H}_2C_6H_5$), 7.41 (s, 5H, OC$H_2C_6\underline{H}_5$).

N-*(Benzyloxycarbonyl)-3-formyl-2,2,4,4-tetramethyl-1-pyrrolidine* (**73**)   To a solution of pyridinium chloroformate (3.2 g, 15 mmol) in methylene chloride (50 mL) was added *N*-(benzyloxycarbonyl)-3-(hydroxymethyl)-2,2,4,4-tetra-methyl-1-pyrrolidine (**72**) (2.9 g, 10 mmol) in methylene chloride (50 mL) in a dropwise fashion so that the temperature of the reaction would not exceed 25°C. After stirring for 3 hr at this temperature, the reaction was diluted with ether (200 mL) and passed over fluorosil (10 gm). The mixture was then washed with a saturated NaCl solution. After separating the layers, the organic solution was dried over anhydrous Na_2SO_4, filtered, and evaporated *in vacuo*, giving a crude residue. Column chromatography, using silica gel (mesh 230–400) and ethyl acetate–hexane (1:7), gave N-(benzyloxycarbonyl)-3-formyl-2,2,4,4-tetramethyl-1-pyrrolidine (73) (2.7 g, 93%) as a white oil (Arya, *et al.*, 1992). NMR (CDCl$_3$) δ: 1.30, 1.31, 1.59, 1.69 (4s, 12H, 4CH$_3$), 2.4 (bs, 1H at H-3 position), 3.2, 3.52 (AB, $J = 10.8$ Hz, 2H, H-5 position), 5.14 (s, 2H, OCH$_2$C$_6$H$_5$), 7.38 (bs, 5H, OCH$_2$C$_6$H$_5$), 9.86 (s, 1H, HC=O).

N-(Benzyloxycarbonyl)-3-[1,2-didehydro-2-(ethoxycarbonyl) ethyl]2,2,4,4-tetramethylpyrrolidine (74)   A solution of N-(benzyloxycarbonyl)-3-formyl-2,2,4,4-tetramethyl-1-pyrrolidine (73) (2.9 g, 10 mmol) and (carbethoxymethylene)triphenylphosphorane (5.2 g, 15 mmol) in benzene (50 mL) was stirred at 60°C for 1 hr, followed by the supplementation of additional benzene (50 mL). At this point, the mixture was washed with a saturated NaCl solution, dried over anhydrous Na$_2$SO$_4$, filtered, and evaporated *in vacuo*, giving a crude solid. Purification by column chromatography, using silica gel (mesh 230–400) and ethyl acetate–hexane (1:10), gave the desired tetramethylpyrrolidine (74) (2.74 g, 76%) as a white solid from ethyl acetate–hexane (1:3), mp 79–80°C (Arya, *et al.*, 1992). NMR (CDCl$_3$) δ: 1.01, 1.06, 1.44, 1.51 (4s, 12H, 4CH$_3$), 1.25 (t, $J = 7.1$ Hz, 3H, OCH$_2$CH$_3$), 2.35 (bs, $J = 2.9$ Hz and 10.7 Hz, 1H at H-3 position), 3.21, 3.57 (AB, $J = 10.8$ Hz, 2H, H-5 position), 4.26 (q, $J = 7.2$ Hz, 2H, OCH$_2$CH$_3$), 5.15 (s, 2H, OCH$_2$C$_6$H$_5$), 5.94 (dd, $J = 2.9$ Hz and 15.4 Hz, 1H, CH=C), 7.05 (dd, $J = 10.7$ Hz and 15.4 Hz, 1H, C=CHCO$_2$R) 7.39 (bs, 5H, OCH$_2$C$_6$H$_5$).

3-[2-(Ethoxycarbonyl)ethyl]-2,2,4,4-tetramethyl-1-pyrrolidine (75)   A solution of N-(benzyloxycarbonyl)-3-[1,2-didehydro-2-(ethoxycarbonyl)ethyl]-2,2,4,4-tetramethylpyrrolidine (74) (3.6 g, 10 mmol) in absolute ethanol (30 mL) and palladium on charcoal (10%, 200 mg) was hydrogenated under atmospheric pressure for 10 hr. The mixture was filtered over fluorosil (5 g) and the remaining solution was rotary evaporated to dryness, yielding 3-[2-(ethoxycarbonyl)ethyl]-2,2,4,4-tetramethyl-1-pyrrolidine (75) (2.1 g, 95%) as a white oil (Arya, *et al.*, 1992). NMR (CDCl$_3$) δ: 0.97, 1.06, 1.09, 1.21 (4s, 12H, 4CH$_3$), 1.28 (t, $J = 7.1$ Hz, 3H, OCH$_2$CH_3), 1.66 (m, 3H, H-3 and H-3′ position), 2.37 (t, $J = 7.8$ Hz, 2H at CH$_2$CO$_2$), 2.80 (bs, 2H, H-5 position), 4.16 (q, $J = 7.1$ Hz, 2H, OCH$_2$CH$_3$).

4-[2-(Ethoxycarbonyl)ethyl]-3,3,5,5-tetramethyl-1-pyrroline-N-oxide (76)   To a stirred solution of 3-[2-(ethoxycarbonyl)ethyl]-2,2,4,4-tetramethyl-1-pyrrolidine (75) (1.13 g, 5 mmol) and sodium tungstate dihydrate (0.82 g, 20 mmol) in methanol (25 mL), under anaerobic conditions at 0°C under argon, was added H$_2$O$_2$ (30%, 2 mL, 17.33 mmol). The reaction was continually stirred anaerobically at room temperature for 3 hr. Methanol was then removed *in vacuo*, at which point a saturated NaCl solution (20 mL) was added. After extraction with chloroform, the organic solution was dried over anhydrous Na$_2$SO$_4$, filtered, and rotary evaporated to dryness, giving an oil. Purification by column chromatography, using silica gel (mesh 230–400) and methylene chloride/methanol (20:1), gave 4-[2-(ethoxycarbonyl)ethyl]-3,3,5,5-tetramethyl-1-pyrroline-N-oxide (76) (1 g, 85%) as a yellow oil, which became a solid at temperatures below 0°C (Arya, *et al.*, 1992). NMR (CDCl$_3$) δ: 0.96, 1.05, 1.18, 1.27 (4s, 12H, 4CH$_3$), 1.10 (t, $J = 7.1$ Hz, 3H,

OCH$_2$CH$_3$), 1.70 (m, 3H, H-4, H-3′, and H-4′), 2.25 (t, J = 7.7 Hz, 2H, CH$_2$CO$_2$), 3.99 (q, J = 7.1 Hz, 2H, OCH$_2$CH$_3$), 6.48 (s, 1H, HC=N–O).

N-(Benzyloxycarbonyl)-3-(N-maleimidobutyryloxymethyl)-2,2,4,4-tetramethylpyrrolidine (**78**)   To a solution of *N*-maleimidobutyric acid (**77**) (439 mg, 2.4 mmol) and triethylamine (300 mg, 0.42 mL, 3 mmol) in tetrahydrofuran (25 mL) at room temperature under argon was added freshly distilled 2,4,6-trichlorobenzoyl chloride (583 mg, 2.4 mmol) in tetrahydrofuran (15 mL). The mixture was stirred for 1.5 hr and the triethylamine hydrochloride was filtered. The remaining solution was added to a mixture of *N*-(benzyloxycarbonyl)-3-(hydroxymethyl)-2,2,4,4-tetramethyl-1-pyrrolidine (**72**) (582 mg, 2 mmol) and, 4-dimethylaminopyridine (268 mg, 2.2 mmol) in tetrahydrofuran (20 mL) under argon. After addition was completed, the reaction was stirred for 30 min, diluted with methylene chloride (100 mL), and quenched with the addition of sodium phosphate buffer, pH 7.4. After separation, the organic layer was dried over anhydrous MgSO$_4$, and evaporated *in vacuo*. The remaining residue was chromatographed using silica gel (mesh 230–400) and ethyl acetate/hexane (1:3) to give *N*-(benzyloxycarbonyl)-3-(*N*-maleimidobutyryloxymethyl)-2,2,4,4-tetramethylpyrrolidine (**78**) (848 mg, 93%) as an oil (Arya, 1996). NMR (CDCl$_3$) δ: 1.01, 1.09, 1.45, and 1.48 (s, 12H, CH$_3$), 1.95 (m, 3H, OCOCH$_2$CH$_2$– and H-3 on nitrone), 2.33 (t, 2H, J = 7.26 Hz, –NCH$_2$–), 3.12 and 3.43 (dd, J = 10.7 Hz each, H-5 on nitrone), 3.57 (t, 2H, J = 6.6 Hz, OCOCH$_2$–), 4.18 (m, 2H at 3′ position), 5.08 (s, 2H, –COOCH$_2$C$_6$H$_5$), 6.69 (s, 2H, maleimido-H), 7.34 (bs, 5H, –COOCH$_2$C$_6$H$_5$).

3-(N-Maleimidobutyryloxymethyl)-2,2,4,4-tetramethyl-pyrrolidine (**79**)   Hydrogen bromide (1 mL, 30% in acetic acid) was added dropwise to a mixture of *N*-(benzyloxycarbonyl)-3-(*N*-maleimido-butyryloxymethyl)-2,2,4,4-tetramethylpyrrolidine (**78**) (456 mg, 1 mmol) in methylene chloride (5 mL) at room temperature. After 30 min, the reaction was quenched with dilute Na$_2$CO$_3$ (10%, 10 mL), the organic layer was separated, dried over anhydrous MgSO$_4$, and evaporated *in vacuo*. Purification by pouring a methylene chloride/methanol (40:3) solution of the oil through a silica gel (mesh 230–400) column afforded 3-(*N*-maleimidobutyryloxymethyl)-2,2,4,4-tetramethylpyrrolidine (**79**) (276 mg, 85%) (Arya, 1996). NMR (CDCl$_3$) δ: 1.12, 1.25, 1.47, and 1.65 (s, 12H, 4CH$_3$), 1.85 (t, 2H, J = 6.9 Hz, OCOCH$_2$CH$_2$–), 2.09 (t, 1H, H-3 on nitrone), 2.27 (t, J = 7.2 Hz, 2H, –NCH$_2$–), 3.12 (bs, 2H, H-5 on nitrone), 3.51 (t, 2H, J = 6.7 Hz, OCOCH$_2$–), 4.07 (ABX, 2H, J = 6.4, 8.5, 11.4 Hz 2H, H-3′ on nitrone), 5.27 (s, 1H, NH), 6.68 (s, 2H, maleimido-H).

4-(N-Maleimidobutyryloxymethyl)-3,3,5,5-tetramethyl-1-pyrroline-N-oxide (**80**)   To a solution of 3-(*N*-maleimidobutyryloxymethyl)-2,2,4,4-tetramethylpyrrolidine (**79**) (312 mg, 1 mmol) in dry tetrahydrofuran (15 mL) was added Davis' reagent, trans-2-(phenylsulfonyl)-3-phenyloxaziridine (574 mg, 2.2 mmol), prepared as described in the literature (Vishwakarma, *et al.*, 1988) in dry tetrahydrofuran (15 mL). The reaction was stirred at room temperature for 20 min, filtered, and the remaining solution was rotary evaporated to afford an oil.

Purification through column chromatography with silica gel (230–400 mesh) and methylene chloride/methanol (40:1.5) gave 4-(N-maleimidobutyryloxymethyl)-3,3,5,5-tetramethyl-1-pyrroline-N-oxide (**80**) (292 mg, 87%) as an oil (Arya, 1996). NMR (CDCl$_3$) δ: 1.08, 1.15, 1.25, and 1.40 (s, 12H, CH$_3$), 1.97 (m, 3H, OCOCH$_2$C$\underline{H}_2$– and H-3 on nitrone), 2.39 (t, $J = 7.4$ Hz, 2H, –NCH$_2$–), 3.63 (t, 2H, $J = 6.8$ Hz, OCOCH$_2$–), 4.24 (d, 2H, $J = 8.1$ Hz at position 3′), 6.75 (s, 2H, maleimido-H), 7.16 (s, 1H, HC=N–O).

3,3-Dimethyl-3,4-dihydroisoquinoline (**83**)   To a mechanically stirred solution of N-(1,1-dimethyl-2-phenylethyl)formamide (**82**) (33.4 g, 188 mmol), prepared as described in the literature from β,β-dimethylphenethyl alcohol (**81**) (Shetty, 1961; Ritter and Kalish, 1964; Lavin, *et al.*, 1981) in dry toluene (325 mL) at room temperature, was added phosophorous pentoxide (100 g, 705 mmol). The reaction was refluxed for 6 hr and allowed to stand overnight at room temperature. After toluene is decanted off, ice-water was added to the remaining residue, which was followed by extraction with ether. The aqueous layer was basified with NaOH (50%) to pH 8 and extracted with ether. After drying the etheral filtrate with anhydrous MgSO$_4$, the remaining solution was evaporated *in vacuo*, affording 3,3-dimethyl-3,4-dihydroisoquinoline (**83**) as a colorless oil, bp 71–75°C at 0.2 mm Hg (Carr, *et al.*, 1994).

4,8b-Dihydro-3,3-dimethyl-3H-oxazirino[3,2a]isoquinoline (**84**)   To a stirred solution of 3,3-dimethyl-3,4-dihydroisoquinoline (**83**) (1 g, 6.3 mmol) in methylene chloride (50 mL) at 0°C was added *m*-chloroperbenzoic acid (1.5 g, 7.0–7.4 mmol, 80–85%), portionwise over several minutes. After stirring for 4 hr at 0°C, the mixture was washed with a saturated solution of Na$_2$CO$_3$, dried over anhydrous MgSO$_4$, filtered, and rotary evaporated to dryness. The remaining oil was chromatographed using silica gel (mesh 230–400). Eluting the column with a solvent mixture of hexane/ethyl acetate (20:80) gave 4,8b-dihydro-3,3-dimethyl-3H-oxazirino[3,2a]isoquinoline (**84**) as a colorless oil, bp 105°C at 0.1 mm Hg (Carr, *et al.*, 1994).

3,3-dimethyl-3,4-dihydroisoquinoline-N-oxide (**85**)   To a methanol/H$_2$O (30:6 mL) solution of 4,8b-dihydro-3,3-dimethyl-3H-oxazirino[3,2a]isoquinoline (**84**) (657 mg, 3.75 mmol) was added concentrated sulfuric acid (24 mL). After stirring at room temperature overnight, the solution was carefully poured into an ice-cold saturated solution of Na$_2$CO$_3$, and extracted with ether. The ethereal solution was washed with aqueous dihydrogen phosphate, dried over anhydrous MgSO$_4$, filtered, and evaporated to dryness. The remaining liquid was distilled, giving 3,3-dimethyl-3,4-dihydroisoquinoline-N-oxide (**85**) as a colorless oil, bp 156°C at 0.05 mm Hg, which solidified upon standing, mp 70–72°C (Carr, *et al.*, 1994).

7-Trifluoromethyl-1,2,3,4-tetrahydro-3,3-dimethyl-oxazolo[2,3-a]isoquinoline-2, 3-dione (**87**)   To a solution of N-formyl-1-(4-trifluorophenyl)-2-methyl-2-aminopropane (**86**) (8.35 g, 34 mmol) in methylene chloride was added oxalyl chloride (3.26 mL, 37.4 mmol). Gas soon evolved and, after 1 hr, the reaction was cooled in an ice bath and treated with ferric chloride (6.62 g, 40.8 mmol). The reaction was allowed to warm to room temperature. After 16 hr, the reaction was treated with dilute hydrochloric acid (2M, 340 mL) and stirred for 1 hr. After separation, the aqueous mixture was extracted with methylene chloride. The combined methylene chloride solutions were dried over anhydrous Na_2SO_4, filtered, and evaporated *in vacuo* to give a viscous, foaming oil. This material was dissolved in hot ethyl acetate:hexane (3:7) and the product rapidly crystallized. Hexanes were added to improve recovery of 7-trifluoromethyl-1,2,3,4-tetrahydro-3,3-dimethyl-oxazolo[2,3-a]isoquinoline-2,3-dione (**87**) (8.42 g, 83%) as a white solid, mp 123.5–124.5°C (Bernotas, *et al.*, 1996b). IR (KBr) 1716 (C=O) cm^{-1}. NMR (CDCl$_3$) δ: 1.39 (s, 3H), 1.77 (s, 3H), 2.99 (m, 2H), 6.42 (s, 1H), 7.44 (d, 1H, $J = 7.7$ Hz), 7.71 (d, 1H, $J = 8.2$ Hz), 7.77 (s, 1H).

7-Trifluoromethyl-3,3-dimethyl-1,2,3,4-tetrahydroisoquinoline (**89**)   7-Trifluoro-methyl-1,2,3,4-tetrahydro-3,3-dimethyl-oxazolo[2,3-a]isoquinoline-2,3-dione (**87**) (2.99 g, 10 mmol) stirred in a flask was immersed in an oil bath at 155–160°C. As the compound melted, gas evolved. After 30 min, gas evolution ceased. At this point, the reaction was cooled in an ice bath, methanol was introduced (20 mL), and sodium borohydride (0.76 g, 20 mmol) was carefully added. The reaction became warm and gas evolved. After 16 hr, sodium hydroxide (1M, 20 mL) was added to the reaction. After an additional 20 min, the reaction was diluted with water (50 mL), extracted with methylene chloride, and dried over anhydrous Na_2SO_4. The remaining solution was filtered and evaporated *in vacuo* to give a viscous oil. Upon flash chromatographic separation using silica gel (mesh 230–400) ethyl acetate followed by ethanol/ethyl acetate (2:8), 7-trifluoromethyl-3,3-dimethyl-1,2,3,4-tetrahydroisoquinoline (**89**) (2.1 g, 92%) was isolated as an off-white solid, mp 41–42°C (Bernotas, *et al.*, 1996b). NMR (CDCl$_3$) δ: 1.47 (s, 6H), 3.14 (s, 2H), 7.31 (bs, 1H), 7.34 (bs, 1H), 7.52 (bm, 1H), 7.73 (s, 1H).

7-Trifluoromethyl-3,3-dimethyl-3,4-dihydroisoquinoline-N-oxide (**90**)   To a solution of 7-trifluoromethyl-3,3-dimethyl-1,2,3,4-tetrahydroisoquinoline (**89**) (2 g, 8.7 mmol) in ethanol (20 mL) was added sodium tungstate dihydrate (0.14 gm, 0.44 mmol) in water (10 mL), followed by the addition of hydrogen peroxide (30%, 2.2 mL, 22 mmol). An additional 1 mL of the peroxide was added after 1 hr, followed by a subsequent 0.5 mL within 30 min. After 2 hr, the reaction was completed as judged by TLC analyses. The reaction was diluted with water (70 mL) and extracted with methylene chloride. The combined extracts were dried over anhydrous Na_2SO_4, concentrated *in vacuo*, and flash chromatographed with silica gel (mesh 230–400) and ethyl acetate to afford 7-trifluoromethyl-3,3-dimethyl-3,4-dihydroisoquinoline-*N*-oxide (**90**) as an oil, which solidified upon trituration with hexane (1.95 g, 92%), mp 51.5–53°C (Bernotas, *et al.*, 1996b). NMR (CDCl$_3$) δ: 1.19 (s, 6H), 2.68 (s, 2H), 4.09 (s, 2H), 7.16 (bd, 1H, $J = 8.2$ Hz), 7.30 (bs, 1H), 7.38 (bd, 1H, $J = 8.2$ Hz).

1,2,4,5-Tetrahydro-3,3-dimethyl-oxazolo[2,3-a]benzazepine-2,3-dione (**92**)   To a solution of *N*-(1,1-dimethyl-3-phenylpropyl)formamide (**91**) (13.37 g, 70 mmol), prepared as described in the literature (Moed, *et al.*, 1955), in methylene chloride

(600 mL) under a nitrogen atmosphere at room temperature, was added oxalyl chloride (9.8 g, 6.72 mL, 77 mmol). Upon addition of oxalyl chloride, gas evolution ensued. After 1 hr at ambient temperature, the reaction was cooled in an ice bath and anhydrous ferric chloride (13.6 g, 84 mmol) was added. After 10 min, the ice bath was removed and the reaction was stirred for 16 hr at room temperature, at which point aqueous hydrochloric acid (2 M, 600 mL) was added to this rapidly mixed reaction. After 1.5 hr, the organic layer was separated, washed with a saturated NaCl solution, dried over anhydrous Na_2SO_4, filtered, and evaporated *in vacuo* to dryness. Purification by flash chromatography with silica gel (230–400 mesh) and ethyl acetate/hexane (1:1) gives 1,2,4,5-tetrahydro-3,3-dimethyl-oxazolo[2,3-a]benzazepine-2,3-dione (**92**) (12.56 g, 73%) as an off-white solid, mp 119–121°C (Carr, *et al.*, 1994; Bernotas, *et al.*, 1996a). IR (CHCl₃) 1817 (C=O), 1734 (C=O) cm⁻¹. NMR (CDCl₃) δ: 1.59 (s, 3H), 1.87 (s, 3H), 1.91 (ddd, 1H, $J = 4.0, 5.4, 15.6$ Hz), 2.31 (ddd, 1H, $J = 4.0, 11.9, 15.7$ Hz), 3.15–3.25 (b, 2H), 6.85 (s, 1H), 7.19–7.35 (b, 3H), 7.53 (dd, 1H, $J = 1.2, 7.2$ Hz).

3,3-Dimethyl-4,5-dihydro-2-benzazepine-N-oxide (**95**)   1,2,4,5-Tetrahydro-3,3-dimethyl-oxazolo[2,3-a]benzazepine-2,3-dione (**92**) (2.45 g, 10 mmol) was heated under a nitrogen atmosphere at 140–150°C, during which time the solid melted, releasing a gas. After 15 min, gas evolution had ceased, yielding 3,3-dimethyl-4,5-dihydro-2-benzazepine (**93**). The crude (**93**) was then immersed into an ice bath and methanol (20 mL) was poured into the flask. Sodium borohydride (0.76 g, 20 mmol) was added over 5 min, leading to the release of a gas and an increase in the temperature of the reaction. After 5 min, the ice bath was removed and the mixture was stirred for 2 hr, whereupon, dilute NaOH (1 M, 20 mL) was added. Following 20 min of mixing, the reaction was extracted with methylene chloride, and the organic layer was dried over Na_2SO_4, filtered, and evaporated *in vacuo* to dryness, yielding an orange-brown oil. Purification was achieved by flash chromatography with silica gel (230–400 mesh) and diethylamine. Changing to diethylamine/ethyl acetate (5:95), 3,3-dimethyl-1,2,4,5-tetrahydro-2-benzazepine (**94**) (1.09 g, 62%) was afforded with R_f of 0.1 (with streaking) with diethylamine.

To amine (**94**) (1.09 g, 62 mmol) dissolved in ethanol (95%, 8 mL) at 0°C under argon was added sodium tungstate dihydrate (102 mg, 310 μmol) and H_2O_2 (30%, 1.41 mL, 12.21 mmol). The reaction was continually stirred anaerobically at room temperature for 6 hr. At this point, the reaction was diluted with H_2O and extracted with methylene chloride. The organic solution was dried over Na_2SO_4, filtered, and rotary evaporated to dryness, giving a yellow oil. Purification of the crude benzazepine-N-oxide (**95**) is accomplished by flash chromatography with silica gel (230–400 mesh) and ethyl acetate. Changing to ethyl acetate/ethanol (80:20), 3,3-dimethyl-4,5-dihydro-2-benzazepine-N-oxide (**95**) is isolated as a colorless oil, which solidified upon storing at 0°C, mp 62–65°C (Carr, *et al.*, 1994; Bernotas, *et al.*, 1996a). IR (KBr) 1543 (C=N–O) cm⁻¹. NMR (CDCl₃) δ: 1.63 (s, 6H), 2.17–2.21 (m, 2H), 3.01–3.05 (m, 2H), 7.15–7.28 (m, 4H), 7.91 (s, 1H).

2,2-Dimethyl-4-hydroximinomethyl-2H-imidazole-N-oxide (**98**)   A mixture of 1,3-dioximinoacetone (**96**), prepared as described by Geissman, *et al.* 1946), (4 g, 34 mmol), ammonium acetate (15.7 g, 20 mmol), acetone (**97**) (47.4 g, 60 mL,

817 mmol), and acetic acid (24 g, 23 mL, 400 mmol) was stirred at room temperature for 65 hr, during which time the reaction turned black. After removing the excess ammonium acetate through filtration, the remaining solution was evaporated *in vacuo*, yielding a viscous black oil. The oil was taken up in brine (50 mL) and extracted with ethyl acetate. After drying the mixture with anhydrous Na_2SO_4, the solution was rotary evaporated, leaving a brown solid. Upon recrystallization with ethyl acetate, 2,2-dimethyl-4-hydroximinomethyl-2*H*-imidazole-*N*-oxide (**98**) is obtained as a brown solid (3.3 g, 63%), mp 186–188°C (Kirilyuk, *et al.*, 1991; Pou,[4] 1997). IR (KBr) 1610, 1590 (C=N–OH), 1530 (C=N–O) cm^{-1}. NMR [$(CD_3)_2SO$] δ: 1.41 (s, 6H), 7.88 (s, 1H), 8.07 (s, 1H), 12.51 (s, 1H).

5,5-Dimethyl[2,3,3-²H₃]-1-pyrroline-N-oxide (**99**)   A solution of 5,5-dimethyl-1-pyrroline-*N*-oxide (**5**) (1 g, 8.62 mmol) in D_2O (10 mL) and NaOD (310 mg, 7.56 mmol) was gently heated at 70°C for 12 hr. At this point, the reaction mixture was rotary evaporated to near dryness. The remaining residue was taken up into chloroform (50 mL), dried over anhydrous K_2CO_3, and evaporated to dryness, affording a brown oil. With this procedure a yield > 90% for d_3/d_2 ratio was found. Kugelrohr distillation (20°C at 0.05 mm Hg) afforded 910 mg (90%) of 5,5-dimethyl[2,3,3-²H₃]-1-pyrroline-*N*-oxide (**99**) (Pou, *et al.*, 1990). IR (Neat Film) 1535 (C=N–O) cm^{-1}. NMR (CDCl$_3$) δ: 1.42 (s, 6H), 2.10 (s, 2H); MS $C_6H_8D_3NO$, *m/e* 116 (M⁺).

5-Nitro(1,1,1,3,3-²H₅]pentan-2-one (**101**)   A two-phase emulsion consisting of D_2O (45 mL), concentrated DCl (9 mL), and 5-nitropentan-2-one (**100**) (45 g, 0.34 mol) was stirred at 55°C for 48 hr. The organic layer was separated and the aqueous solution was extracted with ether. The combined organic solutions were dried over anhydrous $MgSO_4$, filtered, and evaporated to dryness, giving partial deuterium ($\approx$ 80%) incorporation into the pentanone. This exchange reaction was repeated several more times until incorporation of deuterium was complete, resulting in 5-nitro(1,1,3,3-²H₅]pentan-2-one (**101**) (32.4 g, 70%) as a colorless liquid, bp 46–50°C at 0.3 mm Hg (Pou, *et al.*, 1990). NMR (CDCl$_3$) δ: 2.21 (t, 2H), 4.42 (t, 2H).

1-([²H₃]Methyl)[2,2-²H₃]-1-pyrroline-N-oxide (**102**)   To a solution of 5-nitro(1,1,3,3-²H₅]pentan-2-one (**101**) (15 g, 0.11 mol) and NH_4Cl (6.36 g, 119 mmol) in D_2O (45 mL) at −20°C was added zinc dust (36 g) in small portions over 3 hr, maintaining the temperature of the reaction at 5°C. After the addition was completed, the reaction was stirred for another 0.5 hr at 25°C. The mixture was filtered and the remaining solid was washed with methanol. The combined filtrates were concentrated and the remaining oil was extracted with chloroform, dried over anhydrous $MgSO_4$, filtered, and evaporated to dryness, yielding 1-([²H₃]methyl)[2,2-²H₃]-1-pyrroline-*N*-oxide (**102**) (7.05 g, 62%) as a white oil, bp 65–67°C at 0.3 mm Hg (Pou, *et al.*, 1990). NMR (CDCl$_3$) δ: 2.08 (t, 2H), 3.99 (t, 2H).

5,5-Di([²H₃]methyl)[4,4-²H₂]-1-pyrroline-N-oxide (**103**)   To a stirred solution of methyl-d$_3$-magnesium bromide (100 mmol, 100 mL) in ether was added a solution of 1-([^{2}H$_3$]methyl)[2,2-^{2}H$_3$]-1-pyrroline-*N*-oxide (**102**) (5.2 g, 50 mmol) in dry ether (50 mL) at a rate sufficient to maintain a gentle reflux of the ether. After the addition was completed, the reaction was stirred for 0.5 hr at 25°C. Then, the reaction was cooled to 0°C, and a saturated solution of NH$_4$Cl (20 mL) was added. The ether layer was decanted, and the remaining aqueous solution was extracted with ether. The combined filtrates were evaporated to dryness, leaving a yellow oil (2.89 g). To this oil was added methanol (50 mL), concentrated NH$_4$OH (5 mL), and copper acetate (500 mg) in the presence of a stream of air. Within 15 min, a blue color persisted, at which point the solution was evaporated to dryness. The blue oil was taken up in ether, washed with saturated sodium hydrogen carbonate (10 mL) and brine (10 mL), dried over anhydrous MgSO$_4$, filtered, and evaporated to dryness, affording 5,5-di([^{2}H$_3$]methyl)[4,4-^{2}H$_2$]-1-pyrroline-*N*-oxide (**103**) (1 g, 17%) as a white oil after distillation, bp 35–37°C at 0.03 mm Hg (Pou, *et al.*, 1990). NMR (CDCl$_3$) δ: 2.55 (br s, 2H), 6.79 (t, 1H).

5,5-Di([^{2}H$_3$]methyl)[2,3,3,4,4-^{2}H$_5$]-1-pyrroline-N-oxide (**104**)   A D$_2$O solution (10 mL) of 5,5-di([^{2}H$_3$]methyl)[4,4-^{2}H$_2$]-1-pyrroline-*N*-oxide (**103**) (1 g, 8.26 mmol) and NaOD (310 mg, 7.56 mmol) was heated to 70°C for 12 hr. The solution was evaporated to dryness. The brown residue was extracted with chloroform, dried over anhydrous MgSO$_4$, filtered, and evaporated to dryness, giving 5,5-di([^{2}H$_3$]methyl)[2,3,3,4,4-^{2}H$_5$]-1-pyrroline-*N*-oxide (**104**), after distillation, bp 35–37°C at 0.03 mm Hg, as a colorless oil, which solidified at temperatures below 0°C (900 mg, 88%) (Pou, *et al.*, 1990). GC/MS indicated a mixture of isotopically labeled nitrones of which 62% was 5,5-di([^{2}H$_3$]methyl)[2,3,3,4,4-^{2}H$_5$]-1-pyrroline-*N*-oxide (**104**).

4-([^{2}H$_3$]Methyl)-4-[^{15}N]nitro)[5,5,5-^{2}H$_3$]pentanal (**108**)   To a D$_2$O solution (5 mL) of (^{15}N)hydroxylamine hydrochloride (1.5 g, 15.6 mmol) was added NaOD (960 mg, 23.4 mmol) in D$_2$O (2 mL). The reaction mixture was stirred for 15 min at 25°C. To this was added acetone-d_6 (**105**) (2 mL, 27.3 mmol) dropwise. After 2 hr, the reaction was saturated with NaCl, followed by extensive extraction with ether. The combined organic solutions were dried over anhydrous K$_2$CO$_3$, filtered, and the remaining filtrate was evaporated to dryness, giving the oxime (**106**) as a white solid. Chlorine was bubbled into a solution of (**106**) dissolved in methylene chloride (20 mL) at −10°C until a blue-green color persisted. This was followed by bubbling ozone into the above reaction until the blue solution was discharged. The solvent was removed *in vacuo* and the residual oil was taken up in NaOH (1.8 g, 45 mmol in 20 mL) and hydrogenated using Pd/C (10%, 150 mg) and H$_2$ under 60 psi for 10 hr. At that point, the reaction was filtered, and the pH of the remaining solution was adjusted to 8 with 2 N HCl. Upon evaporation of this solution to dryness, a white powder, consisting of NaCl and the anion of 2-nitro[1,1,1,3,3,3-^{2}H$_3$]propane (**107**), remained. This mixture was further dried through azeotropic distillation with ethanol (99.5%), followed by subjection to high vacuum. Once completed, the powder was suspended in ethanol (99.5%, 30 mL) and cooled to −60°C, at which point freshly distilled acrolein (2.60 g, 46.4 mmol) was carefully added such that the temperature did not exceed −60°C. After the addition, the temperature was allowed to rise to −15°C over 4 hr. To the resulting yellow suspension was added glacial acetic acid (1.5 g, 25.1 mmol) and the mixture was stirred for 15 min, resulting in a white suspension. Evaporation to dryness gave a sticky light yellow residue, which was taken up in methylene chloride (20 mL), filtered, and again evaporated to dryness. The resulting liquid was distilled, yielding 4-([²H₃]methyl)-4-[¹⁵N]nitro)[5,5,5-²H₃]pentanal (**108**) (2.91 gm, 90%) as a colorless liquid, bp 59–65°C at 0.1 mm Hg (Warner and Moe, 1952; Pou, *et al.*, 1990). NMR (CDCl₃) δ: 2.24 (t, 2H), 2.51 (t, 2H), 9.78 (s, 1H).

4-([²H₃]Methyl)-4-[¹⁵N]nitro)[5,5,5-²H₃]pentanal ethylene acetal (**109**)   A benzene solution (20 mL) of 4-([²H₃]methyl)-4-[¹⁵N]nitro)[5,5,5-²H₃]pentanal (**108**) (2.91 g, 19.3 mmol) ethylene glycol (1.8 g, 29 mmol), and *p*-toluenesulfonic acid monohydrate (4 mg) was refluxed for 3 hr while water was azeotropically removed in a Dean-Stark trap containing CaSO₄ (2 g). Upon cooling, the solutions were washed with a saturated NaHCO₃ solution (2 mL) and brine (2 mL), dried over anhydrous K₂CO₃, and evaporated to dryness. The remaining oil was distilled, giving 4-([²H₃]methyl)-4-[¹⁵N]nitro)[5,5,5-²H₃]pentanal ethylene acetal (**109**) (2.64 g, 70%) as a colorless liquid, bp 85–86°C at 0.05 mm Hg (Bonnett, *et al.*, 1959a; Pou, *et al.*, 1990). NMR (CDCl₃) δ: 1.60–1.67 (m, 2H), 1.9–2.03 (m, 2H), 3.85–3.98 (m, 4H), 4.86 (t, 1H).

5,5-Di([²H₃]methyl)[¹⁵N]-1-pyrroline-N-oxide (**110**)   To an aqueous solution (10 mL) of 4-([²H₃]methyl)-4-[¹⁵N]nitro)[5,5,5-²H₃]pentanal ethylene acetal (**108**) (2.64 g, 13.5 mmol) and NH₄Cl (850 mg, 16 mmol) at 0°C was added zinc dust (2.2 g, 33.6 mmol) portionwise over 2 hr at such a rate that the temperature of the reaction remained ≈ 0°C. After being stirred at this temperature for an additional 4 hr, the mixture was filtered and washed with warm water. The combined filtrates were acidified with 12 N HCl (1.8 mL) and warmed to 70°C for 1 hr. The volume of the reaction was then reduced on a rotary evaporator until a yellow-white precipitate appeared. The solution was neutralized with NaOH (1 g in 3 mL of H₂O) at 0°C. The resultant mixture was saturated with sodium borate (2 g) and extensively extracted with chloroform. The combined organic solutions were dried over anhydrous K₂CO₃, filtered, and evaporated to dryness, yielding 5,5-di([²H₃]methyl)[¹⁵N]-1-pyrroline-*N*-oxide (**110**) (543 mg, 34%) as a colorless liquid, bp 45–47°C at 0.1 mm Hg (Pou, *et al.*, 1990). NMR (CDCl₃) δ: 2.09 (t, 2H) 2.14 (m, 2H), 6.78 (t, 1H). HRMS (70 eV, *m/z*) calcd for C₆H₅D₆¹⁵NO, 120.1187; observed, 120.1190.

5,5-Di([²H₃]methyl)[¹⁵N,2,3,3-²H₃]-1-pyrroline-N-oxide (**111**)   A D₂O solution (7 mL) of 5,5-di([²H₃]methyl)[¹⁵N]-1-pyrroline-*N*-oxide (**110**) (443 mg, 3.69 mmol) and NaOD (133 mg, 3.24 mmol) was heated at 70°C for 20 hr. The solvent was evaporated to near dryness. The remaining brown residue was taken up in chloroform, dried over anhydrous K₂CO₃, evaporated to dryness, and distilled, giving 5,5-di([²H₃]methyl)[¹⁵N,2,3,3-²H₃]-1-pyrroline-*N*-oxide (**111**) (346 mg, 76%) as a colorless liquid, which solidified at temperatures below 0°C (Pou, *et al.*, 1990). IR (CHCl₃) 1572 (C=N–O) cm⁻¹. NMR (CDCl₃) δ: 2.03 (s, 2H); HRMS (70 eV, *m/z*) calcd for C₆H₂D₉¹⁵NO, 123.1376; observed, 123.1376.

5,5-Dimethyl-2-([^{13}C]methyl-1-pyrroline-N-oxide (**112**)   A solution of 5,5-methyl-1-pyrroline-*N*-oxide (**5**) (1.2 g, 10.6 mmol in 20 mL of dry ether) was added to an ethereal (40 mL) solution of [^{13}C]methylmagnesium iodide, prepared by the reaction of [^{13}C]iodomethane (2 g, 14 mmol) and magnesium turnings (0.34 g, 14 mmol) in anhydrous ether (10 mL) and cooled to 0°C. After addition was completed, the reaction mixture was stirred at room temperature for 30 min. The reaction was then cooled in an ice bath and a saturated NH$_4$Cl solution (15 mL) was carefully added, with stirring. The aqueous mixture was extracted several times with ether, and the combined organic solutions were dried over anhydrous Na$_2$SO$_4$, filtered, and rotary evaporated to dryness to give an oil, 1-hydroxy-5,5-dimethyl-2-([^{13}C]methylpyrrolidine (9 g, 65%) (Barasch, *et al.,* 1994). IR (Neat film) 3340 (br, N–OH) cm^{-1}. ^{13}C NMR (CDCl$_3$) δ19.3 (s, 2-CH$_3$), 19.8 (strong s, 5-CH$_3$), 26.2 (s, C$_4$), 27.7 (s, 2-CH$_3$), 34.3 (s, C$_3$), 59.2 (d, $J = 41$ Hz, C$_5$), 63.5 (s, C$_2$).

Continual aeration of this crude oil (0.9 g, 6.9 mmol) in absolute ethanol (10 mL) with copper acetate (3 mg, 16.6 µmol) and ammonium hydroxide (25%, 2 mL) for 3 hrs, when the blue color persisted, gave, after standard isolation procedures, crude nitrone (0.54 g). Fractional distillation afforded 5,5-dimethyl-2-([^{13}C]methyl-1-pyrroline-*N*-oxide (**112**) as a colorless oil, bp 56–58°C at 0.5 mm Hg (Barasch, *et al.*, 1994). IR (Neat film) 1600 (C=N–O) cm^{-1}. ^{13}C NMR (CDCl$_3$) δ: 13.0 (strong s, 2-CH$_3$), 25.4 (s, 2 × 5-CH$_3$), 29.0 (s, C$_3$), 32.2 (s, C$_4$), 73.1 (s, C$_5$).

4-(Diethylamino)-[2-^{13}C]-2-butanone (**114**)   To an ice-cooled solution of diiodomethane (2 mL, 25 mmol) and [2-^{13}C]acetone (**113**) (1 gm, 17 mmol) in dry tetrahydrofuran (15 mL) was added dropwise diethylamine (5.3 mL, 51 mmol) in anhydrous tetrahydrofuran (5 mL). After addition was completed, the reaction was refluxed for 3 hr, cooled, filtered, and the solid was washed several times with tetrahydrofuran. The combined solutions were evaporated *in vacuo*. The resulting residue was taken up in dilute hydrochloric acid (1 M) and was extracted with petroleum ether (bp 40–60°C). The remaining aqueous solution was made basic with NaOH (10%), saturated with K$_2$CO$_3$, and extracted with ether. After drying with anhydrous K$_2$CO$_3$, the ether solution was rotary evaporated to give 4-(diethylamino)-[2-^{13}C]-2-butanone (**114**) as a brown liquid (2.25 g) (Barasch, *et al.*, 1994). IR (Neat film) 1713 (C=O) cm^{-1}. ^{13}C NMR (CDCl$_3$) δ: 11.5 {s, N(CH$_2$CH$_3$)$_2$}, 30.1 (d, $J = 40$ Hz, C$_1$), 41.3 (d, $J = 40$ Hz, C$_3$), 46.7 {s, N(CH$_2$CH$_3$)$_2$}, 47.2 (s, C$_4$), 208.4 (strong s, C$_2$). This material was used for the synthesis of 5-methyl-5-nitro-[2-^{13}C]-2-hexanone without further purification.

5-Methyl-5-nitro-[2-^{13}C]-2-hexanone (**115**)   To a solution of 2-nitropropane (27.6 g, 28 mL, 310 mmol) and 4-(diethylamino)-[2-^{13}C]-2-butanone (**114**) (2.2 g, 12.2 mmol) in dry acetonitrile (50 mL) was added dry potassium fluoride (0.2 g, 3.4 mmol) and 18-crown-6 (0.24 g, 0.9 mmol). The reaction was refluxed for 5 hr. After rotary evaporation to dryness, the remaining residue was partitioned between methylene chloride and dilute hydrochloric acid (1 M). The organic phase was washed with water, dried over anhydrous K$_2$CO$_3$, and rotary evaporated to dryness, yielding 5-methyl-5-nitro-[2-^{13}C]-2-hexanone (**115**) as a brown liquid (Barasch, *et al.*, 1994). IR (Neat film) 1718 (C=O), 1538 (N–O), 1348 (N–O) cm^{-1}. ^{13}C NMR (CDCl$_3$) δ: 25.8 (s, C$_6$ and 5-CH$_3$), 29.9 (d, $J = 40$ Hz, C$_1$), 33.9 (s, C$_4$),

38.1 (d, $J = 40$ Hz, C_3), 83.7 (s, C_5), 206.3 (strong s, C_2). It was recommended that this nitro-ketone be distilled prior to reductive cyclization.

2,5,5-Trimethyl-[2-^{13}C]-1-pyrroline-N-*oxide* **(116)**   To a solution of 5-methyl-5-nitro-[2-^{13}C]-2-hexanone **(115)** (1.4 g, 8.6 mmol) and ammonium chloride (0.66 g, 124 mmol) dissolved in water (18 mL) was added zinc dust (1.9 g, 29 mmol) in small portions. The mixture was stirred at room temperature for 2 hr, filtered, and the filter cake was washed several times with warm water (70°C). The combined solutions were rotary evaporated to dryness, taken up in methylene chloride, and dried over anhydrous K_2CO_3. After evaporation to dryness, the remaining liquid was distilled, giving 2,5,5-trimethyl-[2-^{13}C]-1-pyrroline-*N*-oxide **(116)** (0.4 g, 25%) as a colorless liquid, bp 52–54°C at 0.4 mm Hg (Barasch, *et al.*, 1994). IR (Neat film) 1600 (C=N–O) cm^{-1}. ^{13}C NMR (CDCl$_3$) δ: 13.0 (d, $J = 49$ Hz, 2-CH$_3$), 25.4 (s, 5,5-CH$_3$), 29.0 (d, $J = 45$ Hz, C_3), 32.2 (s, C_4), 73.1 (d, $J = 11$ Hz, C_5), 141.0 (strong s, C_2).

Phenylvinyl-[2-^{13}C]-ketone **(119)**   A solution of benzoyl-[2-^{13}C]-chloride **(117)** (2.1 g, 15 mmol) and benzyl(chloro)bis-(triphenylphosphine)palladium (II) (20 mg, 26 μmol) in chloroform (5 mL) was treated with tributyl(vinyl)tin (4.95 g, 15.6 mmol) in chloroform (15 mL). The solution was heated at 65°C, with stirring, under an air atmosphere in a sealed tube until the palladium metal precipitated, which occurred in about 35 min. The reaction was then cooled to room temperature, poured into ether (100 mL), extracted with H_2O, dried over anhydrous MgSO$_4$, and evaporated to dryness. The remaining residue was chromatographed using silica gel (mesh 230–400) and eluted with a mixture of pentane/methylene chloride (50:50), affording phenylvinyl-[2-^{13}C]-ketone **(119)** (1.6 g, 80%) (Labadie, *et al.*, 1983; Janzen, *et al.*, 1994). NMR (CDCl$_3$) δ: 5.94 (ddd, $J_H = 10.5$ Hz, 1.4 Hz, J^{13}C $= 10.5$ Hz, 1H trans CH=), 6.44 (ddd, $J_H = 17.1$, 1.8 Hz, J^{13}C $= 6.8$ Hz, 1H, cis CH=), 7.16 (ddd, $J_H = 17.1$, 10.5 Hz, J^{13}C $= 5.4$ Hz, 1H, 2-CH=), 7.46–7.55 (m, 3H, H–Ar), 7.93–7.97 (m, 2H, H–Ar).

4-Methyl-4-nitro-[1-^{13}C]valerophenone **(120)**   To a solution of 2-nitropropane (4.68 g, 52.6 mmol) and benzyltrimethylammonium hydroxide (40%, 0.15 mL, 0.3 mmol) in ether (50 mL) was added phenylvinyl-[2-^{13}C]-ketone **(119)** (1.4 g, 10.5 mmol) in ether (20 mL). After the addition was completed, the reaction was refluxed, with stirring, for 20 hr, neutralized with concentrated HCl (37%), and washed with a saturated NaCl solution. The remaining ether mixture was dried over anhydrous Na$_2$SO$_4$, filtered, and rotary evaporated at 1 Torr to give the crude product **(120)**. Purification by column chromatography with silica gel (mesh 230–400) and pentane/methylene chloride (1:1) afforded 4-methyl-4-nitro-[1-^{13}C]valerophenone **(120)** (1.9 g, 82%) as a clear liquid (Janzen, *et al.*, 1994). NMR (CDCl$_3$) δ: 1.65 (s, 6H, CH$_3$), 2.38 (dt, 2H, CH$_2$, $J_H = 5.2$, J^{13}C $= 2.7$ Hz), 2.98 (q, 2H, CH$_2$, $J_H = J^{13}$C $= 7.2$ Hz), 7.44–7.49 (t, 2H, C$_6$H$_5$), 7.55–7.58 (t, 1H, C$_6$H$_5$), 7.91–7.95 (m, 2H, C$_6$H$_5$).

2-Phenyl-5,5-dimethyl-[2-^{13}C]-1-pyrroline-N-*oxide* **(121)**   To a well-stirred and cooled solution (−5°C) of 4-methyl-4-nitro-[1-^{13}C]valerophenone **(120)** (1.9 g,

8.56 mmol) in 95% ethanol (40 mL) was added zinc powder (1.2 g, 18.3 mmol). After the addition was completed, glacial acetic acid (2.05 g, 1.95 mL, 34.2 mmol) in 95% ethanol (40 mL) was added dropwise. The reaction was stirred at ≈ 0°C for 3 hr and then was placed in a refrigerator overnight at 3–5°C. After filtering the reaction mixture, it was rotary evaporated to dryness. The remaining residue was taken up in chloroform (100 mL), washed with an aqueous saturated NaCl solution, dried over anhydrous Na_2SO_4, filtered, and rotary evaporated to give crude nitrone (**121**). Purification by column chromatography with silica gel (mesh 230–400) and ethyl acetate, followed by sublimation, afforded 2-phenyl-5,5-dimethyl-[2-^{13}C]-1-pyrroline-*N*-oxide (**121**) (1 g, 60%), mp 99°C (Janzen, *et al.*, 1994). NMR ($CDCl_3$) δ: 1.50 (s, 6H, CH_3), 2.12 (dt, 2H, CH_2, $J_H = 7.4$ Hz, $J^{13}C = 2.6$ Hz), 3.04 (q, 2H, CH_2, $J_H = J^{13}C = 7.4$ Hz), 7.42–7.44 (m, 3H, C_6H_5), 8.36–8.40 (m, 2H, C_6H_5).

2,4-Dihydroxy-1,1,1-trifluorobutane (**123**)   To a solution of ethyl 4,4,4-trifluoroacetoacetate (**122**) (18.4 g, 100 mmol) in dry ether (200 mL) at 0–5°C was added sodium borohydride (4 g, 105 mmol) in small portions to maintain the reaction at this temperature. This mixture was stirred for an additional hour at this temperature and then raised to ambient temperature, where it remained overnight. Dilute hydrochloric acid (10%, 100 mL) was then added carefully and the solid was filtered. The aqueous layer was extracted with ether. The combined ether solutions were dried over anhydrous Na_2SO_4, filtered, and rotary evaporated to dryness to give a liquid (18 gm). This liquid was placed into dry ether (50 mL) and added slowly, over 80 min, to a suspension of lithium aluminum hydride (6 g, 158 mmol) in dry ether (50 mL) at 0–5°C. After stirring this mixture overnight at room temperature, the reaction was cooled in an ice bath and dilute hydrochloric acid (10%, 100 mL) was carefully added to decompose the unreacted lithium aluminum hydride. The remaining solid was filtered and the aqueous solution was extracted with ether. The combined ether solutions were dried over anhydrous Na_2SO_4, filtered, and rotary evaporated to dryness to yield 2,4-dihydroxy-1,1,1-trifluorobutane (**123**) (12.6 g, 88%), which was purified by fractional distillation, bp 51.5–54°C at 32 mm Hg (Janzen, *et al.*, 1995). NMR ($CDCl_3$) δ: 1.59 (m, 2H), 3.56 (s, 2H), 4.03 (m, 1H), 4.60 (s, 1H), 6.05 (d, 1H, $J = 6.6$ Hz).

2-Hydroxy-4-(tosyloxy)-1,1,1-trifluorobutane (**124**)   To a solution of 2,4-dihydroxy-1,1,1-trifluorobutane (**123**) (12.6 g, 87.5 mmol) and pyridine (30 mL) in methylene chloride, cooled to 4°C, was added *p*-toluenesulfonyl chloride (20.5 g, 108 mmol) over 30 min. The reaction was stirred at this temperature for 4 days and then poured onto crushed ice containing dilute hydrochloric acid (10%, 100 mL). The organic layer was washed several times with dilute hydrochloric acid (10%, 100 mL) and then with a saturated NaCl/dilute Na_2CO_3 solution (10%, 100 mL). The remaining organic solution was dried over anhydrous Na_2SO_4, filtered, and rotary evaporated to dryness to give crude (**124**). Purification by column chromatography with silica gel (mesh 230–400) and methylene chloride afforded 2-hydroxy-4-(tosyloxy)-1,1,1-trifluorobutane (**124**) (14.5 g, 55%) (Janzen, *et al.*, 1995). NMR ($CDCl_3$) δ: 1.84 and 2.06 (m, 2H), 2.44 (s, 3H), 3.20 (d, 1H,

$J = 5.7\,\text{Hz}$), 4.15 (m, 2H), 4.30 (dt, 1H, $J_1 = 3.9\,\text{Hz}$, $J_2 = 10.2\,\text{Hz}$), 7.38 (d, 2H, $J = 8.4\,\text{Hz}$), 7.80 (d, 2H, $J = 8.4\,\text{Hz}$).

4-(Tosyloxy)-1,1,1-trifluoro-2-butanone (**125**)   Dry N_2 was passed through a flask containing Dess-Martin reagent (Dess and Martin, 1983) to remove trace amounts of acetic anhydride. After several minutes, methylene chloride (100 mL) was added to dissolve the oxidant. A solution of 2-hydroxy-4-(tosyloxy)-1,1,1-trifluorobutane (**124**) (1.9 g, 63 mol) in methylene chloride (40 mL) was then introduced and the flask was sealed. After stirring at room temperature for 4.5 hr, the reaction was poured into a stirred aqueous $NaHSO_3$ solution. At this point, an aqueous saturated Na_2CO_3 solution (100 mL) was slowly added and the aqueous layer was extracted with methylene chloride. The combined methylene chloride extractants were washed with H_2O and dried over anhydrous $MgSO_4$. After filtration, the remaining solution was evaporated *in vacuo*, giving 4-(tosyloxy)-1,1,1-trifluoro-2-butanone (**125**) (1.9 g, 100%) as a clear liquid (Janzen, *et al.*, 1995). NMR ($CDCl_3$) δ: 2.45 (s, 3H), 3.11 (t, 2H, $J = 5.8$), 4.32 (t, 2H, $J = 5.8$), 7.36 (d, 2H, $J = 8, 5$), 7.76 (d, 2H, $J = 8.5$).

5-Methyl-5-nitro-1,1,1-trifluoro-2-hexanone (**126**)   A solution of 2-nitropropane (17.5 g, 196.5 mmol) and sodium ethoxide (42.4 g, 21 wt% EtOH solution, 131 mmol) in ethanol (150 mL) was cooled to 0–5°C under a N_2 atmosphere. To this solution was added 4-(tosyloxy)-1,1,1-trifluoro-2-butanone (**125**) (3.9 g, 13.1 mmol) in absolute ethanol (50 mL) over a 2-hr period at such a rate as to maintain the temperature of the reaction at 0–5°C. Once completed, the reaction was raised to room temperature and stirred for an additional 6 hr. At this point, the reaction was cooled in an ice bath and concentrated hydrochloric acid (37%, 15 mL) was slowly added, with stirring. The solid, which formed during the addition of the acid, was filtered and washed with absolute ethanol. The remaining solution was evaporated *in vacuo* to dryness. The residue was taken up in chloroform, washed with a saturated NaCl solution and dried over anhydrous $MgSO_4$. Upon filtration, the resultant solution was rotary evaporated to dryness, leaving a brown liquid (2.7 g, 96%). Purification was achieved by distillation at bp 46°C at 0.18 mm Hg, affording 5-methyl-5-nitro-1,1,1-trifluoro-2-hexanone (**126**) (2.1 g, 75%) as a colorless liquid (Janzen, *et al.*, 1995). NMR ($CDCl_3$) δ: 1.61 (s, 6H), 2.28 (t, 2H, $J = 7.5\,\text{Hz}$), 2.77 (t, 2H, 7.5).

5,5-Dimethyl-2-(trifluoromethyl)-1-pyrroline-N-oxide (**127**)   To a well-stirred solution of 5-methyl-5-nitro-1,1,1-trifluoro-2-hexanone (**126**) (2.05 g, 9.6 mmol) in ethanol (95%, 60 mL) at 3°C was added zinc dust (1.26 g, 19.2 mmol). Once completed, glacial acetic acid (2.31 g, 2.20 mL, 38.5 mmol) in ethanol (95%, 20 mL) was added at such a rate as to maintain the reaction temperature at ≤ 7°C. The reaction was stirred at this temperature for 3 hr and then filtered. The filter cake was washed with ethanol (95%, 50 mL) and evaporated to dryness. The remaining residue was taken up in chloroform (150 mL), washed with a saturated NaCl solution, and dried over anhydrous $MgSO_4$. After filtration, the remaining solution was rotary evaporated to dryness, leaving a clear liquid (1.22 g, 70%). Fractional distillation afforded 5,5-dimethyl-2-(trifluoromethyl)-1-pyrroline-*N*-

oxide (**127**) (1.1 g, 63%) as a colorless liquid, bp 44°C at 0.18 mm Hg (Janzen, *et al.*, 1995). NMR (CDCl$_3$) δ: 1.31 (s, 3H), 1.40 (s, 3H), 2.13 (t, 2H, $J = 7.4$ Hz), 2.78 (tm, 2H, $J_t = 7.4$ Hz).

2-(Diethylphosphono)-5,5-dimethyl-1-pyrroline-N-oxide (**128**)   The general procedure for the addition of diethylphosphonium anion to nitrones was adapted for this synthetic scheme (Huber, *et al.*, 1985). Under N$_2$, lithium diisopropropylamide (10 mL, 2-M solution, 20 mmol) was added carefully to a solution of diethyl phosphite (5 gm, 36.2 mmol) in methylene chloride (40 mL) at −20°C. After addition was completed, this mixture was stirred for 15 min at this temperature and then cooled to −60°C. At this point, freshly distilled 5,5-dimethyl-1-pyrroline-*N*-oxide (**5**) (2 g, 17.7 mmol) in methylene chloride (4 mL) was added to the reaction at −60°C. After stirring for several minutes at this temperature, the reaction was slowly warmed over 3.5 hr to −20°C. At this point, the reaction was quenched with H$_2$O (5 mL), and warmed to room temperature. The reaction was diluted with methylene chloride (100 mL), washed with a saturated NaCl solution, dried over anhydrous Na$_2$SO$_4$, and filtered. Upon rotary evaporation, a residue was isolated, which, upon fractional distillation, gives 5-(diethylphosphono)-2,2-dimethyl-1-hydroxylpyrrolidine (2.45 g), bp 99–109°C at 1 mm Hg. To an ethanol solution (95%, 20 mL) of the above hydroxylamine (2.45 g, 9.8 mmol) was added copper acetate (100 mg, 0.55 mmol) and ammonium hydroxide (29%, 0.3 mL). This reaction was aerated until a blue color persisted. After rotary evaporation, the remaining oil was chromatographed with silica gel (mesh 230–400) and ethyl acetate, affording 2-(diethylphosphono)-5,5-dimethyl-1-pyrroline-*N*-oxide (**128**) (1.3 g, 29%) as a clear liquid (Janzen and Zhang, 1995). NMR (CDCl$_3$) δ: 1.29 (t, 6H, CH$_3$), 1.35 (s, 6H, CH$_3$), 2.06 (t, 2H, CH$_2$, $J = 7.2$ Hz), 2.74 (dt, 2H, CH$_2$, $J_H = 7.2$ Hz, $J_P = 3.1$ Hz), 4.22 (quintet, 4H, 2OCH$_2$, $J_H = J_P = 7.7$ Hz).

Diethyl (2-methyl-2-pyrrolidinyl)phosphonate (**130**)   Gaseous ammonia was bubbled through a solution of 5-chloropentan-2-one (**129**) (11.8 g, 100 mmol) and diethyl phosphite (14 g, 100 mmol) in absolute ethanol (50 mL) for 4 hr at 50°C. After filtration and removal of the solvent *in vacuo*, the residue was poured into dilute hydrochloric acid (2 M, 40 mL). After washing with methylene chloride, the remaining aqueous layer was basified to pH 10 with Na$_2$CO$_3$. Following extraction with chloroform, the organic layer was dried over anhydrous Na$_2$SO$_4$, filtered, and rotary evaporated to dryness, giving diethyl (2-methyl-2-pyrrolidinyl)phosphonate (**130**) (13.7 g, 62%) as a colorless oil (Fréjaville, *et al.*, 1995). IR (Neat film) 3315 (N–H), 1235, 1055 (O–P=O) cm^{-1}. NMR (C$_6$D$_6$) δ: 1.10 (t, $J = 7$ Hz, 6H, CH_3CH$_2$), 1.28 (d, $J = 15$ Hz, 3H, CH$_3$), 1.2–1.8 (m, 3H, ring, 2CH$_2$ with one near the phosphate), 2.0–2.25 (m, 1H, ring, CH$_2$ with one near the phosphate), 2.7–3.0 (m, 2H, ring, N–CH$_2$), 3.7–4.5 (m, 4H, CH$_2$O).

5-(Diethoxyphosphoryl)-5-methyl-1-pyrroline-N-oxide (**131**)   A solution of *m*-chloroperbenzoic acid (70%, 4.4 g, 18 mmol) in chloroform (40 mL) was added over 1 hr to a stirred solution of diethyl (2-methyl-2-pyrrolidinyl)phosphonate (**130**) (2 g, 9 mmol) at −10°C. The solution was washed sequentially with saturated solutions of NaHCO$_3$ and NaCl. After drying the organic layer with anhydrous Na$_2$SO$_4$, the solution was filtered and rotary evaporated to dryness, giving an oil. After column chromatography with silica gel and ethanol/methylene chloride (15:85), 5-(diethoxyphosphoryl)-5-methyl-1-pyrroline-*N*-oxide (131) was obtained as an oil (600 mg, 30%). (Fréjaville, *et al.*, 1995). IR (Neat film) 1574 (C=N–O) cm^{-1}. NMR (CDCl$_3$) δ: 1.35 (t, 3H, $J = 7.1$ Hz), 1.36 (t, 3H, $J = 7.1$ Hz), 1.70 (d, 3H, $J = 14.9$ Hz), 2.0–2.2 (m, 1H), 2.5–2.7 (m, 1H), 2.7–2.9 (m, 2H), 4.1–4.4 (m, 4H), 6.9–7.0 (m, 1H, HC=N–O).

5-(Diethoxyphosphoryl)-5-methyl[2,3,-^{2}H$_2$]-1-pyrroline-N-oxide (132) *and 5-(Diethoxyphosphoryl)-5-methyl[2,3,3,-^{2}H$_3$]-1-pyrroline-N-oxide* (133)   NaOD (0.18 M, 10 mL) was slowly added under nitrogen atmosphere to a stirred solution of 5-(diethoxyphosphoryl)-5-methyl-1-pyrroline-*N*-oxide (131) (500 mg) in D$_2$O (3 mL) at 5°C. The reaction was monitored by following the disappearance of H–C=N–O by NMR. Upon neutralization with D$_2$SO$_4$ and extraction with DMF, the solvent was removed *in vacuo* to give a brown oil. Molecular distillation at 10^{-6} mm Hg gave 80 mg (16%) of a colorless oil, consisting of 5-(diethoxyphosphoryl)-5-methyl[2,3,-^{2}H$_2$]-1-pyrroline-*N*-oxide (132) and 5-(diethoxyphosphoryl)-5-methyl[2,3,3,-^{2}H$_3$]-1-pyrroline-*N*-oxide (133) with a d_3/d_2 ratio of 80:20 (Fréjaville, *et al.*, 1995). NMR (C$_6$D$_6$) δ: 1.02 (t, 3H, $J = 7.1$ Hz), 1.12 (dt, 3H, $J = 7.1$ Hz and 0.5 Hz), 1.5–1.7 (m, 1H), 1.58 (d, 3H, $J = 14.6$ Hz), 2.4–2.8 (m, 1.2H, 0.2H excess due to d_2), 3.9–4.0 (m, 2H), 4.2–4.5 (m, 2H).

N-tert-Butyl-α-phenylnitrone (136)   A mixture of freshly distilled benzaldehyde (134) (5.3 g, 50 mmol) and *N*-(*tert*-butyl)hydroxylamine hydrochloride (135) (4.5 g, 50 mmol), prepared as described in the literature (Calder, *et al.*, 1972), was heated to 45°C. At reaching this temperature, the reaction became exothermic, which subsided within a few minutes. The mixture was then kept at 50–60°C for 1 hr, until an aqueous layer began to separate. The reaction was cooled in an ice bath and methylene chloride was added (50 mL). After the organic layer was separated, it was dried over anhydrous Na$_2$SO$_4$, filtered, and rotary evaporated to dryness, giving N-*tert*-butyl-α-phenylnitrone (136) (5.5 g, 62%). Recrystallization from hexane afforded nitrone (136) as a white solid, mp 75–76°C (Emmons, 1957). IR (KBr) 1585 (C=N–O) cm^{-1}.

2-tert-Butyl-3-phenyloxazirane (138)   To vigorously stirring and ice-cooled methylene chloride (50 mL) was added hydrogen peroxide (90%, 15 mL, 550 mmol) and two drops of sulfuric acid. This reaction should be carried out in a hood behind a safety screen as such high concentrations of hydrogen peroxide are unstable and can explode without warning. Acetic anhydride (67.2 g, 660 mmol) was slowly added to the above reaction and cooled in an ice bath. At this point, the reaction was stirred at ≈ 0°C for 15 min and for an additional 30 min at room temperature. The clear solution of peracetic acid was added dropwise to a mixture of *N*-benzylidene-*tert*-butylamine (137) (80.5 g, 500 mmol), prepared as described in Emmons (1957), in methylene chloride (100 mL) at ≈ 0°C. During the addition, the solution became blue, presumably due to the formation of *tert*-nitrosobutane. After the addition, the reaction was allowed to remain in the ice bath until the ice had melted, and then it was stirred at room temperature overnight. The solution was washed with water, followed by dilute ammonium hydroxide (15%). The remaining solution was dried over anhydrous Na_2SO_4, filtered, and evaporated *in vacuo*, giving, after distillation in a spinning band column, 2-*tert*-butyl-3-phenyloxazirane (**138**) (63.1 g, 71%), bp 61–63°C at 0.3 mm Hg (Emmons, 1957).

N-tert-*Butyl-α-phenylnitrone* (**136**).   A solution of 2-*tert*-butyl-3-phenyloxazir-ane (**138**) (8.8 g, 50 mmol) in dry acetonitrile (100 mL) was refluxed for 3 days. Evaporation of this reaction *in vacuo* gave *N-tert*-butyl-α-phenylnitrone (**136**) (8 g, 95%). Recrystallization from hexane gave nitrone (**136**) as a white solid, mp 75–76°C (Emmons, 1957). IR (KBr pellet) 1580 (C=N–O) cm^{-1}.

α-(4-Pyridyl 1-oxide)-N-tert-*butylnitrone* (**140**)   To a methanol (20 mL) solution containing N-(*tert*-butyl)hydroxylamine hydrochloride (**135**) (1.25 g, 10 mmol) and triethylamine (1 g, 1.4 mL, 10 mmol) was added 4-pyridinecarboxy-aldehyde N-oxide (**139**) (1.23 g, 10 mmol). The reaction was refluxed overnight and the solvent was removed *in vacuo*. The remaining residue was recrystallized from ethylene glycoldimethyl ether, giving α-(4-pyridyl 1-oxide)-*N-tert*-butylnitrone (**140**) (1.4 g, 72%) as a white solid, mp 181–183°C (Janzen, *et al.*, 1978). IR (KBr pellet) 1590 (C=N–O) cm^{-1}.

N-tert-*Butyl-α-(4-dodecyloxy)phenylnitrone* (**144**)   To a rapidly stirred mixture consisting of 4-dodecyloxybenzadehyde (**141**) (11.7 g, 40.3 mmol), prepared as described in Huie and Cherry (1985), 2-methyl-2-nitropropane (8.31 g, 80.6 mmol), and zinc dust (7.91 g, 121 mmol) in ethanol (95%, 300 mL) at 10°C was added glacial acetic acid (14.5 g, 242 mmol) in a dropwise fashion over 1 hr. The mixture was mixed for 2 hr at 10°C and then stored in a cold room, $T \approx 6°C$, for 48 hr. The sample was subsequently filtered to remove $Zn(OAc)_2$ and the filter cake was washed with ether. The remaining solution was rotary evaporated to dryness. This afforded a solid, which was taken up in ether (150 mL) and washed with water (100 mL). The ether solution was dried over anhydrous Na_2SO_4, filtered, and evaporated *in vacuo*, giving N-*tert*-butyl-α-(4-dodecyloxy)phenylnitrone (**144**) as white fluffy crystals from acetone/water (11.1 g, 76%), mp 62–64°C (Janzen and Coulter, 1984; Huie and Cherry, 1985). IR (KBr pellet) 1585 (C=N–O) cm^{-1}. NMR ($CDCl_3$) δ: 0.67–1.40 (m, 23H), 1.57 (s, 9H), 3.98 (t, 2H, $J = 6.5$ Hz), 6.88 (d, 2H, $J = 9$ Hz), 7.42 (s, 1H, HC=N-O), 8.24 (d, 2H, $J = 9$ Hz).

α-(4-Methylpyridyl)-N-tert-*butylnitrone methyl sulfate* (**149**)   A mixture of 4-pyridinecarboxaldehyde (**146**) (1 g, 8.1 mmol) and dimethyl sulfate (1.02 g, 8.1 mmol) in dry benzene (20 mL) was refluxed for 12 hr. The precipitate that formed was filtered and washed with dry benzene. This aldehyde, without further purification, was added to a methanol solution (20 mL) containing N-(*tert*-butyl)hydroxylamine hydrochloride (**135**) (1.25 g, 10 mmol) and triethylamine (1 g, 1.4 mL, 10 mmol). This reaction was refluxed for 20 hr. At this point, the solvent was removed by rotary evaporation. The remaining residue was recrystallized from a mixture of chloroform/anhydrous ether, giving α-(4-methylpyridyl)-*N-tert*-butylnitrone methyl sulfate (**149**) (1.8 g, 71%) as a white solid, mp 127–128°C

(Janzen, *et al.*, 1979). NMR (CDCl₃) δ: 1.6 (s, 9H, N–C(CH₃)₃), 3.65, 4.50 (s, 3H, CH₃), 8.36 (s, 1H, HC=N–O), 8.74–9.06 (m, 4H, C₅H₄N).

Sodium N-tert-*butyl*-α-*(* o-*sulfonato)phenylnitrone* (**151***)   A solution of *o*-benzaldehyde sulfonic acid sodium salt (**150**) (2.1 g, 10 mmol) in absolute ethanol (40 mL) was added to a solution of *N*-(*tert*-butyl)hydroxylamine hydrochloride (**135**) (1.5 g, 12 mmol) and triethylamine (1.21 g, 1.67 mL, 12 mmol) in absolute ethanol (50 mL). The reaction was stirred under reflux for 2 days, at which point the mixture was rotary evaporated to dryness. The remaining residue was recrystallized from absolute ethanol/ether, affording sodium *N-tert*-butyl-α-(*o*-sulfonato)phenylnitrone (**151**) (2.2 g, 78%) as a white solid, mp 253–255°C (with decomposition) (Janzen and Shetty, 1979). NMR (D₂O) δ: 1.52 (s, 9H, N–C(CH₃)₃), 7.52–7.80 (m, 2H, C₆H₄), 7.9–8.12 (m, 1H, C₆H₄), 8.55–8.70 (m, 1H, C₆H₄), 8.72 (s, 1H, HC=N–O).

α-*Phenyl*-N-*[(1-hydroxy-2-methyl)*-2-*propyl]nitrone* (**155**)   To a cooled solution of 2-methyl-2-nitropropanol (**152**) (5.95 g, 50 mmol) in ethanol (95%, 150 mL) and ammonium chloride (3.25 g, 61 mmol) dissolved in distilled water (50 mL) was added zinc dust (13 g, 200 mmol) over a period of 15 min, while the temperature of the reaction was maintained below 30°C. After stirring for 4 hr, the reaction was initially filtered and then washed with hot ethanol (95%, 2 × 100 mL), followed by hot chloroform (2 × 100 mL). The light green filtrate was evaporated *in vacuo* to 50 mL and extracted with chloroform. To this organic solution was added benzaldehyde (**134**) (5.25 g, 50 mmol) and the reaction was refluxed for 3.5 hr. After cooling in an ice bath, the reaction was dried over anhydrous MgSO₄ and rotary evaporated, affording a residual paste. Crystallization with cold carbon tetrachloride gave α-phenyl-*N*-[(1-hydroxy-2-methyl)-2-propyl]nitrone (**155**) (5.2 g, 55%) as a white solid, mp 75–76°C (Janzen and Zawalski, 1978). IR (KBr pellet) 3180 (O–H), 1591 (C=N–O) cm⁻¹, NMR (CDCl₃) δ: 1.55 (s, 6H, 2 CH₃), 3.73 (s, 2H, CH₂OH), 4.45 (s, broad, 1H, CH₂OH), 7.40 (m, 4H, C₆H₅—2 meta, 1 para, HC=N–O), 8.23 (m, 2H, C₆H₅—2 ortho).

N-tert-*Butyl*-α-*(4-nitro)phenylnitrone* (**174**)   A mixture of *p*-nitrobenzaldehyde (**173**) (1 g, 3.6 mmol), sodium acetate (500 mg, 3.6 mmol) and *N*-(*tert*-butyl)hydroxylamine hydrochloride (**135**) (830 mg, 6.6 mmol) was stirred in 95% ethanol (100 mL) at room temperature for 24 hr. The solid, which separated from the yellow solution, was filtered and washed with additional ethanol (15 mL). The combined filtrate and washings were evaporated to dryness, and the residue was stirred with ether (50 mL) and filtered to remove inorganic salts. The remaining solution was decreased in volume, resulting in the precipitation of a yellow solid, *N-tert*-butyl-α-(4-nitro)phenylnitrone (**174**), which recrystallized from benzene/petroleum ether (bp 30–60°C), mp 134–136°C (1 g, 70%) (Pou, *et al.*, 1995). IR (KBr) 1595 (C=N–O), 1555 and 1345 (NO₂) cm⁻¹. NMR (CDCl₃) δ: 1.6 (s, 9H), 7.6 (s, 1H), 8.5–8.1 (m, 4H).

N-tert-*Butyl*-α-*(4-amino)phenylnitrone* (**175**)   A mixture of *N-tert*-butyl-α-(4-nitro)phenylnitrone (**174**) (1 g, 4.5 mmol) and platinum (IV) oxide (100 mg) in methanol (150 mL) was shaken in a Parr hydrogenator at 30 psi at room temperature for 30 min. The mixture was filtered through Celite and the filtrate was evaporated to dryness, affording *N-tert*-butyl-α-(4-amino)phenylnitrone (**175**) as a yellow solid, mp 145–148°C, from ether (750 mg, 85%) (Pou, *et al.*, 1995). IR (KBr) 3200 (NH$_2$), 1600 (C=N–O) cm^{-1}. NMR (CDCl$_3$) δ: 1.5 (s, 9H), 3.9 (br, s, 2H, exchangeable with D$_2$O, NH$_2$), 6.6 (d, $J = 8.8$ Hz, 2H), 7.3 (s, 1H), 8.1 (d, $J = 8.8$ Hz, 2H).

N-tert-*Butyl-α-4-[5-((2-carboxy)phenyl)-5-hydroxy-4-oxo-3-phenyl)-2-pyrrolin-1-yl]phenylnitrone sodium salt* (**176**)   A solution of *N-tert*-butyl-α-(4-amino)phenylnitrone (**175**) (100 mg, 0.52 mmol) in methanol (2 mL) was added to a sodium phosphate buffer (pH 9, 50 mL) and the mixture was stirred. A solution of fluorescamine (200 mg, 0.71 mmol) in methanol (5 mL) was added dropwise, as the color changed to fluorescent yellow. The reaction mixture was stirred at room temperature for 8 hr, at which point TLC (silica gel, CHCl$_3$/MeOH, 8:1) shows the formation of a UV-absorbing product. After concentration of the solution to 10 mL, the reaction was filtered to remove insoluble material, and the filtrate was desalted using Sep-Pak C18 cartridges with 1-mL injections at a time and washing with H$_2$O. Elution of the cartridges with 50% aqueous acetonitrile and evaporation of the solvent left a bright yellow fluorescent material, which, upon trituration with ether, afforded *N-tert*-butyl-α-4-[5-((2-carboxy)phenyl)-5-hydroxy-4-oxo-3-phenyl)-2-pyrrolin-1-yl]phenylnitrone sodium salt (**176**) as a yellow solid (200 mg, 78%), mp 230°C with decomposition (Pou, *et al.*, 1995). (KBr) 3400 (NH$_2$), 1670 (br, C=O), 1570 (br, C=C, C=N–O) cm^{-1}. NMR (DMSO-d$_6$) δ: 1.4 (s, 9H), 6.7 (d, $J = 8.0$ Hz, 1H), 7.0 (t, $J = 7.5$ Hz, 1H), 7.1 (m, 1H), 7.3 (t, $J = 7.5$ Hz, 2H), 7.4 (d, $J = 9.0$ Hz, 2H), 7.6 (s, 1H, HC=N-O), 7.9 (d, $J = 8.0$ Hz, 2H), 8.0 (d, $J = 7.5$ Hz, 1H), 8.2 (d, $J = 9.0$ Hz, 2H), 9.66 (s, 1H), 15.3 (s, 1H, exchangeable with D$_2$O).

8-Formyl-6-methyl-1′,3′,3′-trimethylspiro[2H-1 benzopyran 2,2′-indoline] (**179**) A mixture of 1,2,3,3-tetramethylindolenimium perchlorate (2.5 g, 9.1 mmol) and 2,6-diformyl-4-methylphenol (1.25 g, 7.6 mmol) in ethanol (50 mL) was refluxed, while piperidine (0.65 g, 7.6 mmol) was carefully added. Once the addition was completed, the reaction was refluxed for 1.5 hr. Upon removal of the solvent, the residue was chromatographed using silica gel (mesh 230–400) and eluted with pentane/ether (99:1), affording 8-formyl-6-methyl-1′,3′,3′-trimethylspiro-[2*H*-1 benzopyran 2,2′-indoline] (**179**) as a solid (850 mg, 36%) (Luccioni-Houzé, *et al.*, 1996). NMR (CDCl$_3$) δ: 1.21 (s, 3H, CH$_3$ on C-3′), 1.34 (s, 3H, CH$_3$ on C-3′), 2.28 (s, 3H, CH$_3$ on C-6), 2.76 (s, 3H, CH$_3$ on N), 5.80 (d, 1H, H-3), 6.53 (d, 1H, H-7′), 6.86 (t, 1H, H-5′), 6.89 (d, 1H, H-4), 7.10 (d, 1H, H-4′), 7.12 (s, 1H, H-5), 7.15 (t, 1H, H-6'), 7.45 (s, 1H, H-7), 10.113 (s, 1H, CH=O).

*8-[2-(*N-tert-*Butylmethanalnitrone)]-6-methyl-1′,3′,3′,-trimethylspiro[2H-1 benzopyran 2,2′-indoline]* (**180**)   A solution of 8-formyl-6-methyl-1′,3′,3′-trimethylspiro[2H-1 benzopyran 2,2′-indoline] (**179**) (400 mg, 1.25 mmol) in ethanol (25 mL) was added under argon to a mixture of *N-(tert*-butyl)hydroxylamine hydrochloride (**135**) (160 mg, 1.25 mmol) and pyridine (100 mg, 1.25 mmol). This mixture was refluxed for 2 hr and then the solvent was removed *in vacuo*. The remaining residue was chromatographed using silica gel (mesh 230–400) and eluted with pentane/ether (99:1), yielding 8-[2-(N-*tert*-butylmethanalnitrone)]-6-methyl-1′,3′,3′,-trimethylspiro[2H-1 benzopyran 2,2′-indoline] (**180**) as a solid (50 mg, 10%) (Luccioni-Houzé, *et al.*, 1996). NMR (CDCl₃) δ: 1.19 (s, 3H, CH₃ on C-3′), 1.25 (s, 3H, CH₃ on C-3′), 1.28 (s, 9H, *t*-Bu), 2.25 (s, 3H, CH₃ on C-6), 2.61 (s, 3H, CH₃ on N), 5.73 (d, 1H, H-3), 6.48 (d, 1H, H-7′), 6.82 (t, 1H, H-5′), 6.84 (d, 1H, H-4), 6.90 (s, 1H, H-5), 7.05 (d, 1H, H-4′), 7.12 (t, 1H, H-6′), 7.55 (s, 1H, H-9), 8.87 (s, 1H, H-7).

N-tert-*Butyl[²H₉]-α-phenyl[2,3,4,5,6-²H₅]nitrone* (**195**)   A mixture of benz [2,3,4,5,6-²H₅]aldehyde (**190**) (0.991 g, 8.92 mmol), perdeuterated 2-methyl-2-nitrosopropane (**193**) (2 g, 17.83 mmol), prepared as described in the literature (Haire and Janzen, 1994), and zinc dust (1.75 g, 26.8 mmol) was dissolved in 95% ethanol (65 mL) and cooled to 10°C. Acetic acid (3.21 g, 53.5 mmol) was added dropwise over 30 min with brisk stirring. Thereafter, the mixture was vigorously stirred for an additional 2 hr and then stored in a cold room at 4°C for 48 hr. After filtering the mixture to remove zinc acetate, the remaining liquid was evaporated *in vacuo*. The resulting residue, containing zinc acetate and crude product, was washed with ether. The organic washes were combined and evaporated to dryness, leaving the title compound as a crude solid. Purification was two-fold. Initially, the nitrone was sublimated at 60°C at (0.05 mm Hg). This product was recrystallized from diethyl ether/light petroleum ether (1:9) (bp 90–110°C) followed by further recrystallization from diethyl ether/hexanes (2:8) to give N-*tert*-butyl[²H₉]-α-phenyl[2,3,4,5,6-²H₅]nitrone (**195**) (1 g, 60%) as a white solid, mp 72–73°C (Haire and Janzen, 1994). ¹³C NMR (CDCl₃) δ: 28.30 (3CD₃), 70.77 (C[CD₃]₃), 129.69 (HC=N–O), 128–131 (C₆D₅).

N-tert-*Butyl-α-phenyl[¹⁵N]nitrone* (**201**)   The initial step in the preparation of the title compound is the synthesis of trichloroamine[¹⁵N] (**198**). In a three-necked flask equipped with a mechanical stirrer and cooled to 2°C, calcium hypochlorite (70%, 12.35 g, 58.7 mmol), water (70 mL), and methylene chloride (41 mL) were mixed. With vigorous stirring, a solution of [¹⁵N]ammonium chloride (2 g, 36.7 mmol) in dilute hydrochloric acid (10%, 30 mL) was added dropwise, maintaining the temperature of the reaction at at −5°C. The solution was stirred for another 25 min with cooling. At this point, the two layers were separated and the aqueous mixture was washed with methylene chloride. The combined organic solutions were washed with water, dried over anhydrous Na₂SO₄, and filtered, giving a methylene chloride solution of trichloroamine[¹⁵N] (**198**).

The above solution, used immediately after filtration, was added to a precooled flask at −10°C under a N₂ atmosphere, containing aluminum chloride (7.33 g, 55 mmol) and methylene chloride (15 mL). With stirring, a solution of 2-chloro-2-methylpropane (10.18 g, 110 mmol) in methylene chloride (10 mL) was added dropwise while the temperature of the reaction was maintained at −10°C. After the reaction was stirred at this temperature for an additional hour, the mixture was poured over crushed ice containing hydrochloric acid (37%, 5 mL). The reaction was stirred for several minutes and then left to stand overnight. The layers were separated and the organic solution was washed with water. The combined aqueous solutions were extracted with ether and bubbled with N₂ for 10 min. This solution contained the hydrochloric salt of 2-[^{15}N]amino-2-methylpropane (199). The solution was cooled to 0–5°C and a concentrated solution of sodium hydroxide in water (50%, 30 g) was added. The remaining solution was steam distilled (100 mL) and carefully collected in a precooled flask containing potassium permanganate (10 g, 63.3 mmol). Additional permanganate (10 g) was added and the solution was stirred for 2 hr at 0–5°C and overnight at 20°C. More potassium permanganate (10 g) was added to the reaction and it was stirred for 3 hr at 20°C. At this point, the mixture was steam distilled, collecting 100 mL of solution in a cooled flask at 0–5°C. The distillate was extracted with ether, dried over anhydrous MgSO$_4$, filtered, and distilled, affording 2-methyl-2[^{15}N]nitropropane (200) (1.6 g).

This crude nitropropane (200), without further purification, was added to a solution of ethanol (95%, 25 mL) containing benzaldehyde (3.26 g, 30.76 mmol) at 0–5°C. Zinc dust (2 g, 30.76 mmol) was introduced into the well-stirred reaction. Thereafter, acetic acid (3.67 g, 61.52 mmol) was added dropwise, while the temperature of the reaction did not exceed 9°C. After addition of acetic acid was completed, the reaction was stirred for an additional 8 hr at < 3°C and then stored in a refrigerator overnight at this temperature. The next day, the reaction was filtered, the zinc salts were washed with ethanol, and the resulting solution was evaporated *in vacuo*. The remaining mixture was taken up in chloroform, washed with water, dried over anhydrous MgSO$_4$, and evaporated to dryness. The residue was flash chromatographed with silica gel (200–300 mesh) and methylene chloride, followed by chloroform, affording N-*tert*-butyl-α-phenyl[^{15}N]nitrone (201) (1.18 g), as a sticky solid. Purification was achieved by sublimation at 60–65°C/ 0.1 mm Hg. Additional repetitions of this sublimation step at 50–55°C/0.1 mm Hg led to pure nitrone (201) as a white crystalline solid, mp 70–71°C (Zhang, 1996). NMR (CDCl$_3$) δ: 1.62 (d, 9H, $J = 2.4$ Hz, C(CH$_3$)$_3$), 7.40–7.43 (m, 3H, Ar–H), 7.56 (d, 1H, $J = 2.4$ Hz, HC=N–O).

N-*Hydroxy*-N-*(1,1-dimethylethyl)benzamide* (209) The preparation of this hydroxamic acid is a two-step process. Initially, N-(*tert*-butyl)hydroxylamine hydrochloride (135) (1.26 g, 10 mmol) was dibenzoylated with benzoyl chloride (4.2 g, 40 mmol) in dilute sodium hydroxide (10%, 30 mL). After standing for 1 day, a solid formed. After collecting and washing this compound with water, N,O-dibenzoyl-N-*tert*-butylhydroxylamine (208) (1.97 g, 66%) was isolated from ethanol as white crystals, mp 101°C.

To a solution of sodium ethoxide, prepared by the addition of sodium metal (120 mg, 5.2 mmol) to absolute ethanol (5 mL), was added N,O-dibenzoyl-N-*tert*-butylhydroxylamine (208) (1.49 g, 10 mmol). When addition was completed, the reaction was brought to the boil and then cooled in an ice bath, whereupon, water (15 mL) was added to quench the reaction. This mixture was extracted with ether. The remaining aqueous solution was then saturated with CO$_2$, resulting in the precipitation of N-hydroxy-N-(1,1-dimethylethyl)benzamide (209) (670 mg, 68%) from cyclohexane, mp 112–114°C (Exner and Kakáč, 1963; Warshaw, *et al.,* 1989). IR (CH$_3$Cl) 3239 broad (N–OH), 1464 (C=O) cm^{-1}. NMR (CDCl$_3$) δ: 1.31 (s, 9H, N–C(CH$_3$)$_3$), 7.41 (s, 5H, C$_6$H$_5$).

α-*Methoxy*-α-*phenyl*-N-tert-*butylnitrone* (**211**).   To a solution of *N*-hydroxy-*N*-(1,1-dimethylethyl)benzamide (**209**) (5.1 g, 26 mmol) in P_2O_5-dried methylene chloride (100 mL) was added methyl trifluoromethanesulfonate (methyl triflate) (5.2 g, 32 mmol) in dry methylene chloride (65 mL). After standing at room temperature for 2 days, the reaction was evaporated *in vacuo* to give the nitrone hydrotriflate as a white solid. Extensive washing with carbon tetrachloride under N_2 gave 97% pure hydrotriflate (**210**), mp 45–50°C (Warshaw, *et al.*, 1989). IR (CH_3Cl) 3500–2500 (N–OH), 1619 (C=N) cm^{-1}. NMR ($CDCl_3$) δ: 1.71 (s, 9H, N–C(CH_3)$_3$), 4.00 (s, 3H OCH_3), 7.62–7.67 (s, 5H, C_6H_5).

Deprotonation using either preparative TLC, triethylamine, or basic resin afforded α-methoxy-α-phenyl-*N*-*tert*-butylnitrone (**211**) in reasonable yields. For example, hydrotriflate (**210**) (170 mg, 97% pure, 460 μmol) was loaded onto a silica gel TLC plate and eluted with methylene chloride/methanol (9:1). After scrapping the nitrone band from the glass plate, the silica gel was washed with dry methanol and evaporated *in vacuo* under a stream of either argon or nitrogen at 0°C, giving α-methoxy-α-phenyl-*N*-*tert*-butylnitrone (**211**), 98% pure by NMR (Warshaw, *et al.*, 1989). IR ($CHCl_3$) 1600 (C=N–O) cm^{-1}, NMR *E* isomer ($CDCl_3$) δ: 1.63 (s, 9H, N–C(CH_3)$_3$), 3.63 (s, 3H, OCH_3), 7.37–7.47 (s, 3H, C_6H_5), 7.87–7.93 (s, 2H, C_6H_5). Elemental analysis was not possible due to the unstable nature of nitrone (**211**). Attempts to prepare gram quantitites of this nitrone by means of flash chromatography with silica gel and methylene chloride have proven to be unsuccessful, resulting in the isolation of only methyl benzoate.

2-Methyl-2-nitrosopropane (**212**)   The preparation of the title compound is dependent upon the initial formation of *N*-(*tert*-butyl)hydroxylamine (**135**). In a three-neck flask (fitted with a dropping funnel, two condensors in series, and a mechanical stirrer) was added an ether (1.5 L) and H_2O (15 mL) solution containing aluminum–mercury amalgam, prepared as described by Calder, *et al.* (1972). This reaction should be run in a well-ventilated fume hood. To this flask was added 2-methyl-2-nitropropane, dropwise at such a rate so that the ether solution was lightly refluxing. If the reaction begins to move up the condensors, the heat should be removed and immediately cooled in an ice bath. After the addition was completed, the reaction was stirred for 30 min, at which point the stirring was stopped, allowing the gelatinous precipitate to settle. The liquid mixture was carefully but rapidly decanted through a glass-wool filter into a separatory filter containing dilute NaOH (2 M). The precipitate was washed with ether. The ether solution was dried over anhydrous Na_2SO_4, filtered, and evaporated *in vacuo*, giving *N-(tert*-butyl)hydroxylamine (**135**) (33 g, 65%) as a solid. The crude hydroxylamine (**135**) was sufficiently pure to be used in the oxidation step. Nevertheless, an analytic sample can be obtained by recrystallization from pentane, mp 64–65°C (Calder, *et al.*, 1972).

A suspension of *N*-(*tert*-butyl)hydroxylamine (**135**) (26.7 g, 900 mmol) in H_2O (50 mL) is added to a solution of sodium hypobromite at −20°C, prepared by adding bromine (57.5 g, 18.5 mL, 360 mmol) to a solution of NaOH (36 g, 900 mmol) in 225 mL of H_2O. The rate of hydroxylamine (**135**) addition was such that the reaction temperature does not exceed 0°C. Then, the solution was cooled to −20°C, at which point the cooling bath was removed. The reaction was stirred for 4 hr while it warmed to room temperature. The nitroso dimer, which precipitated from the solution, was filtered and washed with H_2O to remove trace amounts of base. The remaining solid was dried under reduced pressure, affording the dimer of 2-methyl-2-nitrosopropane (212) (19 g, 75%), mp 80–81°C (Wajer, *et al.*, 1967; Stowell, 1971; Calder, *et al.*, 1972). IR (CCl_4) 1565 (N=O) cm^{-1}, in equilibrated mixture; IR (KBr) of dimer lacks 1565 cm^{-1} peak. NMR (CCl_4) δ: 1.24 (s, 9H, N–C(CH$_3$)$_3$) for monomer; 1.57 (s, 18H, N–C(CH$_3$)$_3$) for dimer.

2-(Hydroxymethyl)-2-nitrosopropane (213)   2-Methyl-2-nitropropanol (25 g, 210 mmol) was added to an aqueous solution (250 mL) of ammonium chloride (9 g, 170 mmol). Once this nitroalkane was dissolved, zinc dust (45 g, 692 mmol) was added over 1 hr, while the temperature of the well-mixed solution was maintained at about 60°C. Once the addition was completed, the reaction was stirred for an additional hour at this temperature. Then, the reaction was filtered and the filter cake was washed with warm water. The aqueous solution was rotary evaporated *in vacuo*, maintaining the temperature below 35°C. The remaining residue was taken up in methanol (≈ 10 mL) and rotary evaporated to dryness, giving *N*-(2-hydroxymethyl-2-propane)hydroxylamine (221) as an oil. This material was used in the next step without further purification.

To a suspension of silver carbonate (17 g, 61.6 mmol), prepared as described in the literature (Maassen and De Boer, 1971) in ether (150 mL), was added an aqueous (50 mL) solution of *N*-(2-hydroxylmethyl-2-propane)hydroxylamine (221) (2.1 g, 20 mmol). The reaction was stirred for 2 min, during which time the color of the solution changed from yellow to black. The mixture was filtered, the Celite was rinsed with additional ether, and the combined ether solutions were washed with dilute sulfuric acid (1 M) and then water. The remaining ether solution was dried over anhydrous MgSO$_4$, filtered, and evaporated *in vacuo*, giving 2-(hydroxymethyl)-2-nitrosopropane (213) as a solid (De Groot, *et al.*, 1973). IR (KBr) 1545 (N=O) cm^{-1}, UV$_{ethanol}$ λ$_{max}$ at 215, 294, and 675 nm.

Light should be eliminated as much as possible. Therefore, it is recommended that the reaction flask be covered with aluminum foil. Additionally, the reaction, washings, and rotary evaporation should be undertaken in dim light to avoid homolytic cleavage, which generates di-*tert*-butylnitroxide.

Sodium 3,5-dibromo-4-nitrosobenzenesulfonate (214)   To a solution of sodium 3,5-dibromosulfanilate (222) (3.53 g, 10 mmol) in glacial acetic acid (30 mL) was added hydrogen peroxide (30%, 7.94 mL, 70 mmol) and anhydrous sodium acetate (820 mg, 10 mmol). The mixture was gently warmed at 30–40°C until the salts went into solution. The reaction was allowed to stand at room temperature for 14 days, during which time crystals formed. After filtering, the solid was washed sequentially with acetic acid (5 mL) and then ether (50 mL). Recrystallization from ethanol gave sodium 3,5-dibromo-4-nitrosobenzenesulfonate (214) as a pale yellow powder, mp > 300°C (Kaur, *et al.*, 1981). IR (KBr) 1545 (N=O) cm^{-1}.

1-Methyl-4-nitroso-3,5-diphenylpyrazole (216)   To an ice-cooled solution of isonitrosodibenzoylmethane (223) (10 g, 40 mmol) in ethanol (50 mL) was added methylhydrazine (2.6 g, 56 mmol) over 15 min. The reaction was then stirred in an ice bath for an additional hour. At this point, the mixture was poured into dilute hydrochloric acid (1 M, 400 mL) and was extracted with methylene chloride. After drying with anhydrous MgSO₄, the solution was evaporated *in vivo*, leaving a gum. Crystallization using ethanol afforded 1-methyl-4-nitroso-3,5-diphenylpyrazole (**216**) (6 g, 58%) as green crystals, mp 96–97°C (Boyd and Norris, 1974). Additional purification raised the mp to 117–118°C (Kaur and Perkins, 1982). IR (KBr pellet) 1485 (N–O) cm⁻¹. NMR (CDCl₃) δ: 3.77 (s, 3H, NCH₃), 7.42 (m, 10H, C₆H₅).

Sodium 3-(4-nitroso-3,5-diphenyl-1-pyrazolyl)propane-1-sulfonate (**217**)   To a stirred solution of isonitrosodibenzoylmethane (**223**) (4.2 g, 16.8 mmol) in methanol (120 mL) was added 3-hydrazinopropane-4-sulfonic acid (2.6 g, 16.8 mmol), prepared as described by Schindler and Pietrzok (1965). The reaction was refluxed for 7 hr, at which point most of the solvent was removed by rotary evaporation. The remaining green aqueous solution was washed with ether, made acidic with dilute sulfuric acid (3 M, 150 mL), and continually extracted with ether. After drying the combined ether extractions with anhydrous MgSO₄, the solvent was evaporated *in vacuo*, giving 3-(4-nitroso-3,5-diphenyl-1-pyrazolyl)propane-1-sulfonic acid as a brown solid (2.3 g, 38%), mp 100–104°C, although the material softens at 73°C. The sodium salt was prepared by adding sodium hydrogen carbonate (240 mg, 2.9 mmol) to an aqueous (10 mL) solution of the sulfonic acid (1 g, 2.7 mmol). Over the next 30 min, green crystals were separated from the solution, affording sodium 3-(4-nitroso-3,5-diphenyl-1-pyrazolyl)propane-1-sulfonate (**217**) (Kaur and Perkins, 1982). NMR (D₂O) δ: 2.0–2.45 (broad m, 2H), 2.55–2.95 (broad t, 2H), 3.95–4.20 (broad t, 2H), 7.3–7.8 (m, 10H, C₆H₅).

1,1,3,3-Tetramethyl-2-indanone (**227**)   In a three-necked flask, fitted with a reflux condensor, a mechanical stirrer, and dropping funnel, a mixture of 2-indanone (**226**) (12.2 g, 100 mmol, prepared as described by Horan and Schiessler (1961) and methyl iodide (114 g, 50 mL, 800 mmol) was added to a suspension of potassium hydroxide (112 g, 200 mmol) in dimethyl sulfoxide (190 mL), with stirring, at 50–60°C. Initially, external heating was required, but as the reaction proceeds the exothermal reaction maintains the temperature in this range. After the addition was completed, the reaction was stirred for an hour, the slurry was poured into ice/water (500 mL) and then extracted with pentane. After washing the pentane solution with water, the organic layer was dried over MgSO₄ and evaporated to dryness, giving 1,1,3,3-tetramethyl-2-indanone (**227**) (14 g, 75%) as a white solid, mp 76°C, from methanol (Langhals and Langhals, 1990; Korth, *et al.*, 1994). NMR (CDCl₃) δ: 1.34 (s, 12 H), 7.28 (m, 2H, *J* = 1.8 Hz), 7.29 (m, 2H, *J* = 1.8 Hz).

5-Bromo-1,1,3,3-tetramethyl-2-indanone (**230**)   Bromination of 1,1,3,3-tetra-methyl-2-indanone (**227**) (10 g, 53.1 mmol) was undertaken using the standard iron powder and bromine method as described in *Vogel's Textbook of Practical Organic Chemistry* (Furniss, *et al.*, 1978). After purification by sublimation (80°C at 10⁻³ mbar), 5-bromo-1,1,3,3-tetramethyl-2-indanone (**230**) (5.1 g, 34%) was isolated as a white solid, mp 76°C (Korth, *et al.*, 1994). IR (KBr pellet) 1742 (C=O) cm⁻¹.

NMR (CDCl$_3$) δ: 1.32 (s, 6H), 1.34 (s, 6H), 7.15 (dd, 1H, J = 8.0, 0.7 Hz), 7.40 (dd, 1H, J = 2.0, 0.7 Hz), 7.42 (dd, 1H, J = 8.0, 2.0 Hz).

5-Carboxy-1,1,3,3-tetramethyl-2-indanone (**231**)   Carboxylation of 5-bromo-1,1,3,3-tetramethyl-2-indanone (**230**) (1 gm, 5.4 mmol) was accomplished by reacting the corresponding Grignard, prepared as described in *Vogel's Textbook of Practical Organic Chemistry*, with gaseous carbon dioxide (Furniss, *et al.*, 1978). After work-up following standard procedures, 5-carboxy-1,1,3,3-tetramethyl-2-indanone (**231**) was obtained as a white solid (510 mg, 51%), mp 215°C (Korth, *et al.*, 1994). IR (KBr pellet) 3000–2500 (OH), 1750 (C=O) and 1682 (C=O) cm^{-1}. NMR (CDCl$_3$) δ: 1.36 (s, 6H), 1.37 (s, 6H), 7.37 (dd, 1H, J = 7.8, 0.6 Hz), 8.01 (dd, 1H, J = 1.3, 0.6 Hz), 8.07 (dd, 1H, J = 7.8, 1.3 Hz).

1,3-Dihydroxy-2-phenyl-4,4,5,5-tetramethylimidazolidine (**237**)   An aqueous solution of 2,3-bis(hydroxylamino)2,3-dimethylbutane (**236**) (4.44 g, 300 mmol), prepared by dissolving the sulfate of this hydroxylamine in dilute NaOH (10%) and adjusting the pH to 9.0 (Joseph, *et al.*, 1993), was added to benzaldehyde (**134**) (3.5 g, 330 mmol) dissolved in methanol (150 mL). The reaction was stirred at room temperature for 24 hr, at which point the desired product (**237**) precipitated and was removed by filtration. By concentrating the remaining liquid, additional product could be obtained. The combined 1,3-dihydroxy-2-phenyl-4,4,5,5-tetramethylimidazolidine (**237**) (5.25 g, 74%) was recrystallized from benzene/ether, mp 168–169°C (Ullman, *et al.*, 1972). IR (KBr pellet) 3260 (OH) cm^{-1}, NMR (DMSO) δ: 1.03 (s, 2CH$_3$), 1.07 (s, 2CH$_3$), 4.51 (s, CH), 7.20–7.55 (m, C$_6$H$_5$), 7.71 (2OH).

2-Phenyl-4,4,5,5-tetramethylimidazolidine 3-oxide 1-oxyl (**234**)   Lead dioxide (20 g, 83.6 mmol) was added to a solution of 1,3-dihydroxy-2-phenyl-4,4,5,5-tetramethylimidazolidine (**237**) (1.45 g, 6.1 mmol) in benzene (250 mL) at room temperature. After the reaction was stirred for 45 min, the mixture was filtered and the remaining benzene solution was evaporated *in vacuo*, affording dark blue crystals of 2-phenyl-4,4,5,5-tetramethylimidazolidine 3-oxide 1-oxyl (**234**) (1 g, 74%) mp 85°C, from ether (Ullman, *et al.*, 1972). IR (CHCl$_3$) 1140, 1371 cm^{-1}. UV λ$_{max}$ (EtOH) 238 mμ(ε 9400), 263 (12,200), 260 (13,300), 588 (685).

References

Acken, B.J., Warshaw, J.A., Gallis, D.E., and Crist, D.R. (1989). Acyclic α-alkoxynitrones. A new class of spin-trapping agents. *J. Org. Chem.* **54**: 1743–1745.

Akaike, T., and Maeda, H. (1996). Quantitation of nitric oxide using 2-phenyl-4,4,5,5-tetramethylimidazoline-1-oxyl-3-oxide (PTIO). *Methods Enzymol.* **268**: 211–221.

Akaike, T., Yoshida, M., Miyamoto, Y., Sato, K., Kohno, M., Sasamoto, K., Miyazaki, K., Ueda, S., and Maeda, H. (1993). Antagonistic action of imidazolineoxyl N-oxides against endothelium-derived relaxing factor/·NO through a radical reaction. *Biochemistry* **32**: 827–832.

Archer, S. (1993). Measurement of nitric oxide in biologic models. *FASEB J.* **7**: 346–360.

Arya, P. (1996). New derivatives of pyrroline N-oxides as spin traps. *Heterocycles* **43**: 397–407.

Arya, P., Stephens, J.C., Griller, D., Pou, S., Ramos, C.L., Pou, W.S., and Rosen, G.M. (1992). Design of modified pyrroline N-oxide derivatives as spin traps specific for hydroxyl radical. *J. Org. Chem.* **57**: 2297–2301.

Ashburn, S.P., and Coates, R.M. (1985). Preparation of oxazoline N-oxides and imidate N-oxides by amide acetal condensation and their [3 + 2] cycloaddition reactions. *J. Org. Chem.* **50**: 3076–3081.

Bancroft, E.E., Blount, H.N., and Janzen, E.G. (1980). Spin trapping with covalently immobilized α-phenyl-N-[(1-hydroxy-2-methyl)-2-propyl]nitrone. *J. Phys. Chem.* **84**: 557–558.

Bapat, J.B. and Black, D.St.C. (1968). Nitrones and oxazirans. I. Preparation of 1-pyrroline 1-oxides and 1-pyrrolines by reductive cyclization of γ-nitro carbonyl compounds. *Aust. J. Chem.* **21**: 2483–2495.

Barasch, D., Krishna, M.C., Russo, A., Katzhendler, J., and Samuni, A. (1994). Novel DMPO-derived ^{13}C-labeled spin traps yield identifiable stable nitroxides. *J. Am. Chem. Soc.* **116**: 7319–7324.

Barber, M.J., Rosen, G.M., and Rauckman, E.J. (1983). Studies of the mobility of maleimide spin labels within the erythrocyte membrane. *Biochim. Biophys. Acta* **732**: 126–132.

Barker, P., Beckwith, A.L.J., Cherry, W.R., and Huie, R. (1985). Characterization of spin adducts obtained with hydrophobic nitrone spin traps. *J. Chem. Soc., Perkin Trans. 2*, 1147–1150.

Barnes, M.W., and Patterson, J.M. (1976). An oxime to nitro conversation. A superior synthesis of secondary nitroparaffins. *J. Org. Chem.* **41**: 733–735.

Becker, D.A. (1996). Highly sensitive colorimetric detection and facile isolation of diamagnetic free radical adducts of novel chromotropic nitrone spin trapping agents readily derived from guaiazulene. *J. Am. Chem. Soc.* **118**: 905–906.

Belsky, I. (1977). "Naked" fluoride-catalysed Michael-additions. *J. Chem. Soc., Chem. Commun.* 237.

Bernotas, R.C., Adams, G., and Carr, A.A. (1996c). Synthesis of benzazepine-based nitrones as radical traps. *Tetrahedron* **52**: 6519–6526.

Bernotas, R.C., Adams, G., and Nieduzak, T.R. (1996b). Thermal cleavage of oxazolidine-4,5-diones to imines: A short synthesis of 3,4-dihydro-3,3-dimethyl-7-trifluoromethylisoquinoline-2-oxide. *Synth. Commun.* **26**: 3471–3477.

Bernotas, R.C., Thomas, C.E., Carr, A.A., Nieduzak, T.R., Adams, G., Ohlweiler, D.F., and Hay, D.A. (1996a). Synthesis and radical scavenging activity of 3,3-dialkyl-3,4-dihydro-isoquinoline 2-oxides. *Biorg. Med. Chem. Lett.* **6**: 1105–1110.

Beth, A.H., Venkataramu, S.D., Balasubramanian, K., Dalton, L.R., Robinson, B.H., Pearson, D.E., Park, C.R., and Park, J.H. (1981). [15]N- and [2]H- substituted maleimide spin labels: Improved sensitivity and resolution for biological EPR studies. *Proc. Natl. Acad. Sci. USA* **78**: 967–971.

Black, D.St.C., and Boscacci, A.B. (1976). Nitrones and oxaziridines. XV. Approaches to 3-oxo-1-pyrroline 1-oxides by oxidation of 1-pyrroline 1-oxides. *Aust. J. Chem.* **29**: 2511–2524.

Blough, N.V., and Simpson, D.J. (1987). Chemically mediated fluorescence yield switching in nitroxide-fluorophore adducts: Optical sensors of radical/redox reactions. *J. Am. Chem. Soc.* **110**: 1915–1917.

Bonnett, R., Brown, R.F.C., Clark, V.M., Sutherland, I.O., and Todd, A. (1959a). Experiments towards the synthesis of corrins. Part II. The preparation and reactions of Δ^1-pyrroline 1-oxides. *J. Chem. Soc.* 2094–2102.

Bonnett, R., Clark, V.M., Giddy, A., and Todd, A. (1959b). Experiments towards the synthesis of corrins. Part I. The preparation and reactions of some Δ^1-pyrrolines. A novel proline synthesis. *J. Chem. Soc.* 2087–2093.

Boyd, G.V., and Norris, T. (1974). Mesoionic compounds derived from pyrazole, isothiazole and isoxazole. *J. Chem. Soc., Perkin Trans. 1* 1028–1033.

Brandman, H.A., and Conley, R.T. (1973). Unambiguous synthesis of a monocyclic 5,6-dihydro-1,2-oxazine. *J. Org. Chem.* **38**: 2236–2238.

Buettner, G.R., and Oberley, L.W. (1978). Considerations in the spin trapping of superoxide and hydroxyl radical in aqueous systems using 5,5-dimethyl-1-pyrroline-1-oxide. *Biochem. Biophys. Res. Commun.* **83**: 69–74.

Buettner, G.R., Oberley, L.W., and Leuthauser, C.W.H. (1978). The effect of iron on the distribution of superoxide and hydroxyl radicals as seen by spin trapping and on the superoxide dismutase assay. *Photochem. Photobiol.* **28**: 693–695.

Butterfield, D.A., Roses, A.D., Appel, S.H., and Chesnut, D.B. (1976). Electron spin resonance studies on membrane proteins in erythrocytes in myotonic dystrophy. *Arch. Biochem. Biophys.* **177**: 226–234.

Calder, A., Forrester, A.R., and Hepburn, S.P. (1972). 2-Methyl-2-nitropropane and its dimer. *Org. Synth.* **52**: 77–82.

Carney, J.M., Starke-Reed, P.E., Oliver, C.N., Landum, R.W., Cheng, M.S., Wu, J.F., and Floyd, R.A. (1991). Reversal of age-related increase in brain protein oxidation, decrease in enzyme activity, and loss in temporal and spatial memory by chronic administration of the spin trapping compound N-*tert*-butyl-α-phenylnitrone. *Proc. Natl. Acad. Sci. USA* **88**: 3633–3636.

Carr, A.A., Thomas, G.E., Bernotas, R.C., and Ku, G. (1994). Cyclic nitrones, pharmaceutical composition thereof and their use in treating shock. U.S. Patent Number 5,292,746.

Dage, J.L., Ackermann, B.L., Barbuch, R.J., Bernotas, R.C., Ohlweiler, D.F., Haegele, K.D., and Thomas, C.E. (1997). Evidence for a novel pentyl radical adduct of the cyclic nitrone spin trap MDL 101,002. *Free Radical Biol. Med.* **22**: 807–812.

Dalton, J.C., and Stokes, B.G. (1975). Formation of a rearranged β, γ-enone from reaction of an acyclic β,γ-unsaturated carboxylic acid with methyl lithium. *Tetrahedron Lett.* 3179–3182.

De Groot, J.J.M.C., Garssen, G.J., Vliegenthart, J.F.G., and Boldingh, J. (1973). The detection of linoleic acid radicals in the anerobic reaction of lipoxygenase. *Biochim. Biophys. Acta* **326**: 279–284.

Dehnel, A., and Lavielle, G. (1980). Anions lithiens derives d'aza-2 allylphosphonates. Synthese regio et stereospecifique de nouvelles Δ^1-pyrrolines. *Tetrahedron Lett.* **21**: 1315–1318.

Dehnel, A., Griller, D., and Kanabus-Kaminska, J.M. (1988). Designer spin traps with a cyclic nitrone structure. *J. Org. Chem.* **53**: 1566–1567.

Dess, D.B., and Martin, J.C. (1983). Readily accessible 12-I-5 oxidant for the conversion of primary and secondary alcohols to aldehydes and ketones. *J. Org. Chem.* **48**: 4155–4156.

Dikalov, S., Kirilyuk, I., and Grigor'ev, I. (1996). Spin trapping of O-, C-, and S-centered radicals and peroxynitrite by 2H-imidazole-1-oxides. *Biochem. Biophys. Res. Commun.* **218**: 616–622.

Dikalov, S.I., Kirilyuk, I.A., Grigor'ev, I.A., and Volodarskii, L.B. (1992). 2H-Imidazole-N-oxides as spin traps. *Bull. Russ. Acad. Sci.* **41**: 834–837.

Elsworth, J.F., and Lamchen, M. (1970). Nitrones. Part X. Tautomerism in cyclic nitrones. *J. S. Afr. Chem. Inst.* **23**: 61–70.

Emmons, W.D. (1957). The preparation and properties of oxaziranes. *J. Am. Chem. Soc.* **79**: 5739–5754.

Exner, O., and Kakáč (1963). Acyl derivatives of hydroxylamine. VIII. A spectroscopic study of tautomerism of hydroxamic acids. *Collect. Czech. Chem. Commun.* **28**: 1656–1663.

Finkelstein, E., Rosen, G.M., and Rauckman, E.J. (1980). Spin trapping of superoxide and hydroxyl radical: Practical aspects. *Arch. Biochem. Biophys.* **200**: 1–16.

Finkelstein, E., Rosen, G.M., and Rauckman, E.J. (1982). Production of hydroxyl radical by decomposition of superoxide spin-trapped adducts. *Mol. Pharmacol.* **21**: 262–265.

Finkelstein, E., Rosen, G.M., Rauckman, E.J., and Paxton, J. (1979). Spin trapping of superoxide. *Mol. Pharmacol.* **16**: 676–685.

Freeman, B.A., Rosen, G.M., and Barber, M.J. (1986). Superoxide perturbation of the organization of vascular endothelial cell membranes. *J. Biol. Chem.* **261**: 6590–6593.

Fréjaville, C., Karoui, H., Tuccio, B., Le Moigne, F., Culcasi, M., Pietri, S., Lauricella, R., and Tordo, P. (1994). 5-Diethoxyphosphoryl-5-methyl-1-pyrroline N-oxide (DEPMPO): A new phosphorylated nitrone for efficient *in vitro* and *in vivo* spin trapping of oxygen-centred radicals. *J. Chem. Soc., Chem. Commun.* 1793–1794.

Fréjaville, C., Karoui, H., Tuccio, B., Le Moigne, F., Culcasi, M., Pietri, S., Lauricella,   R., and Tordo, P. (1995). 5-Diethoxyphosphoryl-5-methyl-1-pyrroline N-oxide (DEPMPO): A new efficient phosphorylated nitrone for the *in vitro* and *in vivo* spin trapping of oxygen-centered radicals. *J. Med. Chem.* **38**: 258–265.

French, J.F., Thomas, G.E., Downs, T.R., Ohlweiler, D.F., Carr, A.A., and Dage, R.C. (1994). Protective effects of a cyclic nitrone antioxidant in animal models of endotoxic shock and chronic bacteremia. *Circ. Shock* **43**: 130–136.

Furniss, B.S., Hannaford, A.J., Rogers, V., Smith, P.W.G., and Tatchell, A.R. (1978). *Vogel's Textbook of Practical Organic Chemistry*, 4th edition, Longman House, Essex, England.

Gabr, A.M., Rai, U.S., and Symons, M.C.R. (1993). Conversion of nitric oxide into a nitroxide radical using 2,3-dimethylbutadiene and 2,5-dimethylhexadiene. *J. Chem. Soc., Chem. Commun.* 1099–1100.

Geissman, T.A., Schlatter, M.J., and Webb, I.D. (1946). The preparation of 1,3-diamino-2-methylaminopropane and 1,3-diamino-2-amino-methylpropane. *J. Org. Chem.* **11**: 736–740.

Green, S.A., Simpson, D.J., Zhou, G., Ho, P.S., and Blough, N.V. (1990). Intramolecular quenching of excited states by stable nitroxyl radicals. *J. Am. Chem. Soc.* **112**: 7337–7346.

Haire, D.L., and Janzen, E.G. (1982). Synthesis and spin trapping kinetics of new alkyl substituted cyclic nitrones. *Can J. Chem.* **60**: 1514–1522.

Haire, D.L., and Janzen, E.G. (1994). Enhanced   diagnostic   EPR   and ENDOR spectroscopy of radical spin adducts of deuterated α-phenyl N-*tert*-butyl nitrone. *Magn. Reson. Chem.* **32**: 151–157.

Haire, D.L., Hilborn, J.W., and Janzen, E.G. (1986). A more efficient synthesis of DMPO-type (nitrone) spin traps. *J. Org. Chem.* **51**: 4298–4300.

Haire, D.L., Oehler, U.H., Goldman, H.D., Dudley, R.L., and Janzen, E.G. (1988a). The first ^{1}H and ^{14}N ENDOR spectra of an oxygen-centered radical adduct of DMPO-type nitrones. *Can. J. Chem.* **66**: 2395–2402.

Haire, D.L., Oehler, U.H., Krygsman, P.H., and Janzen, E.G. (1988b). Correlation of radical structure with EPR spin adduct parameters: Utility of ^{1}H, ^{13}C, and ^{14}N hyperfine splitting constants of aminoxyl adducts of PBN-*nitronyl*-^{13}C for three parameter scatter plots. *J. Org. Chem.* **53**: 4535–4542.

Halpern, H.J., Pou, S., Peric, M., Yu, C., Bath, E., and Rosen, G.M. (1993). Detection and imaging of oxygen-centered free radicals with low-frequency electron paramagnetic resonance and signal-enhancing deuterium-containing spin traps. *J. Am. Chem. Soc.* **115**: 218–223.

Halpern, H.J., Yu, C., Barth, E., Peric, M., and Rosen, G.M. (1995). *In situ* detection, by spin trapping of hydroxyl radical markers produced from ionizing radiation in the tumor of a living mouse. *Proc. Natl. Acad. Sci. USA* **92**: 796–800.

Hamer, J., and Macaluso, A. (1964). Nitrones. *Chem. Rev.* **64**: 474–495.

Harbour, J.R., and Bolton, J.R. (1975). Superoxide formation in spinach chloroplasts: Electron spin resonance detection by spin trapping. *Biochem. Biophys. Res. Commun.* **64**: 803–807.

Harbour, J.R., Chow, V., and Bolton, J.R. (1974). An electron spin resonance study of the spin adducts of OH and HO_2 radicals with nitrones in the ultraviolet photolysis of aqueous hydrogen peroxide solutions. *Can. J. Chem.* **52**: 3549–3553.

Holman, R.J., and Perkins, M.J. (1971). A probe for homolytic reactions in solution. Part V. Perdeuterionitrosobutane. An improved spin trap. *J. Chem. Soc. C*: 2324–2326.

Horan, J.E., and Schiessler, R.W. (1961). 2-Indanone. *Org. Synth.* **41**: 52–55.

Huang, C., Charlton, J.P., Shyr, C.I., and Thompson, J.E. (1970). Studies on phosphatidylcholine vesicles with thiocholesterol and thiocholesterol-linked spin label incorporated in the vesicle wall. *Biochemistry* **9**: 3422–3426.

Huber, R., Knierzinger, A., Obrecht, J.-P., and Vasella, A. (1985). Nucleophilic additions to N-glucosylnitrones. Asymmetric synthesis of α-aminophosphonic acids. *Helv. Chim. Acta* **68**: 1730–1747.

Huie, R., and Cherry, W.R. (1985). Facile one-step synthesis of phenyl-*tert*-butylnitrone (PBN) and its derivatives. *J. Org. Chem.* **50**: 1531–1532.

Ignarro, L.J, Buga, G.M., Wood, K.S., Byrns, R.E., and Chaudhuri, G. (1987). Endothelium-derived relaxing factor produced and released from artery and vein is nitric oxide. *Proc. Natl. Acad. Sci. USA* **84**: 9265–9269.

Ingold, K.U. (1984). Reactions with a different radical, in: *Landolt-Börnstein. New Series. Radical Reaction Rates in Liquids* (Fischer, H., ed.), Vol. 13, Part c, pp. 181–200, Springer-Verlag, Berlin.

Iwamura, M., and Inamoto, N. (1967). Novel formation of nitroxide radicals by radical addition to nitrones. *Bull. Chem. Soc. Japan* **40**: 703.

Iwamura, M., and Inamoto, N. (1970a). Reactions of nitrones with free radicals. I. Radical 1,3-addition to nitrones. *Bull. Chem. Soc. Japan* **43**: 856–860.

Iwamura, M., and Inamoto, N. (1970b). Reactions of nitrones with free radicals. II. Formation of nitroxides. *Bull. Chem. Soc. Japan* **43**: 860–863.

Janzen, E.G. (1971). Spin Trapping. *Acc. Chem. Res.* **4**: 31–40.

Janzen, E.G., and Coulter, G.A. (1984). Spin trapping in SDS micelles. *J. Am. Chem. Soc.* **106**: 1962–1968.

Janzen, E.G., and Liu, J.I.-P. (1973). Radical addition of 5,5-dimethyl-1-pyrroline-1-oxide. ESR spin trapping with a cyclic nitrone. *J. Magn. Reson.* **9**: 510–512.

Janzen, E.G., and Shetty, R.V. (1979). Spin trapping chemistry of sodium 2-sulfonatophenyl-*t*-butyl nitrone (Na$^+$ 2-SPBN$^-$). A negatively charged water-soluble spin trap. *Tetrahedron Lett.* 3229–3232.

Janzen, E.G., and Zawalski, R.C. (1978). Synthesis of nitronyl alcohols and their benzoate esters. *J. Org. Chem.* **43**: 1900–1903.

Janzen, E.G., and Zhang, Y.-K. (1995). Identification of reactive free radicals with a new ^{31}P-labeled DMPO spin trap. *J. Org. Chem.* **60**: 5441–5445.

Janzen, E.G., Dudley, R.L., and Shetty, R.V. (1979). Synthesis and electron spin resonance chemistry of nitronyl labels for spin trapping. α-Phenyl N-[5-(5-methyl-2,2-dialkyl-1,3-dioxanyl)] nitrones and α-(N-alkylpyridinium N-(*tert*-butyl)nitrones. *J. Am. Chem. Soc.* **101**: 243-245.

Janzen, E.G., Jandrisits, L.T., Shetty, R.V., Haire, D.L., and Hilborn, J.W. (1989). Synthesis and purification of 5,5-dimethyl-1-pyrroline-N-oxide for biological applications. *Chem.-Biol. Interact.* **70**: 167–172.

Janzen, E.G., Kotake, Y., and Hinton, R.D. (1992). Stabilities of hydroxyl radical spin adducts of PBN-type spin traps. *Free Radical Biol. Med.* **12**: 169–173.

Janzen, E.G., Lopp, I.G., and Morgan, T.V. (1973). Detection of fluoroalkyl and acyl radicals in the gas-phase photolysis of ketones and aldehydes by electron spin resonance gas-phase spin trapping techniques. *J. Phys. Chem.* **77**: 139–141.

Janzen, E.G., Wang, Y.Y., and Shetty, R.V. (1978). Spin trapping with α-pyridyl 1-oxide N-*tert*-butyl nitrones in aqueous solutions. A unique electron spin resonance spectrum for the hydroxyl radical adduct. *J. Am. Chem. Soc.* **100**: 2923–2925.

Janzen, E.G., Zhang, Y.-K., and Arimura, M. (1995). Synthesis and spin-trapping chemistry of 5,5-dimethyl-2-(trifluoromethyl)-1-pyrroline N-oxide. *J. Org. Chem.* **60**: 5434–5440.

Janzen, E.G., Zhang, Y.-K., and Haire, D.L. (1994). Synthesis of a novel nitrone, 2-phenyl-5,5-dimethyl-1-pyrroline N-oxide (nitronyl-^{13}C), for enhanced radical addend recognition and spin adduct persistence. *J. Am. Chem. Soc.* **116**: 3738–3743.

Jiang, J.J., Liu, K.J., Jordan, S.J., Swartz, H.M., and Mason, R.P. (1996). Detection of free radical metabolite formation using *in vivo* EPR spectroscopy: Evidence of rat hemoglobin thiyl formation following administration of phenylhydrazine. *Arch. Biochem. Biophys.* **330**: 266–270.

Jiang, J.J., Liu, K.J., Shi, X., and Swartz, H.M. (1995). Detection of short-lived free radicals by low-frequency electron paramagnetic resonance spin trapping in whole living animals. *Arch. Biochem. Biophys.* **319**: 570–573.

Joseph, J., Kalyanaraman, G., and Hyde, J.S. (1993). Trapping of nitric oxide by nitronyl nitroxides: An electron spin resonance investigation. *Biochem. Biophys. Res. Commun.* **192**: 926–934.

Kalyanaraman, B., Perez-Reyes, E., and Mason, R.P. (1979). The reduction of nitroso-spin traps in chemical and biological systems. A cautionary note. *Tetrahedron Lett.* 4809–4812.

Kaur, H., and Perkins, M.J. (1982). From early spin-trapping experiments to acyl nitroxide chemistry: Synthesis of and spin trapping with a water-soluble nitrosopyrazole derivative. *Can. J. Chem.* **60**: 1587–1593.

Kaur, H., Leung, K.H.W., and Perkins, M.J. (1981). A water-soluble, nitrosoaromatic spin-trap. *J. Chem. Soc., Chem. Commun.* 142–143.

Kaur, H., Perkins, M.J., Scheffer, A., and Vennor-Morris, D.C. (1982). Spin trapping with 9-nitrosotriptycene. *Can. J. Chem.* **60**: 1594–1596.

Keana, J.F., and Pou, S. (1989). Synthesis of two photolabile benzyl manganese derivatives incorporating a geometrically inaccessible radical trap. *Org. Prep. Proced.* **21**: 303–308.

Kieber, D.J., and Blough, N.V. (1990). Fluorescence detection of carbon-centered radicals in aqueous solution. *Free Radical Res. Commun.* **10**: 109–117.

Kirilyuk, I.A., Grigor'ev, I.A., and Volodarskii, L.B. (1991). Synthesis of 2H-imidazole 1-oxides and stable nitroxyl radicals based on them. *Bull. Russ. Acad. Sci.* **40**: 1871–1879.

Kloetzel, M.C. (1947). Reactions of nitroparaffins. I. Synthesis and reduction of some γ-nitroketones. *J. Am. Chem. Soc.* **69**: 2271–2275.

Kon, G.A.R., and Linstead, R.P. (1925a). Chemistry of the three-carbon system. III. The α,β–β, γ change in unsaturated acids. *J. Chem. Soc.* **127**: 616–624.

Kon, G.A.R., and Linstead, R.P. (1925b). Chemistry of the three-carbon system. IV. Case of retarded mobility. *J. Chem. Soc.* **127**: 815–821.

Konaka, R., Kawai (Nee Abe), M., Noda, H., Kohno, M., and Niwa, R. (1995). Synthesis and evaluation of DMPO-type spin traps. *Free Radical Res.* **23**: 15–25.

Konorev, E.A., Tarpey, M.M., Joseph, J., Baker, J.E., and Kalyanaraman, B. (1995). Nitronyl nitroxides as probes to study the mechanism of vasodilatory action of nitrovasodilators, nitrone spin traps, and nitroxides: Role of nitric oxide. *Free Radical Biol. Med.* **18**: 169–177.

Kornblum, N., Blackwood, R.K., and Powers, J.W. (1957). A new synthesis of α-nitroesters. *J. Am. Chem. Soc.* **79**: 2507–2509.

Korth, H.-G., Ingold, K.U., Sustmann, R., de Groot, H., and Sies, H. (1992). Tetramethyl-ortho-quinodimethane. First member of a family of custom-tailored cheletropic spin traps for nitric oxide. *Angew. Chem., Int. Ed. Engl.* **31**: 891–893.

Korth, H.-G., Sustmann, R., Lommes, P., Paul, T., de Groot, H., Hughes, L., and Ingold, K.U. (1994). Nitric oxide cheletropic traps (NOCTs) with improved thermal stability and water solubility. *J. Am. Chem. Soc.* **116**: 2767–2777.

Kotake, Y., and Janzen, E.G. (1991). Decay and fate of the hydroxyl radical adduct of α-phenyl-N-*tert*-butylnitrone in aqueous media. *J. Am. Chem. Soc.* **113**: 9503–9506.

Labadie, J.W., Tueting, D., and Stille, J.K. (1983). Synthetic utility of the palladium-catalyzed coupling reaction of acid chlorides with organotins. *J. Org. Chem.* **48**: 4634–4642.

Lai, C.-S., and Komarov, A.M. (1994). Spin trapping of nitric oxide produced *in vivo* in septic-shock in mice. *FEBS Lett.* **345**: 120–124.

Langhals, E., and Langhals, H. (1990). Alkylation of ketones by use of sold KOH in dimethyl sulfoxide. *Tetrahedron Lett.* 859–862.

Lavin, T.N., Heald, S.L., Jeffs, P.W., Shorr, R.G.L., Lefkowitz, R.J., and Caron, M.G. (1981). Photoaffinity labeling of the β-adrenergic receptor. *J. Biol. Chem.* **256**: 11944–11950.

Linderman, R.J., and Graves, D.M. (1989). Oxidation of fluoroalkyl-substituted carbinols by the Dess-Martin reagent. *J. Org. Chem.* **54**: 661–668.

Lui, K.J., Miyake, M., Panz, T., and Swartz, H. (1999). Evaluation of DEPMPO as a spin trapping agent in biological systems. *Free Radical Biol. Med.* **26**: 714–721.

Luccioni-Houzé, B., Nakache, P., Campredon, M., Guglielmetti, R., and Giusti, G. (1996). Synthesis of new photochromic compounds containing a spin-trap or a spin-label. *Res. Chem. Intermed.* **22**: 449–457.

Lunt, E. (1964). Synthesis of tetraalkyl-pyrrolidines from γ-nitroketones, in: *Nitro Compounds* (Urban'ski, T., ed.), pp. 291–315, Pergamon Press, Oxford, England.

Maassen, J.A., and De Boer, Th.J. (1971). Silver carbonate. A convenient reagent for preparing C-nitroso compounds from hydroxylamines. *Recl. Trav. Chim.* **90**: 373–376.

Makino, K., Hagiwara, T., Hagi, A., Nishi, M., and Murakami, A. (1990). Cautionary note for DMPO spin trapping in the presence of iron ion. *Biochem. Biophys. Res. Commun.* **172**: 1073–1080.

March, J. (1992). *Advanced Organic Chemistry. Reactions, Mechanisms, and Structure*, 4th edition, pp. 795–797, John Wiley & Sons, New York.

Moed, H.D., van Dijk, J., and Niewind, H. (1955). Synthesis of phenethylamine derivatives. III. Bronchodilators. *Recl. Trav. Chim.* **74**: 919–936.

Mordvintcev, P., Mülsch, A., Busse, R., and Vanin, A. (1991). On-line detection of nitric oxide formation in liquid aqueous phase by electron paramagnetic resonance spectroscopy. *Anal. Biochem.* **199**: 142–146.

Murahashi, S.-I., Mitsui, H., Shiota, T., Tsuda, T., and Watanabe, S. (1990). Tungstate-catalyzed oxidation of secondary amines to nitrones. α-Substitution of secondary amines via nitrones. *J. Org. Chem.* **55**: 1736–1744.

Novelli, G.P. (1992). Oxygen radicals in experimental shock: Effects of spin trapping nitrones in ameliorating shock pathophysiology. *Crit. Care Med.* **20**: 499–507.

Okazaki, R., Hosogai, T., Iwadare, E., Hashimoto, M., and Inamoto, N. (1969). Preparation of sterically hindered nitrosobenzenes. *Bull. Chem. Soc. Japan* **42**: 3611–3612.

Osiecki, J.H., and Ullman, E.F. (1968). Studies of free radicals. I. α-Nitronyl nitroxides, a new class of stable radicals. *J. Am. Chem. Soc.* **90**: 1078–1079.

Pogrebniak, H.W., Merino, M.J., Hahn, S.M., Mitchell, J.B., and Pass, H.I. (1992). Spin trap salvage from endotoxemia: The role of cytokine down-regulation. *Surgery* **112**: 130–139.

Pou, S., Bhan, A., Bhadti, V.S., Wu, S.Y., Hosmane, R.S., and Rosen, G.M. (1995). The use of fluorophore-containing spin traps as potential probes to localize free radicals in cells with fluorescence imaging methods. *FASEB J.* **9**: 1085–1090.

Pou, S., Huang, Y.-I., Bhan, A., Bhadti, V.S., Hosmane, R.S., Wu, S.Y., Cao, G.-L., and Rosen, G.M. (1993). A fluorophore-containing nitroxide as a probe to detect superoxide and hydroxyl radical generated by stimulated neutrophils. *Anal. Biochem.* **212**: 85–90.

Pou, S., Keaton, L., Surichamorn, W., Frigillana, P., and Rosen, G.M. (1994b). Can nitric oxide be spin trapped by nitrone and nitroso compounds? *Biochim. Biophys. Acta* **1201**: 118–124.

Pou, S., Ramos, C.L., Gladwell, T., Renks, E., Centra, M., Young, D., Cohen, M.S., and Rosen, G.M. (1994a). A kinetic approach to the selection of a sensitive spin trapping system for the detection of hydroxyl radical. *Anal. Biochem.* **217**: 76–83.

Pou, S., Rosen, G.M., Wu, Y., and Keana, J.F.W. (1990). Synthesis of deuterium- and ^{15}N-containing pyrroline 1-oxides: A spin trapping study. *J. Org. Chem.* **55**: 4438–4443.

Ratcliffe, R.W., and Christensen, B.G. (1973). Total synthesis of β-lactam antibiotics I. α-Thioformamidodiethylphosphonoacetates. *Tetrahedron Lett.* 4645–4648.

Ritter, J.J., and Kalish, J. (1964). α,α-Dimethyl-β-phenethylamine. *Org. Synth.* **44**: 44–47.

Rosen, G.M., and Freeman, B.A. (1984). Detection of superoxide by endothelial cells. *Proc. Natl. Acad. Sci. USA*, **81**: 7269–7273.

Rosen, G.M., and Rauckman, E.J. (1980). Spin trapping of the primary radical involved in the activation of the carcinogen N-hydroxy-2-acetylaminofluorene by cumene hydroperoxide-hematin. *Mol. Pharmacol.* **17**: 233–238.

Rosen, G.M., and Rauckman, E.J. (1984). Spin trapping of superoxide and hydroxyl radicals. *Methods Enzymol.* **105**: 198–209.

Rosen, G.M., and Turner, M.J., III. (1988). Synthesis of spin traps specific for hydroxyl radical. *J. Med. Chem.* **31**: 428–432.

Rosen, G.M., Barber, M.J., and Rauckman, E.J. (1983). Disruption of erythrocyte membranal organization by superoxide. *J. Biol. Chem.* **258**: 2225–2228.

Rosen, G.M., Halpern, H.J., Brunsting, L.A., Spencer, D.P., Strauss, K.E., Bowman, M.J., and Wechsler, A.S. (1988). Direct measurement of nitroxide pharmacokinetics in isolated hearts situated in a low-frequency electron spin resonance spectrometer: Implications for spin trapping and *in vivo* oxymetry. *Proc. Natl. Acad. Sci. USA* **85**: 7772–7776.

Rosen, G.M., Pou, S., Cohen, M.S., Hassett, D.J., Britigan, B.E., Barber, M.J., Cao, G.-L., Cosby, K., Sturgeon, B.E., and Halpern, H.J. (1999). ^{14}N-Spin trapping of free radicals in the presence of ^{15}N-spin labeled *Neisseria gonorrhoeae*. *J. Chem. Soc., Perkin Trans.* 2: 297–300.

Roubaud, V., Lauricella, R., Tuccio, B., Bouteiller, J.-C., and Tordo, P. (1996). Decay of superoxide spin adducts of new PBN-type phosphorylated nitrones. *Res. Chem. Intermed.* **22**: 405–416.

Sankuratri, N., Janzen, E.G., West, M.S., and Poyer, J.L. (1997). Spin trapping with 5-methyl-5-phenylpyrroline N-oxide. A replacement for 5,5-dimethylpyrroline N-oxide. *J. Org. Chem.* **62**: 1176–1178.

Scaiano, J.C. (1982). Laser flash photolysis studies of the reactions of some 1,4-biradicals. *Acc. Chem. Res.* **15**: 252–258.

Schindler, W., and Pietrzok, H. (1965). Hydrazine compounds. German (East) Patent Number 44,067; *Chem. Abst.* **64**: 17426.

Sealy, R.C., Swartz, H.M., and Olive, P.L. (1978). Electron spin resonance-spin trapping. Detection of superoxide formation during aerobic microsomal reduction of nitro-compounds. *Biochem. Biophys. Res. Commun.* **82**: 680–684.

Shechter, H., Ley, D.E., and Zeldin, L. (1952). Addition reactions of nitroalkanes with acrolein and methyl vinyl ketone. Selective reduction of nitrocarbonyl compounds to nitrocarbinols. *J. Am. Chem. Soc.* **74**: 3664–3668.

Shetty, B.V. (1961). Preparation of 1-phenyl-2-methyl-2-hydrazino-propane hydrochloride and its isopropyl derivative. *J. Org. Chem.* **26**: 3002–3004.

Stork, G., Brizzolara, A., Landesman, H., Szmuskovicz, J., and Terrell, R. (1963). The enamine alkylation and acylation of carbonyl compounds. *J. Am. Chem. Soc.* **85**: 207–222.

Stowell, J.C. (1971). *tert*-Alkylnitroso compounds. Synthesis and dimerization equilibria. *J. Org. Chem.* **36**: 3055–3056.

Terabe, S., and Konaka, R. (1969). Electron spin resonance studies on oxidation with nickel peroxide. Spin trapping of free-radical intermediates. *J. Am. Chem. Soc.* **91**: 5655–5657.

Terabe, S., and Konaka, R. (1971). Spin trapping of short-lived free radicals by use of 2,4,6-tri-*tert*-butylnitrosobenzene. *J. Am. Chem. Soc.* **93**: 4306–4307.

Terabe, S., Kuruma, K., and Konaka, R. (1973). Spin trapping by use of nitroso-compounds. Part VI. Nitrosodurene and other nitrosobenzene derivatives. *J. Chem. Soc., Perkin II* 1252–1258.

Tordeux, M., and Wakselman, C. (1982). Synthese de la trifluoromethylvinyl-cetone. *J. Fluorine Chem.* **20**: 301–306.

Tuccio, B., Zeghdaoui, A., Finet, J.-P., Cerri, V., and Tordo, P. (1996). Use of new β-phosphorylated nitrones for the spin trapping of free radicals. *Res. Chem. Intermed.* **22**: 393–404.

Turner, M.J., III, and Rosen, G.M. (1986). Spin trapping of superoxide and hydroxyl radicals with substituted pyrroline 1-oxides. *J. Med. Chem.* **29**: 2439–2444.

Ullman, E.F., Osiecki, J.H., Boocock, G.B., and Darcy, R. (1972). Studies of stable free radicals. X. Nitronyl nitroxide monoradicals and biradicals as possible molecule spin labels. *J. Am. Chem. Soc.* **94**: 7049–7059.

Vishwakarma, L.C., Stringer, O.D., and Davis, F.A. (1988). trans-2-(Phenylsulfonyl)-3-phenyloxaziridine. *Org. Synth.* **66**: 203–210.

Wajer, Th.A.J.W., Mackor, A., De Boer, Th.J., and Van Voorst, J.D.W. (1967). C-Nitroso compounds. II. On the photochemical and thermal formation of nitoxides from nitroso compounds, as studied by E.S.R. *Tetrahedron* **23**: 4021–4026.

Warner, D.T., and Moe, O.A. (1952). The reaction of α,β-unsaturated aldehydes with nitro compounds. *J. Am. Chem. Soc.* **74**: 1064–1066.

Warshaw, J.A., Gallis, D.E., Acken, B.J., Gonzalez, O.J., and Crist, D.R. (1989). α-Heteroatom-substituted nitrones. Synthesis and reactions of acyclic α-alkoxynitrones. *J. Org. Chem.* **54**: 1736–1743.

Yoshimura, T., Yokoyama, H., Fujii, S., Takayama, F., Oikawa, K., and Kamada, H. (1996). *In vivo* EPR detection and imaging of endogenous nitric oxide in lipopolysaccharide-treated mice. *Nat. Biotechnol.* **14**: 992–994.

Zeghdaoui, A., Tuccio, B., Finet, J.-P., Cerri, V., and Tordo, P., (1995). β-Phosphorylated   α-phenyl-*N-tert*-butylnitrone (PBN) analogues: A new series of spin traps for oxyl radicals. *J. Chem. Soc., Perkin Trans. 2*, 2087–2089.

Zhang, Y.-K. (1996). First synthesis of C-phenyl   *N-tert*-butyl[^{15}N]nitrone (PBN-^{15}N) for the EPR spin trapping methodology. *Z. Naturforsch.* **51b**: 139–143.

Zhang, Y., and Xu, G. (1989). ESR evidence for the stereospecific spin trapping of 5-alkyl-5-methyl-1-pyrroline N-oxides. *Magn. Reson. Chem.* **27**: 846–851.

Zhang, Y.-K., Lu, D.-H., and Xu, G.-Z. (1990). Synthesis and plane selective spin trapping of a novel trap 5,5-dimethyl-3-(2-ethoxycarbonylethyl)-1-pyrroline N-oxide. *Z. Naturforsch.* **45b**: 1075–1083.

Zhang, Y.-K., Lu, D.-H., and Xu, G.-Z. (1991). Synthesis and radical additional stereochemistry of two trimethyl-1-pyrroline 1-oxides as studied by EPR spectroscopy. *J. Chem. Soc., Perkin Trans. 2*, 1855–1860.

7

Electron Paramagnetic Resonance Spectroscopy

Principles and Practical Aspects

We succeeded to show, that for these frequencies there is a maximum of absorption in the region of weak fields H; this maximum being shifted towards the region of strong fields H; when the frequency of the oscillating field increases.

Zavoiskii, 1946

The spectroscopy used to detect spin trapped adducts is electron paramagnetic resonance (EPR) or electron spin resonance (ESR). As with all spectroscopy, these names refer to a pattern of energy absorption or radiation—in this case, principally absorption. For EPR, such energy is from electromagnetic radiation of frequencies between low (~ 1 megahertz, MHz, 10^6 Hz) and high (several terahertz, THz, 10^{12} Hz) frequencies, although the most common is the microwave X-band at approximately 9.5 GHz.

Energy is absorbed through the stimulation of transitions between quantized (*vide infra*) magnetic moment or angular momentum states. Since most transitions are between quantized spin angular momentum states, the term electron spin resonance—ESR—was adopted by some scientists to recognize that the major focus of the technique is on spin angular momentum (Wertz and Bolton, 1986). In contrast, the term electron paramagnetic resonance—EPR—recognizes that orbital angular momentum can contribute to the magnetic moment of a state (Abragam and Bleaney, 1986). Therefore, the two designations apply, under most circumstances, to the same spectroscopy.

Theoretical Background

Intrinsic spin is a unique quantum property of electrons, nuclei, and nuclear and subnuclear particles. Charged particles with spin or orbital angular momentum will have magnetic moments, a vector quantity denoted as such by the bold character μ. We discussed in chapter 5 the fact that μ is quantized. We will describe this further below. As noted, the magnetic moment μ interacts with the static vector magnetic field **B** to produce an energy E through the following scalar product relationship:

$$E = -\mu \cdot \mathbf{B} \qquad [7.1]$$

This component of the electron energy is designated as the Zeeman interaction, after Pieter Zeeman, who, in 1896, discovered the effect of magnetic fields on atomic spectra (Zeeman, 1913). This component of the energy is often written E_z with a subscript capital Z to distinguish it from the z-component.

Static magnetic fields generate an energy-level structure. Magnetic fields that vary rapidly with time stimulate transitions between energy levels or states, giving spectral lines. The available energy states are quantitatively described by a matrix energy function called the spin Hamiltonian. From the Hamiltonian, the energy-level structure can be derived. Usually, simplifications of the terms and approximations are used to express the Hamiltonian so that salient features can be understood and so that solving for the energy-level structure in the Hamiltonian is made easier or, in some cases, made possible. For this discussion, we will consider a sample composed of a single type of molecule with one unpaired electron that has an intrinsic spin. This spin is described by the vector of spin matrices **S** ($= S_x, S_y, S_z$, each of which represents a spatial component of the observable spin angular momentum). The intrinsic spin gives the electron its intrinsic spin magnetic moment, as one might expect from a spinning charged object. This spinning charge will create a current loop which will, in turn, create a magnetic moment. This is described through the relationship

$$\mu = -g\beta\mathbf{S} \qquad [7.2]$$

where g is a dimensionless factor, the electron Zeeman factor. This was originally introduced to correct and increase, by a factor of 2, the relationship between magnetic moment and spin from what would be expected from the relationship between magnetic moment and orbital angular momentum. The apparent discrepancy arises from relativistic effects which reduce the magnetic moment associated with the orbital angular momentum of an electron in an atom, the Thomas term (Messiah, 1966). For a free electron, g is 2.002319 (Cohen and Taylor, 1988). The g value can vary considerably for bound electrons, depending on the atom or molecule in which the unpaired electron principally resides. The Bohr magneton, β, is the quantity which relates the units of angular momenta, here taken to be an integer or half-integer number, to the electron magnetic moment. In the standard system of units MKS (meter-kilogram-second), β is 9.274×10^{-24} amperes-meter^2 or joules/tesla (Cohen and Taylor, 1988). Expressed in terms of fundamental constants, β is equal to

$$\beta = |e|h/4\pi m \qquad [7.3]$$

where e is the quantum of charge (the charge of one electron, 1.602×10^{-19} coulombs), h is Planck's constant (6.63×10^{-34} J-s), and m is the mass of the electron (9.11×10^{-31} kg). The units of angular momentum are those of h, joule-seconds or kg-m^2/s but for simplicity the factor of h is incorporated into β leaving **S** dimensionless (Messiah, 1966).

From equation [7.1], we can write the Zeeman Hamiltonian expression as

$$H_Z = g\beta \mathbf{S} \cdot \mathbf{B} = g\beta S_z B_o \qquad [7.4]$$

where the main magnetic field B_o is directed in the z-direction.

A fundamental aspect of quantum mechanics is the necessity, in certain circumstances, that measurable quantities have a finite or infinite number of discrete values. These measurable quantities are described as quantized. A second fundamental aspect of quantum mechanics is the inability to measure or define all possible aspects of a physical system simultaneously. A generic example of this is the spin of the electron. If the z-component of the electron spin is defined at a particular time, then the x- and y-components are undefined. Furthermore, the defined z-component of spin of the electron, S_z, may take on only two values, $+1/2$ and $-1/2$. As a result, the energy E, described in equation [7.4], can take on one of two values. Figure 7.1 shows this simple set of energy levels.

A measurable aspect of a physical system, as in the case of the quantum mechanical spin component of the electron S_z, is represented as a mathematical object that operates on a system state; for instance, an electron spin state $|m_s\rangle$. The state is mathematically referred to as an eigenstate. The result of this operation is to give back the spin state $|m_s\rangle$ multiplied by its spin (eigen-)value m_s:

$$S_z|m_s\rangle = m_s|m_s\rangle \qquad [7.5]$$

This eigenvalue is the value of the measurement of the spin — the spin value. The Hamiltonian H_z acts on the spin state $|m_s\rangle$, as well through the S_z terms. This is denoted $H_z|m_s\rangle$. If the state $|m_s\rangle$ is an eigenstate of the Hamiltonian, this is equal to the Zeeman energy multiplying the spin state:

$$H_z|m_s\rangle = g\beta B_o S_z|m_s\rangle = E_z|m_s\rangle = g\beta B_o m_s|m_s\rangle \qquad [7.6]$$

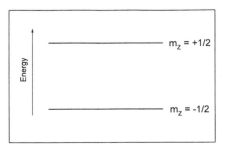

Figure 7.1 Energy-level diagram for spin 1/2 particle in a magnetic field.

In general, the effect of the operation by the Hamiltonian on an eigenstate is the multiplication of that state by its energy. In this case, the energy can take one of two values, $g\beta B_0/2$ or $-g\beta B_0/2$, through the two values available for $m_S + 1/2$ and $-1/2$. For simplicity, we will not refer to the electronic spin states or the Hamiltonian but will write down the energy of the electron as sum products of spin quantum number values with other factors or other spin quantum numbers. Thus, equation [7.6] can be simplified to

$$E_z = g\beta B_o m_s \qquad [7.7]$$

A more general descripton of the interaction of an electron that has a spin **S** with an applied magnetic field **B** involves the g tensor. This is referred to as the g tensor interaction. The g factor is replaced by a tensor indicated by the bold character **g**. The tensor nature of **g** allows components of **S**, not parallel to **B**, to contribute to the Hamiltonian energy matrix function. Equations involving these quantities can, therefore, be quite complex. As we will discuss below, expressions involving these terms may often be simplified to expose the basic principles they illustrate. It is generally represented as

$$H_z = \beta \mathbf{B} \cdot \mathbf{g} \cdot \mathbf{S} \qquad [7.8]$$

The **g** tensor allows the combination of all components of the vectors **B** and **S** to yield at a scalar quantity. This Hamiltonian can be used to derive the energy levels described by the Zeeman Hamiltonian, H_z. The process by which this can be derived is described below for a simplified Hamiltonian.

The unpaired electron spin **S** also interacts with local magnetic fields produced by the intrinsic spin of nuclei in the molecule in which the electron resides. The atomic nuclei of the molecule, situated specifically with respect to the unpaired electron, provide a molecular signature through this interaction with the unpaired electron. The interaction is called the hyperfine interaction. Each compound provides a specific hyperfine interaction that allows the spectroscopic distinction between quite similar molecular species, although there are classes of molecules that are difficult to distinguish with this spectroscopy. The hyperfine interaction is described mathematically in terms of direct interaction between the electron spin **S** and a nuclear spin **I**. In general, there may be a total of N such nuclei. Each nucleus can be described by the subscript (k) with nuclear spins represented by the vector of matrices $\mathbf{I}_k$. The nuclear spin $\mathbf{I}_k$ interacts with the electron spin **S** through the hyperfine interactions as described by the N hyperfine tensors or matrices $\mathbf{A}^{(k)}$. The tensors $\mathbf{A}^{(k)}$ are referred to as the hyperfine tensors. The hyperfine interaction spin Hamiltonian H_{HF} can thus be written as

$$H_{HF} = \sum_{k=1,N} \mathbf{S} \cdot \mathbf{A}^{(k)} \cdot \mathbf{I}_k \qquad [7.9]$$

Because the electron spin **S** and the nuclear spins $\mathbf{I}_k$ are quantized, the effect of the hyperfine interaction is to split the energy levels of the electron spin into multiple levels, clustering about the major electron spin energy levels defined by the Zeeman interaction. For spin trapping, the hyperfine interaction and its effect on the

Hamiltonian are the key elements. It is this interaction that provides EPR spectroscopy with the ability to distinguish between nitroxides derived from the reaction of different free radicals with the same spin trap.

The full spin Hamiltonian of relevance to this discussion is

$$H = \beta \mathbf{B} \cdot \mathbf{g} \cdot \mathbf{S} + \sum_{k=1,N} \mathbf{S} \cdot \mathbf{A}^{(k)} \cdot \mathbf{I}_k \qquad [7.10]$$

ignoring nuclear spin interaction with the magnetic field or nuclear hyperfine, quadrupole interactions, spin orbit interactions, crystal field interactions, interactions with other electron spins, and the basic electron-nuclear electrostatic intractions that produce the atomic and molecular orbital energies (Nordio, 1976; Poole, 1983). The heavy dots indicate inner product multiplication of a vector with a second-rank tensor. The components of the vectors and tensors are defined with respect to directions in the coordinate system in which the experiment is carried out. The z-direction is usually taken to be in the direction of the imposed static magnetic field. When applied to a proper set of electron and nuclear spin states, these elements of the spin matrix function provide a list of energies available to the spin system, otherwise known as energy states. The use of these vector products of matrices is required when studying crystals whose molecules are aligned. The matrix energy function will depend, therefore, on the relative orientation of the magnetic field $\mathbf{B}$, the electron spin $\mathbf{S}$, and the kth nuclear spin $\mathbf{I}_k$, and the pseudo-tensors $\mathbf{A}$ and $\mathbf{g}$, which will depend on the molecular orientation. This Hamiltonian is more complicated than is necessary for the discussion of high-field spin trapping EPR spectroscopy. It is necessary to understand, however, the origins of low-field effects, particularly the irregular positions of an electron line split by the effects of nuclear spins.

For low-viscosity solution measurements (viscosity $\sim$ 1 cP), the molecule carrying the unpaired spin is tumbling rapidly with respect to the decay of an excited spin. The rotation of the molecule averages the different orientations of the $\mathbf{g}$ tensor, fixed to the molecular orientation, with respect to the static magnetic field. It thereby averages the various energies produced by the different orientations, giving a single number. The tensor characteristics of $\mathbf{g}$, in the determination of the energy of the spin states of the electron, are lost through this averaging. Thus, the Zeeman component of the Hamiltonian can be written, as initially presented:

$$H_z = g\beta \mathbf{B} \cdot \mathbf{S} \qquad [7.11a]$$

$$E_z = g\beta B_o S_z \qquad [711b]$$

The energy E_z represents the high-field approximation in which the imposed magnetic field (in the z-direction) is to such a degree the largest magnetic field that no other component need be considered.

As we will see in a majority of cases relevant to spin trapping, EPR spectral lines derive from energy absorbed or emitted when an electron spin is stimulated to make a transition between these two energy levels. At the frequencies commonly used for spin trapping, the Zeeman interaction is usually the largest of the interaction terms in EPR spectroscopy and sets the scale for the magnetic field at which experiments are performed for a given electromagnetic radiation frequency.

The rapid tumbling of a molecule in a low-viscosity solvent will average the relative orientations of nuclear and electron spins and hyperfine tensor in much the same way in which it averages the relative orientations of the electron spin, main imposed magnetic field, and **g** tensor. This will modify the hyperfine component of the Hamiltonian similar to the modification of the Zeeman term. One other condition needs to apply for this simplifcation. The magnetic field B_o must be large enough so that the Zeeman term is much larger than the hyperfine term. This, again, is the high-field approximation. In the following, the terms $A^{(k)}$ are simple scalar numbers often referred to as the hyperfine coupling constants or hyperfine couplings, I_{kz} are the z-components of the nuclear spin for each nucleus k, m_s is the electron spin magnetic quantum number, and m_{kI} is the nuclear spin magnetic quantum number for the nucleus k. The operation of nuclear spin operators I_{kz} on nuclear spin (eigen-)states $|m_{kI}\rangle$ produces the same result as with the electron spins:

$$I_{kz}|m_{kI}\rangle = m_{kI}|m_{kI}\rangle \qquad [7.12]$$

The hyperfine component of the Hamiltonian can then be written as

$$H_{\mathrm{HF}}|m_s\rangle|m_{kI}\rangle = \beta\beta_{\mathrm{N}} \sum_{k=1,N} A^{(k)} S_z I_{kz}|m_s\rangle|m_{kI}\rangle =$$
$$\beta\beta_{\mathrm{N}} \sum_{k=1,N} A^{(k)} M_s M_{kI}|m_s\rangle|m_{kI}\rangle \qquad [7.13a]$$

$$E_{\mathrm{HF}} = \beta\beta_{\mathrm{N}} \sum_{k=1,N} A^{(k)} m_s m_{kI} \qquad [7.13b]$$

Here, β_{N} is the nuclear magneton. The hyperfine component of the Hamiltonian is reduced to a list of scalar couplings that multiply the number that specifies the magnetic substate value of the electron and the magnetic substate value of nucleus k.

The final simplified spin energy is the sum of the Zeeman and hyperfine terms:

$$E_s = E_z + E_{\mathrm{HF}} = \beta g B_o m_s + \beta\beta_N \sum_{k=1,N} A^{(k)} m_s m_{kI} \qquad [7.14]$$

This defines the energies at which resonant transitions will occur. It determines the positions of the spectral lines. The distances between spectral lines will be defined by the "splitting constants," $A^{(k)}$. These constants, in turn, are species dependent; they help to identify the free radical which has been spin trapped.

Selection Rules and Transitions for an Unpaired Electron in a Molecule with a Nitrogen Atom and No Other Nonzero Nuclear Spin

The Hamiltonian matrix energy function is dependent on the magnetic field as described in equation [7.10]. Since the energy states are determined (from the first term in equation [7.10]—the Zeeman term) by the imposed magnetic field B, the energy of the states defined by the Hamiltonian matrix energy function is written as $E(B)$. Transitions between an initial energy state $E_i(B)$ and a final energy state $E_f(B)$ result in the absorption or emission of a quantum of energy from or to the electromagnetic wave of frequency v to which the spin sample is being subjected. The energy of the quantum of radiation hv matches the energy change of the electron spin energy levels, the resonance condition. This occurs at an imposed magnetic field B such that

$$\Delta E = hv = E_f(B) - E_i(B) \qquad [7.15]$$

where ΔE is the difference between the energies E of the unpaired electron in the initial (i) and final (f) states of an allowed transition as a function of B, and h is Planck's constant. The majority of events involve single photon absorption (or emission) by the electron. The angular momentum of the photon is either plus or minus one angular momentum unit, corresponding to circular polarization parallel to or antiparallel to its direction of motion. The spin of the electron will change either from $S_z = m_s = +1/2$ to $S_z = m_s = -1/2$ or from $S_z = m_s = -1/2$ to $S_z = m_s = +1/2$. This is necessary to conserve the angular momentum. The angular momentum absorbed (emitted) by the electron in the form of a photon will just balance the change in spin angular moment in the electron itself. This is a selection rule, so-called because it selects the allowed transitions. Thus, the absolute value of the change in the z quantum number of the electron spin z-component must be 1:

$$|\Delta S_z| = |\Delta m_s| = 1 \qquad [7.16]$$

In the approximation that a single photon is absorbed by the electron and that no other absorption or emission takes place, all other spins—the nuclear spins—must not change energy; that is they remain in the same orientation. Were this not true for first-order single photon emission or absorption, conservation of angular momentum would be violated. Thus, there is no change in the z-component of the nuclear angular momenta, I_{zk}:

$$\Delta m_{Ik} = 0 \qquad [7.17]$$

This is a second selection rule that is derived from the same conservation principle. It should be noted that this situation characterizes most interactions, although there are a number of instances where processes taking place in this approximation are not the only significant or even dominant ones.

Energy Level Diagrams and Spectra

For the energy-level diagram that corresponds to the nitroxide shown in Figure 7.2, having a single nucleus with nonzero spin, the nitrogen nucleus (denoted N) will have $m_N = -1$, 0, or 1. An example of such a nitroxide is peroxylamine disulfonate, Fremy's salt. The energy E from the spin Hamiltonian will contain a Zeeman term and a single nitrogen hyperfine term with nitrogen coupling A_N:

$$E = \beta g B_o m_s + \beta \beta_N A_N m_s m_N \qquad [7.18]$$

The energy of the state with the electron spin z-component equal to $+1/2$ will be

$$E_{+1/2} = \beta g B(1/2) + \beta \beta_N A_N (1/2) m_N \qquad [7.19]$$

This corresponds to the upper three energy levels in Figure 7.2. The energy of an electron with spin z-component equal to $-1/2$ will be

$$E_{-1/2} = \beta g B(-1/2) + \beta \beta_N A_N (-1/2) m_N \qquad [7.20]$$

This corresponds to the lower three energy levels in Figure 7.2. Note the inversion of the order of increasing energy of the series $I_Z = m_N = -1$, 0, 1 in the lower triplet of levels compared with the upper triplet of levels. This is generated by the multiplication of m_N in equation [7.20] by the electron angular momentum z-component equal to $-1/2$ rather than $+1/2$, as it is in equation [7.19]. We have, therefore, assumed that A_N is positive. Because the m_N must remain the same for a single photon transition ($|\Delta m_N| = 0$), only three transitions are allowed. These are shown by the arrows in Figure 7.2.

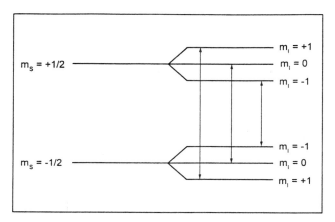

Figure 7.2 Energy levels for an unpaired electron in a magnetic field interacting with a nitrogen nucleus, $m_I = -1, 0, +1$

From equations [7.15], [7.19], and [7.20], the energy-level difference between $E_{-1/2}$ and $E_{+1/2}$, as ΔE, can be written for the transition from the lower lying $E_{-1/2}$ state to the higher lying $E_{+1/2}$ state and expressed as an energy absorption:

$$\Delta E = h\nu = \beta g B - \beta \beta_N A_N m_N \qquad [7.21]$$

When this condition is fulfilled, the frequency of transitions with emission and absorption of photons greatly increases. The lower lying state is more heavily populated. Thus, there will be more transitions involving absorption of a photon rather than emission of a photon. As noted in chapter 5 and above, this induces absorption of energy by the spin system. This is seen as an absorption spectral line. For a nitrogen nucleus, because it has total nuclear angular momentum (I) of 1, there will be a total of $2I + 1$ quantized z-components of the nuclear angular momentum, or three values of $m_N = -1$, 0, 1. Thus, as the magnetic field of the spectrometer is increased at constant frequency, the condition described by equation [7.15] is fulfilled three times. This is shown in Figure 7.3. The single absorp-

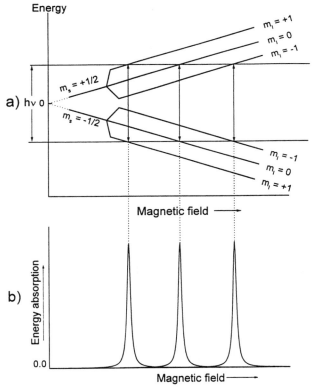

Figure 7.3 (a) Energy-level diagram as a function of magnetic field for an electron in an increasing magnetic field B whose energy levels are split by a nitrogen nucleus. Also shown is the energy-stimulating electromagnetic field. (b) Absorption of energy from stimulating electromagnetic field versus magnetic field B.

tion line that would occur with the Zeeman interaction alone is replaced by a triplet of lines. The difference in the magnetic field between the spectral lines is known as the hyperfine splitting. In the high-field approximation described in equation [7.14] and the equations leading to it, the splitting between lines corresponding to the z-component of the nuclear angular momentum −1 and 0, as well as 0 and 1, is equal. This is not necessarily the case at lower fields or with other Hamiltonians whose non-diagonal elements cannot be ignored. At these low fields, the hyperfine coupling A_N does not correspond precisely to the observed splittings and a distinction between splittings and coupling must be made. The magnetic field splittings shown in Figure 7.3 are characteristic of a nitrogen nucleus. The precise value of the nitrogen splittings will, however, differ significantly among various molecular species, depending on the specific electron spin density at the nitrogen nucleus.

Conventional EPR spectrometers operate at a frequency of approximately 9.5 $\times 10^9$ hertz or 9.5 gigahertz (GHz). This region of the microwave spectrum is called the X-band. Other commonly used spectral bands include K-band (25 GHz), Q-band (35 GHz), S-band (2–4 GHz), L-band (1–2 GHz) and, more recently, W-band (~100 GHz). The magnetic field for a free electron with $g = 2.0023$ will be approximately 3400 gauss (G) at X-band. Common values for the nitrogen splitting of nitroxides are approximately 15 G.

EPR Spectrum of the α-Hydroxyethyl Spin Trapped Adduct of α-(4-Pyridyl 1-Oxide)-N-tert-Butylnitrone [4-POBN]

The EPR spectrum of the α-hydroxyethyl spin trapped adduct of α-(4-pyridyl 1-oxide)-N-tert-butylnitrone [4-POBN–CH(CH$_3$)OH] (Figure 7.4) illustrates the effect of various hyperfine interactions required to produce a molecular signature. This nitroxide has an unpaired electron whose major localization is at the N–O bond, giving substantial coupling to the nitrogen nucleus. The Hamiltonian will resemble that described in the previous section in that it will have a Zeeman term and a hyperfine term due to the nitrogen nucleus. Neither the oxygen nor the [^{12}C]carbon nuclei, which have $I = 0$, will contribute to the hyperfine splitting. Now, as we march about the molecule, proceeding away from the primary locus of the unpaired electron, distance, particularly through single bonds, reduces the overlap probability of the unpaired electron with a nucleus. The next atom that will contribute a hyperfine splitting is the hydrogen on the β-carbon adjacent to the nitrogen. The unpaired electron will have considerable spin density at the hydro-

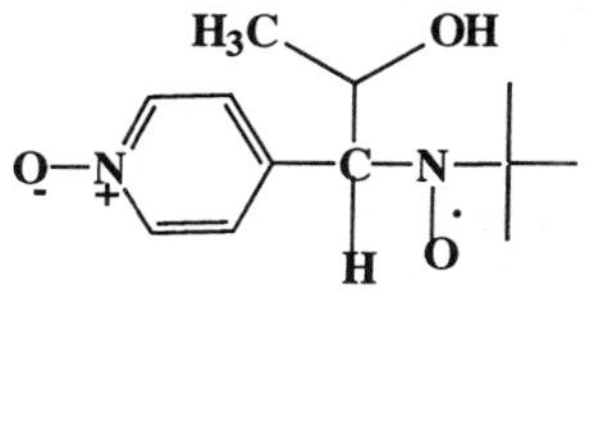

Figure 7.4 Structure of the α-hydroxyethyl spin trapped adduct of α-(4-pyridyl 1-oxide)-N-tert-butylnitrone [4-POBN–CH(CH$_3$)OH].

gen of the β-carbon. An electronegative group attached to the β-carbon will attract spin density to itself. This will decrease the coupling of this hydrogen, A_H, to the unpaired electron, decreasing the spectral splitting. Moving to the *tert*-butyl group, there are 9-quivalent hydrogen atoms. At such distances from the site of the unpaired electron, these hydrogen atoms exhibit modest couplings that, under normal spectroscopic conditions, are not resolvable. They contribute, nevertheless, to the inhomogeneous linewidth described below. The four pyridyl hydrogen atoms have similar weak couplings to the unpaired electron.

Thus, we can write a spin energy E_s equation for 4-POBN-CH(CH$_3$)OH:

$$E_s = \beta g B_o m_s + \beta \beta_N A_N m_s m_N + \beta \beta_N A_H m_s m_H \qquad [7.22]$$

Here, we have relabeled the magnetic quantum numbers of the nitrogen nucleus as m_N and the magnetic quantum number of the hydrogen of the β-carbon as m_H. The extra term, $\beta \beta_N A_H m_s m_H$ will introduce a further splitting of the nitrogen lines. This is shown in Figure 7.5, where the energy levels of the Hamiltonian are plotted as a function of small variations of the magnetic field. As with the simple Fremy's salt diagram, (Figure 7.3), the transitions must conserve the nuclear magnetic quantum numbers. Similarly, the lower lying energy-state splittings are inverted with respect to the upper level set of energies. The allowed transitions are shown by arrows. A fixed frequency transition is shown as two horizontal lines between which the energy transition may take place. The magnetic field at which the fixed frequency line intersects each energy level is the field at which the resonance condition is satisfied. The EPR spectrum of 4-POBN-CH(CH$_3$)OH is shown in Figure 7.5. The spectral lines correspond to those shown in the energy diagram.

Lower Field EPR

At lower fields and frequencies, the Zeeman term becomes comparable to the hyperfine term. The scalar approximation to the **g** tensor in the determination of the electron spin energy states remains a good approximation. At low field, however, the approximation breaks down for the hyperfine interaction. In the Hamiltonian, the interaction $\mathbf{A}^{(k)}$ of the electron spin with the nuclear spin cannot be characterized as a simple scalar quantity. A description of the tensor must be included. At low magnetic fields, matrix equations need to be solved to determine energy levels; the levels will no longer be equally spaced. Since spectral lines are generated by transitions between these levels, the positions of these lines at low fields will have irregular spacing, as recognized initially by Breit and Rabi (1931). This irregularity can be seen at 250 MHz in an approximately 90 G magnetic field.

Spin Density and Molecular Orbital Modeling

The magnitude of the splitting of the electron spin energy levels by the nonzero spin nuclei reflected in the constants $\mathbf{A}_{(k)}$, depends, to the first order, on the magnetic moment of the nucleus and the spin density of the electron at the nucleus (McConnell, 1956; McConnell and Chesnut, 1958; Pople, *et al.*, 1968; Weil, *et al.*, 1994). The spin density is the spin of the electron multiplied by the probability that

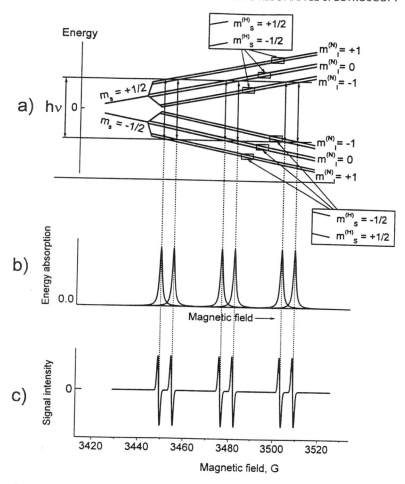

Figure 7.5 (a) Energy diagram of 4-POBN–CH(CH$_3$)OH versus magnetic field. (b) 4-POBN–CH(CH$_3$)OH absorption spectral lines versus magnetic field. (c) 4-POBN–CH(CH$_3$)OH absorption spectra with field modulation at X-band frequency versus magnetic field.

the electron is located at a position in space. This probability is the square of the electron wave function which, in turn, is the spatial distribution (the squared modulus) of the molecular orbital occupied by the unpaired electron. The nuclei that induce the splitting select points in space at which to sample the squared modulus of the orbital wave function. This splitting then reports the electron spin density at nuclear loci, which are thereby the identifiers of the molecule bearing the unpaired spin.

Spin trapping offers the possibility of detecting and identifying short-lived free radicals based on EPR spectral splitting. This has important implications for the identification of free radicals in the biologic milieu (Janzen and Blackburn, 1969;

Janzen and Liu, 1973; Janzen, *et al.*, 1973). The resultant hyperfine splitting constants of a spin trapped adduct have been found to be dependent on both the nature of the free radical and the spin trap (Janzen and Blackburn, 1969; Janzen and Liu, 1973; Janzen, *et al.*, 1973). Introduction of a hetero atom, for example, at the β-carbon, with increasing electronegativity, will withdraw spin density from both the nitrogen and the β-hydrogen. This results in an observed reduction in both A_N and $A_H{}^\beta$. In contrast, the γ-hydrogens, $A_H{}^\gamma$ exhibit enhanced splitting. These findings have been attributed to increased ring puckering centered at the β-carbon in association with the electronegativity of the atom. Evidence in support of this comes, in part, from the observation that the EPR spectrum of DMPO–OOH (Figure 7.6) demonstrates substantial nitrogen and β-hydrogen splitting reduction (relative to an alkane), as well as an increased γ-splitting with less β-hydrogen splitting than has been found with alkoxyl spin trapped adducts (Harbour, *et al.*, 1974; Finkelstein, *et al.*, 1979). As has been noted in chapter 5, the splitting constants for different spin trapping adducts may be similar. This will

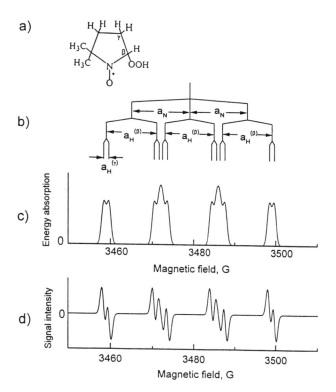

Figure 7.6 (a) Structure of 2-hydroperoxy-5,5-dimethyl-1-pyrrolidinoxyl (DMPO–OOH). (b) Stick diagram of the splitting pattern for DMPO–OOH. (c) DMPO–OOH absorption spectral lines versus magnetic field. (d) DMPO–OOH absorption spectra with field modulation at X-band frequency *versus* magnetic field.

make identification of the initial transient free radical, based on spectral distinction of the spin trapped adduct, difficult and, in some cases, nearly impossible.

Lineshapes—Simple Absorption Profiles

The position of the absorption line is determined by the energy values from the Hamiltonian as described above. For an idealized situation of an isolated unpaired electron in a magnetic field, the Hamiltonian will be simply the Zeeman term described in the above discussion. We may write for this transition,

$$\Delta E_o = h\nu_o = \beta g B_o \quad \text{or} \quad \nu_o = (\beta g/h)B_o = \gamma B_o \qquad [7.23]$$

where the constant $\gamma = \beta g/h$ is equal to $2.802 \times 10^6 \text{Hz/G}$ or 28.02×10^9 Hz/T.

Absorption will not take place at a single frequency or magnetic field due at least to the uncertainty principle. The absorption will have some functional dependence on frequency or magnetic field. The most common of these functional dependences are simple solutions of the Bloch equations (Bloch, 1946; Abragam, 1961a). These empirical equations govern the behavior of a macroscopic magnetization precessing in a fixed magnetic field, driven by an oscillating magnetic field. The basic form of a steady-state solution to these equations corresponds to a mathematical construct in complex number space: the imaginary part of a simple pole. We consider a function f of a complex number $z = x + iy$, where x and y are real numbers and $i = \sqrt{(-1)}$ is the unit imaginary. The simplest form of f that approaches infinity at a particular complex value z_o with nonzero imaginary part is called a simple pole and is written

$$f(z) = 1/(z - z_o) = [(x - x_o) - i(y - y_o)]/[(x - x_o)^2 + (y - y_o)^2] \qquad [7.24]$$

Because the value z_o is displaced from the real axis, z_o cannot be accessed for real values of z. We assume that physical values of z, those values that are accessible, are real with $y = 0$. The real part of z is denoted $x = \text{Re}(z)$. Mathematically, the absorption line is described as the imaginary part of a simple pole, $\text{Im}(f(z))$ with z real $(y = 0)$. This is referred to as the Lorentzian shape. As the real, accessible part of z approaches closest to z_o, as x approaches x_o, the value of the imaginary part of $f(z)$ will increase. In its full generality, the Lorentzian functional form $f(x)$ is written

$$f(x) = a/(1 + b^2(x - x_o)^2) \qquad [7.25]$$

where a ($= 1/y_o$ in equation [7.24]) and b ($1/y_o^2$ in equation [7.24]) are arbitrary constants (Figure 7.7).

Although this basic form gives a general sense of the absorption shape as a function of either frequency or magnetic field, the constants in this expression may be related to physically relevant parameters, characterizing the absorption and the conditions of the experiment. The sample is irradiated with electromagnetic radiation of frequency ν_o, with a resonance condition imposed by the static magnetic field B_o. The amplitude of the time-varying magnetic field is B_1. Low values of B_1

transitions between energy levels do not substantially alter the energy level populations, an experimental condition referred to as nonsaturating. The power absorbed by the sample subjected to an external magnetic field B, $P(B)$, can be written as a modification of an earlier expression (Kittel, 1967):

$$P(B) = [\chi_o/\mu_o](2\pi\gamma)^2 B_o BB_1^{\,2} T_2/[1 + \{2\pi T_2(B_o - B)\gamma\}^2] \qquad [7.26]$$

The basic form of equation [7.26] is that of equation [7.27] with the resonance line centered at the magnetic field B_o. Here, χ_o is the static susceptibility of the sample and μ_o is the permeability of free space; T_2, the transverse relaxation time, determines the linewidth since γ is the electron gyromagnetic ratio (defined in cycles per gauss), the constant defined above. T_2 is inversely related to the homogenous linewidth (absorption full-width half-maximum) $\Delta B_{1/2}$ by the relationship

$$\Delta B_{1/2} = 1/(\pi T_2\gamma) \qquad [7.27]$$

When B moves from the value B_o to $B_o + 1/2\Delta B_{1/2}$ or to $B_o - 1/2\Delta B_{1/2}$, a total range of $\Delta B_{1/2}$, the denominator in equation [7.26] increases from the value 1 to the value 2, reducing the value of $P(B)$ to approximately $1/2$ of its value at B_o.

The water content of living organisms greatly affects the T_2 of a spin trapped adduct. This isolated linewidth can be referred to as the intrinsic linewidth or the spin packet linewidth. Thus, care must be exercised when comparing signal heights of certain nitroxides or spin trapped adducts with minimal inhomogeneous broadening, as reduction in the linewidth for the same number of spins will increase the signal height. Other paramagnetic species, usually dissolved oxygen in buffers, collide with the spin trapped adduct. Through Heisenberg spin exchange, the intrinsic linewidth increases (Eastman, et al., 1969). Another factor that will affect the intrinsic linewidth is the viscosity of the solvent. Increased viscosity will increase the intrinsic linewidth (McConnell, 1958). These effects make data interpretation more complicated. When these broadening effects are substantial, the height or even relative height of a spectral line will no longer be directly related to the concentration of the spin trapped adduct. Measurement of the spectral linewidths may be necessary. Even more subtle differences may require spectral fitting to extract the doubly integrated spectral intensity (Halpern, et al., 1993). These considerations become even more important when attempting to quantitate the distribution of the spin trapped adduct between intra- and extracellular compartments of isolated organs and in vivo (Rosen, et al., 1988; Halpern, et al., 1995). A further factor that modifies the linewidth but which is conceptually different from the intrinsic linewidth, inhomogeneous broadening, is discussed in the next section.

Unresolved Splitting and Inhomogeneous Broadening

As discussed above, nuclear spins interact with an unpaired electron via the hyperfine interaction. They add or subtract a discrete amount from the imposed magnetic field to shift the position (the B_o in equation [7.26]) of the resonant absorption

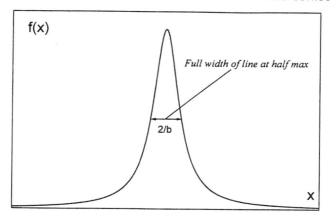

Figure 7.7 Lorentzian profile. The term b refers to the parameter in equation [7.25].

line. When the magnitude of the splitting is significantly less than the linewidth of a single isolated line, the effect is seen as a broadening of the spectral line. The principal determinant of the splitting induced by a nucleus is the electron spin density at the position of the nucleus. Thus, nuclei that are distant in the molecule from the major locus of the unpaired electron spin will have small splittings (McConnell, 1956; McConnell and Chesnut, 1958). The width of the isolated line will also determine what is resolved and what is not resolved. The major contribution to these small unresolved splittings comes from hydrogen atoms (Windle, 1981; Bales, 1989; Bales, et al., 1992). Hydrogen atoms that are positioned with respect to the unpaired electron with sufficient symmetry cause identical splitting in the unpaired electron spectrum. Symmetrically disposed protons are referred to as equivalent. The splitting induced by n equivalent protons will add

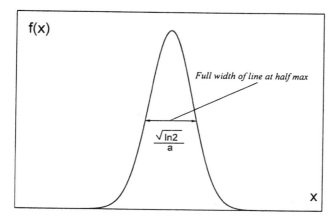

Figure 7.8 Gaussian profile. The term a refers to the parameter in equation [7.28].

and subtract to give $n + 1$ different unresolved lines. The number of possible combinations of spins combining to give any one of the unresolved lines gives the relative intensity of that line. This distribution of intensities is a binomial distribution. When unresolved, this splitting appears as a smearing of an otherwise sharp line. A line dominated by unresolved splitting will have a shape approximated by the following function, known as a Gaussian shape (Bales, 1989):

$$f(x) = exp(-a^2 x^2) \qquad [7.28]$$

We can equate x with $B - B_0$. The parameter a is inversely related to the linewidth and is proportional first to the individual hydrogen nucleus splitting and second to the square root of the number of equivalent protons.

When a line with a Lorentzian shape is smeared by unresolved splitting, this can be modeled mathematically as a convolution of the Lorentzian lineshape of equation [7.25] with a Gaussian distribution (Voigt, 1912). The equivalent linewidth due to unresolved splitting can be larger than the intrinsic Lorentzian linewidth. Under these circumstances, the Gaussian line will dominate both the linewidth and the lineshape. This form of line broadening, in contrast with the Lorentzian lineshape, is known as inhomogeneous broadening (Figure 7.8). The principal difference between the Gaussian and Lorentzian shape can be seen in Figures 7.7 and 7.8 where the Lorentzian has a sharper peak but diminishes toward baseline more slowly than the Gaussian.

Instrumentation

The First EPR Spectrometer

As indicated in chapter 5, EPR spectroscopy was discovered in 1944–1945, based on the historic experiments by Zavoiskii in Kazan, Tartarstan, former USSR (Zavoiskii, 1945a, 1945b). Zavoiskii's initial experiments involved the placement of crystalline and aqueous solution samples of cupric chloride inside the loops of an inductor of an LC (inductor–capacitor) resonant circuit tuned to the frequency of 133-MHz. The leads of the LC circuit were then attached to the grids of a vacuum tube. A magnetic field was applied to the sample and gradually swept. At resonance, the sample absorbed energy from the 133-MHz electromagnetic field. The absorption of energy made the LC circuit appear to have an increased resistance. This, in turn, changed the current output of the vacuum tube. Both the LC circuit and the vacuum tube amplified the effective change in resistance to give a large signal. The universal grid current method of Zavoiskii (1945a) was further modified as described by El'sting (1960). With the microwave technology[1] spurred by the development of radar during World War II and the higher signal-to-noise available at higher frequencies, a shift to microwave frequencies soon followed. This breakthrough technology has led to the modern EPR spectrometer (Poole, 1983).

Basic Continuous-Wave EPR Spectrometer

The experimental paradigm for the basic continuous-wave EPR spectrometer is no different than the instrumentation developed by Zavoiskii over 50 years ago. A sample is placed in a resonant structure and subjected to fixed frequency electromagnetic radiation. A second magnetic field is imposed on the sample perpendicular to the magnetic component of the fixed frequency electromagnetic radiation. This second high magnetic field is slowly swept through the resonance conditions of the spins in the sample. A voltage change in the resonator output is detected as resonance is approached. The plot of voltage versus the slowly swept main magnetic field gives the lineshapes discussed above. The continuous-wave EPR spectrometer consists of a source of microwave, far infrared or radiofrequency electromagnetic energy, a resonant structure containing the sample and, between the sample resonator and the electromagnetic energy source, a device to extract the resonance signal, often referred to as the bridge. The bridge can be subdivided into a sensing element and an element acting like a Wheatsone bridge as discussed below. This is schematically shown in Figure 7.9.

Sources of Electromagnetic Energy

Radiofrequency (RF) Sources For radiofrequency (RF) applications, the electromagnetic energy is usually provided by standard RF signal generators, either resonator tuned or synthesized from a crystal oscillator. The basic design is usually a variant on the LC tuned Hartley oscillator (Horowitz and Hill, 1989).

Klystrons In the late 1940s and early 1950s, EPR spectrometers rapidly moved into microwave frequencies. This was the result of two events. The first was the availability of one of several microwave devices, called the klystron, that could be designed to operate reliably at approximately 10-GHz frequencies, easily propagated by a hollow air-filled waveguide. The second was the relative ease of fabrication and convenience of a waveguide with inner dimensions of approximately 2.2 cm by 1 cm. Thereafter, X-band frequencies, between 8 and 12 GHz, propagated by a waveguide of these dimensions, became the standard frequencies of operation.

The reflex klystron, shown diagramatically in Figure 7.10, has been the standard power source for X-band spectrometers until recently. Electrons, boiled off from a heated metal cathode, are accelerated toward the two positively charged anode grids. A small alternating electric field between the two anodes bunches the electrons. The bunched electron beam passes the anode grid toward the repeller, whose potential is more negative than that of the cathode. The bunched electrons are further focused as they proceed near to and then back from the repeller. The faster, more energetic electrons will move closer to the repeller, then back to the anode. The slower, less energetic electrons will reverse their direction in a shorter path length than the faster more energetic electrons. A side-coupled cavity is present near the anode. The distances and voltages are arranged so that the returning electrons converge and maximally bunch as they again pass through the anodes. This gives a repetition of the bunches at the resonant frequency of the side-coupled

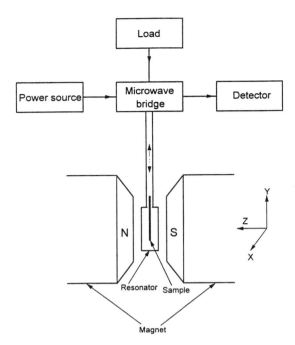

Figure 7.9 Block diagram of an EPR spectrometer.

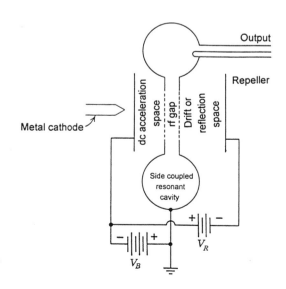

Figure 7.10 Structure of a reflex klystron. (Taken from Montgomery, 1947.)

cavity. The electromagnetic pulse created by the passage of a tight electron bunch excites resonant modes in the resonant cavity. Pulsing at the resonant frequency of the cavity maximizes the transfer of energy from the electrons to the resonator. This, in turn, is coupled to the waveguide, making available electromagnetic energy for transmission to the cavity and the bridge. The klystron is a durable power source with low-phase noise, usually the major limiting source of noise in an EPR spectrometer.

Gunn Diode   Another microwave source is the Gunn diode (Strangeway, *et al.*, 1988; Oles, *et al.*, 1992). These were introduced originally at higher frequencies, but recent designs have brought phase noise down to that of a klystron at X-band. They are now standard on new Bruker EPR X-band spectrometers. Solid-state diodes consist of semiconductor materials adherent to the surface layer into which impurity atoms have been introduced. This produces a tendency to give up or attract electrons at the surface. Materials with two different impurities, placed in contact, will have special conductive properties. In particular, certain materials will easily conduct current in response to a voltage difference in one direction perpendicular to the contact surface, but not in the opposite direction. The response of current to applied voltage is, therefore, highly nonlinear. For certain contact junctions, however, the nonlinear response of the current to the applied voltage has an unusual behavior. As the voltage is increased in the direction of current conduction, the current can, at a certain voltage, decrease. This is, effectively, a region where the conductivity of the diode is negative. As the voltage increases, the current will eventually increase. However, a region is formed where, for a given voltage, there are two currents. Between these currents, the voltage diminishes to a local minimum and then increases. Like a marble in a round-bottomed cup, the voltage will oscillate between these states, converting direct current energy from the impressed voltage and current into alternating or oscillatory electromagnetic energy. The oscillatory current can be coupled to the waveguide of an EPR spectrometer and used as a power source.

Transmission of RF or Microwave Energy

Waveguide   Higher frequency, microwave electromagnetic energy is transmitted over substantial distances (several meters) with minimal loss by means of hollow, conductive tubes called waveguides. These allow the propagation of electromagnetic fields along their axis within a range of frequencies that depend on the transverse dimensions of the tube. For a rectangular cross-section tube of 2.2 cm by 1 cm inner dimension, X-band frequencies propagate. For tubing with this aspect ratio, the upper cut-off frequency will be less than, but near to, that corresponding to the free space electromagnetic wavelength of the long dimension of the waveguide. Thus, a 2.2-cm long dimension waveguide will propagate frequencies at 12 GHz and below.

There are two modes by which radiation can propagate through a hollow waveguide. One mode involves the magnetic field transverse to the direction of propagation in the waveguide, the TM mode.[2] The other involves the electric field transverse to the direction of propagation in the wave, known as the TE mode[2]

(Poole, 1967a). A number will accompany the mode. This refers to the number of half-wavelengths in the long transverse dimension and the short transverse dimension, respectively. Thus, a TE_{10} mode corresponds to an electromagnetic wave propagating with one-half of a full wavelength variation along the long transverse dimension of the waveguide and no variation along the short dimension of the waveguide (Marcuvitz, 1951a; Poole, 1967b).

Circular waveguides can be used with similar transverse electric and transverse magnetic modes. The subscripts (e.g., TE_{mn}), describing cylindrical waveguide, refer to the variation of the transverse electric field along the azimuth of the cylinder (m) and the radial direction of the cylinder (n). Although the azimuthal variation can be characterized in terms of a simple sinusiodal variation, the radial behavior is more complex.

Coaxial and Multiaxial Cables

For lower frequencies, the size of waveguide becomes larger. For laboratory measurements, this can become cumbersome. Furthermore, at lower frequencies, the amplitude of reflections induced by connectors diminishes. This allows the use of flexible cable structures with rigid connections that have small impedance discontinuities. At lower frequencies, one or more center conductors can be added to the conductive tube configuration. This changes the electromagnetic boundary conditions and allows a much smaller transverse diameter to transmit the electromagnetic energy. Thus, RG174, approximately 2.5-mm diameter semirigid coaxial cable, is rated from 26 GHz to DC. There is no cut-off as the wavelength increases or the frequency decreases. Losses, however, are higher than those of waveguides. Coaxial cable propagates electromagnetic energy with both the magnetic and electric fields transverse to the direction of propagation, the TEM mode (Marcuvitz, 1951b; Poole, 1967c).

Resonant Sample Holders

Standard Cavities   Standard resonant cavities are hollow, highly conductive structures with both limited longitudinal (along the direction of microwave propagation) and transverse dimensions. Into such a confined box, microwave energy enters through a small variable opening known as the iris. The energy reflects from the wall of the box opposite the iris and again from the wall containing the iris. If a half-integral multiple of the wavelength of the microwave energy just matches the length of the highly conductive box, a standing wave will develop. This will have a much larger amplitude than the microwave entering the box through the iris. The most important of the box dimensions is its length, since this determines the frequency of the cavity resonance. Resonant cavities are characterized by the particular transverse field: TE, transverse electric and TM, transverse magnetic. There is a third subscript for the specification of a cavity, relating to the number of half-integral wavelengths to which the length of the cavity corresponds.

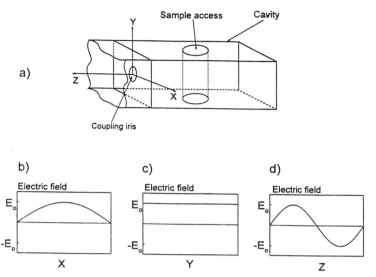

Figure 7.11 (a) Diagram of a TE_{102} EPR cavity variation of transverse electric field in (b) the x-direction, (c) the y-direction, (d) the z-direction.

There are a number of resonant modes that can be excited by the microwave energy that depend on the geometry of the cavity. The most commonly used resonator is the rectangular cavity operating in a TE_{102} mode, as shown in Figure 7.11. This can be a resonant structure fabricated from a TE_{10} waveguide with a length corresponding to two half-wavelengths. Figure 7.11 also shows the variation of the transverse electric field in the three orthogonal directions. In the x-, y-, and z-directions, the TE_{102} has, respectively, 1, 0, and 2 half-wavelengths.

The sharpness of the resonance of the cavity without sample or external connections is quantitated as the unloaded Q. It is determined, in part, by the conductivity of the cavity walls. To achieve high Q, the walls are internally silvered and polished. The amplitude of the microwave electromagnetic field in the resonator will depend on the balance between delivered power and that lost due to dissipation either in the walls of the cavity, the heating of the sample, or radiation from the cavity. One definition of Q is

$$Q = \text{Energy stored in the cavity electromagnetic field/energy}$$
$$\text{dissipated in one cycle}$$
$$= 1/2(\varepsilon_o E_1^2 + [1/\mu_o]B_1^2)dV/(P/v_o) \qquad [7.29]$$

where E_1 and B_1 are, respectively, the microwave electric field and magnetic induction in the cavity at resonance, which are integrated over the cavity volume V; ε_o is the permittivity of free space; μ_o is the permeability of free space; P is the power lost in the cavity; and v_o is the resonant frequency. The unloaded Q, as a measurement of resonance sharpness, is defined as the center frequency of the resonance, ν_o, divided by the difference in frequency between the upper frequency and lower frequency at half power points, $\Delta\nu_{1/2}$:

$$Q = \nu_o/\Delta\nu_{1/2} \qquad [7.30]$$

These definitions can be shown to be equivalent. The loaded Q, Q_L, is the same parameter with the cavity coupled to external connections. The Q for a standard TE_{102} cavity with a nonlossy sample is typically approximately 5000.

Sensitivity Considerations   Many factors contribute to the sensitivity of the measurement and to a large extent are dependent on many aspects of experimental conditions (Feher, 1957; Abragam, 1961b; Poole, 1967d). For instance, the fractional change in the measured parameter, X, which can be denoted as $\Delta X/X$—the fractional change in voltage, $\Delta V/V$, or the fractional change in power, $\Delta P/P$—is related to three factors: the resonator Q, the filling factor η, and the imaginary component of the sample susceptibility, χ'' (Feher, 1957; Abragam, 1961b). As it turns out, the susceptibility is dependent on the number of spins in the sample. Thus, the fractional magnitude of the signal becomes

$$\Delta X/X \propto \chi''\eta Q \qquad [7.31]$$

From equation [7.31], it is seen that the signal from a resonator is proportional to the Q. Most relevant to the experimenter is the loaded Q with a sample present. Many samples absorb microwave energy, particularly those with aqueous and conductive solvents. This can exceed the importance of cavity wall resistance and the radiated energy. Whatever factors contribute to energy loss, they will diminish Q and reduce the signal. It is, therefore, important to minimize the volume of lossy sample at X-band in order to maintain the amplitude ΔX of the EPR signal.

Another parameter to which signal is proportional is the filling factor, η. This term is derived from the ratio of the squared magnetic field of the resonator integrated over the volume of the sample to the squared magnetic field of the resonator integrated over all space. It is a measure of the fraction of the volume capable of delivering electromagnetic energy to the sample, weighted by the electromagnetic energy. Although this definition is nearly universal, it ignores the fact that the linearly polarized magnetic field in the resonator must be regarded as the superposition of two circularly polarized magnetic fields. Only one of these, the in-phase component, will contribute to magnetic resonance. Thus, the often computed value of filling factor is, in fact, half as large. For aqueous samples, a quartz flat cell is often used. The volume of this cell inserted into the resonator is small, typically 0.3 mm thick, 1 cm wide, and 2.5 cm long, affording a low filling factor of less than 0.01.

Other considerations affecting the sensitivity of a spectrometer involve sources of noise. A common source of noise is the variation of the phase/frequency of the signal from the klystron. Changes in phase will appear as a signal (see below).

Standard Cavity Sensitivity Considerations   The parameter commonly used to define sensitivity is the minimum number of spins detectable for a particular resonator. This must be interpreted carefully because it can depend on the solvent of the sample. The solvent can contribute to the absorption of electromagnetic energy which will reduce Q and, by definition, diminish sensitivity. Samples dissolved in water, for example, will have a high electromagnetic loss, making this solvent less desirable than others to conduct measurements. Minimum number of spins may not be germane for experiments where the paramagnetic substrate is copious. At low concentrations of parmagnetic species, a larger sample volume may allow accurate measurements to be acquired. A paramagnetic compound with a narrow spectral line will have a higher amplitude signal—the signal from its spins will be concentrated at more nearly a single field value. Likewise, lower numbers of spectral lines will invariably increase sensitivity, since the spins will be divided among fewer lines. Individual lines will, therefore, have greater amplitude.

Sensitivity is usually expressed as the number of detectable spins per gauss linewidth per line. For ease of convenience, sensitivity is designated as a 1:1 signal-to-noise ratio. The general-purpose TE_{102} cavity, for example, can measure relatively large volumes of dry samples, like coal or the Varian standard weak pitch dispersed on KCl, with a sensitivity to relatively low numbers of spins, on the order of 10^{12} (Poole, 1967d). However, for aqueous samples, this cavity is limited to measurements of small volumes in a flat cell, since, as noted above, water has a high dielectric absorption at X-band frequencies. Thus, as the water sample extends into portions of the resonator with higher electric fields, dielectric losses increase rapidly and the Q diminishes. To minimize these problems, samples are often placed in a quartz flat cell, which is oriented in the $x-y$ plane of Figure 7.11 to occupy the center plane of the TE_{102} cavity with high magnetic field and lowest electric field. This allows for a good Q, ranging in the thousands. Larger sample thicknesses decreases the Q more rapidly than the increase in volume. For standard cavities at X-band, there is a tradeoff between sample size and filling factor, on one hand, and resonator Q, on the other.

The Loop-Gap Resonator   The potential of the loop-gap resonator was realized by Froncisz and Hyde (1982) and thereafter applied to EPR spectroscopy. In its simplest form, the loop-gap is merely an inductive–capacitive LC circuit with the inductive element serving as the sample holder and generator of the magnetic field. In general, the Q of loop-gap resonators is lower than that of microwave cavity structures, typically 10–20% that of other cavities. However, the loop-gap is not restricted to a size that is an integral number of half-wavelengths. Thus, very small resonators can be constructed relative to the microwave or radiofrequency employed. The loss of Q is more than compensated by an increase in the filling factor. Values near the theoretical maximum of 1/2 (for linearly polarized radiation) can be attained. A major capability of loop-gap resonators is their ability to provide high-sensitivity measurements of small volumes of sample, such as micrograms of spin-labeled proteins. It is enabling for radiofrequencies where the free space wavelength is in meters requiring meter-length resonant cavities, which would be unacceptibly large with impossibly small filling factors.

Froncisz and Hyde define the parameter Λ as

$$\Lambda = B_1/\sqrt{P} \qquad\qquad [7.32]$$

where B_1 is the RF or microwave magnetic field at the sample and P is the incident power. In the loop-gap, Λ is much higher than for standard microwave cavities (Froncisz and Hyde, 1982). This gives a much higher magnetic field at the sample with lower incident power. For nonsaturating samples, this gives a potential improvement in sensitivity that can more than compensate for the lower Q. The loop-gap allows the radiation of the sample at two distinct frequencies, as in an electron–electron double resonance (ELDOR) experiment. Since the Q is lower, the frequency interval or bandwidth of the resonator is larger. Pulsed applications can take advantage of the lower Q, with reduced ring down-time (*vide infra*, Hyde and Froncisz, 1986).

Microwave Bridges

Magic T Bridge   The term microwave bridge is applied to the portion of the EPR spectrometer which is attached to both the source of the microwave energy and the resonating sample holder or cavity. Classically, the bridge was connected to a dummy load that acted electrically like the cavity at resonance without a resonant sample. A fourth microwave port led to the signal detector. A common configuration for this four-way connection employs a microwave structure referred to as a magic T. Figure 7.12 shows a schematic of a magic T. Microwave power entering the magic T through the E arm will divide equally between arms 1 and 2 but will not transmit power to the H arm. Microwave power entering the T from the H arm will divide equally between arms 1 and 2 but will not transmit power to the E arm. Structures can be added to ensure impedance matching and eliminate reflected power. When the sample spins are not at resonance, the dummy load just balances the load presented by the cavity/sample. The bridge behaves like a microwave

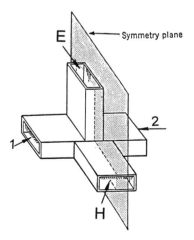

Figure 7.12 A magic T. (Taken from Montgomery, *et al.*, 1947.)

analog of the Wheatstone bridge, another four-port device, from which it derives its name. Microwaves entering the bridge are split between the dummy load and the cavity, but do not enter the detector. At balance between the dummy and cavity loads, the microwave energy reflected from the dummy load and the cavity is transmitted into the detector arm but not into the arm from which the klystron source microwaves emerge. In the ideal circumstance, the signal from the dummy load and the resonator cancel at the output to the signal detector. In practice, however, a small imbalance is necessary to bias the detector diode, bringing it into the region of square law response. Electron spin resonance in the cavity sample increases the apparent resistance of the cavity. This changes the cavity load and creates an increased imbalance between the dummy and cavity loads. This is seen as an increase in the diode current proportional to the resonance signal.

Disadvantages of this kind of bridge include both the loss of one-half of the klystron power into the dummy load and one-half of the signal power. In addition, it is necessary to apply a bias power to the diode detector to bring the diode into the proper operating region (see the discussion of Crystal Detectors below). This is a problem with a magic T. The stability of the imbalance can be noisy and may overwhelm the signal at low signal amplitudes. This will result in a lower signal-to-noise ratio.

Circulator Bridge with Reference Arm   Modern EPR spectrometer designs avoid the loss of microwave power in the dummy load. Instead, a three-port device known as a circulator is employed, as depicted in Figure 7.13. Here, the cavity/sample holder is tuned to appear as a termination with the same impedance as the waveguide. With the impedance of the load matching that of the waveguide, all the energy transmitted by the waveguide will be absorbed by the cavity. There will be no reflected signal; nothing is transmitted to the detector. The absence of a dummy load avoids a loss of half of the microwave power. The EPR absorption of electromagnetic energy by the spins of the sample change the cavity impedance and create a reflection from the cavity. This is seen as a signal transmitted to the detector.

Many quality modern spectrometers use a reference arm. A portion of the microwave power (approximately one-hundredth of the klystron power) is diverted from the circulator. Instead, it passes into the reference arm. Power from the reference arm combines with the output from the circulator. The

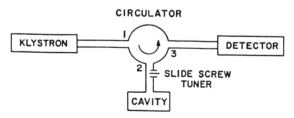

Figure 7.13 A circulator-based bridge in an EPR spectrometer. (Taken from Poole, 1967a.)

reference arm power can, therefore, be used to directly bias the detector crystal without depending on the magic T bridge balance. This results in better stability and avoids the waste of half of the microwave power.

Microwave Detectors

Crystal Detectors The electromagnetic wave from the reference arm is combined with the output of the circulator. The superposed electromagnetic waves proceed to a load where their energy is dissipated. Just before the load is a wire loop that converts the time-varying magnetic fields to currents I_1 and I_2 and feeds this current to the crystal detector or diode. The diode has a nonlinear current response to the sum of both of the inputs, I_1 and I_2. The nonlinear response of this output voltage, V_{out}, to the currents input to the crystal can be expressed as a mathematical power series in the sum of the input. This, in turn, will contain terms that are various products of the two inputs:

$$V_{out} = F(I_1 + I_2) = a_0 + a_1(I_1 + I_2) + a_2(I_1 + I_2)^2 + a_3(I_1 + I_2)^2 + \ldots \quad [7.33]$$

For typical crystal bias, the lowest order and highest amplitude product will be the third term. This is referred to as square law detection. There will be terms proportional to I_1^2 and I_2^2. The I_2^2 will usually be small compared with the other two terms and can thus be ignored. There will also be a term proportional to the simple product of the two input currents I_1 and I_2, equal to $2 \cdot a_2 \cdot I_1 \cdot I_2$. The trigonometric identity

$$\sin(a) \cdot (b) = 1/2 \cdot (\cos(a - b) - \cos(a + b)) \quad [7.34]$$

allows the expression of the product of two currents, which vary in time in a sinusoidal fashion with frequencies ω_1 and ω_2, as a group of terms that have simple sinusoidal variations with the sum and difference of the two frequencies ω_1 and ω_2:

$$I_1 = A_1 \sin(\omega_1 t) \quad [735a]$$

$$I_2 = A_2 \sin(\omega_2 t) \quad [7.35b]$$

$$I_1 \cdot I_2 = A_1 \sin(\omega_1 t) \cdot A_2 \cdot \sin(\omega_2 t)$$
$$= 1/2 \cdot A_1 \cdot A_2 \cdot (\cos((\omega_1 - \omega_2)t) - \cos((\omega_1 + \omega_2)t)) \quad [7.36]$$

Thus, this term can be expressed as the sum of two terms: one containing the sum and the other containing the difference of the input frequencies. These are often referred to as sidebands. A general rule can be derived from this expression. A device whose output is nonlinear in the input current will give a frequency spectrum in response to the input current or currents that is a sum of voltages consisting of harmonics of the input frequencies. These harmonics are sums and differences of multiples of the input frequencies.

In most examples, the dominant term will be that proportional to I_1^2. In this illustration, the frequencies of the two terms involved in the multiplication are the same. Thus, using the simple trigonometric identity derived from equation [7.34], the square law response V_{out} to the sine wave-induced current $I_{in} \cdot \sin(\omega t)$ will be

$$V_{out} = (I_{in} \cdot \sin(\omega t))^2 = 1/2 I_{in}^2 (\cos((\omega - \omega)t) - \cos((\omega + \omega)t))$$
$$= 1/2 I_{in}^2 - 1/2 I_{in}^2 \cdot \cos(2\omega t) \qquad [7.37]$$

Thus, the sum frequency harmonic will be double the microwave frequency. This is easily filtered out by components with limited frequency responses. It is, for instance, well beyond the cut-off frequency of the waveguide and will be substantially attenuated by short pieces of most coaxial cables. The remaining voltage proportional to the difference frequency will be a constant voltage, proportional to the cavity signal squared.

The most common microwave detector is a microwave diode called the crystal detector. The propagated electromagnetic wave induces a sinusoidal voltage in a conducting loop which protrudes into the detection arm in the bridge. This is connected to the crystal detector. As noted above, this has a nonlinear voltage response to the current induced in the conducting loop by the microwaves propagating into the detection arm. Most of the commercial diodes use, under the operating conditions, square law detectors: the dominant response is proportional to the square of the generated voltage[3] (Poole, 1967e). Although this relationship is assumed in the discussion above, typically the response for small microwave power is, however, not square law. Therefore, as noted above, a bias power must be applied to the crystal in order to achieve the predominant square law response.

The output will contain the first term in equation [7.37], a term which does not vary rapidly with time. This portion of the output from the square wave detector can be measured with relatively standard electronic circuitry, amplified and fed into either a chart recorder in older spectrometers, or a digitizer, and stored for later spectral analysis and display.

Modulation

If the absorption by the sample is varied by changing the magnetic field in a sinusoidal fashion with angular frequency, ω_{mod}, the EPR signal from the sample will be amplitude modulated. Poole's book (1967f) has an extensive discussion of the effect of this variation on the detected lineshape as do analyses by Wahlquist (Walhquist, 1961), and Smith (Smith, 1964). From the short discussion above, the absorption lineshape may appear as primarily Lorentzian. Modulation of the magnetic field with a sinusoidally varying modulation field of amplitude B_m will give a crystal output that will vary with magnetic field. The mathematical description of a line with field modulation, for the example of a pure Lorentzian shape, is

$$P(B - B_m(\sin(\omega_{mod}t))) = [\chi_o/\mu_o](2\pi\gamma)^2 B_o BB_1 P^2 T_2/[1 + \{2\pi T_2$$
$$(B_o - B - B_m \sin(\omega_{mod}t))/\gamma\}^2] \qquad [7.38]$$

For B_m magnitude much less than the half-width at half-maximum of the spectral line, the perturbation of the lineshape introduced by the modulation can be approximated by

$$P(B - B_m(\sin(\omega_{mod}t))) = P(B) - [dP(B)/dt]B_m \sin(\omega_{mod}t) \qquad [7.39]$$

For slow time variation of the main magnetic field B, the effect of the modulation is, to a reasonable approximation, to multiply the derivative of the spectral shape by a sinusoidally varying term. Modulation, then, will convert the spectral shape from a monophasic spectral line, varying in one direction above or below the baseline, into the familiar biphasic EPR spectral line. Figure 7.14 shows the shape of the derivative spectra for the Lorentzian and Gaussian lineshapes. Traditionally, good definition of the lineshape without distortion of the derivative or first harmonic line shape requires that the modulation field amplitude be less than one-tenth the peak-to-peak linewidth. Recent studies with spectral fitting indicate that the Lorentzian component can be extracted reliably from an inhomogeneously broadened line with substantially larger modulation fields (Peric and Halpern, 1994).

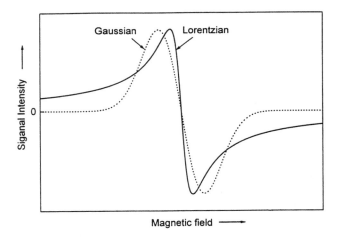

Figure 7.14 Lorentzian (solid line) and Gaussian (dotted line) absorption lines with magnetic field modulation.

The reason for the use of a technique that results in this distortion of the line-shape lies in its ability to suppress noise. The current induced in a loop protruding into the waveguide can be shown to be well represented by the expression

$$I = I_0(1 + m \cdot \sin(\omega_{mod}t)) \cdot \sin(\omega t) = I_0(\sin(\omega t) + 1/2m \cos(\omega - \omega_{mod})$$
$$- 1/2m \cos(\omega + \omega_{mod})) \tag{7.40}$$

while the main magnetic field is held constant or is only slowly varying. The term m is proportional to the modulation amplitude and the line shape derivative. The microwave frequency is ω. Typically, ω_{mod} is between 1 and 500 kHz, at least several orders of magnitude lower than ω. The expression in the parentheses of equation [7.40] is a simplification of the right-hand side of equation [7.39]. As was described above, the trigonometric identity equation [7.34] shows that this time variation gives harmonics or sidebands in the EPR cavity signal with the modulation at frequencies $\omega + \omega_{mod}$ and $\omega - \omega_{mod}$. This will create time variation in the square law crystal detector output with frequency $\omega_{mod} = (\omega + \omega_{mod}) - \omega$, where the subtracted ω comes from the first, predominant, sinusoid in the rightmost expression in equation [7.40], as well as other combinations of the fundamental and sideband frequencies. The frequency response of other elements in the system will suppress other, much higher frequencies near the microwave frequency or multiples of the microwave frequency. Much as the microwave carrier frequency is eliminated in the crystal detector or demodulated, the modulation time variation is removed from the signal by a lock-in amplifier. It is synchronized with the modulation frequency and designed to pass signal only at the modulation frequency plus or minus a small pass band. The pass band is set through the lock-in amplifier time constant. Voltage variation at any other frequency than ω_{mod} is suppressed by the lock-in amplifier. This places a strong constraint on the voltages to which the system is sensitive. This constraint eliminates noise with random time variation. It can be shown that a major component of randomly varying circuit noise has a power that varies as 1/frequency and is clearly maximal at zero frequency or at constant voltage (Hooge, 1994). Modulation shifts the detection frequency from zero to the modulation frequency, reducing the noise dependent on 1/frequency. By operating with modulation frequencies in the kilohertz range, this major noise component is substantially reduced. The strong noise suppression provided by modulation justifies, in most situations, the potential distortion that it introduces.

Other Techniques

Pulse or Time Domain Spectroscopy   The experimental paradigm underlying the discussion above involves a sample whose electron (and nuclear) spins have been prepared or equilibrated in a constant magnetic field. This gives an energy level structure and a difference in populations of the spins of various energies according to the Boltzmann distribution. Those electron spins in the higher energy levels will have a reduced occupation number or population relative to those in the lower lying energy levels. These spins are then subjected to an electromagnetic radiation field that stimulates transitions between the spin orientation-dependent energy levels. The stimulating electromagnetic field remains sufficiently weak to avoid equalizing the energy-level populations, a situation known as saturation. The basic experiment involves the sweeping of the preparative magnetic field and holding the frequency constant. This is, in essence, equivalent to slowly sweeping the electromagnetic radiation frequency; it simply entails variation of one of the variables in the resonance condition, equation [7.15]. This basic experiment involves radiation of the sample with a well-defined or fixed frequency over a large time interval. The sweep time is sufficiently slow relative to the relaxation times of the spins that the frequency or magnetic field is considered constant. This paradigm is often referred to as the continuous-wave data acquisition or CW spectroscopy.

The same information—the same pattern of energy absorption—can be obtained from an experiment where the time at which the electromagnetic energy is applied to the spin system is well defined and, therefore, of short duration, but where the frequency range of the electromagnetic radiation is very broad. That is, one may derive the same data by applying a strong short pulse of radiation to the spin system. For very short pulses, this will coherently rotate the spin system. The individual spins will add and behave as a single vector quantity. The resultant magnetization will precess like a spun top. It will also relax to its original direction. The magnetization will induce currents in the resonant circuit—in the cavity walls—which can be monitored with the bridge described above. A form of the pulsed experiment is known as the Fourier transform EPR spectrosopy. The equivalence of these data from the two experiments derives from the general complementarity of information from systems whose descriptions are related through Fourier transforms (Bracewell, 1978). The frequency dependence of the magnetization may be recovered by Fourier transforming the time dependence of the magnetization. The specific pattern of oscillations as the magnetization relaxes to equilibrium, under Fourier transform, yields the absorption or dispersion spectrum. For this reason, the name given to this aspect of the technique is Fourier transform EPR spectroscopy.

Sequences of pulses can be used to focus on various aspects of the response of the spin system magnetization to electromagnetic radiation. The stationary magnetic field, as discussed above, prepares the electron spins with a net magnetic moment or magnetization pointing antiparallel to the magnetic field. Over the short time intervals of the pulse, the electron spins respond together or coherently. Thus, the pulse can be used to rotate the net magnetic moment $180°$ to parallel the magnetic field. This corresponds to an inversion of the spin population. The subsequent free induction decay can be used to study the rate at which the energy deposited in the spin system is dispersed into the environment of the spins, the inverse of the longitudinal relaxation time or $1/T_1$. Rotating the magnetization $90°$ from its equilibrium direction results in a short time interval for the spins to rotate or precess in the plane perpendicular to the stationary magnetic field. By subsequently applying a $180°$ pulse, the spins reverse their direction of precession. Spins which have precessed more or less rapidly in respectively higher or lower regions of magnetic field will retrace their angular rotation. At a time equal to that between the 90° and the 180° pulse, the individual spin directions coincide again, much as they had coincided when the 90° pulse was applied, and a spin echo is detected. The reduction in the amplitude of the echo from the initial transverse magnetization is, classically, a measure of the transverse relaxation time, T_2. Other mechanisms that move spins off resonance will also contribute to the reduction of the amplitude of the echo (Slichter, 1988).

The duration of such a pulsed electromagnetic radiation is in the range of 10 ns. This short pulse duration is necessary because the detector cannot distinguish signal from the electromagnetic pulse and the precessing magnetization that it stimulates. A short pulse also contains a broad band of frequencies. The stimulating magnetic field must be eliminated before the induced magnetization in the spin sample diminishes by much more than $1/e$ in the time T_1 or T_2. The T_1 or T_2 for paramagnetic species in solution can vary from > 10 μs to < 1 ns, although, in practice, it is rare to use pulsed spectroscopy for spins with relaxation times less than 100 ns. This requires much faster electronic detection than the corresponding nuclear magnetic resonance experiment.

The short times over which pulsed EPR experiments can be conducted may be exploited to probe rapidly decaying free radical events. Such a process has been used by radiation chemists for years (Shkrob and Trifunac, 1997). Short pulses of radiation can, for instance, be delivered in < 1 ns time intervals. Multiple repeated pulses of radiation, therefore, can stimulate free radical production, allowing the averaging of many spectra. This, in turn, gives excellent signal-to-noise ratios, particularly if the kinetics of the chemical reactions that follow have time constants consistent with the relaxation times of the EPR-detected magnetization.

For longer lived species with relaxation times on the order of 1 μs, pulsed EPR spectroscopy at low frequencies can give signal-to-noise ratios comparable to continuous wave spectroscopy. It has the further advantage that pulsed spectra can be correlated with animal motion to reduce motion-induced artifacts. Using pulse EPR spectroscopy, high quality *in vivo* spectra and images at very low frequency have recently been published as discussed below (Murugesan, *et al.*, 1997).

There are, nevertheless, major impediments to the use of pulse techniques in living systems. These include, for instance, the lack of convenient and well-defined stimuli for the evolution of free radicals in animal models that can be generated repetitively with a response time consistent with the relaxation times of the relevant free radicals. Often, these free radicals are diatomic. In liquid solutions, they exhibit spectral lines whose breadth makes them difficult to measure with Fourier transform EPR spectroscopy, other pulse techniques, or more conventional continuous-wave techniques. Thus, spin traps have been introduced to "trap" the short-lived free radical as spin trapped adducts, which can be detected with continuous-wave spectroscopy.

Multiquantum Detection   The initial impetus to use multiquantum detection came from problems associated with the distortion of the absorption spectral profile and sensitivity limitations induced by magnetic field modulation (Hyde, *et al.*, 1989). The underlying concepts for multiquantum detection recapitulate the theme of using a nonlinear response to detect sidebands, as developed above (Redfield, 1955; Anderson, 1956; Froncisz, et al., 1989; Sczaniecki, et al., 1990, 1991). The response of the magnetization in the spin sample to an oscillatory applied electromagnetic field is effectively nonlinear, particularly at relatively large magnetic component electromagnetic field amplitudes, near resonance (Shirley, 1965; Mchaourab and Hyde, 1993a). Thus, when two electromagnetic fields irradiate the spin sample, each with a well-defined frequency near resonance, the spin system mixes the frequencies of the electromagnetic radiation. This was discussed above as occurring through the simple trigonometric identity equation [7.34] and the consequence for the square law response equation [7.37]. The nonlinear response of the spins (e.g., square law response) produces harmonics of the two frequencies. The frequencies of the two fields must be close to each other, differing by less than the intrinsic width of a single spectral line or spin packet linewidth. They must be locked to each other so that the difference between the frequency of the two fields is well defined and does not vary significantly. Among the harmonics produced will be one varying with the difference between the frequencies of the two electromagnetic fields. If this difference is conveniently low, it can be used as an input to a phase-sensitive detector and used to accept signal that varies with a specific frequency, in much the same way that a field-modulated spin system will give a signal that varies with the modulation frequency. Here, however, the signal will be displayed with a classical absorption profile similar to that of the basic Lorentzian form of equation [7.25]. Thus, modulation-associated distortion is avoided. The height of various harmonics will have specific functional (power law) dependence on relaxation times T_1 and T_2 (Mchaourab and Hyde, 1993b). Relaxation times may be directly evaluated by comparing spectral heights of several harmonics. This is achieved with a continuous-wave technique. It must be remembered, moreover, that the technique demands high-quality, low-noise signal sources, and the locking of one frequency to another through the use of common oscillators or phase-locking techniques. Components must be of sufficient quality to avoid or minimize the the generation of spurious harmonics, referred to as intermodulation sidebands, that will distort or add a spurious component to the harmonics generated by the spin system. To achieve sensitivity and spectral purity comparable to that of the usual field-modulated spin system, a continuous-wave spectrometer requires significant technical capability, not to say virtuosity.

Lower Frequency: *In Vivo* EPR Spectroscopy

Lower frequencies, other than X-band EPR spectroscopy, are necessary for the measurement of free radicals in viable biological systems larger than spheroids (largest diameter $\sim$1 mm) (Dobrucki, et al., 1990). This is the result of the absorption of short-wavelength, high-frequency electromagnetic radiation by water through its substantial complex dielectric constant at high frequencies, and by conductive losses associated with electrolytes dispersed in living aqueous solutions (Poole, 1983). As the frequency diminishes, the depth of penetration increases (Bottomley and Andrew, 1978). The depth of penetration is usually characterized by the quantity referred to as the skin depth, whereas absorption of electromag-

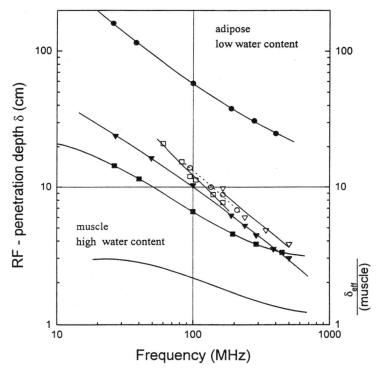

Figure 7.15 The penetration depth of the RF radiation in the sample as a function of frequency of the EPR spectrometer. (Modified from Roschmann, 1987).

netic radiation by the medium is known as the skin effect absorption. A medium with substantial skin effect absorption can be described as lossy. Clearly, biological aqueous fluids are lossy. Figure 7.15 shows the increase of skin depth with reduction of frequency, presented in graphic form, based on human data from Roschmann (1987). Higher frequency electromagnetic radiation has even smaller skin depth. At a frequency of 9.5 GHz (X-band), maximal skin depth absorption is < 1 mm, which increases to several millimeters at 2–4 GHz (S-band) and to approximately 1 cm at 1–2 GHz (L-band). By using lower frequencies, localized spectroscopic measurements of free radicals can be taken in the tissues of living organisms in real time (Berliner and Fujii, 1985; Berliner, *et al.*, 1987; Bacic, *et al.*, 1989a; Berliner and Wan, 1989; Lukiewicz, *et al.*, 1989; Ferrari, *et al.*, 1990, 1994; Utsumi, *et al.*, 1990; Komarov, *et al.*, 1993; Fujii, *et al.*, 1994, 1997; Lai and Komarov, 1994; Chzhan, *et al.*, 1995; Halpern, *et al.*, 1995; Hiramatsu, *et al.*,

1995; Jiang, *et al.*, 1995; Yoshimura, *et al.*, 1996). At still lower frequencies, such as 250 MHz and beyond, skin depth absorption in tissues approaches 7 cm and, in some situations, the depth may be even larger. At these low frequencies, measurements in larger animals become possible without the need for tissue-invasive catheter placement.

Sensitivity at Lower Frequencies

The principal reason that measurements at lower frequencies have not been commonly made is that the sensitivity or signal-to-noise ratio is lessened. The degree of this reduction and the conditions under which this occurs are controversial. To a large extent, it is dependent upon experiment configuration. The geometry of the resonator also affects the reduction in sensitivity with frequency. The scaling of sensitivity with frequency from a cylindrical sample-holder enclosing the sample is different than a loop applied to a slab-shaped object due to a different frequency dependence of skin depth. The cylindrical resonator geometry focuses electromagnetic energy at its center, increasing the effective skin depth. Although simple skin depth considerations, applicable to an appositional loop geometry, indicate that skin depth varies as $(1/v)^{1/2}$, data (Roschmann, 1987) and theoretical calculations solving Maxwell's equations in a cylindrical geometry (Bottomley and Andrews, 1978; Halpern and Bowman, 1991) indicate the scaling of effective skin depth with frequency for samples in a cylindrical enclosing resonator holder to be much closer to $(1/v)^1$.

This issue is further complicated by an error in the otherwise meticulous and excellent treatise by Poole (1983). This has resulted in an unsubstantiated concern about the loss of signal at very low frequencies. For example, the linear dimension of a sample and the loop-gap resonator for measuring it might increase as $1/v$, where v is the frequency at which the EPR spectroscopy experiment is carried out. Under these circumstances, it can be shown that the signal-to-noise ratio will increase as $(1/v)^{1/4}$ (Gareth Eaton, private communication; see also Hoult and Richards, 1976; Hoult and Lauterbur, 1979). This does not take into consideration the change in tissue conductivity with frequency. This will further increase the sensitivity of the low-frequency *in vivo* measurement relative to what would otherwise be predicted at higher frequencies (Foster, 1992).

L-Band Spectroscopic Measurements

While the detection of nitroxides in living animals has been made at 9.5 GHz, this was achieved through the implantation of a helix coil into a rat. Direct coupling to the spectrometer allowed spectra to be recorded (Feldman, *et al.*, 1975). With the development of lower frequency EPR spectrometers in the range of 1-2 GHz, the penetrability of the electromagnetic radiation to a depth of nearly 1 cm of skin has made it possible to record spectra from a more varied anatomic cross-section of a small animal. As such, imaging and localized spectroscopy have been attained at this frequency. The experimental paradigm involves the injection of a concentrated solution of a nitroxide into the animal, measuring the distribution of the spin label by EPR spectroscopy as it apportions into different tissues (Lukiewicz and Lukiewicz, 1984; Nishikawa, *et al.*, 1985; Subczynski, *et al.*, 1986; Bacic, *et al.*, 1989a, 1989b; Ishida, *et al.*, 1989; Utsumi, *et al.*, 1990). Metabolic activity and other aspects of tissue physiology were deduced from the rate of nitroxide reduction to the corresponding hydroxylamine (Feldman, *et al.*, 1975; Bacic, *et al.*, 1989b; Utsumi, *et al.*, 1990; Takeshita, *et al.*, 1993).

Measurements from which tissue O_2 concentration can be inferred were made from an EPR spectrum of a model nitroxide. Oxygen broadens spectral lines, particularly the Lorentzian or homogeneous components of the spectral lines. By measuring the width of the spectral lines, O_2 concentrations could be estimated. In this way, the concentration of O_2 was measured in model systems (Backer, *et al.*, 1977; Sarna, *et al.*, 1980; Lai, *et al.*, 1982) and in isolated organ preparations (Zweier and Kuppusamy, 1988; Zweier, *et al.*, 1991; Kuppusamy, *et al.*, 1995a). Subsequent discoveries of the influence of O_2 on the width of spectral lines from high spin density sugar chars (Boyer and Clarkson, 1994), various powdered coals such as fusinite (Vahidi, *et al.*, 1994), and crystals of lithium phthalocyanine (Liu, *et al.*, 1993) have made possible a wide array of sensitive O_2 probes. The dramatic response of these substances to very small changes in O_2 concentration make it feasible to estimate regional tissue differences by as little as a fraction of a torr. As these particulate carbon-centered radicals can be implanted into specific animal tissues, repeated O_2 measurements of the same area can be made over periods of up to 2 weeks (Vahidi, *et al.*, 1994). Several small regions of anatomic interest can be measured simultaneously by imposing a magnetic field gradient on the animal. The gradient causes the magnetic field to change with position. Thus, several regions can report spectra with their lines shifted to characteristic positions separated from one another, allowing independent measurements of their widths. This technique has been used to measure changes in O_2 concentrations following a dose of radiation (O'Hara, *et al.*, 1995). Likewise, the recovery of O_2 tension following spinal cord injury has been monitored using this method (Liu, *et al.*, 1995).

Radiofrequency EPR Spectroscopy

At frequencies lower than the L-band, EPR spectra can be recorded from larger animals. The skin depth sensitivity for a cylindrical geometric resonator, such as a conventional NMR unit or a loop-gap resonator at 250 MHz, is 7 cm (Roschmann, 1987). This is the frequency at which a 6 tesla NMR spectrometer operates. In fact, current 4.4-tesla NMR spectrometers with operating frequencies of 185 MHz are in current use as diagnostic and physiologic tools for human subjects. Thus, measurements acquired at these frequencies are relevant to the design of low-frequency EPR spectrometers. As will be discussed below, sufficient EPR signal intensity has been recorded at these very low frequencies, allowing for *in vivo* imaging of nitroxides (Halpern, *et al.*, 1996a).

A radiofrequency spectrometer specifically designed for animal measurements was of the standard continuous-wave construction, which incidentally included features of an NMR spectrometer (Halpern, *et al.*, 1989). Early experimentation explored the *in situ* pharmacokinetics of structurally diverse nitroxides infused into the vasculature of isolated perfused hearts housed in a Plexiglas cuvette, which is then inserted into the resonator of this spectrometer (Rosen, *et al.,* 1988). Other radiofrequency spectrometers have achieved similar success with whole-body measurements of nitroxide metabolism and images (see below) measuring distribution in rats and mice (Colacicchi, *et al.,* 1989, 1993; Ferrari, *et al.,* 1990).

One of the consequences of operating at these very low frequencies is that a given absolute tolerance for magnetic field homogeneity or stability translates into far less restrictive fractional tolerances. Thus, the requirement of a 1-mG variation at 250 MHz (90-G field) is one part in 10^5 instead of one part in 3×10^6 needed at X-band. This results in reliable spectral shape acquisition and analyses. This concept is illustrated by measurements of O_2 concentrations in mouse tumors using spectral shape analyses (Halpern, *et al.,* 1994, 1996b).

EPR Imaging

The very fact that EPR, as is the case for NMR, is a magnetic resonance technique means that a paramagnetic species within a sample can be localized spectroscopically and imaged (Karthe and Wehrsdorfer, 1979; Halpern and Bowman, 1991). This can be achieved if, in addition to a constant applied magnetic field, a field which changes with position in a paramagnetic sample is added. Then, the resonance condition for a specific spectral line will be satisfied at different values of the applied magnetic field. The value of the magnetic field of the spectral line in the sweep of the main magnetic field reflects its locus within the sample and an image of the distribution of the paramagnetic species can be obtained. The fact that there is a 10^6 difference in magnitude between EPR and NMR relaxation times means that pulsed imaging techniques are not as extensively used in the former as in the latter; equipment with response rates of nanoseconds rather than milliseconds are necessary for EPR images to be acquired in this fashion. Early examples include the imaging of nitroxides in celery (Berliner and Fujii, 1985) and the application of EPR imaging as an alternative to stopped-flow kinetics (Ohno, 1982a, 1982b).

A crucial development for both NMR and EPR imaging of biologic samples was the development of spectral–spatial imaging (Lauterbur, *et al.,* 1984; Maltempo, 1986; Maltempo, *et al.,* 1987). To conceptualize this imaging modality, consider an EPR spectrum as another dimension of the image. Figures 7.16 and 7.17 illustrate this. A spectrum obtained in the presence of a gradient represents a mixture of spectral and spatial information, much like a rotation in a two-dimensional space mixes information from two orthogonal spatial directions. The spectrum so obtained can be viewed as a projection of this mixed spectral and spatial information. Just as back-projection reconstruction allows the reconstruction of a two-dimensional image from a series of projections of the image in different directions, a series of spectral-spatial projections can give an image consisting of one spatial dimension and one spectral dimension. A series of such spectral–spatial images in different spatial directions will yield a full 4-dimensional spectral–spatial image. For each subvolume of the sample that is imaged (voxel), a spectrum can

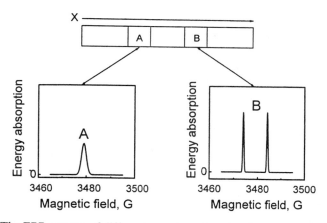

Figure 7.16 The EPR spectra of different sample regions in a homogeneous magnetic field.

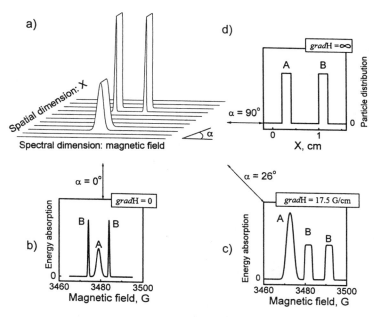

Figure 7.17 (a) A two-dimensional spectral–spatial EPR image of the same sample as that in Figure 7.16. This shows the projections at the different magnetic field gradient directions: (b) $gradH = 0$, spectral–spatial rotation angle $\alpha = 0°$; (c) $gradH = 17.5\,G - cm^{1}$, spectral–spatial rotation angle $\alpha = 26°$, (d) $gradH = \infty$, spectral–spatial rotation $\alpha = 90°$. The spectral line A is assumed to have width 2G, and line B is assumed to have width 0.5 G. Lines A and B have the same number of spins.

be derived. This image is, therefore, a spectral image and all of the information contained within the spectrum can be imaged. Thus, the spectral height can be mapped over the sample with the resolution of the image. In a similar fashion, the width of the spectral line within each voxel can be imaged. Spectral–spatial imaging allows EPR to map this spectral parameter, which, in turn, can reflect O_2 concentration. For example, O_2 content in small volumes (in certain cases, a < 1 mm linear dimension) within a tumor, can be mapped by using spectral–spatial imaging of a paramagnetic substrate that is infused into an animal. This may be essential for the appropriate radiation treatment of the cancerous tissue, and the diagnosis and treatment of stroke and myocardial infarction. Introduction of a narrowed-line paramagnetic species, whose linewidth is sensitive to O_2 concentration, would, therefore, permit the corresponding image to provide a quantitative measure of O_2 content within a region of the imaged tissue.

Images have been obtained at X-band in studies of spheroids bathed in nitroxides, with a microscopic resolution approaching 200 μm (Dobrucki, et al., 1990). Similarly, EPR imaging at this frequency has been employed to distinguish between fatty- and muscle-rich areas of subcutaneous adipose tissue containing well-differentiated fat and muscle regions (Bacic, et al., 1988). Other imaging examples include a scanning technique in which a sample is passed over a 1 mm aperture and rotated, producing an image of a heavily radiated human tooth (Hochi, et al., 1993), and the evaluation of the diffusibility of structurally diverse nitroxides into mouse epidermis (Mäder, et al. 1992; Fuchs, et al., 1992; Freisleben, et al., 1994).

The L-band, with its increased depth sensitivity, opens the possibility of imaging paramagnetic species in organ preparations and, to a lesser extent, in small animals. The potential of EPR imaging can readily be seen from two-dimensional image studies of a phantom model of a beating heart at different times in the heart cycle (Testa, et al., 1993). More recent high-resolution three-dimensional images of nitroxides in isolated perfused hearts have enhanced our understanding of myocardial metabolism, not easily obtained from other modalities (Kuppusamy, et al., 1994, 1995b, 1996a, 1996b). These accomplishments are so impressive that further improvements will, undoubtedly, allow for the monitoring of heart function in disease states.

One of the earliest investigations of L-band imaging explored the uptake of nitroxides into a murine tail tumor (Berliner, et al., 1987). These images used only the highest gradients available, applied to the sample from different directions. The image was derived by deconvolution of the spectral shape. Soon thereafter, three-dimensional images of nitroxides in rats were achieved (Colacicchi, et al., 1988; Alecci, et al., 1990; Ishida, et al., 1992). By using EPR spectrometers operating at even lower frequencies, such as radiofrequency, outstanding in vivo images of paramagnetic species in real time have been observed (Yoshimura, et al., 1996). Early longitudinal spectral–spatial images at 250 MHz were, for instance, generated from nitroxide images of the torso of a mouse (Halpern, et al., 1991). Additionally, one- and two-dimensional images of the compartmentation dynamics of different nitroxides in mice has allowed a more accurate prediction of their in vivo distribution, which was based, to a large extent, on the structural characteristics of these spin labels (Halpern, *et al.*, 1996). More recent high-resolution two-dimensional images of a narrow-lined trityl free radical (Andersson, *et al.*, 1996) were produced from a mouse tail using a pulsed EPR spectrometer operating at 300 MHz (Murugesan, *et al.*, 1997). Here, stepped gradients were used to acquire the image, but the spectral acquisition was obtained using Fourier transform pulsed EPR spectroscopy.

Concluding Thoughts

In the 50 years since the development of EPR spectroscopy, novel innovations have spurred its development in many fields. The explosive developments detailed here have been motivated by the desire to study human physiology and the role of free radicals in modulating biological function. Narrow-linewidth stable free radicals providing high spatial resolution (Andersson, *et al.*, 1996; Radner, *et al.*, 1996) appear to be a crucial development for *in vivo* EPR spectroscopy and imaging (Murugesan, *et al.*, 1997; Kuppusamy, *et al.*, 1997). In the illustration shown in Figure 7.16, the spectral–spatial image of a one-spatial-dimensional object, such as a single hair, is shown. For a wide spectral line, large angles are necessary to "look around" the spectral line. This corresponds to large magnetic field gradients or large changes of magnetic field with position. If the spectral line narrows, it is easier to "look around" the spectral line and smaller spatial separations can be distinguished with more modest magnetic field gradients. The narrow-line probes further enhance sensitivity, since, for a given concentration, the spectral line has a larger amplitude, improving the signal-to-noise ratio. Finally, the narrower intrinsic linewidth increases the fractional change in spectral parameters, such as linewidth to physiologic change. This increases physiologic sensitivity. For this reason, narrow-line spin probes promise much greater spatial resolution and sensitivity to physiologic conditions, including the detection of small regional variations in O_2 concentration.

The *in vivo* pharmacokinetics and pharmacodynamics of spin traps and spin trapped adducts will invariably dictate the success of *in vivo* EPR spectroscopy and imaging. In the next chapter, we develop kinetic models aimed at evaluating the potential to detect and characterize a specific free radical in various biological systems, including animals by the use of low-frequency EPR spectroscopy in real time.

References

Abragam, A. (1961a). *The Principles of Nuclear Magnetisms*, pp. 46–48, Oxford University Press, New York.

Abragam, A. (1961b). *The Principles of Nuclear Magnetisms*, pp. 71–93, Oxford University Press, New York.

Abragam, A., and Bleany, B. (1986). *Electron Paramagnetic Resonance of Transition Ions*, Dover, New York.

Alecci, M., Colacicohi, S., Indovina, P.L., Momo, F., Pavone, P., and Sotgiu, A. (1990). Three-dimensional *in vivo* ESR imaging in rats. *Magn. Reson. Imaging* **8**: 59–63.

Anderson, W.A. (1956). Nuclear magnetic resonance spectra of some hydrocarbons. *Phys. Rev.* **102**: 151–167.

Andersson, S., Radner, F., Rydbeek, A., Servin, R., and Wistrand, L.-G. (1996). Free radicals. U.S. Patent Number 5,530,140.

Bacic, G., Nilges, M.J., Magin, R.L., Walczak, T., and Swartz, H.M. (1989a). *In vivo* localized ESR spectroscopy reflecting metabolism. *Magn. Reson. Med.* **10**: 266–272.

Bacic, G., Nilges, M.J., Walczak, T., and Swartz, H.M. (1989b). The use of *in vivo* EPR to follow the pharmacokinetics of drugs. *Phys. Med.* **5**: 307–313.

Bacic, G., Walczak, T., Demsar, F., and Swartz, H.M. (1988). Electron spin resonance imaging of tissues with lipid-rich areas. *Magn. Reson. Med.* **8**: 209–219.

Backer, J.M., Budker, V.G., Eremenko, S.I., and Molin, Yu.N. (1977). Detection of the kinetics of biochemical reactions with oxygen using exchange broadening in the ESR spectra of nitroxide radicals. *Biochim. Biophys. Acta* **460**: 152–156.

Bales, B.L. (1989). Inhomogenously broadened spin-label spectra, in: *Spin Labeling. Theory and Applications* (Berliner, L.J., and Reuben, J., eds.), Vol. 8, pp. 77–130, Plenum Press, New York.

Bales, B.L., Blum, R.A., Mareno, D., Peric, M., and Halpern, H.J. (1992). Solvent and temperature dependence of the hyperfine coupling constants in CTPO. *J. Magn. Reson.* **98**: 299–307.

Berliner, L.J., and Fujii, H. (1985). Magnetic resonance imaging of biological specimens by electron paramagnetic resonance of nitroxide spin labels. *Science* **227**: 517–519.

Berliner, L.J., and Wan, X.M. (1989). *In vivo* pharmacokinetics by electron magnetic resonance spectroscopy. *Magn. Reson. Med.* **9**: 430–434.

Berliner, L.J., Fujii, H., Wan, X., and Lukiewicz, S.J. (1987). Feasibility study of imaging a living murine tumor by electron paramagnetic resonance. *Magn. Reson. Med.* **4**: 380–384.

Bloch, F. (1946). Nuclear induction. *Phys. Rev.* **70**: 460–474.

Bottomley, P.A., and Andrew, E.R. (1978). RF magnetic field penetration, phase shift and power dissipation in biological tissues: Implications for NMR imaging. *Phys. Med. Biol.* **23**: 630–643.

Boyer, S.J., and Clarkson, R.B. (1994). Electron paramagnetic resonance studies of an active carbon: The influence of preparation procedure on the oxygen response of the linewidth. *Colloids Surf. A, Physiochem. Eng. Asp.* **82**: 217–224.

Bracewell, R.N. (1978). *The Fourier Transform and Its Applications*, 2nd edition, McGraw-Hill, New York.

Breit, G., and Rabi, I.I. (1931). Measurement of nuclear spin. *Phys. Rev.* **38**: 2082–2083.

Chzhan, M., Kuppusamy, P., and Zweier, J.L. (1995). Development of an electronically tunable L-band resonator for EPR spectroscopy and imaging of biological samples. *J. Magn. Reson. B.* **108**: 67–72.

Cohen, E.R., and Taylor, B.N. (1988). The 1986 CODATA recommended values of the fundamental physical constants. *J. Phys. Chem. Ref. Data* **17**: 1795.

Colacicchi, S., Allecci, M., Gualtieri, G., Quaresima, V., Ursini, C.L., Ferrari, M., and Sotgiu, A. (1993). New experimental procedures for *in vivo* L-band and radio frequency EPR spectroscopy/imaging. *J. Chem. Soc., Perkin Trans. 2* 2077–2082.

Colacicchi, S., Ferrari, M., Gualtieri, G., Santini, M.T., and Sotgiu, A. (1989). Spatial localization and oximetry of EPR signal in mice. *Phys. Med.* **5**: 297–305.

Colacicchi, S., Indovina, P.L., Momo, F., and Sotgiu, A. (1988). Low-frequency three-dimensional ESR imaging of large samples. *J. Appl. Phys. E: Sci. Instrum.* **21**: 910–913.

Dobrucki, J.W., Demsar, F., Walczak, T., Woods, R.K., Bacic, G., and Swartz, H.M. (1990). Electron spin resonance microscopy of an *in vitro* tumour model. *Br. J. Cancer* **61**: 221–224.

Eastman, M.P., Kooser, R.G., Das, M.R., and Freed, J. (1969). Studies of Heisenberg spin exchange in ESR spectra. I.

Linewidth and saturation effects. *J. Chem. Phys.* **51**: 2690–2709.

El'sting, O.G. (1960). A device for investigating paramagnetic electronic resonance at meter-band frequencies. *Prib. Tekh. Eksp.* **5**: 64–66; Translation in: *Instrum. Exp. Tech.* **4**: 753–755.

Feher, G. (1957). Sensitivity considerations in microwave paramagnetic resonance absorption techniques. *Bell System Tech. J.* **36**: 449–484.

Feldman, A., Wildman, E., Bartolini, G., and Piette, L.H. (1975). *In vivo* electron spin resonance in rats. *Phys. Med. Biol.* **20**: 602–612.

Ferrari, M., Colacicchi, S., Gualtieri, G., Santini, M.T., and Sotgiu, A. (1990). Whole mouse nitroxide free radical pharmacokinetics by low frequency electron paramagnetic resonance. *Biochem. Biophys. Res. Commun.* **166**: 168–173.

Ferrari, M., Quaresima, V., Ursini, C.L., Alecci, M., and Sotgiu, A. (1994). *In vivo* electron paramagnetic resonance spectroscopy-imaging in experimental oncology: The hope and the reality. *Int. J. Radiat. Oncol. Biol. Phys.* **29**: 421–425.

Finkelstein, E., Rosen, G.M., Rauckman, E.J., and Paxton J. (1979). Spin trapping of superoxide. *Mol. Pharmacol.* **16**: 676–685.

Foster, T.H. (1992). Tissue conductivity modifies the magnetic resonance intrinsic signal-to-noise ratio at high frequencies. *Magn. Reson. Med.* **23**: 383–385.

Freisleben, H.-J., Groth, N., Fuchs, J., Rudolph, P., and Herrling, T. (1994). Penetration of spin-labeled dihydrolipoate into the skin of hairless mice. Modification of epidermal and dermal polarity. *Arzneim.-Forsch./Drug Res.* **44**(II): 1047–1050.

Froncisz, W., and Hyde, J.S. (1982). The loop-gap resonator: A new microwave lumped circuit ESR sample structure. *J. Magn. Reson.* **47**: 515–521.

Froncisz, W., Sczaniecki, P.B., and Hyde, J.S. (1989). Electron paramagnetic rotary resonance spectroscopy of spin labels. *Phys. Med.* **5**: 163–175.

Fuchs, J., Groth, N., Herrling, T., Milbradt, R., Zimmer, G., and Packer, L. (1992). Electron paramagnetic resonance (EPR) imaging in skin: Biophysical and biochemical microscopy. *J. Invest. Dermatol.* **98**: 713–719.

Fujii, H., Koscielniak, J., and Berliner, L.J. (1997). Determination and characterization of nitric oxide generation in mice by *in vivo* L-band EPR spectroscopy. *Magn. Reson. Med.* **38**: 565–568.

Fujii, H., Zhao, B., Koscielniak, J., and Berliner, L.J. (1994). *In vivo* EPR studies of the metabolic fate of nitrosobenzene in the mouse. *Magn. Reson. Med.* **31**: 77–80.

Halpern, H.J., and Bowman, M.K. (1991). Low-frequency EPR spectrometers: MHz range, in: *EPR Imaging and In Vivo EPR* (Eaton, G.R., Eaton, S.S., and Ohno, K., eds.), pp. 45–63, CRC Press, Boca Raton, FL.

Halpern, H.J., Bowman, M.K., Spencer, D.P., Van Polen, J., Dowey, E.M., Massoth, R.J., Nelson, A.C., and Teicher, B.A. (1989). An imaging radio-frequency electron spin resonance spectrometer with high resolution and sensitivity for *in vivo* measurements. *Rev. Sci. Instrum.* **60**: 1040–1050.

Halpern, H.J., Peric, M., Nguyen, T.D., Bowman, M.K., Lin, Y.J., and Teicher, B.A. (1991). *In vivo* O_2 sensitive imaging at low frequencies. *Phys. Med.* **7**: 39–45.

Halpern, H.J., Peric, M., Yu, C., and Bales, B.L. (1993) Rapid evaluation of parameters from inhomogeneously broadened EPR spectra. *J. Magn. Reson.* **103**: 13–22.

Halpern, H.J., Peric, M., Yu, C., Barth, E.D., Chandramouli, G.V.R., Makinen, M.W., and Rosen, G.M. (1996a). *In vivo* spin-label murine pharmacodynamics using low-frequency electron paramagnetic resonance imaging. *Biophys. J.* **71**: 403–409.

Halpern, H.J., Yu, C., Barth, E., Peric, M., and Rosen, G.M. (1995). *In situ* detection, by spin trapping of hydroxyl radical markers produced from ionizing radiation in the tumor of a living mouse. *Proc. Natl. Acad. Sci. USA* **92**: 796–800.

Halpern, H.J., Yu, C., Peric, M., Barth, E., Grdina, D.J., and Teicher, B.A. (1994). Oxymetry deep in tissues with low-frequency electron paramagnetic resonance. *Proc. Natl. Acad. Sci. USA* **91**: 13047–13051.

Halpern, H.J., Yu, C., Peric, M., Barth, E.D., Karczmar, G.S., River, J.N., Grdina, D.J., and Teicher, B.A. (1996b). Measurement of differences in pO_2 in response to perfluorocarbon/carbogen in FSa and NFSa murine fibrosarcomas with low-frequency EPR oximetry. *Radiat Res.* **145**: 610–618.

Harbour, J.R., Chow, V., and Bolton, J.R. (1974). An electron spin resonance study of the spin adducts of OH and HO_2 radicals with nitrones in the ultraviolet photolysis of aqueous hydrogen peroxide solutions. *Can. J. Chem.* **52**: 3549–3553.

Hiramatsu, M., Oikawa, K., Noda, H., Mori, A., Ogata, T., and Kamada, H. (1995). Free radical imaging by electron spin resonance computed tomography in rat brain. *Brain Res.* **697**: 44–47.

Hochi, A., Furusawa, M., and Ikeya, M. (1993). Applications of microwave scanning ESR microscope: Human tooth with metal. *Appl. Radiat. Isot.* **44**: 401–405.

Hooge, F.N. (1994) 1/f noise sources. *IEEE Trans. Electron* Devices **41**: 1926–1935.

Horowitz, P., and Hill, W. (1989). *The Art of Electronics*, 2nd edition, pp. 297–300, Cambridge University Press, Cambridge, England.

Hoult, D.I., and Lauterbur, P.C. (1979). The sensitivity of the Zeugmatographic experiment involving human samples. *J. Magn. Reson.* **34**: 425–433.

Hoult, D.I., and Richards, R.E. (1976). The signal-to-noise ratio of the nuclear magnetic resonance experiment. *J. Magn. Reson.* **24**: 71–85.

Hyde, J.S., and Froncisz, W. (1986). Loop-gap resonators, in: *Electron Spin Resonance* (Symons, M.R.C., ed.), pp. 175–184, Royal Society of Chemistry, London.

Hyde, J.S., Sczaniecki, P.B., and Froncisz, W. (1989). The Bruker Lecture: Alternatives to field modulation in electron spin resonance spectroscopy. *J. Chem Soc., Faraday Trans. 1* **85**: 3901–3912.

Ishida, S., Kumashiro, H., Tsuchihashi, N., Ogata, T., Ono, M., Kamada, H., and Yoshida, E. (1989). *In vivo* analysis of nitroxide radicals injected into small animals by L-band ESR technique. *Phys. Med. Biol.* **34**: 1317–1323.

Ishida, S., Matsumoto, S., Yokoyama, H., Mori, N., Kumashiro, H., Tsuchihashi, N., Ogata, T., Yamada, M., Ono, M., Kitajima, T., Kamada, H., and Yoshida, E. (1992). An ESR-CT imaging of the head of a living rat receiving an administration of a nitroxide radical. *Magn. Reson. Imaging* **10**: 109–114.

Janzen, E.G., and Blackburn, B.J. (1969). Detection and identification of short-lived free radicals by electron spin resonance trapping techniques (spin trapping). Photolysis of organolead, -tin, and mercury compounds. *J. Am. Chem. Soc.* **91**: 4481–4490.

Janzen, E.G., and Liu, J.I.-P. (1973). Radical addition reactions of 5,5-dimethyl-1-pyrroline-1-oxide. ESR spin trapping with a cyclic nitrone. *J. Mag. Reson.* **9**: 510–512.

Janzen, E.G., Evans, C.A., and Liu, J.I.-P. (1973). Factors influencing hyperfine splitting in the ESR spectra of five-membered ring nitroxides. *J. Mag. Reson.* **9**: 513–516.

Jiang, J.J., Liu, K.J., Shi, X., and Swartz, H.M. (1995). Detection of short-lived free radicals by low-frequency electron paramagnetic resonance spin trapping in whole living animals. *Arch. Biochem. Biophys.* **319**: 570–573.

Johnson, C.C., and Guy, A.W. (1972). Non-ionizing electromagnetic wave effects in biological materials and systems. *Proc. IEEE* **60**: 692–718.

Karthe, W., and Wehrsdorfer, E. (1979). The measurement of inhomogeneous distribution of paramagnetic centers by means of EPR. *J. Magn. Reson.* **33**: 107–111.

Kittel, C. (1967). *Introduction to Solid State Physics*, 3rd edition, pp. 508, John Wiley & Sons, New York.

Komarov, A., Mattson, D., Jones, M.M., Singh, P.K., and Lai, C.-S. (1993). *In vivo* spin trapping of nitric oxide in mice. *Biochem. Biophys. Res. Commun.* **195**: 1191–1198.

Kuppusamy, P., Chzhan, M., Vij, K., Shteynbuk, M., Lefer, D.J., Giannella, E., and Zweier, J.L. (1994). Three-dimensional spectral-spatial EPR imaging of free radicals in the heart: A technique for imaging tissue metabolism and oxygenation.

Proc. Natl. Acad. Sci. USA **91**: 3388–3392.

Kuppusamy, P., Chzhan, M., Wang, P., and Zweier, J.L. (1996a). Three-dimensional gated EPR imaging of the beating heart: Time-resolved measurements of free radical distribution during the cardiac contractile cycle. *Magn. Reson. Med.* **35**: 323–328.

Kuppusamy, P., Wang, P., Chzhan, M., and Zweier, J.L. (1997). High resolution electron paramagnetic resonance imaging of biological samples with a single line paramagnetic label. *Magn. Reson. Med.* **37**: 479–483.

Kuppusamy, P., Wang, P., and Zweier, J.L. (1995a). Evaluation of nitroxides for the study of myocardial metabolism and oxygenation. *Magn. Reson. Chem.* **33**: S123-S128.

Kuppusamy, P., Wang, P., and Zweier, J.L. (1995b). Three-dimensional spatial EPR imaging of the rat heart. *Magn. Reson. Med.* **34**: 99–105.

Kuppusamy, P., Wang, P., Zweier, J.L., Krishna, M.C., Mitchell, J.B., Ma, L., Trimble, C.E., and Hsia, C.J.C. (1996b). Electron paramagnetic resonance imaging of rat heart with nitroxide and polynitroxyl-albumin. *Bio-chemistry* **35**: 7051–7057.

Lai, C.-S., and Komarov, A.M. (1994). Spin trapping of nitric oxide produced *in vivo* in septic-shock in mice. *FEBS Lett.* **345**: 120–124.

Lai, C.-S., Hopwood, L.E., Hyde, J.S., and Lukiewicz, S. (1982). ESR studies of O_2 uptake by Chinese hamster ovary cells during the cell cycle. *Proc. Natl. Acad. Sci. USA* **79**: 1166–1170.

Lauterbur, P., Levin, D.N., and Marr, R.B. (1984). Theory and simulation of NMR spectroscopic imaging and field plotting by projection reconstruction involving an intrinsic frequency dimension. *J. Magn. Reson.* **59**: 536–541.

Liu, K.J., Bacic, G., Hoopes, P.J., Jiang, J., Du, H., Ou, L.C., Dunn, J.F., and Swartz, H.M. (1995). Assessment of cerebral pO_2 by EPR oximetry in rodents: Effects of anesthesia, ischemia, and breathing gas. *Brain Res.* **685**: 91–98.

Liu, K.J., Gast, P., Moussavi, M., Norby, S.W., Vahidi, N., Walczak, T., Wu, M., and Swartz, H.M. (1993). Lithium phthalocyanine: A probe for electron paramagnetic resonance oximetry in viable biological systems. *Proc.Natl. Acad. Sci. USA* **90**: 5438–5442.

Lukiewicz, S.J., and Lukiewicz, S.G. (1984). *In vivo* ESR spectroscopy of large biological objects. *Magn. Reson. Med.* **1**: 297–298.

Lukiewicz, S.J., Cieszka, K., Wojcik, K., Lackowska, B., Markowska, E., Pajak, S., Elas, M., and Dubis, K. (1989). *In vivo* ESR studies on the influence of hyperthermia on biological half-lifes (BHL) of nitroxides in B-16 murine melanoma. *Phys. Med.* **5**: 315–320.

Mäder, K., Strösser, R., and Borchert, H.-H. (1992). Transcutaneous absorption of nitroxide radicals detected *in vivo* by means of X-band ESR. *Pharmazie* **47**: 946–947.

Maltempo, M.M. (1986). Differentiation of spectral and spatial components in EPR imaging using 2-D image reconstruction algorithms. *J. Magn. Reson.* **69**: 156–161.

Maltempo, M., Eaton, S.S., and Eaton, G.R. (1987). Spectral-spatial two-dimensional EPR imaging. *J. Magn. Reson.* **72**: 449–455.

Marcuvitz, N. (1951a). *Waveguide Handbook*, MIT Radiation Laboratory Series, No. 10, pp. 56–60, McGraw-Hill, New York.

Marcuvitz, N. (1951b). *Waveguide Handbook*, MIT Radiation Laboratory Series, No. 10, pp. 72–80, McGraw-Hill, New York.

McConnell, H.M. (1956). Indirect hyperfine interactions in the paramagnetic resonance spectra of aromatic free radicals. *J. Chem. Phys.* **24**: 764–766.

McConnell, H.M. (1958). Reaction rates by nuclear magnetic resonance. *J. Chem. Phys.* **28**: 430–431.

McConnell, H.M., and Chesnut, D.B. (1958). Theory of isotropic hyperfine interactions in π-electron radicals. *J. Chem. Phys.* **28**: 107–117.

Mchaourab, H.S., and Hyde, J.S. (1993a). Dependence of the multiple-quantum EPR signal on the spin-lattice relaxation time. Effect of oxygen in spin-labeled membranes. *J. Magn. Reson.* **101B**: 178–184.

Mchaourab, H.S., and Hyde, J.S. (1993b). Continuous wave multiquantum electron paramagnetic resonance spectroscopy. III. Theory of inter-modulation sidebands. *J. Chem. Phys.* **98**: 1786–1796.

Messiah, A. (1966). *Quantum Mechanics*, Vol. II, pp. 544 and 936, North Holland, Amsterdam.

Montgomery, C.G. (1947). *Technique of Microwave Measurements*, MIT Radiation Laboratory Series, No. 11, p. 25, McGraw Hill, New York.

Montgomery, C.G., Dicke, R.H., and Purcell, E.M. (1947). *Principles of Microwave Circuits*, MIT Radiation Laboratory Series, No. 8, pp. 306–308, McGraw Hill, New York.

Murugesan, R., Cook, J.A., Devasahayam, N., Afeworki, M., Subramanian, S., Tschudin, R., Larsen, J.H.A., Mitchell, J.B., Russo, A., and Krishna, M.C. (1997). *In vivo* imaging of a stable paramagnetic probe by pulsed-radiofrequency electron paramagnetic resonance spectroscopy. *Magn. Reson. Med.* **38**: 409–414.

Nishikawa, N., Fujii, H., and Berliner, L.J. (1985). Helices and surface coils for low-field *in vivo* ESR and EPR imaging applications. *J. Magn. Reson.* **62**: 79–86.

Nordio, P.L. (1976). General magnetic resonance theory, in: *Spin Labelling: Theory and Applications* (Berliner, L.K., ed.), pp. 5–52, Academic Press, New York.

O'Hara, J.A., Goda, F., Liu, K.J., Bacic, G., Hoopes, P.J., and Swartz, H.M. (1995). The pO$_2$ in a murine tumor after irradiation: An *in vivo* electron paramagnetic resonance oximetry study. *Radiat. Res.* **144**: 222–229.

Ohno, K. (1982a). Application of ESR imaging to a continuous flow method for study on kinetics of short-lived radicals. *J. Magn. Reson.* **49**: 56–63.

Ohno, K. (1982b). ESR imaging: A deconvolution method for hyperfine patterns. *J. Magn. Reson.* **50**: 145–150.

Oles, T., Strangeway, R.A., Luglio, J., Froncisz, W, and Hyde, J. S. (1992). X-band low-phase-noise gunn diode oscillator for EPR spectroscopy. *Rev. Sci. Instrum.* **63**: 4010–4011.

Peric, M., and Halpern, H.J. (1994). Modulation broadening and fitting of the

Voigt ESR line. *J. Magn. Reson.* **109**: 198–202.

Poole, C.P. (1967a). *Electron Spin Resonance, a Comprehensive Treatise on Experimental Techniques*, p. 85, John Wiley & Sons, New York.

Poole, C.P. (1967b). *Electron Spin Resonance, a Comprehensive Treatise on Experimental Techniques*, pp. 66–72, John Wiley & Sons, New York.

Poole, C.P. (1967c). *Electron Spin Resonance, a Comprehensive Treatise on Experimental Techniques*, pp. 85–96, John Wiley & Sons, New York.

Poole, C.P. (1967d). *Electron Spin Resonance, a Comprehensive Treatise on Experimental Techniques*, pp. 523–601, John Wiley & Sons, New York.

Poole, C.P. (1967e). *Electron Spin Resonance, a Comprehensive Treatise on Experimental Techniques*, pp. 430–439, John Wiley & Sons, New York.

Poole, C.P. (1967f). *Electron Spin Resonance, a Comprehensive Treatise on Experimental Techniques*, pp. 387–426, John Wiley & Sons, New York.

Poole, C.P. (1983). *Electron Spin Resonance, a Comprehensive Treatise on Experimental Techniques*, 2nd edition, John Wiley & Sons, New York.

Pople, J.A., Beveridge, D.L., and Dobosh, P.A. (1968). Molecular orbital theory of the electronic structure of organic compounds. II. Spin densities in paramagnetic species. *J. Am. Chem. Soc.* **90**: 4201–4209.

Radner, F., Rassat, A., and Hervsall, C.-J. (1996). Preparation of perdeuteriated 2,5-di-*tert*-butyl-3,4-di(methoxycarbonyl)pyrroloxyl: A stable nitroxide free radical with remarkably narrow intrinsic EPR linewidth. *Acta Chem. Scand.* **50**: 146–149.

Redfield, A.G. (1955). Nuclear magnetic resonance saturation and rotary saturation in solids. *Phys. Rev.* **98**: 1787–1809.

Roschmann, P. (1987). Radiofrequency penetration and absorption in the human body: Limitations to high-field whole-body nuclear magnetic resonance imaging. *Med. Phys.* **14**: 922–928.

Rosen, G.M., Halpern, H.J., Brunsting, L.A., Spencer, D.P., Strauss, K.E., Bowman, M.J., and Wechsler, A.S.

(1988). Direct measurement of nitroxide pharmacokinetics in isolated hearts situated in a low-frequency electron spin resonance spectrometer: Implications for spin trapping and *in vivo* oxymetry. *Proc. Natl. Acad. Sci. USA* **85**: 7772–7776.

Sarna, T., Duleba, A., Korytowski, W., and Swartz, H. (1980). Interaction of melanin with oxygen. *Arch. Biochem. Biophys.* **200**: 140–148.

Sczaniecki, P.B., Hyde, J.S., and Froncisz, W. (1990). Continuous wave multiquantum electron paramagnetic resonance spectroscopy. *J. Chem. Phys.* **93**: 3891–3898.

Sczaniecki, P.B., Hyde, J.S., and Froncisz, W. (1991). Continuous wave multiquantum electron paramagnetic resonance spectroscopy. II. Spin-system generated intermodulation sidebands. *J. Chem. Phys.* **94**: 5907–5916.

Shirley, J.H. (1965). Solution of the Schrödinger equation with a Hamiltonian periodic in time. *Phys. Rev.* **138**: B979-B987.

Shkrob, I.A., and Trifunac, A.D. (1997). Spin and time-resolved magnetic resonance in radiation chemistry. Recent developments and perspectives. *Radiat. Phys. Chem.* **50**: 227–243.

Slichter, C.P. (1988). *Principles of Magnetic Resonance*, 3rd edition, p 38, Springer Verlag, Berlin.

Smith, G.W. (1964). Modulation effects in magnetic resonance: Widths and amplitudes for Lorentzian and Gaussian lines. *J. Appl. Phys.* **33**: 1217–1221.

Strangeway, R.A., Ishii, T.K., and Hyde, J.S. (1988). Low-phase-noise Gunn diode oscillator design. *IEEE Trans. Microwave Theory and Techniques* **36**: 792–794.

Subczynski, W.K., Lukiewicz, S., and Hyde, J.S. (1986). Murine *in vivo* L-band ESR spin-label oximetry with a loop-gap resonator. *Magn. Reson. Med.* **3**: 747–754.

Takeshita, K., Utsumi, H., and Hamada, A. (1993). Whole mouse measurement of paramagnetism—Loss of nitroxide free radical in lung with a L-band ESR spectrometer. *Biochem. Mol. Biol. Int.* **29**: 17–24.

Testa, L., Gualtieri, G., and Sotgui, A. (1993). Electron paramagnetic resonance imaging of a model of a beating heart. *Phys. Med. Biol.* **38**: 259–266.

Utsumi, H., Muto, E., Masuda, S., and Hamada, A. (1990). *In vivo* ESR measurement of free radicals in whole mice. *Biochem. Biophys. Res. Commun.* **172**: 1342–1348.

Vahidi, N., Clarkson, R.B., Liu, K.J., Norby, S.W., Wu, M., and Swartz, H.M. (1994) *In vivo* and *in vitro* EPR oximetry with fusinite: A new coal-derived, particulate EPR Probe. *Magn. Reson. Med.* **31**: 139–146.

Voigt, W. (1912). Uber das gesetz der intensitatsverteilung innerhalb der linien eines gassperktrums. *S.B. Bayer Akad. Wiss.* 603–620.

Wahlquist, H. (1961). Modulation broadening of unsaturated Lorentzian lines. *J. Chem. Phys.* **35**: 1708–1710.

Weil, J.A., Bolton, J.R., and Wertz, J.E. (1994). *Electron Spin Resonance: Elementary Theory and Practical Applications*, pp. 239–259, John Wiley & Sons, New York.

Wertz, J.E., and Bolton, J.R. (1986). *Electron Spin Resonance: Elementary Theory and Practical Applications*, Chapman and Hall, New York.

Windle, J.J. (1981). Hyperfine coupling constants for nitroxide spin probes in water and carbon tetrahydrochloride. *J. Magn. Reson.* **45**: 432–439.

Yoshimura, T., Yokoyama, H., Fujii, S., Takayama, F., Oikawa, K., and Kamada, H. (1996). *In vivo* EPR detection and imaging of endogenous nitric oxide in lipopolysaccharide-treated mice. *Nat. Biotechnol.* **14**: 992–994.

Zavoiskii, E. (1945a). Paramagnetic relaxation of liquid solutions for perpendicular fields. *J. Phys.* **9**: 211–216.

Zavoiskii, E. (1945b). Spin-magnetic resonance in paramagnetic substances. *J. Phys.* **9**: 245.

Zavoiskii, E. (1946). Paramagnetic absorption in some salts in perpendicular magnetic fields. *J. Phys.* **10**: 170–173.

Zeeman, P. (1913). *Researches in Magneto-Optics, With Special Reference to the Magnetic Resolution of Spectrum Lines*, Macmillan, London.

Zweier, J.L., and Kuppusamy, P. (1988). Electron paramagnetic resonance measurements of free radicals in the intact beating heart: A technique for detection and characterization of free radicals in whole bological tissues. *Proc. Natl. Acad. Sci. USA* **85**: 5703–5707.

Zweier, J.L., Thompson-Gorman, S., and Kuppusamy, P. (1991). Measurement of oxygen concentrations in the intact beating heart using electron paramagnetic resonance spectroscopy: A technique for measuring oxygen concentrations *in situ. J. Bioenerg. Biomembr.* **23**: 855–871.

8

Kinetics of Spin Trapping Free Radicals

Failure to observed a spin adduct does not prove the absence of radicals. It may simply reflect low rates of trapping or the formation of a short-lived spin adduct.

Marriott, Perkins, and Griller, 1980

The detection of free radicals, generated from either purified enzyme preparations, isolated cell suspensions, or animal models, is dependent upon the ability of the nitrone, the nitrosoalkane, or the ferro-chelate to compete favorably for the free radical with other biologic targets. The efficiency of spin trapping, most often the limiting factor in the identification of specific free radicals, is, therefore, reflected in the differing rate constants for a variety of similar, yet diverse, reactions (Ingold, 1973; Marriott, *et al.*, 1980). Although, at first glance, the number and complexity of potential interactions may appear to be overwhelming, one should not be deterred by the effort required to obtain reliable estimates of specific rate constants. The scope of the problem can be illustrated by considering the following example. At a constant flux of an initiating free radical (X·), the detection of spin trapped adducts, ST–X· and ST–Y·, is influenced by the rate constants, k_2, k_3, k_4, k_5, and k_6—derived from the rate of secondary free radical formation, the rates of spin trapping the various free radicals, and the rates of different spin trapped adduct decomposition (Rosen and Rauckman, 1981; Janzen, 1984):

$$X \xrightarrow{k_1} X\cdot$$

$$X\cdot + Y \xrightarrow{k_2} Y\cdot + X$$

$$X\cdot + ST \xrightarrow{k_3} ST\text{-}X\cdot$$

$$Y\cdot + ST \xrightarrow{k_4} ST\text{-}Y\cdot$$

$$ST\text{--}X\cdot \xrightarrow{k_5} \text{ nonradical species}$$

$$ST\text{--}Y\cdot \xrightarrow{k_6} \text{ nonradical species}$$

Despite the uncertainty as to whether specific spin trapped adducts can be identified in the biological milieu, success has been achieved either through good fortune or, in some cases, cognizance of the relevant rate constants. In this chapter, we will begin to place these qualitative estimates of reaction rates on sound mathematical foundations by developing kinetic models (Ingold, 1973; Janzen and Evans, 1973; Schmidt and Ingold, 1978) for spin trapping the most frequently studied of the biologically generated free radicals, $O_2\cdot^-$, $HO\cdot$, small carbon-centered and sulfur-centered free radicals, and $NO\cdot$.

Superoxide

The early literature describing the spin trapping of $O_2\cdot^-$ by DMPO in aqueous solutions is confusing. One publication reported that this reaction appeared to be diffusion limited (Harbour, et al., 1974), whereas another communication questioned whether there was any reaction between $O_2\cdot^-$ and DMPO (Lai and Piette, 1977). This disparity is not surprising when one considers known reaction mechanisms—a faster rate might be anticipated if the addition of $O_2\cdot^-$ to a spin trap were by a free radical reaction in contrast to an anionic nucleophilic attack at a double bond. There is some evidence for $O_2\cdot^-$ acting as a strong nucleophile in aqueous solutions (Divisek and Kastening, 1975; Fee and Valentine, 1977), even though there is a rich literature describing its reactivity in aprotic organic solvents (Dietz, et al., 1970; Merritt and Sawyer, 1970; Johnson and Nidy, 1975; San Filippo, et al., 1975; Danen and Warner, 1977; Merritt and Johnson, 1977; Saito, et al., 1979; Afanas'ev, 1989a). At acidic pH, the reaction of $HO_2\cdot$ toward an activated electron rich center has likewise received limited attention (Divisek and Kastening, 1975; Fee and Valentine, 1977).

For spin trapping $O_2\cdot^-$, instead of determining absolute rate constants for these reactions (Chateauneuf, et al., 1988), apparent rate constants using two different competitive kinetic models have been developed: the self-dismutation of $O_2\cdot^-$ and the inhibition by SOD (Finkelstein, et al., 1980). Of the possible sources of this free radical, the oxidative metabolism of xanthine by xanthine oxidase was chosen, due to the lack of enzymic inhibition by various spin traps, such as α-(4-pyridyl 1-oxide)-N-tert-butylnitrone (4-POBN), PBN, DMPO, and 2,5,5-trimethyl-1-pyrroline-N-oxide (TMPO), even at concentrations as high as 200 mM (Finkelstein, et al., 1979). In contrast, other $O_2\cdot^-$ generating systems, such as the reaction of $HO\cdot$ with H_2O_2 (Harbour, et al., 1974), were discarded, since the spin trapping of $HO\cdot$ was found to be considerably faster than was the addition of $O_2\cdot^-$ to these nitrones:

$$HO\cdot + H_2O_2 \rightarrow HO_2\cdot + H_2O$$

The first approach examined the effect of varying the spin trap concentration at a fixed pH and constant flux of $O_2 \cdot^-$. An aqueous solution of TMPO[1] was prepared in the presence of a continued flux of $O_2 \cdot$ (Figure 8.1). Loss of $O_2 \cdot^-$ is either through disproportionation to H_2O_2 or by reaction with this nitrone (Finkelstein, et al., 1979). As the spin trap concentration is enhanced, the amount of $O_2 \cdot^-$ spin trapped is increased, shifting the equilibrium away from the dismutation pathway:

$$HO_2 \cdot \;\rightleftarrows\; H^+ + O_2 \cdot^- \tag{8.1}$$

$$HO_2 \cdot + HO_2 \cdot \rightarrow H_2O_2 + O_2 \tag{8.2}$$

$$O_2 \cdot^- + TMPO \rightarrow TMPO-OO^- + H^+ \rightarrow TMPO-OOH \tag{8.3}$$

$$HO_2 \cdot + TMPO \rightarrow TMPO{-}OOH \tag{8.4}$$

If the rates of spin trapping and of dismutation are known, the rate constant for the reaction of TMPO with $O_2 \cdot^-$ can be determined, assuming the flux of $O_2 \cdot^-$, s, remains uniform and the concentration of TMPO is unchanged over the lifetime of the experiment. If k_d and k_t represent the pH-dependent apparent rate constants for the dismutation and spin trapping of $O_2 \cdot^-$, respectively[2], then, the disappearance of this free radical can be obtained from equation [8.5]:

$$-d[O_2 \cdot^-]/dt = k_d[O_2 \cdot^-]^2 + k_t[TMPO][O_2 \cdot^-] \tag{8.5}$$

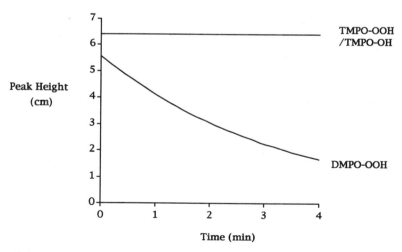

Figure 8.1 Relative stability of TMPO–OOH (and TMPO–OH) and DMPO–OOH, produced by the reaction of the respective spin trap with a DMF solution of tetramethylammonium superoxide in a phosphate buffer at pH 7.8. (Taken from Finkelstein, et al., 1979.)

Applying the steady-state assumption[3] that $d[O_2 \cdot^-]/dt = s$, the following relationship between TMPO and $O_2 \cdot^-$ can be derived:

$$k_d[O_2 \cdot^-]^2 + k_t[TMPO][O_2 \cdot^-] = s \qquad [8.6]$$

If we let the rate of spin trap adduct formation, $d[TMPO–OOH]/dt$ be represented as V, then

$$V = k_t[TMPO][O_2 \cdot^-]. \qquad [8.7]$$

Then, equation [8.6] can be represented as

$$V = s - k_d[O_2 \cdot^-]^2 \qquad [8.8]$$

Rearranging equation [8.7] in terms of $[O_2 \cdot^-]$ leads to

$$[O_2 \cdot^-] = V/k_t[TMPO] \qquad [8.9]$$

Substitution for $O_2 \cdot^-$ in equation [8.8] results in

$$V = s - (k_d V^2 / k_t^2 [TMPO]^2) \qquad [8.10]$$

Equation [8.10] contains terms that are measurable and the expression can be presented graphically. Thus, by plotting V versus $V^2/[TMPO]^2$, a straight line can be obtained, the slope of which is $-k_d/k_t^2$ and the intercept of which is s.

The validity of equation [8.10] was tested by varying the concentration of TMPO at a fixed $O_2 \cdot^-$ flux, generated during the metabolism of xanthine by xanthine oxidase at pH 8.1 (Finkelstein, *et al.*, 1980). As shown in Figure 8.2, the rate of the reaction between $O_2 \cdot^-$ and TMPO was dependent on the concentration of the spin trap. From equation [8.10], we may set the extrapolated value of V at infinite TMPO concentration (see Figure 8.2), equal to the calculated flux of $O_2 \cdot^-$, s, as determined by the SOD-inhibitable reduction of ferricytochrome c (Kuthan, *et al.*, 1982). The slope of the line in Figure 8.2 is $-2.46 \times 10^4 \, M \cdot s$. Since this number is equal to $-k_d/k_t^2$, k_t can be calculated from the k_d, the dismutation rate constant, which at pH 8.1 is $5.1 \times 10^4 \, M^{-1} s^{-1}$ (data taken from Figure 2.3). By solving the equation for k_t, the apparent rate constant for the spin trapping of $O_2 \cdot^-$ by TMPO was determined to be $1.44 \, M^{-1} s^{-1}$. In a similar manner, the apparent rate constant for the reaction of DMPO with $O_2 \cdot^-$ was determined to be $15.7 \, M^{-1} s^{-1}$ (Table 8.1). Considering the enormous disparity in rate constants between competing reaction pathways—dismutation with a k_d of $5.1 \times 10^4 \, M^{-1} s^{-1}$ and spin trapping with a k_t of 1.44 for TMPO and $15.7 \, M^{-1} s^{-1}$ for DMPO—a high concentration of various Δ^1-pyrroline-*N*-oxides is an essential ingredient for spin trapping $O_2 \cdot^-$ in aqueous solutions.

Competition using SOD as the inhibitor provides the second kinetic model for determining the apparent rate constant for spin trapping $O_2 \cdot^-$ (Sawada and Yamazaki, 1973; Asada, *et al.*, 1974). The choice of SOD as the antagonist is supported by the following (Klug, *et al.*, 1972; Fridovich, 1978). First, the enzyme

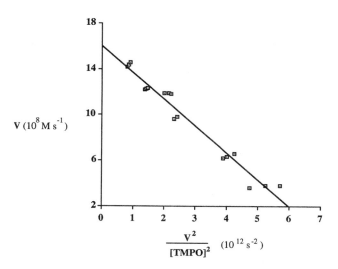

Figure 8.2 Effect of varying concentrations of TMPO on spin trapping $O_2\cdot^-$, generated at a constant flux of 0.16 μMs^{-1}. From the slope of this line and the k_d of $5.1 \times 10^4\,M^{-1}s^{-1}$ at pH 8.1, the apparent rate constant, k_t, was found to be 1.44 $M^{-1}s^{-1}$. (Taken from Finkelstein, *et al.*, 1980.)

Table 8.1 Apparent Rate Constants for the Spin Trapping of Superoxide

Spin Trap	Rate Constant $(M^{-1}s^{-1})$	Experimental Condition
TMPO[a]	7	Xanthine/xanthine oxidase, pH 7.8, competition with SOD
TMPO[a]	1.44	Xanthine/xanthine oxidase, pH 8.1, competition with dismutation of $O_2\cdot^-$
DMPO[a]	10	Xanthine/xanthine oxidase, pH 7.8, competition with TMPO
DMPO[a]	15.7	Light/riboflavin/DTPA, pH 8.0, competition with SOD
DMPO[b]	10.2	Aerobic solution of MPDP[+], pH 10, competition with SOD
DMPO[c]	1.2	NADPH–cytochrome P-450 reductase/paraquat, pH 7.4 competition with dismutation of $O_2\cdot^-$
DMPO[d]	18	Xanthine/xanthine oxidase, pH 7.8, competition with SOD
DMPO[d]	310	Xanthine/xanthine oxidase, pH 6.2, competition with SOD
DMPO[e]	12	PMA-stimulated neutrophils, pH 7.4, competition with SOD
BMPO[f]	4.6	Xanthine/xanthine oxidase, pH 7.8, competition with SOD
DPPO[f]	2.0	Xanthine/xanthine oxidase, pH 7.8, competition with SOD
CPPO[f]	2.1	Xanthine/xanthine oxidase, pH 7.8, competition with SOD
M₄PO[g]	1.0	Xanthine/xanthine oxidase, pH 7.8, competition with SOD
DEMPO[g]	0	Xanthine/xanthine oxidase, pH 7.8, competition with SOD
M₄EEPO[h]	0	Xanthine/xanthine oxidase, pH 7.8, competition with SOD
DEPMPO[i]	60	Xanthine/xanthine oxidase, pH 7.0, competition with DMPO

[a]TMPO = 2,5,5-trimethyl-1-pyrroline-*N*-oxide, DMPO = 5,5-dimethyl-1-pyrroline-*N*-oxide, Finkelstein, *et al.*, 1980. [b]DMPO, Zang and Misra, 1992. [c]DMPO, Yamazaki, *et al.*, 1990. [d]DMPO, Mitsuta, *et al.*, 1994. [e]DMPO, Kotake, *et al.*, 1994. [f]BMPO = 5-butyl-5-methyl-1-pyrroline-*N*-oxide, DPPO = 5,5-dipropyl-1-pyrroline-*N*-oxide, CPPO = 2-aza-2-cyclopentenespirocyclo-pentane 2-oxide, Turner and Rosen, 1986. [g]M₄PO = 3,3,5,5-tetramethyl-1-pyrroline-*N*-oxide, DEDMPO = 3,3-diethyl-5,5-dimethyl-1-pyrroline-*N*-oxide, Rosen and Turner, 1988. [h]M₄EEPO = 4-[(2-ethoxycarbonyl)ethyl]-3,3,5,5-tetramethyl-1-pyrroline-1-oxide, Arya, *et al.*, 1992. DEPMPO = 5-diethylphosphoryl-5-methyl-1-pyrroline-*N*-oxide, Fréjaville, *et al.*, 1994.

has a high degree of specificity towards $O_2 \cdot^-$. Second, the reaction is first order with regard to enzyme and its substrate. Third, the rate constant is uniform and pH independent over the range from 5.0 to 9.5. Finally, the catalytic removal of $O_2 \cdot^-$ does not diminish the effective concentration of the enzyme. To decrease the complexity of the various reaction pathways, as defined by equations [8.2], [8.11], and [8.12], the concentration of DMPO was set at 180 mM.[4] With the concentration of $O_2 \cdot^-$ approximately 1–10 μM, this DMPO concentration is sufficiently high to trap essentially all of the $O_2 \cdot^-$ in the absence of SOD.[5] The dismutation equation [8.2] can, therefore, be dropped from the overall rate expression, since $k_d[O_2 \cdot^-]^2$ is small with respect to $k_t[DMPO][O_2 \cdot^-]$. This leaves two competing schemes to eliminate $O_2 \cdot^-$:

$$O_2 \cdot^- + DMPO \rightarrow DMPO - OO^- + H^+ \rightarrow DMPO - OOH \qquad [8.11]$$

$$O_2 \cdot^- + H^+ + SOD \rightarrow \tfrac{1}{2} H_2O_2 + \tfrac{1}{2} O_2 \qquad [8.12]$$

Thus, from equations [8.11] and [8.12], the rate of $O_2 \cdot^-$ elimination can be represented as

$$V = -d[O_2 \cdot^-]/dt = k_{app}[DMPO][O_2 \cdot^-] + k_{SOD}[SOD][O_2 \cdot^-] \qquad [8.13]$$

In the absence of SOD, equation [8.11] can be described as

$$v = -d[O_2 \cdot^-]/dt = d[DMPO - OOH]/dt = k_{app}[DMPO][O_2 \cdot^-] \qquad [8.14]$$

By substituting equation [8.14] into equation [8.13] and rearranging the terms, competing reactions can be displayed as ratios of reactants and rate constants:

$$V/v = 1 + k_{SOD}[SOD]/k_{app}[DMPO] \qquad [8.15]$$

As the initial rate of spin trapping is first order with respect to $[O_2 \cdot^-]$, $k_{SOD}/k_{app}[DMPO]$ becomes the constant, k', at fixed [DMPO]. Then, equation [8.15] can be simplified to

$$V/v = 1 + k'[SOD] \qquad [8.16]$$

By plotting V/v versus [SOD], a straight line is obtained, the slope of which is k′ (Asada, et al., 1974). From these data, the k_{app} at pH 7.8 for DMPO was found to be $16 \, M^{-1} \, s^{-1}$ and $7 \, M^{-1} \, s^{-1}$ for TMPO (Table 8.1, Figure 8.3). This kinetic model has been verified in other laboratories (Yamazaki, et al., 1990; Zang and Misra, 1992; Fréjaville, et al., 1994, 1995). With this kinetic model, apparent rate constants for the reaction of $O_2 \cdot^-$ with other nitrones have been calculated (Table 8.1).

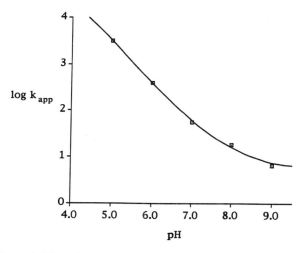

Figure 8.3 Effect of pH on the apparent rate constant for spin trapping $O_2 \cdot^-$ by DMPO. The line, a plot of equation [8.18], provides a reasonable fit to the experimental data of 6.6×10^3 $M^{-1}s^{-1}$, 10 $M^{-1}s^{-1}$, and 4.88 for $k_{HO_2 \cdot}$, $k_{O_2 \cdot^-}$, and $pK_a = 4.88$, respectively. (Taken from Finkelstein, et al., 1980.)

pH Dependence on Spin Trapping Superoxide

The reactiveness of $HO_2 \cdot$ and $O_2 \cdot^-$ is notably different, as illustrated by the breadth of dismutation rate constants found over a broad pH range (Figure 2.3). It is, therefore, essential to determine rate constants for spin trapping $O_2 \cdot^-$ in distinct biologic milieus. In so doing, this approach verifies that the rate constant is dependent upon the pH of the solution. This is particularly important as $O_2 \cdot^-$ has been identified in such dissimilar environments as the acidic phagosomal vacuole of various leukocytes to the more basic fluid of extracellular space. Because the flux of $O_2 \cdot^-$, produced by the action of xanthine oxidase on xanthine, is pH dependent with a maximal $O_2 \cdot$ rate at pH 7.8 (McCord and Fridovich, 1968; McCord, et al., 1977), an alternative source of $O_2 \cdot^-$ was sought for studying the effect of pH on the rate of spin trapping. After investigating several models, a light–riboflavin–DTPA system (Fife and Moore, 1979) was chosen, as it is neither pH sensitive nor limited by spin trap concentration (at levels $\leq$ 200 mM).

Using the kinetic competition model described above with SOD as the competitive inhibitor, a family of apparent rate constants can be calculated over a broad range of pHs. Plotting these numbers as a function of pH results in a graph with a changing slope (Figure 8.3). Between pH 5 and 7, the rate constants decrease precipitously. Thereafter, the values begin to level off. In fact, at pH $\geq$ 8, the rate constant is, for all practical purposes, unchanged (compare data from Figure 8.3 and Table 8.1). This shift in slope as a function of pH can be explained, if we assume that DMPO reacts with $HO_2 \cdot$ and $O_2 \cdot^-$ at different rates.

Equations [8.3] and [8.4] must be modified to take into account the concentrations of $O_2 \cdot^-$ and $HO_2 \cdot$ available to be spin trapped:

$$k_{8.3} = k_{O_2^-}[O_2^{\cdot-}]/([HO_2\cdot] + [O_2^{\cdot-}]) \tag{8.17}$$

$$k_{8.4} = k_{HO_2}[HO_2\cdot]/([HO_2\cdot] + [O_2^{\cdot-}]) \tag{8.18}$$

where $k_{8.3}$ and $k_{8.4}$ are the new pH-dependent rate constants for spin trapping $O_2^{\cdot-}$ and $HO_2\cdot$. From the Henderson-Hasselbach relationship, let

$$x = K_a/[H^+] = 10^{(pH-pK_a)} \tag{8.19}$$

where K_a is the dissociation constant for $HO_2\cdot$ /$O_2^{\cdot-}$. Thus, the apparent rate constant for the reaction of $HO_2\cdot$ with DMPO as a function of pH is $k_{HO_2}[1/(1 + x)]$, while the apparent rate constant for the reaction of $O_2^{\cdot-}$ with DMPO as a function of pH is $k_{O_2^-}[x/(1 + x)]$. Then, the total apparent rate constant, taking into account the contribution of $HO_2\cdot$ and $O_2^{\cdot-}$, becomes

$$k_{app} = [k_{HO_2} + k_{O_2^-}(x)]/(1 + x) \tag{8.20}$$

This equation provides a reasonable fit to the experimental data in Figure 8.3, when $k_{HO_2} = 6.6 \times 10^3 \, M^{-1} \, s^{-1}$, $k_{O_2^-} = 10 \, M^{-1} \, s^{-1}$, and $pK_a = 4.88$ (Behar, *et al.*, 1970). The enormous disparity in rate constants between $HO_2\cdot$ and $O_2^{\cdot-}$ suggests two different mechanisms must undoubtedly account for the formation of DMPO–OOH. At pH values of < 4.88, where the concentration of $HO_2\cdot$ dominates, a free radical reaction is a likely pathway to this spin trapped adduct. Under more basic conditions where pH > 7, a Michael-type addition to the nitrone, with its slower rate of reaction, is an acceptable model. Further evidence for this hypothesis comes from the finding that the rate constant for reaction of DMPO with $(CH_3)_3CO\cdot$ in cyclohexane is $5 \times 10^8 \, M^{-1} \, s^{-1}$ (Janzen and Evans, 1973). The enormous difference in the rate constants between the spin trapping of $(CH_3)_3CO\cdot$ and $O_2^{\cdot-}$ by DMPO cannot be explained by an increase in the polarity of the solvent, moving from cyclohexane to an aqueous buffer.

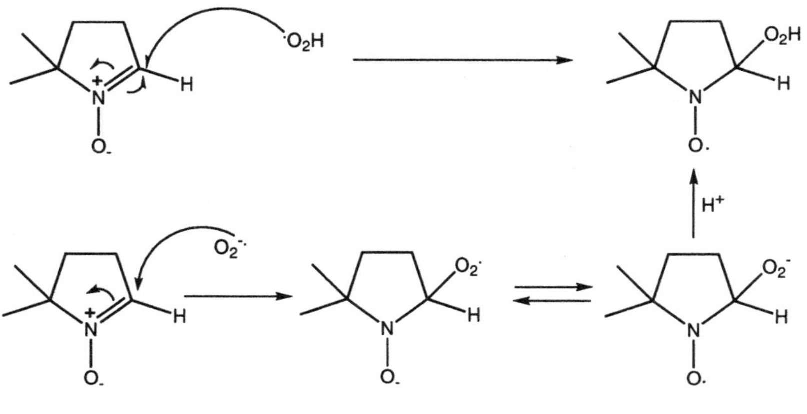

Hydroxyl Radical

A competitive kinetic model has been developed to determine rate constants for the spin trapping of HO· (Finkelstein, *et al.*, 1980; Marriott, *et al.*, 1980; Lown and Chen, 1981; Morehouse and Mason, 1988; Buettner and Mason, 1990; Hanna and Mason, 1992). For a preponderance of these experiments, ethanol was chosen as the competitive inhibitor (Figure 8.4), due to its lack of interference with the optical production of HO· and the publication of the rate constant for the reaction (Buxton, *et al.*, 1988), making the solution for the derived equation possible. Although two different sources of HO·—UV photolysis of H_2O_2 and Fe^{2+} reduction of H_2O_2—have been utilized, for discussion purposes the first model will be illustrated. Consider the photolysis of low concentrations of H_2O_2:

$$H_2O_2 + hv \rightarrow 2HO\cdot \qquad [8.21]$$

In the presence of ethanol, the loss of this free radical is either through an abstraction of a hydrogen atom from this alcohol (Asmus, *et al.*, 1973; Pou, *et al.*, 1994) or reaction with DMPO:

$$HO\cdot + CH_3CH_2OH \rightarrow CH_3 \cdot CHOH + H_2O \qquad [8.22]$$

$$HO\cdot + DMPO \rightarrow DMPO-OH \qquad [8.23]$$

$$CH_3 \cdot CHOH + DMPO \rightarrow DMPO-CH(CH_3)OH \qquad [8.24]$$

At a given steady-state production of HO·, the rate of spin trapped adduct formation, $d[DMPO-OH]/dt$ in the presence of ethanol can be expressed as

$$V = -d[HO\cdot]/dt = k_{app}[DMPO][HO\cdot] + k_e[CH_3CH_2OH][HO\cdot] \qquad [8.25]$$

where k_{app} and k_e are the rates constants for the spin trapping of HO· and the second-order reaction with ethanol. In the absence of ethanol, equation [8.25] can be simplified to:

$$v = -d[HO\cdot]/dt = d[DMPO-OH]/dt = k_{app}[DMPO][HO\cdot] \qquad [8.26]$$

Thus, by dividing equation [8.25] by equation [8.26], the rate of spin trapping HO· in the presence and absence of ethanol can be expressed as a function of the concentration of spin trap and ethanol:

$$V/v = 1 + k_e[CH_3CH_2OH]/k_{app}[DMPO] \qquad [8.27]$$

A typical plot of $(V/v) - 1$ versus $[CH_3CH_2OH]/[DMPO]$ results in a straight line; the slope k_e/k_{app} is 0.52 (Figure 8.5). Assuming k_e equals $1.8 \times 10^9 \, M^{-1} s^{-1}$ (Burton, *et al.*, 1988), then k_{app} becomes $3.4 \times 10^9 \, M^{-1} s^{-1}$. Rate constants for the reaction of various spin traps with HO· are shown in Table 8.2. Not surprisingly, these values are large, with no statistical difference between Δ^1-pyrroline-*N*-oxides and aromatic conjugated alkane-*N*-oxides.

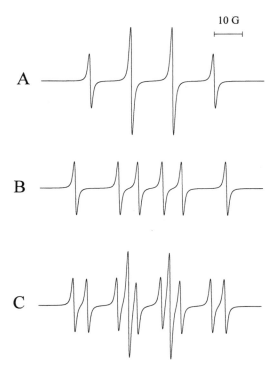

Figure 8.4 Effect of ethanol on the spin trapping of HO· by DMPO. (A) The EPR spectral simulation of DMPO–OH with $A_N = A_H = 14.9\,\mathrm{G}$. (B) The EPR spectral simulation of DMPO–CH(CH₃)OH with $A_N = 15.8\,\mathrm{G}$ and $A_H = 22.8\,\mathrm{G}$. (C) The EPR spectral simulation of an equal intensity of DMPO–OH and DMPO–CH(CH₃)OH.

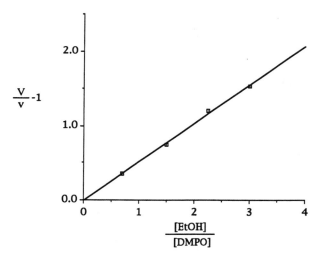

Figure 8.5 Inhibition of DMPO spin trapping HO· by ethanol. Hydroxyl radical was produced by photolysis of H_2O_2 (176 mM) in phosphate buffer, pH 7.4. (Taken from Finkelstein, *et al.*, 1980.)

Table 8.2 Apparent Rate Constants for the Spin Trapping of Hydroxyl Radical

Spin Trap	Rate Constant $(\times 10^9 M^{-1} s^{-1})$	Experimental Condition
DMPO[a]	2.7	Photolysis of H_2O_2, pH 7, competition with methanol
DMPO[b]	2.1	Fenton reaction, pH 7.4, competition with ethanol
DMPO[b]	3.4	Photolysis of H_2O_2, pH 7.4, competition with ethanol
DMPO[c]	2.6	Photolysis of H_2O_2, pH 7, competition with sodium formate
DMPO[d]	5.2	Ferredoxin/NADPH–ferredoxin reductase/NADPH, pH 7.4, H_2O_2, competition with ethanol
DMPO[e]	2.0	Ferredoxin/NADPH–ferredoxin reductase/NADPH, pH 7.4, H_2O_2, competition with sodium formate
DMPO[f]	4.3	Ionizing radiation, aqueous solutions of N_2O
DMPO[g]	3.6	Ionizing radiation, aqueous solutions of N_2O
TMPO[h]	3.8	Fenton reaction, pH 7.4, competition with ethanol
M_4PO[i]	1.2	Photolysis of H_2O_2, pH 7.4, competition with DMPO
M_3PO[i]	8.1	Photolysis of H_2O_2, pH 7.4, competition with DMPO
M_4EEPO[j]	4.3	Photolysis of H_2O_2, pH 7.4, competition with ethanol
M_4CEPO[k]	3.1	Photolysis of H_2O_2, pH 7.4, competition with ethanol
DEPMPO[l]	7.8	Photolysis of H_2O_2, pH 7.0, competition with ethanol
MPPO[m]	14	Photolysis of H_2O_2, competition with PBN
4-POBN[g]	4.0	Ionizing radiation, aqueous solutions of N_2O
4-POBN[n]	1.9	Fenton reaction, pH 7.4, competition with ethanol
4-POBN[o]	3.5	Ionizing radiation, aqueous solutions of N_2O
2-POBN[o]	3.2	Ionizing radiation, aqueous solutions of N_2O
3-POBN[g]	4.5	Ionizing radiation, aqueous solutions of N_2O
3-POBN[o]	4.8	Ionizing radiation, aqueous solutions of N_2O
PBN[g]	6.1	Ionizing radiation, aqueous solutions of N_2O
PBN[p]	8.5	Ionizing radiation, aqueous solutions of N_2O
4-MePyBN[q]	0.5	Photolysis of H_2O_2, pH 6, competition with methanol
4-PyBN[r]	8.3	Ionizing radiation, aqueous solutions of N_2O
3-PyBN[r]	4.6	Ionizing radiation, aqueous solutions of N_2O
2-PyBN[r]	9.6	Ionizing radiation, aqueous solutions of N_2O
4-MePBN[s]	4.7	Ionizing radiation, aqueous solutions of N_2O
4-MeOPBN[s]	6.4	Ionizing radiation, aqueous solutions of N_2O
4-CNPBN[s]	6.5	Ionizing radiation, aqueous solutions of N_2O
4-NO_2PBN[s]	7.4	Ionizing radiation, aqueous solutions of N_2O

[a]DMPO = 5,5-dimethyl-1-pyrroline-N-oxide, Marriott, et al., 1980. [b]DMPO, Finkelstein, et al., 1980. [c]DMPO, Castelhano, et al., 1983. [d]DMPO, Morehouse and Mason, 1988. [e]DMPO, Morehouse and Mason, 1988. [f]DMPO, Neta, et al., 1980. [g]DMPO, Sridhar, et al., 1986; PBN = N-tert-butyl-α-phenylnitrone, Sridhar, et al., 1986; [h]TMPO = 2,5,5-trimethyl-1-pyrroline-N-oxide, Finkelstein, et al., 1980. [i]M_4PO = 3,3,5,5-tetramethyl-1-pyrroline-N-oxide, Nishi, et al., 1992. M_3PO = 2,5,5-trimethyl-1-pyrroline-1-oxide, Nishi, et al., 1992. [j]M_4EEPO = 4-[(2-ethoxycarbonyl)ethyl]-3,3,5,5-tetramethyl-1-pyrroline-1-oxide, Arya, et al., 1992. [k]M_4CEPO = 4-carbethoxy-3,3,5,5-tetramethyl-1-pyrroline-N-oxide, Arya, et al., 1992. [l]DEPMPO = 5-diethylphosphoryl-5-methyl-1-pyrroline-N-oxide, Fréjaville, et al., 1994. [m]MPPO = 5-methyl-5-phenyl-1-pyrroline-N-oxide, Sankuratri, et al., 1996. [n]4-POBN = α-(4-pyridyl 1-oxide)-N-tert-butylnitrone, Finkelstein, et al., 1980. [o]4-POBN = α-(4-pyridyl 1-oxide)-N-tert-butylnitrone, Neta, et al., 1980; 2-POBN = α-(2-pyridyl 1-oxide)-N-tert-butylnitrone, Neta, et al., 1980; 3-POBN = α-(3-pyridyl 1-oxide)-N-tert-butylnitrone, Neta, et al., 1980. [p]PBN, Greenstock and Wiebe, 1982. [q]4-MePyBN = α-(4-methylpyridyl)-N-tert-butylnitrone methyl sulfate, Marriott, et al., 1980. [r]4-PyBN = α-(4-pyridyl)-N-tert-butylnitrone, Sridhar, et al., 1986; 3-PyBN = α-(3-pyridyl)-N-tert-butylnitrone, Sridhar, et al., 1986; 2-PyBN = α-(2-pyridyl)-N-tert-butylnitrone, Sridhar, et al., 1986. [s]4-MePBN = N-tert-butyl-α-(4-methyl)phenylnitrone, Greenstock and Wiebe, 1982; 4-MeOPBN = N-tert-butyl-α-(4-methoxy)phenylnitrone, Greenstock and Wiebe, 1982; 4-CNPBN = N-tert-butyl-α-(4-cyano)phenylnitrone, Greenstock and Wiebe, 1982; 4-NO_2PBN = N-tert-butyl-α-(4-nitro)phenylnitrone, Greenstock and Wiebe, 1982.

Small Carbon-Centered Free Radicals

As with $O_2 \cdot^-$ and $HO\cdot$, rate constants for the spin trapping of small carbon-centered free radicals can be obtained using a kinetic model based on the competition between different free radicals (Janzen, et al., 1975; Finkelstein, et al., 1980; Marriott, et al., 1980; Morehouse and Mason, 1988; Pou, et al., 1994). For illustrative purposes, the spin trapping of $CH_3 \cdot CHOH$[6] by PBN will be described. If the concentration of ethanol in the reaction mixture is in excess compared with the steady-state concentration of $HO\cdot$, then reaction of this free radical with PBN becomes insignificant and equation [8.28] can be eliminated from the overall rate expression.

$$HO\cdot + PBN \rightarrow PBN-OH \qquad [8.28]$$

With this competitive kinetic model, the fate of $HO\cdot$ is intimately linked to the rate equations [8.22], [8.29], and [8.30]:

$$HO\cdot + CH_3CH_2OH \rightarrow CH_3 \cdot CHOH + H_2O \qquad [8.22]$$

$$CH_3 \cdot CHOH + CH_3 \cdot CHOH \rightarrow CH_3-C(OH)H-CH(OH)CH_3 \qquad [8.29]$$

$$CH_3 \cdot CHOH + PBN \rightarrow PBN-CH(CH_3)OH \qquad [8.30]$$

By combining equations [8.22], [8.29], and [8.30], the disappearance of $CH_3 \cdot CHOH$ can be expressed as

$$- d[CH_3 \cdot CHOH]/dt = -k_e[CH_3CH_2OH][HO\cdot] + k_d[CH_3 \cdot CHOH]^2$$
$$+ k_{app}[PBN][CH_3 \cdot CHOH] \qquad [8.31]$$

Applying the steady-state assumption, $d[CH_3 \cdot CHOH]/dt = 0$, then equation [8.31] can be rearranged as a quadratic expression:

$$k_d[CH_3 \cdot CHOH]^2 + k_{app}[PBN][CH_3 \cdot CHOH] - k_e[CH_3CH_2OH][HO\cdot] = 0$$
$$[8.32]$$

The rate of spin trapping $CH_3 \cdot CHOH$ can then be represented as

$$v = d[PBN-CH(OH)CH_3]/dt = k_{app}[PBN][CH_3 \cdot CHOH] \qquad [8.33]$$

Rearranging this expression in terms of $[CH_3 \cdot CHOH]$ gives

$$[CH_3 \cdot CHOH] = v/k_{app}[PBN] \qquad [8.34]$$

Substituting equation [8.34] into equation [8.32], the quadratic equation can be expressed as

$$k_d v^2/(k_{app}[PBN])^2 + v - C = 0 \qquad [8.35]$$

if $k_e[CH_3CH_2OH][HO\cdot]$ is represented by C. Rearranging terms for v as a function of v^2 leads to

$$v = (-k_d/k_{app}^2)(v^2/[PBN]^2) + C \qquad [8.36]$$

By plotting v versus $v^2/[PBN]^2$, a straight line is obtained, the slope of which is $-k_d/k_{app}^2$ and the intercept of which is C (Figure 8.6). Given that $k_d = 5.5 \times 10^8$ M^{-1} s^{-1} (Taub and Dorfman, 1962), k_{app} for the spin trapping of $CH_3\cdot CHOH$ by PBN was found to be 1.5×10^6 $M^{-1} s^{-1}$. Rate constants for the spin trapping of this and other free radicals by various nitrones are presented in Table 8.3. One rather interesting observation points to the nucleophilic character of various α-hydroxyalkyl radicals (Greenstock and Wiebe, 1982). Not surprisingly, electron withdrawing substitutents at the para position of PBN increase the rate of spin trapping (Table 8.3). Likewise, the nature of the α-hydroxyalkyl radical dictates the

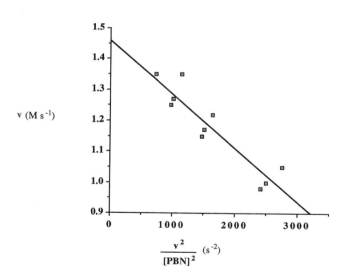

Figure 8.6 Determination of the apparent rate constant for the spin trapping of α-hydroxyethyl radical by PBN. α-Hydroxyethyl radical was generated by photolysis of an ethanol (1.7 M) solution of H_2O_2 (176 mM) in phosphate buffer, pH 7.4. (Taken from Pou, et al., 1994.)

Table 8.3 Apparent Rate Constants for the Spin Trapping of Carbon-Centered and Sulfur-Centered Free Radicals

Spin Trap	Free Radical	Rate Constant ($\times 10^6$ M^{-1} s^{-1})	Experimental Condition
DMPO[a]	$CH_3 \cdot CHOH$	1.1	Photolysis of H_2O_2, ethanol, pH 7.4
DMPO[b]	$(CH_3)_2 \cdot COH$	150	Ionizing radiation, isopropanol, pH 11
DMPO[b]	$CO_2 \cdot^-$	750	Ionizing radiation, formate, pH 11
PBN[a]	$CH_3 \cdot CHOH$	1.5	Photolysis of H_2O_2, ethanol, pH 7.4
PBN[c]	$CH_3 \cdot CHOH$	16	Ionizing radiation, ethanol
PBN[c]	$\cdot CH_2OH$	43	Ionizing radiation, methanol
PBN[c]	$CH_3CH_2 \cdot CHOH$	13	Ionizing radiation, n-propanol
PBN[c]	$(CH_3)_2 \cdot COH$	10	Ionizing radiation, isopropanol
4-POBN[a]	$CH_3 \cdot CHOH$	31	Photolysis of H_2O_2, ethanol, pH 7.4
4-POBN[d]	$CH_3 \cdot CHOH$	0.42	Ionizing radiation, ethanol, pH 7.4
4-POBN[e]	$(CH_3)_2 \cdot COH$	100	Ionizing radiation, isopropanol
4-POBN[b]	$(CH_3)_2 \cdot COH$	380	Ionizing radiation, isopropanol, pH 11
4-POBN[b]	$CO_2 \cdot^-$	610	Ionizing radiation, formate, pH 11
4-MePBN[c]	$\cdot CH_2OH$	25	Ionizing radiation, methanol
4-MePBN[c]	$CH_3 \cdot CHOH$	1.0	Ionizing radiation, ethanol
4-MePBN[c]	$CH_3CH_2 \cdot CHOH$	3.5	Ionizing radiation, n-propanol
4-MePBN[c]	$(CH_3)_2 \cdot COH$	1.0	Ionizing radiation, isopropanol
4-MeOPBN[c]	$\cdot CH_2OH$	29	Ionizing radiation, methanol
4-MeOPBN[c]	$CH_3 \cdot CHOH$	9.0	Ionizing radiation, ethanol
4-MeOPBN[c]	$CH_3CH_2 \cdot CHOH$	1.0	Ionizing radiation, n-propanol
4-MeOPBN[c]	$(CH_3)_2 \cdot COH$	1.5	Ionizing radiation, isopropanol
4-CNPBN[c]	$\cdot CH_2OH$	750	Ionizing radiation, methanol
4-CNPBN[c]	$CH_3 \cdot CHOH$	20	Ionizing radiation, ethanol.
4-CNPBN[c]	$CH_3CH_2 \cdot CHOH$	17	Ionizing radiation, n-propanol
4-NO$_2$PBN[c]	$\cdot CH_2OH$	1200	Ionizing radiation, methanol
4-NO$_2$PBN[c]	$CH_3 \cdot CHOH$	35	Ionizing radiation, ethanol
4-NO$_2$PBN[c]	$CH_3CH_2 \cdot CHOH$	6.5	Ionizing radiation, n-propanol
MNP[f]	$\cdot CH_2OH$	140	Radiolysis of H_2O, methanol
MNP[f]	$CH_3 \cdot CHOH$	320	Radiolysis of H_2O, ethanol
MNP[f]	$CO_2 \cdot^-$	1700	Ionizing radiation, formate, pH 9.2
DBNBS[g]	$CO_2 \cdot^-$	1500	Ionizing radiation, formate, pH 9.2
DMPO[h]	GS$\cdot$	260	Radiolysis of N_2 saturated aqueous solutions of propan-2-ol, propanone, and glutathione, pH 6
M$_4$PO[h]	GS$\cdot$	50	Radiolysis of N_2 saturated aqueous solutions of propan-2-ol, propanone, and glutathione, pH 6
DMPO[h]	CS$\cdot$	210	Radiolysis of N_2 saturated aqueous solutions of propan-2-ol, propanone, and cysteine, pH 6
M$_4$PO[h]	CS$\cdot$	40	Radiolysis of N_2 saturated aqueous solutions of propan-2-ol, propanone, and cysteine, pH 6

[a]DMPO = 5,5-dimethyl-1-pyrroline-N-oxide, Pou, *et al.*, 1994; PBN = N-*tert*-butyl-α-phenylnitrone, Pou, *et al.*, 1994; 4-POBN = α-(4-pyridyl 1-oxide)-N-*tert*-butylnitrone, Pou, *et al.*, 1994. [b]DMPO = Faraggi, *et al.*, 1984; 4-POBN = α-(4-pyridyl 1-oxide)-N-*tert*-butylnitrone, Faraggi, *et al.*, 1994. [c]PBN = Greenstock and Wiebe, 1982; 4-MePBN = N-*tert*-butyl-α-(4-methyl)phenylnitrone, Greenstock and Wiebe, 1982; 4-MeOPBN = N-*tert*-butyl-α-(4-methoxy)phenylnitrone, Greenstock and Wiebe, 1982; 4-CNPBN = N-*tert*-butyl-α-(4-cyano)phenylnitrone, Greenstock and Wiebe, 1982; 4-NO$_2$PBN = N-*tert*-butyl-α-(4-nitro)phenylnitrone, Greenstock and Wiebe, 1982. [d]4-POBN, Santiard, *et al.*, 1995. [e]4-POBN, Sridhar, *et al.*, 1986. [f]MNP = 2-methyl-2-nitrosopropane, Madden and Taniguchi, 1991. [g]DBNBS = sodium 3,5-dibromo-4-nitrosobenzenesulfonate, Bors and Stettmaier, 1992. [h]DMPO, absolute rate constants, Davies, *et al.*, 1987; M$_4$PO = 3,3,5,5-tetramethyl-1-pyrroline-N-oxide, absolute rate constants Davies, *et al.*, 1987.

reaction rate with constants decreasing with increased steric hindrance: $\cdot CH_2OH > CH_3 \cdot CHOH > CH_3CH_2 \cdot CHOH > (CH_3)_2 \cdot COH$ (Table 8.3).

Nitric Oxide

The competitive kinetic models developed for the spin trapping of carbon-centered free radicals can easily be adapted to determine rate constants for the reaction of NO· with different probes. The oxidation of reduced oxyhemoglobin [$Hb(Fe^{2+})O_2$] by NO· has been found to be $3.7 \times 10^7 M^{-1} s^{-1}$ at 25°C (Doyle and Hoekstra, 1981). Thus, reduced oxyhemoglobin can be used as a competitive inhibitor (Figure 8.7). When such experiments are undertaken, apparent rate constants (Table 8.4) for the reaction of either $Fe(DTCS)_2$ or $Fe(MGD)_2$ with NO· have

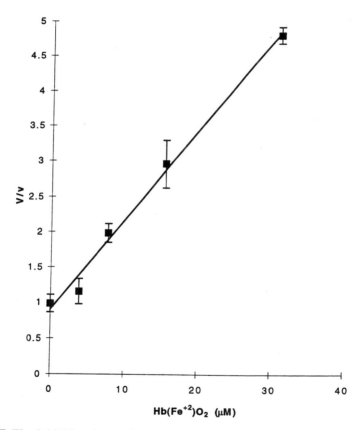

Figure 8.7 The inhibition by oxyhemoglobin of the spin trapping of nitric oxide with $Fe^{2+}(DTCS)_2$. Nitric oxide, in an anaerobic sodium phosphate buffer, pH 7.4, was added to the spin trap in the absence and presence of oxyhemoglobin. From the slope of this line, the concentration of oxyhemoglobin, and the published rate constant for the reaction of nitric oxide with oxyhemoglobin, the apparent rate constant for the spin trapping of this free radical by $Fe^{2+}(DTCS)_2$ can be calculated. (Taken from Pou, *et al.*, 1999.)

Table 8.4 Rate Constants for the Spin Trapping of Nitric Oxide

Spin Trap	Rate Constant ($\times 10^3 \, M^{-1} \, s^{-1}$)	Experimental Condition
PTIO[a]	5.15	Aqueous phosphate buffered solutions of NO· at pH 7.4
COMe-PTIO[a]	5.27	Aqueous phosphate buffered solutions of NO· at pH 7.4
CO-PTIO[a]	10.1	Aqueous phosphate buffered solutions of NO· at pH 7.4
N(Me)$_3$-PTIO[b]	6.0	Aqueous phosphate buffered solutions of NO· at pH 7.4
Fe(MGD)$_2$[c]	1200	Aqueous phosphate buffered solutions of NO· at pH 7.4
Fe(DTCS)$_2$[c]	1700	Aqueous phosphate buffered solutions of NO· at pH 7.4

[a]PTIO = 2-phenyl-4,5,5-tetramethylimidazoline-1-oxyl 3-oxide; COMe-PTIO = 2-(4-carboxymethoxyphenyl)-4,4,5,5-tetramethylimidazoline-1-oxyl 3-oxide; CO-PTIO = 2-(4-carboxyphenyl)-4,4,5,5-tetramethyl- imidazoline-1-oxyl 3-oxide, Akaike, *et al.*, 1993. [b]N(Me)$_3$-PTIO = 2-(4-trimethylaminophenyl)-4,4,5,5-tetramethylimidazoline-1-oxyl 3-oxide methyl sulfate, Woldman, *et al.*, 1994. [c]Fe(MGD)$_2$ = bis-(*N*-methyl-D-glucaminedithiocarbamato)iron; Fe(DTCS)$_2$ = bis-(*N*-(dithiocarboxy) sarcosino)iron, Pou, *et al.*, 1999. The values are absolute rate constants for PTIO, COMe-PTIO and CO-PTIO and apparent rate constants for Fe(MGD)$_2$ and Fe(DTCS)$_2$ where oxyhemoglobin was the competitive inhibitor.

been found to exceed $1 \times 10^6 \, M^{-1} \, s^{-1}$ (Pou, *et al.*, 1999). Not surprisingly, *in vivo* EPR spectroscopy of the corresponding spin trapped adduct, NO–Fe(DTCS)$_2$, in real time has allowed its localization through imaging this paramagnetic species (Yoshimura, *et al.*, 1996). One of the drawbacks with these iron chelates is their propensity to concentrate at interstitial sites, limiting spectroscopic measurements to loci distal from intracellular sources of NO·. Alternatives such as the more lipophilic nitronyl nitroxides, whose concentration within cells should be considerably greater than the iron chelates, are not without their limitations, the most pressing of which are their poor rate constants, ranging from $5 \times 10^3 \, M^{-1} \, s^{-1}$ to $1 \times 10^4 \, M^{-1} \, s^{-1}$ (Akaike, *et al.*, 1993; Woldman, *et al.*, 1994; Akaike and Maeda, 1996).

Singlet Oxygen

Even though singlet oxygen, $^1O_2$ ($^1\Delta_g$), is not a free radical, there are reports of its addition to nitrones (Ching and Foote, 1975), with rate constants ranging from $1.8 \times 10^7 \, M^{-1} \, s^{-1}$ for DMPO to $1.4 \times 10^8 \, M^{-1} \, s^{-1}$ for PBN (Table 8.5) (Harbour, *et al.*, 1980). Of the spin traps that have been studied, the identifiable products are 4-POBN–OH and DMPO–OH. This suggests that formation of these spin trapped adducts requires $^1O_2$ to add across the nitrone, rather than participate in a 1,3-dipolar cycloaddition (Ching and Foote, 1975; Harbour, *et al.*, 1980; Bilski, *et al.*, 1996).

Although the importance of $^1O_2$ in biology remains unresolved, documentation of its secretion by phagocytic cells has stimulated considerable interest in its potential microbicidal activity (Steinbeck, *et al.*, 1992, 1993). As part of that inquiry, spin trapping could play an important role in identifying a myriad of reactive oxygen species, including $^1O_2$. However, caution should be exercised when experimental conditions promote production of both $^1O_2$ and HO·. As each reactant will

Table 8.5 Apparent Rate Constants for the Reaction of Singlet Oxygen with Spin Traps

Spin Trap[a]	Rate Constant ($\times 10^7 \, M^{-1} \, s^{-1}$)
PBN	14
4-PyBN	14
4-POBN	12
4-MePyBN	8.0
2-SSPB	9.4
DMPO	1.8

Source: Data taken from Harbour, *et al.*, 1980.

[a]PBN = *N-tert*-butyl-a phenylnitrone; 4-PyBN = α-(4-pyridyl)-*N-tert*-butylnitrone; 4-POBN = α-(4-pyridyl 1-oxide)-*N-tert*-butylnitrone; 4-MePyBN = α-(4-methylpyridyl) -*N-tert*-butylnitrone methyl sulfate; 2-SSPB = sodium *N-tert*-butyl-α-(o-sulfonato)phenylnitrone; DMPO = 5,5-dimethyl-1-pyrroline-*N*-oxide.

result in the formation of DMPO–OH, determining the source of this spin trapped adduct will not be a trivial task. Study designs that investigate the effect of promoters and inhibitors of $^1O_2$ on the generation of DMPO–OH can greatly assist in discriminating between these alternative reactions. This can be accomplished by conducting experiments in D_2O, which enhances the lifetime of $^1O_2$, and $^1O_2$ quenchers such as 1,4-diazabicyclo[2.2.2]octane (Foote, 1976; Nye, *et al.*, 1987).

Stability of Spin Trapped Adducts

As important as rate constants are to predicting spin trapping efficiency, the stability of the spin trapped adduct can be considered a more important variable in determining the feasibility of observing an EPR spectrum. This point is appropriately illustrated by studying the reaction of HO· with DMPO and 4-POBN. Even though the rate constants are similar, in the range of $2-5 \times 10^9 \, M^{-1} \, s^{-1}$ (Table 8.2), the first-order half-lives of the corresponding spin trapped adducts in physiologic buffer are remarkably different, ranging from 10-23 s for 4-POBN–OH (Finkelstein, *et al.*, 1980; Janzen, *et al.*, 1992) to 12–156 min for DMPO–OH (Table 8.6) (Finkelstein, *et al.*, 1980; Marriott, *et al.*, 1980). Spin trapped adducts of $O_2\cdot^-$ degrade with equal rapidity (Table 8.7).

Surprisingly, the mechanism for this decomposition has not been well characterized. From the few studies available, the decay seems to follow two different reaction pathways (Finkelstein, *et al.*, 1980; Marriott, *et al.*, 1980; Kirino, *et al.*, 1981; Carmichael, *et al.*, 1984; Janzen, 1984; Kotake and Janzen, 1991; Janzen, *et al.*, 1992; Fréjaville, *et al.*, 1994; Tuccio, *et al.*, 1995). Early on, when the concentration of the spin trapped adduct is high, the rate of disappearance is second order, consistent with a disproportionation reaction[7] (Dupeyre and Rassat, 1966; Adamic, *et al.*, 1971; Bowman, *et al.*, 1971a; 1971b; Kirino, *et al.*, 1981; Rauckman, *et al.*, 1982; Keana, 1984; Kotake and Janzen, 1991).

Table 8.6 First Order Half-Life of Hydroxyl Radical Spin Trapped Adducts

Spin Trap	Half-Life[a]	Experimental Condition
DMPO[b]	2.6 hr	Photolysis of H_2O_2, pH 7.4
DMPO[c]	25 min	Photolysis of H_2O_2, pH 6
DMPO[c]	14 min	Photolysis of H_2O_2, pH 7
DMPO[c]	12 min	Photolysis of H_2O_2, pH 8
DMPO[d]	2 min	Ionizing radiation, degassed aqueous solutions, pH 3
DMPO[d]	23 min	Ionizing radiation, degassed aqueous solutions, pH 6.7
DMPO[d]	7 min	Ionizing radiation, degassed aqueous solutions, pH 9.5
MPPO[e]	27 min	Photolysis of H_2O_2, pH 5
MPPO[e]	76 min	Photolysis of H_2O_2, pH 7
MPPO[e]	74 min	Photolysis of H_2O_2, pH 8
PBN[f]	90 s	Photolysis of H_2O_2, pH 6
PBN[f]	46 s	Photolysis of H_2O_2, pH 7
PBN[g]	38 s	Photolysis of H_2O_2, pH 7.4
PBN[g]	11 s	Photolysis of H_2O_2, pH 8
4-PyBN[g]	185 s	Photolysis of H_2O_2, pH 6
4-PyBN[g]	31 s	Photolysis of H_2O_2, pH 7.4
4-PyBN[g]	3 s	Photolysis of H_2O_2, pH 8
3-PyBN[g]	120 s	Photolysis of H_2O_2, pH 6
3-PyBN[g]	28 s	Photolysis of H_2O_2, pH 7.4
3-PyBN[g]	5 s	Photolysis of H_2O_2, pH 8
2-PyBN[g]	23 s	Photolysis of H_2O_2, pH 6
2-PyBN[g]	24 s	Photolysis of H_2O_2, pH 7.4
2-PyBN[g]	3 s	Photolysis of H_2O_2, pH 8
4-POBN[b]	23 s	Photolysis of H_2O_2, pH 7.4
4-POBN[g]	80 s	Photolysis of H_2O_2, pH 6
4-POBN[g]	10 s	Photolysis of H_2O_2, pH 7.4
4-POBN[g]	3 s	Photolysis of H_2O_2, pH 8
3-POBN[g]	56 s	Photolysis of H_2O_2, pH 6
3-POBN[g]	22 s	Photolysis of H_2O_2, pH 7.4
3-POBN[g]	2 s	Photolysis of H_2O_2, pH 8
2-POBN[g]	730 s	Photolysis of H_2O_2, pH 6
2-POBN[g]	610 s	Photolysis of H_2O_2, pH 7.4
2-POBN[g]	43 s	Photolysis of H_2O_2, pH 8
4-MPyBN[g]	430 s	Photolysis of H_2O_2, pH 6
4-MPyBN[g]	120 s	Photolysis of H_2O_2, pH 7.4
4-MPyBN[g]	13 s	Photolysis of H_2O_2, pH 8
3-MPyBN[g]	240 s	Photolysis of H_2O_2, pH 6
3-MPyBN[g]	24 s	Photolysis of H_2O_2, pH 7.4
3-MPyBN[g]	5 s	Photolysis of H_2O_2, pH 8
2-MPyBN[g]	290 s	Photolysis of H_2O_2, pH 6
2-MPyBN[g]	30 s	Photolysis of H_2O_2, pH 7.4
2-MPyBN[g]	2 s	Photolysis of H_2O_2, pH 8

[a]The half-life is based on a first-order decay of the spin trapped adducts. [b]DMPO = 5,5-dimethyl-1-pyrroline-N-oxide, Finkelstein, et al., 1980; 4-POBN = α-(4-pyridyl 1-oxide)-N-tert-butylnitrone, Finkelstein, et al., 1980. [c]DMPO, Marriott, et al., 1980. [d]DMPO, Carmichael, et al., 1984. [e]MPPO = 5-methyl-5-phenyl-1-pyrroline-N-oxide, Sankuratri, et al., 1996. [f]PBN = N-tert-butyl-α-phenylnitrone, Kotake and Janzen, 1991. [g]PBN, 4-PyBN = α-(4-pyridyl)-N-tert-butylnitrone; 3-PyBN = α-(3-pyridyl)-N-tert-butylnitrone; 2-PyBN = α-(2-pyridyl)-N-tert-butylnitrone; 4-POBN, 3-POBN = α-(3-pyridyl 1-oxide)-N-tert-butylnitrone; 4-MPyBN = α-(4-methyl-pyridyl)-N-tert-butylnitrone methyl sulfate; 3-MPyBN = α-(3-methylpyridyl)-N-tert-butylnitrone methyl sulfate; 2-MPyBN = α-(2-methylpyridyl)-N-tert-butylnitrone methyl sulfate, Janzen, et al., 1992.

Table 8.7 First-Order Half-Life of Superoxide Spin Trapped Adducts

Spin Trap	Half-Life[a]	Experimental Condition
DMPO[b]	91 s	Riboflavin/DTPA/light, pH 5
DMPO[b]	80 s	Riboflavin/DTPA/light, pH 6
DMPO[b]	58 s	Riboflavin/DTPA/light, pH 7
DMPO[b]	35 s	Riboflavin/DTPA/light, pH 8
DMPO[c]	60 s	$(CH_3)_4N^+O_2\cdot^-$, pH 7.8
DMPO[d]	60 s	Pea chloroplasts/light, pH 7
DMPO[e]	66 s	NADPH–cytochrome P-450 reductase/paraquat, pH 7.4
DMPO[f]	63 s	PMA-stimulated neutrophils, pH 7.4, competition with SOD
DMPO[g]	60 s	Riboflavin/DTPA/light, pH 7
DMPO[h]	87 s	Riboflavin/DTPA/light, pH 5.6
DMPO[h]	50 s	Riboflavin/DTPA/light, pH 7
DMPO[h]	41 s	Riboflavin/DTPA/light, pH 8.2
DEPMPO[g]	890 s	Riboflavin/DTPA/light, pH 7
DEPMPO[h]	1824 s	Riboflavin/DTPA/light, pH 5.6
DEPMPO[h]	780 s	Riboflavin/DTPA/light, pH 7
DEPMPO[h]	630 s	Riboflavin/DTPA/light, pH 8.2
DEPMPO[i]	1060 s	Riboflavin/DTPA/light, pH 7.4
DEPMPO[l]	640 s	Xanthine/xanthine oxidase, pH 8.2
MPPO[j]	384 s	Riboflavin/DTPA/light, pH 5
MPPO[j]	342 s	Riboflavin/DTPA/light, pH 7.4
MPPO[j]	282 s	Riboflavin/DTPA/light, pH 8.4
M$_4$PO[k]	35 s	Riboflavin/DTPA/light, pH 7.4
4-MePyBN[l]	83 s	Riboflavin/DTPA/light, pH 5.5
4-MePyBN[l]	78 s	Riboflavin/DTPA/light, pH 7
4-MePyBN[l]	65 s	Riboflavin/DTPA/light, pH 8
PPN[m]	307 s	Riboflavin/DTPA/light, pH 5.8
4-PyOPN[m]	425 s	Riboflavin/DTPA/light, pH 5.8

[a]The half-life is based on a first-order decay of the spin trapped adducts. [b]DMPO = 5,5-dimethyl-1-pyrroline-N-oxide, Buettner and Oberley, 1978. [c]DMPO, Finkelstein, *et al.*, 1979. [d]DMPO, Bowyer and Camilleri, 1985. [e]DMPO, Yamazaki, *et al.*, 1990. [f]DMPO, Kotake, *et al.*, 1994. [g]DMPO, Fréjaville, *et al.*, 1994; DEPMPO = 5-diethoxyphosphoryl-5-methyl-1-pyrroline N-oxide, Fréjaville, *et al.*, 1994. [h]DMPO, Tuccio, *et al.*, 1995; DEPMPO, Tuccio, *et al.*, 1995. [i]DEPMPO, Roubaud, *et al.*, 1997. [j]MPPO = 5-methyl-5-phenyl-1-pyrroline-N-oxide, Sankuratri, *et al.*, 1996. [k]M$_4$PO = 3,3,5,5-tetramethyl-1-pyrroline-N-oxide, Buettner and Britigan, 1990. [l]4-MePyBN = α-(4-methylpyridyl)-N-*tert*-butylnitrone methyl sulfate, Sanderson and Chedekel, 1980. [m]PPN = diethyl 1-(N-benzylidene N-oxyamino) 1-methylethyl phosphonate, Roubaud, *et al.*, 1996; 4-PYOPN = diethyl 1-N-oxyl-N-[4(N'-oxypyridinyl)formylene]amino 1-methylethyl phosphonate, Roubaud, *et al.*, 1996.

Disproportionation Reaction

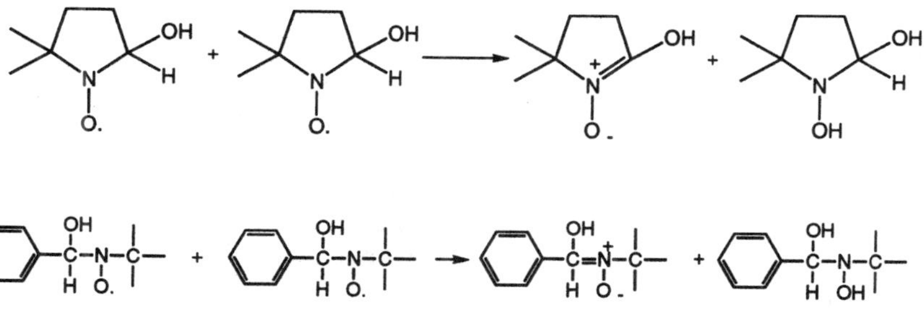

Over time, as the concentration of the spin trapped adduct decreases, the dismutation pathway no longer dominates, shifting instead to a first-order decay—a simple solvolysis reaction (Marriott, *et al.*, 1980; Carmichael, *et al.*, 1984; Kotake and Janzen, 1991):

$$-d[ST - X\cdot]/dt = k'(ST - X\cdot)$$ [8.38]

where k' is the pseudo first-order rate constant. Evidence in support of this scheme comes from the finding of a pH-dependent rate of hydrolysis (Table 8.6) and time-delayed production of *tert*-butylhydroaminoxyl (Kotake and Janzen, 1991), identified by its hyperfine splitting constants, $A_H = A_N = 14.8\,G$ (Kalyanaraman, *et al.*, 1979).

Hydrolysis Reaction

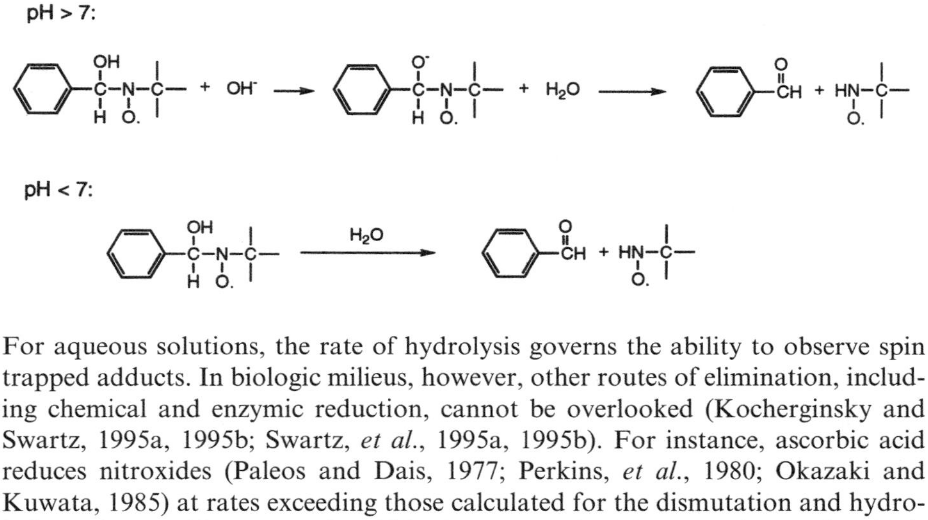

pH > 7:

pH < 7:

For aqueous solutions, the rate of hydrolysis governs the ability to observe spin trapped adducts. In biologic milieus, however, other routes of elimination, including chemical and enzymic reduction, cannot be overlooked (Kocherginsky and Swartz, 1995a, 1995b; Swartz, *et al.*, 1995a, 1995b). For instance, ascorbic acid reduces nitroxides (Paleos and Dais, 1977; Perkins, *et al.*, 1980; Okazaki and Kuwata, 1985) at rates exceeding those calculated for the dismutation and hydrolytic pathways (Couet, *et al.*, 1984; Keana, *et al.*, 1987; Morris, *et al.*, 1991). Alternative mechanisms for nitroxide reduction include the reaction of $O_2\cdot^-$ in the presence (Finkelstein, *et al.*, 1984; Rosen, *et al.*, 1988; Pou, *et al.*, 1995) and absence of thiols (Samuni, *et al.*, 1988, 1989; Krishna, *et al.*, 1992). Since SOD controls intracellular steady-state fluxes of $O_2\cdot^-$, the role that this free radical may play in nitroxide reduction is, therefore, limited to specific experimental conditions (Rosen and Freeman, 1984) and unique cellular models, such as neutrophils (Hawley, *et al.*, 1983; Kleinhans and Barefoot, 1987; Rosen, *et al.*, 1988; Samuni, *et al.*, 1988; Pou, *et al.*, 1989).

Besides ascorbate reduction, enzymic redox cycling is another avenue for nitroxide degradation (Swartz, *et al.*, 1995a, 1995b). First described in 1971 (Stier and Reitz, 1971), it appears that a family of NADPH-dependent cytoplasmic enzymes, as well as the monooxygenase, cytochrome P-450, can catalyze this reduction (Rosen and Rauckman, 1977; Floyd, 1983; Rauckman, *et al.*, 1984).

Similar metabolic pathways have been suggested to account for the *in vivo* loss of nitroxides (Couet, *et al.*, 1984; Griffeth, *et al.*, 1984).

A discussion of the chemistry of nitrones and nitrosoalkanes is presented in the next chapter. Nevertheless, it would be neglectful not to mention the one reaction that is pertinent to the present discussion—hydrolysis of spin traps. The primary synthetic route to α-aryl-*tert*-butylnitrones is the condensation of an aldehyde with *N*-(*tert*-butyl)hydroxylamine, resulting in the desired spin trap and loss of water:

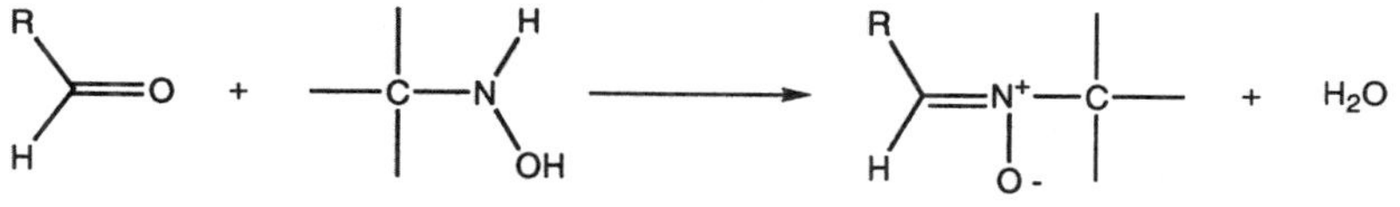

Not surprisingly, the reverse reaction, solvolysis of the nitrone to the corresponding aldehyde and *N*-(*tert*-butyl)hydroxylamine, proceeds at reasonable rates, depending on the structure of the spin trap and pH of the aqueous solution (De La Mare and Coppinger, 1963; Hamer and Macaluso, 1964; Janzen, *et al.*, 1978; Chamulitrat, *et al.*, 1993):

For example, at physiologic pH the concentration of 4-POBN remains constant, even after 32 h in an aqueous media (Janzen, *et al.*, 1978). In fact, based on a series of pharmacodynamic and pharmacokinetic studies, the solvolysis reaction may have only a minimal impact on the ability of acyclic nitrones to spin trap free radicals *in vivo* (Chen, *et al.*, 1990). However, under more acidic conditions, the half-life of 4-POBN steadily decreases, dropping to 14 min at pH 2 (Janzen, *et al.*, 1978). In contrast, Δ^1-pyrroline 1-oxides appear to be considerably less susceptible to hydrolysis than are α-aryl-*tert*-butylnitrones (Hamer and Macaluso, 1964). For instance, sealed aqueous solutions of DMPO kept in a cool dark place have been reported to be stable for 5 months (Elsworth and Lamchen, 1971). Similarly, base-catalyzed deuterium exchange of DMPO at 70°C results in a high yield of the desired spin trap without detectable amounts of ring opening being observed (Pou, *et al.*, 1990).

Experimental Design

The general approach described in this chapter for the calculation of rate constants is based on peak height measurements of spin trapped adducts as a function of time. Competitive kinetics are undertaken by including inhibitors specific for a particular free radical. Although the experiments are simple in concept, there are many places along the way that invite error. In the following sections, we will cover the most salient points.

Spin Trapping Superoxide

General Considerations Nitrones and nitrosoalkanes are synthesized to the purity required to spin trap free radicals as described in chapter 6. Adventitious iron and copper salts are common contaminants of most laboratory buffers. As these metal ions are redox active, it is important to eliminate them prior to initiating spin trapping experiments. Removal of these metal ions can be accomplished by passing the buffer through an ion-exchange resin, such as Chelex 100 (Bio-Rad, Richmond, CA) (Poyer and McCay, 1971). Even though this resin is efficient at extracting copper ions from the buffer, it is less efficient with iron salts (Borg and Schaich, 1984). Furthermore, there have been reported inconsistencies in the metal-removing capability of various batches of this resin (Van Reyk, et al., 1995). For many experiments, conalbumin has proven to be a more reliable method for removing iron salts (Gutteridge, 1987). Verification that redox active metal ions are no longer in the buffer can be accomplished by using a simple ascorbate assay (Buettner, 1988). Inclusion of the metal ion chelator DTPA (diethylenetriaminepentaacetic acid), but not EDTA (ethylenediaminetetraacetic acid), renders Fe^{2+}, Cu^{2+} and Mn^{2+} ineffective SOD mimics (Marklund and Marklund, 1974; Buettner, et al., 1978).

Xanthine–Xanthine Oxidase Superoxide Generating System The oxidation of xanthine by xanthine oxidase is a common enzymatic source of $O_2 \cdot ^-$ (McCord and Fridovich, 1968), as reagents are inexpensive and readily available from a number of vendors. Since the commercial preparations are frequently contaminated with iron salts (Britigan, et al., 1990), the enzyme should be dialyzed against conalbumin prior to commencement of the experiment (Buettner and Mason, 1990). For $O_2 \cdot ^-$ generation, xanthine and hypoxanthine are the substrates most often used, although others—such as acetaldehyde (Arneson, 1970; Archibald and Fridovich, 1982) and lumazine (Nagano and Fridovich, 1985),—are reliable alternatives, especially for higher fluxes of $O_2 \cdot ^-$. Typical reaction conditions include xanthine (400 μM), DTPA (100 μM to 1 mM), sodium phosphate buffer (50 mM), pH 7.8, and xanthine oxidase such that the rate of $O_2 \cdot ^-$ generation, measured as the SOD-inhibitable reduction of ferricytochrome c at 550 nm, using an extinction coefficient of 21 mM^{-1} cm^{-1} (McCord and Fridovich, 1969), is 10 μM min^{-1} at 25°C. By varying the concentration of xanthine oxidase, variable flux rates of $O_2 \cdot ^-$ can be produced.

Prior to initiating kinetic studies, it is important to determine whether the spin trap will inhibit the metabolism of xanthine by xanthine oxidase, thereby attenuating production of $O_2 \cdot ^-$. This can be accomplished by monitoring the conversion of xanthine to uric acid at 292 nm (Della Corte and Stirpe, 1968; Stirpe and Della Corte, 1969). In studies reporting such information, concentrations of nitrones below 200 mM were found not to antagonize the oxidation of xanthine by xanthine oxidase (Finkelstein, et al., 1980; Turner and Rosen, 1986; Arya, et al., 1992).

Light–Riboflavin Superoxide Generating System   The metabolism of xanthine by xanthine oxidase is a dependable $O_2\cdot^-$ generating system within a narrow range around physiologic pH (McCord and Fridovich, 1968; McCord, *et al.*, 1977). For experiments requiring $O_2\cdot^-$ production under acidic or basic conditions, alternative sources of this free radical have been described (Fridovich, 1982; Anfanas'ev, 1989b). Of those methods, the action of light on riboflavin–DTPA appears to be an ideal choice for spin trapping $O_2\cdot^-$ (Buettner and Oberley, 1978). First, $O_2\cdot^-$ flux can be closely controlled, as production of this free radical is dependent upon the visible illumination of an aqueous solution of riboflavin–DTPA. Therefore, formation of the spin trapped adduct can be monitored by irradiating, using a slide projector as a source of light, a sample of the mixture in the cavity of the EPR spectrometer. Second, a constant flux of $O_2\cdot^-$ can be observed over a pH range from 5 to 9 without a change in pH of the solution upon illumination. Finally, the irradiation does not adversely effect the catalytic activity of SOD (Finkelstein, *et al.*, 1980).

Upon illumination of a solution of riboflavin–DPTA and DMPO, EPR spectra are recorded that correspond to DMPO–OOH, DMPO–OH, and a third component with hyperfine splitting constants of $A_N = 15.8\,G$ and $A_H = 22.0\,G$ (Finkelstein, *et al.*, 1980). If cysteine is used as an electron donor, the EPR spectrum of the third species was found to be $A_N = 15.3\,G$ and $A_H = 22.1\,G$. These findings suggest that DMPO spin trapped a free radical derived from DTPA, an observation consistent with similar studies with EDTA (Fife and Moore, 1979). The ratio of spin trapped adducts can be modulated by varying the concentration of the nitrone. At DMPO concentrations $\leq 200\,mM$, the observed EPR spectrum consists primarily of DMPO–OOH and to a lesser extent DMPO–OH. At concentrations of the spin trap $\geq 300\,mM$, DMPO–DTPA is the dominate species, while DMPO–OOH is markedly inhibited. The source of DMPO–OH remains uncertain, although hydrolysis is a likely pathway to this nitroxide, as neither SOD nor HO· scavengers attenuate the formation of this nitroxide (Finkelstein, *et al.*, 1980). For kinetic studies, typical reaction mixtures contain DTPA (200 mM) and, riboflavin (42 µM) in potassium phosphate buffer (50 mM), cleansed of metal ions and adjusted to the desired pH. It has been found that fixing the concentration of DMPO at 180 mM and varying the dose of SOD allowed reliable data to be collected on the spin trapping of $O_2\cdot^-$ (Finkelstein, *et al.*, 1980). Although SOD prevented the formation of DMPO–OOH, it had no effect on the stability of DMPO–OOH once formed. Inactivation of SOD, through either boiling the enzyme or removal of copper by treatment with diethyldithiocarbamate (Heikkila and Cohen, 1977; Misra, 1979; Cocco, *et al.*, 1981), was found not to inhibit the production of DMPO–OOH (Finkelstein, *et al.*, 1980).

Spin Trapping Hydroxyl Radical

Photolysis of Hydrogen Peroxide as a Source of Hydroxyl Radical   There are three general methods for the formation of HO·: the metal ion-catalyzed Haber-Weiss reaction, photolysis of H_2O_2, and radiolysis of H_2O. For kinetic studies, the photolysis method is easier to control, as long as H_2O_2 concentrations are sufficiently low to ensure that there is no $O_2\cdot^-$ production (Harbour, *et al.*, 1974). In contrast, sources of ionizing radiation are not always easily accessible or as convenient to use as is photolysis. Since the ultraviolet light might effect the spin trap, a tandem EPR quartz cell system was employed in which the front cuvette, closest to the light source, contains DTPA (100 µM), chelexed sodium phosphate buffer (50 mM), pH 7.4, and DMPO (90 mM). In the back cuvette, H_2O_2 (290 mM) and ethanol (at various concentrations), when required for competition experiments, are added to the above reactants. Each of these solutions was made anaerobic prior to the start of the experiment. This tandem arrangement was irradiated with ultraviolet light for a period of 1 min. The cell farthest from the light source was placed into the cavity of the EPR spectrometer and spectra recorded (Figure 8.4).

Spin Trapping α-Hydroxyethyl Radical

Photolysis of Hydrogen Peroxide in the Presence of Ethanol as a Source of α-Hydroxyethyl Radical   The experimental design is identical to that of spin trapping HO·, except that the concentration of ethanol is increased to 1.7 M (10%, v/v). Verification that the α-carbon is the site of the unpaired electron can be accomplished using [1-^{13}C]-ethanol, obtained from commercial sources. The additional

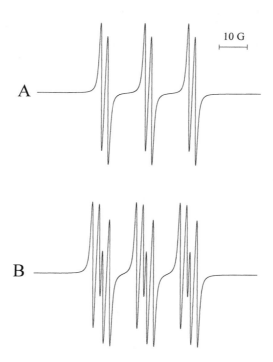

Figure 8.8 Verification of the spin trapping α-hydroxyethyl radical by 4-POBN. (A) The EPR spectral simulation of 4-POBN-CH(CH$_3$)OH with $A_N = 15.75$ G, and $A_H = 2.40$ G. (B) The EPR spectral simulation of 4-POBN–^{13}CH(CH$_3$)OH with $A_N = 15.75$ G, $A_H = 2.40$ G, and $A^{13}C = 3.75$ G.

splitting from the odd-number nucleus of $CH_3 \cdot {}^{13}CHOH$ can be observed in the EPR spectrum of 4-POBN–${}^{13}CH(CH_3)OH$ (Figure 8.8).

Acquisition of Data   With reagents prepared and experiments carefully designed, data collection—measurement of nitroxide peak height as a function of time—can commence. Frequently, in the literature of this field, it is not unusual to find publications containing a spin trap saturation curve. The purpose of such an experiment is to obtain optimal conditions for spin trapping a specific free radical (Lai and Piette, 1978; Floyd and Lewis, 1983). Although these data are often helpful in designing kinetic studies, the results must be viewed, however, with a certain degree of caution. This is particularly so when the stability of the nitroxide is in doubt.

For nitroxides with full substitution at both α-carbons, like TMPO–OOH or TMPO–OH, one may use any one of the three EPR spectral lines for estimating rates of nitroxide formation. As the tumbling of nitroxides in aqueous solutions most affects the high-field line at X-band, either of the first two lines maximizes the accuracy of data collection. For spin trapped adducts with additional hyperfine splitting, such as DMPO–OOH, choosing the line with the least amount of overlap or line broadening will increase the correctness of the measurement. Even under the best of conditions, the short lifetime of many of these nitroxides makes data acquisition difficult. In those cases, initial rates of spin trapped adduct formation should be used for calculating rate constants.

When two free radicals, such as HO· and $CH_3 \cdot CHOH$, compete for the same spin trap, the splitting pattern of each nitroxide must be taken into account when determining peak height ratios. The spin trapping of these free radicals by DMPO illustrates the necessity of these calculations:

$$HO \cdot + DMPO \rightarrow DMPO-OH$$

$$CH_3 \cdot CHOH + DMPO \rightarrow DMPO-CH(CH_3)OH$$

The EPR spectrum of DMPO–$CH(CH_3)OH$ is composed of six lines with hyperfine splitting constants of $A_N = 15.8\,G$ and $A_H = 22.8\,G$ and an equivalent peak height ratio over the complete spectrum (Figure 8.4B). In contrast, the structure of DMPO–OH dictates that the spectral hyperfine splitting constants are equal: $A_N = A_H = 14.9\,G$. This results in overlapping peaks with a signal intensity ratio of 1:2:2:1 (Figure 8.4A).

When the ratio of EPR signal intensities is 1:1, as would be found by choosing the low-field peaks of DMPO–OH and DMPO–$CH(CH_3)OH$, then a direct peak height ratio can be used to determine apparent rate constants. When the signal intensity ratios are 2:1, as would be observed by selecting one of the two central spectral lines of DMPO–OH and any of the six equal lines of DMPO–$CH(CH_3)OH$, then this ratio must be accounted for in calculating apparent rate constants.

Concluding Thoughts

Kinetic experiments provide a window through which the investigator can predict the feasibility of spin trapping a specific free radical. Likewise, the stability of the resultant nitroxide offers the practical reality of estimating the ability to record an EPR spectrum. As useful as these quantitative measurements are, spin trapped adducts may arise from reaction pathways that do not involve free radicals. Therefore, knowledge of the source of this nitroxide may allow experimental designs that avoid these unwanted reactions. In the next chapter, we will discuss the myriad of chemical transformations that may befall spin traps, offering alternatives to minimize these undesirable outcomes.

References

Adamic, K., Bowman, D.F. Gillan, T., and Ingold, K.U. (1971). Kinetic-applications of electron paramagnetic resonance spectroscopy. I. Self-reactions of diethyl nitroxide radicals. *J. Am. Chem. Soc.* **93**: 902–908.

Afanas'ev, I.B. (1989a). Nucleophilic reactions of superoxide ion, in: *Superoxide Ion: Chemistry and Biological Implications*, Vol. I, pp. 107–155, CRC Press, Boca Raton, FL.

Afanas'ev, I.B. (1989b). Nonenzymatic and enzymatic production of oxygen radicals in biological systems, in: *Superoxide Ion: Chemistry and Biological Implications*, Vol. II, pp. 5–25, CRC Press, Boca Raton, FL.

Akaike, T., and Maeda, H. (1996). Quantitation of nitric oxide using 2-phenyl-4,4,5,5-tetramethylimidazoline-1-oxyl 3-oxide (PTIO). *Methods Enzymol.* **268**: 211–221.

Akaike, T., Yoshida, M., Miyamoto, Y., Sato, K., Kohno, M., Sasamoto, K.,Miyazaki, K., Ueda, S., and Maeda, H. (1993). Antagonistic action of imidazolineoxyl N-oxides against endothelium-derived relaxing factor/·NO through a radical reaction. *Biochemistry* **32**: 827–832.

Archibald, F.S., and Fridovich, I. (1982). The scavenging of superoxide radical by manganous complexes: *In vitro*. *Arch. Biochem. Biophys.* **214**: 452–463.

Arneson, R.M. (1970). Substrate-induced chemiluminescence of xanthine oxidase and aldehyde oxidase. *Arch. Biochem. Biophys.* **136**: 352–360.

Arya, P., Stephens, J.C., Griller, D., Pou, S., Ramos, C.L., Pou, W.S., and Rosen, G.M. (1992). Design of modified pyrroline N-oxide derivatives as spin traps specific for hydroxyl radical. *J. Org. Chem.* **57**: 2297–2301.

Asada, K., Takahashi, M.-A., and Nagate, M. (1974). Assay and inhibitors of spinach superoxide dismutase. *Agric. Biol. Chem.* **38**: 471–473.

Asmus, K.-D., Möckel, H., and Henglein, A. (1973). Pulse radiolytic study of the site of ·OH radical attack on aliphatic alcohols in aqueous solution. *J. Phys. Chem.* **77**: 1218–1221.

Behar, D., Czapski, G., Rabani, J., Dorfman, L.M., and Schwarz, H.A. (1970). The acid dissociation constant and decay kinetics of the perhydroxyl radical. *J. Phys. Chem.* **74**: 3209–3212.

Bilski, P., Reszka, K., Bilska, M., and Chignell, C.F. (1996). Oxidation of the spin trap 5,5-dimethyl-1-pyrroline N-oxide by singlet oxygen in aqueous solution. *J. Am. Chem. Soc.* **118**: 1330–1338.

Borg, D.C., and Schaich, K.M. (1984). Cytotoxicity from coupled redox cycling of autoxidizing xenobiotics and metals. A selective critical review and commentary on work-in-progress. *Is. J. Chem.* **24**: 38–53.

Bors, W., and Stettmaier, K. (1992). Determination of rate constants of the spin trap 3,5-dibromo-4-nitrosobenzenesulfonic acid with various radicals by pulse radiolysis and competition kinetics. *J. Chem. Soc., Perkin Trans.* 2 1509–1512.

Bowman, D.F., Brokenshire, J.L., Gillan, T., and Ingold, K.U. (1971a). Kinetic applications of electron paramagnetic resonance spectroscopy. II. Self-reactions of N-alkyl nitroxides and N-phenyl nitroxide. *J. Am. Chem. Soc.* **93**: 6551–6555.

Bowman, D.F., Gillan, T., and Ingold, K.U. (1971b). Kinetic applications of electron paramagnetic resonance spectroscopy. III. Self-reactions of dialkyl nitroxide radicals. *J. Am. Chem. Soc.* **93**: 6555–6561.

Bowyer, J.R., and Camilleri, P. (1985). Spin-trap study of the reactions of ferredoxin with reduced oxygen species in pea chloroplasts. *Biochim. Biophys. Acta* **808**: 235–242.

Britigan, B.E., Pou, S., Rosen, G.M., Lilleg, D.M., and Buettner, G.R. (1990). Hydroxyl radical is not a product of the reaction of xanthine oxidase and xanthine. The confounding problem of adventitious iron bound to xanthine oxidase. *J. Biol. Chem.* **265**: 17533–17538.

Buettner, G.R., and Britigan, B.E. (1990). The spin trapping of superoxide with M_4PO (3,3,5,5-tetramethylpyrroline-N-oxide). *Free Radical Biol. Med.* **8**: 57–60.

Buettner, G.R. (1988). In the absence of catalytic metals ascorbate does not autoxidize at pH 7: Ascorbate as a test for catalytic metals. *J. Biochem. Biophys. Methods* **16**: 27–40.

Buettner, G.R., and Mason, R.P. (1990). Spin-trapping methods for detecting superoxide and hydroxyl free radicals *in vitro* and *in vivo*. *Methods Enzymol.* **186**: 127–133.

Buettner, G.R., and Oberley, L.W. (1978). Considerations in the spin trapping of superoxide and hydroxyl radical in aqueous systems using 5,5-dimethyl-1-pyrroline-1-oxide. *Biochem. Biophys. Res. Commun.* **83**: 69–74.

Buettner, G.R., Oberley, L.W., and Leuthauser, S.W.H.C. (1978). The effect of iron on the distribution of

superoxide and hydroxyl radicals as seen by spin trapping and on the superoxide dismutase assay. *Photochem. Photobiol.* **28**: 693–695.

Buxton, G.V., Greenstock, C.L., Helman, W.P., and Ross, A.B. (1988). Critical review of rate constants for reactions of hydrated electrons, hydrogen atoms and hydroxyl radical ($\cdot OH/\cdot O^-$) in aqueous solutions. *J. Phys. Chem. Ref. Data* **17**: 513–886.

Carmichael, A.J., Makino, K., and Riesz, P. (1984). Quantitative aspects of ESR and spin trapping of hydroxyl radicals and hydrogen atoms in gamma-irradiated aqueous solutions. *Radiat. Res.* **100**: 222–234.

Castelhano, A.L., Perkins, M.J., and Griller, D. (1983). Spin trapping of hydroxyl in water: Decay kinetics for the $\cdot OH$ and $CO_2 \cdot^-$ adducts to 5,5-dimethyl-1-pyrroline-N-oxide. *Can. J. Chem.* **61**: 298–300.

Chamulitrat, W., Jordan, S.J., Mason, R.P., Saito, K., and Cutler, R.G. (1993). Nitric oxide formation during light-induced decomposition of phenyl N-*tert*-butylnitrone. *J. Biol. Chem.* **268**: 11520–11527.

Chateauneuf, J., Lusztyk, J., and Ingold, K.U. (1988). Absolute rate constants for the reactions of some carbon-centered radicals with 2,2,6,6-tetramethyl-piperidine-N-oxyl. *J. Org. Chem.* **53**: 1629–1632.

Chen, G., Bray, T.M., Janzen, E.G., and McCay, P.B. (1990). Excretion, metabolism and tissue distribution of a spin trapping agent, α-phenyl-N-*tert*-butyl-nitrone (PBN) in rats. *Free Radical Res. Commun.* **9**: 317–323.

Ching, T.-Y., and Foote, C. (1975). Chemistry of singlet oxygen. XXII. Photoxidation of nitrones. *Tetra-hedron Lett.* 3771–3774.

Cocco, D., Calabrese, L., Rigo, A., Argese, E., and Rotilio, G. (1981). Re-examination of the reaction of diethyldithiocarbamate with the copper of superoxide dismutase. *J. Biol. Chem.* **256**: 8983–8986.

Couet, W.R., Eriksson, U.G., Tozer, T.N., Tuck, L.D., Wesbey, G.E., Nitecki, D., and Brasch, R.C. (1984). Pharmacokinetics and metabolic fate of two nitroxides potentially useful as contrast agents for magnetic resonance imaging. *Pharm. Res.* **4**: 203–209.

Danen, W.C., and Warner, R.J. (1977). The remarkable nucleophilicity of super-oxide anion radical. Rate constants for reaction of superoxide ion with aliphatic bromides. *Tetrahedron Lett.* 989–992.

Davies, M.J., Forni, L.G., and Shuter, S.L. (1987). Electron spin resonance and pulse radiolysis studies on the spin trapping of sulphur-centered radicals. *Chem.-Biol. Interact.* **61**: 177–188.

De La Mare, H.E., and Coppinger, G.M. (1963). Oxidation of N,N-dialkyl hydroxylamines with *t*-butyl hydroperoxide. A new synthesis of nitrones. *J. Org. Chem.* **28**: 1068–1070.

Della Corte, E., and Stirpe, F. (1968). Regulation of xanthine oxidase in rat liver: Modification of the enzyme activity of rat liver supernatant on storage at −20°C. *Biochem. J.* **108**: 349–351.

Dietz, R., Forno, A.E.J., Larcombe, B.E., and Peover, M.E. (1970). Nucleophilic reactions of electrogenerated super-oxide ion. *J. Chem. Soc.* 816–820.

Divisek, J., and Kastening, B. (1975). Electrochemical generation and reactivity of the superoxide ion in aqueous solutions. *J. Electroanal. Chem.* **65**: 603–621.

Doyle, M.P., and Hoekstra, J.W. (1981). Oxidation of nitrogen oxides by bound dioxygen in hemoproteins. *J. Inorg. Biochem.* **14**: 351–358.

Dupeyre, R.-M., and Rassat, A. (1966). Nitroxides. XIX. Norpseudopelletier-ine-N-oxyl, a new, stable unhindered free radical. *J. Am. Chem. Soc.* **88**: 3180–3181.

Elsworth, J.F., and Lamchen, M. (1971). Nitrones. XI. Spectral studies on the oxidation of cyclic nitrones by iron (III) chloride. *J. S. Afr. Chem. Inst.* **24**: 196–204.

Faraggi, M., Carmichael, A., and Riesz, P. (1984). OH radical formation by photolysis of aqueous porphyrin solutions. A spin trapping and e.s.r. study. *Int. J. Radiat. Biol.* **46**: 703–713.

Fee, J.A., and Valentine, J.S. (1977). Chemical and physical properties of super-oxide, in: *Superoxide and Superoxide Dismutases* (Michelson, A.M., McCord, J.M., and Fridovich, I.,

eds.), pp. 19–60, Academic Press, New York.

Fife, D.J., and Moore, W.M. (1979). The reduction and quenching of photo-excited flavins by EDTA. *Photochem. Photobiol.* **29**: 43–47.

Finkelstein, E., Rosen, G.M., and Rauckman, E.J. (1980). Spin trapping. Kinetics of the reaction of superoxide and hydroxyl radicals with nitrones. *J. Am. Chem. Soc.* **102**: 4994–4999.

Finkelstein, E., Rosen, G.M., and Rauckman, E.J. (1984) Superoxide dependent reduction of nitroxides by thiols. *Biochim. Biophys. Acta* **802**, 90–98.

Finkelstein, E., Rosen, G.M., Rauckman, E.J., and Paxton J. (1979). Spin trapping of superoxide. *Mol. Pharmacol.* **16**: 676–685.

Floyd, R.A. (1983). Hydroxyl free-radical spin-adducts in rat brain synapto-somes: Observations on the reduction of the nitroxide. *Biochim. Biophys. Acta* **756**: 204–216.

Floyd, R.A., and Lewis, C.A. (1983). Hydroxyl free radical formation from hydrogen peroxide by ferrous iron–nucleotide complexes. *Biochemistry* **22**: 2645–2649.

Foote, C.S. (1976). Photosensitized oxidation and singlet oxygen: Consequences in biological systems, in: *Free Radicals in Biology* (Pryor, W.A., ed.), Vol. II, pp. 85–133, Academic Press, New York.

Fréjaville, C., Karoui, H., Tuccio, B., Le Moigne, F., Culcasi, M., Pietri, S., Lauricella, R., and Tordo, P. (1994). 5-Diethoxyphosphoryl-5-methyl-1-pyr-roline N-oxide (DEPMPO): A new phosphorylated nitrone for efficient *in vitro* and *in vivo* spin trapping of oxygen-centred radicals. *J. Chem. Soc., Chem. Commun.* 1793–1794.

Fréjaville, C., Karoui, H., Tuccio, B., Le Moigne, F., Culcasi, M., Pietri, S., Lauricella, R., and Tordo, P. (1995). 5-Diethoxyphosphoryl-5-methyl-1-pyr-roline N-oxide (DEPMPO): A new efficient phosphorylated nitrone for the *in vitro* and *in vivo* spin trapping of oxygen-centered radicals. *J. Med. Chem.* **38**: 258–265.

Fridovich, I. (1978). The biology of oxygen radicals. *Science* **201**: 875–880.

Fridovich, I. (1982). Measuring the activity of superoxide dismutases: An embarrassment of riches, in: *Superoxide Dismutase* (Oberley, L.W., ed.), Vol. I, pp. 69–77, CRC Press, Boca Raton, FL.

Greenstock, C.L., and Wiebe, R.H. (1982). Substituent effects in the kinetic analysis of free radical reactions with nitrone spin traps. *Can. J. Chem.* **60**: 1560–1564.

Griffeth, L.K., Rosen, G.M., Rauckman, E.J., and Drayer, B.P. (1984). Pharmacokinetics of nitroxide NMR contrast agents. *Invest. Radiol.* **19**: 553–562.

Gutteridge, J.M.C. (1987). A method for removal of trace iron contamination from biological buffers. *FEBS Lett.* **214**: 362–364.

Hamer, J., and Macaluso, A. (1964). Nitrones. *Chem. Rev.* **64**: 473–495.

Hanna, P.M., and Mason, R.P. (1992). Direct evidence for inhibition of free radical formation from Cu(I) and hydrogen peroxide by glutathione and other potential ligands using EPR spin-trapping technique. *Arch. Biochem. Biophys.* **295**: 205–213.

Harbour, J.R., Chow, V., and Bolton, J.R. (1974). An electron spin resonance study of the spin adducts of OH and HO$_2$ radicals with nitrones in the ultraviolet photolysis of aqueous hydrogen peroxide solutions. *Can. J. Chem.* **52**: 3549–3553.

Harbour, J.R., Issler, S.L., and Hair, M.L. (1980). Singlet oxygen and spin trapping with nitrones. *J. Am. Chem. Soc.* **102**: 7778–7779.

Hawley, D.A., Kleinhans, F.W., and Biesecker, J.L. (1983). Determination of alternate pathway complement kinetics by electron spin resonance spectroscopy. *Am. J. Clin. Pathol.* **79**: 673–677.

Heikkila, R.E., and Cohen, G. (1977). The inactivation of copper-zinc superoxide dismutase by diethyldithiocarbamate, in: *Superoxide and Superoxide Dismutases* (Michelson, A.M., McCord, J.M., and Fridovich, I., eds.), pp. 367–373, Academic Press, New York.

Ingold, K.U. (1973). Rate constants for free radical reactions in solution, in: *Free Radicals* (Kochi, J.K., ed), Vol. 1, pp. 37–112, John Wiley & Sons, New York.

Janzen, E.G. (1984). Spin trapping. *Methods Enzymol.* **105**: 188–198.

Janzen, E.G., and Evans, C.A. (1973). Rate constants for spin trapping *tert*-butoxy radicals as studied by electron spin resonance. *J. Am. Chem. Soc.* **95**: 8205–8206.

Janzen, E.G., Kotake, Y., and Hinton, R.D. (1992). Stabilities of hydroxyl radical spin adducts of PBN-type spin traps. *Free Radical Biol. Med.* **12**: 169–173.

Janzen, E.G., Nutter, D.E., Jr., and Evans, C.A. (1975). Rate constants for the hydrogen atom abstraction by phenyl radical methanol, ethanol, and 2-propanol as studied by electron spin resonance spin trapping techniques. *J. Phys. Chem.* **79**: 1983–1984.

Janzen, E.G., Wang, Y.Y., and Shetty, R.V. (1978). Spin trapping with α-pyridyl 1-oxide N-*tert*-butyl nitrones in aqueous solutions. A unique electron spin resonance spectrum for the hydroxyl radical adduct. *J. Am. Chem. Soc.* **100**: 2923–2925.

Johnson, R.A., and Nidy, E.G. (1975). Superoxide chemistry: A convenient synthesis of dialkyl peroxides. *J. Org. Chem.* **40**: 1680–1681.

Kalyanaraman, B., Perez-Reyes, E., and Mason, R.P. (1979). The reduction of nitroso-spin traps in chemical and biological systems. A cautionary note. *Tetrahedron Lett.* 4809–4812.

Keana, J.F.W. (1984). Synthesis and chemistry of nitroxide spin labels, in: *Spin Labeling in Pharmacology* (Holtzman, J., ed.), pp. 1–85, Academic Press, New York.

Keana, J.F.W., Pou, S., and Rosen, G.M. (1987). Nitroxides as potential contrast enhancing agents for MRI applications: Influence of structure on the rate of reduction by rat hepatocytes, subcellular fractions, and ascorbate. *Magn. Reson. Med.* **5**: 525–536.

Kleinhans, F.W., and Barefoot, S.T. (1987). Spin trap determination of free radical burst kinetics in stimulated neutrophils. *J. Biol. Chem.* **262**: 12452–12457.

Kirino, Y., Ohkuma, T., and Kwan (1981). Spin trapping with dimethylpyrroline-N-oxide in solution. *Chem. Pharm. Bull.* 34.

Klug, D., Fridovich, I., and (1972). A direct demon catalytic action of sur the use of pulse radiolysis. *J. Biol. Chem.* **247**: 4839–4842.

Kocherginsky, N., and Swartz, H.M. (1995a). Applications of reactions of nitroxides in biophysics, in: *Nitroxide Spin Labels. Reactions in Biology and Chemistry* (Kocherginsky, N., and Swartz, H.M., eds.), pp. 67–94, CRC Press, Boca Raton, FL.

Kocherginsky, N., and Swartz, H.M. (1995b). Biochemical basis of reduction of nitroxides in membrane, in: *Nitroxide Spin Labels. Reactions in Biology and Chemistry* (Kocherginsky, N., and Swartz, H.M., eds.), pp. 94–112, CRC Press, Boca Raton, FL.

Kotake, Y., and Janzen, E.G. (1991). Decay and fate of the hydroxyl radical adduct of α-phenyl-N-*tert*-butylnitrone in aqueous media. *J. Am. Chem. Soc.* **113**: 9503–9506.

Kotake, Y., Reinke, L.A., Tanigawa, T., and Koshida, H. (1994). Determination of the rate of superoxide generation from biological systems by spin trapping: Use of rapid oxygen depletion to measure the decay rate of spin adducts. *Free Radical Biol. Med.* **17**: 215–223.

Krishna, M.C., Grahame, D.A., Samuni, A., Mitchell, J.B., and Russo, A. (1992). Oxoammonium cation intermediate in the nitroxide-catalyzed dismutation of superoxide. *Proc. Natl. Acad. Sci. USA* **89**: 5537–5541.

Kuthan, H., Ullrich, V., and Estabrook, R. (1982). A quantitative test for superoxide radicals produced in biological systems. *Biochem. J.* **203**: 551–558.

Lai, C.S., and Piette, L.H. (1977). Hydroxyl radical production in lipid peroxidation of rat microsomes. *Biochem. Biophys. Res. Commun.* **78**: 51–59.

Lai, C.S., and Piette, L.H. (1978). Spin-trapping studies of hydroxyl radical production involved in lipid peroxidation. *Arch. Biochem. Biophys.* **190**: 27–38.

Lown, J.W., and Chen. H.-H. (1981). Evidence for the generation of free hydroxyl radicals from certain quinone antitumor antibiotics upon reductive activation in solution. *Can. J. Chem.* **59**: 390–395.

Madden, K.P., and Taniguchi, H. (1991). An *in situ* radiolysis time-resolved electron spin resonance study of 2-methyl-2-nitrosopropane spin trapping

kinetics. *J. Am. Chem. Soc.* **113**: 5541–5547.

Marklund, S., and Marklund, G. (1974). Involvement of the superoxide anion radical in the autoxidation of pyrogallol and a convenient assay for superoxide dismutase. *Eur. J. Biochem.* **47**: 469–474.

Marriott, P.R., Perkins, M.J., and Griller, D. (1980). Spin trapping for hydroxyl in water: A kinetic evaluation of two popular traps. *Can. J. Chem.* **58**: 803–807.

McCord, J.M., and Fridovich, I. (1968). The reduction of cytochrome c by milk xanthine oxidase. *J. Biol. Chem.* **243**: 5753–5760.

McCord, J.M., and Fridovich, I. (1969). Superoxide dismutase. An enzymic function for erythrocuprein (hemocuprein). *J. Biol. Chem.* **244**: 6049–6055.

McCord, J.M., Crapo, J.D., and Fridovich, I. (1977). Superoxide dismutase assays: A review of methodology, in: *Superoxide and Superoxide Dismutases* (Michelson, A.M., McCord, J.M., and Fridovich, I., eds.), pp. 11–17, Academic Press, New York.

Merritt, M.V., and Johnson, R.A. (1977). Spin trapping, alkylperoxy radicals, and superoxide–alkyl halide reactions. *J. Am. Chem. Soc.* **99**: 3713–3719.

Merritt, M.V., and Sawyer, D.T. (1970). Electrochemical studies of the reactivity of superoxide ion with alkyl halides in dimethyl sulfoxide. *J. Org. Chem.* **35**: 2157–2159.

Misra, H.P. (1979). Reaction of copper-zinc superoxide dismutase with diethyldithiocarbamate. *J. Biol. Chem.* **254**: 11623–11628.

Mitsuta, K., Hiramatsu, M., Ohya-Nishiguchi, H., Kamada, H., and Fujii, K. (1994). A kinetic analysis of superoxide adduct formation in the presence of typical scavengers. *Bull. Chem. Soc. Jpn.* **67**: 529–538.

Morehouse, K.M., and Mason, R.P. (1988). The transition metal-mediated formation of the hydroxyl radical during the reduction of molecular oxygen by ferredoxin-ferredoxin:NADP+ oxidoreductase. *J. Biol. Chem.* **263**: 1204–1211.

Morris, S., Sosnovsky, G., Hui, B., Huber, C.O., Rao, N.U.M., and Swartz, H.M. (1991). Chemical and electrochemical reduction rates of cyclic nitroxides (nitroxyls). *J. Pharm. Sci.* **80**: 149–152.

Nagano, T., and Fridovich, I. (1985). Superoxide radical from xanthine oxidase and lumazine. *J. Free Radical Biol. Med.* **1**: 39–42.

Neta, P., Steenken, S., Janzen, E.G., and Shetty, R.V. (1980). Pattern of addition of hydroxyl radicals to the spin traps α-pyridyl 1-oxide N-*tert*-butyl nitrone. *J. Phys. Chem.* **84**: 532–534.

Nishi, M., Hagi, A., Ide, H., Murakami, A., and Makino, K. (1992). Comparison of 2,5,5-trimethyl-1-pyrroline-N-oxide (M$_3$PO) and 3,3,5,5-tetramethyl-1-pyrroline-N-oxide (M$_4$PO) with 5,5-dimethyl-1-pyrroline-N-oxide (DMPO) as spin traps. *Biochem. Int.* **27**: 651–659.

Nye, A.C., Rosen, G.M., Gabrielson, E.W., Keana, J.F.W., and Prabhu, V.S. (1987). Diffusion of singlet oxygen into human bronchial epithelial cells. *Biochim. Biophys. Acta* **928**: 1–7.

Okazaki, M., and Kuwata, K. (1985). A stopped-flow ESR study on the reactivity of some nitroxide radicals with ascorbic acid in the presence of β-cyclodextrin. *J. Phys. Chem.* **89**: 4437–4440.

Paleos, C.M., and Dais, P. (1977). Ready reduction of some nitroxide free radicals with ascorbic acid. *J. Chem. Soc., Chem. Commun.* 345–346.

Perkins, R.C., Beth, A.H., Wilkerson, L.S., Serafin, W., Dalton, L.R., Park, C.R., and Park, J.H. (1980). Enhancement of free radical reduction at elevated concentrations of ascorbic acid in avian dystropic muscle. *Proc. Natl. Acad. Sci. USA* **77**: 790–794.

Pou, S., Cohen, M.S., Britigan, B.E., and Rosen, G.M. (1989). Spin-trapping and human neutrophils. Limits of detection of hydroxyl radical. *J. Biol. Chem.* **264**: 12299–12302.

Pou, S., Davis, P.L., Wolf, G.L., and Rosen, G.M. (1995). Use of nitroxides as NMR contrast enhancing agents for joints. *Free Radical Res.* **23**: 353–364.

Pou, S., Ramos, C.L., Gladwell, T., Renks, E., Centra, M., Young, D., Cohen, M.S., and Rosen, G.M. (1994). A kinetic approach to the selection of a sensitive spin trapping system for the detection of hydroxyl radical. *Anal. Biochem.* **217**: 76–83.

Pou, S., Rosen, G.M., Wu, Y., and Keana, J.F.W. (1990). Synthesis of deuterium- and ^{15}N-containing pyrroline-1-oxides: A spin trapping study. *J. Org. Chem.* **55**: 4438–4443.

Pou, S., Tsai, P., Porasuphatana, S., Halpern, H.J., Chandramouli, G.V.R., Barth, E.D., and Rosen, G.M. (1999). Spin trapping of nitric oxide by ferrochelates: Kinetic and *in vivo* pharmacokinetic studies. *Biochim. Biophys. Acta* **1427**: 216–266.

Poyer, J.L., and McCay, P.B. (1971). Reduced triphosphopyridine nucleotide oxidase-catalyzed alterations of membrane phospholipids. IV. Dependence on Fe^{3+}. *J. Biol. Chem.* **246**: 263–269.

Rauckman, E.J., Rosen, G.M., and Griffeth, L.K. (1984). Enzymatic reactions of spin labels, in: *Spin Labeling in Pharmacology* (Holtzman, J., ed.), pp. 175–190, Academic Press, New York.

Rauckman, E.J., Rosen, G.M., and Cavagnaro, J. (1982). Norcocaine nitroxide. A potential hepatotoxic metabolite of cocaine. *Mol. Pharmacol.* **21**: 458–463.

Rosen, G.M., and Freeman, B.A. (1984). Detection of superoxide generated by endothelial cells. *Proc. Natl. Acad. Sci. USA* **81**: 7269–7273.

Rosen, G.M., and Rauckman, E.J. (1977). Formation and reduction of a nitroxide radical by liver microsomes. *Biochem. Pharmacol.* **26**: 675–678.

Rosen, G.M., and Rauckman, E.G. (1981). Spin trapping of free radicals during hepatic microsomal lipid peroxidation. *Proc. Natl. Acad. Sci. USA* **78**: 7346–7349.

Rosen, G.M., and Turner, M.J., III (1988). Synthesis of spin traps specific for hydroxyl radical. *J. Med. Chem.* **31**: 428–432.

Rosen, G.M., Britigan, B.E., Cohen, M.S., Ellington, S.P., and Barber, M.J. (1988). Detection of phagocyte-derived free radicals with spin trapping techniques: Effect of temperature and metabolism. *Biochim. Biophys. Acta* **969**: 236–241.

Roubaud, V., Lauricella, R., Tuccio, B., Bouteiller, J.-C., and Tordo, P. (1996). Decay of superoxide spin adducts of new PBN-type phosphorylated nitrones. *Res. Chem. Intermed.* **22**: 405–416.

Roubaud, V., Sankarapandi, S., Kuppusamy, P., Tordo, P., and Zweier, J.L. (1997). Quantitative measurement of superoxide generation using the spin trap 5-(diethoxyphosphoryl)-5-methyl-1-pyrroline N-oxide. *Anal. Biochem.* **247**: 404–411.

Saito, I., Otsuki, T., and Matsuura, T. (1979). The reaction of superoxide ion with vitamin K_1 and its related compounds. *Tetrahedron Lett.* 1693–1696.

Samuni, A., Black, C.D.V., Krishna, C.M., Malech, H.L., Bernstein, E.F., and Russo, A. (1988). Hydroxyl radical production by stimulated neutrophils reappraised. *J. Biol. Chem.* **263**: 13797–13801.

Samuni, A., Krishna, C.M., Riesz, P., Finkelstein, E., and Russo, A. (1989). Superoxide reaction with nitroxide spin-adducts. *Free Radical* Biol. Med. **6**: 141–148.

Sanderson, D.G., and Chedekel, M.R. (1980). Spin trapping of the superoxide radical by 4-(N-methylpyridinium) *t*-butyl nitrone. *Photochem. Photobiol.* **32**: 573–576.

San Filippo, J., Jr., Chern, C.-I., and Valentine, J.S. (1975). The reaction of superoxide with alkyl haldies and tosylates. *J. Org. Chem.* **40**: 1678–1680.

Sankuratri, N., Kotake, Y., and Janzen, E.G. (1996). Studies on the stability of oxygen radical spin adducts of a new spin trap: 5-Methyl-5-phenylpyrroline-1-oxide (MPPO). *Free Radical Biol. Med.* **21**: 889–894.

Santiard, D., Ribière, C., Nordmann, R., and Houee-Levin, C. (1995). Inactivation of Cu,Zn-superoxide dismutase by free radicals derived from ethanol metabolism: A γ radiolysis study. *Free Radical Biol. Med.* **19**: 121–127.

Sawada, Y., and Yamazaki, I. (1973). One-electron transfer reactions in biochemical systems VIII. Kinetic study of superoxide dismutase. *Biochim. Biophys. Acta* **327**: 257–265.

Schmidt, P., and Ingold, K.U. (1978). Kinetic applications of electron paramagnetic resonance spectroscopy. 31. Rate constants for spin trapping. 1. Primary alkyl radicals. *J. Am. Chem. Soc.* **100**: 2493–2500.

Sridhar, R., Beaumont, P.C., and Powers, E.L. (1986). Fast kinetics of the reactions of hydroxyl radicals with nitrone spin traps. *J. Radioanal. Nucl. Chem.* **101**: 227–237.

Steinbeck, M.J., Khan, A.U., and Karnovsky, M.J. (1992). Intracellular singlet oxygen generation by phagocytosing neutrophils in response to particles coated with a chemical trap. *J. Biol. Chem.* **267**: 13425–13433.

Steinbeck, M.J., Khan, A.U., and Karnovsky, M.J. (1993). Extracellular production of singlet oxygen by stimulated macrophages quantified using 9,10-diphenylanthracene and perylene in a polystyrene film. *J. Biol. Chem.* **268**: 15649–15654.

Stier, A., and Reitz, I. (1971). Radical production in amine oxidation by liver microsomes. *Xenobiotica* **1**: 499–500.

Stirpe, F., and Della Corte, E. (1969). The regulation of rat liver xanthine oxidase. Conversion *in vitro* of the enzyme activity from dehydrogenase (type D) to oxidase (type O). *J. Biol. Chem.* **244**: 3855–3863.

Swartz, H.M., Sentjuric, M., and Kocherginsky, N. (1995a). Metabolism of nitroxides and their products in cells, in: *Nitroxide Spin Labels. Reactions in Biology and Chemistry* (Kocherginsky, N., and Swartz, H.M., eds.), pp. 113–147, CRC Press, Boca Raton, FL.

Swartz, H.M., Sentjuric, M., and Kocherginsky, N. (1995b). Metabolism and distribution of nitroxides *in vivo*, in: *Nitroxide Spin Labels. Reactions in Biology and Chemistry* (Kocherginsky, N., and Swartz, H.M., eds.), pp. 153–173, CRC Press, Boca Raton, FL.

Taub, I.A., and Dorfman, L.M. (1962). Pulse radiolysis studies. II. Transient spectra and rate processes in irradiated ethanol and aqueous ethanol solution. *J. Am. Chem. Soc.* **84**: 4053–4059.

Tuccio, B., Lauricella, R., Fréjaville, C., Bouteiller, J.-C., and Tordo, P. (1995). Decay of the hydroperoxyl spin adduct of 5-diethoxyphosphoryl-5-methyl-1-pyrroline N-oxide: An EPR kinetic study. *J. Chem. Soc., Perkin Trans. 2* 295–298.

Turner, M.J., III, and Rosen, G.M. (1986). Spin trapping of superoxide and hydroxyl radicals with substituted pyrroline 1-oxides. *J. Med. Chem.* **29**: 2439–2444.

Van Reyk, D.M., Brown, A.J., Jessup, W., and Dean, R.T. (1995). Batch-to-batch variation of chelex-100 confounds metal-catalysed oxidation. Leaching of inhibitory compounds from a batch of chelex-100 and their removal by a pre-washing procedure. *Free Radical Res.* **23**: 533–535.

Woldman, Y.Y., Khramtsov, V.V., Grigor'ev, I.A., Kiriljuk, I.A., and Utepbergenov, D.I. (1994). Spin trapping of nitric oxide by nitronyl-nitoxides: Measurement of the activity of NO synthase from rat cerebellum. *Biochem. Biophys. Res. Commun.* **202**: 195–203.

Yamazaki, I., Piette, L.H., and Grover, T.A. (1990). Kinetic studies on spin trapping of superoxide and hydroxyl radicals generated in NADPH–cytochrome P-450 reductase–paraquat systems. Effect of iron chelates. *J. Biol. Chem.* **265**: 652–659.

Yoshimura, T., Yokoyama, H., Fujii, S., Takayama, F., Oikawa, K., and Kamada, H. (1996). *In vivo* EPR detection and imaging of endogenous nitric oxide in lipopolysaccharide-treated mice. *Nat. Biotechnol.* **14**: 992–994.

Zang, L.-Y., and Misra, H.P. (1992). EPR kinetic studies of superoxide radicals generated during the autoxidation of 1-methyl-4-phenyl-2,3-dihydropridinium, a bioactivated intermediate of Parkinsonian-inducing neurotoxin 1-methyl-4-phenyl-1,2,3,6-tetrahydropyridine. *J. Biol. Chem.* **267**: 23601–23608.

9

The Chemistry of Spin Traps and Spin Trapped Adducts

Addition of Grignard reagents to nitrones proceeds smoothly ... giving the 2-alkyl cyclic hydroxylamines in high yields.

Bonnett, Brown, Clark, Sutherland, and Todd, 1959

By most accounts, the chemistry of nitrones is dominated by aldol condensations and addition reactions, as this functional group mirrors the reactivity of aldehydes and ketones (Bonnett, *et al.*, 1959; Hamer and Macaluso, 1964; Torssell, 1988). The similarity of structure to these carbonyl groups plays a major role in the synthesis of this class of spin traps. In other situations, the reactivity of this functional group can initiate unwanted side reactions, masking the ability to decipher whether a biologic event is mediated by a free radical (Forrester and Hepburn, 1971). For instance, one only needs to be mindful of the difficulty in assigning the source of DMPO–OH to appreciate the myriad of complex reactions involving these compounds (Floyd and Wiseman, 1979; Chew and Bolton, 1980; Nohl, *et al.*, 1981; Tero-Kubota, *et al.*, 1982; Rosen and Rauckman, 1984; Chandra and Symons, 1986; Janzen, *et al.*, 1987; Makino, *et al.*, 1989, 1990a, 1990b, 1992; Buettner and Mason, 1990; Hanna, *et al.*, 1992; Nishi, *et al.*, 1992; Rosen, *et al.*, 1994).

The chemistry of nitrosoalkanes is primarily driven by nucleophilic additions, leading to hydroxylamines and, from this, acyclic nitrones can be derived (Huie and Cherry, 1985):

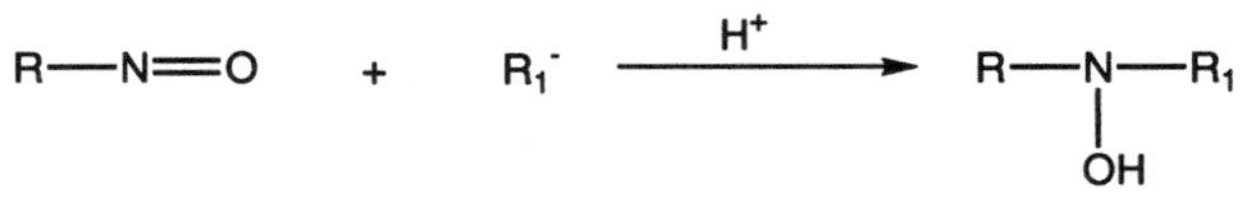

$$R\!-\!N\!\!=\!\!O \quad + \quad R_1^{\cdot} \quad \xrightarrow{\ H^+\ } \quad R\!-\!\underset{\underset{OH}{|}}{N}\!-\!R_1$$

In biological systems, however, reactions of this class of spin traps are considerably more restrictive, involving sequential reductive and/or oxidative steps:

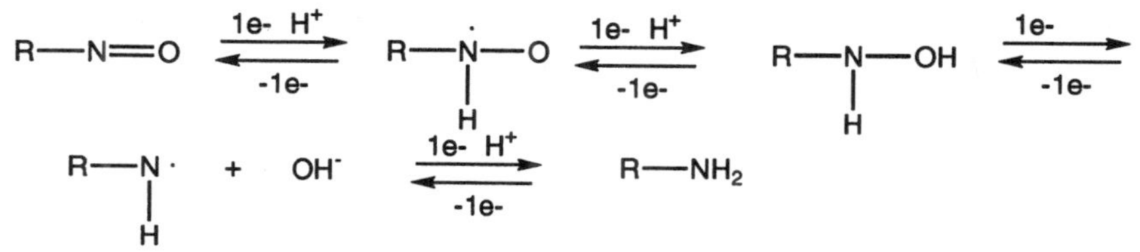

$$R-N{=}O \; \underset{-1e-}{\overset{1e- \; H^+}{\rightleftarrows}} \; R-\underset{\underset{H}{|}}{\overset{\cdot}{N}}-O \; \underset{-1e-}{\overset{1e- \; H^+}{\rightleftarrows}} \; R-\underset{\underset{H}{|}}{N}-OH \; \underset{-1e-}{\overset{1e-}{\rightleftarrows}}$$

$$R-\underset{\underset{H}{|}}{N}\cdot \; + \; OH^- \; \underset{-1e-}{\overset{1e- \; H^+}{\rightleftarrows}} \; R-NH_2$$

Of recent biologic importance is the formation of NO·, arising from the photo-activated cleavage of nitrosoalkanes (Wajer, *et al.*, 1967; Sargent and Gardy, 1976; Pou, *et al.*, 1994a):

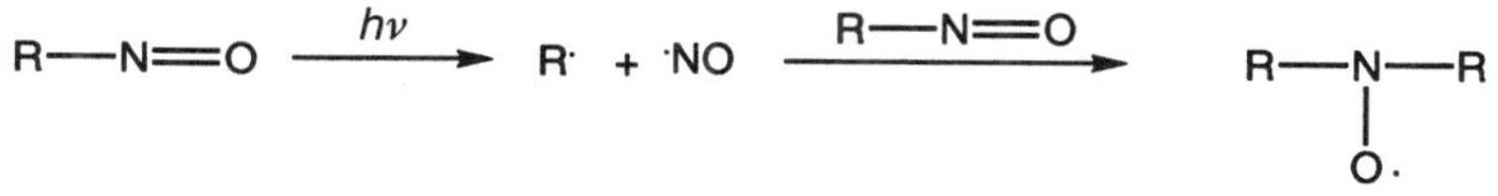

$$R-N{=}O \; \overset{h\nu}{\longrightarrow} \; R\cdot \; + \; \cdot NO \; \overset{R-N=O}{\longrightarrow} \; R-\underset{\underset{O\cdot}{|}}{N}-R$$

In chapter 6, the reader was introduced to synthetic schemes for the preparation of representative Δ^1-pyrroline 1-oxides, aromatic conjugated alkane-*N*-oxides, and nitrosoalkanes. Herein, the emphasis is placed on the chemical nature of these compounds, focusing primarily on reactions that can take place under biologic conditions. An awareness of their reactivity will, we hope, result in decreased misinterpretation of experimental findings.

Chemical Properties of Nitrones

Nitrones, which are tautomers of oximes, display physiochemical properties that readily allow characterization of this functional group.

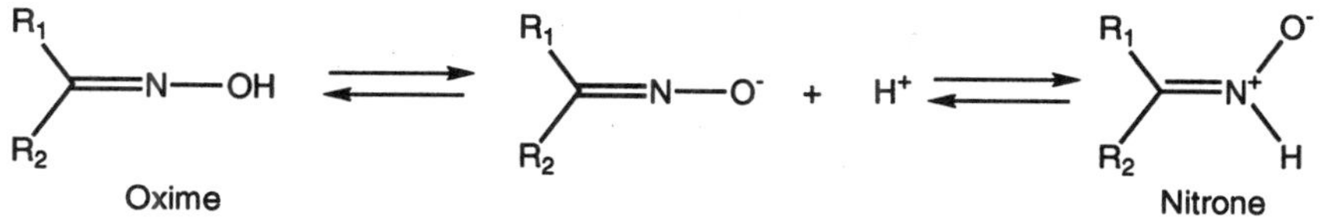

$$\underset{R_2}{\overset{R_1}{>}}{=}N-OH \; \rightleftarrows \; \underset{R_2}{\overset{R_1}{>}}{=}N-O^- \; + \; H^+ \; \rightleftarrows \; \underset{R_2}{\overset{R_1}{>}}{=}\overset{+}{\underset{\underset{H}{|}}{N}}\overset{O^-}{}$$

| Oxime | | Nitrone |

Common to most nitrones is the distinctive infrared spectral absorbance, exhibiting a C=N–O stretching frequency in the range of 1560–1600 cm^{-1} (Bonnett, *et al.*, 1959) and a strong UV absorption at a λ_{max} of 225–320 mµ (Wheeler and Gore, 1956; Bonnett, *et al.*, 1959) with intensities increasing with conjugation. The ^{1}H NMR chemical shift of approximately 7–8 δ is representative of both nitrones and aldehydes, which incidentally defines the similarity of their chemical behavior.

Nitrones are weakly basic with pK_a values, ranging from 7–9 (Bren, *et al.*, 1973). This chemical property undoubtedly contributes to the polar characteristics of these compounds, but is partially offset by the lipophilic nature of aromatic conjugated alkane-*N*-oxides (Pou, *et al.*, 1994b), isoquinoline-*N*-oxides, and benzazepine-*N*-oxides (Thomas, *et al.*, 1996). Not surprisingly, the small partition coefficients for Δ^1-pyrroline-*N*-oxides (Turner and Rosen, 1986) and highly charged aromatic conjugated alkane-*N*-oxides (Pou, *et al.*, 1994b) dictate that these compounds primarily distribute extracellularly (Table 9.1).

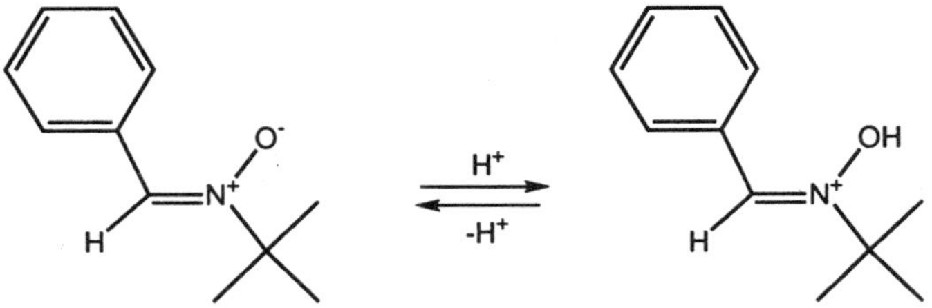

Most often, spin trapping experiments use racemic mixtures of two isomeric nitrones. The ability of each conformeric structure to spin trap a free radical has not, however, been extensively studied. From the few published reports, it appears that the anti form of PBN is, for example, the most stable chiral molecule and the one more easily approached by the free radical (Murofushi, *et al.*, 1987; Abe, *et al.*, 1991; Thomas, *et al.*, 1996). Future studies evaluating the dynamic stereochemistry of the interaction between a nitrone and a specific free radical may allow the design of more efficient spin traps.

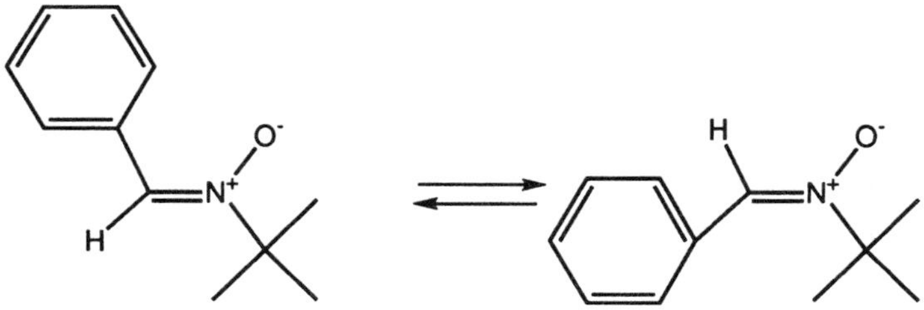

Anti   Syn

Addition Reactions and Aldol Condensations

Nitrones facilitate the formation of and stability of carbanions (Utzinger and Regenass, 1954; Brown, *et al.*, 1959a, 1959b; Lamchen and Mittag, 1966; Keana, 1984; Nazarski and Showronski, 1989; March, 1992). As such, the anion, once formed, can participate in a Michael or Michael-type reaction. One of the earliest reported examples includes the addition of the anion of 2,4,4-tri-methyl-1-pyrroline-*N*-oxide to 4,5,5-trimethyl-1-pyrroline-*N*-oxide (Brown, *et al.*, 1959a):

Table 9.1 Lipophilicity of Spin Traps

Spin Trap	Partition Coefficient	Experimental Condition
DMPO[a]	0.09	n-Octanol/phosphate buffer, pH 7.4
DMPO[b]	0.02	n-Octanol/phosphate buffer, pH 7.4
DMPO[c]	0.03	n-Octanol/HBSS buffer, pH 7.4
DMPO[d]	0.08	n-Octanol/phosphate buffer, pH 7.5
DMPO[e]	0.1	n-Octanol/water
4HDMPO[e]	0.01	n-Octanol/water
PBN[c]	11.4	n-Octanol/HBSS buffer, pH 7.4
PBN[d]	10.4	n-Octanol/phosphate buffer, pH 7.5
PBN[e]	21	n-Octanol/water
4-POBN[c]	0.09	n-Octanol/HBSS buffer, pH 7.4
4-POBN[d]	0.09	n-Octanol/phosphate buffer, pH 7.5
BPPO[b]	1.1	n-Octanol/phosphate buffer, pH 7.4
DPPO[b]	4.9	n-Octanol/phosphate buffer, pH 7.4
CPPO[b]	0.8	n-Octanol/phosphate buffer, pH 7.4
M_4PO^f	0.14	n-Octanol/phosphate buffer, pH 7.4
MNP[d]	8.2	n-Octanol/phosphate buffer, pH 7.5
NMP–OH[d]	0.3	n-Octanol/phosphate buffer, pH 7.5
NB[d]	73	n-Octanol/phosphate buffer, pH 7.5
DBNBS[d]	0.15	n-Octanol/phosphate buffer, pH 7.5
PBN[g]	17	Computer-generated
MDL[g] 101,002	10.23	Computer-generated
MDL[g] 102,389	37.15	Computer-generated

[a]DMPO = 5,5-dimethyl-1-pyrroline-N-oxide, Rosen, et al., 1982. [b]DMPO, Turner and Rosen, 1986; BPPO = 5-butyl-5-methyl-1-pyrroline-N-oxide, Turner and Rosen, 1986; DPPO = 5,5-dipropyl-1-pyrroline-N-oxide, Turner and Rosen, 1986; CPPO = 5-cyclopenyl-1-pyrroline-N-oxide, Turner and Rosen, 1986. [c]DMPO, Pou, et al., 1994b; PBN = N-tert-butyl-α-phenylnitrone, Pou, et al., 1994b; 4-POBN = α-(4-pyridyl 1-oxide)-N-tert-butylnitrone, Pou, et al., 1994b. [d]DMPO, PBN, and 4-POBN, Konorev, et al., 1993; MNP = 2-methyl-2-nitrosopropane, Konorev, et al., 1993; MNP–OH = 2-(hydroxymethyl)-2-nitrosopropane, Konorev, et al., 1993; NB = nitrosobenzene, Konorev, et al., 1993; DBNBS = sodium 3,5-dibromo-4-nitrosobenzenesulfonate, Konorev, et al., 1993. [e]DMPO, PBN, and 4HMDMPO = 5,5-dimethyl-4-hydroxymethyl-1-pyrroline-N-oxide, Konaka, et al., 1995. [f]M₄PO = 3,3,5,5-tetramethyl-1-pyrroline-N-oxide, Rosen and Turner, 1988. [g]PBN, Thomas, et al., 1996; MDL 101,002 = 3,4-dihydro-3,3-dimethylisoquinoline-N-oxide, Thomas, et al., 1996; MDL 102,389 = 3,3-dimethyl-4,5-dihydro-2-benzazepine-N-oxide, Thomas, et al., 1996.

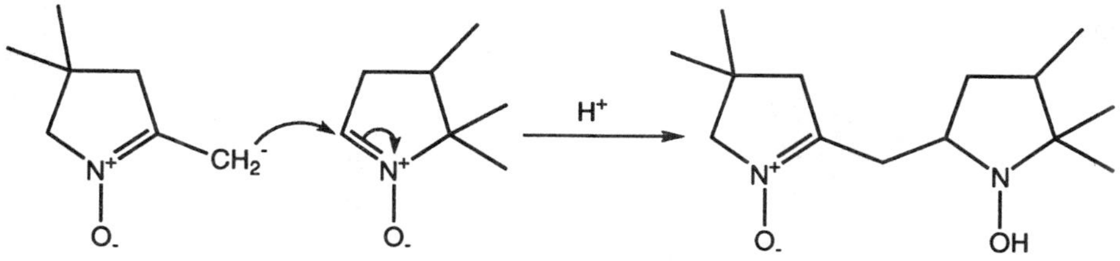

In some situations, self-condensations have been observed during Grignard additions to nitrones (Bonnett, et al., 1959; Black, et al., 1976; Nazarski and Skowronski, 1989). Here, the Grignard, serving as a base, generates an anion of, for instance, 2,5,5-trimethyl-1-pyrroline-N-oxide. Reaction with another molecule of the nitrone results in the corresponding hydroxylamine nitrone:

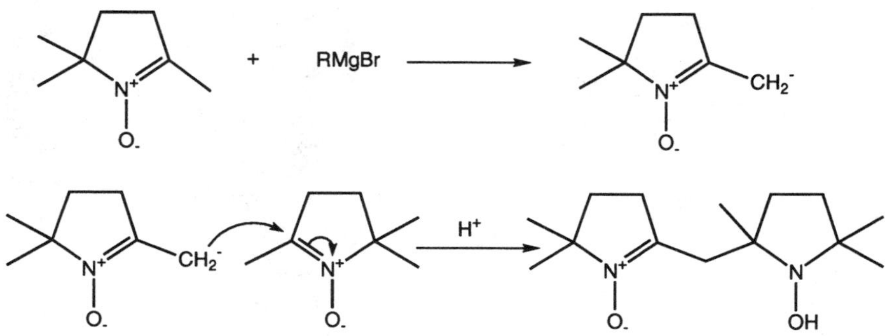

Even during the purification of 2,5,5-trimethyl-1-pyrroline-*N*-oxide under the mildest of conditions—high-vacuum Kugelrohr distillation at room temperature—self-condensation may occur, arising from the addition of the electron-rich enamine to the nitrone. Air oxidation leads to the observed nitroxide triplet (Janzen, *et al.*, 1993):

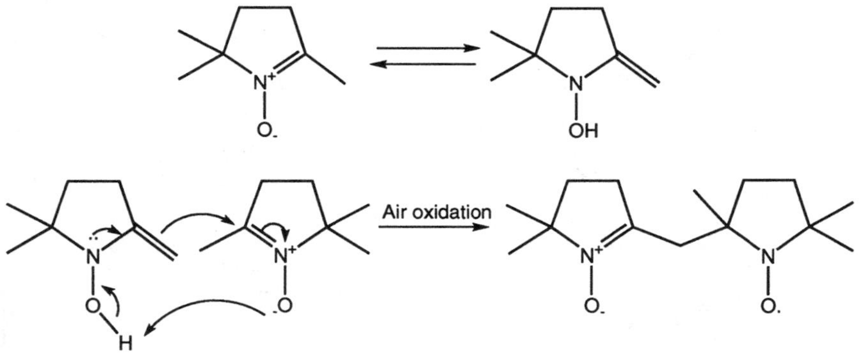

For 5,5-dimethyl-1-pyrroline-*N*-oxide, enolization would promote the dimerization of this nitrone. After aerobic oxidation, the resultant spin trapped adduct would exhibit a six-line EPR spectrum. This self-reaction pathway has been proposed to account for the observed nitroxide that frequently contaminantes aqueous solutions of this spin trap (Janzen, *et al.*, 1993):

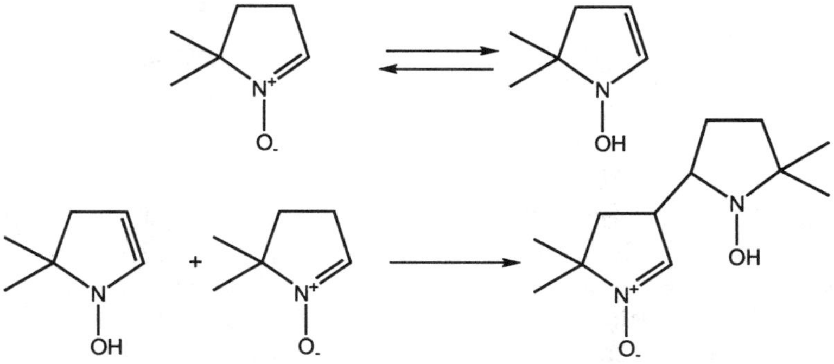

The breadth of aldol condensations is boundless, even though, in some situations, steric hindrance and the electrophilic character of the aldehyde become self-limiting (Nazarski and Showronski, 1989). For example, the anion of 2-methyl–pyrroline-*N*-oxide easily adds to benzaldehyde, giving 2-styryl-1-pyrroline-*N*-oxide (Bonnett, *et al.*, 1959):

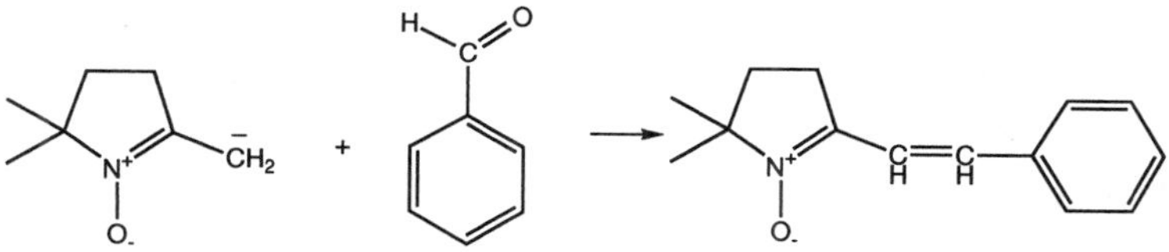

In contrast, the corresponding anion of 5,5-dimethyl-2-ethyl-1-pyrroline-*N*-oxide has been shown not to react with aromatic aldehydes, even with those containing electron withdrawing groups, such as *p*-nitrobenzaldehyde (Bonnett, *et al.*, 1959):

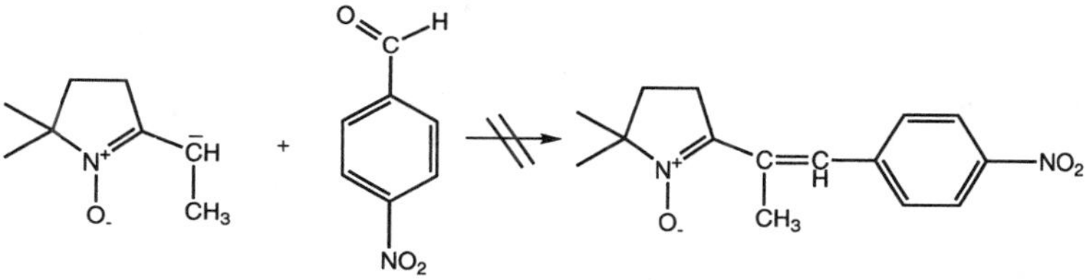

With 5,5-dimethyl-1-pyrroline-*N*-oxide, the failure to observe the desired product may be linked to diminished stability of the corresponding anion, as well as steric factors that make addition to an aldehyde difficult (Bonnett, *et al.*, 1959; Nazarski and Skowronski, 1989). In aqueous solutions, however, deuterium exchange at the 2- and 3-positions of 5,5-dimethyl-1-pyrroline-*N*-oxide attests to the acidity of the α-hydrogen and the delocalization of the charge over several adjacent atoms (Pou, *et al.*, 1990; Janzen, *et al.*, 1993):

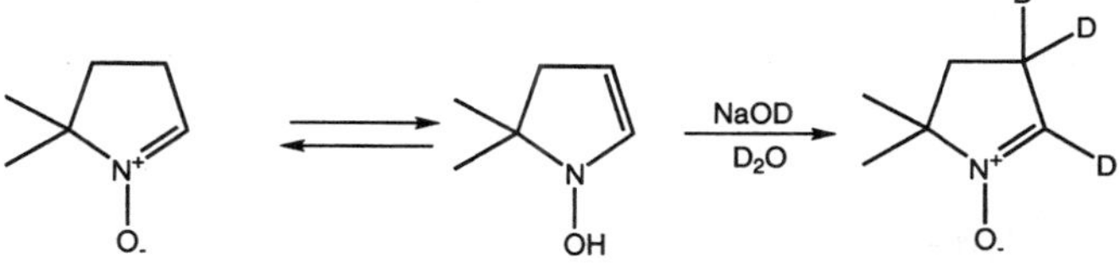

In aprotic solvents, acyclic nitrones, like their cyclic counterparts, display sufficient acidity to initiate an aldol condensation. For instance, the reaction of the anion of α-methyl-*N*-phenylnitrone with benzaldehyde results in α-styryl-*N*-phenylnitrone (Utzinger and Regenass, 1954):

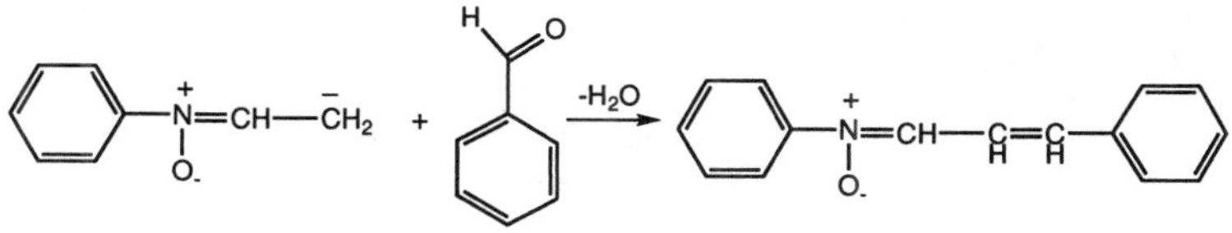

Rearrangement Reactions

One of the best documented reactions of spin trapped adducts of DMPO involves the base-catalyzed intramolecular rearrangement of 2-alkylperoxyl-5,5-dimethyl-1-pyrroldinyloxyl (DMPO–OOR) to the hydroxamic acid nitroxide 5,5-dimethyl-pyrrolidone-(2)-oxyl (DMPOX) (Floyd and Soong, 1977; Rosen and Rauckman, 1980; Hill and Thornalley, 1982; Kalyanaraman, *et al.*, 1983; Thornalley, *et al.*, 1983; Mason, 1984; Mossoba, *et al.*, 1984; Ben-Hur, *et al.*, 1985; Bernofsky, *et al.*, 1988; Davies, 1988; Van der Zee, *et al.*, 1996). Independent synthesis has verified the structural assignment for DMPOX (Bonnett, *et al.*, 1959; Rosen and Rauckman, 1980).

The initial step in the formation of DMPOX is the spin trapping of an alkylperoxyl radical (ROO·). The resultant nitroxide exhibits an EPR spectrum with hyperfine splitting constants that are independent of the alkyl group: ethyl (Kalyanaraman, *et al.*, 1983), *n*-propyl (Greenley and Davies, 1992), *tert*-butyl (Davies, 1988; Van der Zee, *et al.*, 1996) and cumene (Rosen and Rauckman, 1980) (Figure 9.1A). At physiologic pH, however, the rate of DMPOX formation is dependent upon the nature of the leaving group with cumene $\gg$ *t*-butyl (Figure 9.1B). Where R is a poor leaving group, such as ethyl or *n*-propyl, rearrangement to DMPOX has not been observed (Kalyanaraman, *et al.*, 1983; Greenley and Davies, 1992). Finally, in the presence of strong oxidants, whether they be iron salts (Elsworth and Lamchen, 1966, 1971; Makino, *et al.*, 1992; Nishi, *et al.*, 1992; Mao, *et al.*, 1994), sodium hypochlorite (Britigan and Hamill, 1989; Bernofsky, *et al.*, 1990), or acidic solutions containing NO· (Pou, *et al.*, 1994c), DMPO can be converted to DMPOX, even in the absence of alkylhydroperoxides:

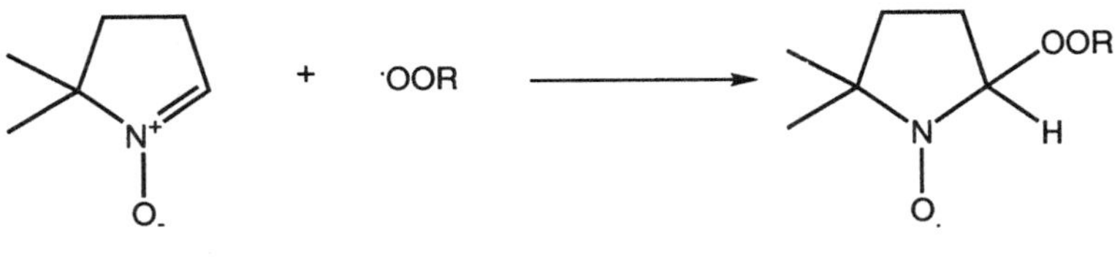

Where R = -CH$_2$CH$_3$,	$A_N = 14.6$ G, $A_H = 11.0$ G, $A_H^\gamma = 1.25$ G
Where R = -CH$_2$CH$_2$CH$_3$,	$A_N = 14.6$ G, $A_H = 10.9$ G, $A_H^\gamma = 1.04$ G
Where R = -C(CH$_3$)$_3$,	$A_N = 14.5$ G, $A_H = 10.7$ G, $A_H^\gamma = 1.28$ G
Where R = -C(CH$_3$)$_2$C$_6$H$_5$,	$A_N = 14.5$ G, $A_H = 10.75$ G, $A_H^\gamma = 1.75$ G

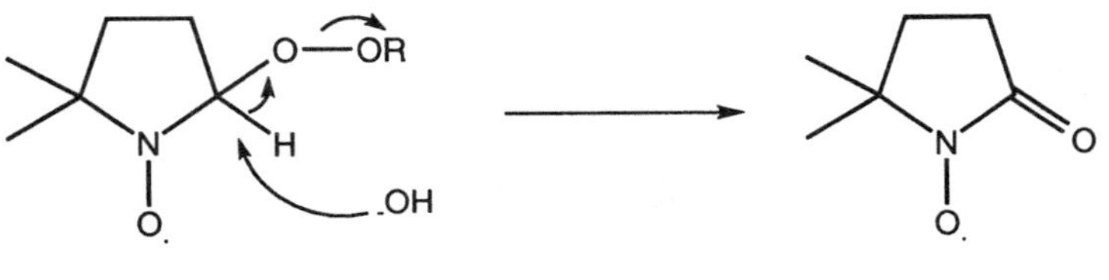

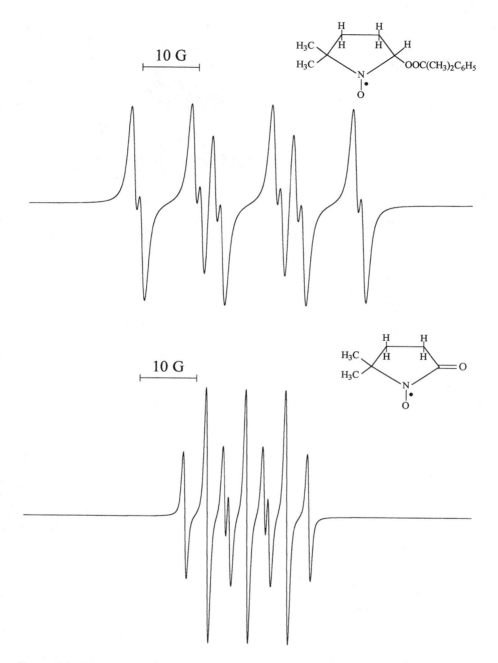

Figure 9.1 Aqueous solution simulation of the cumeneperoxyl spin trapped adduct of DMPO (5,5-dimethyl-1-pyrroline-N-oxide) and rearrangement product DMPOX. (A) DMPO–OOC(CH$_3$)$_2$C$_6$H$_5$ (2,2-dimethyl-5-cumeneperoxy-1-pyrrolinyloxyl). The hyperfine splitting constants at 9.5 GHz are $A_N = 14.54$ G, $A_H = 10.75$ G, and $A_H^{\gamma} = 1.75$ G. (B) DMPOX (5,5-dimethylpyrrolidone-(2)-oxyl). The hyperfine splitting constants at 9.5 GHz are $A_N = 7.1$ G and $A_H = 4.0$ G.

A similar rearrangement has been observed with PBN–OOR, resulting in benzoyl *tert*-butyl aminoxyl radical (PBNOX) (Janzen, *et al.*, 1990, 1992a; Sang, *et al.*, 1996):

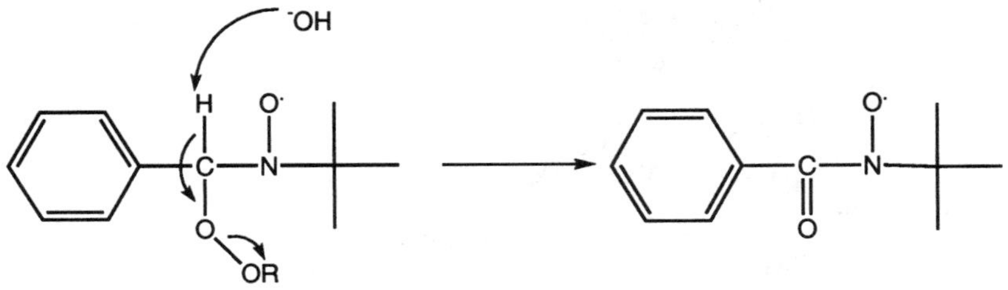

Decomposition of Spin Trapped Adducts

In the early years of spin trapping, it was not uncommon to attribute an EPR spectrum of DMPO–OH to the reaction between HO· and DMPO, even though, in a few of those publications, SOD but not catalase was reported to inhibit the formation of this spin trapped adduct (Rosen and Klebanoff, 1979; Britigan, *et al.*, 1986). It was soon realized that alternative pathways, those emerging from reactions not involving HO·, could give rise to this particular nitroxide (Finkelstein, *et al.*, 1979). Thereafter, the search for the origin of this spin trapped adduct led to many unanswered questions, solutions of which continue to elude even the most persistent of scientists.

There is little doubt that $O_2\cdot^-$ can be spin trapped by DMPO (Harbour, *et al.*, 1974; Buettner and Oberley, 1978; Sealy, *et al.*, 1978; Finkelstein, *et al.*, 1979; Lai, *et al.*, 1979; Kalyanaraman, *et al.*, 1980; Hanna, *et al.*, 1992). Early reports, however, seemed to suggest that the resulting spin trapped adduct, DMPO–OOH, was unstable, decomposing into DMPO–OH and other products (Finkelstein, *et al.*, 1979, 1980), the pathways of which still remain in doubt (Samuni, *et al.*, 1989a; Buettner, 1990; Yamazaki, *et al.*, 1990; Reinke, *et al.*, 1996). Based on mechanistic studies, the rearrangement of DMPO–OOH to 4-methyl-4-nitrosopentanal is an attractive route to the formation of HO· (Finkelstein, *et al.*, 1982). A more recent report proposes the identical rearrangement to account for the degradation of MPPO–OOH (Sankuratri, *et al.*, 1996).

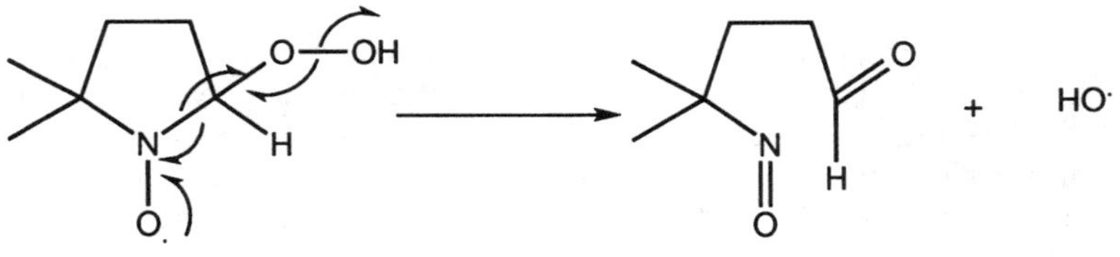

Not surprisingly, this ring opening reaction results in formation of HO· and 4-phenyl-4-nitrosopentanal. Attempts to enhance the stability of the corresponding spin trapped adducts have been marginally successful, dependent upon the site and nature of ring substitution. For instance, the degradation of DEPMPO–OOH, derived from the reaction of 5-(diethyoxyphosphoryl)-5-methyl-1-pyrroline N-oxide (DEPMPO) with $O_2\cdot^-$, does not follow a similar decomposition pathway as DMPO–OOH, since DEPMPO–OH was not observed in aqueous solutions of the parent hydroperoxide (Fréjaville, *et al.*, 1995). In contrast, outstanding leaving groups, such as the trifluoromethyl in TFDMPO–OH, actually promote re-arrangement, resulting in 5-methyl-5-nitroso-1,1,1-trifluoro-2-hexanone (Janzen, *et al.*, 1995).

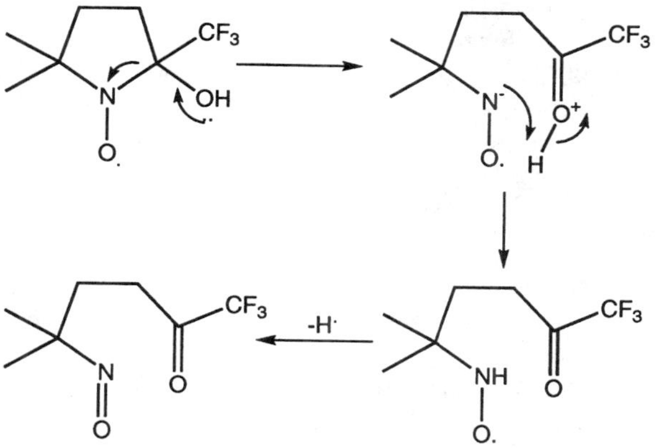

The substituent effect on the stability of acyclic nitrone spin trapped adducts has likewise received considerable attention. From those experiments, it appears that the rate of decomposition of substituted PBN–OHs is similarly linked to the nature of the group and its location on the aromatic ring. Electron withdrawing groups decrease the rate of decay whereas electron donating groups increase the rapidity of decomposition (Janzen, *et al.*, 1992b,1992c).

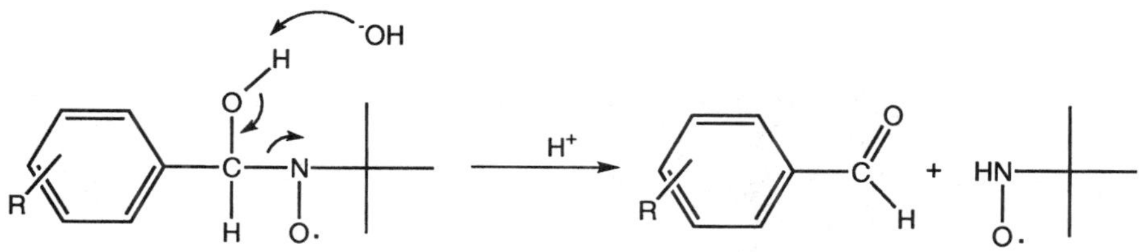

An understanding of the fate of DMPO–OOH in aqueous solutions must unquestionably be tied to the chemistry of hydroperoxides, their stability being defined by the structure of the peroxide and experimental conditions (Hiatt and Irwin, 1968; Hiatt, *et al.*, 1968a, 1968b, 1968c, 1968d). Decomposition of DMPO–OOH can occur through heterolytic, homolytic, and redox cleavage of the –O–O–

bond. From the diversity of products obtained, it appears that this degradation follows different reaction pathways; the importance of each pathway remains to be determined. Nevertheless, a thoughtful discussion of these reactions will test the veracity of these proposed schemes.

Even though thermal degradation of primary alkyl hydroperoxides involves heterolysis (Durham, *et al.*, 1958; Wurster *et al.*, 1958), for secondary alkyl hydroperoxides, like DMPO–OOH, homolytic cleavage of the –O–O– bond (Benson, 1964) is a reasonable route for the breakdown of this spin trapped adduct:

$$DMPO–OOH \rightarrow DMPO–O\cdot + HO\cdot$$

$$DMPO + HO\cdot \rightarrow DMPO–OH$$

Support for this reaction comes from the observation that small amounts of HO· are spin trapped at a time when $O_2\cdot^-$ is no longer present in the reaction mixture (Finkelstein, *et al.*, 1982). Under these conditions, appearance of DMPO–OH is unaffected by catalase and independent of redox active metal ions, which frequently taint aqueous solutions (Finkelstein, *et al.*, 1982).

The fortunes of DMPO–O· are less clear. Even though it is possible to spin trap this alkoxyl radical, there is no EPR spectral evidence that this reaction takes place (Finkelstein, *et al.*, 1979, 1982). Rather, a more likely outcome would be an intramolecular rearrangement of the pyrrolinyloxyl, leading to 4-methyl-4-nitrosopentanal.

Base-catalyzed decomposition of tertiary alkyl hydroperoxides has been reported to yield alcohols along with O_2 (Kharasch, *et al.*, 1952). Even though DMPO–OOH is a secondary alkyl hydroperoxide, steric hindrance may actually promote accumulation of DMPO–OH through this pathway:

$$DMPO–OOH + NaOH \rightarrow DMPO–OO^- + Na^+ + H_2O$$

$$DMP–OO^- + DMPO–OOH + H^+ \rightarrow 2DMPO–OH + O_2$$

Phosphate buffers, which are frequently contaminated with redox active metal ions, can accelerate the degradation of DMPO–OOH in the presence of $O_2\cdot^-$ (Kharasch, *et al.*, 1950, 1952):

$$O_2\cdot^- + Fe^{3+} \rightarrow O_2 + Fe^{2+}$$

$$Fe^{2+} + DMPO–OOH \rightarrow DMPO–O\cdot + Fe^{3+}$$

$$O_2\cdot^- + Fe^{3+} \rightarrow O_2 + Fe^{2+}$$

$$DMPO–O\cdot + Fe^{2+} + H^+ \rightarrow Fe^{3+} + DMPO–OH$$

Finally, DMPO–OOH is a substrate for glutathione peroxidase (Rosen and Freeman, 1984). Similar reductions by other peroxidases undoubtedly occur.

$$DMPO–OOH + 2GSH \xrightarrow{\text{glutathione peroxidase}} DMPO–OH + GSSG + H_2O$$

Oxidation of Nitrones

Thus far, the presentation has been limited in scope to potential mechanisms by which DMPO–OOH decomposition can lead to DMPO–OH. There is, of course, an alternative route to this nitroxide, one intimately linked to the catalytic activity of redox active metal ions, such as the cycling of Fe^{3+}/Fe^{2+}, whether these ions are confined to buffers (Elsworth and Lamchen, 1966, 1971, 1972; Makino, et al., 1990a, 1990b, 1992; Hanna, et al., 1992; Nishi, et al., 1992) or associated with enzyme preparations (Britigan, et al., 1990; Lloyd and Mason, 1990). One very reasonable hypothesis envisions the initial chelation of a ferric ion to DMPO, thereby enhancing the electrophilic character of the β-carbon by shunting electron density towards Fe^{3+}. This shift would promote an S_N2 addition by even as weak a nucleophile as H_2O, resulting in, after further oxidation, DMPO–OH (Makino, et al., 1990a, 1990b, 1992; Nishi, et al., 1992).

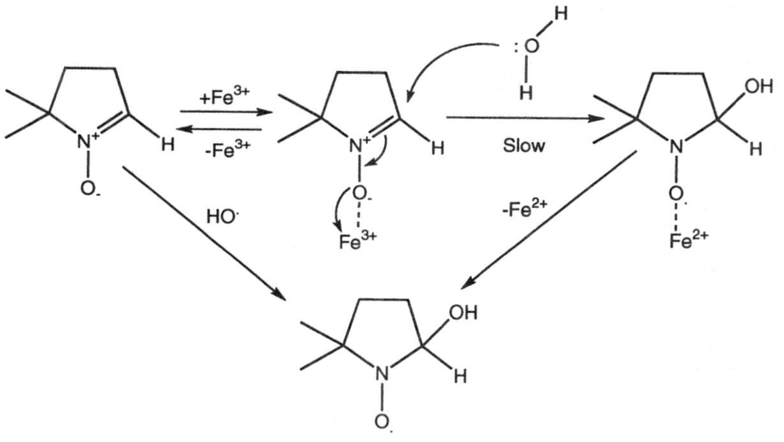

Under some experimental conditions, further oxidation by ferric salts can lead to DMPOX (Makino, et al., 1992). A similar reaction has been proposed for other redox active metal ions such as Cu^{2+}/Cu^{1+} (Hanna, et al., 1992).

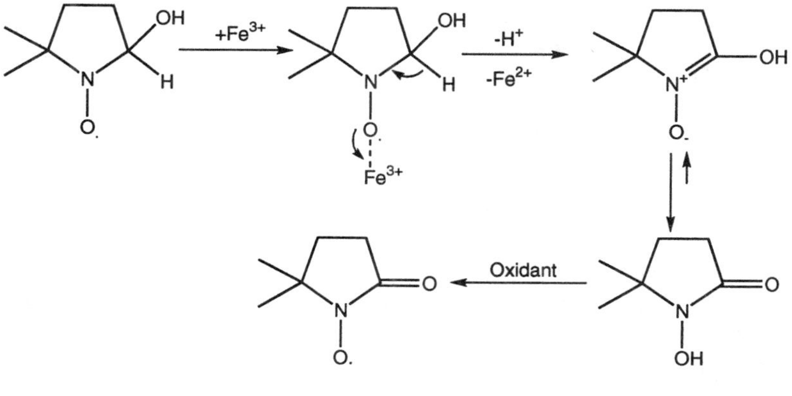

Formation of DMPO–OH in buffers suitable for biologic preparations is often artifact driven, leading to erroneous conclusions and missed opportunities for insightful discoveries. The depth of the problem can be well illustrated by noting that a common, frequently used buffer such as 50 mM potassium phosphate at pH 7.4 contains approximately 0.5 μM Fe^{3+} as a contaminant. Considering that the flux of $O_2\cdot^-$, generated, for example, by stimulated neutrophils, can approach 10 μM/1 × 10^6 cells per min, this level of redox active iron in the buffer is sufficiently great to catalyze the oxidation of the spin trap or promote the rate of spin trapped adduct decomposition. Removal of this and other trace metal ion impurities can be achieved by passing buffers through an ion-exchange resin, such as Chelex-100 (Poyer and McCay, 1971), or dialyzing against conalbumin (Gutteridge, 1987), and inclusion of metal ion chelators, including DTPA (Buettner, et al., 1978).

Chemical Properties of Nitrosoalkanes

The room temperature nitrosation of alkanes with *tert*-butylnitrite results in high yields of nitrosoalkanes with little, if any, oxime formation (Mackor, *et al.*, 1969; Mackor and de Boer, 1970):

$$\text{t-BuONO} \xrightarrow{h\nu} \text{t-BuO}\cdot + \text{NO}\cdot$$

$$\text{t-BuO}\cdot + \text{RH} \rightarrow \text{R}\cdot + \text{t-BuOH}$$

$$\text{R}\cdot + \text{NO}\cdot \rightarrow \text{RNO} \rightarrow \text{dimer}$$

The tendency not to isomerize is the result of the poor catalytic activity of *tert*-BuOH. Under more acidic conditions, this reaction is considerably faster than dimerization, especially when the nitroso compound contains an α-hydrogen (de Boer, 1982).

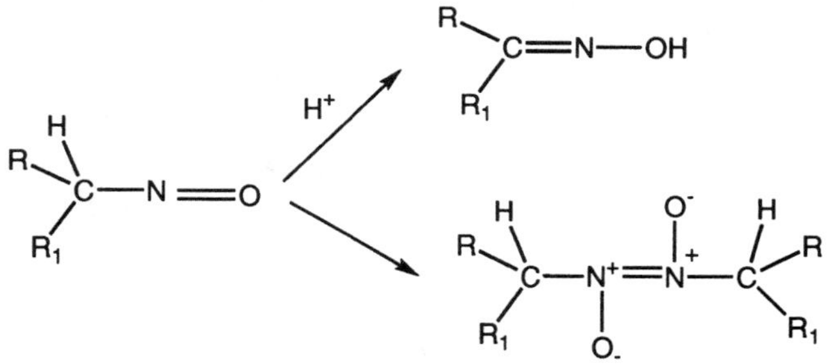

In fact, the isomerization of nitrosoalkanes under acidic conditions is an often used synthetic route to oximes (Metzger and Müller, 1957). More recent studies suggest a similar reaction can be catalyzed under basic conditions (Boyer, 1969):

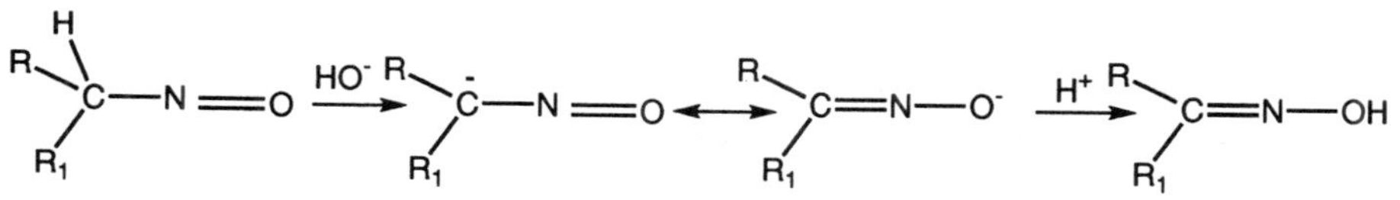

$$\underset{R_1}{\overset{R}{\underset{H}{C}}}\!-\!N\!=\!O \xrightarrow{HO^-} \underset{R_1}{\overset{R}{\overset{..}{C}}}\!-\!N\!=\!O \longleftrightarrow \underset{R_1}{\overset{R}{C}}\!=\!N\!-\!O^- \xrightarrow{H^+} \underset{R_1}{\overset{R}{C}}\!=\!N\!-\!OH$$

One of the most distinctive characteristics of primary and secondary nitrosoalkanes is the propensity to dimerize into colorless *cis*- and *trans*-azodioxy isomers (Sergeev and Leenson, 1978). The tendency of *tertiary*-nitrosoalkanes to undergo this reaction is considerably diminished, due to their dissociation into the monomer, as demonstrated by a blue-tinged solution (Mackor and de Boer, 1970). The origin of the blue, and in some cases green, color results from $n \to \pi^*$ transition with a weak absorption ($\varepsilon = 1$–60) in the 630–830 mµ range (Boyer, 1969).

Nitrosoalkanes exhibit distinctive infrared spectral absorbances, with C–N=O stretching frequencies around 1575 cm^{-1}, whereas for the *trans*-azodioxy isomer, the RN=NR absorption occurs at 1176 and 1290 cm^{-1} for aliphatic groups and slightly higher at 1253 and 1299 cm^{-1} for aromatic groups (Rao and Bhaskar, 1969). With the *cis*-azodioxy isomer, two bands are observed at 1323–1344 and 1330–1420 cm^{-1} and at 1389–1397 and 1409 cm^{-1} for aliphatic and aromatic groups, respectively (Rao and Bhaskar, 1969). A weak absorption maximal of nitrosoalkanes is found in a narrow range of 600–750 mµ, while the azodioxy isomers exhibit a very strong absorption from 290 to 370 mµ (Mackor, *et al.*, 1969).

Photochemical Formation of Nitroxides

From the perspective of spin trapping free radicals, the most important side reaction of nitrosoalkanes is the photodegradation to nitroxides (Mackor, *et al.*, 1967; Wajer, *et al.*, 1967; Doba, *et al.*, 1977; Pfab, 1978; Makino, *et al.*, 1981; Ohkuma, *et al.*, 1981). Although it was originally thought that the dimer was essential to nitroxide formation (Theilacker, *et al.*, 1965; Strom and Bluhm, 1966), more extensive investigations suggest that this hypothesis is only partially correct. At elevated temperatures, in which the equilibrium between the monomer and dimer is shifted toward the monomer (Sergeev and Leenson, 1978), the homolytic cleavage is at a considerably faster rate than was observed with the dimer per se (Mackor, *et al.*, 1966; Wajer, *et al.*, 1967).

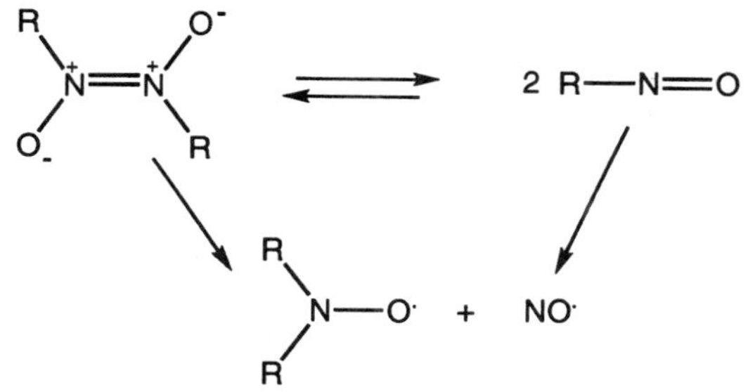

The λ_{max} for this reaction is frequently dependent on the structure of the nitrosoalkane. For example, the formation of diphenylnitroxide from the photolysis of nitrosobenzene requires energy from light in the ultraviolet range, 320–360 mμ (Mackor, et al., 1966). On the other hand, irradiation of 2-methyl-2-nitrosopropane is most efficient at longer wavelengths, around 680 mμ (Wajer, et al., 1967). Knowledge of these requirements can often dictate experimental conditions, allowing spin trapping experiments to be conducted with minimal formation of nitroxides from untoward side reactions.

Redox Cycling of Nitrosoalkanes

Since nitrosoalkanes can participate in either reductive or oxidative processes (Fischer and Mason, 1986), especially as a conduit for enzymatic shunting of electrons, the utility of this class of spin traps in describing specific free radical events in biological systems is greatly limited. For example, early studies with 2-methyl-2-nitrosopropane (MNP) found this spin trap to be susceptible to hepatic microsomal reduction, yielding *tert*-butylhydroxylamine (Kalyanaraman, et al., 1979). Oxidation by $O_2 \cdot {}^-$, produced during the enzymatic cycling of cytochrome P-450 (Kuthan and Ullrich, 1982) and/or FAD-containing monooxygenase (Patton, et al., 1980) resulted in *tert*-butyl hydronitroxide (Kalyanaraman, et al., 1979).

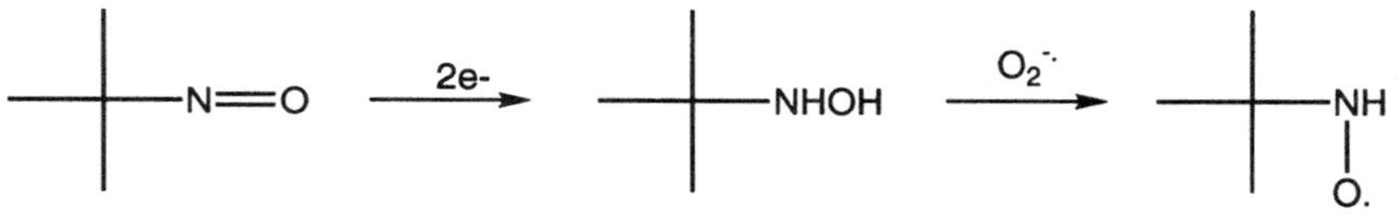

Likewise, endogenous biologic reductants, such as ascorbate, or readily oxidized xenobiotics can initiate similar reactions. This frequently results in the misconception that a free radical was spin trapped, whereas the data actually support the conclusion that the spin trap was a recipient of an electron. For instance, the autoxidation of dithranol in the presence of 2,4,6-tri-*tert*-butylnitrosobenzene resulted in an EPR spectrum, which was attributed to the spin trapping of the 1,8-dihydroxy-9-anthrone free radical (Martinmaa, et al., 1978). On closer examination, it was discovered that the 1,8-dihydroxy-9-anthrone free radical did not directly react with this nitrosobenzene, but, rather, this free radical reduced the spin trap to the corresponding hydronitroxide (Mottley, et al., 1981). Independent synthesis confirmed the structural assignment (Mottley, et al., 1981).

Aromatic nitrosoalkanes are especially susceptible to peroxidase-catalyzed oxidation to iminoxyls. Typical of this one-electron oxidation is the formation of 2-oxo-1,2-dihydronaphthalene-1-iminoxyl from 1-nitroso-2-naphthol (Fischer and Mason, 1986):

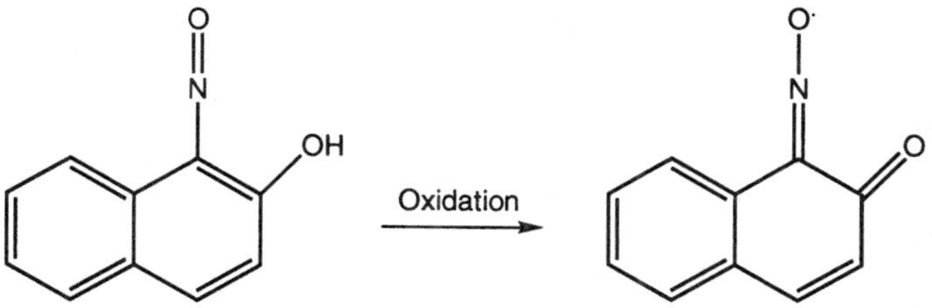

The ease of redox cycling of sodium 3,5-dibromo-4-nitrosobenzenesulfonate (DBNBS) has significantly limited the biologic applications of this spin trap (Mani and Crouch, 1989; Samuni, et al., 1989b; Nazhat, et al., 1990), despite the breadth of its reactions with a wide variety of carbon-centered free radicals and its remarkable stability toward light and heat (Ettinger, et al., 1981; Kaur, et al., 1981; Ozawa and Hanaki, 1987; Smith and Robertson, 1988; Athar, et al., 1989; Kaur, 1996).

Concluding Thoughts

An appreciation of the diversity of reactions that nitrones and nitrosoalkanes can undergo is the first step in understanding the complex chemistry of these classes of spin traps. This is an especially important process to ferret out, as similar biologic events will unquestionably arise and may lead the uninformed to an errant conclusion. In the ensuing chapter, we will describe the spin trapping of biologically generated free radicals in milieus ranging from well-defined homogenous enzyme preparations to the complexity of animal models.

References

Abe, Y., Seno, S.Y., Sakakibara, K., and Hirota, M. (1991). Spin trapping of oxygen-centred radicals by substituted N-benzylidine-*tert*-butylamine N-oxides. *J. Chem. Soc., Perkin Trans. 2*, 897–903.

Athar, M., Mukhtar, H., Bickers, D.R., Khan, I.U., and Kalyanaraman, B. (1989). Evidence for the metabolism of tumor promoter organic hydroperoxides into free radicals by human carcinoma skin keratinocytes: An ESR-spin trapping study. *Carcinogenesis* 10: 1499–1503.

Ben-Hur, E., Carmichael, A., Reisz, P., and Rosenthal, I. (1985). Photochemical generation of superoxide radical and the cytotoxicity of phthalocyanines. *Int. J. Radiat. Biol.* 48: 837–846.

Benson, S.W. (1964). Kinetics of pyrolysis of alkyl hydroperoxides and their O–O bond dissociation energies. *J. Chem. Phys.* 40: 1007–1013.

Bernofsky, C., Bandara, B.M.R., and Hinojosa, O. (1990). Electron spin resonance studies of the reaction of hypochlorite with 5,5-dimethyl-1-pyrroline-N-oxide. *Free Radical Biol. Med.* 8: 231–239.

Bernofsky, C., Olavesen, E.Y., Felix, C.C., and Kalyanaraman, B. (1988). Evidence for a hydroperoxyl radical from hypochlorite-modified adenosine 5′-

monophosphate (AMP). *Biochem. Arch.* **4**: 103–107.

Black, D.St.C., Clark, V.M., Odell, B.G., Suthjerland, I.O., and Todd, A. (1976). Experiments towards the synthesis of corrins. Part XI. Attempted synthesis of β-dinitrones by oxidative routes. *J. Chem. Soc., Perkin Trans. I* 1942–1943.

Bonnett, R., Brown, R.F.C., Clark, V.M., Sutherland, I.O., and Todd, A. (1959). Experiments towards the synthesis of corrins. Part II. The preparation and reactions of Δ^1-pyrroline 1-oxides. *J. Chem. Soc.* 2094–2102.

Boyer, J.H. (1969). Methods of formation of the nitroso group and its reactions, in: *The Chemistry of the Nitro and Nitroso Groups* (Feuer, H., ed.), Part 1, pp. 215–299, John Wiley & Sons and Interscience Publishers, New York.

Bren, V.A., Medyantseva, E.A., Andreeva, I.M., and Minkin, V.I. (1973). Structure and properties of nitrones. IV. Basicity of α-aryl-N-methylaldonitrones. *Zh. Org. Kim.* **9**: 767–770.

Britigan, B.E., and Hamill, D.R. (1989). The interaction of 5,5-dimethyl-1-pyrroline-N-oxide with human myeloperoxidase and its potential impact on spin trapping of neutrophil-derived radicals. *Arch. Biochem. Biophys.* **275**: 72–81.

Britigan, B.E., Pou, S., Rosen, G.M., Lilleg, D.M., and Buettner, G.R. (1990). Hydroxyl radical is not a product of the reaction of xanthine oxidase and xanthine. The confounding problem of adventitious iron bound to xanthine oxidase. *J. Biol. Chem.* **265**: 17533–17538.

Britigan, B.E., Rosen, G.M., Chai, Y., and Cohen, M.S. (1986). Do human neutrophils make hydroxyl radical? Detection of free radicals generated by human neutrophils activated with a soluble or particulate stimulus using electron paramagnetic resonance spectrometry. *J. Biol. Chem.* **261**: 4426–4431.

Brown, R.F.C., Clark, V.M., Lamchen, M., Sklarz, B., and Todd, A. (1959a). Aldol, pinacol, and benzoin-type reactions of Δ^1-pyrroline 1-oxide. *Proc. Chem. Soc.* 169–171.

Brown, R.F.C., Clark, V.M., and Todd, A. (1959b). Experiments towards the synthesis of corrins. Part IV. The oxidation and ring expansion of 2:4:4:-trimethyl-Δ^1-pyrroline 1-oxide. *J. Chem. Soc.* 2105–2108.

Buettner, G.R. (1990). On the reaction of superoxide with DMPO/·OOH *Free Radical Res. Commun.* **10**: 11–15.

Buettner, G.R., and Mason, R.P. (1990). Spin-trapping methods for detecting superoxide and hydroxyl free radicals *in vitro* and *in vivo*. *Methods Enzymol.* **186**: 127–133.

Buettner, G.R., and Oberley, L.W. (1978). Considerations in the spin trapping of superoxide and hydroxyl radical in aqueous systems using 5,5-dimethyl-1-pyrroline-1-oxide. *Biochem. Biophys. Res. Commun.* **83**: 69–74.

Buettner, G.R., Oberley, L.W., and Leuthauser, S.W.H.C. (1978). The effect of iron on the distribution of superoxide and hydroxyl radicals as seen by spin trapping and on the superoxide dismutase assay. *Photochem. Photobiol.* **28**: 693–695.

Chandra, H., and Symons, M.C.R. (1986). Hydration of spin-trap cations as a source of hydroxyl adducts. *J. Chem. Soc., Chem. Commun.* 1301–1302.

Chew, V.S.F., and Bolton, J.R. (1980). Photochemistry of 5-methyl-phenazinium salts in aqueous solution. Products and quantum yield of the reaction. *J. Phys. Chem.* **84**: 1903–1908.

Davies, M.J. (1988). Detection of peroxyl and alkoxyl radicals produced by reaction of hydroperoxides with heme-proteins by electron spin resonance spectroscopy. *Biochim. Biophys. Acta* **964**: 28–35.

de Boer, Th.J. (1982). Spin-trapping in early and some recent nitroso chemistry. *Can. J. Chem.* **60**: 1602–1609.

Doba, T., Ichikawa, T., and Yoshida, H. (1977). Kinetic studies of spin-trapping reactions. I. The trapping of the *t*-butyl radical generated by the photodissociation of 2-methyl-2-nitrosopropane by several spin-trapping agents. *Bull. Chem. Soc. Jpn.* **50**: 3158–3163.

Durham, L.J., Wurster, C.F., Jr., and Mosher, H.S. (1958). Peroxides. VIII. The mechanism for the thermal decomposition of *n*-butyl hydroperoxides and *n*-butyl 1-hydroxybutyl peroxide. *J. Am. Chem. Soc.* **80**: 332–337.

Elsworth, J.F., and Lamchen, M. (1966). Cyclic nitrones. II. The oxidation of cyclic nitrones by iron (III) chloride. *J. Chem. Soc. C* 1477–1479.

Elsworth, J.F., and Lamchen, M. (1971). Nitrones. XI. Spectral studies on the oxidation of cyclic nitrones by iron (III) chloride. *J. S. Afr. Chem. Inst.* **24**: 196–204.

Elsworth, J.F., and Lamchen, M. (1972). Nitrones. Part XII. Kinetic and mechanistic studies on the oxidation of cyclic nitrones and cyclic hydroxyl-amines by iron (II) chloride. *J. S. Afr. Chem. Inst.* **25**: 1–9.

Ettinger, K.V., Forrester, A.R., and Hunter, C.H. (1981). Lyoluminescence and spin trapping. *Can. J. Chem.* **60**: 1549–1559.

Finkelstein, E., Rosen, G.M., and Rauckman, E.J. (1980). Spin trapping. Kinetics of the reaction of superoxide and hydroxyl radicals with nitrones. *J. Am. Chem. Soc.* **102**: 4994–4999.

Finkelstein, E., Rosen, G.M., and Rauckman, E.J. (1982). Production of hydroxyl radical by decomposition of superoxide spin trapped adducts. *Mol. Pharmacol.* **21**: 262–265.

Finkelstein, E., Rosen, G.M., Rauckman, E.J., and Paxton J. (1979). Spin trapping of superoxide. *Mol. Pharmacol.* **16**: 676–685.

Fischer, V., and Mason, R.P. (1986). Formation of iminoxyl and nitroxide free radicals from nitrosonaphthols: An electron spin resonance study. *Chem.-Biol. Interact.* **57**: 129–142.

Floyd, R.A., and Soong, L.M. (1977). Spin trapping in biological systems. Oxidation of the spin trap 5,5-dimethyl-1-pyrroline-1-oxide by a hydroperoxide–hematin system. *Biochem. Biophys. Res. Commun.* **74**: 79–84.

Floyd, R.A., and Wiseman, B.B. (1979). Spin-trapping free radicals in the auto-oxidation of 6-hydroxydopamine. *Biochim. Biophys. Acta* **586**: 196–207.

Forrester, A.R., and Hepburn, S.P. (1971). Spin traps. A cautionary note. *J. Chem. Soc. C* 701–703.

Fréjaville, C., Karoui, H., Tuccio, B., Le Moigne, F., Culcasi, M., Pietri, S., Lauricella, R., and Tordo, P. (1995). 5-(Diethyoxyphosphoryl)-5-methyl-1-pyrroline N-oxide: A new efficient phosphorylated nitrone for the *in vitro* and *in vivo* spin trapping of oxygen-centered radicals. *J. Med. Chem.* **38**: 258–265.

Greenley, T.L., and Davies, M.J. (1992). Detection of radicals produced by reaction of hydroperoxides with rat liver microsomal fractions. *Biochim. Biophys. Acta* **1116**: 192–203.

Gutteridge, J.M.C. (1987). A method for removal of trace iron contamination from biological buffers. *FEBS Lett.* **214**: 362–364.

Hamer, J., and Macaluso, A. (1964). Nitrones. *Chem. Rev.* **64**: 473–495.

Hanna, P.M., Chamulitrat, W., and Mason, R.P. (1992). When are metal ion-dependent hydroxyl and alkoxyl radical adducts of 5,5-dimethyl-1-pyrroline N-oxide artifacts? *Arch. Biochem. Biophys.* **296**: 640–644.

Harbour, J.R., Chow, V., and Bolton, J.R. (1974). An electron spin resonance study of the spin adducts of OH and HO_2 radicals with nitrones in the ultraviolet photolysis of aqueous hydrogen peroxide solutions. *Can. J. Chem.* **52**: 3549–3553.

Hiatt, R., and Irwin, K.C. (1968). Homolytic decomposition of hydroperoxides. V. Thermal decompositions. *J. Org. Chem.* **33**: 1436–1441.

Hiatt, R., Irwin, K.C., and Gould, C.W. (1968a). Homolytic decomposition of hydroperoxides. IV. Metal-catalyzed decompositions. *J. Org. Chem.* **33**: 1430–1435.

Hiatt, R., Mill, T., Irwin, K.C., and Castleman, J.K. (1968b). Homolytic decompositions of hydroperoxides. II. Radical-induced decompositions of *t*-butyl hydroperoxide. *J. Org. Chem.* **33**: 1421–1428.

Hiatt, R., Mill, T., Irwin, K.C., and Castleman, J.K. (1968c). Homolytic decompositions of hydroperoxides. III. Radical-induced decomposition of primary and secondary hydroperoxides. *J. Org. Chem.* **33**: 1428–1430.

Hiatt, R., Mill, T., and Mayo, F.R. (1968d). Homolytic decompositions of hydroperoxides. I. Summary and implications for autoxidation. *J. Org. Chem.* **33**: 1416–1420.

Hill, H.A.O., and Thornalley, P.J. (1982). The oxidative ring opening of the cyclic nitrone spin trap 5,5-dimethyl-1-pyrroline-1-oxide (DMPO): Free radical involvement. *Inorg. Chim. Acta* **67**: L35–L36.

Huie, R., and Cherry, W.R. (1985). Facile one-step synthesis of phenyl-*tert*-butyl-nitrone (PBN) and its derivatives. *J. Org. Chem.* **50**: 1531–1532.

Janzen, E.G., Hinton, R.D., and Kotake, Y. (1992b). Substituent effect on the stability of the hydroxyl radical adduct of α-phenyl N-*tert*-butyl nitrone (PBN). *Tetrahedron Lett.* **33**: 1257–1260.

Janzen, E.G., Jandrisits, L.T., and Barber, D.L. (1987). Studies on the origin of the hydroxyl spin adduct of DMPO produced from the stimulation of neutrophils by phorbol-12-myristate-13-acetate. *Free Radical Res. Commun.* **4**: 115–123.

Janzen, E.G., Kotake, Y., and Hinton, R.D. (1992c). Stabilities of hydroxyl radical spin adducts of PBN-type spin traps. *Free Radical Biol. Med.* **12**: 169–173.

Janzen, E.G., Krygsman, P.H., Lindsay, D.A., and Haire, D.L. (1990). Detection of alkyl, alkoxyl, and alkyperoxyl radicals from the thermolysis of azobis-(isobutyronitrile) by ESR/spin trapping. Evidence for double spin adducts from liquid-phase chromatography and mass spectroscopy. *J. Am. Chem. Soc.* **112**: 8279–8284.

Janzen, E.G., Lin, C.-R., and Hinton, R.D. (1992a). Spontaneous free-radical formation in reaction of *m*-chloroperbenzoic acid with C-phenyl-N-*tert*-butylnitrone (PBN) and 3- or 4-substituted PBNs. *J. Org. Chem.* **57**: 1633–1635.

Janzen, E.G., Zhang, Y.-K., and Arimura, M. (1993). Proposed mechanism for production of stable aminoxyl radical impurities in the synthesis of substituted 5,5-dimethylpyrroline-N-oxide (DMPO) spin traps. *Chem. Lett.* 497–500.

Janzen, E.G., Zhang, Y.-K., and Arimura, M. (1995). Synthesis and spin-trapping chemistry of 5,5-dimethyl-2-(trifluoromethyl)-1-pyrroline N-oxide. *J. Org. Chem.* **60**: 5434–5440.

Kalyanaraman, B., Mottley, C., and Mason, R.P. (1983). A direct electron spin resonance and spin-trapping investigation of peroxyl free radical formation by hematin/hydroperoxide systems. *J. Biol. Chem.* **258**: 3855–3858.

Kalyanaraman, B., Perez-Reyes, E., and Mason, R.P. (1979). The reduction of nitroso-spin traps in chemical and biological systems. A cautionary note. *Tetrahedron Lett.* 4809–4812.

Kalyanaraman, B., Perez-Reyes, E., and Mason, R.P. (1980). Spin-trapping and direct electron spin resonance investigations of the redox metabolism of quinone anticancer drugs. *Biochem. Biophys. Acta* **630**: 119–130.

Kaur, H. (1996). A water soluble C-nitroso-aromatic spin-trap—3,5-dibromo-4-nitrosobenzensulphoic acid. "The Perkins spin-trap." *Free Radical Res.* **24**: 409–422.

Kaur, H., Leung, K.H.W., and Perkins, M.J. (1981). A water-soluble, nitroso-aromatic spin-trap. *J. Chem. Soc., Chem. Commun.* 142–143.

Keana, J.F.W. (1984). Synthesis and chemistry of nitroxide spin labels, in: *Spin Labeling in Pharmacology* (Holtzman, J., ed.), pp. 1–85, Academic Press, New York.

Kharasch, M.S., Fono, A., and Nudenberg, W. (1950). The chemistry of hydroperoxides. III. The free-radical decomposition of hydroperoxides. *J. Org. Chem.* **15**: 763–774.

Kharasch, M.S., Fono, A., Nudenberg, W., and Bischof, B. (1952). The chemistry of hydroperoxides. XI. Hydroperoxides as oxidizing and reducing agents. *J. Org. Chem.* **17**: 207–220.

Konaka, R., Kawai (Nee Abe), M., Noda, H., Kohno, M., and Niwa, R. (1995). Synthesis and evaluation of DMPO-type spin traps. *Free Radical Res.* **23**: 15–25.

Konorev, E.A., Baker, J.E., Joseph, J., and Kalyanaraman, B. (1993). Vasodilator and toxic effects of spin traps on aerobic cardiac function. *Free Radical Biol. Med.* **14**: 127–137.

Kuthan H., and Ullrich, V. (1982). Oxidase and oxygenase function of the microsomal cytochrome P450 monooxygenase system. *Eur. J. Biochem.* **126**: 583–588.

Lai, C.S., Grover, T.A., and Piette, L.H. (1979). Hydroxyl radical production in a purified NADPH-cyrcohrome c

(P-450) reductase system. *Arch. Biochem. Biophys.* **193**: 373–378.

Lamchen, M., and Mittag, T.W. (1966). Nitrones. IV. Synthesis and properties of a monocyclic α-dinitrone. *J. Chem. Soc. C* 2300–2303.

Lloyd, R.V., and Mason, R.P. (1990). Evidence against transition metal-independent hydroxyl radical generation by xanthine oxidase. *J. Biol. Chem.* **265**: 16733–16736.

Mackor, A., and de Boer, Th.J. (1970). C-Nitroso compounds. Part XII. *Cis-* and *trans-*azodioxy compounds (dimeric nitroso compounds) by photochemical nitrosation of hydrocarbons with *tert-*butyl nitrite. *Recl. Trav. Chim. Pays-Bas* **89**: 151–158.

Mackor, A., Veenland, J.U., and de Boer, Th.J. (1969). C-Nitroso compounds. Part XI. *Trans-*azodioxycyclohexane (dimeric nitrosocyclohexane) by photochemical nitrosation of cyclohexane with alkyl nitrites. *Recl. Trav. Chim. Pays-Bas* **88**: 1249–1262.

Mackor, A., Wajer, Th.A.J.W., and de Boer, Th.J. (1966). C-Nitroso compounds. Part I. The formation of nitroxides by photolysis of nitroso compounds as studied by electron spin resonance. *Tetrahedron Lett.* 2115–2123.

Mackor, A., Wajer, Th.A.J.W., de Boer, Th.J., and van Voorst, J.D.W. (1967). C-Nitroso compounds. Part. III. Alkoxy-alkyl nitroxides as intermediates in the reaction of alkoxyl-radicals with nitroso compounds. *Tetrahedron Lett.* 385–390.

Makino, K., Hagi, A., Ide, H., Murakami, A., and Nishi, M. (1992). Mechanistic studies on the formation of aminoxyl radicals from 5,5-dimethyl-1-pyrroline-N-oxide in Fenton systems. Characterization of key precursors giving rise to background ESR signals. *Can. J. Chem.* **70**: 2818–2827.

Makino, K., Hagiwara, T., Hagi, A., Nishi, M., and Murakami, A. (1990a). Cautionary note for DMPO spin trapping in the presence of iron ion. *Biochem. Biophys. Res. Commun.* **172**: 1073–1080.

Makino, K., Hagiwara, T., Imaishi, H., Nishi, M., Fuji, S., Ohya, H., and Murakami, A. (1990b). DMPO spin trapping in the presence of Fe ion. *Free Radical Res. Commun.* **9**: 233–240.

Makino, K., Imaishi, H., Morinishi, S., Hagiwara, T., Takeuchi, T., and Murakami, A. (1989). An artifact in the ESR spectrum obtained by spin trapping with DMPO. *Free Radical Res. Commun.* **6**: 19–28.

Makino, K., Suzuki, N., Moriya, F., Rokushika, S., and Hatano, H. (1981). A fundamental study on aqueous solutions of 2-methyl-2-nitrosopropane as a spin trap. *Radiat. Res.* **86**: 294–310.

Mani, V., and Crouch R.K. (1989). Spin trapping of the superoxide anion: Complications in the use of the water-soluble nitroso-aromatic reagent DBNBS. *J. Biochem. Biophys. Methods* **18**: 91–96.

Mao, G.D., Thomas, P.D., and Poznansky, M.J. (1994). Oxidation of spin trap 5,5-dimethyl-1-pyrroline-N-oxide in an electron paramagnetic resonance study of the reaction of methemoglobin with hydrogen peroxide. *Free Radical Biol. Med.* **16**: 493–500.

March, J. (1992). *Advanced Organic Chemistry. Reactions, Mechanisms, and Structure*, 4th edition, pp. 175–181, John Wiley & Sons, New York.

Martinmaa, J., Vanhala, L., and Mustakallio, K.K. (1978). Free radical intermediates produced by autoxidation of 1,8-dihydroxy-9-anthrone (dianthanol) in pyridine. *Experientia* **34**: 872–873.

Mason, R.P. (1984). Spin trapping free radical metabolites of toxic chemicals, in: *Spin Labeling in Pharmacology* (Holtzman, J., ed.), pp. 87–129, Academic Press, New York.

Metzger, H., and Müller, E. (1957). Über die einwirkung von licht auf die umwandlung von sek. Cycloaliphatischen bis-nitrosoverbindungen zu oximen mit chlorwasserstoff. *Chem. Ber.* **90**: 1185–1188.

Mossoba, M.M., Rosenthal, I., Carmichael, A.J., and Reisz, P. (1984). Photochemistry of porphyrins as studied by spin trapping and electron spin resonance. *Photochem. Photobiol.* **39**: 731–734.

Mottley, C., Kalyanaraman, B., and Mason, R.P. (1981). Spin trapping artifacts due to the reduction of nitroso spin traps. *FEBS Lett.* **130**: 12–14.

Murofushi, K., Abe, K., and Hirota, M. (1987). Substituent effect on the spin trapping reactions of substituted N-benzylidene-*t*-butylamine N-oxides. *J. Chem. Soc., Perkin Trans. II* 1829–1833.

Nazarski, R.B., and Showronski, R. (1989). Sterically crowded five-membered heterocyclic systems. Part 3. Unexpected formation of stable flexible pyrrolidinoxyl biradicals via nitrone aldol dimers: A spectroscopic and mechanistic study. *J. Chem. Soc., Perkin Trans. I* 1603–1610.

Nazhat, N.B., Yang, G., Allen, R.E., Blake, D.R., and Jones, P. (1990). Does 3,5-dibromo-4-nitrosobenzene sulphonate spin trap superoxide radicals? *Biochem. Biophys. Res. Commun.* 166: 807–812.

Nishi, M., Hagi, A., Ide, H., Murakami, A., and Makino, K. (1992). Comparison of 2,5,5-trimethyl-1-pyrroline-N-oxide (M_3PO) and 3,3,5,5-tetramethyl-1-pyrroline-N-oxide (M_4PO) with 5,5-dimethyl-1-pyrroline-N-oxide (DMPO) as spin traps. *Biochem. Int.* 27: 651–659.

Nohl, H., Jordan, W., and Hegner, D. (1981). Identification of free hydroxyl radicals in respiring heart mitochondria by spin trapping with the nitrone DMPO. *FEBS Lett.* 123: 241–244.

Ohkuma, T., Kirino, Y., and Kwan, T. (1981). Some physicochemical properties of 2-methyl-2-nitrosopropne, phenyl N-tert-butyl nitrone, 5,5-dimethyl-1-pyrroline-N oxide and 2,5,5-trimethyl-1-pyrroline-N-oxide and the feasibility of their use as spin traps in aqueous solution. *Chem. Pharm. Bull.* 29: 25–28.

Ozawa, T., and Hanaki, A. (1987). Spin-trapping of alkyl radical by a water-soluble nitroso aromatic spin-trap, 3,5-dibromo-4-nitroso-benzenesulfonate. *Bull. Chem. Soc. Jpn.* 60: 2304–2306.

Patton, S.E., Rosen, G.M., and Rauckman, E.J. (1980). Superoxide production by purified hamster hepatic nuclei. *Mol. Pharmacol.* 18: 588–593.

Pfab, J. (1978). Alkylperoxynitroxides in the photo-oxidation of C-nitrosoalkanes and the "spin trapping" of peroxy radicals by C-nitroso-compounds. *Tetrahedron Lett.* 843–846.

Pou, S., Anderson, D.E., Suricharmorn, W., Keaton, L.L., and Tod, M.L. (1994a). Biological studies of a nitroso compound that releases nitric oxide upon illumination. *Mol. Pharmacol.* 46: 709–715.

Pou, S., Keaton, L., Surichamorn, W., Frigillana, P., and Rosen, G.M. (1994c). Can nitric oxide be spin trapped by nitrone and nitroso compounds? *Biochim. Biophys. Acta* 1201: 118–124.

Pou, S., Ramos, C.L., Gladwell, T., Renks, E., Centra, M., Young, D., Cohen, M.S., and Rosen, G.M. (1994b). A kinetic approach to the selection of a sensitive spin trapping system for the detection of hydroxyl radical. *Anal. Biochem.* 217: 76–83.

Pou, S., Rosen, G.M., Wu, Y., and Keana, J.F.W. (1990). Synthesis of deuterium- and ^{15}N-containing pyrroline-1-oxides: A spin trapping study. *J. Org. Chem.* 55: 4438–4443.

Poyer, J.L., and McCay, P.B. (1971). Reduced triphosphopyridine nucleotide oxidase-catalyzed alterations of membrane phospholipids. IV. Dependence on Fe^{3+}. *J. Biol. Chem.* 246: 263–269.

Rao, G.N.R., and Bhaskar, K.R. (1969). Spectroscopy of the nitroso group, in: *The Chemistry of the Nitro and Nitroso Groups* (Feuer, H., ed.), Part 1, pp. 137–162, John Wiley & Sons, and Interscience Publishers, New York.

Reinke, L.A., Moore, D.R., and McCay, P.B. (1996). Degradation of DMPO adducts from hydroxyl and 1-hydroxyethyl radicals by rat liver microsomes. *Free Radical Res.* 25: 467–474.

Rosen, G.M., and Freeman, B.A. (1984). Detection of superoxide by endothelial cells. *Proc. Natl. Acad. Sci. USA* 81: 7269–7273.

Rosen, G.M., and Rauckman, E.J. (1980). Spin trapping of the primary radical involved in the activation of the carcinogen N-hydroxy-2-acetylaminofluorene by cumene hydroperoxide-hematin. *Mol. Pharmacol.* 17: 233–238.

Rosen, G.M., and Rauckman, E.J. (1984). Spin trapping of superoxide and hydroxyl radicals. *Methods Enzymol.* 105: 198–209.

Rosen, G.M., and Turner, M.J., III (1988). Synthesis of spin traps specific for hydroxyl radical. *J. Med. Chem.* **31**: 428–432.

Rosen, G.M., Finkelstein, E., and Rauckman, E.J. (1982). A method for the detection of superoxide in biological systems. *Arch. Biochem. Biophys.* **215**: 367–378.

Rosen, G.M., Pou, S., Britigan, B.E., and Cohen, M.S. (1994). Spin trapping of hydroxyl radicals in biological systems. *Methods Enzymol.* **233**: 105–111.

Rosen, H., and Klebanoff, S.J. (1979). Hydroxyl radical generation by polymorphonuclear leukocytes measured by electron spin resonance spectroscopy. *J. Clin. Invest.* **64**: 1725–1729.

Samuni, A., Krishna, M., Riesz, P., Finkelstein, E., and Russo, A. (1989a). Superoxide reaction with nitroxide spinadducts. *Free Radical Biol. Med.* **6**: 141–148.

Samuni, A., Samuni, A., and Swartz, H.M. (1989b). Evaluation of dibromonitrosobenzene sulfonate as a spin trap in biological systems. *Free Radical Biol. Med.* **7**: 37–43.

Sang, H., Janzen, E.G., and Lewis, B.H. (1996). Mass spectrometry and electron paramagnetic resonance study of free radicals spontaneously formed in nitrone-peracid reactions. *J. Org. Chem.* **61**: 2358–2363.

Sankuratri, N., Kotake, Y., and Janzen, E.G. (1996). Studies on the stability of oxygen radical spin adducts of a new spin trap: 5-Methyl-5-phenylpyrroline-1-oxide (MPPO). *Free Radical Biol. Med.* **21**: 889–894.

Sargent, F.P., and Gardy, E.M. (1976). Spin trapping of radicals formed during radiolysis of aqueous solutions. Direct electron spin resonance observations. *Can. J. Chem.* **54**: 275–279.

Sealy, R.C., Swartz, H.N., and Olive, P.L. (1978). Electron spin resonance-spin trapping. Detection of superoxide formation during aerobic microsomal reduction of nitro-compounds. *Biochem. Biophys. Res. Commun.* **82**: 680–684.

Sergeev, G.B., and Leenson, I.A. (1978). Spectrophotometric study of the kinetics and the thermodynamics of the monomer–dimer equilibrium in solutions of 2-methyl-2-nitrosopropane. *Russ. J. Phys. Chem.* **52**: 312–315.

Smith, P., and Robertson, J.S. (1988). Electron paramagnetic resonance study of the reaction of the spin trap 3,5-dibromo-4-nitroso-benzenesulfonate. *Can. J. Chem.* **66**: 1153–1158.

Strom, E.T., and Bluhm, A.L. (1966). Photochemical formation of dialkyl nitroxides from nitrosoalkanes. Evidence for fluorine p–π conjugation. *J. Chem. Soc., Chem. Commun.* 115–116.

Tero-Kubota, S., Ikegami, Y., Kurokawa, T., Sasaki, R., Sugioka, K., and Nakano, M. (1982). Generation of free radicals and initiation of radical reactions in nitrones-Fe^{2+}—phosphate buffer systems. *Biochem. Biophys. Res. Commun.* **108**: 1025–1031.

Theilacker, W., Knop, A., and Uffmann, H. (1965). Über den radikcharakter von nitrosoverbindungen. *Angew. Chem.* **77**: 717.

Thomas, C.E., Ohlweiler, D.F., Carr, A.A., Nieduzak, T.R., Hay, D.A., Adams, G., Vaz, R., and Bernotas, R.C. (1996). Characterization of the radical trapping activity of a novel series of cyclic nitrone spin traps. *J. Biol. Chem.* **271**: 3097–3104.

Thornalley, P.J., Trotta, R.J., and Stern, A. (1983). Free radical involvement in the oxidative phenomenona induced by *tert*-butyl hydroperoxide in erythrocytes. *Biochim. Biophys. Acta* **759**: 16–22.

Torssell, K.B.G. (1988). *Nitrile Oxides, Nitrones, and Nitronates in Organic Synthesis. Novel Strategies in Synthesis*, pp. 75–93, VCH, New York.

Turner, M.J., III, and Rosen, G.M. (1986). Spin trapping of superoxide and hydroxyl radicals with substituted pyrroline 1-oxides. *J. Med. Chem.* **29**: 2439–2444.

Utzinger, G.Ed., and Regenass, F.A. (1954). N-Arylnitrone. *Helv. Chim. Acta* **37**: 1892–1901.

Van der Zee, J., Barr, D.P., and Mason, R.P. (1996). ESR spin trapping investigation of radical formation from the reaction between hematin and *tert*-butyl hydroperoxide. *Free Radical Biol. Med.* **20**: 199–206.

Wajer, Th.A.J.W., Mackor, A., De Boer, Th.J., and Van Voorst, J.D.W. (1967). C-Nitroso compounds. II. On the photochemical and thermal formation of nitoxides from nitroso compounds as studied by E.S.R. *Tetrahedron* **23**: 4021–4026.

Wheeler, O.H., and Gore, P.H. (1956). Absorption spectra of azo- and related compounds. II. Substituted phenylnitrones. *J. Am. Chem. Soc.* **78**: 3363–3366.

Wurster, C.F., Jr., Durham, L.J., and Mosher, H.S. (1958). Peroxides. VII. The thermal decomposition of primary hydroperoxides. *J. Am. Chem. Soc.* **80**: 327–331.

Yamazaki, I., Piette, L.H., and Grover, T.A. (1990). Kinetic studies on spin trapping of superoxide and hydroxyl radicals generated in NADPH–cytochrome P-450 reductase-paraquat systems. Effect of iron chelates. *J. Biol. Chem.* **265**: 652–659.

10

Spin Trapping Free Radicals in Biological Systems

We have demonstrated in vivo EPR imaging of endogenous NO·, produced in LPS-treated mice, by employing a new spin-trapping reagent and 700 MHz-microwave EPR spectroscopy.

Yoshimura, Yokoyama, Fujii, Takayama, Oikawa,and Kamada, 1996

With the advent of *in vivo in situ* spin trapping of free radicals in real time (Komarov, *et al.*, 1993; Lai and Komarov, 1994; Halpern, *et al.*, 1995; Jiang, *et al.*, 1995, 1996; Komarov and Lai, 1995; Yoshimura, *et al.*, 1996; Liu, *et al.*, 1999), a new chapter in the exploration of biologically generated free radicals has begun. Such achievements could not have been possible without the pioneering efforts of many scientists, whose visions have allowed stepwise advances to be made.

Spin trapping/EPR spectroscopy is at the interface of biology, chemistry, and physics, each discipline contributing a wealth of knowledge to enhance the success of this technology. From the early years of isolated spinach chloroplasts (Harbour and Bolton, 1975), to purified enzyme preparations (De Groot, *et al.*, 1973; Lai, *et al.*, 1979a; Finkelstein, *et al.*, 1980a), to crude organ homogenates (Lai and Piette, 1977; Poyer, *et al.*, 1978; Kalyanaraman, *et al.*, 1979a; Rosen and Rauckman, 1981), to isolated cell suspensions (Rosen and Klebanoff, 1979; Bannister and Thornally, 1983; Rosen and Freeman, 1984; Horton, *et al.*, 1986; Mansbach, *et al.*, 1986) and, most recently to experiments with real time detection of free radicals in living animals (Komarov, *et al.*, 1993; Mäder, *et al.*, 1993; Fujii, *et al.*, 1994, 1997; Lai and Komarov, 1994; Halpern, *et al.*, 1995; Komarov and Lai, 1995; Jiang, *et al.*, 1995, 1996; Yoshimura, *et al.*, 1996; Liu, *et al.*, 1999), the primary objective has been to characterize specific free radicals produced as the result of an enzymatic process or after exposure to high-energy radiation.

The volume of scientific literature devoted to the identification of biologically generated free radicals is enormous, increasing in an exponential manner. In a text such as this, it is impractical to present all these studies with the depth that one would expect from a series of review articles. Therefore, selected topics, areas of considerable scientific inquiry, will be highlighted in this chapter as an illustration of the benefits and limitations of applying spin trapping to the study of free radical reactions in biology.

Spin Trapping Superoxide and Hydroxyl Radicals

Detection in Crude Cellular Homogenates

Our discussion of spin trapping free radicals will begin with $O_2\cdot^-$ and $HO\cdot$, as their role in biologic events continues to dominate the scientific literature of this discipline. Less than a decade after SOD was discovered to regulate cellular accumulation of $O_2\cdot^-$ (McCord and Fridovich, 1969), the capacity to spin trap this free radical was documented. Initial studies focused primarily on the chemical reactivity of photolytically generated $HO\cdot$ and $O_2\cdot^-$ toward DMPO (Harbour, et al., 1974; Harbour and Bolton, 1975; Harbour and Hair, 1977, 1978). At low concentrations of H_2O_2 (e.g., 1% in H_2O), a four-line EPR spectrum was observed, having hyperfine splitting constants of $A_H = A_N = 15.3\ G^1$ (Figure 10.1)

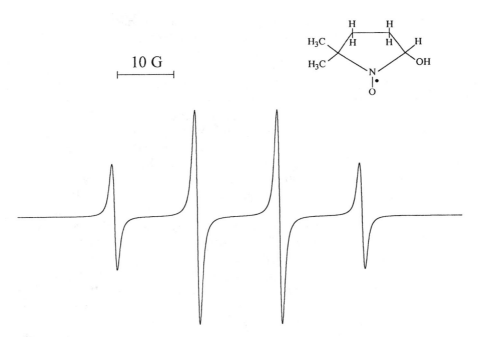

Figure 10.1 Aqueous solution simulation of the $HO\cdot$ spin trapped adducts of DMPO (5,5-dimethyl-1-pyrroline-N-oxide) and DMPO–OH (2,2-dimethyl-5-hydroxy-1-pyrrolinyloxyl). The hyperfine splitting constants at 9.5 GHz are $A_N = 14.9\ G$ and $A_H^\beta = 14.9\ G$. In the case of A_H^γ, it is not detected with a standard EPR spectrometer in aqueous solutions.

(Harbour, *et al.*, 1974). From the known photochemical behavior of this peroxide, it was proposed that the EPR spectrum corresponded to the reaction of HO· with DMPO:

$$H_2O_2 + h\nu \rightarrow 2HO·$$

$$DMPO + HO· \rightarrow DMPO\text{–}OH$$

At concentrations of H_2O_2 approaching 30%, an additional and overlapping 12-lined EPR spectrum was recorded. The species responsible for this new spectrum was suggested to be the hydroperoxide DMPO–OOH with hyperfine splitting constants of $A_N = 14.3$ G, $A_H^\beta = 11.7$ G, and $A_H^\gamma = 1.25$ G (Figure 10.2) (Harbour, *et al.*, 1974). This assignment was based exclusively on established chemical reactions of HO·:

$$H_2O_2 + h\nu \rightarrow 2HO·$$

$$H_2O_2 + HO· \rightarrow HO_2· + H_2O$$

$$DMPO + HO_2· \rightarrow DMPO\text{–}OOH$$

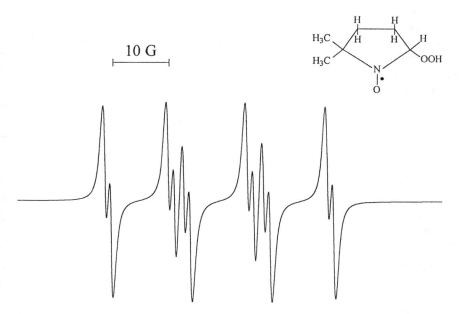

Figure 10.2 Aqueous solution simulation of the $O_2·^-$ spin trapped adducts of DMPO (5,5-dimethyl-1-pyrroline-*N*-oxide) and DMPO–OOH (2,2-dimethyl-5-hydroperoxy-1-pyrrolinyloxyl). The hyperfine splitting constants at 9.5 GHz are $A_N = 14.3$ G, $A_H^\beta = 11.4$ G, and $A_H^\gamma = 1.25$ G.

Even though an early communication questioned whether biologically produced $O_2\cdot^-$ could be spin trapped (Lai and Piette, 1977), a number of subsequent publications laid to rest any doubts as to the activity of this free radical toward cyclic and acyclic nitrones (Buettner and Oberley, 1978; Buettner, *et al.*, 1978; Felix, *et al.*, 1978; Janzen, *et al.*, 1978; Sealy, *et al.*, 1978; Finkelstein, *et al.*, 1979, 1980a; Green, *et al.*, 1979; Kalyanaraman, *et al.*, 1980).

Notwithstanding the success of these initial studies (Harbour, *et al.*, 1974; Harbour and Bolton, 1975; Buettner and Oberley, 1978; Janzen, *et al.*, 1978), unequivocal verification that the 12-lined EPR spectrum ascribed to DMPO–OOH (Figure 10.2) was correctly assigned was surprisingly late in coming, despite the commercial availability of SOD (Finkelstein, *et al.*, 1979, 1980a). Inclusion of this enzyme, catalase or $^{17}O_2$ into various reaction mixtures has allowed hyperfine splitting constants to be accurately calculated for DMPO–OOH, PBN–OOH, and 4-POBN–OOH (Finkelstein, *et al.*, 1979; Kalyanaraman, *et al.*, 1980; Mottley, *et al.*, 1986). In addition to DMPO, PBN and 4-POBN were likewise explored as possible spin traps for $O_2\cdot^-$ (Harbour, *et al.*, 1974; Janzen, *et al.*, 1978; Finkelstein, *et al.*, 1979). Nevertheless, the poor stability of the corresponding spin trapped adducts, 4-POBN–OOH and PBN–OOH (Finkelstein, *et al.*, 1979, 1980b) all but eliminated these acyclic nitrones as useful probes to identify this free radical. Therefore, DMPO and similarly related pyrroline-*N*-oxides have, more by default than by design, emerged as preferred spin traps for $O_2\cdot^-$ (Finkelstein, *et al.*, 1980a; Fréjaville, *et al.*, 1995). However, side reactions involving these compounds have greatly limited their ability to verify the formation of this free radical, especially in complex biological systems (Buettner and Oberley, 1978; Finkelstein, *et al.*, 1979). One encouraging finding is that DEPMPO–OOH has a significantly longer lifetime than DMPO–OOH (Table 8.7), and that the decomposition of DEPMPO–OOH does not result in the formation of DEPMPO–OH (Fréjaville, *et al.*, 1995). This has allowed DEPMPO to be used in a number of experimental paradigms where the use of DMPO is problematic (Karoui, *et al.*, 1996; Rouaud, *et al.*, 1997, 1998; Vásquez-Vivar, *et al.*, 1997; 1998).

With initial reports successfully detecting chemically produced $O_2\cdot^-$ by DMPO, investigators began to apply spin trapping to enzymatically generated free radicals. The best documented examples come from studies investigating the metabolic fate of aromatic nitro-containing compounds, bipyridylium dication salts, and 1,2- and 1,4-quinones (Mason, 1982, 1990; Livertoux, *et al.*, 1996). As each of these classes of chemicals is biotransformed to a free radical intermediate, subsequent reaction with O_2 initiates a cascade of other reactions. In the next few sections, we will explore the pathways by which these redox active chemicals alter enzymic electron transport and how spin trapping has played an integral role in defining the impact that these xenobiotics have on cell function.

Aromatic Nitro-Containing Compounds

Since the 1950s, hepatic microsomes were known to reduce nitroalkanes to amines, a six-electron process (Fouts and Brodie, 1957):

$$RNO_2 \xrightarrow[H_2O]{2e-,\ 2H^+} RN{=}O \xrightarrow{2e-,\ 2H^+} RNHOH \xrightarrow[H_2O]{2e-,\ 2H^+} RNH_2$$

Soon thereafter, this microsomal reaction was attributed to the catalytic activity of cytochrome P-450 (Gillette, *et al.*, 1968; Sasame and Gillette, 1969; Feller, *et al.*, 1971), even though examples of alternative reductive pathways have been described (Kato, *et al.*, 1970). As discussed in chapter 3, cytochrome P-450 accepts electrons from NADPH by means of an intermediate reductase, capable of transferring two electrons, individually, to the terminal oxidase, cytochrome P-450 (Guengerich, 1983; Guengerich and Macdonald, 1984; White, 1994). When an aromatic nitro-containing compound is present, however, the cytochrome P-450 electron transport chain becomes uncoupled, shunting an electron from one of the flavin prosthetic groups on NADPH-cytochrome P-450 reductase to the nitroalkane. This results in the formation of a nitro anion free radical (Figure 10.3) (Geske and Maki, 1960; Mason and Holtzman, 1975a; Rossi, *et al.*, 1996):

$$RNO_2 + 1e^- \rightarrow RNO_2 \cdot{}^-$$

Early on, the fortune of this free radical in an aerobic environment was not well understood. From initial experiments, an inhibitory role for O_2 in this reductive route was envisioned, as only the starting nitroalkane was isolated (Gillette, *et al.*,

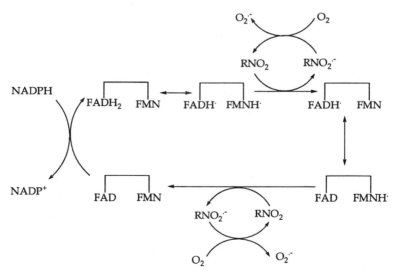

Figure 10.3 A schematic depicting the uncoupling of flavin-containing enzymes by aromatic nitroalkanes. In this case, electrons are shunted from the enzyme to the xenobiotic, forming a nitro anion free radical. In the presence of O_2, $O_2 \cdot{}^-$ is generated at the expense of other cellular functions.

1968). Subsequent studies suggested a mechanism in which an electron is transferred to O_2, generating the original nitro-containing compound concomitant with the formation of $O_2 \cdot^-$ (Figure 10.3) (Kastening, 1964; Russell and Bemis, 1967; Greenstock and Dunlop, 1973; Mason and Holtzman, 1975a; Willson and Searle, 1975; Peterson, *et al.*, 1979):

$$RNO_2 \cdot^- + O_2 \rightarrow O_2 \cdot^- + RNO_2$$

Verifying this latter pathway soon followed, as $O_2 \cdot^-$ was detected using either probes/spectrophotometric or spin trapping/EPR spectroscopic methods (Mason and Holtzman, 1975b; Biaglow, *et al.*, 1976; Wardman and Clarke, 1976; Sealy, *et al.*, 1978; Kalyanaraman, *et al.*, 1979b, 1981; Perez-Reyes, *et al.*, 1980; Docampo, *et al.*, 1981; Moreno, *et al.*, 1982, 1983; Rosen, *et al.*, 1984; Lloyd and Pedersen, 1985). Not surprisingly, other nitrogen-containing compounds, which readily accept and stabilize an electron through delocalization, including azo groups, were found to behave in a similar fashion (Hernandez, *et al.*, 1967; Mason, *et al.*, 1977, 1978). In delineating these enzymatic pathways, spin trapping/EPR spectroscopy has proven to be invaluable in defining these reaction mechanisms.

Bipyridylium Dication Salts

Paraquat cation free radical has a long and storied history. First described more than 60 years ago, this free radical appeared as a deep blue color in reducing solutions of the parent herbicide (Michaelis and Hill, 1933). Although those early experiments did not detect $O_2 \cdot^-$, its existence was inferred from the rapid redox cycling of this N,N'-dimethyl-γ,γ'-dipyridyl salt (Figure 10.4). Several decades would elapse, nevertheless, before $O_2 \cdot^-$ was associated with the cellular metabolism of paraquat and related 1,2- and 1,4-quinones (Homer, *et al.*, 1960; Mees, 1960; Gage, 1968; Epel and Neumann, 1973; Bus, *et al.*, 1974, 1977; Harbour and Bolton, 1975; Mason and Holtzman, 1975; Miller and Macdowall, 1975; Hassan and Fridovich, 1977, 1978, 1979; Ledwith, 1977; Montgomery, 1977; Thor, *et al.*, 1982; Horton, *et al.*, 1986; Rabinowitch, *et al.*, 1987; Zhang, *et al.*, 1995).

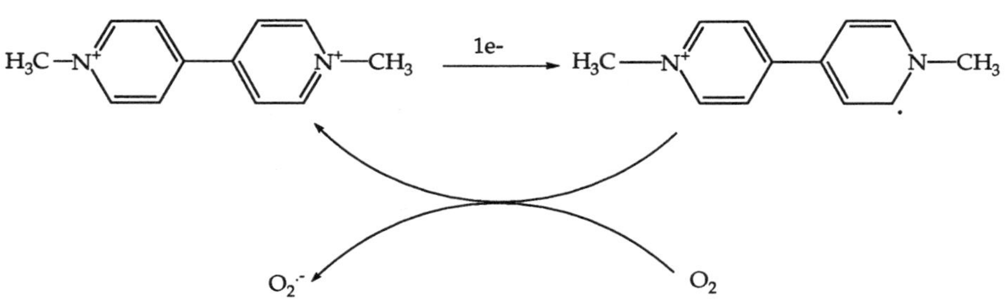

Figure 10.4 The futile redox cycling of paraquat, N,N'-dimethyl-γ,γ'-dipyridyl salt, results in the continued production of $O_2 \cdot^-$. The limiting factors are the availability of reducing equivalents and O_2.

When data supporting cellular formation of $O_2 \cdot^-$ began to surface, it came from many directions, but none were more compelling than the finding of a rapid and profound increase in the rate of MnSOD synthesis in *Escherichia coli* exposed to paraquat (Hassan and Fridovich, 1977, 1978; Hassan, 1984) and the ability of this N,N'-dimethyl-γ,γ'-dipyridyl salt to inhibit bacterial growth in the presence, but not in the absence of O_2 (Fisher and Williams, 1976).

From these and other mechanistic studies, it appears that paraquat (PQ^{2+}), like aromatic nitroalkanes, can uncouple cellular electron transport, leading to the paraquat free radical ($PQ \cdot^+$) (Figure 10.4). Subsequent reaction with O_2 is rapid, with a rate constant of $7.7 \times 10^8 \, M^{-1} \, s^{-1}$ (Farrington, *et al.*, 1973; Ledwith, 1977):

$$PQ^{2+} + 1e^- \rightarrow PQ \cdot^+$$

$$PQ \cdot^+ + O_2 \rightarrow O_2 \cdot^- + PQ^{2+}$$

$$O_2 \cdot^- + M^{n+} \rightarrow M^{(n-1)+} + O_2$$

$$PQ \cdot^+ + M^{n+} \rightarrow M^{(n-1)+} + PQ^{2+}$$

$$O_2 \cdot^- + O_2 \cdot^- + 2H^+ \rightarrow H_2O_2 + O_2$$

$$M^{(n-1)+} + H_2O_2 \rightarrow M^{n+} + HO \cdot + HO^-$$

The continued production of $O_2 \cdot^-$ and products derived from this free radical is dependent upon the availability of O_2 and reducing equivalents (Winterbourn and Sutton, 1984). In the presence of redox active metal ions, such as iron and copper, the metal-catalyzed Haber-Weiss reaction assures the formation of $HO \cdot$ (Sutton and Winterbourn, 1984; Winterbourn and Sutton, 1984; Kohen and Chevion, 1985a, 1985b, 1988; Sandy, *et al.*, 1987; Kadiiska, *et al.*, 1993). Not surprisingly, these events can result in enhanced rates of lipid peroxidation (Mead, 1976; Shu, *et al.*, 1979; Steffen and Netter, 1979; Trush, *et al.*, 1981) with the accompanying toxicity to cellular models (Bagley, *et al.*, 1986; Korbashi, *et al.*, 1986; Krall, *et al.*, 1988) and *in vivo* (Nakagawa, *et al.*, 1995). Further support for the role of $O_2 \cdot^-$ in paraquat-mediated toxicity comes from a series of studies in which elevated levels of O_2 in inhaled air enhanced lung injury after paraquat poisoning in rodent models (Fisher, *et al.*, 1973; Bus, *et al.*, 1976). Administration of SOD attenuated this injury (Autor, 1974).

1,2- and 1,4-Quinones

The metabolic fate of 1,2- and 1,4-quinones is particularly interesting as two competing enzymes contend for these xenobiotics (Thor, *et al.*, 1982). The first, as described above, is NADPH–cytochrome P-450 reductase. This enzyme reduces quinones to the corresponding semiquinone free radicals. In the presence of O_2, the parent quinone is regenerated concomitant with the formation of $O_2 \cdot^-$ (Figure 10.5). The second is DT–diaphorase (Ernster, 1967; Lind, *et al.*, 1990). Unlike

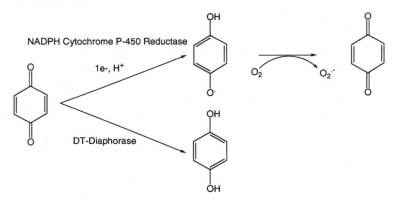

Figure 10.5 The competing metabolic fate of 1,4-benzoquinone. This quinone can shunt an electron from NADPH–cytochrome P-450 reductase to give the semiquinone free radical. In the presence of O_2, $O_2 \cdot^-$ is generated along with 1,4-benzoquinone. Alternatively, the enzyme DT–diaphorase can act upon this quinone by transferring two-electrons to 1,4-benzoquinone, yielding 1,4-dihydroxybenzene. Which of these two competing pathways will dominate is dependent upon the structure of the quinone and the metabolic status of the cell.

NADPH–cytochrome P-450 reductase, this diaphorase transfers two electrons to a quinone, yielding a dihydroquinone (Figure 10.5) (Iyanagi and Yamazaki, 1970; Lind, et al., 1990). Notably, these two biotransformations result in different cellular outcomes. While the one-electron pathway is perceived as cytotoxic (Gant, et al., 1988; Weiner, 1994), the two-electron route is viewed as cytoprotective (Lind, et al., 1990; Cadenas, 1991). As the biotransformation of a specific quinone is dependent on its structure, the catalytic route of reduction, and the metabolic state of the cell, it is difficult, a priori, to predict which of these differing pathways will dominate for a certain xenobiotic.

The redox cycling of doxorubicin and similarly related anticancer anthracyclines, forming intermediate semiquinone free radicals (Malisza, et al., 1996), has been thought by many to be the leading pathway by which these drugs mediate their antitumor activity (Figure 10.6) (Schreiber, et al., 1987; Gutierrez, 1989; Powis, 1989), even though an alternative hypothesis of doxorubicin intercalation into DNA has been considered as a more viable mechanism (Myers, et al., 1977; Kang, et al., 1996). Less controversial is the role of free radicals in doxorubin-induced cardiomyopathy (Myers, et al., 1977; Keizer, et al., 1990; Lee, et al., 1991; Kang, et al., 1996). Evidence in support of the free radical theory comes from studies pointing to enhanced $O_2 \cdot^-$ production by microsomal preparations containing doxorubicin (Handa and Sato, 1975) and the amelioration by free radical scavengers (Myers, et al., 1977; Keizer, et al., 1990; Lee, et al., 1991; Kang, et al., 1996).

Isolated Cell Suspensions

In models as sophisticated as cells, affirming formation of $O_2 \cdot^-$ and $HO \cdot$ is often difficult. Experimental findings must invariably rely on the selectivity of inhibitors,

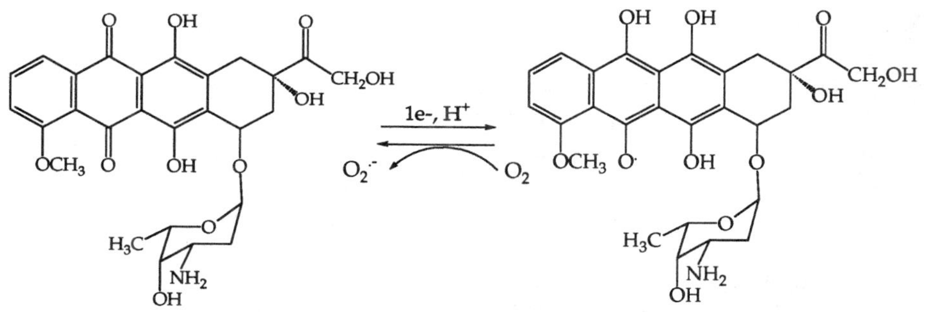

Figure 10.6 The redox cycling of doxorubicin and similarly related anticancer anthracyclines is depicted. As the result of a one-electron reduction, intermediate semiquinone free radical is produced, which can transfer an electron to O_2, resulting in the formation of the parent anthracycline and $O_2 \cdot^-$.

which are frequently known to exhibit such specificity only in homogeneous solutions of purified enzyme preparations. In contrast, EPR spectroscopic identification of specific free radicals can be accomplished by spin trapping these reactive species in real time and at the site of formation, whether they arise from intracellular or extracellular loci. Typical of the early spin trapping studies were those suggesting the production of $O_2 \cdot^-$ and HO· by stimulated neutrophils (Green, et al., 1979; Rosen and Klebanoff, 1979; Hume, et al., 1983). Even though these initial reports focused primarily on the identification of these free radicals, within a decade, fluorescence and spin trapping methods would confirm the phagosomal localization of $O_2 \cdot^-$ and its link to the activation of an unique NADPH-dependent oxidase (Burow and Valet, 1987; Pou, et al., 1989b; Rothe and Valet, 1990; Ryan, et al., 1990).

Interestingly enough, the suggestion that neutrophils generate HO· was based primarily on the recorded EPR spectrum characteristic of DMPO–OH, even though SOD attenuated the signal (Rosen and Klebanoff, 1979). At that time, the rapidity of DMPO–OOH decomposition to DMPO–OH was not appreciated and it resulted in the misconception that activated leukocytes produced HO· through a mechanism not dependent on redox active metals. A number of years would elapse, however, before spin trapping would play a pivotal role in defining an alternative pathway for neutrophil and monocyte derived HO· —an MPO-dependent reaction (Ramos, et al., 1992; Britigan, et al., 1996).

The spin trapping of cellular-derived $O_2 \cdot^-$ and HO· in models other than professional phagocytes has resulted in a myriad of difficulties, not the least of which has centered on the identity of the free radical actually spin trapped, intracellularly (Bannister and Thornalley, 1983; Rosen and Freeman, 1984; Mansbach, et al., 1986; Samuni, et al., 1986; Zweier, et al., 1988, 1994; Rosen, et al., 1989; Turner, et al., 1989; Britigan, et al., 1992a; Hirayama, et al., 1995). The ability to detect free radicals within cells requires the fulfillment of three separate, yet connecting, processes. First, nitrones or nitrosoalkanes have to exhibit sufficient lipophilicity to permit high concentrations of the spin trap to diffuse into the cell. Second, a propitious kinetic profile must allow the spin trap to favorably compete with other biologic targets for the free radical. Third, the spin trapped adduct should demonstrate a sufficiently long lifetime to permit its characterization. When these criteria have been met, the likelihood of success invariably follows.

For illustration purposes, let us consider the metabolic fate of the quinone menadione. As endothelial cells contain cytochrome P-450 reductase and DT–diaphorase, one- and two-electron reductive pathways compete for this quinone. In the case of cytochrome P-450 reductase, univalent reduction of menadione leads to the semiquinone free radical. Transfer of an electron to O_2 results in formation of $O_2 \cdot^-$. Alternatively, divalent reduction of menadione by DT–diaphorase results in 2-methyl-1,4-dihydroquinone. Although conjugation of this hydroquinone by one of several phase II enzymes can inactivate this compound toward further redox cycling (Thor, *et al.*, 1982), a competing disproportionation reaction between menadione and 2-methyl-1,4-dihydroquinone, producing the semiquinone free radical (Figure 10.7), can, under some conditions, dwarf the cytoprotective effects offered by DT–diaphorase.

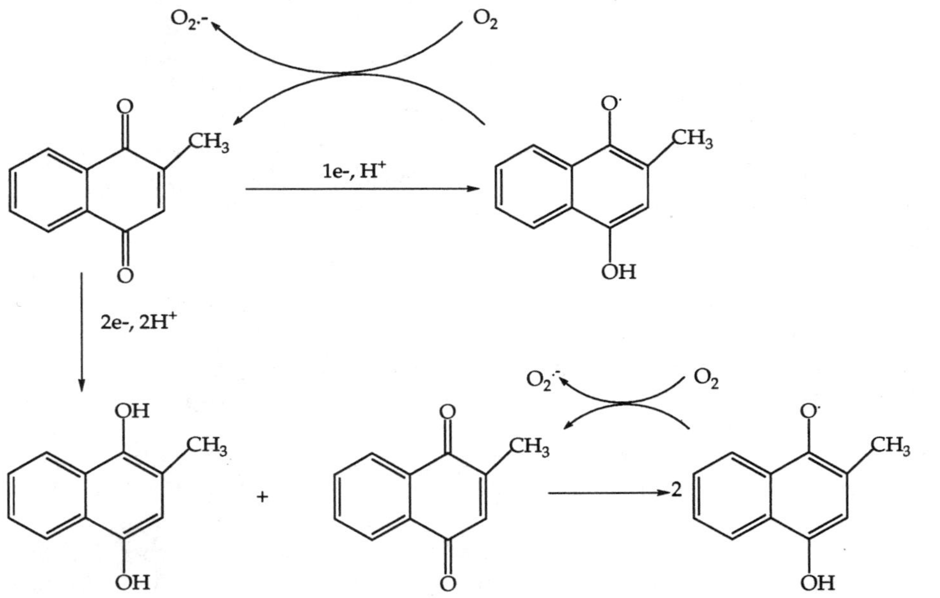

Figure 10.7 Formation of $O_2 \cdot^-$ ocurrs from two competing cellular pathways. In the case of univalent reduction of menadione via cytochrome P-450 reductase, the semiquinone free radical is the initial intermediate. Transfer of an electron to O_2 results in generation of $O_2 \cdot^-$. Alternatively, divalent reduction of menadione by DT–diaphorase leads to 2-methyl-1,4-dihydroquinone. In this case, $O_2 \cdot^-$ production is dependent upon the disproportionation of menadione and 2-methyl-1,4-dihydroquinone. The resultant semiquinone free radical rapidly reacts with O_2, regenerating menadione.

When endothelial cells were incubated with menadione and DMPO, an EPR spectrum was recorded, which was composed of DMPO–OOH and, to a lesser extent, DMPO–OH (Figure 10.8). With the addition of SOD but not catalase to the reaction mixture, a small but distinctive EPR spectrum characteristic of DMPO–OH was recorded (Figure 10.9) (Rosen and Freeman, 1984). Assuming catalase and SOD remained outside the endothelial cell, the spin trapping of $O_2\cdot^-$ had to take place within the confines of the surrounding media. Although there are several theories as to how $O_2\cdot^-$ arrived at this locus, the most logical proposition rests with an extracellular source of this free radical, undoubtedly the result of the disproportionation reaction described above. In this instance, 1,4-dihydroxy-2-methylnaphthalene rather than $O_2\cdot^-$ may have diffused from intracellular sites of formation into the extracellular

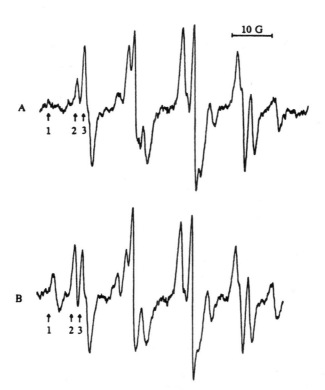

Figure 10.8 The EPR spectra obtained when endothelial cells were incubated with mena-dione, dissolved in DMSO. Scan A is a composite of three different spin trapped adducts: (peak 1) 2,2,5-trimethyl-1-pyrrolidinyloxyl (DMPO–CH$_3$), (peak 2) 5-hydroxy-2,2-dimethyl-1-pyrrolidinyloxyl (DMPO–OH), and (peak 3) 5-hydroperoxy-2,2-dimethyl-1-pyrrolidinyloxyl (DMPO–OOH). Ten minutes later, the spectrum depicted in scan B was recorded. Based on a series of independent experiments, it has been suggested that peaks 1 and 2 were derived from the decomposition of DMPO–OOH. (Spectra taken from Rosen and Freeman, 1984.) For a detailed discussion of the fate of DMPO–OOH, the interested reader should consult chapter 9.

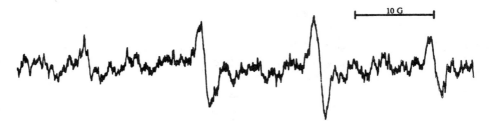

Figure 10.9 The EPR spectrum characteristic of DMPO–OH, obtained from endothelial cells incubated with menadione to which SOD was added. (Spectra taken from Rosen and Freeman, 1984.)

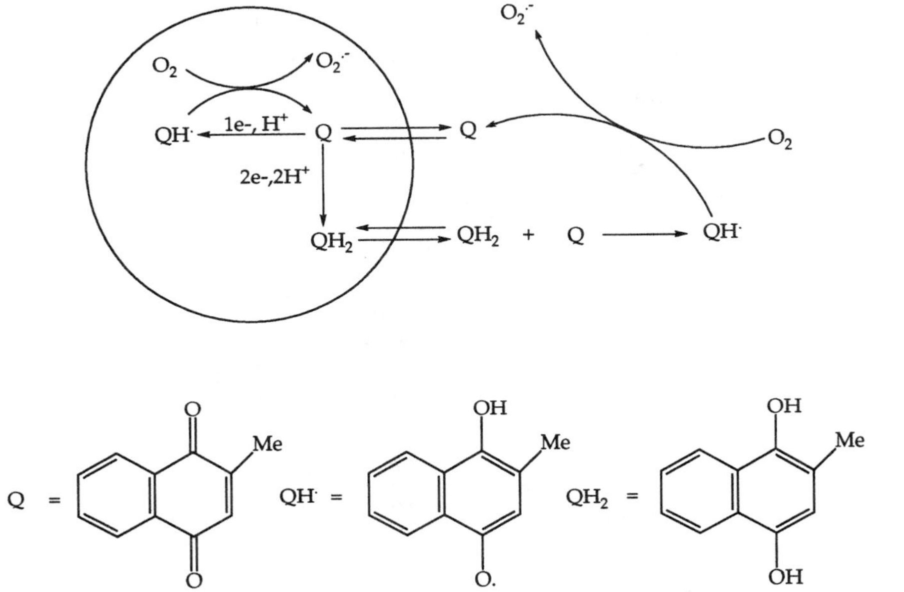

Figure 10.10 The metabolic fate of menadione is shown. Menadione can be reduced by either NADPH–cytochrome P-450 reductase or DT–diaphorase. The former results in the formation of the menadione semiquinone free radical, whereas the latter leads to 1,4-dihydroxy-2-methylnaphthalene. Diffusion of this diol from the cell to the surrounding milieu followed by a disproportionation reaction involving menadione results in extracellularly generated $O_2 \cdot^-$. (Adapted from Rosen, et al., 1989.)

milieu (Figure 10.10). Data in support of this hypothesis remain elusive, even though the indirect evidence of a lack of an EPR spectrum when endothelial cells were either made dysfunctional in the presence of menadione or treated with DT–diaphorase inhibitors, is supportive of this theory.

Discovering the mechanism responsible for the intracellular formation of DMPO–OH is an arduous task. As discussed in chapter 9, DMPO–OH may arise from a number of dissimilar reactions. Even though in homogeneous solutions, well-designed experiments can resolve with reasonable certainty whether DMPO–OH resulted from the spin trapping of $O_2 \cdot^-$ or HO·, studies applying this technique to isolated cell suspensions do not often result in data that lead to a definitive conclusion. One approach is to develop a spin trap unique for HO·. Considering that there is greater than a 10^8-fold difference in rate constants between the spin trapping of HO· and $O_2 \cdot^-$ by DMPO at physiologic pH (Finkelstein, et al., 1980a), synthesis of nitrones with specificity towards HO· seemed a likely means to address the uncertainty as to the source of DMPO–OH. Early experiments with 3,3,5,5-tetramethyl-1-pyrroline-N-oxide (M_4PO) were quite promising and suggested more sterically hindered derivatives of DMPO might lead to a spin trap selective for HO· (Rosen and Turner, 1988; Buettner and Britigan, 1990). This goal was soon realized as 3,3-diethyl-5,5-dimethyl-1-pyrroline-N-oxide (DEDMPO) was found to spin trap only HO·; it was unable to react with $O_2 \cdot^-$.

The singularity of DEDMPO for HO· allowed inquiry into whether HO· or $O_2 \cdot^-$ was the free radical responsible for intracellular formation of DMPO–OH, arising during the cellular metabolism of menadione (Rosen and Freeman, 1984; Rosen and Turner, 1988). Since no EPR spectrum was recorded with DEDMPO under experimental conditions that had previously resulted in an EPR signal with DMPO, it may be inferred that $O_2 \cdot^-$ and not HO· was the source of intracellular DMPO–OH (Rosen and Turner, 1988).

Still, the spin trapping of intracellularly generated O_2-centered free radicals remains an arduous task. Studies using DMPO, for instance, have suggested that the instability of the DMPO spin trapped adducts, rather than the partitioning of the nitrone into intracellular compartments, was the primary hindrance to the detection of $O_2 \cdot^-$ and HO· inside cells (Samuni, et al., 1986). Newer analogs of DMPO, such as 4-[(2-ethoxycarbonyl)ethyl]-3,3,5,5-tetramethyl-1-pyrroline-N-oxide (M_4EEPO), whose spin trapped adducts are considerably more stable than either DMPO–OOH or DMPO–OH may be an attractive alternative (Arya, et al., 1992).

The search for other spin trapping systems specific for HO· followed many paths, the most fruitful of which led to 4-POBN/EtOH (Pou, et al., 1994a). This discovery was based on a series of reports detailing the remarkable stability of alkyl-substituted PBNs, even under a variety of harsh conditions that rapidly destroyed other spin trapped adducts (Connor, et al., 1986b; Keana, et al., 1989; Samuni, et al., 1989a). The success of this spin trapping system is dependent upon the specificity of the reaction and a large rate constant for the spin trapping of α-hydroxyethyl radical by 4-POBN (Table 8.3):

$$HO\cdot + CH_3CH_2OH \rightarrow CH_3\cdot CHOH$$

$$CH_3\cdot CHOH + 4\text{-POBN} \rightarrow 4\text{-POBN--CH(CH}_3)OH$$

In addition, the stability of 4-POBN-CH(OH)CH$_3$ was found to be remarkable, appearing to be relatively unaffected in the presence of a high flux of $O_2\cdot^-$. Finally, 4-POBN/EtOH is about 10-times more sensitive to the detection of HO· than is either DMPO/EtOH or PBN/EtOH (Pou, *et al.*, 1994a). Using 4-POBN/EtOH, but not PBN/DMSO (Britigan, *et al.*, 1990), it was discovered that neutrophils and eosinophils can produce HO· through a peroxidase-driven pathway that is independent of the metal ion-catalyzed Haber-Weiss reaction (Ramos, *et al.*, 1992; McCormick, *et al.*, 1994). The importance of MPO in the production of HO· was evident, as neither neutrophils from a MPO-deficient patient nor macrophages, which lack MPO, would generate this free radical (Ramos, *et al.*, 1992). Since these initial observations, the 4-POBN/EtOH spin trapping system has been successfully used to identify HO· in eucarya and bacteria. These include endothelial cells, epithelial cells, hepatocytes, and *Escherichia coli.* (Britigan, *et al.*, 1992b, 1996, 1997; Hoehler, *et al.*, 1997; McCormick, *et al.*, 1998a).

Detection in Isolated Organ Preparations

As the result of a particular enzymatic process, the detection of cellular-derived $O_2\cdot^-$ and HO· is, by most accounts, a significant advance in our understanding of cell biology. Nevertheless, experiments with cells are limited by the static state of the system; spin traps at a macroscopic level compartmentalize between intra- and extracellular milieus. Until the recent development of lower frequency EPR spectrometers, application of spin trapping to functional tissues was restricted to measurements of the effluent from isolated organ preparations. Under such conditions, the results are frequently skewed, as dynamic factors such as lipophilicity and metabolism of the spin trap and spin trapped adduct, as well as the rate of perfusion, dictate outcomes, leaving the findings often in doubt (Rosen, *et al.*, 1988).

Studies on the role of free radical formation in ischemia/reperfusion injury provide an excellent illustration of the difficulties in applying spin trapping to isolated organ preparations (Davies, 1989). Most of the evidence suggesting that the O_2-centered free radicals, $O_2\cdot^-$ and/or HO·, are responsible for this pathology was indirect and relied on the pharmacologic action of specific free radical scavengers to ameliorate tissue injury (Granger, *et al.*, 1981; Myers, *et al.*, 1985; Bernier, *et al.*, 1986; Przylenk and Kloner, 1986; McCord, 1988; Tamura, *et al.*, 1988; Siems, *et al.*, 1989; Yoshikawa, *et al.*, 1989; Gutteridge and Halliwell, 1990; Hall, *et al.*, 1990; Grisham, 1992; Ma, *et al.*, 1992). Yet, even after a lengthy series of complicated spin trapping experiments, there remain significant doubts as to whether biologically generated free radicals were actually detected during reperfusion of a previously made ischemic tissue, as an isolated perfused preparation or intact within an animal (Blasig, *et al.*, 1986, 1989, 1990; Arroyo, *et al.*, 1987; Garlick, *et al.*, 1987; Kramer, *et al.*, 1987; Bolli, *et al.*, 1988; Culcasi, *et al.*, 1989; Pietri, *et al.*, 1989; Zweier, *et al.*, 1989; Bolli and McCay, 1990; Kuzuya, *et al.*, 1990; Pincemail, *et al.*, 1990; Tosaki, *et al.*, 1990, 1993; Coghlan, *et al.*, 1991;

Mergner, *et al.*, 1991; Zini, *et al.*, 1992; Defraigne, *et al.*, 1993; Sen and Phillis, 1993; Bremer, *et al.*, 1994; Fréjaville, *et al.*, 1995; Kadkhodaee, *et al.*, 1996). For instance, it has been suggested (Pou, *et al.*, 1989a; Nohl, *et al.*, 1991; Boucher, *et al.*, 1992) that EPR spectra reported in many of the aforementioned studies were actually artifacts, arising from the redox cycling of metal ions excreted into the coronary flow following myocardial ischemia (Bolli, *et al.*, 1990; Chevion, *et al.*, 1993; Sekili, *et al.*, 1993). In the presence of O_2, $O_2 \cdot^-$ and $HO \cdot$ would be generated in the vasculature where the spin trap is highly concentrated (Rosen, *et al.*, 1988). Alternatively, iron-catalyzed oxidation of nitrones, particularly pyrroline N-oxides, has been shown to yield nitroxides independent of spin trapping either $O_2 \cdot^-$ or $HO \cdot$ (Makino, *et al.*, 1992). Despite this rather gloomy picture, *in situ* identification of organ-derived free radicals as the result of an ischemic/reperfusion event may be realized by modifying the animal model (Dugan, *et al.*, 1995); by synthesizing newer spin traps that are specific for a particular free radical—less prone to oxidation and more resistant to cellular reduction; and by continuing to develop more sensitive lower frequency EPR spectrometers. Even with the most promising of experimental designs, detection of free radicals may be limited to interstitial and vascular compartments, as measuring intracellularly produced $O_2 \cdot^-$ will, at best, be problematic.

Detection in Living Animals

Traditionally, EPR spectroscopic measurements, made at 9.5 GHz, have had limited application for the detection of free radical evolution in animals. Microwave radiation is absorbed by aqueous and biological samples due to high dielectric and conductive losses at these frequencies (Poole, 1983). Alternative methods of free radical identification with conventional EPR spectrometers have relied on either rapid freezing followed by sample grinding (Bolton, *et al.*, 1972; Zweier, *et al.*, 1987; Baker, *et al.*, 1988; Nakazawa, *et al.*, 1988; Baker and Kalyanaraman, 1989; van der Kraaij, *et al.*, 1989), or chemical extraction of the spin trapped adduct from the homogenated biological sample (Lai, *et al.*, 1979b, 1986; Poyer, *et al.*, 1980; Kubow, *et al.*, 1984). With regard to the latter approach, animals are preloaded with the spin trap, usually by intraperitoneal injection, followed by induction of the experimental manipulation of interest. After a defined period, the animal is sacrificed, and tissue is removed, homogenized, and extracted with a mixture of highly purified organic solvents (Folch, *et al.*, 1957; Bösterling and Trudell, 1981; Trudell, 1987). Careful evaporation under anaerobic conditions affords a residue, which, when taken up in an aqueous buffer, has been found to contain an EPR spectrum (Kalyanaraman, *et al.*, 1984). Identification of the parent free radical has remained, for the most part, unresolved and in some cases it appears to be solvent-derived (Connor, *et al.*, 1994). Despite this suboptimal outcome, there have been some successes, usually realized through careful selection of the appropriate spin trap, *in vivo* administration of isotope-labeled substrates, such as $^{13}CCl_4$, and meticulous handling of materials (Lai, *et al.*, 1979b; Poyer, *et al.*, 1980; Albano, *et al.*, 1982; McCay, *et al.*, 1984). Recent modification of this cumbersome isolation procedure by sampling body fluids such as urine, bile, and blood in place of extracted isolated tissue homogenates, has greatly enhanced the reproducibility and accuracy of the experimental findings (Connor, *et al.*, 1986b, 1994; Knecht and Mason, 1988; Burkitt and Mason, 1991; Hughes, *et al.*, 1991; Reinke and Janzen, 1991; Knecht, *et al.*, 1992; Kadiiska, *et al.*, 1994, 1995).

Nevertheless, even under the best of conditions, conclusions often are skewed due to differences in regional pharmacodynamics and *in vivo* metabolism (Bacic, *et al.*, 1989; Ferrari, *et al.*, 1990; Miura, *et al.*, 1992; Mäder, *et al.*, 1993; Quaresima, *et al.*, 1993; Hiramatsu, *et al.*, 1995; Utsumi, *et al.*, 1995; Gallez, *et al.*, 1996; Halpern, *et al.*, 1996; Kazama, *et al.*, 1996; Yokoyama, *et al.*, 1996). The detection of 4-POBN–CH(CH$_3$)OH, as a marker of *in vivo in situ* production of radiation-produced HO·, illustrates this point (Halpern, *et al.*, 1995). Ethanol and 4-POBN were injected into a mouse extremity tumor to which was delivered a high, toxic dose of radiation to a substantial bulk of the tissue with minimal effect on the physiology of the animal. Based on the irradiation chemistry of H$_2$O (Weiss, 1944) and subsequent reactions thereof, the recording of an EPR spectrum characteristic of 4-POBN–CH(CH$_3$)OH was obtained (Halpern, *et al.*, 1995). It was assumed that 4-POBN and EtOH were apportioned in the same tissue compartment so that α-hydroxyethyl radical (CH$_3$·CHOH) could react with 4-POBN:

$$H_2O + h\nu \rightarrow H_2O\cdot^+ + e^-(\rightarrow e_{aq})$$

$$H_2O\cdot^+ + H_2O \rightarrow HO\cdot + H_3O^+$$

$$HO\cdot + CH_3CH_2OH \rightarrow CH_3\cdot CHOH$$

$$CH_3\cdot CHOH + 4\text{-POBN} \rightarrow 4\text{-POBN–CH(CH}_3)OH$$

Even though EtOH can diffuse into many sites within a tissue, 4-POBN, which has a small *n*-octanol partition coefficient (≈ 0.09), does not passively enter cells (Pou, *et al.*, 1994a). Therefore, the detection of HO· as 4-POBN–CH(CH$_3$)OH was solely limited to interstitial and vascular spaces.

The poor cellular uptake of 4-POBN results in measurements of spin trapped adducts that localize away from the site of relevant intracellular production of free radicals (Rosen, *et al.*, 1988). Whatever free radical events that may take place within this compartment, they would not be discernible with the 4-POBN/EtOH spin trapping system, even though irradiation of H$_2$O within cells has long been known to produce HO· at sensitive cellular sites (Johansen and Howard-Flanders, 1965; Roots and Okada, 1972; Czapski, 1984; Ewing and Walton, 1991).

The *in vivo* identification of O$_2$·$^-$ is not feasible with the present generation of spin traps. Even though cyclic nitrones such as DMPO are excellent reporters of this free radical, generated by enzymatic preparations and by models as sophisticated as cells, the tendency of this family of spin traps to undergo a variety of reactions unrelated to spin trapping O$_2$·$^-$ complicates data interpretation. Even if it were possible to limit these unwanted outcomes, the ease by which spin trapped adducts can be metabolized to diamagnetic species further limits the use of cyclic nitrones in this *in vivo* setting (Swartz, 1990). Changing to acyclic nitrones, including PBN and 4-MePyBN, offers no better opportunity to identify O$_2$·$^-$, as the poor stability of PBN–OOH and 4-MePyBN–OOH (Table 8.6) is an insurmountable obstacle for the *in vivo* characterization of this free radical (Finkelstein, *et al.*, 1979; Sanderson and Chedekel, 1980). Using a cascade of free radical reactions to detect $O_2 \cdot^-$, in a manner similar to HO· with 4-POBN/EtOH or PBN/DMSO (Britigan, *et al.*, 1990; Burkitt and Mason, 1991; Ramos, *et al.*, 1992; Pou, *et al.*, 1994a), has proven to be unsuccessful, since the reactivity of $O_2 \cdot^-$ toward these targets is too slow to support such a reaction, especially so when SOD is present (Adams and Wardman, 1977).

Spin Trapping Trichloromethyl and Other Halogenated Hydrocarbon Radicals

Detection in Crude Homogenates and Isolated Cell Suspensions

Despite the enormous capacity of the liver to detoxify a broad spectrum of drugs and other foreign compounds, there are examples of biologically inert xenobiotics being biotransformed to toxic intermediates (Recknagel and Ghoshal, 1966; Slater, 1966). Carbon tetrachloride is one such chemical (Recknagel, 1967; Recknagel and Glende, 1973; Cheeseman, *et al.*, 1985). In this case, this halogenated hydrocarbon elicits extensive toxicity to the liver by its hepatic conversion to a more reactive metabolite (Ahr, *et al.*, 1980). However, unlike aromatic nitro-containing compounds, bipyridylium dication salts and 1,2- and 1,4-quinones, whose uncoupling of NADPH–cytochrome P-450 reductase results in the formation of a free radical of the parent compound, CCl_4 is directly reduced by the terminal oxidase cytochrome P-450 to $\cdot CCl_3$ (Sipes, *et al.*, 1977; Wolf, *et al.*, 1980):

$$P\text{-}450\text{--}Fe^{2+} + CCl_4 \rightarrow P\text{-}450\text{--}Fe^{3+} + \cdot CCl_3$$

Several lines of evidence support a role for cytochrome P-450 in this toxicity, the most substantial of which are the observations that CCl_4-increased hypersensitivity found in rats can be directly correlated to enhanced hepatic levels of this enzyme after prior *in vivo* administration of phenobarbital (Garner and McLean, 1969).

Biochemical findings for the generation of $\cdot CCl_3$ come from the recovery of $CHCl_3$ (Butler, 1961) and Cl_3CCCl_3 (Fowler, 1969) after the metabolism of CCl_4, the former through hydrogen atom abstraction of lipids and the latter from dimerization of $\cdot CCl_3$. Furthermore, the bond dissociation energy for the formation of $\cdot CCl_3$ from $BrCCl_3$, CCl_4, and $CHCl_3$ is 49 kcal/mol, 68 kcal/mol and 90 kcal/mol, respectively (Recknagel, *et al.*, 1977), reflecting the *in vivo* potency of these halogenated hydrocarbons (Plaa, *et al.*, 1958; Koch, *et al.*, 1974; Ingall, *et al.*, 1978). These data, when taken together, add credence to the theory that CCl_4 mediates hepatotoxicity through the formation of an initiating free radical, which subsequently catalyzes the peroxidation of membrane lipids (Calligaro and Vannini, 1975).

Even as a wealth of biochemical data pointing to involvement of cytochrome P-450 in the one-electron reduction of CCl_4 have accumulated, spin trapping studies

(Janzen, *et al.*, 1985) have played an invaluable role in support of a free radical theory for the bioactivation of CCl_4 (Ingall, *et al.*, 1978; Poyer, *et al.*, 1978). Although these ground-breaking papers differed in their conclusions as to the nature of the free radical spin trapped, they did substantiate the notion that free radicals were produced during hepatic metabolism of this halogenated hydrocarbon.

These articles were likewise remarkable because of their other contributions. Prior to these investigations (Ingall, *et al.*, 1978; Poyer, *et al.*, 1978), there were few publications in which spin trapping was an integral part in solving complicated biochemical mechanisms of enzyme function. Rather, this technique was viewed as an analytic tool to characterize specific free radicals. Through these publications, however, it became clear that spin trapping could be employed to answer fundamental questions on the role of free radicals in biologic events. As a direct outgrowth of these previous studies, the importance of a specific isozyme of cytochrome P-450 in catalyzing the formation of CCl_4-derived free radicals was demonstrated (Noguchi, *et al.*, 1982). Nevertheless, questions still remained as to the identity of the free radicals produced during microsomal and cellular metabolism of CCl_4. Although the initial studies proposed that $\cdot CCl_3$ was the free radical spin trapped by PBN (Poyer, *et al.*, 1978), a lipid dienyl radical was soon thereafter suggested as a more reasonable alternative (Kalyanaraman, *et al.*, 1979a). Confirmatory experiments using either $^{13}CCl_4$ or HPLC and/or GC/MS have, however, swung the pendulum of evidence in favor of $\cdot CCl_3$ (Figure 10.11) (Poyer, *et al.*, 1980; Tomasi, *et al.*, 1980; Albano, *et al.*, 1982; McCay, *et al.*, 1984; Janzen, *et al.*, 1990b). Yet, the underlying correctness of the earlier studies (Kalyanaraman, *et al.*, 1979a) is without question; lipid dienyl free radicals are a consequence of $\cdot CCl_3$ reaction with membrane lipids.

Besides self-condensation, $\cdot CCl_3$ can initiate the peroxidation of lipids through two similar pathways. The first mechanism involves the abstraction of a hydrogen atom from a polyunsaturated fatty acid, resulting in the formation of a lipid dienyl free radical (L·) and chloroform ($HCCl_3$). Subsequent reaction with O_2 gives the corresponding lipid peroxyl radical (LOO·). Aerobic dismutation of these lipid peroxyl radicals leads to lipid oxyl radicals (LO·):

$$LH + \cdot CCl_3 \rightarrow L\cdot + HCCl_3$$

$$L\cdot + O_2 \rightarrow LOO\cdot$$

$$2LOO\cdot + O_2 \rightarrow 2LO\cdot + O_2$$

The second route is linked to the formation of trichloromethyl peroxyl radical ($Cl_3COO\cdot$) (Symons, *et al.*, 1982). The high reactivity of this free radical toward a variety of biologic targets, such as polyunsaturated fatty acids, amino acids, and thiols (Tomasi, *et al.*, 1980; Packer, *et al.*, 1981), makes identification of $Cl_3COO\cdot$ in this setting problematic, even though this peroxyl radical may be an important propagator of lipid peroxidation.

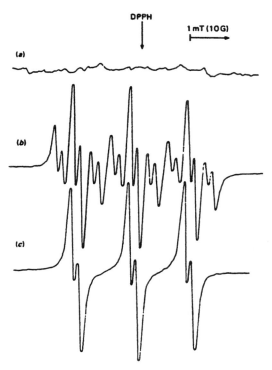

DPPH

1 mT (10 G)

(a)

(b)

(c)

Figure 10.11 The EPR spectra obtained from *in vivo* administration of CCl_4 and PBN. After 15 minutes, animals were sacrificed and the liver homogenated, extracted with organic solvents, evaporated to dryness and taken up in aqueous buffer. Scan (a) is a control – PBN in the absence of CCl_4. Scan (b) animals were administered PBN and $^{13}CCl_4$. Verification that the recorded EPR spectrum results from the spin trapping of $\cdot CCl_3$ comes from the additional hyperfine splitting associated with $^{13}CCl_3$. Scan (c) is after rats had received CCl_4 and PBN. (Spectra taken from Albano, *et al.*, 1982.)

$$\cdot CCl_3 + O_2 \rightarrow Cl_3COO\cdot$$

$$LH + Cl_3COO\cdot \rightarrow L\cdot + CCl_3COOH$$

$$L\cdot + O_2 \rightarrow LOO\cdot$$

$$2LOO\cdot + O_2 \rightarrow 2LO\cdot + O_2$$

With such a medley of free radicals generated subsequent to the biologic formation of $\cdot CCl_3$, spin trapping cannot easily distinguish among the various reactions. One only needs to consider the nearly 500-fold difference in rate constants between the spin trapping of $Cl_3COO\cdot$ and $\cdot CCl_3$ by PBN to appreciate the problem (Packer, *et al.*, 1978; Tomasi, *et al.*, 1980). It seems unlikely, therefore, that $\cdot CCl_3$ could be spin trapped in biologic milieus. Yet, in a number of *in vitro* and *in vivo* models, PBN has, surprisingly, been found to be specific toward ·CCl$_3$ at the expense of both Cl$_3$COO· and LOO· (Poyer, *et al.*, 1980; Tomasi, *et al.*, 1980; Albano, *et al.*, 1982; McCay, *et al.*, 1984), even though line broadening in aerobic environments make unequivocal EPR spectral assignments difficult, in the absence of ^{13}·CCl$_3$ (Janzen, *et al.*, 1990b). One method of overcoming this limitation in specificity is to use deuterium-labeled spin traps, whose corresponding nitroxides after reaction with free radicals exhibit decreased EPR spectral linewidths (Haire, *et al.*, 1988; Pou, *et al.*, 1990; Halpern, *et al.*, 1993). When such experiments were undertaken, carbon-centered lipid radicals were detected from biologic mixtures containing PBN(d$_{14}$) and CCl$_4$ (Janzen, *et al.*, 1987, 1990b).

As ·CCl$_3$ is formed by a membrane-associated hemoprotein, succeeding reaction with the spin trap is often dictated by the lipophilic character of the nitrone or the nitrosoalkane. For instance, hydrophilic nitrones such as 4-POBN compartmentalize in the aqueous milieu surrounding a lipid membrane. As such, it would prefer to spin trap more polar lipid peroxyl free radicals at the membrane surface (Rosen and Rauckman, 1982). In the case of hydrophobic nitrosoalkanes like MNP, whose lipophilic nature ensures a large distribution into lipid matrices, lipid dienyl free radicals have been identified (Albano, *et al.*, 1982).

Even though the discussion thus far has focused exclusively on CCl$_4$, there are many structural analogs of this halogenated hydrocarbon whose toxicity is dependent upon cytochrome P-450-mediated reduction to carbon-centered free radicals (Hake, *et al.*, 1960; Ross, *et al.*, 1979, McLain, *et al.*, 1979; Callen, *et al.*, 1980; Reynolds and Moslen, 1980; Cheeseman, *et al.*, 1985). Some of the well-documented xenobiotics, which have been found to undergo similar biotransformations, include halothane (Poyer, *et al.*, 1981; Trudell, *et al.*, 1981, 1982; Plummer, *et al.*, 1982; Mönig, *et al.*, 1983; Tomasi, *et al.*, 1983b; Fujii, *et al.*, 1984; Janzen, *et al.*, 1990a; Hughes, *et al.*, 1991; Knecht, *et al.*, 1992), trihalomethanes (Tomasi, *et al.*, 1985), 1,2-dibromoethane (Tomasi, *et al.*, 1983a; Albano, *et al.*, 1984), 1,1,2,2-tetrachloroethane (Paolini, *et al.*, 1992), and 1,1,1-trichloroethane (Dürk, *et al.*, 1992; Steel-Goodwin, *et al.*, 1996).

Detection in Living Animals

Even with nearly ideal experimental designs, *in vitro* identification of specific free radicals is not always a predicator of similar events *in vivo*. Pharmacodynamic, pharmacokinetic, and metabolic factors may greatly influence both the locale and nature of the free radical spin trapped (Knecht and Mason, 1991). For instance, the detection of carbon dioxide anion radical (CO$_2$·$^-$), arising during the *in vivo* transformation of CCl$_4$, was a surprising finding, as this metabolite had not previously been documented in liver homogenates (Connor, *et al.*, 1986b; Reinke, *et al.*, 1992). In some situations, the *in vivo* metabolism of the spin trap has created an environment where alternative free radical reactions can be identified. Such is the case with MO$_3$PBN (*N-tert*-butyl-α-(2,4,6-trimethoxy)phenylnitrone). Administration of MO$_3$PBN and CCl$_4$ resulted in the spin trapping of a large carbon-centered free radical, other than ·CCl$_3$, attached to HO(MO)$_2$PBN (*N-tert*-butyl-α-(2-hydroxy-4,6-dimethoxy)phenylnitrone), an *o*-demethylated metabolite of MO$_3$PBN (McCay, *et al.*, 1984). Similar findings have been reported after rats were challenged with endotoxin. The EPR spectra from heart and liver extracts suggested that a carbon-centered spin trapped adduct was derived from $HO(MO)_2PBN$ (Brackett, *et al.*, 1989).

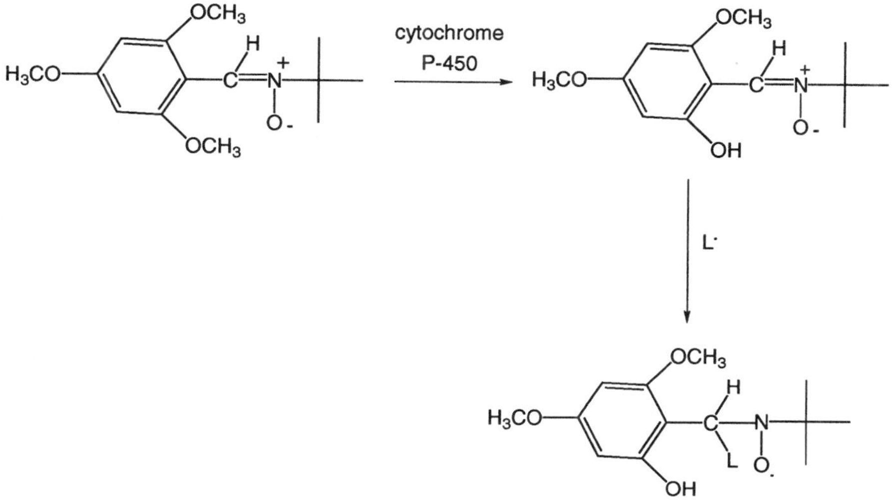

In contrast, PBN was found to be specific for $\cdot CCl_3$ (McCay, *et al.*, 1984). This difference in selectivity may reflect the more polar character of $HO(MO)_2PBN$ and the ability of this nitrone to partition at a site distal from the locus of cytochrome P-450-mediated reduction of CCl_4 (Janzen, *et al.*, 1985). Despite these and other uncertainties, the identification of free radicals during the *in vivo* biotransformation of CCl_4 has opened new vistas into human metabolism and the role that bioactivation plays in subsequent toxic events.

Based on data from many laboratories, the triggering event in CCl_4-induced hepatotoxicity is the one-electron reduction of this halogenated hydrocarbon by cytochrome P-450. As cellular distribution of this enzyme is compartmentalized within the endoplasmic reticulum and the nucleus (Bresnick, *et al.*, 1977; Rogan and Cavalieri, 1978), studies have focused on these sites of free radical formation. After administration of CCl_4 and PBN, the characteristic EPR spectrum of PBN–CCl_3 was found to be exclusively limited to the microsomal fraction, suggesting a prominent role for the endoplasmic reticulum in the bioactivation of CCl_4 (Lai, *et al.*, 1979b).

The *in vivo* detection of PBN–CO_2^- in the bile and urine, but not from liver extracts, of rats during metabolism of either CCl_4 or $CBrCl_3$ has stimulated debate as to the source of this spin trapped adduct (Conner, *et al.*, 1986a; LaCagnin, *et al.*, 1988). Along the way, surprising findings have led to new insights into the role of GSH in detoxifying CCl_4 (Reiter and Burk, 1988; Connor, *et al.*, 1990; Sentjurc and Mason, 1992). Initial studies suggested that the *in vivo* formation of PBN–CO_2^- was intimately linked to the production of $Cl_3COO\cdot$. Although this peroxyl radical undoubtedly propagates lipid peroxidation, a significant concentration of $Cl_3COO\cdot$ must unquestionably further dehalogenate, as phosgene (Kubic and

Anders, 1980) and carbon dioxide (Paul and Rubinstein, 1963) are known meta-bolites of CCl_4. From these early studies (Connor, et al., 1986b), it was unclear, however, whether $PBN-CO_2^-$ was derived from either the spin trapping of $CO_2 \cdot^-$ or another intermediate free radical such as $\cdot C(Cl)O$:

$$CCl_4 + \text{P-450-Fe}^{2+} \rightarrow \cdot CCl_3 + \text{P-450-Fe}^{3+}$$

$$\cdot CCl_3 + PBN \rightarrow PBN-CCl_3$$

$$\cdot CCl_3 + O_2 \rightarrow Cl_3COO \cdot \rightarrow \cdot C(Cl)O \rightarrow CO_2 \cdot^-$$

$$CO_2 \cdot^- + PBN \rightarrow PBN-CO_2^-$$

$$\cdot C(Cl)O + PBN \rightarrow PBN-C(Cl)O \rightarrow PBN-CO_2^-$$

In the case of PBN-C(Cl)O, cytochrome P-450 catalyzed dehalogenation would have to proceed at the expense of other reductive sequences, such as nitroxyl reduction. Despite the propensity of most spin labels to be bioreduced by small-molecular-weight reductants such as ascorbate or cellular reductases, including cytochrome P-450 (Stier and Reitz, 1971; Rosen and Rauckman, 1977; Couet, et al., 1984; Griffeth, et al., 1984; Iannone, et al., 1990a; Fuchs, et al., 1993; Janzen, et al., 1993; Kocherginsky and Swartz, 1995) and cytoplasmic reductases (Griffeth, et al., 1984; Iannone, et al., 1990b; Kocherginsky and Swartz, 1995), $PBN-CO_2^-$ is remarkably resistant to this reaction (Keana, et al., 1987). As $PBN-C(Cl)O$ has not been synthesized, its relative stability, compared with $PBN-CO_2^-$, cannot be evaluated at this time.

Even though there is general agreement that $\cdot CCl_3$ is the proximal free radical to which $PBN-CO_2^-$ owes its presence, questions remain as to the pathway by which this spin trapped adduct was derived. Considering the ubiquity of GSH, one possible link to $PBN-CO_2^-$ is through this thiol. It has been known for some time that GSH reacts with phosgene, forming a diglutathionyl dithiocarbonate (Pohl, et al., 1981); however, its reaction with $\cdot CCl_3$, producing an unidentified free radical, possibly $\cdot GSH-CCl_3$, has only recently been reported (Reiter and Burk, 1988). Evidence in support of $\cdot GSH-CCl_3$ as the key intermediate in the formation of $PBN-CO_2^-$ comes from a series of in vitro and in vivo experiments in which the EPR spectrum for $PBN-GSH-CCl_3$ was recorded (Connor, et al., 1990; Sentjurc and Mason, 1992). Rapid decomposition resulted in the generation of $PBN-CO_2^-$ (Connor, et al., 1990; Sentjurc and Mason, 1992). Even though GSH is known to detoxify a broad spectrum of foreign compounds, the critical role that this thiol plays in the formation of $PBN-CO_2^-$ is an additional example of its physiologic control of xenobiotic-derived free radicals.

Cellular models have often played an important role in testing theories of xenobiotic-induced hepatotoxicity. Carbon tetrachloride is no exception to this general rule. Along the way, mechanistic studies have provided valuable insights into cellular production of free radicals (Rau, et al., 1990). Interestingly, these experiments have exposed one of the intrinsic weaknesses of spin trapping—diffi-culties in characterizing intracellular free radicals. For instance, an EPR spectrum unique to PBN–CO_2^- was directly recorded from an isolated hepatocyte preparation aerobically exposed to CCl_4 (Rau, *et al.*, 1990). In contrast, identification of PBN-CCl_3 required a toluene extraction of this cell suspension (Rau, *et al.*, 1990). Based on available evidence, it is clear that PBN–CCl_3 and PBN–CO_2^- were formed intracellularly (Connor, *et al.*, 1986b, 1990; LaCagnin, *et al.*, 1988; Rau, *et al.*, 1990; Sentjurc and Mason, 1992). Yet, the detection of these spin labels required vastly different preparative methods, thereby elevating the potential for error along the way. For PBN–CCl_3, its concentration at lipophilic sites within the cell resulted in decreased motion of the nitroxide. This made it impossible to identify PBN–CCl_3 intracellularly, requiring its removal from this milieu. In the case of PBN–CO_2^-, diffusion to aqueous locales, including extracellular space wherein this nitroxide was free to tumble, allowed for easy measurement of this spin trapped adduct.

Spin Trapping Lipid-Derived Free Radicals

Despite the ease by which nitrones such as DMPO can distinguish between $O_2 \cdot^-$ and HO·, this family of spin traps has not been able to readily discriminate between lipid dienyl (L·), lipid peroxyl (LOO·), or lipid oxyl (LO·) radicals. This lack of specificity has been amply illustrated from experiments carried out in biologic milieus where line-broadening often makes calculations of hyperfine splitting constants difficult (Connor, *et al.*, 1986a). The use of deuterium-containing spin traps such as PBN(d_{14}) has, for instance, allowed identification of carbon-centered lipid radicals during the *in vitro* metabolism of CCl_4 and ethanol; such spectra were not easily observable in the absence of isotope-labeled PBN (Janzen, *et al.*, 1987, 1990a; Reinke, *et al.*, 1991).

Hepatic microsomal lipid peroxidation is one area in which spin trapping has been applied to the study of lipid-derived free radicals. Early experiments suggested that LOO· could be spin trapped by 4-POBN (Rosen and Rauckman, 1981). This conclusion was based on the finding that, in model systems, EPR spectra were tied to the presence of O_2 and inversely correlated to the spin trapping of L· by MNP (De Groot, *et al.*, 1973; Rosen and Rauckman, 1981). More recent studies (Connor, *et al.*, 1986a) drew into question the identity of the free radical spin trapped by 4-POBN during this microsomal peroxidation of lipids. Spiking the reaction mixtures with $^{17}O_2$, which has a nuclear spin of $5/2$, did not result in EPR spectra with additional hyperfine splitting, as would have been expected if $L^{17}O_2 \cdot$ were to react with 4-POBN (Connor, *et al.*, 1986a). This finding, which was in contrast to the earlier study (Rosen and Rauckman, 1981) was unexpected, considering that rate constants for the formation of lipid peroxyl radicals range around $10^9 \, M^{-1} s^{-1}$ (Hasegawa and Patterson, 1978; Salvador, *et al.*, 1995), whereas the reaction of 4-POBN with small carbon-centered free radicals is substantially smaller (Table 8.3). In an attempt to resolve this dilemma, careful analysis of the spin trapped adducts of 4-POBN derived during the *in vitro* peroxidation of lipid acids with HPLC equipped with an EPR spectrometer in tandem with a mass spectrometer has been undertaken (Iwahashi, *et al.*, 1991, 1996). Despite the inability to detect 4-POBN–L, 4-POBN–OOL, or 4-POBN–OL, it was concluded, nevertheless, that L·, LOO· and LO· were probably spin trapped. Their stability, however, was not sufficient to allow for their identification (Iwahashi, *et al.*, 1991, 1996).

In many regards, these findings are not unexpected as they reflect the diversity of reactions, arising after the initiation of lipid peroxidation, and point to the site where the spin trap, depending on its structure (Yamada, *et al.*, 1984; Dage, *et al.*, 1997), interrupts the propagation trail. For instance, given a specific class of nitrones or nitrosoalkanes (Howard and Tait, 1978), differences in reactivity and lipophilicity will undoubtedly dictate the nature of the free radical spin trapped (Dage, *et al.*, 1997). If, for example, the spin trap compartmentalizes in close proximity to the site of L· formation, as would be found with PBN, it is likely that this free radical would be spin trapped within this peroxidizing lipid matrix. Alternatively, if highly charged spin traps, such as 4-POBN, are incubated with peroxidizing membranes, their concentration in the extracellular milieu would result in the spin trapping of more hydrophilic lipids, such as LOO·. Finally, unequivocally demonstrating whether L·, LOO·, or LO· was actually spin trapped by a nitrone or a nitrosoalkane does little more than describe experimental conditions required to inhibit a specific propagation step. Despite these limitations, spin trapping continues to play an important role in the detection, if not in the characterization, of lipid free radicals generated in liposomes, whole cells and tissues (Feix and Kalyanaraman, 1989; North, *et al.*, 1992, 1994; Alexander-North, *et al.*, 1994; Grune, *et al.*, 1997; Numagami, *et al.*, 1998; Wagner, *et al.*, 1998). In the end, knowing that lipid peroxidation has taken place may not only answer the fundamental question posed, but may open insights into important mechanisms of biological oxidation of cell membranes.

Spin Trapping Nitric Oxide

Unlike the free radicals described thus far, the unexpectedly long lifetime of NO·, an essential prerequisite for a physiologic transmitter whose cellular target is often distal from the site of original (Moncada, *et al.*, 1991), presents unique challenges in the design of spin trapping approaches to detect this free radical. In fact, until recently, the fate of this free radical in aqueous solutions was not understood to a degree sufficient to predict whether it was even possible to spin trap NO·. Illustrative of this point is the fact that even though nitrite is the end product of its autoxidation (Ignarro, *et al.*, 1993; Kharitonov, *et al.*, 1994; Lewis, *et al.*, 1995), other intermediates, which may have a direct impact on the spin trapping of NO·, have only recently been cataloged (Goldstein and Czapski, 1995). They include such reactive species as nitrosyloxyl radical (ONOO·), nitrosyl radical (NO$_2$·), and dinitrogen trioxide (N$_2$O$_3$) (Goldstein and Czapski, 1995):

$$NO\cdot + O_2 \rightleftarrows ONOO\cdot$$

$$ONOO\cdot + NO\cdot \rightarrow ONOONO$$

$$ONOONO \rightarrow 2NO_2\cdot$$

$$NO_2\cdot + NO\cdot \rightleftarrows N_2O_3$$

$$N_2O_3 + H_2O \rightarrow 2NO_2^- + 2H^+$$

Considering the complexity of aerobic reactions involving NO·, it is not surprising that experiments designed to spin trap this free radical have been, for the most part, unrewarding. For example, it has been reported that NO·, either dissolved in buffer, or generated by neuroblastoma cells or platelets, can be spin trapped by sodium 3,5-dibromo-4-nitrosobenzenesulfonic acid (DBNBS) (Arroyo and Forray, 1991; Arroyo and Kohno, 1991; Prónai, et al, 1991; Ichimori, et al., 1993; Kaur, 1996). Other data appear to contradict these conclusions. For instance, experiments with ^{15}N-labeled NO· (Arroyo and Kohno, 1991) were not consistent with the spin trapping of this free radical by DBNBS (Korth, et al., 1992, Pou, et al., 1994b). Similarly, neither DMPO, PBN, 4-POBN, nor DBNBS were found to react with NO· generated during the oxidation of L-arginine by purified brain nitric oxide synthase (Pou, et al., 1994b). In fact, the enzymic conversion of L-arginine to L-citrulline and, by implication, the generation of NO· was almost entirely inhibited at concentrations of DBNBS routinely used to spin trap free radicals (Pou, et al., 1994b). What then may be the origin of EPR spectra obtained from aerobic solutions of NO· in the presence of DBNBS? There are several possibilities. One school of thought suggests that redox reactions involving DBNBS account for the observed nitroxide (Mani and Crouch, 1989; Samuni, et al., 1989b; Nazhat, et al., 1990). Although this is an attractive hypothesis, alternative mechanisms cannot be ruled out, as the appearance of similar spectra in neutral, less oxidant-prone buffers have been observed (Pou, et al., 1994b). A final problem with nitrones and nitrosoalkanes is their propensity to generate NO· in the presence of UV-light (Chamulitrat, et al., 1993, 1995; Pou, et al., 1994c; Saito, et al., 1998) or in certain oxidative environments. This, of course, can confound an already complicated biologic experiment and can lead to erroneous conclusions about the presence of NO·.

As the current generation of nitrones and nitrosoalkanes are unable to unambiguously identify NO· in biologic milieus, a number of alternatives have been explored. One of the most innovative schemes draws upon the ease with which "activated" cis-conjugated dienes react with free radicals (Korth, et al., 1992, 1994; Paul, et al., 1996). In fact, the term NO· cheletropic trap (NOCT) was coined to described this reaction (Korth, et al., 1992). The simplest of this new class of spin traps, 2,5-dimethylhexadiene,[2] when introduced into degassed organic solutions of NO·, yielded a three-line EPR spectrum, tentatively assigned to 2,2,5,5-tetramethyl-1-pyrrolinyloxyl (Gabr, et al., 1993).

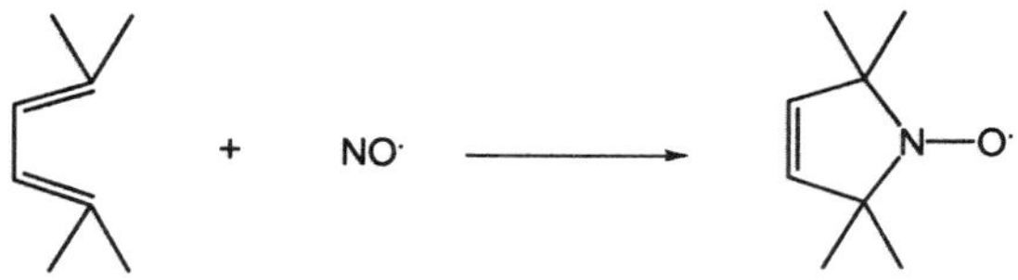

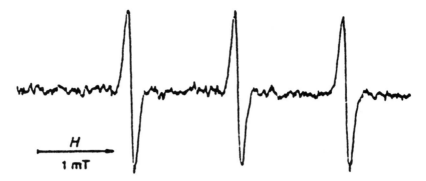

Figure 10.12 The EPR spectrum of 1,1,3,3-tetramethylisoindolin-2-oxyl, arising from the reaction of NO· with the cheletropic trap 7,7,8,8-tetramethyl-*o*-quinodimethane (NOCT). Assignment of the EPR spectrum to that of 1,1,3,3-tetramethylisoindolin-2-oxyl is based on the characteristic nitroxide triplet $A_N = 14.5$ G. (Spectrum taken from Korth, *et al.*, 1992.)

Structural verification[2] was based on the additional hyperfine splitting observed when NO· was added into a solution of 2,3-dimethylbutadiene (Gabr, *et al.,* 1993).

Locking in the cis-isomer as would arise with 7,7,8,8-tetramethyl-*o*-quinodi-methane (NOCT), produced during the photolysis of 1,1,3,3-tetramethyl-2-indan-one, has resulted in the formation of 1,1,3,3-tetramethylisoindolin-2-oxyl after reaction with NO· (Figure 10.12) (Korth, *et al.*, 1992). Detection of activated macrophage-generated NO· was a surprising finding, considering the poor hydro-philicity of this *o*-quinodimethane (Korth, *et al.*, 1992) and the reported low rate of NO· production from similarly stimulated phagocytic cells ($\approx 0.5\,\mu$M/hr per 1×10^6 cells) (Green, *et al.*, 1994). Even though NOCT is a notable advance over 2,5-dimethylhexadiene and similarly related *cis/trans*-dienes, the prerequisite photo-activation of this *o*-quinodimethane, concomitant with poor aqueous solubility, a short half-life of the corresponding nitroxide in biologic systems, and a lack of specificity towards NO· have dampened the enthusiasm for what would otherwise be a most remarkable accomplishment (Korth, *et al.*, 1992).

NOCT

For instance, NOCT reacts with NO_2· at a rate that is at least 200-times faster than its reaction with NO· (Korth, *et al.*, 1994). This is a serious limitation considering the likelihood that NO_2· may be the most significant product from aerobic reac-tions of NO· (Goldstein and Czapski, 1995). Despite these deficiencies, success has been realized as the spin trapping of NO·, generated by stimulated Kupffer cells, has been achieved (Korth, *et al.*, 1992, 1994).

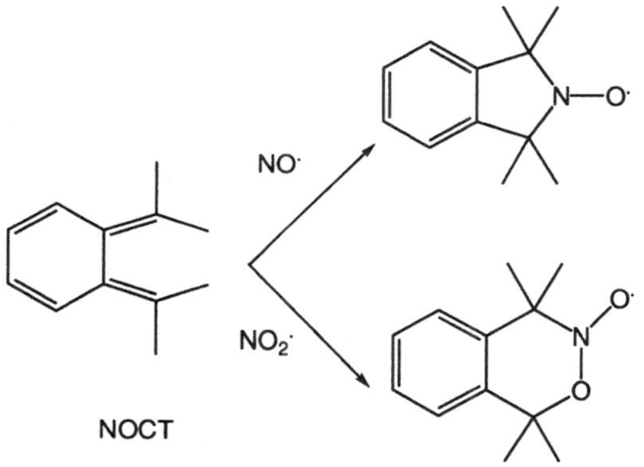

Recent modifications of NOCT have, nevertheless, begun to address many of the inadequacies of this *o*-quinodimethane (Korth, *et al.*, 1994; Paul, *et al.*, 1996). Yet, the depth of the problem may be insurmountable. For illustrative purposes, it is worth comparing experimental results with NOCT-6 and NOCT-7. The water-insoluble NOCT-6 is stable in organic solvents for at least 3 days, and it reacts specifically with NO·, yielding a nitroxide whose stability is nearly a month in length. In contrast, the water-soluble NOCT-7 has such a short lifetime that its ability to report cellular secretion of NO· remains in doubt (Korth, *et al.*, 1994).

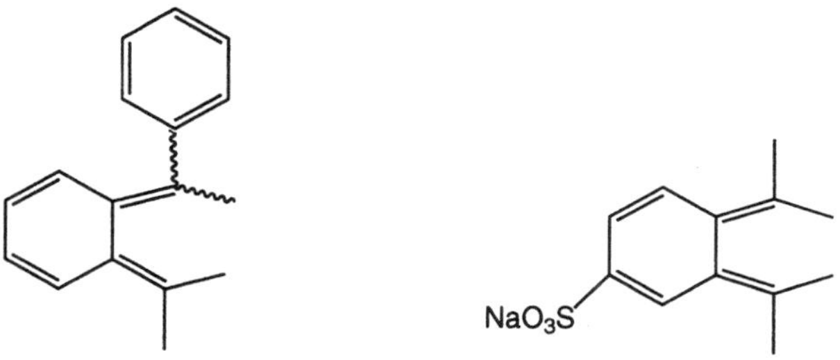

A variant of the "activated-ene" approach has recently been described in which the aci-form of nitromethane, generated at pH ≥ 11, was shown to react with NO·, resulting in a nitroanion free radical (Reszka, *et al.*, 1994, 1996). Loss of a proton in the basic environment leads to a dianion radical:

$$CH_2=NO_2^- + NO· \rightarrow O=NCH_2NO_2·^-$$

$$O=NCH_2NO_2·^- \xrightarrow{-H^+} {}^-ONCHNO_2·^-$$

Not surprisingly, the stability of the dianion radical at neutral pH is marginal, as pH values in excess of the physiologic range are a prerequisite for a reasonable lifetime for this anion radical species in aqueous solutions (Reszka, *et al.*, 1994).

As NO· is known to react either with ferrous salts, giving intense EPR spectra in aqueous solutions at ambient temperatures (McDonald, *et al.*, 1965; Goodman, *et al.*, 1969), or with ferrous hemoglobin at near-diffusion-controlled rates, leading to nitrosylhemoglobin (Murphy and Noack, 1994; Radi, 1996), there has been much interest in developing synthetic iron chelates as spin traps for this free radical:

$$\text{hemoglobin-Fe}^{2+} + \text{NO·} \rightarrow \text{hemoglobin-Fe}^{2+}\text{–NO}$$

One of the better defined complexes in biological milieus is iron(II) diethyl-dithiocarbamate (DETC) (Vanin, *et al.*, 1984, 1993; Mülsch, *et al.*, 1992). Its reaction with NO· is characterized by the low-temperature EPR spectrum at $g_\perp = 2.035$ and $g_{||} = 2.02$, with a triplet hyperfine structure at $g_\perp$ (Mordvintcev, *et al.*, 1991).

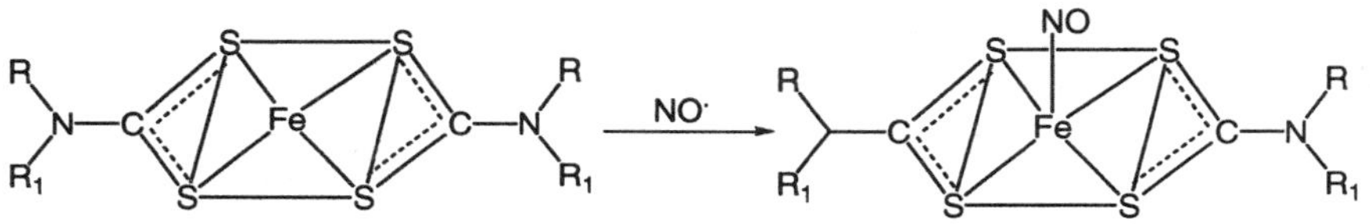

At elevated temperatures, approaching 37°C, the unresolved $g_\perp$ in the frozen state changes to an isotropic triplet with $g_{av} = 2.03$. Nevertheless, there are serious limitations with the use of $\text{Fe}^{2+}(\text{DETC})_2$ to identify NO· *in vivo*. While DETC is water soluble, $\text{Fe}^{2+}(\text{DETC})_2$ is not nearly as miscible in physiologic buffers. Thus, the identification of NO· in animal models relies on the *in situ* formation of $\text{Fe}^{2+}(\text{DETC})_2$ after independent administration of a ferrous salt and DETC. At a defined period of time, the tissue of interest is removed, frozen, and prepared for low-temperature EPR spectroscopy (Kubrina, *et al.*, 1993; Sato, *et al.*, 1993; Tominaga, *et al.*, 1993; Tominaga, *et al.*, 1994; Bune, *et al.*, 1995; Akaike, *et al.*, 1996; Hamada, *et al.*, 1996; Tsuchiya, *et al.*, 1996; Komarov, *et al.*, 1997; Mikoyan, *et al.*, 1997; Shutenko, *et al.*, 1999). These series of manipulations, required to characterize NO–$\text{Fe}^{2+}(\text{DETC})_2$, makes it difficult to quantify NO· production and questions the general applicability of this spin trapping system. However, one recent study has, surprisingly, reported the measurement, although weak, of an EPR spectrum of NO–$\text{Fe}^{2+}(\text{DETC})_2$ from the head of a living rat (Suzuki, *et al.*, 1998).

Nevertheless, the search for alternatives to $\text{Fe}^{2+}(\text{DETC})_2$ has led to the development of ferrous chelates coupled to either *N*-methyl-D-glucamine dithiocarbamate (MGD) (Shinobu, *et al.*, 1984), giving $\text{Fe}^{2+}(\text{MGD})_2$ or *N*-(dithiocarboxy)sarcosine (DTCS), yielding $\text{Fe}^{2+}(\text{DTCS})_2$ (Yoshimura, *et al.*, 1995, 1996). Upon reaction with NO·, generated from either viable cell suspensions, isolated tissue preparations, or rodent models, a three-lined EPR spectrum is recorded, which corresponds to NO–$\text{Fe}^{2+}(\text{MGD})_2$ or NO–$\text{Fe}^{2+}(\text{DTCS})_2$

(Komarov, *et al.*, 1993; Lai and Komarov, 1994; Komarov and Lai, 1995; Kotake, *et al.*, 1995, 1996; Zweier, *et al.*, 1995; Kuppusamy, *et al.*, 1996; Reinke, *et al.*, 1996; Xia, *et al.*, 1996; Yoshimura, *et al.*, 1996; Fujii, *et al.*, 1997; Norby, *et al.*, 1997; Fujii, *et al.*, 1998; Giulivi, *et al.*, 1998).

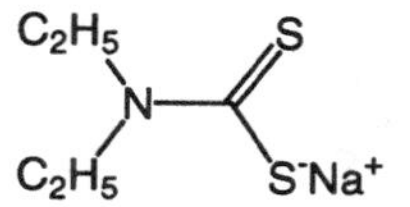

Diethyl dithiocarbamate
DETC

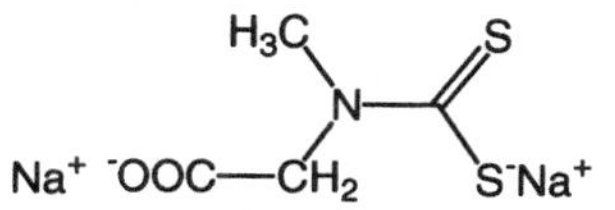

N-(dithiocarboxy)sarcosine
DTCS

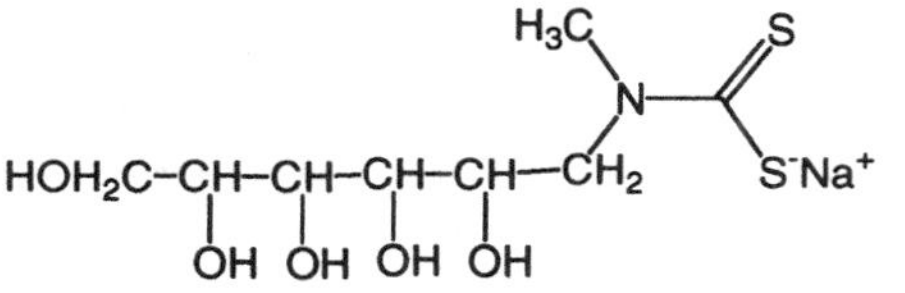

N-methyl-D-glucamine dithiocarbamate
MGD

Purified NOS I, NOS II and NOS III have been shown to produce NO· and $O_2 \cdot^-$ in which the flux rates of these free radicals can be modulated through manipulation of L-arginine levels (Pou, *et al.*, 1992, 1999a, Vásquez-Vivar, *et al.*, 1998; Xia, *et al*, 1998a, 1998b). Yet, cellular consequences have remained ambiguous, as unequivocal identification of NOS-derived free radicals had not, until recently, been forthcoming. In the first series of such studies, cellular secretion of NO· and $O_2 \cdot^-$, as monitored by the spin traps $Fe^{2+}(MGD)_2$ and DMPO, were identified from control and L-arginine-depleted cells (Xia, *et al.*, 1996; Xia and Zweier, 1997) (Figure 10.13). Simultaneous generation of these free radicals resulted in formation of $ONOO^-$ concomitant with enhanced cytotoxicity (Xia, *et al.*, 1996). One interesting observation not dwelled upon was the site where NO· and $O_2 \cdot^-$ were spin trapped. Based on the highly charged nature of $Fe^{2+}(MGD)_2$ and DMPO, one can infer that NO· and $O_2 \cdot^-$ were detected extracellularly. This suggests that the most likely cellular site of NOS distribution in this model is near or within the plasma membrane of the cell, as diffusion of NO· and $O_2 \cdot^-$ from this locus to the surrounding milieu was a prerequisite to spin trapping these free radicals.

The high reactivity of NO· toward other functional groups (Fontecave and Pierre, 1994) hints of its reaction with nitronyl nitroxides (Ullman, *et al.*, 1972). In this case, the loss of $NO_2 \cdot$ leads to an imino nitroxide and the accompanying shift in the EPR spectrum (Figure 10.14) (Akaike, *et al.*, 1993; Joseph, *et al.*, 1993; Hogg, *et al.*, 1995; Konorev, *et al.*, 1995).

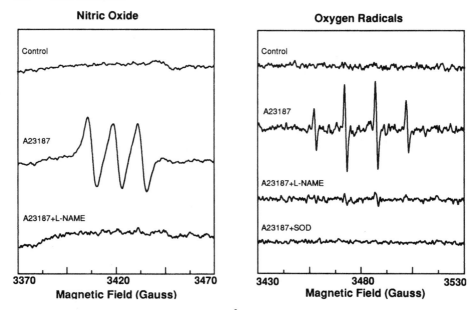

Figure 10.13 Detection of NO·, as NO–Fe^{2+}(MGD)$_2$ and O$_2$·$^-$, as DMPO–OH from the decomposition of DMPO–OOH, derived from cells activated under experimental conditions that selectively promote the formation of these free radicals. (Spectra taken from Xia, *et al.*, 1996.)

Even though this reaction, at first glance, appears to be unique to NO·, in models as sophisticated as isolated tissue preparations, comparable findings (Akaike, *et al.*, 1993; Lilley and Gibson, 1996) should be viewed with caution, as reductases are known to catalyze similar reactions. For instance, cytochrome P-450 has been shown to reduce amine *N*-oxides to the corresponding amines (Iwasaki, *et al.*, 1977; Kloss, *et al.*, 1983). Similarly, nitronyl nitroxides, like other nitroxides, are susceptible to reduction, leading to the corresponding hydroxylamine (Woldman, *et al.*, 1994; Akaike and Maeda, 1996; Haseloff, *et al.*, 1997).

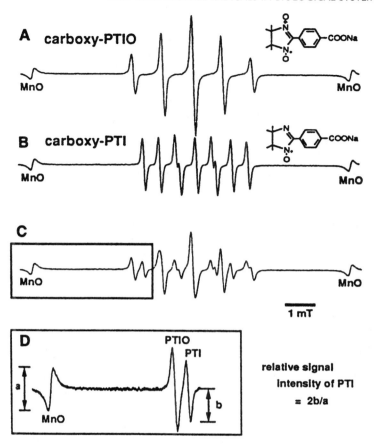

Figure 10.14 The EPR spectral measurement of NO·, which was based on the conversion of 2-(p-carboxy)phenyl-4,4,5,5-tetramethylimidazoline 3-oxide 1-oxyl (scan A) to 2-(p-carboxy)phenyl-4,4,5,5-tetramethylimidazoline 3-oxide (scan B). Scan C is a composite of 2-(p-carboxy)phenyl-4,4,5,5-tetramethylimidazoline 3-oxide 1-oxyl and 2-(p-carboxy) phenyl-4,4,5,5-tetramethylimidazoline 3-oxide. Scan D is an enlargement of the low-field peaks in scan C, allowing a more accurate determination of the ratio of each nitroxide in the reaction mixture. (Spectra taken from Akaike and Maeda, 1996.)

Experiments using the photolysis of nitroprusside as a source of NO· (Bates, *et al.*, 1991) were found to require nearly 10 min to initially observe the conversion of the nitronyl nitroxide to the imino nitroxide and several hours to complete this reaction (Joseph, *et al.*, 1993). In light of these observations and kinetic studies that demonstrate the rapidity of their bioreduction, these spin traps may not be suitable for the *in vivo* detection of NO· in real time and at the site of its evolution (Akaike and Maeda, 1996).

Of the spin traps for NO·, the ferrous chelates such as $Fe^{2+}(MGD)_2$ and $Fe^{2+}(DTCS)_2$ are the most attractive of reporters, as these complexes allow for the spectral identification and the imaging of this free radical (Komarov, *et al.*,

1993; Lai and Komarov, 1994; Komarov and Lai, 1995; Kotake, *et al.*, 1995; Miyajima and Kotake, 1995, 1997; Yoshimura, *et al.*, 1996; Norby, *et al.*, 1997). Most recent whole animal pharmacokinetic and metabolic studies of NO–$Fe^{2+}(MGD)_2$ and NO–$Fe^{2+}(DTCS)_2$ have suggested that $Fe^{2+}(DTCS)_2$ holds great promise for quantifying NO· at sites of evolution (Pou, *et al.*, 1999b). Other inquiries will undoubtedly be forthcoming, thereby allowing detailed *in vivo* mapping of this free radical, generated under a variety of physiological and/ or pathological states.

Spin Trapping of Other Free Radicals of Biological Interest

In addition to successfully detecting $O_2\cdot^-$, HO·, L· (and secondary free radicals derived from L·), and NO·, spin trapping has characterized other free radicals, whose importance in biology has just begun to be defined. For instance, thiyl radicals (RS·), principally derived from a secondary free radical reaction, have been successfully spin trapped (Harman, *et al.*, 1984; Ross, *et al.*, 1985; Eling, *et al.*, 1986; Schreiber, *et al.*, 1989; Anthroline, *et al.*, 1991; Romero, *et al.*, 1992; Kwak, *et al.*, 1995; Karoui, *et al.*, 1996; Singh, *et al.*, 1996; Stoyanovsky, *et al.*, 1996; Pou and Rosen, 1998). In the case of GS·, the EPR spectrum of DMPO–SG is so similar to that of DMPO–OH that confusion could arise as to whether GS· or HO· is actually spin trapped, unless care is exercised in measuring the hyperfine splitting constants and thoughtfully designed experiments are undertaken to verify the source of the spin trapped adduct (Schreiber, *et al.*, 1989; Kwak, *et al.*, 1995, Pou and Rosen, 1998). Another example is the tyrosyl free radical, whose identification has been made possible by spin trapping. Here, formation of a protein-derived free radical, from H_2O_2-activated metalloenzymes, was found to react with nitrones and nitrosoalkanes (Barr, *et al.*, 1996; Ostdal, *et al.*, 1997; Gunther, *et al.*, 1998; McCormick, *et al.*, 1998b).

Experimental Design

In chapter 8, we designed kinetic experiments for the spin trapping of free radicals in homogeneous solutions. These general procedures can easily be adapted to more complex systems, such as microsomal preparations, without the necessity to drastically alter the general approach previously described. For more sophisticated models, however, additional care can often result in valuable data that could not have been garnered from other methods of measuring free radicals. In the following sections, we will present some of the more insightful observations where, without such attention to detail, the identification of free radicals would have been difficult, at best.

Isolated Cell Suspensions   As important as crude homogenates have been to the study of free radical events, isolated cells preparations have greatly advanced our mechanistic understanding of how these reactive species integrate into the biochemical machinery of life. One should be aware, nevertheless, that the physiologic status of these mammalian cells, obtained from enzymic digestion of the tissue of interest, may be significantly changed as the result of the isolation procedure. This, of course, may dictate the outcome of the experiment. In the most severe case, the inability of freshly isolated, but poorly viable, enterocytes to participate in xenobiotic-dependent production of $O_2 \cdot^-$ has been directly linked to diminished ability of this compromised cell suspension to metabolize quinones and aromatic nitro-containing compounds to the prerequisite free radical intermediates (Mansbach, *et al.*, 1986).

Spin trapping is adaptable to the study of free radicals derived from other isolated cells, such as endothelial or epithelial cells, cultured as an adherent monolayer. As such, they are less amenable to the requirements of spin trapping than are cells grown in suspension, such as bacteria or freshly isolated enterocytes or leukocytes. Three general approaches for the study of adherent cells have been used. The simplest method is to add the spin trap to the media overlaying the monolayer and, at defined times, transfer an aliquot of this solution to a quartz flat cell for subsequent EPR spectroscopic studies (Lafon-Cazal, *et al.*, 1993). Although this method has the advantage of leaving the monolayer undisturbed, there are significant limitations with this procedure. First, only extracellular events are reliably measured. Second, monitoring of intracellular reactions is, however, dependent on diffusion of the spin trapped adduct from the intracellular site of formation to the extracellular environment where it can be sampled.

Besides phagocytes, there are only a few examples of cellular-secreted free radicals (Horton, *et al.*, 1986; Rabinowitch, *et al.*, 1987; Rosen, *et al.*, 1989; Britigan, *et al.*, 1992a; Lafon-Cazal, *et al.*, 1993; Xia, *et al.*, 1996; Xia and Zweier, 1997). Most often, these reactive species are produced intracellularly and react therein with a variety of acceptor molecules. What enters the extracellular milieu, and is therefore the content of the sample, is undoubtedly a limited and perhaps an inaccurate profile of free radical events taking place intracellularly. Second, the opportunity to continually monitor spin trapped adduct production over time is greatly compromised.

Another experimental design requires the disruption of the cells from the monolayer through either physical means or protease treatment. The remaining cell suspension can then be introduced into the EPR quartz flat cuvette, where *in situ* free radical events can be studied (Rosen and Freeman, 1984). The most apparent limitation with this procedure is that removal of the attached cells, by whatever method chosen, may lead to unintended alterations in cell function by disrupting the plasma membrane.

An alternative technique is to create a "suspension of adherent cells" by culturing them on microcarrier beads (Busch, *et al.*, 1982; Britigan, *et al.*, 1992a) or other particulate material, such as the three-dimensional matrix, Gelfoam® (Centra, *et al.*, 1992). Experimental manipulations are readily carried out with the "bead" suspension, including the ease of introduction into the EPR flat cell. This approach has successfully been used to study the intra- and extracellular localization of $O_2 \cdot^-$ produced during the metabolism of paraquat by endothelial cells cultured on microcarrier beads (Britigan, *et al.*, 1992a).

Typically, spin trapping experiments involving cell suspensions are conducted in an aqueous flat sample cell in which the flow of O_2 is frequently restricted. Most often, this presents little, if any, difficulty, as EPR spectra are recorded for short periods of time. However, for more lengthy experiments or those with metaboli-cally active cells, such as stimulated neutrophils—whose consumption of O_2 is at a rapid rate, introduction of O_2 into the cuvette may be a necessity to prevent anoxia, as in its absence the catalog of free radical reactions would more closely reflect the anaerobic state than the desired homeostatic condition (De Groot, et al., 1973).

Growth media can greatly impact the nature of the lipid-derived free radical that is spin trapped. Culturing cells in the presence of various polyunsaturated fatty acids has been found, for instance, to enhance the formation of carbon-centered spin trapped adducts of 4-POBN upon cellular exposure to oxidants (North, et al., 1992; Wagner, et al., 1993, 1994). This presumably relates to result-ing alterations in the plasma membrane lipid profile, leading to increased forma-tion of lipid radicals. Similarly, growth media deficient in α-tocoperol, selenium, and glutathione can significantly alter the antioxidant status of the cell under study, thereby artificially increasing the peroxidative potential of membrane lipids (Kelley, et al., 1995; Leist, et al., 1996).

Interrupting a free radical cascade by the introduction of a spin trap does not guarantee a successful outcome. One should be mindful that the verification of cellular-derived free radicals is an arduous task, made more complicated by high concentrations of metal ions that frequently contaminate the growth media cover-ing the monolayer of cells. Dialyzing the media against conalbumin prior to its addition to the monolayer may provide a reasonable method of removing metal ions, which all too often can lead to spurious nitroxide signals (Gutteridge, 1987).

Study design, independent of cell status, can greatly influence the experimental result. By way of illustration, consider the continual recording of EPR spectra over prolonged periods of time. Under such conditions, cell aggregation and settling to the bottom of the quartz aqueous flat sample cuvette can frequently result in uneven data collection. This limitation can be minimized by shifting the cavity by 90°. As such, the cuvette enters the spectrometer in a position that allows cells to distribute uniformly over the 0.5-mm thickness of the flat quartz cell, as opposed to over its 60-mm length that would occur if using the standard configured cavity (Kotake, et al., 1994).

The localization of spin trapped adducts has been of overriding interest to the understanding of free radical events at a cellular level. As the spin trap is added to cells, partitioning between intracellular and extracellular pools is dependent upon the lipophilic characteristics of the spin trap. Although this is primarily governed by passive diffusion, in some circumstances an active transport mechanism might promote intracellular entry of charged nitrones or nitrosoalkanes. Alternatively, the addition of enzymes, such as SOD, which by virtue of their size remain outside the cell, has often been used to discriminate between intra- and extracellular spin trapping of $O_2 \cdot^-$. The literature is replete with such studies. In other situations where it is less clear whether a specific inhibitor resides solely within the extra-cellular milieu or can diffuse into the cell, the use of line-broadening agents, including chromium oxalate or ferricyanide, has been a useful tool to discriminate between these compartments (Samuni, et al., 1986; Britigan, et al., 1992a; Clapp, et al., 1994). These highly charged metal salts will induce a spin-relaxation in the extracellular environment surrounding the spin trapped adduct. This results in a loss of EPR spectral intensity outside the cell with no apparent impact on intracellular production of free radicals, the rate of spin trapping, or the distribution of the corresponding nitroxide (Samuni, *et al.*, 1986; Britigan, *et al.*, 1992a; Clapp, *et al.*, 1994).

Intact Tissue Preparations   The detection of free radicals from isolated perfused organs has, until recently, been conducted by rapidly freezing a previously ground organ in liquid N_2 or sampling tissue effluent at defined times. Each procedure is not without limitations, which make accurate identification of free radicals problematic. Many of the underlying issues surrounding tissue preparation for low-temperature EPR spectroscopy and time-dependent sample collection from perfusate were introduced earlier in this chapter. Nevertheless, a few additional salient points are worth noting.

As a consequence of an ischemic insult, reperfusion of that tissue has resulted in the generation of free radicals. Identification of specific species has relied upon the localization of the spin trap at defined sites. For highly lipophilic agents, such as PBN, the reporting of intracellular events is a likely outcome, whereas for hydrophobic spin traps, like DMPO, free radical cascades arising in the vasculature will be described by this spin trap. Even under the best of experimental conditions, the source of corresponding spin trapped adduct (Blasig, *et al.*, 1986; Arroyo, *et al.*, 1987; Zweier, 1988) must be viewed with skepticism, as a variety of reactions that do not involve free radicals dominate the chemistry of many of these spin traps. Consider, for example, the formation of DMPO–OH, recorded after addition of DMPO to the effluent of a reperfused isolated heart previously subjected to a brief period of ischemia (Blasig, *et al.*, 1989, 1990). Reflecting on the short lifetime of $O_2 \cdot^-$ and HO· (see Table 2.3), it would be unlikely for the generation of DMPO–OH to arise from the diffusion of these free radicals from intracellular stores to the perfusate, where, with time, the spin trap will be introduced after this fluid exits the tissue. A more probable scenario is that DMPO–OH arose from pathways involving addition and oxidative reactions of DMPO (Makino, *et al.*, 1992).

Often, when the time interval between samples is short, and does not allow for immediate EPR spectroscopic measurements between collections, experimental design includes the rapid freezing in liquid N_2 and analyses made at a subsequent point. The rationale behind this procedure is based on the theory that by slowing the degradation of the spin trapped adduct at these low temperatures, the integrity of the sample might remain intact. Although this appears to hold for DMPO–OH, the freezing–thawing procedure has been found to hasten the decomposition of DMPO–OOH to DMPO–OH (Pou and Rosen, 1990), thereby eliminating the unique EPR spectral properties that can distinguish between these nitroxides.

Ex vivo tissue slices have become of considerable interest (Fisher, *et al.*, 1995; Kane and Thohan, 1996; Stolc, *et al.*, 1996), as these preparations can allow *in situ* detection of free radicals in models closely aligned with whole-animal preparations. Mounting thin slices of tissue onto a tissue sample cell allows, after entry into the cavity of an EPR spectrometer, direct measurement of free radicals (Timmins and Davies, 1993; Jurkiewicz and Buettner, 1994). The inability to properly bathe the slice in buffered solutions, essential to maintaining physiologic conditions, greatly limits the experimental design to short periods of free radical reporting.

Whole-Animal Experiments   With the development of lower frequency EPR spectrometers, many of the limitations and artifacts attributed to sampling tissue perfusates can be minimized by *in vivo* detection of free radicals in real time and at the site of their evolution. In chapter 7, we explored many of the essential elements required to *in vivo in situ* spin trap free radicals. We will not further elaborate on these important issues here, except to embellish the difficult task of quantitating spin trapped adduct formation in whole-animal preparations. To begin with, glass rather than plastic syringes should be used for *in vivo* administration of the spin trap, as aqueous solutions of DMPO, for example, have been reported to rapidly undergo reactions that result in spurious nitroxide signals with plastic ware (Buettner, *et al.*, 1991). After infusion of the nitrone, nitrosoalkane, or iron chelate, the ability to identify a specific free radical is governed by the pharmacodynamic properties of the spin trap, as well as by the efficiency of competing reactions for the free radical. Even under the most ideal of conditions in which the formation of the spin trapped adduct is favored, identification of this stable free radical within a tissue compartment is dependent primarily upon dynamic factors, which dictate the diffusion from the site of formation and, to a lesser extent, the rates of bioreduction (Bacic, *et al.*, 1989; Miura, *et al.*, 1992, 1995; Halpern, *et al.*, 1995, 1996, Pou, *et al.*, 1999b). Even when these factors have not been taken into consideration, quantitative estimation of the spin trapped adduct can still be reliably assessed by whole-animal spectroscopy and imaging (Yoshimura, *et al.*, 1996). For more localized determinations, such as the identification of free radicals in appendages, rapid dilution of the spin trapped adduct from the site of reaction can all too often dictate the feasibility of detecting the corresponding nitroxide, as only the extremity is placed into the cavity of the spectrometer. In this case, extensive pharmacokinetics is an essential prerequisite to the successful identification of specific spin trapped adducts at localized tissue sites (Halpern, *et al.*, 1995).

Concluding Thoughts

Over the years, the characterization of specific free radicals produced during cellular metabolism has been a complex and frequently unrewarding task. Despite limited insight as to how cells generate and regulate formation of these reactive intermediates, identifying unique free radicals has become an important step in the discovery of cellular function. The previous sections have outlined, in some detail, the successes and limitations in the application of spin trapping to the study of free radicals in biology. Successful inquiries generally reflect careful experimental design, taking into account the model, the nature and reactivity of the free radical of interest, its potential site(s) of formation, and competing reactions—all of which influence data interpretation.

Defining the appropriate model is often a difficult task, which can, on occasion, lead to a misinterpretation of the resultant data. For instance, it is clear that in open-chested dogs, postischemic dysfunction and the accompanied concentration of the spin trapped adducts upon reperfusion is greatly exaggerated when compared with identical experiments conducted in conscious animals (Triana, *et al.*, 1991; Li, *et al.*, 1993; Sekili, *et al.*, 1993). In fact, the formation of spin trapped adducts in the stunned myocardium is variable, linked to the severity of the ischemic period (Li, *et al.*, 1993). These findings clearly have an impact on our perception of the importance of free radicals in this disease state.

Choice of spin traps must constantly be matched to the relative reactivity of a particular free radical and the dynamics of the biological system under investigation. This is especially relevant to models in which cellular metabolism may significantly attenuate the steady-state concentration of a spin trapped adduct, and thereby influence the sensitivity of the EPR spectroscopic measurement. In some cases, synthesis of new spin traps will be required to meet these ever increasing demands imposed by the development of more sophisticated instrumentation. This will become more critical when lower frequency EPR spectrometers become commercially available with the capability to image, thereby allowing the localization and quantitation of free radicals, as their spin trapped adducts.

Successfully implemented experiments are dependent on the ability of the spin trap to reach the site where the free radical originated. This is less of a problem with homogenous enzymes than it is with isolated cell suspensions, perfused organ preparations, and animal models. Knowledge of the hydrophobicity and other physical characteristics of the spin trap that may drastically affect the ability to reach intracellular sites of limited accessibility is a key aspect of any study design. Even under the best of conditions, competing reactions, particularly those involving cellular systems, can alter many of the processes expected to produce specific free radicals and the accompanying spin trapped adducts. For illustrative purposes, consider the impact that growth media has had on the differential formation of cellular-derived NO· and $O_2 \cdot^-$ (Xia, *et al.*, 1996). Similarly, potential confounding effects of $ONOO^-$ have greatly influenced the spin trapping of HO· (Pou, *et al.*, 1995; Goldstein, *et al.*, 1996). In other settings, the spin trap can alter the biochemical reaction of interest (Britigan, *et al.*, 1991). In the end, it is critical to authenticate, through independent experiments, the source of the spin trapped adduct, confirming its assignment from literature reports. This task has been made easier by the existence of a database that catalogs spin trapped adducts, their hyperfine splitting constants, and the experimental conditions through which they were derived. This database is available through the Internet at two sites: http://hippo.niehs.-nih.gov/stdbl.html or http://mole.chm.bris.ac.uk/stdb/.

The application of spin trapping to biology is an ever expanding and fruitful inquiry into the understanding of free radical events. With the development of lower frequency EPR spectrometers, the *in vivo in situ* spin trapping of free radicals in real time has become a reality, allowing, for the first time, a glimpse into the role of free radicals in mammalian physiology and disease processes. With the administration of spin traps to animal models, the pharmacologic activity of these agents will need to be assessed. In the next chapter, we will begin to explore the therapeutic action of nitrones and nitrosoalkanes, as their *in vivo* effects may significantly influence the nature of the spin trapped free radical.

References

Adams, G.E., and Wardman, P. (1977). Free radicals in biology: The pulse radiolysis approach, in: *Free Radicals in Biology* (Pryor, W.A., ed.), Vol. III, pp. 53–95, Academic Press, New York.

Ahr, H.J., King, L.J., Nastainczyk, W., and Ullrich, V. (1980). The mechanism of chloroform and carbon monoxide formation from carbon tetrachloride by microsomal cytochrome P-450. *Biochem. Pharmacol.* **29**: 2855–2861.

Akaike, T., and Maeda, H. (1996). Quantitation of nitric oxide using 2-phenyl-4,4,5,5-tetramethylimidazoline-1-oxyl 3-oxide (PTIO). *Methods Enzymol.* **268**: 211–221.

Akaike, T., Noguchi, Y., Ijiri, S., Setoguchi, K., Suga, M., Zheng, Y.M., Dietzschold, B., and Maeda, H. (1996). Pathogenesis of influenza virus-induced pneumonia: Involvement of both nitric oxide and oxygen radicals. *Proc. Natl. Acad. Sci. USA* **93**: 2448–2453.

Akaike, T., Yoshida, M., Miyamoto, Y., Sato, K., Kohno, M., Sasamoto, K., Miyazaki, K., Ueda, S., and Maeda, H. (1993). Antagonistic action of imidazolineoxyl N-oxides against endothelium-derived relaxing factor/·NO through a radical reaction. *Biochemistry* **32**: 827–832.

Albano, E., Lott, K.A.K., Slater, T.F., Stier, A., Symons, M.C.R., and Tomasi, A. (1982). Spin-trapping studies on the free-radical products formed by metabolic activation of carbon tetrachloride in rat liver microsomal fractions isolated hepatocytes and *in vivo* in the rat. *Biochem. J.* **204**: 593–603.

Albano, E., Poli, G., Tomasi, A., Bini, A., Vannini, V., and Dianzani, M.U. (1984). Toxicity of 1,2-dibromoethane in isolated hepatocytes: Role of lipid peroxidation. *Chem.-Biol. Interact.* **50**: 255–265.

Alexander-North, L.S., North, J.A., Kiminyo, K.P., Buettner, G.R., and Spector, A.A. (1994). Polyunsaturated fatty acids increase lipid radical formation by oxidant stress in endothelial cells. *J. Lipid Res.* **35**: 1773–1785.

Anthroline, W.E., Kalyanaraman, B., Templin, J.A., Byrnes, R.W., and Petering, D.H. (1991). Spin-trapping studies of the oxidation-reduction reactions of iron bleomycin in the presence of thiols and buffer. *Free Radical Biol. Med.* **10**: 119–123.

Arroyo, C.M., and Forray, C. (1991). Activation of cyclic GMP formation in mouse neuroblastoma cells by a labile nitroxyl radical. An electron paramagnetic resonance/spin trapping study. *Eur. J. Pharmacol.* **208**: 157–161.

Arroyo, C.M., and Kohno, M. (1991). Difficulties encountered in the detection of nitric oxide (NO) by spin trapping techniques: A cautionary note. *Free Radical Res. Commun.* **14**: 145–149.

Arroyo, C.M., Kramer, J.H., Dickens, B.F., and Weglicki, W.B. (1987). Identification of free radicals in myocardial ischemia/reperfusion by spin trapping with nitrone DMPO. *FEBS Lett.* **221**: 101–104.

Arya, P., Stephens, J.C., Griller, D., Pou, S., Ramos, C.L., Pou, W.S., and Rosen,G.M. (1992). Design of modified pyrroline N-oxide derivatives as spin traps specific for hydroxyl radical. *J. Org. Chem.* **57**: 2297–2301.

Autor, A.P. (1974). Reduction of paraquat toxicity by superoxide dismutase. *Life Sci.* **14**: 1309–1319.

Bacic, G., Nilges, M.J., Magin, R.L., Walczak, T., and Swartz, H.M. (1989). *In vivo* localized ESR spectroscopy reflecting metabolism. *Magn. Reson. Med.* **10**: 266–272.

Bagley, A.C., Krall, J., and Lynch, R.E. (1986). Superoxide mediates the toxicity of paraquat for Chinese hamster ovary cells. *Proc. Natl. Acad. Sci. USA* **83**: 3189–3193.

Baker, J.E., and Kalyanaraman, B. (1989). Ischemia-induced changes in myocardial paramagnetic metabolites: Implications for intracellular oxy-radical generation. *FEBS Lett.* **244**: 311–314.

Baker, J.E., Felix, C.C., Olinger, G.N., and Kalyanaraman, B. (1988). Myocardial ischemia and reperfusion: Direct evidence for free radical generation by electron spin resonance spectroscopy.

Proc. Natl. Acad. Sci. USA **85**: 2786–2789.

Bannister, J.V., and Thornalley, P.J. (1983). The production of hydroxyl radicals by adriamycin in red blood cells. *FEBS Lett.* **157**: 170–172.

Barr, D.P., Gunther, M.R., Deterding, L.J., Tomer, K.B., and Mason, R.P. (1996). ESR spin-trapping of a protein-derived tyrosyl radical from the reaction of cytochrome c with hydrogen peroxide. *J. Biol. Chem.* **271**: 15498–15503.

Bates, J.N., Baker, M.T., Guerra, R., Jr., and Harrison, D.G. (1991). Nitric oxide generation from nitroprusside by vascular tissue. Evidence that reduction of the nitroprusside anion and cyanide loss are required. *Biochem. Pharmacol.* **42S**: S157-S165.

Bernier, M., Hearse, D.J., and Manning, A.S. (1986). Reperfusion-induced arrhythmias and oxygen-derived free radicals. Studies with "anti-free radical" intervention and a free radical-generating system in the isolated perfused rat heart. *Circ. Res.* **58**: 331–340.

Biaglow, J., Nygaard, O.F., and Greenstock, C.L. (1976). Electron transfer in Ehrlich ascites tumor cells in the presence of nitrofurans. *Biochem. Pharmacol.* **25**: 393–398.

Blasig, I.E., Ebert, B., Hennig, C., Pali, T., and Tosaki, A. (1990). Inverse relationship between ESR spin trapping of oxyradicals and degree of functional recovery during myocardial reperfusion in isolated working rat heart. *Cardiovasc. Res.* **24**: 263–270.

Blasig, I.E., Ebert, B., and Love, H. (1986). Identification of free radicals trapped during myocardial ischemia *in vitro* by ESR. *Stud. Biophys.* **116**: 35–42.

Blasig, I.E., Ebert, B., Wallukat, G., and Löwe, H. (1989). Spin trapping evidence for radical generation by isolated hearts and cultured heart cells. *Free Radical Res. Commun.* **6**: 303–310.

Bolli, R., and McCay, P.B. (1990). Use of spin traps in intact animals undergoing myocardial ischemia/reperfusion: A new approach to assessing the role of oxygen radicals in myocardial "stunning." *Free Radical Res. Commun.* **9**: 169–180.

Bolli, R., Patel, B.S., Jeroudi, M.O., Lai, E.K., and McCay, P.B. (1988). Demon-stration of free radical generation in "stunned" myocardium of intact dogs with the use of the spin trap α-phenyl N-tert-butyl nitrone. *J. Clin. Invest.* **82**: 476–485.

Bolli, R., Patel, B.S., Jeroudi, M.O., Li, X.-Y., Triana, J.F., Lai, E.K., and McCay, P.B. (1990). Iron-mediated radical reactions upon reperfusion contribute to myocardial "stunning." *Am. J. Physiol.* **259**: H1901–H1911.

Bolton J.R., Borg, D.C., and Swartz, H.M. (1972). Experimental aspects of biological electron spin resonance studies, in: *Biological Applications of Electron Spin Resonance* (Swartz, H.M., Bolton, J.R., and Borg, D.C. eds.), pp. 64–118, Wiley-Interscience, New York.

Bösterling, B., and Trudell, J.R. (1981). Spin trap evidence for production of superoxide radical anions by purified NADPH-cytochrome P-450 reductase. *Biochem. Biophys. Res. Commun.* **98**: 569–575.

Boucher, F., Pucheu, S., Coudray, C., Favier, A., and de Leiris, J. (1992). Evidence of cytosolic iron release during post-ischemic reperfusion of isolated rat hearts. Influence on spin trapping experiments with DMPO. *FEBS Lett.* **302**: 261–264.

Brackett, D.J., Lai, E.K., Lerner, M.R., Wilson, M.F., and McCay, P.B. (1989). Spin trapping of free radicals produced *in vivo* in heart and liver during endotoxemia. *Free Radical Res. Commun.* **7**: 315–324.

Bremer, C., Bradford, B.U., Hunt, K.J., Knecht, K.T., Connor, H.D., Mason, R.P., and Thurman, R.G. (1994). Role of Kupffer cells in the pathogenesis of hepatic reperfusion injury. *Am. J. Physiol.* **267**: G630–G636.

Bresnick, E., Vaught, J.B., Chuang, A.H.L., Stoming, T.A., Bockman, D., and Mukhtar, H. (1977). Nuclear aryl hydrocarbon hydroxylase and interaction of polycyclic hydrocarbons with nuclear components. *Arch. Biochem. Biophys.* **181**: 257–269.

Britigan, B.E., Coffman, T.J., and Buettner, G.R. (1990). Spin trapping evidence for the lack of significant hydroxyl radical production during the respiration burst of human phagocytes using a spin adduct resistant to superoxide-mediated destruction. *J. Biol. Chem.* **265**: 2650–2656.

Britigan, B.E., Rasmussen, G.T., and Cox, C.D. (1997). Augmentation of oxidant injury to human pulmonary epithelial cells by the *Pseudomonas aeruginosa* siderophore pyochelin. *Infect. Immun.* **65**: 1071–1076.

Britigan, B.E., Ratcliffe, H.R., Buettner, G.R., and Rosen, G.M. (1996). Binding of myeloperoxidase to bacteria: Effect on hydroxyl radical formation and susceptibility to oxidant-mediated killing. *Biochim. Biophys. Acta* **1290**: 231–240.

Britigan, B.E., Roeder, T.L., and Buettner, G.R. (1991). Spin traps inhibit formation of hydrogen peroxide via the dismutation of superoxide: Implications for spin trapping the hydroxyl radical. *Biochim. Biophys. Acta* **1075**: 213–222.

Britigan, B.E., Roeder, T.L., Rasmussen, G.T., Shasby, D.M., McCormick, M.L., and Cox, C.D. (1992b). Interaction of the *Pseudomonas aeruginosa* secretory products pyocyanin and pyochelin generates hydroxyl radical and causes synergistic damage to endothelial cells: Implications for *Pseudomonas*-associated tissue injury. *J. Clin. Invest.* **90**: 2187–2196.

Britigan, B.E., Roeder, T.L., and Shasby, D.M. (1992a). Insight into the nature and site of oxygen-centered free radical generation by endothelial cell monolayers using a novel spin trapping technique. *Blood* **79**: 699–707.

Buettner, G.R., and Britigan, B.E. (1990). The spin trapping of superoxide with M₄PO (3,3,5,5-tetramethylpyrroline-N-oxide). *Free Radical Biol. Med.* **8**: 57–60.

Buettner, G.R., and Oberley, L.W. (1978). Considerations in the spin trapping of superoxide and hydroxyl radical in aqueous systems using 5,5-dimethyl-1-pyrroline-1-oxide. *Biochem. Biophys. Res. Commun.* **83**: 69–74.

Buettner, G.R., Oberley, L.W., and Leuthauser, S.W.H.C. (1978). The effect of iron on the distribution of superoxide and hydroxyl radicals as seen by spin trapping and on the superoxide dismutase assay. *Photochem. Photobiol.* **28**: 693–695.

Buettner, G.R., Scott, B.D., Kerber, R.E., and Mügge, A. (1991). Free radicals from plastic syringes. *Free Radical Biol. Med.* **11**: 69–70.

Bune, A.J., Shergill, J.K., Cammack, R., and Cook, H.T. (1995). L-Arginine depletion by arginase reduces nitric oxide production in endotoxic shock: An electron paramagnetic resonance study. *FEBS Lett.* **366**: 127–130.

Burkitt, M.J., and Mason, R.P. (1991). Direct evidence for *in vivo* hydroxyl-radical generation in experimental iron overload: An ESR spin-trapping investigation. *Proc. Natl. Acad. Sci. USA* **88**: 8440–8444.

Burow, S., and Valet, G. (1987). Flow-cytometric characterization of stimulation, free radical formation, peroxidase activity and phagocytosis of human granulocytes with 2,7-dichlorofluorescein (DCF). *Eur. J. Cell Biol.* **43**: 128–133.

Bus, J.S., Aust, S.D., and Gibson, J.E. (1974). Superoxide- and singlet oxygen-catalyzed lipid peroxidation as a possible mechanism for paraquat (methyl viologen) toxicity. *Biochem. Biophys. Res. Commun.* **58**: 749–755.

Bus, J.S., Aust, S.D., and Gibson, J.E. (1977). Lipid peroxidation as a proposed mechanism for paraquat toxicity, in: *Biochemical Mechanisms of Paraquat Toxicity* (Autor, A.P., ed.), pp. 157–174, Academic Press, New York.

Bus, J.S., Cagen, S.Z., Olgaard, M., and Gibson, J.E. (1976). A mechanism of paraquat toxicity in mice and rats. *Toxicol. Appl. Pharmacol.* **35**: 501–513.

Busch, C., Cancilla, P.A., DeBault, L.E., Goldsmith, J.C., and Owen, W.G. (1982). Use of endothelium cultured on microcarriers as a model for the microcirculation. *Lab. Invest.* **47**: 498–504.

Butler, T.C. (1961). Reduction of carbon tetrachloride *in vivo* and reduction of carbon tetrachloride and chloroform *in vitro* by tissue and tissue constituents. *J. Pharmacol. Exp. Ther.* **134**: 311–319.

Cadenas, E. (1991). Two-electron bioreductive activation of 1,4-naphthoquinones: Prooxidant and antioxidant aspects, in: *Oxidative Damage and Repair. Chemical, Biological and Medical Aspects*

(Davies, K.J.A., ed.), pp. 606–611, Pergamon Press, Oxford, England.

Callen, D.F., Wolf, C.R., and Philphot, R.M. (1980). Cytochrome P-450 mediated genetic activity and cytotoxicity of seven halogenated aliphatic hydrocarbons in *Saccharomyces cerevisiae*. *Mutat. Res.* **77**: 55–63.

Calligaro, A., and Vannini, V. (1975). Electron spin resonance study of homolytic cleavage of carbon tetrachloride in rat liver: Trichloromethyl free radicals. *Pharmacol. Res. Commun.* **7**: 323–329.

Centra, M., Ratych, R.E., Cao, G.-L., Li, J., Williams, E., Taylor, R.M., and Rosen, G.M. (1992). Culture of bovine pulmonary artery endothelial cells on Gelfoam blocks. *FASEB J.* **6**: 3117–3121.

Chamulitrat, W., Jordan, S.J., Mason, R.P., Saito, K., and Cutler, R.G. (1993). Nitric oxide formation during light-induced decomposition of phenyl N-tert-butylnitrone. *J. Biol. Chem.* **268**: 11520–11527.

Chamulitrat, W., Parker, C.E., Tomer, K.B., and Mason, R.P. (1995). Phenyl N-*tert*-butyl nitrone forms nitric oxide as a result of its FE(III)-catalyzed hydrolysis or hydroxyl radical adduct formation. *Free Radical Res.* **23**: 1–14.

Cheeseman, K.H., Albano, E.F., Tomasi, A., and Slater, T.F. (1985). Biochemical studies on the metabolic activation of halogenated alkanes. *Environ. Health Perspect.* **64**: 85–101.

Chevion, M., Jiang, Y., Har-El, R., Berenshtein, E., Uretzky, G., and Kitrossky, N. (1993). Copper and iron are mobilized following myocardial ischemia: Possible predictive criteria for tissue injury. *Proc. Natl. Acad. Sci. USA* **90**: 1102–1106.

Clapp, P.A., Davies, M.J., French, M.S., and Gilbert, B.C. (1994). The bactericidal action of peroxides; An E.P.R. spin-trapping study. *Free Radical Res.* **21**: 147–167.

Coghlan, J.G., Flitters, W.D., Holley, A.E., Norell, M., Mitchell, A.G., Ilsley, C.D., and Slater, T.F. (1991). Detection of free radicals and cholesterol hydroperoxides in blood taken from the coronary sinus of man during percutaneous transluminal coronary angioplasty. *Free Radical Res. Commun.* **14**: 409–417.

Connor, H.D., Fischer, V., and Mason, R.P. (1986a). A search for oxygen-centered free radicals in the lipoxygenase/linoleic acid system. *Biochem. Biophys. Res. Commun.* **141**: 614–621.

Connor, H.D., Gao, W., Mason, R.P., and Thurman, R.G. (1994). New reactive oxidizing species causes formation of carbon-centered radical adducts in organic extracts of blood following liver transplantation. *Free Radical Biol. Med.* **16**: 871–875.

Connor, H.D., LaCagnin, L.B., Knecht, K.T., Thurman, R.G., and Mason, R.P. (1990). Reaction of glutathione with a free radical metabolite of carbon tetrachloride. *Mol. Pharmacol.* **37**: 443–451.

Connor, H.D., Thurman, R.G., Galizi, M.D., and Mason, R.P. (1986b). The formation of a novel free radical metabolite from CCl₄ in the reperfused rat liver and *in vivo*. *J. Biol. Chem.* **261**: 4542–4548.

Couet, W.R., Eriksson, U.G., Tozer, T.N., Tuck, L.D., Wesbey, G.E., Nitecki, D., and Brasch, R.C. (1984). Pharmacokinetics and metabolic fate of two nitroxides potentially useful as contrast agents for magnetic resonance imaging. *Pharm. Res.* **4**: 203–209.

Culcasi, M., Pietri, S., and Cozzone, P.J. (1989). Use of 3,3,5,5-tetramethyl-1-pyrroline-1-oxide spin trap for the continuous flow ESR monitoring of hydroxyl radical generation in the ischemic and reperfused myocardium. *Biochem. Biophys. Res. Commun.* **164**: 1274–1280.

Czapski, G. (1984). On the use of OH· scavenger in biological systems. *Is. J. Chem.* **24**: 29–32.

Dage, J.L., Ackermann, B.L., Barbuch, R.J., Bernotas, R.C., Ohlweiler, D.F., Haegele, K.D., and Thomas, C.E. (1997). Evidence for a novel pentyl radical adduct of the cyclic nitrone spin trap MDL 101,002. *Free Radical Biol. Med.* **22**: 807–812.

Davies, M.J. (1989). Direct detection of radical production in the ischaemic and reperfused myocardium: Current status. *Free Radical Res. Commun.* **7**: 275–284.

Defraigne, J.O., Pincemail, J., Franssen, C., Meurisse, M., Defechereux, T., Philippart, C., Serteyn, D., Lamy, M., Deby, C., and Limet, R. (1993). *In vivo* free radical production after cross-clamping and reperfusion of the renal artery in the rabbit. *Cardiovasc. Surg.* **1**: 343–349.

De Groot, J.J.M.C., Garssen, G.J., Vliegenthart, J.F.G., and Boldingh, J. (1973). The detection of linoleic acid radicals in the anaerobic reaction of lipoxygenase. *Biochim. Biophys. Acta* **326**: 279–284.

Docampo, R., Mason, R.P., Mottley, C., and Muniz, R.P.A. (1981). Generation of free radicals induced by nifurtimox in mammalian tissues. *J. Biol. Chem.* **256**: 10930–10933.

Dugan, L.L., Lin, T.-S., He, Y.-Y., Hsu, C.Y., and Choi, D.W. (1995). Detection of free radicals by microdialysis/spin trapping EPR following focal cerebral ischemia-reperfusion and a cautionary note on the stability of 5,5-dimethyl-1-pyrroline N-oxide (DMPO). *Free Radical Res.* **23**: 27–32.

Dürk, H., Poyer, J.L., Klessen, C., and Frank, H. (1992). Acetylene, a mammalian metabolite of 1,1,1-trichloroethane. *Biochem. J.* **286**: 353–362.

Eling, T.E., Curtis, J.F., Harman, L.S., and Mason, R.P. (1986). Oxidation of glutathione to its free radical metabolite by prostaglandin H synthase. A potential endogenous substrate for the hydroperoxidase. *J. Biol. Chem.* **261**: 5023–5028.

Epel, B.L., and Neumann, J. (1973). The mechanism of the oxidation of ascorbate and Mn^{2+} by chloroplasts. The role of the radical superoxide. *Biochim. Biophys. Acta* **325**: 520–529.

Ernster, L. (1967). DT diaphorase. *Methods Enzymol.* **10**: 309–317.

Ewing, D., and Walton, H.L. (1991). Radiation protection of *in vitro* mammalian cells: Effects of hydroxyl radical scavengers on the slopes and shoulders of survival curves. *Radiat. Res.* **126**: 187–197.

Farrington, J.A., Ebert, M., Land, E.J., and Fletcher, K. (1973). Bipyridylium quaternary salts and related compounds. V. Pulse radiolysis studies of the reaction of paraquat radical with oxygen. Implications for the mode of action of bipyridyl herbicides. *Biochim. Biophys. Acta* **314**: 372–381.

Feix, J.B., and Kalyanaraman, B. (1989). Spin trapping of lipid-derived radicals in liposomes. *Biochim. Biophys. Acta* **992**: 230–235.

Felix, C.C., Hyde, J.S., Sarna, T., and Sealy, R.C. (1978). Melanin photoreactions in aerated media: Electron spin resonance evidence for production of superoxide and hydrogen peroxide. *Biochem. Biophys. Res. Commun.* **84**: 335–341.

Feller, D.R., Morita, M., and Gillette, J.R. (1971). Enzymatic reduction of niridazole by rat liver microsomes. *Biochem. Pharmacol.* **20**: 203–215.

Ferrari, M., Colacicchi, S., Gualtieri, G., Santini, M.T., and Sotgiu, A. (1990). Whole mouse nitroxide free radical pharmacokinetics by low frequency electron paramagnetic resonance. *Biochem. Biophys. Res. Commun.* **166**: 168–173.

Finkelstein, E., Rosen, G.M., and Rauckman, E.J. (1980a). Spin trapping. Kinetics of the reaction of superoxide and hydroxyl radicals with nitrones. *J. Am. Chem. Soc.* **102**: 4994–4999.

Finkelstein, E., Rosen, G.M., and Rauckman, E.J. (1980b). Spin trapping of superoxide and hydroxyl radical: Practical aspects. *Arch. Biochem. Biophys.* **200**: 1–16.

Finkelstein, E., Rosen, G.M., Rauckman, E.J., and Paxton J. (1979). Spin trapping of superoxide. *Mol. Pharmacol.* **16**: 676–685.

Fisher, H.K., and Williams, G. (1976). Paraquat is not bacteriostatic under anaerobic conditions. *Life Sci.* **19**: 421–426.

Fisher, H.K., Clements, J.A., and Wright, R.R. (1973). Enhancement of oxygen toxicity by the herbicide paraquat. *Am. Rev. Respir. Dis.* **107**: 246–252.

Fisher, R.L., Shaughnessy, R.P., Jenkins, P.M., Austin, M.L., Roth, G.L., Gandolfi, A.J., and Brendel, K. (1995). Dynamic organ culture is superior to multiwell plate culture for maintaining precision-cut tissue slices: Optimization of tissue slice culture, Part. 1. *Toxicol. Methods*, **5**: 99–113.

Folch, J., Lees, M., and Stanley, G.H.S. (1957). A simple method for the isolation and purification of total lipids from animal tissue. *J. Biol. Chem.* **226**: 497–509.

Fontecave M., and Pierre, J.-L. (1994). The basic chemistry of nitric oxide and its possible biologic reactions. *Bull. Soc. Chim. Fr.* **131**: 620–631.

Fouts, J.R., and Brodie, B.B. (1957). The enzymatic reduction of chloramphenicol, p-nitrobenzoic acid and other aromatic nitro compounds in mammals. *J. Pharmacol. Exp. Ther.* **119**: 197–207.

Fowler, J.S.L. (1969). Carbon tetrachloride metabolism in the rabbit. *Br. J. Pharmacol.* **37**: 733–737.

Fréjaville, C., Karoui, H., Tuccio, B., Le Moigne, F., Culcasi, M., Pietri, S., Lauricella, R., and Tordo, P. (1995). 5-Diethoxyphosphoryl-5-methyl-1-pyrroline N-oxide: A new efficient phosphorylated nitrone for the *in vitro* and *in vivo* spin trapping of oxygen-centered radicals. *J. Med. Chem.* **38**: 258–265.

Fuchs, J., Freisleben, H.J., Podda, M., Zimmer, G., Milbradt, R., and Packer, L. (1993). Nitroxide radical biostability in skin. *Free Radical Biol. Med.* **15**: 415–423.

Fujii, H., Koscielniak, J., and Berliner, L.J. (1997). Determination and characterization of nitric oxide generation in mice by *in vivo* L-band EPR spectroscopy. *Magn. Reson. Med.* **38**: 565–568.

Fujii, H., Zhao, B., Koscielniak, J., and Berliner, L.J. (1994). *In vivo* EPR studies of the metabolic fate of nitrosobenzene in the mouse. *Magn. Reson. Med.* **31**: 77–80.

Fujii, K., Morio, M., Kikuchi, H., Ishihara, S., Okida, M., and Ficor, F. (1984). *In vivo* spin trapping study on anaerobic dehalogenation of halothane. *Life Sci.* **35**: 463–468.

Fujii, S., Suzuki, Y., Yoshimura, T., and Kamada, H. (1998). *In vivo* three-dimensional EPR imaging of nitric oxide production from isosorbide dinitrate in mice. *Am. J. Physiol.* **274**: G857–G862.

Gabr, A.M., Rai, U.S., and Symons, M.C.R. (1993). Conversion of nitric oxide into a nitroxide radical using 2,3-dimethylbutadiene and 2,5-dimethylhexadiene. *J. Chem. Soc., Chem. Commun.* 1099–1100.

Gage, J.C. (1968). The action of paraquat and diquat on the respiration of liver cell fractions. *Biochem. J.* **109**: 757–761.

Gallez, B., Mäder, K., and Swartz, H.M. (1996). Noninvasive measurement of the pH inside the gut by using pH-sensitive nitroxides. An *in vivo* EPR Study. *Magn. Reson. Med.* **36**: 694–697.

Gant, T.W., Rao, D.N.R., Mason, R.P., and Cohen, G.M. (1988). Redox cycling and sulphydryl arylation: Their relative importance in the mechanism of quinone cytotoxicity to isolated hepatocytes. *Chem.-Biol. Interact.* **65**: 157–173.

Garlick, P.B., Davies, M.J., Hearse, D.J., and Slater, T.F. (1987). Direct detection of free radicals in the reperfused rat heart using electron spin resonance spectroscopy. *Circ. Res.* **61**: 757–760.

Garner, R.C., and McLean, A.E.M. (1969). Increased susceptibility to carbon tetrachloride poisoning in the rat after pretreatment with oral phenobarbitone. *Biochem. Pharmacol.* **18**: 645–650.

Geske, D.H., and Maki, A. H. (1960). Electrochemical generation of free radicals and their study by electron spin resonance spectroscopy; The nitrobenzene anion radical. *J. Am. Chem. Soc.* **82**: 2671–2676.

Gillette, J.R., Kamm, J.J., and Sasame, H.A. (1968). Mechanism of p-nitrobenzoate reduction in liver: The possible role of cytochrome P-450 in liver microsomes. *Mol. Pharmacol.* **4**: 541–548.

Giulivi, C., Poderoso, J.J., and Boveris, A. (1998). Production of nitric oxide by mitochondria. *J. Biol. Chem.* **273**: 11038–11043.

Goldstein, S., and Czapski, G. (1995). Kinetics of nitric oxide autoxidation in aqueous solution in the absence and presence of various reductants. The nature of the oxidizing intermediates. *J. Am. Chem. Soc.* **117**: 12078–12084.

Goldstein, S., Squadrito, G.L., Pryor, W.A., and Czapski, G. (1996). Direct and indirect oxidations by peroxynitrite, neither involving the hydroxyl radical. *Free Radical Biol. Med.* **21**: 965–974.

Goodman, B.A., Raynor, J.B., and Symons, M.C.R. (1969). Electron spin resonance of bis(N,N-diethylthiocarbamato)nitrosyliron. *J. Chem. Soc.* 2572–2575.

Granger, D.N., Rutili, G., and McCord, J.M. (1981). Superoxide radicals in feline intestinal ischemia. *Gastroenterology* **81**: 22–29.

Green, M.R., Hill, H.A.O., Okolow-Zubkowska, M.J., and Segal, A.W. (1979). The production of hydroxyl and superoxide radicals by stimulated human neutrophils—measurements by EPR spectroscopy. *FEBS Lett.* **100**: 23–26.

Green, S.J., Scheller, L.F., Marletta, M.A., Seguin, M.C., Klotz, F.W., Slayter, M., Nelson, B.J., and Nacy, C.A. (1994). Nitric oxide: Cytokine-regulation of nitric oxide in host resistance to intracellular pathogens. *Immunol. Lett.* **43**: 87–94.

Greenstock, C.L., and Dunlop, I. (1973). Electron-transfer studies of nitrofurans using pulse radiolysis. *Int. J. Radiat.* **23**: 197–199.

Griffeth, L.K., Rosen, G.M., Rauckman, E.J., and Drayer, B.P. (1984). Pharmacokinetics of nitroxide NMR contrast agents. *Invest. Radiol.* **19**: 553–562.

Grisham, M.B. (1992). *Reactive Metabolites of Oxygen and Nitrogen in Biology and Medicine*, pp. 56–69, R.L. Landes Co., Austin, TX.

Grune, T., Müller, K., Zöllner, S., Haseloff, R., Blasig, I.E., David, H., and Siems, W. (1997). Evaluation of purine nucleotide loss, lipid peroxidation and ultrastructural alterations in post-hypoxic hepatocytes. *J. Physiol.* **498**: 511–522.

Guengerich, F.P. (1983). Oxidation-reduction properties of rat liver cytochrome P-450 and NADPH-cytochrome P-450 reductase related to catalysis in reconstituted systems. *Biochemistry* **22**: 2811–2820.

Guengerich, F.P., and Macdonald, T.L. (1984). Chemical mechanisms of catalysis by cytochrome P-450: A unified view. *Acc. Chem. Res.* **17**: 9–16.

Gunther, M.R., Tschirret-Guth, R.A., Witkowska, H.E., Fann, Y.C., Barr, D.P., Ortiz De Montellano, P.R., and Mason, R.P. (1998). Site-specific spin trapping of tyrosine radicals in the oxidation of metmyoglobin by hydrogen peroxide. *Biochem. J.* **330**: 1293–1299.

Gutierrez, P. (1989). Mechanism(s) of bioreductive activation. The example of diaziquone (AZQ). *Free Radical Biol. Med.* **6**: 405–445.

Gutteridge, J.M.C. (1987). A method for removal of trace iron contamination from biological buffers. *FEBS Lett.* **214**: 362–364.

Gutteridge, J.M.C., and Halliwell, B. (1990). Reoxygenation injury and antioxidant protection: A tale of two paradoxes. *Arch. Biochem. Biophys.* **283**: 223–226.

Haire, D.L., Oehler, U.M., Krygsman, P.H., and Janzen, E.G. (1988). Correlation of radical structure with EPR spin adduct parameters: Utility of the ^{1}H, ^{13}C, and ^{14}N hyperfine splitting constants of aminoxyl adducts of PBN-*nitronyl*-^{13}C for three parameter scatter plots. *J. Org. Chem.* **53**: 4535–4542.

Hake, C.L., Waggoner, T.B., Robertson, D.N., and Rowe, V.K. (1960). The metabolism of 1,1,1-trichloroethane by the rat. *Arch. Environ. Health* **1**: 101–107.

Hall, E.D., Pazara, K.E., Braughler, J.M., Linseman, K.L., and Jacobsen, E.J. (1990). Nonsteroidal lazaroid U78517F in models of focal and global ischemia. *Stroke* **21** (suppl. III): III-83–III-87.

Halpern, H.J., Peric, M., Yu, C., Barth, E.D., Chandramouli, G.V.R., Makinen, M.W., and Rosen, G.M. (1996). *In vivo* spin-label murine pharmacodynamics using low-frequency electron paramagnetic resonance imaging. *Biophys. J.* **71**: 403–409.

Halpern, H.J., Pou, S., Peric, M., Yu, C., Barth, E., and Rosen, G.M. (1993). Detection and imaging of oxygen-centered free radicals with low-frequency electron paramagnetic resonance and signal-enhancing deuterium-containing spin traps. *J. Am. Chem. Soc.* **115**: 218–223.

Halpern, H.J., Yu, C., Barth, E., Peric, M., and Rosen, G.M. (1995). *In situ* detection, by spin trapping of hydroxyl radical markers produced from ionizing radiation in the tumor of a living mouse. *Proc. Natl. Acad. Sci. USA* **92**: 796–800.

Hamada, Y., Ikata, T., Katoh, S., Tsuchiya, K., Niwa, M., Tsutsumishita, Y., and Fukuzawa, K. (1996). Roles of nitric oxide in compression injury of rat spinal cord. *Free Radical Biol. Med.* **20**: 1–9.

Handa, K., and Sato, S. (1975). Generation of free radicals of quinone group-containing anticancer chemicals in NADPH-microsome system as evidenced by initiation of sulfite oxidation. *Gann* **66**: 43–47.

Harbour, J.R., and Bolton, J.R. (1975). Superoxide formation in spinach chloroplasts: Electron spin resonance detection by spin trapping. *Biochem. Biophys. Res. Commun.* **64**: 803–807.

Harbour, J.R., and Hair, M.L. (1977). Superoxide generation in the photolysis of aqueous cadmium sulfide dispersions. Detecting by spin trapping. *J. Phys. Chem.* **81**: 1791–1793.

Harbour, J.R., and Hair, M.L. (1978). Detection of superoxide ions in nonaqueous media. Generation by photolysis of pigment dispersions. *J. Phys. Chem.* **82**: 1397–1399.

Harbour, J.R., Chow, V., and Bolton, J.R. (1974). An electron spin resonance study of the spin adducts of OH and HO$_2$ radicals with nitrones in the ultraviolet photolysis of aqueous hydrogen peroxide solutions. *Can. J. Chem.* **52**: 3549–3553.

Harman, L.S., Mottley, C., and Mason, R.P. (1984). Free radical metabolites of L-cysteine oxidation. *J. Biol. Chem.* **259**: 5606–5611.

Hasegawa, K., and Patterson, L.K. (1978). Pulse radiolysis studies in model lipid systems: Formation and behavior of peroxy radicals in fatty acids. *Photochem. Photobiol.* **28**: 817–823.

Haseloff, R.F., Zöllner, S., Kirilyuk, I.A., Grigor'ev, I.A., Reszka, R., Bernhardt, R., Mertsch, K., Roloff, B., and Blasig, I.E. (1997). Superoxide-mediated reduction of the nitroxide group can prevent detection of nitric oxide by nitronyl nitroxides. *Free Radical Res.* **26**: 7–17.

Hassan, H.M. (1984). Exacerbation of superoxide radical formation by paraquat. *Methods Enzymol.* **105**: 523–553.

Hassan, H.M., and Fridovich, I. (1977). Regulation of the synthesis of superoxide dismutase in *Escherichia coli*. Induction by methyl viologen. *J. Biol. Chem.* **252**: 7667–7672.

Hassan, H.M., and Fridovich, I. (1978). Superoxide radical and the oxygen enhancement of the toxicity of paraquat in *Escherichia coli*. *J. Biol. Chem.* **253**: 8143–8148.

Hassan, H.M., and Fridovich, I. (1979). Intracellular production of superoxide radical and of hydrogen peroxide by redox active compounds. *Arch. Biochem. Biophys.* **196**: 385–395.

Hernandez, P.H., Gillette, J.R., and Mazel, P. (1967). Studies on the mechanism of action of mammalian hepatic azoreductase-I. Azoreductase activity of reduced nicotinamide adenine dinucleotide phosphate–cytochrome c reductase. *Biochem. Pharmacol.* **16**: 1859–1875.

Hiramatsu, M., Oikawa, K., Noda, H., Mori, A., Ogata, T., and Kamada, H. (1995). Free radical imaging by electron spin resonance computed tomography in rat brain. *Brain Res.* **697**: 44–47.

Hirayama, S., Ueda, R., and Sugata, K. (1995). Detection of hydroxyl radical in intact cells of *Chlorella vulgaris*. *Free Radical Res.* **23**: 51–59.

Hoehler, D., Marquardt, R.R., McIntosh, A.R., and Hatch, G.M. (1997). Induction of free radicals in hepatocytes, mitochondria and microsomes of rats by ochratoxin A and its analogs. *Biochim. Biophys. Acta* **1357**: 225–233.

Hogg, N., Singh, R.J., Joseph, J., Neese, F., and Kalyanaraman, B. (1995). Reactions of nitric oxide with nitronyl nitroxides and oxygen: Prediction of nitrite and nitrate formation by kinetic stimulation. *Free Radical Res.* **22**: 47–56.

Homer, R.F., Mees, G.C., and Tomlinson, T.E. (1960). Mode of action of dipyridyl quaternary salts as herbicides. *J. Sci. Food Agric.* **11**: 309–315.

Horton, J.K., Brigelius, R., Mason, R.P., and Bend, J.R. (1986). Paraquat uptake into freshly isolated rabbit lung epithelial cells and its reduction to the paraquat radical under anaerobic conditions. *Mol. Pharmacol.* **29**: 484–488.

Howard, J.A., and Tait, J.C. (1978). Electron paramagnetic resonance spectra of the *tert*-butylperoxy and *tert*-butoxy adducts of phenyl *tert*-butyl nitrone and 2-methyl-2-nitrosopropane. Oxygen-17 hyperfine coupling constants. *Can. J. Chem.* **56**: 176–178.

Hughes, H.M., George, I.M., Evans, J.C., Rowland, C.C., Powell, G.M., and Curtis, C.G. (1991). The role of the liver in the production of free radicals during halothane anaesthesia in the rat. Quantification of N-*tert*-butyl-α-(4-

nitrophenyl)nitrone (PBN)-trapped adducts in bile from halothane as compared to carbon tetrachloride. *Biochem. J.* **277**: 795–800.

Hume, D.A., Gordon, S., Thornalley, P.J., and Bannister, J.V. (1983). The production of oxygen-centred radicals by Bacillus-Calmette-Guerin-activated macrophages. An electron paramagnetic resonance study of the response to phorbol myristrate acetate. *Biochim. Biophys. Acta* **763**: 245–250.

Iannone, A., Tomasi, A., Vannini, V., and Swartz, H.M. (1990a). Metabolism of nitroxide spin labels in subcellular fraction of rat liver. I. Reduction by microsomes. *Biochim. Biophys. Acta* **1034**: 285–289.

Iannone, A., Tomasi, A., Vannini, V., and Swartz, H.M. (1990b). Metabolism of nitroxide spin labels in subcellular fractions of rat liver. II. Reduction in the cytosol. *Biochim. Biophys. Acta* **1034**: 290–293.

Ichimori, K., Arroyo, C.M., Pronai, L., Fukahori, M., and Nakazawa, H. (1993). The reactions of 3,5-dibromo-4-nitrosobenzene sulphonate and its biological applications. *Free Radical Res. Commun.* **19S**: S129–S139.

Ignarro, L.J., Fukuto, J.M., Griscavage, J.M., Rogers, N.E., and Byrns, R.E. (1993). Oxidation of nitric oxide in aqueous solution to nitrite but not nitrate: Comparison with enzymatically formed nitric oxide from L-arginine. *Proc. Natl. Acad. Sci. USA* **90**: 8103–8107.

Ingall, A., Lott, K.A.K., Slater, T.F., Finch, S., and Stier, A. (1978). Metabolic activation of carbon tetrachloride to a free-radical product: Studies using a spin trap. *Biochem. Soc. Trans.* **6**: 962–964.

Iwahashi, H., Albro, P.W., McGown, S.R., Tomer, K.B., and Mason, R.P. (1991). Isolation and identification of α-(4-pyridyl-1-oxide)-N-*tert*-butylnitrone radical adducts formed by the decomposition of the hydroperoxides of linoleic acid, linolenic acid and arachidonic acid by soybean lipoxygenase. *Arch. Biochem. Biophys.* **285**: 172–180.

Iwahashi, H., Deterding, L.J., Parker, C.E., Mason, R.P., and Tomer, K.B. (1996). Identification of radical adducts formed in the reactions of unsaturated fatty acids with soybean lipoxygenase using continuous flow fast atom bombardment with tandem mass spectrometry. *Free Radical Res.* **25**: 255–274.

Iwasaki, K., Noguchi, H., Kato, R., Imai, Y., and Sato, R. (1977). Reduction of tertiary amine N-oxide by purified cytochrome P-450. *Biochem. Biophys. Res. Commun.* **77**: 1143–1149.

Iyanagi, T., and Yamazaki, I. (1970). One-electron-transfer reactions in biochemical systems. V. Difference in the mechanism of quinone reduction by the NADH dehydrogenase and the NAD(P)H dehydrogenase (DT-diaphorase). *Biochim. Biophys. Acta* **216**: 281–294.

Janzen, E.G., Nutter, D.E., Jr., Davis, E.R., Blackburn, B.J., Poyer, J.L., and McCay, P.B. (1978). On spin trapping hydroxyl and hydroperoxyl radicals. *Can. J. Chem.* **56**: 2237–2242.

Janzen, E.G., Stronks, H.J., Dubose, C.M., Poyer, J.L., and McCay, P.B. (1985). Chemistry and biology of spin-trapping radicals associated with halocarbon metabolism *in vitro* and *in vivo*. *Environ. Health Perspect.* **64**: 151–170.

Janzen, E.G., Towner, R.A., and Haire, D.L. (1987). Detection of free radicals generated from the *in vitro* metabolism of carbon tetrachloride using improved ESR spin trapping techniques. *Free Radical Res. Commun.* **3**: 357–364.

Janzen, E.G., Towner, R.A., Krygsman, P.H., Haire, D.L., and Poyer, J.L. (1990a). Structure identification of free radicals by ESR and GC/MS of PBN spin adducts from the *in vitro* and *in vivo* rat liver metabolism of halothane. *Free Radical Res. Commun.* **9**: 343–351.

Janzen, E.G., Towner, R.A., Krygsman, P.H., Lai, E.K., Poyer, J.L., Brueggemann, G., and McCay, P.B. (1990b). Mass spectroscopy and chromatography of trichloromethyl radical adduct of phenyl *tert*-butyl nitrone. *Free Radical Res. Commun.* **9**: 353–360.

Janzen, E.G., Zhdanov, R.I., and Reinke, L.A. (1993). Metabolism of phenyl and alkyl spin adducts of PBN in rat hepatocytyes. Rate dependence on size and type of addend group. *Free Radical Res. Commun.* **19S**: S157- S162.

Jiang, J.J., Liu, K.J., Jordan, S.J., Swartz, H.M., and Mason, R.P. (1996). Detection of free radical metabolite formation using *in vivo* EPR spectroscopy: Evidence of rat hemoglobin thiyl formation following administration of phenylhydrazine. *Arch. Biochem. Biophys.* **330**: 266–270.

Jiang, J.J., Liu, K.J., Shi, X., and Swartz, H.M. (1995). Detection of short-lived free radicals by low-frequency electron paramagnetic resonance spin trapping in whole living animals. *Arch. Biochem. Biophys.* **319**: 570–573.

Johansen, I., and Howard-Flanders, P. (1965). Macromolecular repair and free radical scavenging in the protection of bacteria against X-rays. *Radiat. Res.* **24**: 184–200.

Joseph, J., Kalyanaraman, G., and Hyde, J.S. (1993). Trapping of nitric oxide by nitronyl nitroxides: An electron spin resonance investigation. *Biochem. Biophys. Res. Commun.* **192**: 926–934.

Jurkiewicz, B.A., and Buettner, G.R. (1994). Ultraviolet light-induced free radical formation in skin: An electron paramagnetic resonance study. *Photochem. Photobiol.* **59**: 1–4.

Kadiiska, M.B., Burkitt, M.J., Xiang, Q.-H., and Mason, R.P. (1995). Iron supplementation generates hydroxyl radical *in vivo*. An ESR spin-trapping investigation. *J. Clin. Invest.* **96**: 1653–1657.

Kadiiska, M.B., Hanna, P.M., and Mason, R.P. (1993). *In vivo* ESR spin trapping evidence for hydroxyl radical-mediated toxicity of paraquat and copper in rats. *Toxicol. Appl. Pharmacol.* **123**: 187–192.

Kadiiska, M.B., Xiang, Q.-H., and Mason, R.P. (1994). *In vivo* free radical generation by chromium (VI): Electron spin resonance spin-trapping investigation. *Chem. Res. Toxicol.* **7**: 800–805.

Kadkhodaee, M., Hanson, G.R., Towner, R.A., and Endre, A.H. (1996). Detection of hydroxyl and carbon-centred radicals by EPR spectroscopy after ischaemia and reperfusion of the rat kidney. *Free Radical Res.* **25**: 31–42.

Kalyanaraman, B., Mason, R.P., Perez-Reyes, E., Chignell, C.F., Wolf, C.R., and Philpot, R.M. (1979a). Character-
ization of the free radical formed in aerobic microsomal incubations containing carbon tetrachloride and NADPH. *Biochem. Biophys. Res. Commun.* **89**: 1065–1072.

Kalyanaraman, B., Mason, R.P., Rowlett, R., and Kispert, L.D. (1981). An electron spin resonance investigation and molecular orbital calculation of the anion radical intermediate in the enzymatic *cis–trans* isomerization of furylfuramide, a nitrofuran derivative of ethylene. *Biochim. Biophys. Acta* **660**: 102–109.

Kalyanaraman, B., Mottley, C., and Mason, R.P. (1984). On the use of organic extraction in the spin-trapping technique as applied to biological systems. *J. Biochem. Biophys. Methods* **9**: 27–31.

Kalyanaraman, B., Perez-Reyes, E., and Mason, R.P. (1980). Spin-trapping and direct electron spin resonance investigations of the redox metabolism of quinone anticancer drugs. *Biochem. Biophys. Acta* **630**: 119–130.

Kalyanaraman, B., Perez-Reyes, E., Mason, R.P., Peterson, F.J., and Holtzman, J.L. (1979b). Electron spin resonance evidence for a free radical intermediate in the cis–trans isomerization of furylfuramide by oxygen-sensitive nitroreductases. *Mol. Pharmacol.* **16**: 1059–1064.

Kane, A.S., and Thohan, T. (1996). Dynamic culture of fish hepatic tissue slices to assess phase I and phase II biotransformation, in: *Techniques in Aquatic Toxicology* (Ostrander, G.K., ed.), pp. 371–391, CRC Press, Boca Raton, FL.

Kang, Y.J., Chen, Y., and Epstein, P.N. (1996). Suppression of doxorubicin cardiotoxicity by overexpression of catalase in the heart of transgenic mice. *J. Biol. Chem.* **271**: 12610–12616.

Karoui, H., Hogg, N., Fréjaville, C., Tordo, P., and Kalyanaraman, B. (1996). Characterization of sulfur-centered radical intermediates formed during the oxidation of thiols and sulfite by peroxynitrite. ESR-spin trapping and oxygen uptake studies. *J. Biol. Chem.* **271**: 6000–6009.

Kastening, B. (1964). Elektrochemische Bildung, Reaktivität und Eigenschaften des Nitrobenzol-Radikalanions. *Electrochim. Acta* **9**: 241–254.

Kato, R., Takahashi, A., and Oshima, T. (1970). Characteristics of nitro reduction of the carcinogenic agent, 4-nitroquinolone N-oxide. *Biochem. Pharmacol.* **19**: 45–55.

Kaur, H. (1996). A water soluble C-nitrosoaromatic spin-trap—3,5-dibromo-4-nitrosobenzenesulphonic acid. "The Perkins spin-trap". *Free Radical Res.* **24**: 409–420.

Kazama, S., Takashige, G., Yoshioka, H., Tanizawa, H., Ogata, T., Koscielniak, J., and Berliner, L.J. (1996). Dynamic electron spin resonance (ESR) imaging of the distribution of spin labeled dextran in a mouse. *Magn. Reson. Med.* **36**: 547–550.

Keana, J.F.W., Pou, S., and Rosen, G.M. (1987). Nitroxides as potential contrast enhancing agents for MRI application: Influence of structure on the rate of reduction by rat hepatocytes, whole liver homogenate, subcellular fractions and ascorbate. *Magn. Reson. Med.* **5**: 525–536.

Keana, J.F.W., Pou, S., and Rosen, G.M. (1989). Synthesis and properties of some nitroxide α-carboxylate salts. *J. Org. Chem.* **54**: 2417–2420.

Keizer, H.G., Pinedo, H.M., Schuurhuis, G.J., and Joenje, H. (1990). Doxorubicin (admiamycin): A critical review of free radical-dependent mechanisms of cytotoxicity. *Pharmacol. Ther.* **47**: 219–231.

Kelley, E.E., Buettner, G.R., and Burns, C.P. (1995). Relative α-tocopherol deficiency in cultured cells: Free radical-mediated lipid peroxidation, lipid oxidizability, and cellular polyunsaturated fatty acid content. *Arch. Biochem. Biophys.* **319**: 102–109.

Kharitonov, V.G., Sundquist, A.R., and Sharma, V.S. (1994). Kinetics of nitric oxide autoxidation in aqueous solution. *J. Biol. Chem.* **269**: 5881–5883.

Kloss, M.W., Rosen, G.M., and Rauckman, E.J. (1983). N-Demethylation of cocaine to norcocaine. Evidence for participation by cytochrome P-450 and FAD-containing monooxygenase. *Mol. Pharmacol.* **23**: 482–485.

Knecht, K.T., and Mason, R.P. (1988). *In vivo* radical trapping and biliary secretion of radical adducts of carbon tetrachloride-derived free radical metabolites. *Drug Metabol. Dispos.* **16**: 813–817.

Knecht, K.T., and Mason, R.P. (1991). The detection of halocarbon-derived radical adducts in bile and liver of rats. *Drug Metabol. Dispos.* **19**: 325–331.

Knecht, K.T., DeGray, J.A., and Mason, R.P. (1992). Free radical metabolism of halothane *in vivo*: Radical adducts detected in bile. *Mol. Pharmacol.* **21**: 943–949.

Koch, R.P., Glende, E.A., Jr., and Recknagel, R.O. (1974). Hepatotoxicity of bromotrichloromethane—bond dissociation energy and lipoperoxidation. *Biochem. Pharmacol.* **23**: 2907–2915.

Kocherginsky, N., and Swartz, H.M. (1995). Chemical reactivity of nitroxides, in: *Nitroxide Spin Labels. Reactions in Biology and Chemistry* (Kocherginsky, N., and Swartz, H.M., eds.), pp. 27–66, CRC Press, Boca Raton, FL.

Kohen, R., and Chevion, M. (1985a). Transition metals potentiate paraquat toxicity. *Free Radical Res. Commun.* **1**: 79–88.

Kohen, R., and Chevion, M. (1985b). Paraquat toxicity is enhanced by iron and reduced by desferrioxamine in laboratory mice. *Biochem. Pharmacol.* **34**: 1841–1843.

Kohen, R., and Chevion, M. (1988). Cytoplasmic membrane is the target organelle for transition metal mediated damage induced by paraquat in *Escherichia coli*. *Biochemistry* **27**: 2597–2603.

Komarov, A.M., and Lai, C.-S. (1995). Detection of nitric oxide production in mice by spin-trapping electron paramagnetic resonance spectroscopy. *Biochim. Biophys. Acta* **1272**: 29–36.

Komarov, A.M., Mak, I.T., and Weglicki, W.B. (1997). Iron potentiates nitric oxide scavenging by dithiocarbamates in tissue of septic shock mice. *Biochim. Biophys. Acta* **1361**: 229–234.

Komarov, A., Mattson, D., Jones, M.M., Singh, P.K., and Lai, C.-S. (1993). *In vivo* spin trapping of nitric oxide in mice. *Biochem. Biophys. Res. Commun.* **195**: 1191–1198.

Konorev, E.A., Tarpey, M.M., Joseph, J., Baker, J.E., and Kalyanaraman, B.

(1995). Nitronyl nitroxides as probes to study the mechanism of vasodilatory action of nitrovasodilators, nitrone spin traps, and nitroxides: Role of nitric oxide. *Free Radical Biol. Med.* **18**: 169–177.

Korbashi, P., Kohen, R., Katzhendler, J., and Chevion, C. (1986). Iron mediates paraquat toxicity in *Escherichia coli. J. Biol. Chem.* **261**: 12472–12476.

Korth, H.-G., Ingold, K.U., Sustmann, R., De Groot, H., and Sies, H. (1992). Tetramethyl-ortho-quinodimethane. First member of a family of custom-tailored cheletropic spin traps for nitric oxide. *Angew. Chem., Int. Ed. Engl.* **31**: 891–893.

Korth, H.G., Sustmann, R., Lommes, P., Paul, T., de Groot, H., Hughes, L., and Ingold, K.U. (1994). Nitric oxide cheletropic traps (NOCTs) with improved thermal stability and water solubility. *J. Am. Chem. Soc.* **116**: 2767–2777.

Kotake, Y., Reinke, L.A., Tanigawa, T., and Koshida, H. (1994). Determination of the rate of superoxide generation from biological systems by spin trapping: Use of rapid oxygen depletion to measure the decay rate of spin adducts. *Free Radical Biol. Med.* **17**: 215–223.

Kotake, Y., Tanigawa, T., Tanigawa, M., and Ueno, I. (1995). Spin trapping isotopically-labeled nitric oxide produced from [^{15}N]L-arginine and [^{17}O]dioxygen by activated macrophages using a water soluble Fe^{++}-dithiocarbamate spin trap. *Free Radical Res.* **23**: 287–295.

Kotake, Y., Tanigawa, T., Tanigawa, M., Ueno, I., Allen, D.R. and Lai, C.-S. (1996). Continuous monitoring of cellular nitric oxide generation by spin trapping with an iron–dithiocarbamate complex. *Biochim. Biophys. Acta* **1289**: 362–368.

Krall, J., Bagley, A.C., Mullenbach, G.T., Hallewell, R.A., and Lynch, R.E. (1988). Superoxide mediates the toxicity of paraquat for mammalian cells. *J. Biol. Chem.* **263**: 1910–1914.

Kramer, J.H., Arroyo, C.M., Dickens, B.F., and Weglicki, W.B. (1987). Spin-trapping evidence that graded myocardial ischemia alters post-ischemic superox-ide production. *Free Radical Biol. Med.* **3**: 153–159.

Kubic, V.L., and Anders, M.W. (1980). Metabolism of carbon tetrachloride to phosgene. *Life Sci.* **26**: 2151–2155.

Kubow, S., Janzen, E.G., and Bray, T.M. (1984). Spin-trapping of free radicals formed during *in vitro* and *in vivo* metabolism of 3-methylindole. *J. Biol. Chem.* **259**: 4447–4451.

Kubrina, L.N., Caldwell, W.S., Mordvint-cev, P.I., Malenkova, I.V., and Vanin, A.F. (1993). EPR evidence for nitric oxide production from guandino nitrogens of L-arginine in animal tissues *in vivo. Biochim. Biophys. Acta* **1099**: 233–237.

Kuppusamy, P., Wang, P., Samouilov, A., and Zweier, J.L. (1996). Spatial mapping of nitric oxide generation in the ischemic heart using electron paramagnetic resonance imaging. *Magn. Reson. Med.* **36**: 212–218.

Kuzuya, T., Hoshida, S., Kim, Y., Nishida, M., Fuji, H., Kitabatake, A., Tada, M., and Kamada, T. (1990). Detection of oxygen-derived free radical generation in the canine postischemic heart during late phase of reperfusion. *Circ. Res.* **66**: 1160–1165.

Kwak, H.S., Yim, H.S., Chock, P.B., and Yim, M.B. (1995). Endogenous intracellular glutathionyl radicals are generated in neuroblastoma cells under hydrogen peroxide oxidative stress. *Proc. Natl. Acad. Sci. USA* **92**: 4582–4586.

LaCagnin, L.B., Connor, H.D., Mason, R.P., and Thurman, R.G. (1988). The carbon dioxide anion radical in the perfused rat liver: Relationship to halocarbon-induced toxicity. *Mol. Pharmacol.* **33**: 351–357.

Lafon-Cazal, M., Pietri, S., Culcasi, M., and Bockaert, J. (1993). NMDA-dependent superoxide production and neurotoxicity. *Nature* **364**: 535–537.

Lai, C.-S., and Komarov, A.M. (1994). Spin trapping of nitric oxide produced *in vivo* in septic-shock in mice. *FEBS Lett.* **345**: 120–124.

Lai, C.-S., and Piette, L.H. (1977). Hydroxyl radical production involved in lipid peroxidation of rat liver microsomes. *Biochem. Biophys. Res. Commun.* **78**: 51–59.

Lai, C.-S., Grover, T.A., and Piette, L.H. (1979a). Hydroxyl radical production in a purified NADPH-cytochrome c (P-450) reductase system. *Arch. Biochem. Biophys.* **193**: 373–378.

Lai, E.K., Crossley, C., Sridhar, R., Misra, H.P., Janzen, E.G., and McCay, P.B. (1986). *In vivo* spin trapping of free radicals generated in brain, spleen and liver during γ radiation of mice. *Arch. Biochem. Biophys.* **244**: 156–160.

Lai, E.K., McCay, P.B., Noguchi, T., and Fong, K.-L. (1979b). *In vivo* spin-trapping of trichloromethyl radicals formed from CCl₄. *Biochem. Pharmacol.* **28**: 2231–2235.

Ledwith, A. (1977). Electron transfer reaction of paraquat, in: *Biochemical Mechanisms of Paraquat Toxicity* (Autor, A.P., ed.), pp. 21–38, Academic Press, New York.

Lee, V., Randhawa, A.K., and Singal, P.K. (1991). Adriamycin-induced myocardial dysfunction *in vitro* is mediated by free radicals. *Am. J. Physiol.* **261**: H989–H995.

Leist, M., Raab, B., Maurer, S., Rösick, U., and Brigelius-Flohé, R. (1996). Conventional cell culture media do not adequately supply cells with antioxidants and thus facilitate peroxide-induced genotoxicity. *Free Radical Biol. Med.* **21**: 297–306.

Lewis, R.S., Tannenbaum, S.R., and Deen, W.M. (1995). Kinetics of N-nitrosation in oxygenated nitric oxide solutions at physiological pH: Role of nitrous anhydride and effects of phosphate and chloride. *J. Am. Chem. Soc.* **117**: 3933–3939.

Li, X.-Y., McCay, P.B., Zughaib, M., Jeroudi, M.O., Triana, J.F., and Bolli, R. (1993). Demonstration of free radical generation in the "stunned" myocardium in the conscious dog and identification of major differences between conscious and open-chest dogs. *J. Clin. Invest.* **92**: 1025–1041.

Lilley, E., and Gibson, A. (1996). Antioxidant protection of NO-induced relaxation of the mouse anococcygeus against inhibition by superoxide anions, hydroquinone and carboxy-PTIO. *Br. J. Pharmacol.* **119**: 432–438.

Lind, C., Cadenas, E., Hochstein, P., and Ernster, L. (1990). DT-diaphorase: Purification, properties and function. *Methods Enzymol.* **186**: 287–301.

Liu, K.J., Miyake, M., Panz, T., and Swartz, H. (1999). Evaluation of DEPMPO as a spin trapping agent in biological systems. *Free Radical Biol. Med.* **26**: 714–721.

Livertoux, M.-H., Lagrange, P., and Minn, A. (1996). The superoxide production mediated by the redox cycling of xenobiotics in rat brain microsomes is dependent on their reduction potential. *Brain Res.* **725**: 207–216.

Lloyd, D., and Pedersen, J.Z. (1985). Metronidazole radical anion generation *in vivo* in *Trichomonas vaginalis*: Oxygen quenching is enhanced in a drug-resistant strain. *J. Gen. Microbiol.* **131**: 87–92.

Ma, X.-L., Johnson, G., III, and Lefer, A.M. (1992). Low doses of superoxide dismutase and a stable prostacyclin analogue protect in myocardial ischemia and reperfusion. *J. Am. Coll. Cardiol.* **19**: 197–204.

Mäder, K., Stösser, R., and Borchert, H.-H. (1993). Detection of free radicals in living mice after inhalation of DTBN by X-band ESR. *Free Radical Biol. Med.* **14**: 339–342.

Makino, K., Hagi, A., Ide, H., Murakami, A., and Nishi, M. (1992). Mechanistic studies on the formation of aminoxyl radicals from 5,5-dimethyl-1-pyrroline-N-oxide in Fenton systems. Characterization of key precursors giving rise to background ESR signals. *Can. J. Chem.* **70**: 2818–2827.

Malisza, K.L., Mcintosh, A.R., Sveinson, S.E., and Hasinoff, B.B. (1996). Semiquinone free radical formation by daunorubin aglycone incorporated into the cellular membranes of intact Chinese hamster ovary cells. *Free Radical Res.* **24**: 9–18.

Mani, V., and Crouch R.K. (1989). Spin trapping of the superoxide anion: Complications in the use of the water-soluble nitroso-aromatic reagent DBNBS. *J. Biochem. Biophys. Methods* **18**: 91–96.

Mansbach, C.M., II, Rosen, G.M., Rahn, C.A., and Strauss, K.E. (1986). Detection of free radicals as a consequence of rat intestinal cellular drug metabolism. *Biochim. Biophys. Acta* **888**: 1–9.

Mason, R.P. (1982). Free-radical intermediates in the metabolism of toxic chemicals, in: *Free Radicals in Biology* (Pryor, W.A., ed.), Vol. V, pp. 161–222, Academic Press, New York.

Mason, R.P. (1990). Redox cycling of radical anion metabolites of toxic chemicals and drugs and the Marcus theory of electron transfer. *Environ. Health Perspect.* **87**: 237–243.

Mason, R.P., and Holtzman, J.L. (1975a). The mechanism of microsomal and mitochondrial nitroreductase. Electron spin resonance evidence for nitroaromatic free radical intermediates. *Biochemistry* **14**: 1626–1632.

Mason, R.P., and Holtzman, J.L. (1975b). The role of catalytic superoxide formation in the O_2 inhibition of nitroreductase. *Biochem. Biophys. Res. Commun.* **67**: 1267–1274.

Mason, R.P., Peterson, F.J., and Holtzman, J.L. (1977). The formation of an azo anion free radical metabolite during microsomal azo reduction of sulfonazo III. *Biochem. Biophys. Res. Commun.* **75**: 532–540.

Mason, R.P., Peterson, F.J., and Holtzman, J.L. (1978). Inhibition of azo-reductase by oxygen. The role of the azo anion free radical metabolite in the reduction of oxygen to superoxide. *Mol. Pharmacol.* **14**: 665–671.

McCay, P.B., Lai, E.K., Poyer, J.L., DuBose, C.M., and Janzen, E.G. (1984). Oxygen- and carbon-centered free radical formation during carbon tetrachloride metabolism. Observation of lipid radicals *in vivo* and *in vitro*. *J. Biol. Chem.* **259**: 2135–2143.

McCord, J.M. (1988). Free radicals and myocardial ischemia: Overview and outlook. *Free Radical Biol. Med.* **4**: 9–14.

McCord, J.M., and Fridovich, I. (1969). Superoxide dismutase. An enzymic function for erythrocuprein (Hemocuprein). *J. Biol. Chem.* **244**: 6049–6055.

McCormick, M.L., Buettner, G.R., and Britigan, B.E. (1998a). Endogenous superoxide dismutase levels regular iron-dependent hydroxyl radical formation in *Escherichia coli* exposed to hydrogen peroxide. *J. Bacteriol.* **180**: 622–625.

McCormick, M.L., Gaut, J.P., Lin, T.S., Britigan, B.E., Buettner, G.R., and Heinecke, J.W. (1998b). Electron paramagnetic resonance detection of free tyrosyl radical generated by myeloperoxidase, lactoperoxidase, and horseradish peroxidase. *J. Biol. Chem.* **273**: 32030–32037.

McCormick, M.L., Roeder, T.L., Railsback, M.A., and Britigan, B.E. (1994). Eosinophil peroxidase-dependent hydroxyl radical generation by human eosinophils. *J. Biol. Chem.* **269**: 27914–27919.

McDonald, C.C., Philips, W.D., and Mower, H.F. (1965). An electron spin resonance study of some complexes of iron, nitric oxide and anionic ligands. *J. Am. Chem. Soc.* **87**: 3319–3326.

McLain, G.E., Sipes, I.G., and Brown, B.R. (1979). An animal model of halothane hepatotoxicity: Roles of enzyme induction and hypoxia. *Anesthesiology* **51**: 321–326.

Mead, J.F. (1976). Free radical mechanisms of lipid damage and consequences for cellular membranes, in: *Free Radicals in Biology* (Pryor, W.A., ed.), Vol. I, pp. 51–68, Academic Press, New York.

Mees, G.C. (1960). Experiments on the herbicidal action of 1,1′-ethylene-2,2′-dipyridylium dibromide. *Ann. Appl. Biol.* **48**: 601–612.

Mergner, G.W., Weglicki, W.B., and Kramer, J.H. (1991). Postischemic free radical production in the venous blood of the regionally ischemic swine heart. Effect of deferoxamine. *Circulation* **84**: 2079–2090.

Michaelis, L., and Hill, E.S. (1933). Potentiometric studies on semiquinones. I. General and theoretical part. *J. Am. Chem. Soc.* **55**: 1481–1494.

Mikoyan, V.D., Kubrina, L.N., Serezhenkov, V.A., Stukan, R.A., and Vanin, A.F. (1997). Complexes of Fe^{2+} with diethyldithiocarbamate or N-methyl-D-glucamine dithiocarbamate as traps of nitric oxide in animal tissues: Comparative investigations. *Biochim. Biophys. Acta* **1336**: 225–234.

Miller, R.W., and Macdowall, F.D.H. (1975). The tiron free radical as a sensitive indicator of chloroplastic photoautoxidation. *Biochim. Biophys. Acta* **387**: 176–187.

Miura, Y., Hamada, A., and Utsumi, H. (1995). *In vivo* ESR studies of antioxidant activity on free radical reaction in living mice under oxidative stress. *Free Radical Res.* **22**: 209–214.

Miura, Y., Utsumi, H., and Hamada, A. (1992). Effects of inspired oxygen concentration on *in vivo* redox reaction of nitroxide radicals in whole mice. *Biochem. Biophys. Res. Commun.* **182**: 1108–1114.

Miyajima, T., and Kotake, Y. (1995). Spin trapping agent, phenyl N-*tert*-butyl nitrone, inhibits induction of nitric oxide synthase in endotoxin-induced shock in mice. *Biochem. Biophys. Res. Commun.* **215**: 114–121.

Miyajima, T., and Kotake, Y. (1997). Optimal time and dosage of phenyl N-*tert*-butyl nitrone (PBN) for the inhibition of nitric oxide synthase induction in mice. *Free Radical Biol. Med.* **22**: 463–470.

Moncada, S., Palmer, R.M.J., and Higgs, E.A. (1991). Nitric oxide: Physiology pathophysiology, and pharmacology. *Pharmacol. Rev.* **43**: 109–142.

Mönig, J., Krischer, K., and Asmus, K.-D. (1983). One-electron reduction of halothane and formation of halide ions in aqueous solutions. *Chem.-Biol. Interact.* **45**: 43–52.

Montgomery, M.R. (1977). Paraquat toxicity and pulmonary superoxide dismutase: An enzymic deficiency of lung microsomes. *Res. Commun. Chem. Path. Pharmacol.* **16**: 155–158.

Mordvintcev, P., Mülsch, A., Busse, R., and Vanin, A. (1991). On-line detection of nitric oxide formation in liquid aqueous phase by electron paramagnetic resonance spectroscopy. *Anal. Biochem.* **199**: 142–146.

Moreno, S.N.J., Docampo, R., Mason, R.P., Leon, W., and Stoppani, A.O.M. (1982). Different behaviors of benznidazole as free radical generator with mammalian and *Trypanosoma cruzi* microsomal preparations. *Arch. Biochem. Biophys.* **218**: 585–591.

Moreno, S.N.J., Mason, R.P., Muniz, R.P.A., Cruz, S.F., and Docampo, R. (1983). Generation of free radicals from metronidazole and other nitroimidazoles by *Tritrichomonas foetus*. *J. Biol. Chem.* **258**: 4051–4054.

Mottley, C., Connor, H.D., and Mason R.P. (1986). [^{17}O]Oxygen hyperfine structure for the hydroxyl and superoxide radical adducts of the spin traps, DMPO, PBN and 4-POBN. *Biochem. Biophys. Res. Commun.* **141**: 622–628.

Mülsch, A., Vanin, A., Mordvintcev, P., Hauschild, S., and Busse, R. (1992). NO accounts completely for the oxygenated nitrogen species generated by enzymic L-arginine oxygenation. *Biochem. J.* **288**: 597–603.

Murphy, M.E., and Noack, E. (1994). Nitric oxide assay using hemoglobin method. *Methods Enzymol.* **233**: 240–250.

Myers, C., McGuire, W.P., Liss, R.H., Ifrim, I., Grotzinger, K., and Young, R.C. (1977). Adriamycin: The role of lipid peroxidation in cardiac toxicity and tumor response. *Science* **197**: 165–167.

Myers, M.L., Bolli, R., Lekich, R.F., Hartley, C.J., and Roberts, R. (1985). Enhancement of recovery of myocardial function by oxygen free-radical scavengers after reversible regional ischemia. *Circulation* **72**: 915–921.

Nakagawa, I., Suzuki, M., Imura, N., and Naganuma, A. (1995). Enhancement of paraquat toxicity by glutathione depletion in mice *in vivo* and *in vitro*. *J. Toxicol. Sci.* **20**: 557–564.

Nakazawa, H., Ichimori, K., Shinozaki, Y., Okino, H., and Hori, S. (1988). Is superoxide demonstration by electron-spin resonance spectroscopy really superoxide? *Am. J. Physiol.* **255**: H213–H215.

Nazhat, N.B., Yang, G., Allen, R.E., Blake, D.R., and Jones, P. (1990). Does 3,5-dibromo-4-nitrosobenzene sulphonate spin trap superoxide radicals? *Biochem. Biophys. Res. Commun.* **166**: 807–812.

Noguchi, T., Fong, K.-L., Lai, E.K., Alexander, S.S., King, M.M., Olson, L., Poyer, J.L., and McCay, P.B. (1982). Specificity of a phenobarbital-induced cytochrome P-450 for metabolism of carbon tetrachloride to the trichloromethyl radical. *Biochem. Pharmacol.* **31**: 615–624.

Nohl, H., Stolze, K., Napetschnig, S., and Ishikawa, T. (1991). Is oxidative stress primarily involved in reperfusion injury of the ischemic heart? *Free Radical Biol. Med.* **11**: 581–588.

Norby, S.W., Weyhenmeyer, J.A., and Clarkson, R.B. (1997). Stimulation and inhibition of nitric oxide production in macrophages and neural cells as observed by spin trapping. *Free Radical Biol. Med.* **22**: 1–9.

North, J.A., Spector, A.A., and Buettner, G.R. (1992). Detection of lipid radicals by electron paramagnetic resonance spin trapping using intact cells enriched with polyunsaturated fatty acids. *J. Biol. Chem.* **267**: 5743–5746.

North, J.A., Spector, A.A., and Buettner, G.R. (1994). Cell fatty acid composition affects free radical formation during lipid peroxidation. *Am. J. Physiol.* **267**: C177–C188.

Numagami, Y., Marro, P.J., Mishra, O.P., and Delivoria-Papadopoulos, M. (1998). Effect of propentofylline on free radical generation during cerebral hypoxia in the newborn piglet. *Neuroscience* **84**: 1127–1133.

Ostdal, H., Skibsted, L.H., and Andersen, H.J. (1997). Formation of long-lived protein radicals in the reaction between H_2O_2-activated metmyoglobin and other proteins. *Free Radical Biol. Med.* **23**: 754–756.

Packer, J.E., Mahood, J.S., Willson, R.L., and Wolfenden, B.S. (1981). Reactions of the trichloromethyl peroxy free radical ($Cl_3COO\cdot$) with tryptophan, tryptophanyl-tyrosine and lysozyme. *Int. J. Radiat. Biol.* **39**: 135–141.

Packer, J.E., Slater, T.F., and Willson, R.L. (1978). Reactions of the carbon tetrachloride-related peroxy free radical ($CCl_3O_2\cdot$) with amino acids: Pulse radiolysis evidence. *Life Sci.* **23**: 2617–2620.

Paolini, M., Sapigni, E., Mesirca, R., Pedulli, G.F., Corongiu, F.P., Dessi, M.A., and Cantelli-Forti, G. (1992). On the hepatoxicity of 1,1,2,2-tetrachloroethane. *Toxicology* **73**: 101–115.

Paul, B.B., and Rubinstein, D. (1963). Metabolism of carbon tetrachloride and chloroform by the rat. *J. Pharmacol. Exp. Ther.* **141**: 141–148.

Paul, T., Hassan, M.A., Korth, H.-G., Sustmann, R., and Avila, D.V. (1996). Reaction of phenyl-substituted o-quinodimethanes with nitric oxide. Are benzo-

cyclobutenes suitable precursors for nitric oxide cheletropic traps? *J. Org. Chem.* **61**: 6835–6848.

Perez-Reyes, E., Kalyanaraman, B., and Mason, R.P. (1980). The reductive metabolism of metronidazole and ronidazole by aerobic liver microsomes. *Mol. Pharmacol.* **17**: 239–244.

Peterson, F.J., Mason, R.P., Hovsepian, J., and Holtzman, J.L. (1979). Oxygensensitive and -insensitive nitroreduction by *Escherichia coli* and rat hepatic microsomes. *J. Biol. Chem.* **254**: 4009–4014.

Pietri, S., Culcasi, M., and Cozzone, P.J. (1989). Real-time continuous-flow spin trapping of hydroxyl radical in the ischemic and post-ischemic myocardium. *Eur. J. Biochem.* **186**: 163–173.

Pincemail, J., Defraigne, J.O., Franssen, C., Defechereux, T., Canivet, J.-L., Philippart, C., and Meurisse, M. (1990). Evidence of *in vivo* free radical generation by spin trapping with α-phenyl N-*tert*-butyl nitrone during ischemia/reperfusion in rabbit kidneys. *Free Radical Res. Commun.* **9**: 181–186.

Plaa, G.L., Evans, E.A., and Hine, C.H. (1958). Relative hepatotoxicity of seven halogenated hydrocarbons. *J. Pharmacol. Exp. Ther.* **123**: 224–229.

Plummer, J.L., Beckwith, A.L.J., Bastin, F.N., Adams, J.F., Cousins, M.J., and Hall, P. (1982). Free radical formation *in vivo* and hepatotoxicity due to anesthesia and halothane. *Anesthesiology* **57**: 160–166.

Pohl, L.R., Branchflower, R.V., Highet, R.J., Martin, J.L., Nunn, D.S., Monks, T.J., George, J.W., and Hinson, J.A. (1981). The formation of diglutathionyl dithiocarbonate as a metabolite of chloroform, bromotrichloromethane and carbon tetrachloride. *Drug Metab. Dispos.* **9**: 334–339.

Poole, C.P. (1983). *Electron Spin Resonance: A Comprehensive Treatise on Experimental Techniques,* 2nd edition, John Wiley & Sons, New York.

Pou, S., and Rosen, G.M. (1990). Spin-trapping of superoxide by 5,5-dimethyl-1-pyrroline N-oxide: Application to isolated perfused organs. *Anal. Biochem.* **190**: 321–325.

Pou, S., and Rosen, G.M. (1998). Generation of thiyl radical by nitric oxide:

A spin trapping study. *J. Chem. Soc., Perkin Trans.* 2 1507–1512.

Pou, S., Anderson, D.E., Suricharmorn, W., Keaton, L.L., and Tod, M.L. (1994c). Biological studies of a nitroso compound that releases nitric oxide upon illumination. *Mol. Pharmacol.* **46**: 709–715.

Pou, S., Hassett, D.J., Britigan, B.E., Cohen, M.S. and Rosen, G.M. (1989a). Problems associated with spin trapping oxygen-centered free radicals in biological systems. *Anal. Biochem.* **177**: 1–6.

Pou, S., Keaton, L., Surichamorn, W., Frigillana, P., and Rosen, G.M. (1994b). Can nitric oxide be spin trapped by nitrone and nitroso compounds? *Biochim. Biophys. Acta* **1201**: 118–124.

Pou, S., Keaton, L., Suricharmorn, W., and Rosen, G.M. (1999a). Mechanism of superoxide generation by neuronal nitric oxide synthase. *J. Biol. Chem.* **274**: 9573–9580.

Pou, S., Nguyen, S.Y., Gladwell, T., and Rosen, G.M. (1995). Does peroxynitrite generate hydroxyl radical. *Biochim. Biophys. Acta* **1244**: 62–68.

Pou, S., Pou, W.S., Bredt, D.S., Snyder, S.H., and Rosen, G.M. (1992). Generation of superoxide by purified brain nitric oxide synthase. *J. Biol. Chem.* **267**: 24173–24176.

Pou, S., Ramos, C.L., Gladwell, T., Renks, E., Centra, M., Young, D., Cohen, M.S., and Rosen, G.M. (1994a). A kinetic approach to the selection of a sensitive spin trapping system for the detection of hydroxyl radical. *Anal. Biochem.* **217**: 76–83.

Pou, S., Rosen, G.M., Britigan, B.E., and Cohen, M.S. (1989b). Intracellular spin-trapping of oxygen-centered radicals generated by human neutrophils. *Biochim. Biophys. Acta* **991**: 459–464.

Pou, S., Rosen, G.M., Wu, Y., and Keana, J.F.W. (1990). Synthesis of deuterium- and ^{15}N-containing pyrroline-1-oxides: A spin trapping study. *J. Org. Chem.* **55**: 4438–4443.

Pou, S., Tsai, P., Porasuphatana, S., Halpern, H.J., Chandramouli, G.V.R., Barth, E.D., and Rosen, G.M. (1999b). Spin trapping of nitric oxide by ferrochelates: Kinetic and *in vivo*

pharmacokinetic studies. *Biochim. Biophys. Acta*, **1427**: 216–266.

Powis, G. (1989). Free radical formation by antitumor quinones. *Free Radical Biol. Med.* **6**: 63–101.

Poyer, J.L., Floyd, R.A., McCay, P.B., Janzen, E.G., and Davis, E.R. (1978). Spin-trapping of the trichloromethyl radical produced during enzymic NADPH oxidation in the presence of carbon tetrachloride or bromotrichloromethane. *Biochim. Biophys. Acta* **539**: 402–409.

Poyer, J.L., McCay, P.B., Lai, E.K., Janzen, E.G., and Davis, E.R. (1980). Confirmation of assignment of the trichloromethyl radical spin adduct detected by spin trapping during ^{13}C-carbon tetrachloride metabolism *in vitro* and *in vivo*. *Biochem. Biophys. Res. Commun.* **94**: 1154–1160.

Poyer, J.L., McCay, P.B, Weddle, C.C., and Downs, P.E. (1981). *In vivo* spin-trapping of radicals formed during halothane metabolism. *Biochem. Pharmacol.* **30**: 1517–1519.

Prónai, L., Ichimori, K., Nozaki, H., Nakazawa, H., Okino, H., Carmicheal, A.J., and Arroyo, C.M. (1991). Investigation of the existence and biological role of L-arginine/nitric oxide pathway in human platelets by spin trapping/EPR studies. *Eur. J. Biochem.* **202**: 923–930.

Przylenk, K., and Kloner, R.A. (1986). Superoxide dismutase plus catalase improve contractile function in the canine model of the "stunned myocardium." *Circ. Res.* **58**: 148–156.

Quaresima, V., Ursini, C.L., Gualtieri, G., Sotgiu, A., and Ferrari, M. (1993). Oxygen-dependent reduction of a nitroxide free radical by electron paramagnetic resonance monitoring of circulating rat blood. *Biochim. Biophys. Acta* **1182**: 115–118.

Rabinowitch, H.D., Rosen, G.M., and Fridovich, I. (1987). Contrasting fates of paraquat monocation radical in *Escherichia coli* B and in *Danaliella salina*. *Arch. Biochem. Biophys.* **257**: 352–356.

Radi, R. (1996). Reaction of nitric oxide with metalloproteins. *Chem. Res. Toxicol.* **9**: 828–835.

Ramos, C.L., Pou, S., Britigan, B.E., Cohen, M.S., and Rosen, G.M. (1992). Spin trapping evidence for myeloperoxidase-dependent hydroxyl radical formation by human neutrophils and monocytes. *J. Biol. Chem.* **267**: 8307–8312.

Rau, J.M., Reinke, L.A., and McCay, P.B. (1990). Direct observation of spin-trapped carbon dioxide radicals in hepatocytes exposed to carbon tetrachloride. *Free Radical Res. Commun.* **9**: 197–204.

Recknagel, R.O. (1967). Carbon tetrachloride hepatotoxicity. *Pharmacol. Rev.* **19**: 145–208.

Recknagel, R.O., and Ghoshal, A.K. (1966). Lipoperoxidation as a vector in carbon tetrachloride hepatotoxicity. *Lab. Invest.* **15**: 132–148.

Recknagel, R.O., and Glende, E.A., Jr. (1973). Carbon tetrachloride hepatotoxicity. An example of lethal cleavage. *CRC Crit. Rev. Toxicol.* **2**: 263–297.

Recknagel, R.O., Glende, E.A., Jr., and Hruszkewycz, A.M. (1977). Chemical mechanisms in carbon tetrachloride toxicity, in: *Free Radicals in Biology* (Pryor, W.A., ed.), Vol. III, pp. 97–132, Academic Press, New York.

Reinke, L.A., and Janzen, E.G. (1991). Detection of spin adducts in blood after administration of carbon tetrachloride to rats. *Chem.-Biol. Interact.* **78**: 155–165.

Reinke, L.A., Kotake, Y., McCay, P.B., and Janzen, E.G. (1991). Spin-trapping studies of hepatic free radicals formed following the acute administration of ethanol to rats: *In vivo* detection of 1-hydroxyethyl radicals with PBN. *Free Radical Biol. Med.* **11**: 31–39.

Reinke, L.A., Moore, D.R., and Kotake, Y. (1996). Hepatic nitric oxide formation: Spin trapping detection in biliary efflux. *Anal. Biochem.* **243**: 8–14.

Reinke, L.A., Towner, R.A., and Janzen, E.G. (1992). Spin trapping of free radical metabolites of carbon tetrachloride *in vitro* and *in vivo*: Effect of acute ethanol administration. *Toxicol. Appl. Pharmacol.* **112**: 17–23.

Reiter, R., and Burk, R.F. (1988). Formation of glutathione adducts of carbon tetrachloride metabolites in a rat liver microsomal incubation system. *Biochem. Pharmacol.* **37**: 327–331.

Reszka, K.J., Bilski, P., and Chignell, C.F. (1996). EPR and spin trapping investi-

gation of nitric oxide (·NO) from UV irradiated nitrite anions in alkaline aqueous solutions. *J. Am. Chem. Soc.* **118**: 8719–8720.

Reszka, K.J., Chignell, C.F., and Bilski, P. (1994). Spin trapping of nitric oxide (·NO) by *aci*-nitromethane in aqueous solutions. *J. Am. Chem. Soc.* **116**: 4119–4120.

Reynolds, E.S., and Moslen, M.T. (1980). Free-radical damage in liver, in: *Free Radicals in Biology* (Pryor, W.A., ed.), Vol. IV, pp. 49–94, Academic Press, New York.

Rogan, E., and Cavalieri, E. (1978). Differences between nuclear and microsomal cytochrome P-450 in uninduced and induced rat liver. *Mol. Pharmacol.* **14**: 215–219.

Romero, F.J., Ordoñez, I., Arduini, A., and Cadenas, E. (1992). The reactivity of thiols and disulfides with different redox states of myoglobin. Redox and addition reactions and formation of thiyl radical intermediates. *J. Biol. Chem.* **267**: 1680–1688.

Roots, R., and Okada, S. (1972). Protection of DNA molecules of cultured mammalian cells from radiation-induced single strand scissions by various alcohols and SH compounds. *Int. J. Radiat. Biol.* **21**: 329–342.

Rosen, G.M., and Freeman, B.A. (1984). Detection of superoxide generated by endothelial cells. *Proc. Natl. Acad. Sci. USA* **81**: 7269–7273.

Rosen, G.M., and Rauckman, E.J. (1977). Formation and reduction of a nitroxide radical by liver microsomes. *Biochem. Pharmacol.* **26**: 675–678.

Rosen, G.M., and Rauckman, E.G. (1981). Spin trapping of free radicals during hepatic microsomal lipid peroxidation. *Proc. Natl. Acad. Sci. USA*, **78**: 7346–7349.

Rosen, G.M., and Rauckman, E.G. (1982). Carbon tetrachloride-induced lipid peroxidation: A spin trapping study. *Toxicol. Lett.* **10**: 337–344.

Rosen, G.M., and Turner, M.J., III. (1988). Synthesis of spin traps specific for hydroxyl radical. *J. Med. Chem.* **31**: 428–432.

Rosen, G.M., Halpern, H.J., Brunsting, L.A., Spencer, D.P., Strauss, K.E., Bowman, M.J., and Wechsler, A.S.

(1988). Direct measurement of nitroxide pharmacokinetics in isolated hearts situated in a low-frequency electron spin resonance spectrometer: Implications for spin trapping and *in vivo* oxymetry. *Proc. Natl. Acad. Sci. USA* **85**: 7772–7776.

Rosen, G.M., Hassett, D.J., Yankaskas, J.R., and Cohen, M.S. (1989). Detection of free radicals as a consequence of dog tracheal epithelial cellular xenobiotic metabolism. *Xenobiotica* **19**: 635–643.

Rosen, G.M., Rauckman, E.J., Wilson, R.L., Jr., and Tschanz, C. (1984). Production of superoxide during the metabolism of nitrazepam. *Xenobiotica*, **14**: 785–794.

Rosen, H., and Klebanoff, S.J. (1979). Hydroxyl radical generation by polymorphonuclear leukocytes measured by electron spin resonance spectroscopy. *J. Clin. Invest.* **64**: 1725–1729.

Ross, D., Norbeck, K., and Moldéus, P. (1985). The generation and subsequent fate of glutathionyl radicals in biological systems. *J. Biol. Chem.* **260**: 15028–15032.

Ross, W.T., Daggy, B.P., and Cardell, R.R. (1979). Hepatic necrosis caused by halothane and hypoxia in phenobarbital-treated rats. *Anesthesiology* **51**: 327–333.

Rossi, L., De Angelis, I., Pedersen, J.Z., Marchese, E., Stammati, A., Rotilio, G., and Zucco, F. (1996). N-[5-nitro-2-furfurylidene]-3-amino-2-oxazolidinone activation by human intestinal cell line caco-2 monitored noninvasive electron spin resonance spectroscopy. *Mol. Pharmacol.* **49**: 547–555.

Rothe, G., and Valet, G. (1990). Flow cytometric analysis of respiratory burst activity in phagocytes with hydroethidine and 2′,7′-dichlorofluorescin. *J. Leukocyte Biol.* **47**: 440–448.

Rouband, V., Sankarapandi, S., Kuppusamy, P., Tordo, P., and Zweier, J.L. (1997). Quantitative measurement of superoxide generation using the spin trap 5-(diethoxyphosphoryl)-5-methyl-1-pyrroline N-oxide. *Anal. Biochem.* **247**: 404–411.

Roubaud, V., Sankarapandi, S., Kuppusamy, P., Tordo, P., and Zweier, J.L. (1998). Quantitative measurement of

superoxide generation and oxygen consumption from leukocytes using electron paramagnetic resonance spectroscopy. *Anal. Biochem.* **257**: 210–217.

Russell, G.A., and Bemis, A.G. (1967). Preparation and reactions of potassium nitrobenzenide. *Inorg. Chem.* **6**: 403–406.

Ryan, T.C., Weil, G.J., Newburger, P.E., Haugland, R., and Simons, E.R. (1990). Measurement of superoxide release in the phagovacuoles of immune complex-stimulated human neutrophils. *J. Immunol. Methods* **130**: 223–233.

Saito, K., Yoshioka, H., Kazama, S., and Cutler, R.G. (1998). Release of nitric oxide from a spin trap, N-*tert*-butyl-α-phenylnitrone, under various oxidative conditions. *Biol. Pharmac. Bull.* **21**: 401–404.

Salvador, A., Antunes, F., and Pinto, R.E. (1995). Kinetic modelling of *in vitro* lipid peroxidation experiments—"low level" validation of a model of *in vivo* lipid peroxidation. *Free Radical Res.* **23**: 151–172.

Samuni, A., Carmichael, A.J., Russo, A., Mitchell, J.B., and Reisz, P. (1986). On the spin trapping and ESR detection of oxygen-derived free radicals generated inside cells. *Proc. Natl. Acad. Sci. USA* **83**: 7593–7597.

Samuni, A., Krishna, C.M., Reisz, P., Finkelstein, E., and Russo, A. (1989a). Superoxide reaction with nitroxide spin adducts. *Free Radical Biol. Med.* **6**: 141–148.

Samuni, A., Samuni, A., and Swartz, H.M. (1989b). Evaluation of dibromonitrosobenzene sulfonate as a spin trap in biological systems. *Free Radical Biol. Med.* **7**: 37–43.

Sanderson, D.G., and Chedekel, M.R. (1980). Spin trapping of the superoxide radical by 4-(N-methylpyridinium) *t*-butyl nitrone. *Photochem. Photobiol.* **32**: 573–576.

Sandy, M.S., Moldeus, P., Ross, D., and Smith, M.T. (1987). Cytotoxicity of the redox cycling compound diquat in isolated hepatocytes: Involvement of hydrogen peroxide and transition metals. *Arch. Biochem. Biophys.* **259**: 29–37.

Sasame, H.A., and Gillette, J.R. (1969). Studies on the relationship between the effects of various substances on absorption spectrum of cytochrome P-450 and the reduction of p-nitrobenzoate by mouse liver microsomes. *Mol. Pharmacol.* **5**: 123–130.

Sato, S., Tominaga, T., Ohnishi, T., and Ohnishi, S.T. (1993). EPR spin-trapping study of nitric oxide formation during bilateral carotid occlusion in the rat. *Biochim. Biophys. Acta* **1181**: 195–197.

Schreiber, J., Foureman, G.L., Hughes, M.F., Mason, R.P., and Eling, T.E. (1989). Detection of glutathione thiyl free radical catalyzed by prostaglandin H synthase present in keratinocytes. Study of cooxidation in a cellular system. *J. Biol. Chem.* **264**: 7936–7943.

Schreiber, J., Mottley, C., Sinha, B.K., Kalyanaraman, B., and Mason, R.P. (1987). One-electron reduction of daunomycin, daunomycinone, and 7-deoxydaunomycinone by the xanthine/xanthine oxidase system: Detection of semiquinone free radicals by electron spin resonance. *J. Am. Chem. Soc.* **109**: 348–351.

Sealy, R.C., Swartz, H.M., and Olive, P.L. (1978). Electron spin resonance-spin trapping. Detection of superoxide formation during aerobic microsomal reduction of nitro-compounds. *Biochem. Biophys. Res. Commun.* **82**: 680–684.

Sekili, S., McCay, P.B., Li, X.-Y., Zughaib, M., Sun, J.-Z., Tang, L., Thornby, J.I., and Bolli, R. (1993). Direct evidence that the hydroxyl radical plays a pathogenetic role in myocardial "stunning" in the conscious dog and demonstration that stunning can be markedly attenuated without subsequent adverse effects. *Circ. Res.* **73**: 705–723.

Sen, S., and Phillis, J.W. (1993). Alpha-phenyl-tert-butyl-nitrone (PBN) attenuates hydroxyl radical production during ischemia-reperfusion injury of rat brain: An EPR study. *Free Radical Res. Commun.* **19**: 255–265.

Sentjurc, M., and Mason, R.P. (1992). Inhibition of radical adduct reduction and reoxidation of the corresponding hydroxylamine in *in vivo* spin trapping of carbon tetrachloride-derived radi-

cals. *Free Radical Biol. Med.* **13**: 151–160.

Shinobu, L.A., Jones, S.G., and Jones, M.M. (1984). Sodium N-methyl-D-glucamine dithiocarbamate and cadmium intoxication. *Acta Pharmacol. Toxicol.* **54**: 189–194.

Shu, H., Talcott, R.E., Rice, S.A., and Wei, E.T. (1979). Lipid peroxidation and paraquat toxicity. *Biochem. Pharmacol.* **28**: 327–331.

Shutenko, Z., Henry, Y., Pinard, E., Seylaz, J., Potier, P., Berthet, F., Girard, P., and Sercombe, R. (1999). Influence of the antioxidant quercetin *in vivo* on the level of nitric oxide determined by electron paramagnetic resonance in rat brain during global ischemia and reperfusion. *Biochem. Pharmacol.* **57**: 199–208.

Siems, W., Kowalewski, J., Werner, A., Schimke, I., and Gerber, G. (1989). Radical formation in the rat small intestine during and following ischemia. *Free Radical Res. Commun.* **7**: 347–353.

Singh, R.J., Hogg, N., Joseph, J., and Kalyanaraman, B. (1996). Mechanism of nitric oxide release from S-nitrosothiols. *J. Biol. Chem.* **271**: 18596–18603.

Sipes, I.G., Krishna, G., and Gillette, J.R. (1977). Bioactivation of carbon tetrachloride, chloroform and bromotrichloromethane. Role of cytochrome P-450. *Life Sci.* **20**: 1541–1548.

Slater, T.F. (1966). Necrogenic action of carbon tetrachloride in the rat. A speculative mechanism based on activation. *Nature* **209**: 36–40.

Steel-Goodwin, L., Pravecek, T.L., and Carmichael, A.J. (1996). Trichloroethylene metabolism *in vitro*: An EPR/spin trapping study. *Hum. Exp. Toxicol.* **15**: 878–884.

Steffen, C., and Netter, K.J. (1979). On the mechanism of paraquat action on microsomal oxygen reduction and its relation to lipid peroxidation. *Toxicol. Appl. Pharmacol.* **47**: 593–602.

Stier, A., and Reitz, I. (1971). Radical production in amine oxidation by liver microsomes. *Xenobiotica* **1**: 499–500.

Stolc, S., Valko, L., Valko, M., and Lombardi, V. (1996). A technique for the fast sampling of biological tissues for electron paramagnetic resonance spectroscopy. *Free Radical Biol. Med.* **20**: 89–91.

Stoyanovsky, D.A., Goldman, R., Jonnalagadda, S.S., Day, B.W., Claycamp, H.G., and Kagan, V.E. (1996). Detection and characterization of the electron paramagnetic resonance-silent glutathionyl-5,5-dimethylpyrroline N-oxide adduct derived from redox cycling of phenoxyl radicals in model systems and HL-60 cells. *Arch. Biochem. Biophys.* **330**: 3–11.

Sutton, H.C., and Winterbourn, C.C. (1984). Chelated iron-catalyzed HO· formation from paraquat radicals and H_2O_2: Mechanism of formate oxidation. *Arch. Biochem. Biophys.* **235**: 106–115.

Suzuki, Y., Fujii, S., Numagami, Y., Tominaga, T., Yoshimoto, T., and Yoshimura, T. (1998). *In vivo* nitric oxide detection in the septic rat brain by electron paramagnetic resonance. *Free Radical Res.* **28**: 293–299.

Swartz, H.M. (1990). Principles of the metabolism of nitroxides and their implications for spin trapping. *Free Radical Res. Commun.* **9**: 399–405.

Symons, M.C.R., Albano, E., Slater, T.F., and Tomasi, A. (1982). Radiolysis of tetrachloromethane. *J. Chem. Soc., Faraday Trans. 1* **78**: 2205–2214.

Tamura, Y., Chi, L., Driscoll, E.M., Jr., Hoff, P.T., Freeman, B.A., Gallagher, K.P., and Lucchesi, B.R. (1988). Superoxide dismutase conjugated to polyethylene glycol provides sustained protection against myocardial ischemia/reperfusion injury in canine heart. *Circ. Res.* **63**: 944–959.

Thor, H., Smith, M.T., Hartzell, P., Bellomo, G., Jewell, S.A., and Orrenius, S. (1982). The metabolism of menadione (2-methyl-1,4-naphthoquinone) by isolated hepatocytes. A study of the implications of oxidative stress in intact cells. *J. Biol. Chem.* **257**: 12419–12425.

Timmins, G.S., and Davies, M.J. (1993). Free radical formation in murine skin treated with tumour promoting organic peroxides. *Carcinogenesis* **14**: 1499–1503.

Tomasi, A., Albano, E., Biasi, F., Slater, T.F., Vannini, V., and Dianzani, M.U. (1985). Activation of chloroform and related trihalomethanes to free radical intermediates in isolated hepatocytes and in the rat *in vivo* as detected by the ESR-spin trapping technique. *Chem.-Biol. Interact.* **55**: 303–316.

Tomasi, A., Albano, E., Diazani, M.U., Slater, T.F., and Vannini, V. (1983a). Metabolic activation of 1,2-dibromoethane to a free radical intermediate by rat liver microsomes and isolated hepatocytes. *FEBS Lett.* **160**: 191–194.

Tomasi, A., Albano, E., Lott, K.A.K., and Slater, T.F. (1980). Spin trapping of free radicals products of CCl_4 activation using pulse radiolysis and high energy radiation procedures. *FEBS Lett.* **122**: 303–306.

Tomasi, A., Billing, S., Garner, A., Slater, T.F., and Albano, E. (1983b). The metabolism of halothane by hepatocytes: A comparison between free radical spin trapping and lipid peroxidation in relation to cell damage. *Chem.-Biol. Interact.* **46**: 353–368.

Tominaga, T., Sato, S., Ohnishi, T., and Ohnishi, S.T. (1993). Potentiation of nitric oxide formation following bilateral carotid occlusion and focal cerebral ischemia in the rat. *In vivo* detection of the nitric oxide radical by electron paramagnetic resonance spin trapping. *Brain Res.* **614**: 342–346.

Tominaga, T., Sato, S., Ohnishi, T., and Ohnishi, S.T. (1994). Electron paramagnetic resonance (EPR) detection of nitric oxide produced during forebrain ischemia of the rat. *J. Cereb. Blood Flow Metab.* **14**: 715–722.

Tosaki, A., Bagchi, D., Pali, T., Cordis, G.A., and Das, D.K. (1993). Comparisons of ESR and HPLC methods for the detection of ·OH radicals in ischemic/reperfused hearts. A relationship between the genesis of free radicals and reperfusion arrhythmias. *Biochem. Pharmacol.* **45**: 961–969.

Tosaki, A., Blasig, I.E., Pali, T., and Ebert, B. (1990). Heart protection and radical trapping by DMPO during reperfusion in isolated working hearts. *Free Radical Biol. Med.* **8**: 363–372.

Triana, J.F., Li, X.-Y., Jamaluddin, U., Thornby, J.I., and Bolli, R. (1991). Postischemic myocardial "stunning." Identification of major differences between the open-chest and the conscious dog and evaluation of the oxygen radical hypothesis in the conscious dog. *Circ. Res.* **69**: 731–747.

Trudell, J.R. (1987). Ethyl acetate extraction of spin-trapped free radicals: A reevaluation. *Free Radical Biol. Med.* **3**: 133–136.

Trudell, J.R., Bösterling, B., and Trevor, A.J. (1981). 1-Chloro-2,2,2-trifluoroethyl radical: Formation from halothane by human cytochrome P-450 in reconstituted vesicles and binding to phospholipids. *Biochem. Biophys. Res. Commun.* **102**: 372–377.

Trudell, J.R., Bösterling, B., and Trevor, A.J. (1982). Reductive metabolism of halothane by human and rabbit cytochrome P-450. Binding of 1-chloro-2,2,2-trifluoroethyl radical to phospholipids. *Mol. Pharmacol.* **21**: 710–717.

Trush, M.A., Mimnaugh, E.G., Ginsburg, E., and Gram, T.E. (1981). *In vitro* stimulation by paraquat of reactive oxygen-mediated lipid peroxidation in rat lung microsomes. *Toxicol. Appl. Pharmacol.* **60**: 279–286.

Tsuchiya, K., Takasugi, M., Minakuchi, K., and Fukuzawa, K. (1996). Sensitive quantitation of nitric oxide by EPR spectroscopy. *Free Radical Biol. Med.* **21**: 733–737.

Turner, M.J., III, Bozarth, C.H., and Strauss, K.E. (1989). Evidence for intracellular superoxide formation following the exposure of guinea pig enterocytes to bleomycin. *Biochem. Pharmacol.* **38**: 85–90.

Ullman, E.F., Osiecki, J.H., Boocock, G.B., and Darcy, R. (1972). Studies of stable free radicals. X. Nitronyl nitroxide monoradicals and biradicals as possible molecule spin labels. *J. Am. Chem. Soc.* **94**: 7049–7059.

Utsumi, H., Ichikawa, K., and Takeshita, K. (1995). *In vivo* ESR measurements of free radical reactions in living mice. *Toxicol. Lett.* **82/83**: 561–565.

van der Kraaij, A.M.M., Koster, J.F., and Hagen, W.R. (1989). Reappraisal of the e.p.r. signals in (post)-ischaemic cardiac tissue. *Biochem. J.* **264**: 687–694.

Vanin, A.F., Mordvintcev, P.I., Hauschildt, S., and Mülsch, A. (1993). The relationship between L-arginine-dependent nitric oxide synthesis, nitrite release and dinitrosyl-iron complex formation by activated macrophages. *Biochim. Biophys. Acta* **1177**: 37–42.

Vanin, A.F., Mordvintcev, P.I., and Kleschev, A.L. (1984). Nitric oxide incorporation in animal tissues *in vivo*. *Stud. Biophys.* **102**: 135–143.

Vásquez-Vivar, J., Hogg, N., Pritchard, K.A. Jr., Martasek, P., and Kalanaraman, B. (1997). Superoxide anion formation from lucigenin: An electron spin resonance spin-trapping study. *FEBS Lett.* **403**: 127–130.

Vásquez-Vivar, J., Kalyanaraman, B., Martásek, P., Hogg, N., Masters, B.S.S., Karoui, H., Tordo, P., and Pritchard, K.A., Jr. (1998). Superoxide generation by endothelial nitric oxide synthase: The influence of cofactors. *Proc. Natl. Acad. Sci. USA* **95**: 9220–9225.

Wagner, B.A., Buettner, G.R., and Burns, C.P. (1993). Increased generation of lipid-derived and ascorbate free radicals by L1210 cells exposed to the ether lipid edelfosine. *Cancer Res.* **53**: 711–713.

Wagner, B.A., Buettner, G.R., and Burns, C.P. (1994). Free radical-mediated lipid peroxidation in cells: Oxidizability is a function of cell lipid *bis*-allylic hydrogen content. *Biochemistry* **33**: 4449–4453.

Wagner, B.A., Buettner, G.R., Oberley, L.W., and Burns, C.P. (1998). Sensitivity of K562 and HL-60 cells to edelfosine, an ether lipid drug, correlates with production of reactive oxygen species. *Cancer Res.* **58**: 2809–2816.

Wardman, P., and Clarke, E.D. (1976). Oxygen inhibition of nitroreductase: Electron transfer from nitro radical-anions to oxygen. *Biochem. Biophys. Res. Commun.* **69**: 942–949.

Weiner, L.M. (1994). Oxygen radicals generation and DNA scission by anticancer and synthetic quinones. *Methods Enzymol.* **233**: 92–105.

Weiss, J. (1944). Radiochemistry of aqueous solutions. *Nature* **153**: 748–750.

White, R.E. (1994). The importance of one-electron transfers in the mechanism of cytochrome P-450, in: *Cytochrome P450 Biochemistry, Biophysics and Molecular Biology* (Lechner, M.C., ed.), pp. 333–340, John Libbey Eurotext, Paris.

Willson, R.L., and Searle, A.J.F. (1975). Metronidazole (Flagyl): Iron catalysed reaction with sulphydryl groups and tumour radiosensitisation. *Nature* **255**: 498–500.

Winterbourn, C.C., and Sutton, H.C. (1984). Hydroxyl radical production from hydrogen peroxide and enzymatically generated paraquat radicals: Catalytic requirements and oxygen dependence. *Arch. Biochem. Biophys.* **215**: 116–126.

Woldman, Y.Y., Khramtsov, V.V., Grigor'ev, I.A., Kiriljuk, I.A., and Utepbergenov, D.I. (1994). Spin trapping of nitric oxide by nitronyl-nitroxides: Measurement of the activity of NO synthase from rat cerebellum. *Biochem. Biophys. Res. Commun.* **202**: 195–203.

Wolf, C.R., Harrelson, W.G., Jr., Nastainczyk, W.M., Philphot, R.M., Kalyanaraman, B., and Mason, R.P. (1980). Metabolism of carbon tetrachloride in hepatic microsomes and reconstituted monooxygenase systems and its relationship to lipid peroxidation. *Mol. Pharmacol.* **18**: 553–558.

Xia, Y., and Zweier, J.L. (1997). Superoxide and peroxynitrite generation from inducible nitric oxide synthase in macrophages. *Proc. Natl. Acad. Sci. USA* **94**: 6954–6958.

Xia, Y., Dawson, V.L., Dawson, T.M., Snyder, S.H., and Zweier, J.L. (1996). Nitric oxide synthase generates superoxide and nitric oxide in arginine-depleted cells leading to peroxynitrite-mediated cellular injury. *Proc. Natl. Acad. Sci. USA* **93**: 6770–6774.

Xia, Y., Roman, L.J., Masters, B.S.S., and Zweier, J.L. (1998a). Inducible nitric-oxide synthase generates superoxide from the reductase domain. *J. Biol. Chem.* **273**: 22635–22639.

Xia, Y., Tsai, A.-L., Berka, V., and Zweier, J.L. (1998b). Superoxide generation from enthothelial nitric-oxide synthase. A Ca^{2+}/calmodulin-dependent and tetrahydrobiopterin regulatory process. *J. Biol. Chem.* **273**: 25804–25808.

Yamada, T., Niki, E., Yokoi, S., Tsuchiya, J., Yamamoto, Y., and Kamiya, Y. (1984). Oxidation of lipids. XI. Spin

trapping and identification of peroxy and alkoxy radicals of methyl linoleate. *Chem. Phys. Lipids* **36**: 189–196.

Yokoyama, H., Ogata, T., Tsuchihashi, N., Hiramatsu, M., and Mori, N. (1996). A spatiotemporal study on the distribution of intraperitoneally injected nitroxide radical in the rat head using an *in vivo* ESR imaging system. *Magn. Reson. Imaging* **14**: 559–563.

Yoshikawa, T., Ueda, S., Naito, Y., Takahashi, S., Oyamada, H., Morita, Y., Yoneta, T., and Kondo, M. (1989). Role of oxygen-derived free radicals in gastric mucosal injury induced by ischemia or ischemia-reperfusion in rats. *Free Radical Res. Commun.* **7**: 285–291.

Yoshimura, T., Fujii, S., Yokoyama, H., and Kamada, H. (1995). *In vivo* electron paramagnetic resonance imaging of NO-bound iron complex in a rat head. *Chem. Lett.* 309–310.

Yoshimura, T., Yokoyama, H., Fujii, S., Takayama, F., Oikawa, K., and Kamada, H. (1996). *In vivo* EPR detection and imaging of endogenous nitric oxide in lipopolysaccharide-treated mice. *Nat. Biotechnol.* **14**: 992–994.

Zhang, L.-Y., van Kuijk, F.J.G.M., and Misra, H.P. (1995). EPR studies of spin-trapped free radicals in paraquat-treated lung microsomes. *Biochem. Mol. Biol. Int.* **37**: 255–262.

Zini, I., Tomasi, A., Grimaldi, R., Vannini, V., and Agnati, L.F. (1992). Detection of free radicals during brain ischemia and reperfusion by spin trapping and microdialysis. *Neurosci. Lett.* **138**: 279–282.

Zweier, J.L. (1988). Measurement of superoxide-derived free radicals in the reperfused heart. Evidence for a free radical mechanism of reperfusion injury. *J. Biol. Chem.* **263**: 1353–1357.

Zweier, J.L., Broderick, R., Kuppusamy, P., Thompson-Gorman, S., and Lutty, G.A. (1994). Determination of the mechanism of free radical generation in human aortic endothelial cells exposed to anoxia and reoxygenation. *J. Biol. Chem.* **269**: 24156–24162.

Zweier, J.L., Flaherty, J.T., and Weisfeldt, M.L. (1987). Direct measurement of free radical generation following reperfusion of ischemic myocardium. *Proc. Natl. Acad. Sci. USA* **84**: 1404–1407.

Zweier, J.L., Kuppusamy, P., and Lutty, G.A. (1988). Measurement of endothelial cells free radical generation: Evidence for a central mechanism of free radical injury in postischemic tissues. *Proc. Natl. Acad. Sci. USA* **85**: 4046–4050.

Zweier, J.L., Kuppusamy, P., Williams, R., Rayburn, B.K., Smith, D., Weisfeldt, M.L., and Flaherty, J.T. (1989). Measurement and characterization of postischemic free radical generation in the isolated perfused heart. *J. Biol. Chem.* **264**: 18890–18895.

Zweier, J.L., Wang, P., and Kuppusamy, P. (1995). Direct measurement of nitric oxide generation in the ischemic heart using electron paramagnetic resonance spectroscopy. *J. Biol. Chem.* **270**: 304–307.

11

The Pharmacological Activity of Spin Traps

The role of PBN as an effective antioxidant has been well established in the recent literature. Our studies demonstrate PBN has a direct reversible dose-dependent effect on diaphragm force production that is not typical of other antioxidants.

Andersen, Diaz, Wright, and Clanton, 1996

An underlying theme of this book is that spin traps, by reacting with free radicals, compete with biologic targets for these reactive species. As such, nitrones and nitrosoalkanes can be perceived as inhibitors, interrupting the cascade of reactions at specific sites (Kalyanaraman, *et al.*, 1991, 1993; Thomas, *et al.*, 1994a, 1994b). Consider, for example, the competition between nitrosoalkanes and O_2 for the carbon-centered free radical derived from either linoleic acid (Aoshima, *et al.*, 1977) or arachidonic acid (Mason, *et al.*, 1980). The efficiency of these processes is, however, not only dependent upon differing rates of reactions, but is also linked to the efficacious behavior of the spin trap toward a particular biologic target (Aoshima, *et al.*, 1977; Mason, *et al.*, 1980; McCormick, *et al.*, 1995). Whether spin traps ameliorate pathologic conditions by scavenging free radicals or by attenuating specific biologic actions, these compounds are now being considered as therapeutic agents about to take their place in the pharmacopoeia of drug therapy (Carney and Floyd, 1991a, 1995; Floyd and Carney, 1991a; Floyd, *et al.*, 1996). In this chapter, we will begin to investigate these rather sparse suggestions of pharmacological activity by exploring the various roles that spin traps are purported to play in modulating physiologic responses.

Effect of Spin Traps on *In Vitro* Enzymatic Activity

Notwithstanding the importance of enzymes as model free radical-generating systems, there is a dearth of data documenting the effects of spin traps on the enzymatic formation of these reactive species and subsequent reactions. Of the few examples that do exist, one of the earliest suggests that nitrones such as DMPO, TMPO, and 4-POBN, even at concentrations in excess of 100 mM, do not adversely affect the ability of xanthine oxidase to convert xanthine to uric acid (Finkelstein, *et al.*, 1979, 1980). Similar observations have been noted for NADPH–cytochrome P-450 reductase (Kalyanaraman, *et al.*, 1980) and prostaglandin synthase (Mottley, *et al.*, 1982). In contrast, DBNBS, but not DMPO, PBN, or 4-POBN, was found to be a potent inhibitor of purified brain nitric oxide synthase (NOS I) (Pou, *et al.*, 1994a).

There have been a number of studies aimed at determining whether spin traps are inhibitors of cytochrome P-450. For instance, DMPO has been found to antagonize hepatic microsomal oxidation of ethanol, whereas this Δ^1-pyrroline-*N*-oxide not did not inhibit the demethylation of aminopyrine or the hydroxylation of aniline (Cederbaum and Cohen, 1980). Similarly, this family of nitrones has been shown to exhibit marked effects directly on cytochrome P-450 (Albano, *et al.*, 1982, 1986, 1989; Augusto, *et al.*, 1982; Ortiz de Montellano, *et al.*, 1983), as these compounds display type I and type II binding to this hemoprotein (Albano, *et al.*, 1986). Notably, the degree of inhibition is dependent upon the substrate and the lipophilic character of the spin trap (Hill and Thornalley, 1983, Thomas, *et al.*, 1994). Since cytochrome P-450 catalyzes reactions in a lipid matrix (Graham-Lorence and Peterson, 1996), it is striking that the hydrophilic spin traps DMPO and 4-POBN are considerably weaker antagonists of this enzyme than is the more lipophilic nitrone PBN (Augusto, *et al.*, 1982). In contrast to potential concerns about inhibitory effects of enzyme function, 4-POBN has been found to enhance the oxidation of deferoxamine by a number of different peroxidases (McCormick, *et al.*,1995).

Spin traps, by virtue of their reaction with free radicals, can be considered functional inhibitors of pathologic events. By way of illustration, PBN has been found to significantly arrest CCl_4-induced lipid peroxidation, even though this nitrone did not alter the rate of NADPH oxidation and presumably the formation of $\cdot CCl_3$ (Wolf, *et al.*, 1980). Similar results have been noted *in vivo* (Janzen, *et al.*, 1990; Towner, *et al.*, 1993). More recent studies suggest that antioxidant characteristics of PBN are the result of the ability of this nitrone to prevent, rather than inhibit, peroxidation of lipids (Janzen, *et al.*, 1994). In some cases, decomposition of the nitrone might result in by-products that exhibit pharmacological activity. Such is the case with PBN, which upon photolysis or oxidation can lead to NO· (Chamulitrat, *et al.*, 1993, 1995; Saito, *et al.*, 1998a).

Effect of Spin Traps on Cellular Function

Even though spin traps have been shown to alter enzymic responses in models as sophisticated as isolated organelles from a variety of eucarya, efforts to catalog similar reactions in whole-cell preparations have been surprisingly lacking in the literature of this discipline. Few of these inaugural studies explored in depth whether spin traps impinge on cellular function (Hill and Thornalley, 1982, 1993; Bannister and Thornalley, 1983; Thornalley, *et al.*, 1983; Rosen and Freeman, 1984; Albano, *et al.*, 1985; Morgan, *et al.*, 1985; Thornalley and Stern,

1985; Mansbach, *et al.*, 1986; Samuni, *et al.*, 1986; Britigan, *et al.*, 1992; Reinke, *et al.*, 1992; Anderson, *et al.*, 1993). There are occasional inquiries into specific pharmacological and toxicological activity attributed to nitrosoalkanes and nitrones with DMPO, for instance, displaying the least toxicity among the group tested (Morgan, *et al.*, 1985; Konorev, *et al.*, 1993; Chen, *et al.*, 1995a; Janzen, *et al.*, 1995; Schaefer, *et al.*, 1996; Haseloff, *et al.*, 1997), despite the general lack of mutagenicity associated with spin traps (Hampton, *et al.*, 1981). A few cellular and *in situ* distributional studies have correlated pharmacological protection with compartmentalization (Hill and Thornalley, 1983; Cova, *et al.*, 1992). And infrequent investigations have described the sensitivity of spin traps toward enzymic metabolism (Samuni, *et al.*, 1989b; Pou, *et al.*, 1994a). But, for the most part, the impact of spin traps on cellular function remains unexplored territory.

Surprisingly, given the fact that phagocytes, by way of their physiologic role in host immunity, generate $O_2 \cdot^-$ and other free radicals (Bannister and Bannister, 1985; Rosen, *et al.*, 1995), there is a scarcity of papers which investigate the influence of spin traps on leukocyte responses to various stimuli. Rather, the prevailing view has been that spin traps are benign probes, acting merely as reporters of free radical events. Until recently, most publications reflected this narrow view, and focused almost exclusively on characterizing the spin trapped adduct and the kinetics of this and subsequent free radical reactions, independent of the mechanism of granulocyte activation (Green, *et al.*, 1979; Rosen and Klebanoff, 1979; Arthur, *et al.*, 1981; Docampo, *et al.*, 1983; Hawley, *et al.*, 1983; Hume, *et al.*, 1983; Britigan, *et al.*, 1986a, 1986b, 1988, 1989, 1990; Cheung, *et al.*, 1986; Kleinhans and Barefoot, 1987; Rosen, *et al.*, 1988; Samuni, *et al.*, 1988; Pou, *et al.*, 1989a, 1989b, 1993; Ueno, *et al.*, 1989; Black, *et al.*, 1991; Cohen, *et al.*, 1991; Ramos, *et al.*, 1992; Candeias, *et al.*, 1993; Tanigawa, *et al.*, 1993; Seawright, *et al.*, 1995). A few contemporary studies have begun, however, to address potential pharmacological activity associated with DMPO and related spin traps. On those rare occurrences, surprising results must invariably draw into question the long-standing perceptions of phagocytic free radical chemistry. For instance, is the inability to detect neutrophil-dependent HO· solely the result of the instability of spin trapped adducts in the presence of neutrophil-derived $O_2 \cdot^-$ (Rosen, *et al.*, 1988; Samuni, *et al.*, 1988; Pou, *et al.*, 1989a; Ramos, *et al.*, 1992) and insensitive spin trapping systems (Pou, *et al.*, 1994b), or can it be attributed to spin trap inhibition of the NADPH-oxidase, the enzyme responsible for neutrophilic production of $O_2 \cdot^-$?

As different stimuli activate neutrophils through distinct pathways, the effects of spin traps on subsequent formation of $O_2 \cdot^-$ may vary greatly. By way of illustration, consider that DMPO marginally attenuates PMA-stimulated neutrophilic generation of $O_2 \cdot^-$, whereas this nitrone prevents the secretion of this free radical after phagocytic activation by either fMLP or concanavalin A (Britigan and Hamill, 1990). In contrast, other studies found that DMPO does not drastically alter the spin trapping of $O_2 \cdot^-$ by fMLP-stimulated neutrophils (Tanigawa, *et al.*, 1993). Reconciling differences through mechanistic studies may resolve questions regarding the exact pathway by which DMPO impinges on neutrophilic production of $O_2 \cdot^-$. Whether such events occur prior to the assemblage of the NADPH-oxidase remains to be addressed. Despite the minimal effect of DMPO on O_2 consumption and $O_2 \cdot^-$ generation, even at concentrations as high as 100 mM

(Britigan, *et al.*, 1986a; Rosen, *et al.*, 1988), this nitrone has been suggested to retard neutrophil phagocytosis of zymosan particles and to enhance cell lysis (Ueno, *et al.*, 1989). More recent data, however, offer an opposing view (Britigan and Hamill, 1990).

Over the last several years, there has become an increasing recognition of the role of free radicals and products derived from these reactive species, such as $O_2 \cdot^-$, $HO\cdot$, H_2O_2, $LOOH$, and $NO\cdot$, in signal transduction pathways. Among the mechanisms involved are: activation of various kinases, increases in intracellular Ca^{2+}, and regulation of several transcription activator systems including NF-κB and AP-1 (Schilling and Elliott, 1992; Bootman and Berridge, 1995; Chen, *et al.*, 1995b; Clapham, 1995; Baeuerle and Baltimore, 1996; Guyton, *et al.*, 1996; Kyriakis and Avruch, 1996; Lo, *et al.*, 1996; Mendelson, *et al.*, 1996; Sun and Oberley, 1996; Aikawa, *et al.*, 1997; Suzuki, *et al.*, 1997). The ability of spin traps to scavenge intracellular-generated free radicals raises the possibility that many of their biologic effects may be attributed to their ability to modulate cell signal transduction pathways. If this were found to be correct, then spin traps might become experimental tools to study the role of oxidant-regulated signal transduction mechanisms in human physiology and pathology as well as therapeutic agents designed to regulate cellular function through controlling cell signaling (Marterre, *et al.*, 1991; Pogrebniak, *et al.*, 1992; Joseph, *et al.* 1996a, 1996b; Li and Smith, 1996; Kashiwakura, *et al.*, 1997; Mastronarde, *et al.*,1998).

Effect of Spin Traps on Isolated Tissue Preparations

In many pathologic states thought to be initiated by free radicals, spin trapping has played an integral role in understanding an underlying mechanism of a specific disease. During the investigative process, important issues, such as the impact of spin traps on organ preparations, have often not been adequately explored (Culcasi, *et al.*, 1989). Typically, it is assumed that a lack of cytotoxicity in one model can then be extrapolated to others, even though a well-designed experiment could provide the essential control data. When such studies have been completed, surprising findings have frequently affected the choice of spin trap. For instance, PBN was found to exhibit a significant dose-dependent toxicity in isolated perfused rat hearts (Charlon and de Leiris, 1988; Li, *et al.*, 1993b), which commences at levels approaching the detection limits for spin trapping free radicals in aqueous solutions (Pou, *et al.*, 1994b). In contrast, DMPO, even at concentrations as high as 40 mM, was devoid of measurable injury to the isolated tissue preparation (Bradamante, *et al.*, 1993; Nakazawa, *et al.*, 1991), whereas at higher concentrations of this nitrone, toxicity became apparent (Nakazawa, *et al.*, 1991). More extensive pharmacological studies have, until recently, not been available, especially those detailing the impact of spin traps on arterial flow. It now appears that nitrones and nitrosoalkanes, independent of whether these spin traps are lipophilic or hydrophobic, exhibit dose-dependent coronary vasodilative activity (Konorev, *et al.*, 1993). The consequence this may have on spin trapping free radicals remains uncertain, as extensive follow-up inquiries have not appeared.

Spin traps, by virtue of their "specificity" toward free radicals, have been utilized in evaluating secondary drug therapy with isolated tissue preparations as *in vitro* models. For example, adriamycin-induced free radicals, whether a cause or an outcome, adversely impair myocardial contractility (Rajagopalan, *et al.*, 1988). Not suprisingly, spin traps have been shown to impart significant cardioprotective effects (Paracchini, *et al.*, 1993; Monti, *et al.*, 1995; Piccinini, *et al.*, 1995). The efficacy of these compounds parallels the lipophilic character of the nitrone, with PBN demonstrating considerably greater pharmacological activity than either 4-POBN or DMPO (Monti, *et al.*, 1991).

Spin Traps as Pharmaceutical Agents

Molecular O_2 undergoes a series of univalent reductions on its way to H_2O, giving $O_2 \cdot^-$, H_2O_2, and $HO \cdot$ as intermediates. These reactive O_2 species are notable as proposed mediators of a variety of toxic events, not the least of which is reperfusion injury. This pathologic state, although recognized for several decades, was first associated with free radicals in the early 1980s. In a classic paper, it was observed that when cat intestine was made regionally ischemic for 1 hr and then reperfused, capillary permeability increased (Granger, *et al.*, 1981). Pretreatment with SOD markedly inhibited the leakage, thereby demonstrating the importance of $O_2 \cdot^-$ in this pathologic event (Granger, *et al.*, 1981; Parks, *et al.*, 1982). Subsequently termed ischemia/reperfusion injury, this disease appears to be the underlying cause of a wide variety of cardiovascular disorders (McCord, 1985; Bernier, *et al.*, 1986; Bulkley, 1987; Cao, *et al.*, 1988; Darley-Usmar, *et. al.*, 1989; Green, *et al.*, 1989; Downey, 1990; Goldhaber and Weiss, 1992; Korthuis, *et al.*, 1992; Lefer and Lefer, 1993; Grisham, 1994).

During the past decade, mechanistic inquires have concentrated principally on dysfunction of the endothelium, as this cell, lining the vascular bed, appears to be the focal point of the ischemic insult. From these studies, it appears that the initial ischemic insult results in diminished release of endothelial-derived cytoprotective agents, including prostacyclin, $NO \cdot$ and adenosine, concomitant with enhanced production of proinflammatory mediators such as $O_2 \cdot^-$, platelet-activating factor, leukotriene B_4, and endothelin (Simpson, *et al.*, 1987; Werns and Lucchesi, 1988; Reimer, *et al.*, 1989; Hempel, *et al.*, 1990; Johnson, *et al.*, 1990; Kubes, *et al.*, 1990; Ko, *et al.*, 1991; Korthuis, *et al.*, 1992; Ma, *et al.*, 1992; Davenpeck, *et al.*, 1993; Hirafuji and Shinoda, 1993; Lefer and Lefer, 1993; Grisham, 1994). A recent study with isolated perfused rat hearts also suggests enhanced production of $NO \cdot$ after reperfusion injury (Wang and Zweier, 1996).

As the result of this change in homeostasis, neutrophils are recruited to the site of injury. These stimulated phagocytes adhere to the endothelium and secrete cytotoxic agents, including free radicals, proximal to these cells. Evidence in support of a central role for neutrophils in reperfusion injury, primarily through the production of $O_2 \cdot^-$ and $HO \cdot$, comes, in part, from the protective effects afforded by treatment with SOD and/or other free radical scavengers (Craddock, *et al.*, 1979; Hernandez, *et al.*, 1987; Crawford, *et al.*, 1988; Simpson, *et al.*, 1988; Lucchesi, *et al.*, 1989; Suzuki, *et al.*, 1989; Lucchesi, 1990; Korthuis, *et al.*, 1992;

Siminiak and Ozawa, 1993). Based on these and other reports, free radicals have generally been associated with this inflammatory response. In contrast, local delivery of NO· serves to ameloriate reperfusion injury by attenuating neutrophil adherance (Kubes, *et al.*, 1991, 1993; Lefer, *et al.*, 1991,1993; Kubes and Granger, 1992; Ma, *et al.*, 1993; Kurose, *et al.*, 1995), and by acting as a cytoprotective agent[1] through elimination of $O_2\cdot^-$ (Treinin and Hayon, 1970; Seddon, *et al.*, 1973; Overend, *et al.*, 1976; Strehlow and Wagner, 1982; Blough and Zafiriou, 1985; Gryglewski, *et al.*, 1986; Moncada, *et al.*, 1986; Rubanyi and Vanhoutte, 1986; Johnson, *et al.*, 1990; Saran, *et al.*, 1990; Kanner, *et al.*, 1991; Rubanyi, *et al.*, 1991; Siegfried, *et al.*, 1992a, 1992b; Davenpeck, *et al.*, 1993; Gaboury, *et al.*, 1993; Huie and Padmaja, 1993; Wink, *et al.*, 1993).

Considering the increased number of pathologic conditions imputed to be mediated by free radicals, it is not unexpected that spin traps have been regarded as potential therapeutic agents for the control of those disease states. Yet, the thread connecting pharmacological activity with ameloriation of free radical events is thin, lacking a substantial scientific foundation. Rather, it has been assumed that spin traps, by virtue of their ability to react with these reactive species, would prevent cellular toxicity by simply removing the causitive agent— the free radical—from the site of injury. Clearly, there are numerous publications with this view in mind (Table 11.1). For instance, free radicals have been associated with the consequences of traumatic shock in which PBN has been found to reverse many of the pathological indices of this injury (Novelli, *et al.*, 1985; Phillis and Clough-Helfman, 1990a Novelli, 1992).

Administration of DMPO in a dose-dependent manner has been reported to dramatically diminish reperfusion-induced arrhythmias in isolated perfused mammalian hearts (Tosaki and Braquet, 1990; Tosaki, *et al.*, 1990). Similar findings have been established with PBN and HMIO (3,3,4,5,5-hexamethyl-1-imidazoline-*N*-oxide) (Bradamante, *et al.*, 1992; Blasig, *et al.*, 1993, 1994). Given the well-accepted association between free radicals and reperfusion injury, it was surprising that in one model of ischemia/reperfusion injury, PBN and DMPO were found to display a more profound effect on the isolated perfused heart by merely prolonging the ischemic time required for perfusion injury than through the ability of these spin traps to scavenge free radicals (Hearse and Tosaki, 1987a, 1987b). In the face of this unexpected consequence—nitrones ameloriating reperfusion injury through pathways unrelated to free radicals—a number of years elapsed before new avenues of investigation began to question long-held prejudices (Tosaki, *et al.*, 1992; Anderson, *et al.*, 1993). For example, PBN has been shown to afford protection against endotoxemia in a manner that may not be associated with free radicals, but, rather, through mechanisms that interrupt the cytokine cascade that is initiated after exposure to lipopolysaccharide (Pogrebniak, *et al.*, 1992).

Despite the most tenuous of connections between reperfusion-induced production of free radicals and the pharmacologic role of spin traps in ameloriating this pathology, there has been an explosion of reports that detail the therapeutic efficacy of these agents. One of the earliest and, without question, most extensive intact-animal studies correlated free radical production (as assessed by measuring the EPR spectral peak height of the spin trapped adduct PBN–X) with the course and severity of injury in the ischemic/reperfused myocardium (Bolli, *et al.*, 1988).

Table 11.1 *In Vivo* Pharmacological Activity of Spin Traps

Spin Trap[a]	Animal Model	References
PBN	Reperfusion-induced arrhythmias in dogs	1
PBN	Stunned myocardium in anesthetized dogs	2
PBN	Stunned myocardium in open-chested dogs	3
PBN	Traumatic shock in rats	4, 5
PBN	Occlusion shock in rats	5
PBN	Endotoxin shock in rats	5-7
PBN	Endotoxin shock in mice	8
PBN	Cerebral ischemia in gerbils	9–15
PBN	Cerebral ischemia in rats	16–22
PBN	Traumatic brain injury in rats	23
PBN	Age-related changes in gerbil CNS function	24–26
PBN	Cognitive response in aging rats	27
PBN	Prolonged lifespan in mice	28, 29
PBN	Liver protection against halocarbon toxicity	30, 31
PBN	Cardiotoxicity of adriamycin in rats	32
PBN	Cardiotoxicity of doxorubicin in rats	33
PBN	Copper-induced hepatitis in rats	34
PBN	Condition taste aversion in rats	35
PBN	Depletion of striatal dopamine in rats	36
PBN	Increase in cerebral blood flow in rats	37
PBN	Inhibition of induction of NOS in mice	38, 39
PBN	Drug-induced insulin-dependent diabetes	40
PBN	Inhibit excitotoxicity in rats	41
PBN	Muscle fatigue in mice	42
4-POBN	Muscle fatigue in mice	42
4-POBN	Endotoxin shock in rats	7
4-POBN	Occlusion shock in rats	5
DMPO	Endotoxin shock in rats	7
DMPO	Occlusion shock in rats	5
DMPO	Muscle fatigue in mice	42
S-PBN	Neuroprotection in rats	43, 44
2-Ph-DMPO	Cardiotoxicity of adriamycin in rats	45
MDL 101,002	Endotoxin shock and bacteremia in rats	46
MDL 101,002	Inhibition of spontaneous locomotion in rats	47
MDL 101,002	Inhibition of Fe^{2+}-induced seizures in mice	48
MDL 101,002	Cerebral ischemia in rats	49

[a]PBN = *N-tert*-butyl-α-phenylnitrone; 4-POBN = α-(4-pyridyl 1-oxide)-*N-tert*-butylnitrone; DMPO = 5,5,-dimethyl-1-pyrroline-*N*-oxide; S-PBN = sodium *N-tert*-butyl-α-(*o*-sulfonato)phenylnitrone; 2-Ph-DMPO = 2-phenyl-5,5-dimethyl-1-pyrroline-*N*-oxide; MDL 101,002 = 3,4-dihydro-3,3-dimethyl-isoquinoline-*N*-oxide.

1 = Parratt and Wainwright, 1987; 2 = Bolli, *et al.*, 1989a; 3 = Li, *et al.*, 1993b; 4 = Novelli, *et al.*, 1985; 5 = Novelli, 1992; 6 = McKechnie, *et al.*, 1986; 7 = Hamburger and McCay, 1989; 8 = Pogrebniak, *et al.*, 1992; 9 = Phillis and Clough-Helfman, 1990a; 10 = Phillis and Clough-Helfman, 1990b; 11 = Clough-Helfman and Phillis, 1991; 12 = Howard, *et al.*, 1996; 13 = Oliver, *et al.*, 1990; 14 = Yue, *et al.*, 1992; 15 = Hall, *et al.*, 1995; 16 = Sen and Phillis, 1993; 17 = Cao and Phillis, 1994; 18 = Zhao, *et al.*, 1994; 19 = Gidö, *et al.*, 1997; 20 = Pahlmark and Siesjö, 1996; 21 = Aronowski, *et al.*, 1997; 22 = He, *et al.*, 1997; 23 = Sen, *et al.*, 1994; 24 = Carney, *et al.*, 1991; 25 = Floyd and Carney, 1991b; 26 = Carney and Floyd, 1991b; 27 = Sack, *et al.*, 1996; 28 = Edamatsu, *et al.*, 1995; 29 = Saito, *et al.*, 1998b; 30 = Janzen, *et al.*, 1990; 31 = Towner, *et al.*, 1993; 32 = Paracchini, *et al.*, 1993; 33 = Jotti, *et al.*, 1992; 34 = Yamashita, *et al.*, 1996; 35 = Rabin, 1996 ; 36 = Cappon, *et al.*, 1996; 37 = Inanami and Kuwabara, 1995; 38 = Miyajima and Kotake, 1995; 39 = Miyajima and Kotake, 1997; 40 = Tabatabaie, *et al.*, 1997; 41 = Lancelot, *et al.*, 1997; 42 = Novelli, *et al.*, 1990; 43 = Schulz, *et al.*, 1995a; 44 = Schulz, *et al.*, 1995b; 45 = Piccinini, *et al.*, 1995; 46 = French, *et al.*, 1994; 47 = Thomas, *et al.*, 1996; 48 = Thomas, *et al.*, 1997; 49 = Johnson, *et al.*, 1998.

At defined times following ischemia, blood was removed and carefully centrifuged. The resultant plasma was extracted with highly purified organic solvents using well-defined procedures (Poyer, *et al.*, 1981; McCay, *et al.*, 1984; Reinke, *et al.*, 1987). Samples, after appropriate preparation, were immediately frozen at −70°C for evaluation at a subsequent time (Bolli, *et al.*, 1988). Determining the nature of the free radical spin trapped in this ischemic model is, without question, nearly impossible. Yet, from our knowledge of the *in vitro* stability of a variety of spin trapped adducts of PBN (Connor, *et al.*, 1986; Keana, *et al.*, 1989; Samuni, *et al.*, 1989a; Britigan, *et al.*, 1990), lipid dienyl (L·) or secondary alkyl (R·) free radicals (derived from rearrangement reactions involving L·) are the most likely of candidates. More recent studies, however, have begun to address this issue by linking improvements in contractility to the inhibitory action of SOD and catalase or the antioxidant N-(2-mercaptopropionyl)glycine (MPG) with diminution in the EPR spectral intensity of PBN-X (Bolli, *et al.*, 1989a, 1989b; Bolli and McCay, 1990; Triana, *et al.*, 1991; Li, *et al.*, 1993a; Sekili, *et al.*, 1993). Of interest was the finding that PBN, although not nearly as efficacious as SOD and catalase or MPG at enhancing recovery from myocardial stunning, still exhibited significant pharmacological activity (Bolli, *et al.*, 1989a, 1989b; Bolli and McCay, 1990; Sekili, *et al.*, 1993). These findings suggest that the identify of the terminal free radical spin trapped by PBN had its origins in $O_2\cdot^-$ and/or HO·.

In spite of a lack of data demonstrating the precise mechanism by which spin traps elicit their pharmacological activity, localization of these compounds must invariably have an effect on therapeutic outcomes. What little is known about the uptake of spin traps into tissues comes from *in vitro* cell studies in which unique EPR spectra were assigned to intracellular reactions (Bannister and Thornalley, 1983; Rosen and Freeman, 1984; Morgan, *et al.*, 1985; Samuni, *et al.*, 1986; Britigan, *et al.*, 1992). Whether these spin traps are actively transported into cells or enter through passive diffusion has not been the focal point of these or other investigations.

On a macroscopic scale, there have been few reports providing detailed information about the *in vivo* distribution of nitrones or nitrosoalkanes (Chen, *et al.*, 1990a, 1990b; Cheng, *et al.*, 1993; Liu, *et al.*, 1999). What is clear from these and other investigations is that *in vivo* entry of spin traps into cells of specific tissues is complicated, and is dependent upon factors such as route of administration, blood flow dynamics and lipophilic characteristics of the probe. With the advent of *in vivo* EPR spectroscopy and imaging, a detailed pharmacodynamic picture will undoubtedly surface in the coming years.

Concluding Thoughts

The search for new drugs to address many of the diseases that afflict humans is a continual process, in which many roads to discovery often result in unrewarding endpoints. In recent years, with the purported variety of maladies assigned to free radicals, it is not surprising that antioxidant therapy has surfaced as a new class of potential pharmaceutical agents. For instance, in the early years of this decade, age-related macular degeneration was shown to be associated with light-induced production of $O_2\cdot^-$ (Gottsch, *et al.*, 1990). More recently, clinical trials were undertaken to evaluate the effects of the antioxidant vitamins A, C, and E (found in specific foods abundant in these nutrients, as well as in dietary supplementation) in controlling the progression of this disease (Seddon, *et al.,* 1994). Even though there was found to be a significant degree of protection afforded by eating vegetables rich in these vitamins, further clinical studies will undoubtedly be required to convince a skeptic public to change their food consumption patterns.

Despite the numerous array of animal studies (Table 11.1) demonstrating the pharmacological activity of spin traps toward a wide variety of diseases thought to be associated with free radicals, the mechanism by which these compounds actually elicit their therapeutic effectiveness remains less certain. One recent study suggests, for instance, that PBN can improve recovery of energy metabolism after middle cerebral artery occlusion (Folbergrová, *et al.*, 1995). Yet, the pathway by which the spin trap evokes this therapeutic action has not been addressed (Folbergrová, *et al.*, 1995).

Nitrones have been reported to be effective calcium channel blockers (Anderson, *et al.*, 1993). As reperfusion of a previous ischemic myocardium and cardiac arrhythmias can significantly be improved by calcium channel blockers (Henry, 1983; Flameng, 1988; Nademanne and Singh, 1988), one must not lose sight of the fact that the medicinal action of spin traps in these settings may not be linked to the interruption of a free radical cascade. Finally, as spin traps are effective vasodilators (Konorev, *et al.*, 1993), these agents would be contraindicated in patients in certain hypotensive states, such as shock (Pogrebniak, *et al.*, 1992), even though some antioxidant effects would have been beneficial in other situations. In the end, therapeutic benefits of spin traps remain to be determined. Reservations about acute and chronic toxicity will unquestionably surface as these potential drugs weave their way through the drug regulatory process.

References

Aikawa, R., Komuro, I., Yamazaki, T., Zou, Y., Kudoh, S., Tanaka, M., Shiojima, I., Hiroi, Y., and Y. Yazaki, Y. (1997). Oxidative stress activates extracellular signal-regulated kinases through Src and Ras in cultured myocytes in neonatal rats. *J. Clin. Invest.* **100**: 1813–1821.

Albano, E., Cheeseman, K.H., Tomasi, A., Carini, R., Dianzani, M.U., and Slater, T.F. (1986). Effect of spin traps in isolated rat hepatocytes and liver microsomes. *Biochem. Pharmacol.* **35**: 3955–3960.

Albano, E., Lott, K.A.K., Slater, T.F., Stier, A., Symons, M.C.R., and Tomasi, A. (1982). Spin-trapping studies on the free-radical products formed by metabolic activation of carbon

tetrachloride in rat liver microsomal fractions isolated hepatocytes and *in vivo* in the rat. *Biochem. J.* **204**: 593–603.

Albano, E., Tomasi, A., Goria-Gatti, L., and Iannone, A. (1989). Free radical activation of monomethyl and dimethyl hydrazines in isolated hepatocytes and liver microsomes. *Free Radical Biol. Med.* **6**: 3–8.

Albano, E., Tomasi, A., Vannini, V., and Dianzani, M.U. (1985). Detection of free radical intermediates during isoniazid and iproniazid metabolism by isolated rat hepatocytes. *Biochem. Pharmacol.* **34**: 381–382.

Andersen, K.A., Diaz, P.T., Wright, V.P., and Clanton, T.L. (1996). N-*tert*-butyl-α-phenylnitrone: A free radical trap with unanticipated effects on diaphragm function. *J. Appl. Physiol.* **80**: 862–868.

Anderson, D.E., Yuan, X.-J., Tseng, C.-M., Rubin, L.J., Rosen, G.M., and Tod, M.L. (1993). Nitrone spin-traps block calcium channels and induce pulmonary artery relaxation independent of free radicals. *Biochem. Biophys. Res. Commun.* **193**: 878–885.

Aoshima, H., Kajiwara, T., Hatanaka, A., and Hatano, H. (1977). Electron spin resonance studies on the lipoxygenase reaction by spin trapping and spin labelling methods. *J. Biochem.* **82**: 1559–1565.

Aronowski, J., Strong, R., and Grotta, J.C. (1997). Reperfusion injury: Demonstration of brain damage produced by reperfusion after transient focal ischemia in rats. *J. Cereb. Blood Flow Metab.* **17**: 1048–1056.

Arthur, J.R., Boyne, R., Hill, H.A.O., and Okolow-Zubkowska, M.J. (1981). The production of oxygen-derived radicals by neutrophils from selenium-deficient cattle. *FEBS Lett.* **135**: 187–190.

Augusto, O., Beilan, H.S., and Ortiz de Montellano, P.R. (1982). The catalytic mechanism of cytochrome P-450. Spin-trapping evidence for one-electron substrate oxidation. *J. Biol. Chem.* **257**: 11288–11295.

Baeuerle, P.A., and Baltimore, D. (1996). NF-κB: Ten years after. *Cell* **87**: 13–20.

Bannister, J.V., and Bannister, W.H. (1985). Production of oxygen-centered radicals by neutrophils and macrophages as studied by electron spin resonance (ESR). *Environ. Health Perspect.* **64**: 37–43.

Bannister, J.V., and Thornalley, P.J. (1983). The production of hydroxyl radicals by adriamycin in red blood cells. *FEBS Lett.* **157**: 170–172.

Bernier, M., Hearse, D.J., and Manning, A.S. (1986). Reperfusion-induced arrhythmias and oxygen-derived free radicals. Studies with "anti-free radical" intervention and a free radical-generating system in the isolated perfused rat heart. *Circ. Res.* **58**: 331–340.

Black, C.D.V., Cook, J.A., Russo, A., and Samuni, A. (1991). Superoxide production by stimulated neutrophils: Temperature effect. *Free Radical Res. Commun.* **12–13**: 27–37.

Blasig, I.E., Shuter, S., Garlick, P., and Slater, T. (1994). Relative time-profiles for free radical trapping, coronary flow, enzyme leakage, arrhythmias, and function during myocardial reperfusion. *Free Radical Biol. Med.* **16**: 35–41.

Blasig, I.E., Volodarski, L.B., and Tosaki, A. (1993). Nitrone spin trap compounds. Mode of cardioprotective action. *Pharm. Pharmacol. Lett.* **3**: 135–138.

Blough, N.V., and Zafiriou, O.C. (1985). Reaction of superoxide with nitric oxide to form peroxynitrite ion alkaline aqueous solution. *Inorg. Chem.* **24**: 3502–3504.

Bolli, R., and McCay, P.B. (1990). Use of spin traps in intact animals under-going myocardial ischemia/reperfusion: A new approach to assessing the role of oxygen radicals in myocardial "stunning." *Free Radical Res. Commun.* **9**: 169–180.

Bolli, R., Jeroudi, M.O., Patel, B.S., Aruoma, O.I., Halliwell, B., Lai, E.K., and McCay, P.B. (1989a). Marked reduction of free radical generation and contractile dysfunction by antioxidant therapy begun at the time of reperfusion. Evidence that myocardial "stunning" is a manifestation of reperfusion injury. *Circ. Res.* **65**: 607–622.

Bolli, R., Jeroudi, M.O., Patel, B.S., DuBose, C.M., Lai, E.K., Roberts, R., and McCay, P.B. (1989b). Direct evidence that oxygen-derived free radicals contribute to postischemic myocardial dysfunction in the intact dog. *Proc. Natl. Acad. Sci. USA* **86**: 4695–4699.

Bolli, R., Patel, B.S., Jeroudi, M.O., Lai, E.K., and McCay, P.B. (1988). Demonstration of free radical generation in "stunned" myocardium of intact dogs with the use of the spin trap α-phenyl N-tert-butyl nitrone. *J. Clin. Invest.* **82**: 476–485.

Bootman, M. D., and M. J. Berridge, M.J. (1995). The elemental principles of calcium signaling. *Cell* **83**: 675–678.

Bradamante, S., Jotti, A., Paracchini, L., and Monti, E. (1993). The hydrophilic spin trap, 5,5-dimethyl-1-pyrroline-1-oxide, does not protect the rat heart from reperfusion injury. *Eur. J. Pharmacol.* **234**: 113–116.

Bradamante, S., Monti, E., Paracchini, L., Lazzarini, E., and Piccinini, F. (1992). Protective activity of the spin trap tert-butyl-α-phenyl nitrone (PBN) in reperfused rat heart. *J. Mol. Cell. Cardiol.* **24**: 375–386.

Britigan, B.E., and Hamill, D.R. (1990). Effect of the spin trap 5,5-dimethyl-1-pyrroline N-oxide (DMPO) on human neutrophil function: Novel inhibition of neutrophil stimulus-response coupling? *Free Radical Biol. Med.* **8**: 459–470.

Britigan, B.E., Coffman, T.J., Adelberg, D.R., and Cohen, M.S. (1988). Mononuclear phagocytes have the potential for sustained hydroxyl radical production. Use of spin-trapping techniques to investigate mononuclear phagocyte free radical production. *J. Exp. Med.* **168**: 2367–2372.

Britigan, B.E., Coffman, T.J., and Buettner, G.R. (1990). Spin trap evidence for the lack of significant hydroxyl radical production during the respiration burst of human phagocytes using a spin adduct resistant to superoxide mediated destruction. *J. Biol. Chem.* **265**: 2650–2656.

Britigan, B.E., Hassett, D.J., Rosen, G.M., Hamill, D.R., and Cohen, M.S. (1989). Neutrophil degranulation inhibits potential hydroxyl radical formation: Differential impact of myeloperoxidase and lactoferrin release on hydroxyl radical production by iron supplemented neutrophils assessed by spin trapping. *Biochem. J.* **264**: 447–455.

Britigan, B.E., Roeder, T.L., and Shasby, D.M. (1992). Insight into the nature and site of oxygen-centered free radical generation by endothelial cell monolayers using a novel spin trapping technique. *Blood* **79**: 699–707.

Britigan, B.E., Rosen, G.M., Chai, Y., and Cohen, M.S. (1986a). Do human neutrophils make hydroxyl radical? Detection of free radicals generated by human neutrophils activated with a soluble or particulate stimulus using electron paramagnetic resonance spectrometry. *J. Biol. Chem.* **261**: 4426–4431.

Britigan, B.E., Rosen, G.M., Thompson, B.Y., Chai, Y., and Cohen, M.S. (1986b). Stimulated neutrophils limit iron-catalyzed hydroxyl radical formation as detected by spin trapping techniques. *J. Biol. Chem.* **261**: 17026–17032.

Bulkley, G.B. (1987). Free radical-mediated reperfusion injury: A selective review. *Brit. J. Cancer* **55** (Suppl. VIII): 66–70.

Candeias, L.P., Patel, K.B., Stratford, M.R.L., and Wardman, P. (1993). Free hydroxyl radicals are formed on reaction between the neutrophil-derived species superoxide anion and hypochlorous acid. *FEBS Lett.* **333**: 151–153.

Cao, W., Carney, J.M., Duchon, A., Floyd, R.A., and Chevion, M. (1988). Oxygen free radical involvement in ischemia and reperfusion injury to brain. *Neurosci. Lett.* **88**: 233–238.

Cao, X., and Phillis, J.W. (1994). α-Phenyl-tert-butyl nitrone reduces cortical infarct and edema in rats subjected to focal ischemia. *Brain Res.* **664**: 267–272.

Cappon, G.D., Broening, H.W., Pu, C., Morford, L., and Vorhees, C.V. (1996). α-Phenyl-N-*tert*-butyl nitrone attenuates methamphetamine-induced depletion of striatal dopamine without altering hyperthermia. *Synapse* **24**: 173–186.

Carney, J.M., and Floyd, R.A. (1991a). Phenyl butyl nitrone compositions and methods for treatment of oxidative tissue damage. *U.S. Patent* 5,025,032.

Carney, J.M., and Floyd, R.A. (1991b). Protection against oxidative damage to CNS by α-phenyl-*tert*-butyl nitrone (PBN) and other spin-trapping agents: A novel series of nonlipid free radical scavengers. *J. Mol. Neurosci.* **3**: 47–57.

Carney, J.M., and Floyd, R.A. (1995). PBN, DMPO, and POBN compositions and methods of use thereof for inhibition of age-associated oxidation. *U.S. Patent* 5,405,874.

Carney, J.M., Starke-Reed, P.E., Oliver, C.N., Landum, R.W., Cheng, M.S., Wu, J.F., and Floyd, R.A. (1991). Reversal of age-related increase in brain protein oxidation, decrease in enzyme activity, and loss in temporal and spatial memory by chronic administration of the spin-trapping compound N-*tert*-butyl-α-phenylnitrone.

Proc. Natl. Acad. Sci. USA **88**: 3633–3636.

Cederbaum, A.I., and Cohen, G. (1980). Inhibition of the microsomal oxidation of ethanol and 1-butanol by the free radical, spin-trapping agent 5,5-dimethyl-1-pyrroline-1-oxide. *Arch. Biochem. Biophys.* **204**: 397–403.

Chamulitrat, W., Jordan, S.J., Mason, R.P., Saito, K., and Cutler, R.G. (1993). Nitric oxide formation during light-induced decomposition of phenyl N-tert-butylnitrone. *J. Biol. Chem.* **268**: 11520–11527.

Chamulitrat, W., Parker, C.E., Tomer, K.B., and Mason, R.P. (1995). Phenyl N-*tert*-butyl nitrone forms nitric oxide as a result of its FE(III)-catalyzed hydrolysis or hydroxyl radical adduct formation. *Free Radical Res.* **23**: 1–14.

Charlon, V., and de Leiris, J. (1988). Ability of N-tert-butyl alpha phenylnitrone (PBN) to be used in isolated perfused heart spin trapping experiments: Preliminary studies. *Basic Res. Cardiol.* **83**: 306–313.

Chen, G., Bray, T.M., Janzen, E.G., and McCay, P.B. (1990a). Excretion, metabolism and tissue distribution of a spin trapping agent, α-phenyl-N-*tert*-butyl-nitrone (PBN) in rats. *Free Radical Res. Commun.* **9**: 317–323.

Chen, G., Griffin, M., Poyer, J.L., McCay, P.B., and Bourne, D.W.A. (1990b). HPLC procedure for the pharmacokinetic study of the spin-trapping agent, α-phenyl-*N*-tert-butyl nitrone (PBN). *Free Radical Biol. Med.* **8**: 93–98.

Chen, Q., Fisher, A., Reagan, J.D., Yan, L.-J., and Ames, B.N. (1995a). Oxidative DNA damage and senescence of human diploid fibroblast cells. *Proc. Natl. Acad. Sci. USA* **92**: 4337–4341.

Chen, Q., Olashaw, N., and Wu, J (1995b). Participation of reactive oxygen species in the lysophosphatidic acid-stimulated mitogen-activated protein kinase kinase activation pathway. *J. Biol. Chem.* **270**: 28499–28502.

Cheng, H.-Y., Liu, T., Feuerstein, G., and Barone, F.C. (1993). Distribution of spin-trapping compounds in rat blood and brain: *In vivo* microdialysis determination. *Free Radical Biol. Med.* **14**: 243–250.

Cheung, K., Lark, J., Robinson, M.F., Pomery, P.J., and Hunter, S. (1986). The production of hydroxyl radical by human neutrophils stimulated by arachidonic acid - Measurements by ESR spectroscopy. *Aust. J. Exp. Biol. Med. Sci.* **64**: 157–164.

Clapham, D. E. (1995). Calcium signaling. *Cell* **80**: 259–268.

Clough-Helfman, C., and Phillis, J.W. (1991). The free radical trapping agent N-*tert*-butyl-α-phenylnitrone (PBN) attenuates cerebral ischaemic injury in gerbils. *Free Radical Res. Commun.* **15**: 177–186.

Cohen, M.S., Britigan, B.E., Chai, Y.S., Pou, S., Roeder, T.L., and Rosen, G.M. (1991). Phagocyte-derived free radicals stimulated by ingestion of iron-rich *Staphylococcus aureus*: A spin trapping study. *J. Infect. Dis.* **163**: 819–824.

Connor, H.D., Thurman, R.G., Galizi, M.D., and Mason, R.P. (1986). The formation of a novel free radical metabolite of CCl₄ in the perfused rat liver and *in vivo*. *J. Biol. Chem.* **261**: 4542–4548.

Cova, D., De Angelis L., Monti, E., and Piccinini, F. (1992). Subcellular distribution of two spin trapping agents in rat heart: Possible explanation for their different protective effects against doxorubicin-induced cardiotoxicity. *Free Radical Res. Commun.* **15**: 353–360.

Craddock, P.R., Hammerschmidt, D.E., Moldow, C.F., Yamada, O., and Jacob, H.S. (1979). Granulocyte aggregation as a manifestation of membrane interactions with complement: Possible role in leukocyte margination, microvascular occlusion and endothelial damage. *Seminars Hematol.* **16**: 140–147.

Crawford, M.H., Grover, F.L., Kolb, W.P., McMahan, C.A., O'Rourke, R.A., McManus, L.M., and Pinckard, R.N. (1988). Complement and neutrophil activation in the pathogenesis of ischemic myocardial injury. *Circulation* **78**: 1449–1458.

Culcasi, M., Pietri, S., and Cozzone, P.J. (1989). Use of 3,3,5,5-tetramethyl-1-pyrroline-1-oxide spin trap for the continuous flow ESR monitoring of hydroxyl radical generation in the ischemic and reperfused myocardium. *Biochem. Biophys. Res. Commun.* **164**: 1274–1280.

Darley-Usmar, V.M., Stone, D., and Smith, D.R. (1989). Oxygen and reperfusion damage: An overview. *Free Radical Res. Commun.* **7**: 247–254.

Davenpeck, K.L., Guo, J.-P, and Lefer, A.M. (1993). Pulmonary artery endothelial dysfunction following ischemia and reperfusion of the rabbit lung. *J. Vasc. Res.* **30**: 145–153.

Docampo, R., Cassellas, A.M., Madeira, E.D., Cardoni, R.L., Moreno, S.N.J., and Mason, R.P. (1983). Oxygen-derived radicals from *Trypanosoma cruzi*-stimulated human neutrophils. *FEBS Lett.* **155**: 25–30.

Downey, J.M. (1990). Free radicals and their involvement during long-term myocardial ischemia and reperfusion. *Annu. Rev. Physiol.* **52**: 487–504.

Edamatsu, R., Mori, A., and Packer, L. (1995). The spin-trap N-*tert*-butyl-α-phenyl-butylnitrone prolongs the life span of the senescence accelerated mouse. *Biochem. Biophys. Res. Commun.* **211**: 847–849.

Finkelstein, E., Rosen, G.M., and Rauckman, E.J. (1980). Spin trapping. Kinetics of the reaction of superoxide and hydroxyl radicals with nitrones. *J. Am. Chem. Soc.* **102**: 4994–4999.

Finkelstein, E., Rosen, G.M., Rauckman, E.J., and Paxton J. (1979). Spin trapping of superoxide. *Mol. Pharmacol.* **16**: 676–685.

Flameng, W. (1988). Myocardial protection. *Ann. N.Y. Acad. Sci.* **522**: 600–610.

Floyd, R.A., and Carney, J.M. (1991a). Phenylbutyl nitrone compositions and methods for prevention of gastric ulceration. *U.S. Patent* 5,035,097.

Floyd, R.A., and Carney, J.M. (1991b). Age influence on oxidative events during brain ischemia/reperfusion. *Arch. Gerontol. Geriatr.* **12**: 155–177.

Floyd, R.A., Liu, G.-J., and Wong, P.K. (1996). Nitrone radical traps as protectors of oxidative damage in the central nervous system, in: *Handbook of Synthetic Antioxidants* (Cadenas, C., and Packer, L., eds.), pp. 339–350, Marcel Dekker, New York.

Folbergrová, J., Zhao, Q., Katsura, K.-I., and Siesjö, B.K. (1995). *N-tert*-Butyl-α-phenylnitrone improves recovery of brain energy state in rats following transient focal ischemia. *Proc. Natl. Acad. Sci. USA* **92**: 5057–5061.

French, J.F., Thomas, G.E., Downs, T.R., Ohlweiler, D.F., Carr, A.A., and Dage, R.C. (1994). Protective effects of a cyclic nitrone antioxidant in animal models of endotoxic shock and chronic bacteremia. *Circ. Shock* **43**: 130–136.

Gaboury, J., Woodman, R.C., Granger, D.N., Reinhardt, P., and Kubes, P. (1993). Nitric oxide prevents leukocyte adherence: Role of superoxide. *Am. J. Physiol.* **265**: H862-H867.

Gidö, G., Kristián, T., and Siesjö, B.K. (1997). Extracellular potassium in a neocortical core area after transient focal ischemia. *Stroke* **28**: 206- 210.

Goldhaber, J.I., and Weiss, J.N. (1992). Oxygen free radicals and cardiac reperfusion abnormalities. *Hypertension* **20**: 118–127.

Gottsch, J.D., Pou, S., Bynoe, L.A., and Rosen, G.M. (1990). Hematogenous photosensitization. A mechanism for the development of age-related macular degeneration. *Invest. Ophthalmol. Vis. Sci.* **31**: 1674–1682.

Graham-Lorence, S., and Peterson, J.A. (1996). P450s: Structural similarities and functional differences. *FASEB J.* **10**: 206–214.

Granger, D.N., Rutili, G., and McCord, J.M. (1981). Superoxide radicals in feline intestinal ischemia. *Gastroenterology* **81**: 22–29.

Green, C.J., Gower, J.D., Healing, G., Cotterill, L.A., Fuller, B.J., and Simpkin, S. (1989). The importance of iron, calcium and free radicals in reperfusion injury: An overview of studies in ischaemic rabbit kidneys. *Free Radical Res. Commun.* **7**: 255–264.

Green, M.R., Hill, H.A.O., Okolow-Zubkowska, M.J., and Segal, A.W. (1979). The production of hydroxyl and superoxide radicals by stimulated human neutrophils—measurements by EPR spectroscopy. *FEBS Lett.* **100**: 23–26.

Grisham, M.B. (1994). *Reactive Metabolites of Oxygen and Nitrogen in Biology and Medicine,* pp. 56–69, R.G. Landes Co., Austin, TX.

Gryglewski, R.J., Palmer, R.M.J., and Moncada, S. (1986). Superoxide anion is involved in the breakdown of endothelial-derived relaxing factor. *Nature* **320**: 454–456.

Guyton, K. Z., Liu, Y.S., Gorospe, M., Xu, Q., and Holbrook, N.J. (1996). Activation of mitogen-activated protein kinase by H_2O_2. Role in cell survival following oxidant injury. *J. Biol. Chem.* **271**: 4138–4142.

Hall, N.C., Carney, J.M., Cheng, M., and Butterfield, D.A. (1995). Prevention of ischemia/reperfusion-induced alterations in synaptosomal membrane-associated proteins and lipids by *N-tert*-butyl-α-phenylnitrone and difluoromethylornithine. *Neuroscience* **69**: 591–600.

Hamburger, S.A., and McCay, P.B. (1989). Endotoxin-induced mortality in rats is reduced by nitrones. *Circ. Shock* **29**: 329–334.

Hampton, M.J., Floyd, R.A., Janzen, E.G., and Shetty, R.V. (1981). Mutagenicity of free-radical spin-trapping compounds. *Mutat. Res.* **91**: 279–283.

Haseloff, R.F., Mertsch, K., Rohde, E., Baeger, I., Grigor'ev, I.A., and Basig, I.E. (1997). Cytotoxicity of spin trapping compounds. *FEBS Lett.* **418**: 73–75.

Hawley, D.A., Kleinhans, F.W., and Biesecker, J.L. (1983). Determination of alternate pathway complement kinetics by electron spin resonance spectroscopy. *Am. J. Clin. Pathol.* **79**: 673–677.

He, Q.P., Smith, M.L., Li, P.A., and Siesjö, B.K. (1997). Necrosis of the substantia nigra, pars reticulate, in flurothyl-induced status epilepticus is ameliorated by the spin trap a phenyl-N-*tert*-butyl nitrone. *Free Radical Biol. Med.* **22**: 917–922.

Hearse, D.J., and Tosaki, A. (1987a). Free radicals and reperfusion-induced arrhythmias: Protection by spin trap agent PBN in the rat heart. *Circ. Res.* **60**: 375–383.

Hearse, D.J., and Tosaki, A. (1987b). Reperfusion-induced arrhythmias and free radicals: Studies in the rat heart with DMPO. *J. Cardiovasc. Pharmacol.* **9**: 641–650.

Hempel, S.L., Haycraft, D.L., Hoak, J.C., and Spector, A.A. (1990). Reduced prostacyclin formation after reoxygenation of anoxic endothelium. *Am. J. Physiol.* **259**: C738–C745.

Henry, P.D. (1983). Mechanisms of action of calcium antagonists in cardiac and smooth muscle, in: *Calcium Channel Blocking Agents in the Treatment of Cardiovascular Disorders* (Stone, P.H., and Antman, E.M., eds.), pp. 107–154, Futura Publishing Co., Mount Kisco, NY.

Hernandez, L.A., Grisham, M.B., Twohig, B., Arfors, K.E., Harlan, J.M., and Granger, D.N. (1987). Role of neutrophils in ischemia-reperfusion-induced microvascular injury. *Am. J. Physiol.* **253**: H699–H703.

Hill, H.A.O., and Thornalley, P.J. (1982). Free radical production during phenylhydrazine-induced hemolysis. *Can. J. Chem.* **60**: 1528–1531.

Hill, H.A.O., and Thornalley, P.J. (1983). The effect of spin traps on phenylhydrazine-induced haemolysis. *Biochim. Biophys. Acta* **762**: 44–51.

Hirafuji, M., and Shinoda, H. (1993). Roles of prostacyclin, EDRF and active oxygens in leukocyte-dependent platelet adhesion to endothelial cells induced by platelet-activating factor *in vitro*. *Br. J. Pharmacol.* **109**: 524–529.

Howard, B.J., Yastin, S., Hensley, K., Allen, K.L., Kelly, J.P., Carney, J., and Butterfield, D.A. (1996). Prevention of hyperoxia-induced alterations in synaptosomal membrane-associated proteins by *N-tert*-butyl-α-phenylnitrone and 4-hydroxy-2,2,6,6-tetramethylpiperidin-1-oxyl (Tempol). *J. Neurochem.* **67**: 2045–2050.

Huie, R.E., and Padmaja, S. (1993). The reaction of NO with superoxide. *Free Radical Res. Commun.* **18**: 195–199.

Hume, D.A., Gordon, S., Thornally, P.J., and Bannister, J.V. (1983). The production of oxygen-centered radicals by Bacillus-Calmette-Guerin-activated macrophages. An electron paramagnetic resonance study of the response to phorbol myristrate acetate. *Biochim. Biophys. Acta* **763**: 245–250.

Inanami, O., and Kuwabara, M. (1995). α-Phenyl N-tert-butyl nitrone (PBN) increases the cortical cerebral blood flow by inhibiting the breakdown of nitric oxide in anesthetized rats. *Free Radical Res.* **23**: 33–39.

Janzen, E.G., Poyer, J.L., Schaefer, C.F., Downs, P.E., and DuBose, C.M. (1995). Biological spin trapping. II. Toxicity of nitrone spin traps: Dose-ranging in the rat. *J. Biochem. Biophys. Methods* **30**: 239–247.

Janzen, E.G., Towner, R.A., and Yamashiro, S. (1990). The effect of phenyl *tert*-butyl nitrone (PBN) on CCl₄-induced rat liver injury detected by proton magnetic resonance imaging (MRI) *in vivo* and electron microscopy (EM). *Free Radical Res. Commun.* **9**: 325–335.

Janzen, E.G., West, M.S., and Poyer, J.L. (1994). Comparison of antioxidant activity of PBN with hindered phenols in initiated rat liver microsomal lipid peroxidation, in: *Frontiers of Reactive Oxygen Species in Biology and Medicine* (Asada, K., and Toshikawa, T., eds.), pp. 431–434, Elsevier Science, Amsterdam.

Johnson, G., III, Tsao, P.S., Mulloy, D., and Lefer, A.M. (1990). Cardioprotective effects of acidified sodium nitrite in myocardial ischemia with reperfusion. *J. Pharmacol. Exp. Ther.* **252**: 35–41.

Johnson, M.P., McCarty, D.R., Velayo, N.L., Markgraf, C.G., Chmielewski, P.A., Ficorilli, J.V., Cheng, H.C., and Thomas, C.E. (1998). MDL 101,002, a free radical spin trap, is efficacious in permanent and transient focal ischemia models. *Life Sci.* **63**: 241–253.

Joseph, J.A., Denisova, N., Villalobos-Molina, R., Erat, S., and Strain, J. (1996a). Oxidative stress and age-related neuronal deficits. *Mol. Chem. Neuropathol.* **28**: 35–40.

Joseph, J.A., Villalobos-Molina, R., Denisova, N., Erat, S., Jimenez, N., and Strain, J. (1996b). Increased sensitivity to oxidative stress and the loss of muscarinic receptor responsiveness in senescence. *Ann. N.Y. Acad. Sci.* **786**: 112–119.

Jotti, A., Paracchini, L., Perletti, G., and Piccinini, F. (1992). Cardiotoxicity induced by doxorubicin *in vivo*: Protective activity of the spin trap alpha-phenyl-tert-butyl nitrone. *Pharmacol. Res.* **26**: 143–150.

Kalyanaraman, B., Joseph, J., and Parthasarathy, S. (1991). The spin trap, α-phenyl N-*tert*-butylnitrone, inhibits the oxidative modification of low density lipoproteins. *FEBS Lett.* **280**: 17–20.

Kalyanaraman, B., Joseph, J., and Parthasarathy, S. (1993). Site-specific trapping of reactive species in low-density lipoprotein oxidation: Biological implications. *Biochim. Biophys. Acta* **1168**: 220–227.

Kalyanaraman, B., Perez-Reyes, E., and Mason, R.P. (1980). Spin-trapping and direct electron spin resonance investigations of the redox metabolism of quinone anticancer drugs. *Biochem. Biophys. Acta* **630**: 119–130.

Kanner, J., Harel, S., and Granit, R. (1991). Nitric oxide as an antioxidant. *Arch. Biochem. Biophys.* **289**: 130–136.

Kashiwakura, I., Kuwabara, M., Murakami, M., Hayase, Y., and Takagi, Y. (1997). Effects of alpha-phenyl N-tert-butylnitrone, a spin trap reagent, on the proliferation of murine hematopoietic progenitor cells *in vitro*. *Res. Commun. Molec. Pathol. Pharmacol.* **98**: 67–76.

Keana, J.F.W., Pou, S., and Rosen, G.M. (1989). Synthesis and properties of some nitroxide α-carboxylate salts. *J. Org. Chem.* **54**: 2417–2420.

Kleinhans, F.W., and Barefoot, S.T. (1987). Spin trap determination of free radical burst kinetics in stimulated neutrophils. *J. Biol. Chem.* **262**: 12452–12457.

Ko, W., Hawes, A.S., Lazenby, W.D., Calvano, S.E., Shin, Y.T., Zelano, J.A., Antonacci, A.C., Isom, O.W., and Krieger, K.H. (1991). Myocardial reperfusion injury. Platelet-activating factor stimulates polymorphonuclear leukocyte hydrogen peroxide production during myocardial reperfusion. *J. Thorac. Cardiovasc. Surg.* **102**: 297–308.

Konorev, E.A., Baker, J.E., Joseph, J., and Kalyanaraman, B. (1993). Vasodilator and toxic effects of spin traps on aerobic cardiac function. *Free Radical Biol. Med.* **14**: 127–137.

Korthuis, R.J., Carden, D.L., and Granger, D.N. (1992). Cellular dysfunction induced by ischemia/reperfusion: Role of reactive oxygen metabolites and granulocytes, in: *Biological Consequences of Oxidative Stress* (Spatz, L., and Bloom, D. eds.), pp. 50–77, Oxford University Press, London.

Kubes, P., and Granger, D.N. (1992). Nitric oxide modulates microvascular permeability. *Am. J. Physiol.* **262**: H611-H615.

Kubes, P., Ibbotson, G., Russell, J., Wallace, J.L., and Granger, D.N. (1990). Role of platelet-activating factor in ischemia/reperfusion-induced leukocyte adherence. *Am. J. Physiol.* **259**: G300-G305.

Kubes, P., Kanwar, S., Niu, X.F., and Gaboury, J.P. (1993). Nitric oxide synthesis inhibition induces leukocyte adhesion via superoxide and mast cells. *FASEB J.* **7**: 1293–1299.

Kubes, P., Suzuki, M., and Granger, D.N. (1991). Nitric oxide: An endogenous modulator of leukocyte adhesion. *Proc. Natl. Acad. Sci. USA* **88**: 4651–4655.

Kurose, I., Wolf, R., Grisham, M.B., Aw, T.Y., Specian, R.D., and Granger, D.N. (1995). Microvascular responses to inhibition of nitric oxide production. Role of active oxidants. *Circ. Res.* **76**: 30–39.

Kyriakis, J. M., and Avruch, J. (1996). Sounding the alarm: Protein kinase cascades activated by stress and inflammation. *J. Biol. Chem.* **271**: 24313–24316.

Lancelot, E., Revaud, M.-L., Boulu, R.G., Plotkine, M., and Callebert, J. (1997). Alpha-phenyl-*N-tert*-butylnitrone attenuates excitotoxicity in rat striatum by preventing hydroxyl radical accumulation. *Free Radical Biol. Med.* **23**: 1031–1034.

Lefer, A.M., and Lefer, D.J. (1993). Pharmacology of the endothelium in ischemia-reperfusion and circulatory shock. *Annu. Rev. Pharmcol. Toxicol.* **33**: 71–90.

Lefer, D.J., Nakanishi, K., Johnston, W.E., and Vinten-Johansen, J. (1993). Antineutrophil and myocardial protecting actions of a novel nitric oxide donor after acute myocardial ischemia and reperfusion in dogs. *Circulation* **88**: 2337–2350.

Lefer, A.M., Tsao, P.S., Lefer, D.J., and Ma, X.-L. (1991). Role of endothelial dysfunction in the pathogenesis of reperfusion injury after myocardial ischemia. *FASEB J.* **5**: 2029–2034.

Li, M., and Smith, C.P. (1996). β-amyloid$_{1-40}$ inhibits electrically stimulated release of [^{3}H]norepinephrine and enhances the internal calcium response to low potassium in rat cortex: Prevention with a free radical scavenger. *Brain Res. Bull.* **39**: 299–303.

Li, X.-Y., McCay, P.B., Zughaib, M., Jeroudi, M.O., Triana, J.F., and Bolli, R. (1993a). Demonstration of free radical generation in the "stunned" myocardium in the conscious dog and identification of major differences between conscious and open-chest dogs. *J. Clin. Invest.* **92**: 1025–1041.

Li, X.-Y., Sun, J.-Z., Bradamante, S., Piccinini, F., and Bolli, R. (1993b). Effects of the spin trap α-phenyl N-*tert*-butyl nitrone on myocardial function and flow: A dose-dependent response study in the open-chest dog and in the isolated rat heart. *Free Radical Biol. Med.* **14**: 277–285.

Liu, K.J., Kotake, Y., Lee, M., Miyake, M., Sugden, K., Yu, Z., and Swartz, H.M. (1999). HPLC study of the pharmacokinetics of various spin traps for application to *in vivo* spin trapping. *Free Radical Biol. Med.* in press.

Lo, Y.Y.C., Wong, J.M.S., and Cruz, T.F. (1996). Reactive oxygen species mediate cytokine activation of c-Jun NH$_2$-terminal kinases. *J. Biol. Chem.* **271**: 15703–15707.

Lucchesi, B.R. (1990). Modulation of leukocyte-mediated myocardial reperfusion injury. *Annu. Rev. Physiol.* **52**: 561–576.

Lucchesi, B.R., Werns, S.W., and Fantone, J.C. (1989). The role of the neutrophil and free radicals in ischemic myocardial injury. *J. Mol. Cell. Cardiol.* **21**: 1241–1251.

Ma, X.-L., Weyrich, A.S., Krantz, S., and Lefer, A.M. (1992). Mechanisms of the cardioprotective actions of WEB-2170, bepafant, a platelet activating factor antagonist, in myocardial ischemia and reperfusion. *J. Pharmacol. Exp. Ther.* **260**: 1229–1236.

Ma, X.-L., Weyrich, A.S., Lefer, D.J., and Lefer, A.M. (1993). Diminished basal nitric oxide release after myocardial ischemia and reperfusion promotes neutrophil adherence to coronary endothelium. *Circ. Res.* **72**: 403–412.

Mansbach, C.M., II, Rosen, G.M., Rahn, C.A., and Strauss, K.E. (1986). Detection of free radicals as a consequence of rat intestinal cellular drug metabolism. *Biochim. Biophys. Acta* **888**: 1–9.

Marterre, W.F., Jr., Kindy, M.S., Carney, J.M., Landrum, R.W., and Strodel, W.E. (1991). Induction of the protoon-cogene *c-fos* and recovery of cytosolic adenosine triphosphate in reperfused liver after transient warm ischemia: Effect of nitrone free-radical spin-trap agents. *Surgery* **110**: 184–191.

Mason, R.P., Kalyanaraman, B., Tainer, B.E., and Eling, T.E. (1980). A carbon-centered free radical intermediate in the prostaglandin synthetase oxidation of arachidonic acid. Spin trapping and oxygen uptake studies. *J. Biol. Chem.* **255**: 5019–5022.

Mastronarde, J.G.; Monick, M.M., Mukaida, N., Matsushima, K., and Hunninghake, G.W. (1998). Activator protein-1 is the preferred transcription factor for cooperative interaction with nuclear factor-kB respiratory syncytial virus-induced interleukin-8 gene expression in airway epithelium. *J. Infect. Dis.* **177**: 1275–1281.

McCay, P.B., Lai, E.K., Poyer, J.L., DuBose, C.M., and Janzen, E.G. (1984). Oxygen- and carbon-centered free radical formation during carbon tetrachloride metabolism. Observation of lipid radicals *in vivo* and *in vitro*. *J. Biol. Chem.* **259**: 2135–2143.

McCord, J.M. (1985). Oxygen-derived free radicals in postischemic tissue injury. *N. Engl. J. Med.* **312**: 159–163.

McCormick, M.L., Buettner, G.R., and Britigan, B.E. (1995). The spin trap α-(pyridyl-1-oxide)-N-*tert*-butylnitrone stimulates peroxidase-mediated oxidation of deferoxamine. Implications for pharmacological use of spin-trapping agents. *J. Biol. Chem.* **270**: 29265–29269.

McKechnie, K., Furman, B.L., and Parratt, J.R. (1986). Modification by oxygen free radical scavengers of the metabolic and cardiovascular effects of endotoxin infusion in conscious rats. *Circ. Shock* **19**: 429–439.

Mendelson, K. G., Contois, L.-R., Tevosian, S.G., Davis, R.J., and Paulson, K.E. (1996). Independent regulation of JNK/p38 mitogen-activated kinases by metabolic oxidative stress in the liver. *Proc. Natl. Acad. Sci. USA* **93**: 12908–12913.

Miyajima, T., and Kotake, Y. (1995). Spin trapping agent, phenyl N-*tert*-butyl nitrone, inhibits induction of nitric oxide synthase in endotoxin-induced shock in mice. *Biochem. Biophys. Res. Commun.* **215**: 114–121.

Miyajima, T., and Kotake, Y. (1997). Optimal time and dosage of phenyl N-*tert*-butyl nitrone (PBN) for the inhibition of nitric oxide synthase induction in mice. *Free Radical Biol. Med.* **22**: 463–470.

Moncada, S., Palmer, R.M.J., and Gryglewski, R.J. (1986). Mechanism of action of some inhibitors of endothelium-derived relaxing factor. *Proc. Natl. Acad. Sci. USA* **83**: 9164–9168.

Monti, E., Paracchini, L., Perletti, G., and Piccinini, F. (1991). Protective effects of spin-trapping agents on adriamycin-induced cardiotoxicity in isolated rat atria. *Free Radical Res. Commun.* **14**: 41–45.

Monti, E.. Prosperi, E., Supino, R., and Bottiroli, G. (1995). Free radical-dependent DNA lesions are involved in the delayed cardiotoxicity induced by adriamycin in the rat. *Anticancer Res.* **15**: 193–198.

Morgan, D.D., Mendenhall, C.L., Bobst, A.M., and Rouster, S.D. (1985). Incorporation of the spin trap DMPO into cultured fetal mouse liver cells. *Photochem. Photobiol.* **42**: 93–94.

Mottley, C., Mason, R.P., Chignell, C.F., Sivarajah, K., and Eling, T.E. (1982). The formation of sulfur trioxide radical anion during the prostaglandin hydroperoxidase-catalyzed oxidation of bisulfite (hydrated sulfur dioxide). *J. Biol. Chem.* **257**: 5050–5055.

Nademanne, K., and Singh, B.N. (1988). Control of cardiac arrhythmias by calcium antagonism. *Ann. N.Y. Acad. Sci.* **522**: 536–552.

Nakazawa, H., Arroyo, C.M., Ichimori, K.J., Saigusa, Y., Minezaki, K.K., and Pronai, L. (1991). The demonstration of DMPO superoxide adduct upon reperfusion using a low non-toxic concentration. *Free Radical Res. Commun.* **14**: 297–302.

Novelli, G.P. (1992). Oxygen radicals in experimental shock: Effects of spin-trapping nitrones in ameliorating shock pathophysiology. *Crit. Care Med.* **20**: 499–507.

Novelli, G.P., Angiolini, P., Tani, R., Consales, G., and Bordi, L. (1985). Phenyl-T-butyl-nitrone is active against traumatic shock in rats. *Free Radical Res. Commun.* **1**: 321–327.

Novelli, G.P., Bracciotti, G., and Falsini, S. (1990). Spin-trappers and vitamin E prolong endurance to muscle fatigue in mice. *Free Radical Biol. Med.* **8**: 9–13.

Oliver, C.N., Starke-Reed, P.E., Stadtman, E.R., Liu, G.J., Carney, J.M., and Floyd, R.A. (1990). Oxidative damage to brain proteins, loss of glutamine synthetase activity, and production of free radicals during ischemia/reperfusion-induced injury to gerbil brain. *Proc. Natl. Acad. Sci. USA* **87**: 5144–5147.

Ortiz de Montellano, P.R., Augusto, O., Viola, F., and Kunze, K.L. (1983). Carbon radicals in the metabolism of alkyl hydrazines. *J. Biol. Chem.* **258**: 8623–8629.

Overend, R., Paraskevopoulos, G., and Black, C. (1976). Rates of OH radical reactions. II. The combination reaction OH + NO + M. *J. Chem. Phys.* **64**: 4149–4154.

Pahlmark, K., and Siesjö, B.K. (1996). Effects of the spin trap-α-phenyl-N-tert-butyl nitrone (PBN) in transient forebrain ischaemia in the rat. *Acta Physiol. Scand.* **157**: 41–51.

Paracchini, L., Jotti, A., Bottiroli, G., Prosperi, E., Supino, R., and Piccinini, F. (1993). The spin trap alpha-phenyl-tert-butyl nitrone protects against myelotoxicity and cardiotoxicity of adriamycin while preserving the cytotoxicity activity. *Anticancer Res.* **13**: 1607–1612.

Parks, D.A., Bulkley, G.B., Granger, D.N., Hamilton, S.R., and McCord, J.M. (1982). Ischemic injury in the cat small intestine: Role of superoxide radicals. *Gastroenterology* **82**: 9–15.

Parratt, J.R., and Wainwright, C.L. (1987). Failure of allopurinol and a spin trapping agent N-t-butyl-α-phenyl nitrone to modify significantly ischaemia and reperfusion-induced arrhythmias. *Br. J. Pharmac.* **91**: 49–59.

Phillis, J.W., and Clough-Helfman, C. (1990aa). Protection from cerebral ischemic injury in gerbils with the spin trap agent N-tert-butyl-α-phenylnitrone (PBN). *Neurosci. Lett.* **116**: 315–319.

Phillis, J.W., and Clough-Helfman, C. (1990b). Free radicals and ischaemic brain injury: Protection by the spin trap agent PBN. *Med. Sci. Res.* **18**: 403–404.

Piccinini, F., Bradamante, S., Monti, E., Zhang, Y.-K., and Janzen, E.G. (1995). Pharmacological action of a new spin trapping compound, 2-phenyl DMPO, in the adriamycin-induced cardiotoxicity. *Free Radical Res.* **23**: 81–87.

Pogrebniak, H.W., Merino, M.J., Hahn, S.M., Mitchell, J.B., and Pass, H.I. (1992). Spin trap salvage from endotoxemia: The role of cytokine down-regulation. *Surgery* **112**: 130–139.

Pou, S., Cohen, M.S., Britigan, B.E., and Rosen, G.M. (1989a). Spin trapping and human neutrophils: Limits of detection of hydroxyl radical. *J. Biol. Chem.* **264**: 12299–12302.

Pou, S., Huang, Y.-I., Bhan, A., Bhadti, V.S., Hosmane, R.S., Wu, S.Y., Cao, G.-L., and Rosen, G.M. (1993). A fluorophore-containing nitroxide as a probe to detect superoxide and hydroxyl radical generated by stimulated neutrophils. *Anal. Biochem.* **212**: 85–90.

Pou, S., Keaton, L., Surichamorn, W., Frigillana, P., and Rosen, G.M. (1994a). Can nitric oxide be spin trapped by nitrone and nitroso compounds? *Biochim. Biophys. Acta* **1201**: 118–124.

Pou, S., Ramos, C.L., Gladwell, T., Renks, E., Centra, M., Young, D., Cohen, M.S., and Rosen, G.M. (1994b). A kinetic approach to the selection of a sensitive spin trapping system for the detection of hydroxyl radical. *Anal. Biochem.* **217**: 76–83.

Pou, S., Rosen, G.M., Britigan, B.E., and Cohen, M.S. (1989b). Intracellular spin-trapping of oxygen centered radicals generated by human neutrophils. *Biochim. Biophys. Acta* **991**: 459–464.

Poyer, J.L., McCay, P.B, Weddle, C.C., and Downs, P.E. (1981). *In vivo* spin-trapping of radicals formed during halothane metabolism. *Biochem. Pharmacol.* **30**: 1517–1519.

Rabin, B.M. (1996). Free radicals and taste aversion learning in the rat: Nitric oxide, radiation and dopamine. *Prog. Neuro-Psychopharmacol. Biol. Psychiat.* **20**: 691–707.

Rajagopalan, S., Politi, P.M., Sinha, B.K., and Myers, C.E. (1988). Adriamycin-induced free radical formation in the perfused rat heart: Implications for cardiotoxicity. *Cancer Res.* **48**: 4766–4769.

Ramos, C.L., Pou, S., Britigan, B.E., Cohen, M.S., and Rosen, G.M. (1992). Spin trapping evidence for myeloperoxidase-dependent hydroxyl radical formation by human neutrophils and monocytes. *J. Biol. Chem.* **267**: 8307–8312.

Reimer, K.A., Murry, C.E., and Richard, V.J. (1989). The role of neutrophils and free radicals in the ischemic-reperfused heart: Why the confusion and controversy. *J. Mol. Cell. Cardiol.* **21**: 1225–1239.

Reinke, L.A., Lai, E.K., DuBose, C.M., and McCay, P.B. (1987). Reactive free radical generation *in vivo* in heart and liver of ethanol-fed rats: Correlation with radical formation *in vitro*. *Proc. Natl. Acad. Sci. USA* **84**: 9223–9227.

Reinke, L.A., Towner, R.A., and Janzen, E.G. (1992). Spin trapping of free radical metabolites of carbon tetrachloride *in vitro* and *in vivo*: Effect of acute ethanol administration. *Toxicol. Appl. Pharmacol.* **112**: 17–23.

Rosen, G.M., and Freeman, B.A. (1984). Detection of superoxide generated by endothelial cells. *Proc. Natl. Acad. Sci. USA* **81**: 7269–7273.

Rosen, G.M., Britigan, B.E., Cohen, M.S., Ellington, S.P., and Barber, M.J. (1988). Detection of phagocyte-derived free radicals with spin trapping techniques: Effects of temperature and cellular metabolism. *Biochim. Biophys. Acta,* **969**: 236–241.

Rosen, G.M., Pou, S., Ramos, C.L., Cohen, M.S., and Britigan, B.E. (1995). Free radicals and phagocytic cells. *FASEB J.* **9**: 200–209.

Rosen, H., and Klebanoff, S.J. (1979). Hydroxyl radical generation by polymorphonuclear leukocytes measured

by electron spin resonance spectroscopy. *J. Clin. Invest.* **64**: 1725–1729.

Rubanyi, G.M., and Vanhoutte, P.M. (1986). Superoxide anions and hyperoxia inactivate endothelial derived relaxing factor. *Am. J. Physiol.* **250**: H822-H827.

Rubanyi, G.M., Ho, E.H., Cantor, E.H., Lumma, W.C., and Parker-Botelho, L.H. (1991). Cytoprotective function of nitric oxide: Inactivation of superoxide radicals produced by human leukocytes. *Biochem. Biophys. Res. Commun.* **181**: 1392–1397.

Sack, C.A., Socci, D.J., Crandall, B.M., and Arendash, G.W. (1996). Antioxidant treatment with phenyl-α-*tert*-butyl nitrone (PBN) improves the cognitive performance and survival of aging rats. *Neurosci. Lett.* **205**: 181–184.

Saito, K., Yoshioka, H., and Cutler, R.G. (1998b). A spin trap, N-*tert*-butyl- α-phenylnitrone extends the life span of mice. *Biosci. Biotechnol. Biochem.* **62**: 792–794.

Saito, K., Yoshioka, H., Kazama, S., and Cutler, R.G. (1998a). Release of nitric oxide from a spin trap, N-*tert*-butyl-α-phenylnitrone, under various oxidative conditions. *Biol. Pharmac. Bull.* **21**: 401–404.

Samuni, A., Black, C.D.V., Krishna, C.M., Malech, H.L., Bernstein, E.F., and Russo, A. (1988). Hydroxyl radical production by stimulated neutrophils reappraised. *J. Biol. Chem.* **263**: 13797–13801.

Samuni, A., Carmichael, A.J., Russo, A., Mitchell, J.B., and Reisz, P. (1986). On the spin trapping and ESR detection of oxygen-derived free radicals generated inside cells. *Proc. Natl. Acad. Sci. USA* **83**: 7593–7597.

Samuni, A., Krishna, C.M., Reisz, P., Finkelstein, E., and Russo, A. (1989a). Superoxide reaction with nitroxide spin adducts. *Free Radical Biol. Med.* **6**: 141–148.

Samuni, A., Samuni, A., and Swartz, H.M. (1989b). Evaluation of dibromonitrosobenzene sulfonate as a spin trap in biological systems. *Free Radical Biol. Med.* **7**: 37–43.

Saran, M., Michel, C., and Bors, W. (1990). Reaction of nitric oxide with superoxide. Implications for the action of endothelial-derived relaxing factor (EDRF). *Free Radical Res. Commun.* **10**: 221–226.

Schaefer, C.F., Janzen, E.G., West, M.S., Poyer, J.L., and Kosanke, S.D. (1996). Blood chemistry changes in the rat induced by high doses of nitronyl free radical spin traps. *Free Radical Biol. Med.* **21**: 427–436.

Schilling, W. P., and Elliott, S.J. (1992). Ca^{2+} signaling mechanisms of vascular endothelial cells and their role in oxidant-induced endothelial cell dysfunction. *Am. J. Physiol.* **262**: H1617-H1630.

Schulz, J.B., Henshaw, D.R., Siwek, D., Jenkins, B.G., Ferrante, R.J., Cipolloni, P.B., Kowall, N.W., Rosen, B.R., and Beal, M.F. (1995a). Involvement of free radicals in excitotoxicity *in vivo. J. Neurochem.* **64**: 2239–2247.

Schulz, J.B., Matthews, R.T., Jenkins, B.G., Brar, P., and Beal, M.F. (1995b). Improved therapeutic window for treatment of histotoxic hypoxia with a free radical spin trap. *J. Cereb. Blood Flow Metab.* **15**: 948–952.

Seawright, L., Tanigawa, M., Tanigawa, T., Kotake, Y., and Janzen, E.G. (1995). Can spin trapping compounds like PBN protect against self-inflicted damage in polymorphonuclear leukocytes? *Free Radical Res.* **23**: 73–80.

Seddon, J.M., Ajani, U.A., Sperduto, R.D., Hiller, R., Blair, N., Burton, T.C., Farber, M.D., Gragoudas, E.S., Haller, J., Miller, D.T., Yannuzzi, L.A., and Willett, W. (1994). Dietary carotenoids, vitamins A, C, and E, and advanced age-related macular degeneration. *JAMA* **272**: 1413–1420.

Seddon, W.A., Fletcher, J.W., and Sopchyshyn, F.C. (1973). Pulse radiolysis of nitric oxide in aqueous solution. *Can. J. Chem.* **51**: 1123–1130.

Sekili, S., McCay, P.B., Li, X.-Y., Zughaib, M., Sun, J.-Z., Tang, L., Thornby, J.I., and Bolli, R. (1993). Direct evidence that the hydroxyl radical plays a pathogenetic role in myocardial "stunning" in the conscious dog and demonstration that stunning can be markedly attenuated without subsequent adverse effects. *Circ. Res.* **73**: 705–723.

Sen, S., and Phillis, J.W. (1993). Alpha-phenyl-tert-butyl-nitrone (PBN) atten-uates hydroxyl radical production during ischemia-reperfusion injury of rat brain: An EPR study. *Free Radical Res. Commun.* **19**: 255–265.

Sen, S., Goodman, H., Morehead, M., Murphy, S., and Phillis, J.W. (1994). α-Phenyl-*tert*-butyl-nitrone inhibits free radical release in brain concussion. *Free Radical Biol. Med.* **16**: 685–691.

Siegfried, M.R., Carey, C., Ma, X.-L., and Lefer, A.M. (1992a). Beneficial effects of SPM-5185, a cysteine-containing NO donor in myocardial ischemia-reperfusion. *Am. J. Physiol.* **263**: H771-H777.

Siegfried, M.R., Erhardt, J., Rider, T., Ma, X.-L., and Lefer, A.M. (1992b). Cardioprotection and attenuation of endothelial dysfunction by organic nitric oxide donors in myocardial ischemia-reperfusion. *J. Pharmacol. Exp. Ther.* **260**: 668–675.

Siminiak, T., and Ozawa, T. (1993). Neutrophil mediated myocardial injury. *Int. J. Biochem.* **25**: 147–156.

Simpson, P.J., Fantone, J.C., Mickelson, J.K., Gallagher, K.P., and B.R. Lucchesi, B.R. (1988). Identification of a time window for therapy to reduce experimental canine myocardial injury: Suppression of neutrophil activation during 72 hours of reperfusion. *Circ. Res.* **63**: 1070–1079.

Simpson, P.J., Mitsos, S.E., Ventura, A., Gallagher, K.P., Fantone, J.C., Abrams, G.D., Schork, M.A., and Lucchesi, B.R. (1987). Prostacyclin protects ischemic reperfused myocardium in the dog by inhibition of neutrophil activation. *Am. Heart J.* **113**: 129–137.

Strehlow, H., and Wagner, I. (1982). Flash photolysis in aqueous nitrite solutions. *Z. Phys. Chem.* **132**: 151–160.

Sun, Y., and L. W. Oberley, L.W. (1996). Redox regulation of transcriptional activators. *Free Radical Biol. Med.* **21**: 335–348.

Suzuki, M., Inauen, W., Kvietys, P.R., Grisham, M.B., Meininger, C., Schelling, M.E., Granger, H.J., and Granger, D.N. (1989). Superoxide mediates reperfusion-induced leukocyte-endothelial cell interactions. *Am. J. Physiol.* **257**: H1740-H1745.

Suzuki, Y. J., Forman, H.J., and Sevanian, A. (1997). Oxidants as stimulators of signal transduction. *Free Radical Biol. Med.* **22**: 269–285.

Tabatabaie, T., Kotake, Y., Wallis, G., Jacob, J.M., and Floyd, R.A. (1997). Spin trapping agent phenyl *N-tert*-butylnitrone protects against the onset of drug-induced insulin-dependent diabetes mellitus. *FEBS Lett.* **407**: 148–152.

Tanigawa, T., Kotake, Y., and Reinke, L.A. (1993). Spin trapping of superoxide radicals following stimulation of neutrophils with fMLP is temperature dependent. *Free Radical Biol. Med.* **15**: 425–433.

Thomas, C.E., Bernardelli, P., Bowen, S.M., Chaney, S.F., Friedrich, D., Janowick, D.A., Jones, B.K., Keeley, F.J., Kehne, J.H., Kettler, B., Ohlweiler, D.F., Paquette, L.A., Robke, D.J., and Fevig, T.L. (1996). Cyclic nitrone free radical traps: Isolation, identification and synthesis of 3,3-dimethyl-3,4-dihydroisoquinone-4-ol N-oxide, a metabolite with reduced side effects. *J. Med. Chem.* **39**: 4997–5004.

Thomas, C.E., Ku, G., and Kalyanaraman, B. (1994a). Nitrone spin trap lipophilicity as a determinant for inhibition of low density lipoprotein oxidation and activation of interleukin-1β release from human monocytes. *J. Lipid Res.* **35**: 610–619.

Thomas, C.E., Ohlweiler, D.F., and Kalyanaraman, B. (1994b). Multiple mechanisms for inhibition of low density lipoprotein oxidation by novel cyclic nitrone spin traps. *J. Biol. Chem.* **269**: 28055–28061.

Thomas, C.E., Ohlweiler, D.F., Taylor, V.L., and Schmidt, C.J. (1997). Radical trapping and inhibition of iron-dependent CNS damage by cyclic nitrone spin traps. *J. Neurochem.* **68**: 1173–1182.

Thornalley, P.J., and Stern, A. (1985). The effect of nitrone spin trapping agents on red cell glucose metabolism. *Free Radical Res. Commun.* **1**: 111–117.

Thornalley, P.J., Stern, A., and Bannister, J.V. (1983). A mechanism for primaquine mediated oxidation of NADPH in red blood cells. *Biochem. Pharmacol.* **32**: 3571–3575.

Tosaki, A., and Braquet, P. (1990). DMPO and reperfusion injury: Arrhythmia, heart function, electron spin resonance, and nuclear magnetic resonance studies in isolated working guinea pig hearts. *Am. Heart J.* **120**: 819–830.

Tosaki, A., Blasig, I.E., Pali, T., and Ebert, B. (1990). Heart protection and radical trapping by DMPO during reperfusion in isolated working hearts. *Free Radical Biol. Med.* **8**: 363–372.

Tosaki, A., Haseloff, R.F., Hellegouarch, A., Schoenheit, K., Martin, V.V., Das, D.K., and Blasig, I.E. (1992). Does the antiarrhythmic effect of DMPO originate from its oxygen radical trapping property or the structure of the molecule itself? *Basic Res. Cardiol.* **87**: 536–547.

Towner, R.A., Janzen, E.G., Zhang, Y.-K., and Yamashiro, S. (1993). MRI study of the inhibitory effect of new spin traps on *in vivo* CCl$_4$-induced hepatotoxicity in rats. *Free Radical Biol. Med.* **14**: 677–681.

Treinin, A., and Hayon, E. (1970). Absorption spectra and reaction kinetics of NO$_2$, N$_2$O$_3$, and N$_2$O$_4$ in aqueous solution. *J. Am. Chem. Soc.* **92**: 5821–5828.

Triana, J.F., Li, X.-Y., Jamaluddin, U., Thornby, J.I., and Bolli, R. (1991). Postischemic myocardial "stunning." Identification of major differences between the open-chest and the conscious dog and evaluation of the oxygen radical hypothesis in the conscious dog. *Circ. Res.* **69**: 731–747.

Ueno, I., Kohno, M., Mitsuta, K., Mizuta, Y., and Kanegasaki, S. (1989). Reevaluation of the spin-trapped adduct formed from 5,5-dimethyl-1-pyrroline-1-oxide during the respiratory burst in neutrophils. *J. Biochem.* **105**: 905–910.

Wang, P., and Zweier, J.L. (1996). Measurement of nitric oxide and peroxynitrite generation in the postischemic heart. Evidence for peroxynitrite-mediated reperfusion injury. *J. Biol. Chem.* **271**: 29223–29230.

Werns, S.W., and Lucchesi, B.R. (1988). Leukocytes, oxygen radicals, and myocardial injury due to ischemia and reperfusion. *Free Radical Biol. Med.* **4**: 31–37.

Wink, D.A., Hanbauer, I., Krishna, M.C., DeGraff, W., Gamson, J., and Mitchell, J.B. (1993). Nitric oxide protects against cellular damage and cytotoxicity from reactive oxygen species. *Proc. Natl. Acad. Sci. USA* **90**: 9813–9817.

Wolf, C.R., Harrelson, W.G., Jr., Nastainczyk, W.M., Philpot, R.M., Kalyanaraman, B., and Mason, R.P. (1980). Metabolism of carbon tetrachloride in hepatic microsomes and reconstituted monooxygenase systems and its relationship to lipid peroxidation. *Mol. Pharmacol.* **18**: 553–558.

Yamashita, T., Ohshima, H., Asanuma, T., Inukai, N., Miyoshi, I., Kasai, N., Kon, Y., Watanabe, T., Sato, F., and Kuwabara, M. (1996). The effects of α-phenyl-*tert*-butyl nitrone (PBN) on copper-induced rat fulminant hepatitis with jaundice. *Free Radical Biol. Med.* **21**: 755–761.

Yue, T.-L., Gu, J.-L., Lysko, P.G., Cheng, H.-Y., Barone, F.C., and Feuerstein, G. (1992). Neuroprotective effects of phenyl-t-butyl-nitrone in gerbil global brain ischemia and in cultured rat cerebellar neurons. *Brain Res.* **674**: 193–197.

Zhao, Q., Pahlmark, K., Smith, M.-L., and Siesjö, B.K. (1994). Delayed treatment with the spin trap α-phenyl-N-*tert*-butyl nitrone (PBN) reduces infarct size following transient middle cerebral artery occlusion in rats. *Acta Physiol. Scand.* **152**: 349–350.

12

Future Directions

The possibility that a radical addition reaction might provide a means of detecting short-lived radicals had initially been considered during mechanistic studies of the dehydrogenation of hydroaromatics with hot nitrobenzene and thermal decomposition of nitro aromatics.

Janzen, 1971

Notwithstanding Professor Janzen's success in foretelling the birth of spin trapping, such prognostications are more often than not incorrect. Yet, some generalizations can be gleaned from previous chapters that may directly impact on prospective research in this discipline. Advances in spin trapping may well depend on the development of newer spin traps with enhanced specificity toward selective free radicals and the commercialization of lower frequency EPR spectrometers.

Spin Traps

Despite considerable efforts put forth to prepare nitrones whose reactivity toward $O_2 \cdot^-$ is kinetically competitive with other reactions involving this free radical, progress has been minimal. Certainly, DEPMPO, with its nearly five-fold enhancement in the rate of reaction toward $O_2 \cdot^-$ as compared with DMPO (Table 8.1), is an impressive first step; however, at $60 \, M^{-1} \, s^{-1}$ (Fréjaville, et al., 1994; Tuccio, et al., 1995) such a rate constant is unlikely to result in any significant improvement in the detection of $O_2 \cdot^-$ in complex biologic systems. There are, nevertheless, synthetic avenues yet to be explored that may still bear fruit.

A careful examination of the reaction of $O_2 \cdot^-$ toward Δ^1-pyrroline-N-oxides raises new possibilities. At low pH, the spin trapping of $HO_2 \cdot / O_2 \cdot^-$ by DMPO is nearly a thousand-fold faster than its reaction with this nitrone at physiologic pH (Finkelstein, et al., 1980). This disparity in rate constants probably reflects the difference in the reactive species at each pH. Under acidic conditions, $HO_2 \cdot$ dominates, whereas under more basic experimental environments, the anion radical

$O_2 \cdot^-$ prevails. If this theory were to be correct, then by enhancing the electrophilic nature of the β-carbon on DMPO, through inclusion of electron withdrawing groups, the rate of spin trapping of $O_2 \cdot^-$ at physiologic pH should increase, approaching the level seen at lower pH. Even though such a theory awaits testing, the 2-trifluoromethyl analog of DMPO, 5,5-dimethyl-2-(trifluoromethyl)-1-pyrroline-N-oxide (TFDMPO) meets this criterion (Janzen, *et al.*, 1995). Unfortunately, the poor stability of the corresponding spin trapped adduct may limit the use of this nitrone to specific experimental designs (Janzen, *et al.*, 1995). The synthesis of TFDMPO is a first step, awaiting refinements that may, for *in vitro* models as complex as cells, justify such efforts.

A more serious problem will arise when attempting to identify $O_2 \cdot^-$ *in vivo*. Here, even with the best of experimental schemes, the possibility of spin trapping this free radical in living animals in real time seems remote. Consider, for example, the *in vivo* spin trapping of $O_2 \cdot^-$. Upon administration of the nitrone into the animal, this spin trap must diffuse from the extracellular space of its entry point to the intracellular compartments (except in the case of stimulated phagocytes) where this free radical would most likely be generated. Since the preponderance of readily available spin traps is hydrophilic, localization of the nitrone at the site of interest in such a dynamic system would be difficult to accomplish. Assuming these limitations could somehow be overcome, resulting in the recording of an EPR spectrum after the introduction of the nitrone, could one rely on unique hyperfine splitting constants as verification of the free radical actually spin trapped in a living animal? As the half-lives of spin trapped adducts of $O_2 \cdot^-$ in static systems are short (Table 8.7), it may not be possible to make such an assignment. For other free radicals, however, the picture is considerably brighter.

Consider, for instance, the *in vivo in situ* spin trapping of NO· after the activation of murine macrophages (Komarov, *et al.*, 1993; Lai and Komarov, 1994; Komarov and Lai, 1995; Yoshimura, *et al.*, 1996). In each of these studies, either $Fe^{2+}(MGD)_2$ or $Fe^{2+}(DTCS)_2$ was injected into the mouse several days after treatment with lipopolysaccharide. Soon thereafter, EPR spectra were recorded in real time and at the site of NO· evolution (Yoshimura, *et al.*, 1996). Since these ferrous chelates are highly charged molecules, it would be reasonable to assume, in lieu of any definitive experimental data, that the EPR spectrum recorded arose from the extracellular spin trapping of NO·. In the future, one might hope that more lipophilic spin traps would be developed, thereby allowing the identification of this free radical at intracellular sites of formation. As such, verification that each of the isozymes of NOS generates NO· can be documented in animals in real time. Further, these findings can assist in elucidating the mechanism by which NO·regulates specific physiologic functions.

The *in vivo in situ* spin trapping of radiation-induced HO· further illustrates the potential of detecting free radicals in animals in real time (Halpern, *et al.*, 1995). Here, a mixture of 4-POBN/EtOH was injected into a leg tumor of a mouse; a tourniquet was wrapped around the tumor to retard the diffusion of the spin trap from the vascular bed to which the highly charged nitrone was introduced during the 20-min irradiation period when the animal was not restrained within the confines of the cavity of the spectrometer. Based on the hydrophobic characteristics of 4-POBN (Pou, *et al.*, 1994) and the *in vivo* elimination kinetics of 4-POBN–

$CH(CH_3)OH$, it was hypothesized that 4-POBN reported HO· production only in the vasculature and interstitial space of the tumor (Halpern, *et al.*, 1995). Although it is probable that radiolytic generation of HO· occurs intracellularly, data indicate that it is in the nucleus where this free radical exerts its primary toxicity. Development of spin traps that localize in this organelle would allow a more quantitative study of the modulation of radiation damage. Once this development is completed, the hunt to capture and image this free radical, as a spin trapped adduct, in animals in real time can begin in earnest.

EPR Spectroscopy

As with other instruments, it would be naive not to assume that future developments in lower frequency EPR spectrocopy would not open new and exciting research opportunities. However, considerable effort will be required before low-frequency EPR spectroscopy is routinely available to the scientific community. As discussed in chapter 7, measurements and images in tissues require the use of radiofrequencies. Superficial structures in small mammals, for example, can be accomplished at L-band frequencies (1–2 GHz). Although the sacrifice of sensitivity with frequency reduction is not as great as has been assumed, there will be, nonetheless, some loss in signal quality. A major effort needs to be undertaken to estimate the limits of signal acquisition under these circumstances. Exploration of new modalities of signal detection, including longitudinally detected EPR (Nicholson, *et al.*, 1996), multiquantum detection (Mchaourab and Hyde, 1993a, 1993b), and a variety of other methods, need to be further investigated. New resonant structures in the context of EPR (which have been available in NMR) must be defined, such as the bird cage resonator (Hayes, *et al.*, 1985). Finally, novel, rapid signal acquisition methods using continuous-wave or pulsed EPR spectroscopy (Murugesan, *et al.*, 1997) must likewise be studied to determine the optimum approach of extracting weak and moderately broad-lined signals from *in vivo* tissues.

Isotope-labeled spin traps designed to narrow the spectral line of the corresponding spin trapped adduct will further enhance the ability to use EPR spectroscopy for *in vivo in situ* detection and imaging of endogenous free radicals (Halpern, *et al.*, 1993). At this juncture, it is difficult to predict which of the above advancements will have the most significant impact on the capability for *in vivo* imaging of free radicals. Perhaps, the most likely scenario is one where advances in the design of spin traps as outlined above will selectively improve one, or several, of the spectroscopic techniques. This, in turn, may suggest synthetic paths to define unique spin traps for specific free radicals. The advances will undoubtedly take place interactively.

The promise that emerges from spin trapping and low-frequency EPR spectroscopy is the possibility of a noninvasive or minimally invasive means to detect and image free radicals *in vivo* and in real time in humans. This will have far-reaching consequences for the study of tissue function in healthy and diseased states. Spin trapping in combination with low-frequency EPR spectroscopy will have a profound effect on monitoring the efficacy of a pharmacologic intervention for a number of disease processes, including the minimization of myocardial infarction, reduction of vascular tone in hypertens. itoring of the toxicity of radiation in tumor and normal tissues. advances necessary to fulfill this promise.

References

Finkelstein, E., Rosen, G.M., and Rauckman, E.J. (1980). Spin trapping. Kinetics of the reaction of superoxide and hydroxyl radicals with nitrones. *J. Am. Chem. Soc.* **102**: 4994–4999.

Fréjaville, C., Karoui, H., Tuccio, B., Le Moigne, F., Culcasi, M., Pietri, S., Lauricella, R., and Tordo, P. (1994). 5-Diethoxyphosphoryl-5-methyl-1-pyrroline N-oxide (DEPMPO): A new phosphorylated nitrone for efficient *in vitro* and *in vivo* spin trapping of oxygen-centred radicals. *J. Chem. Soc., Chem. Commun.* 1793–1794.

Halpern, H.J., Pou, S., Peric, M., Yu, C., Barth, E., and Rosen, G.M. (1993). Detection and imaging of oxygen-centered free radicals with low-frequency electron paramagnetic resonance and signal-enhancing deuterium-containing spin traps. *J. Am. Chem. Soc.* **115**: 218–223.

Halpern, H.J., Yu, C., Barth, E., Peric, M., and Rosen, G.M. (1995). *In situ* detection, by spin trapping of hydroxyl radical markers produced from ionizing radiation in the tumor of a living mouse. *Proc. Natl. Acad. Sci. USA* **92**: 796–800.

Hayes, C.E., Edelstein, W.A., Schenk J.F., Mueller, O.M., and Eash, M. (1985). An efficient, highly homogeneous radiofrequency coil for whole-body NMR imaging at 1.5 T. *J. Magn. Reson.* **63**: 622–628.

Janzen, E.G. (1971). Spin trapping. *Acc. Chem. Res.* **4**: 31–40.

Janzen, E.G., Zhang, Y.-K., and Arimura, M. (1995). Synthesis and spin-trapping chemistry of 5,5-dimethyl-2-(trifluoromethyl)-1-pyrroline N-oxide. *J. Org. Chem.* **60**: 5434–5440.

Komarov, A.M., and Lai, C.-S. (1995). Detection of nitric oxide production in mice by spin-trapping electron paramagnetic resonance spectroscopy. *Biochim. Biophys. Acta* **1272**: 29–36.

Komarov, A., Mattson, D., Jones, M.M., Singh, P.K., and Lai, C.-S. (1993). *In vivo* spin trapping of nitric oxide in mice. *Biochem. Biophys. Res. Commun.* **195**: 1191–1198.

Lai, C.-S., and Komarov, A.M. (1994). Spin trapping of nitric oxide produced *in vivo* in septic-shock in mice. *FEBS Letts.* **345**: 120–124.

Mchaourab, H.S., and Hyde, J.S. (1993a). Dependence of the multiple- quantum EPR signal on the spin-lattice relaxation time. Effect of oxygen in spin-labeled membranes. *J. Magn. Reson.* **101B**: 178–184.

Mchaourab, H.S., and Hyde, J.S. (1993b). Continuous wave multiquantum electron paramagnetic resonance spectroscopy. III. Theory of intermodulation sidebands. *J. Chem. Phys.* **98**: 1786–1796.

Murugesan, R., Cook, J.A., Devasahayam, N., Afeworki, M., Subramanian, S., Tschudin, R., Larsen, J.H.A., Mitchell, J.B., Russo, A., and Krishna, M.C. (1997). *In vivo* imaging of a stable paramagnetic probe by pulsed-radio-frequency electron paramagnetic resonance spectroscopy. *Magn. Reson. Med.* **38**: 409–414.

Nicholson, I., Foster, M.A., Robb, F.J.L., Hutichison, J.M.S., and Lurie, D.J. (1996). *In vivo* imaging of nitroxide-free-radical clearance in the rat using radiofrequency longitudinally detected ESR imaging. *J. Magn. Reson.* **113**: 256–261.

Pou, S., Ramos, C.L., Gladwell, T., Renks, E., Centra, M., Young, D., Cohen, M.S., and Rosen, G.M. (1994). A kinetic approach to the selection of a sitive spin trapping system for the detection of hydroxyl radical. *Anal. Biochem.* **217**: 76–83.

uccio, B., Lauricella, R., Fréjaville, C., Bouteiller, J.-C., and Tordo, P. (1995). Decay of the hydroperoxyl spin adduct of 5-diethoxyphosphoryl-5-methyl-1-pyrroline N-oxide: An EPR kinetic study. *J. Chem. Soc., Perkin Trans. 2* 295–298.

Yoshimura, T., Yokoyama, H., Fujii, S., Takayama, F., Oikawa, K., and Kamada, H. (1996). *In vivo* EPR detection and imaging of endogenous nitric oxide in lipopolysaccharide-treated mice. *Nat. Biotechnol.* **14**: 992–994.

Notes

Chapter 1

1. There are a number of excellent sources on the origin of life. Several recent texts are especially helpful:

Alberts, B., Bray, D., Lewis, J., Raff, M., Roberts, K., and Watson, J.D. (1994). *Molecular Biology of the Cell*, 3rd edition, Garland Publishing, New York.

Campbell, N.A. (1993). *Biology*, 3rd edition, The Benjamin/Cummings Publishing Co., Redwood City, CA.

Darnell, J., Lodish, H., and Baltimore, D. (1990). *Molecular Cell Biology*, 2nd edition, Scientific American Books, New York.

DeDuve, C. (1991). *Blueprint for a Cell. The Nature and Origin of Life*, Neil Patterson Publishers, Carolina Biological Supply Co., Burlington, NC.

Lehninger, A.L., Nelson, D.L., and Cox, M.M. (1993). *Principles of Biochemistry*, Worth Publishers, New York.

Moran, L.A., Scrimgeour, K.G., Horton, H.R., Ochs, R.S., and Rawn, J.D. (1994). *Biochemistry*, 2nd edition, Neil Patterson Publishers, Prentice Hall, Englewood Cliffs, NJ.

Weinberg, S. (1994). Life in the Universe, *Sci. Am.* **271(4)**: 44–49.

2. The discovery of possible biologic-derived chemicals in a Martian meteorite dating from the time life is thought to have emerged on Earth may suggest the existence of life on other planets in the universe. See, David S. McKay, Everett K. Gibson, Jr., Kathie L. Thomas-Keprta, Hojatollah Vali, Christopher S. Romanek, Simon J. Clemett, Xavier D.F. Chillier, Claude R. Maechling, and Richard N. Zare, "Search for past life on Mars: Possible relic biogenic activity in Martian meteorite ALH84001," *Science* **273** (1996): 924–930.

3. In the past several years, there has been much interest in complex vesicle formation. Typical of these studies is a recent report in which a surfactant was prepared that could spontaneously assemble and undergo an autocatalytic self-reproduction process (Andrea Veronese and Pier Luigi Luisi, "An autocatalytic reaction leading to spontaneously assembled phosphatidyl nucleoside giant vesicles," *J. Am. Chem. Soc.* **120** (1998): 2662–2663).

Chapter 2

Portions of this chapter were adopted from a review article (Gerald M. Rosen, Sovitj

Pou, Carroll L. Ramos, Myron S. Cohen, and Bradley E. Britigan, "Free radicals and phagocytic cells," *FASEB J.* **9** (1995): 200–209).

1. It is uncanny that the discovery of O_2 by Scheele in Sweden and Priestley in London in the early 1770s occurred at a time of great economic and political upheavel throughout Europe. Considering these unsettling times, with frequent disruption in communications and travel, it is surprising that both Priestley and Scheele were in frequent contact with Lavoisier through letters and a visit by Priestley in 1774. The influence that these scientists had on Lavoisier is apparent from his experimental designs and his publications at the time. (see Henry Guerlac, *Lavoisier—The Crucial Year. The Background and Origin of His First Experiments on Combustion in 1772,* Cornell University Press, Ithaca, NY, 1961). For a more extensive review on the history of O_2, the interested reader should entertain the fine account of the discovery of this gas by Daniel L. Gilbert, "Perspective on the history of oxygen and life," in *Oxygen and Living Processes. An Interdisciplinary Approach,* ed. by D.L. Gilbert, Springer-Verlag, New York, 1981, pp. 1–43.

2. While investigating the nature of cathode rays, Röntgen found that when a Crookes tube was covered and enclosed in a box, serendipitously placed at a distance from barium–platinum–cyanide crystals, and charged, the crystals fluoresced. Upon placing his hand between the Crookes tube and the crystalline alloy, he could see an outline of his bones, even as the material glowed in the dark. This suggested to Röntgen that this tube was emitting a new more penetrating form of radiation, which he called the X-ray (see Wilhelm C. Röntgen, "Über eine neue Art von Strahlen," *Würzberger Physik.-Med. Ges.* **9** (1895): 132–141).

At about the same time, Grubbé in Chicago, the manufacturer of Crookes tubes, observed severe dermatitis on his hands, which he attributed to the constant exposure to the newly discovered X-rays, described by Röntgen, during repeated testing of these tubes. Soon thereafter, Grubbé consulted Dr. Cobb and his colleagues at the Hahnemann Medical College in Chicago, who confirmed his suspicions. One of the physicians, Dr. Gilman, thought this radiation might be useful in the treatment of pathologic conditions, including cancer. On January 29, 1896, using the expertise of Grubbé, a patient with carcinoma of the left breast was treated with X-rays. For a personal account, see Emil H. Grubbé, "Priority in the therapeutic use of X-rays," *Radiology,* **21** (1933): 156–162. Independent of the studies conducted in Chicago, in February of 1896, Dr. Voigt of Zürich likewise treated a cancer patient with this newly discovered radiation. Thus began the field of radiation oncology.

3. In 1952, Rebeca Gerschman was a research associate in the Department of Physiology at the University of Rochester. As she was preparing a seminar on "Adrenals and Hyperoxia," she became intrigued by the apparent similarities in toxicity associated with elevated levels of O_2 and ionizing radiation. Over the next several years, the pursuit of this connection led to her novel theory that free radicals were the causal species responsible for O_2 toxicity. For a more insightful discussion, the reader should consult Rebeca Gerschman, "Historial introduction to the free radical theory of oxygen toxicity," in *Oxygen and Living Processes. An Interdisciplinary Approach,* ed. by D.L. Gilbert, Springer-Verlag, New York, 1981, pp 44–46 and Daniel L. Gilbert, "Rebeca Gerschman: A personal remembrance," *Free Radical Biol. Med.* **21** (1996): 1–4.

4. In 1968, Joe McCord, a graduate student with Irwin Fridovich, demonstrated that $O_2 \cdot^-$ was not bound to xanthine oxidase, but, rather, $O_2 \cdot^-$ was released into the reaction medium during oxidative metabolism of xanthine. This observation was soon followed by the discovery of specific superoxide dismutase-like properties associated with a copper-containing protein, erythrocuprein, later to be known as superoxide dismutase. The universality of univalent O_2 reduction pathways in biological systems is confirmed by the wide distribution of SOD throughout the various domains, archaea, bacteria and eucarya. For a personal perspective on the discovery of $O_2 \cdot^-$ and SOD, see Irwin Fridovich, "The discovery of superoxide dismutases: A history," in *Superoxide Dismutases,* Vol. I, ed. by L.W. Oberley, CRC Press, Boca Raton, FL, 1982, pp. 1–9 and Joe M. McCord and

Irwin Fridovich, "Superoxide dismutase: The first twenty years (1968–1988)," *Free Radical Biol. Med.* **5** (1988): 363–369.

5. In 1876, Fenton described the formation of a violet color when tartaric acid was mixed with H_2O_2 and ferrous salts. Over the next decade, Fenton continued his studies, exploring the limits of this reaction and, by 1898, he proposed a mechanism for the oxidation. It was not until several years after Fenton's death, however, that HO· was suggested as the oxidant produced during the reduction of H_2O_2 by Fe^{2+}. In contrast, Haber and Weiss thought that HO· could be formed, independent of the presence of Fe^{2+}, by the mere reaction of H_2O_2 with $HO_2·/O_2·^-$.

$$O_2·^- + H_2O_2 \rightarrow HO· + HO^- + O_2$$

Subsequently, this reaction has been shown to be too slow to be of biologic significance in the absence of iron salts. For thoughtful discussions of this topic, see W. Hugh Melhuish and Harry C. Sutton, "Study of the Haber-Weiss reaction using a sensitive method for detection of OH radicals," *J. Chem. Soc., Chem. Commun.* (1978): 970–971; Judith Weinstein and Benon H.J. Bielski, "Kinetics of the interaction of HO_2 and O_2^- radicals with hydrogen peroxide. The Haber-Weiss reaction," *J. Am. Chem. Soc.* **101** (1979): 58–62; William H. Koppenol, "The centennial of the Fenton reaction", *Free Radical Biol. Med.* **15** (1993): 645–651.

6. It is very difficult to obtain accurate cellular levels for the $[O_2·^-]_{ss}$. For the purposes of this discussion, however, we have chosen high and low fluxes of $O_2·^-$ to emphasize the importance of SOD in regulating $[O_2·^-]_{ss}$. Several assumptions have been made. First, for PMA-stimulated neutrophils, the initial rate of 18 nM/min $O_2·^-$ secreted from 5×10^6 cells was calculated from the SOD-inhibitable reduction of cytochrome c. A similar value for single cells was obtained from chemiluminescence video microsocopy. Second, for nonphagocytic cellular production of $O_2·^-$, the rate of H_2O_2 production in the perfused liver was used to estimate $[O_2·^-]_{ss}$. This was based on the fact that the preponderence of H_2O_2 is generated from $O_2·^-$.

7. Estimation of the K_m for SOD is derived from the inability to saturate the enzyme at 25°. Nevertheless, several reasonable values have been calculated, ranging from 3.5 mM at pH 9.3 and 5.5°C to 0.35 mM at pH 9.9 and 25°C. For a discussion of this point, see Christopher Bull and James A. Fee, "Steady-state kinetics of superoxide dismutases: Properties of the iron containing protein from *Escherichia coli*," *J. Am. Chem. Soc.* **107** (1985): 3295–3304; James A. Fee and Christopher Bull, "Steady-state kinetic studies of superoxide dismutase. Saturative behavior of the copper- and zinc-containing proteins," *J. Biol. Chem.* **261** (1986): 13000–13006; A. Rigo, P. Viglino, and G. Rotilio, "Kinetic study of O_2^- dismutation by bovine superoxide dismutase. Evidence for saturation of the catalytic sites by O_2, *Biochem. Biophys. Res. Commun.* **63** (1975): 1013–1018.

8. For the purpose of completeness, aging has been included, as $O_2·^-$ and oxidants derived from this free radical have been associated with the lifespan of many species. For a review of this topic, see Rajindar S. Sohal and Richard Weindruch, "Oxidative stress, caloric restriction, and aging," *Science* **273** (1996): 59–63.

9. Whether $O_2·^-$ is directly responsible for initiation of lipid peroxidation is superfluous, considering that known mechanisms for generation of HO· require the presence of $O_2·^-$. For a discussion of this point, see Gerald M. Rosen, Sovitj Pou, Carroll L. Ramos, Myron S. Cohen, and Bradley E. Britigan, "Free radicals and phagocytic cells," *FASEB J.* **9** (1995): 200–209.

Chapter 3

Portions of this chapter were adopted from a review article (Gerald M. Rosen, Christine U. Eccles, and Sovitj Pou, "Nitric oxide in the brain: Looking at two sides of the same coin," *Neurologist* **1**: (1995) 311–325). More extensive discussions on the chemical reactions and physiological functions of NO· can be found in a number of recent books on this topic. Of particular relevance to the theme of this chapter are the following references: *Methods in Nitric Oxide Research*, ed. by M. Feelish and J.S. Stamler, John Wiley & Sons, New York, 1996; *Methods in Enzymology*, Vols. 268 and 269, ed. by L. Packer, Academic Press, San Diego, CA, 1996; *Nitric Oxide, Principles and Actions*, ed. by J. Lancaster, Jr., Academic Press, San Diego, CA, 1996.

Robert F. Furchgott, Louis J. Ignarro, and Ferid Murad received the 1998 Nobel Prize in Physiology or Medicine for their research on the biological role of nitric oxide.

1. We would be insensitive not to include the years 1954 (Rebeca Gerschman, Daniel L. Gilbert, Sylvanus W. Nye, Peter Dwyer, and Wallace O. Fenn, "Oxygen poisoning and x-irradiation: A mechanism in common," *Science* **119**: (1954) 623–226), 1968 (Joe M. McCord and Irwin Fridovich, "The reduction of cytochrome c by milk xanthine oxidase," *J. Biol. Chem.* (1968) **243**: 5753–5760), and 1969 (Joe M. McCord and Irwin Fridovich, "Superoxide dismutase—an enzymic function for erythrocuprein (hemocuprein)," *J. Biol. Chem.* **244** (1969): 6049–6055) as milestones in free radical research. For this reason, we use the phrase "another watershed year."

2. At the 4th International Symposium on Mechanisms of Vasodilatation, held in Rochester, MN in July 1986, Drs. Robert Furchgott and Lewis Ignarro presented data documenting that the pharmacological properties associated with endothelium-derived relaxation factor could be duplicated by NO·. However, the compilation of their lectures into a book was not completed until 1988, a year after Palmer, *et al.*, (1987) and Ignarro, *et al.* (1987) had disclosed their findings through publication in scientific journals.

3. A reliable source of the purified human isoform of NOS II has recently become available using different transfection procedures. For details, see Tzeng, *et al.* (1995), Calaycay, *et al.* (1996), and Laubach, *et al.* (1996).

4. The role that tetrahydro-L-biopterin plays in the NOS-catalyzed formation of NO· from L-arginine remains unresolved. For an insightful discussion of this topic, see Masters, *et al.* (1996).

Chapter 4

Portions of this chapter were adopted from a review article (Sharon P. Ellington, Karyn E. Strauss, and Gerald M. Rosen, "Spin trapping of free radicals in whole cells, isolated organs and *in vivo*," *Cellular Antioxidant Defense Mechanisms*, Vol. I,

ed. by C.K. Chow, CRC Press, Boca Raton, FL; 1988, pp. 42–57).

1. Several recent books present detailed analytic methods for the detection of NO·. Of particular interest are *Methods in Nitric Oxide Research*, ed. by M. Feelish and J.S. Stamler, John Wiley & Sons, 1996 and *Methods in Enzymology*, Vol. 268, ed. by L. Packer, Academic Press, San Diego, CA, 1996.

2. Analytic assays for the quantification of ONOO$^-$ have been compiled in *Methods in Enzymology*, Vol. 269, ed. by L. Packer, Academic Press, San Diego, CA, 1996.

Chapter 6

1. Sir Alexander Todd, who recently died at the age of 89, received the 1957 Nobel prize in chemistry for his research on the structure and chemistry of nucleic acids. His later work on vitamin B$_{12}$ was certainly instrumental in the development of spin trapping as a method to identify unstable free radicals.

2. The authors are grateful to Dr. William Cherry for providing a sample of, and synthetic scheme for, 5-hexadecyl-3,3,5-trimethyl-1-pyrroline-N-oxide (**65**).

3. The ester hydrolysis described is typical for the preparation of Δ^1-pyrroline-*N*-oxides, such as 4-carboxyl-3,3,5,5-tetramethyl-1-pyrroline-N-oxide (**70**). Although not published, the synthesis of nitrone (**70**) is a simple hydrolysis of nitrone (**69**) (Dr. Rosen and Dr. Prabhat Arya of the National Research Council of Canada, Ottawa, Canada).

4. The synthesis of 2,2-dimethyl-4-hydroximinomethyl-2H-imidazole-N-oxide (**98**) is based on the preparative scheme of Kirilyuk, *et al.*, (1991) with modifications described herein.

Chapter 7

1. For a discussion of the development of microwave technology and its role in the development of radar during World War II, see the Massachusetts Institute of Technology Radiation Laboratory Series, Vols. 1–28, ed. by L.N. Ridenour, McGraw-Hill, New York, 1948.

2. The TM mode and the TE mode have confusingly been referred to as an E mode and an H mode, respectively. For a discus-

sion on this point, see N. Marcuvitz, *Waveguide Handbook*, MIT Radiation Laboratory Series, No. 10, McGraw Hill, New York, 1951,.

3. For an example Henry C. Torrey and Charles A. Whitmer, *Crystal Rectifiers* MIT Radiation Laboratory Series, No. 15, McGraw-Hill, New York, 1948, pp. 1–5.

Chapter 8

Portions of this chapter were adopted from earlier publications on the kinetics of spin trapping $O_2\cdot^-$ and HO· (Eli Finkelstein, Gerald M. Rosen, and Elmer J. Rauckman, "Spin trapping. Kinetics of the reaction of superoxide and hydroxyl radicals with nitrones," *J. Am. Chem. Soc.* **102** (1980): 4994–4999; Sovitj Pou, Carroll L. Ramos, Tim Gladwell, Erik Renks, Michelle Centra, David Young, Myron S. Cohen, and Gerald M. Rosen, "A kinetic approach to the selection of a sensitive spin trapping system for the detection of hydroxyl radical," *Anal. Biochem.* **217** (1994): 76–83).

1. This nitrone was chosen to simplify the model due primarily to the long half-life of TMPO–OOH and/or TMPO–OH versus DMPO–OOH (see Figure 8.1).

2. The apparent rate constants include the contribution from $O_2\cdot^-$ and $HO_2^{\cdot}$ to the rate equations. The rate of TMPO-OOH formation is, therefore, a summation of the rate of spin trapping $O_2\cdot^-$ and $HO_2\cdot$:

$$d[\text{TMPO}-\text{OOH}]dt =$$

$$k_{O_2\cdot^-}[\text{TMPO}][O_2\cdot^-] + k_{HO_2}[\text{TMPO}][HO_2\cdot]$$

$$[HO_2\cdot] = [O_2\cdot^-][H^+]/K_a$$

$$d[\text{TMPO}-\text{OOH}]/dt =$$

$$(k_{O_2\cdot^-} + k_{HO_2}[H^+]/K_a)][\text{TMPO}][O_2\cdot^-]$$

or

$$d[\text{TMPO}-\text{OOH}]/dt = k_t[\text{TMPO}][O_2\cdot^-]$$

where

$$k_t = (k_{O_2\cdot^-} + k_{HO_2}[H^+]/K_a)$$

A similar derivation can be developed to account for the pH-dependent dismutation of $O_2\cdot^-$, $k_d[O_2\cdot^-]^2$, where k_d is a function of pH. By using apparent rate constants, the spin trapping and dismutation can be expressed in terms of $O_2\cdot^-$, thereby solving the combined rate equation explicitly.

3. The validity of using a steady-state assumption for the concentration of $O_2\cdot^-$ from a xanthine/xanthine oxidase system is supported by the earlier work of McCord and Fridovich. At a given flux of $O_2\cdot^-$ under these experimental conditions, this free radical can either disproportionate or react with the nitrone. For a review on this subject, see Joe M. McCord and Irwin Fridovich, "Superoxide dismutases: A history," in *Superoxide and Superoxide Dismutases*, ed. by A.M. Michelson, J.M., McCord, and I. Fridovich, Academic Press, New York; 1977, pp. 1–10.

4. In developing the kinetic model, the question becomes what concentration of the nitrone is necessary to essentially spin trap all of the available $O_2\cdot^-$.

$$k_d[O_2\cdot^-]^2 = k_t[O_2\cdot^-][\text{DMPO}]$$

Assuming a $k_d \approx 10^5\,\text{M}^{-1}\text{s}^{-1}$ and $k_t \approx 10\,\text{M}^{-1}\text{s}^{-1}$ at a flux of $O_2\cdot^-$ of 1 µM at pH 7.8, then a DMPO concentration of 180 mM would meet this criteria, resulting in an error of no more than 5%, whch is well within an acceptable range.

5. For purposes of normalizing data, the solution of SOD is standardized against cytochrome c. One unit of activity is, therefore, defined as the amount of SOD required to inhibit by 50% the rate of $O_2\cdot^-$-mediated reduction of cytochrome c, as described by Joe M. McCord and I. Fridovich, "Superoxide dismutase. An enzymic function for erythrocuprein (hemocuprein)," *J. Biol. Chem.* **244** (1969): 6049–6055.

6. Although the reaction of HO· with CH_3CH_2OH can occur at three different sites—the α-, β- or HO-position—α-abstraction is the preferred location (Asmus, *et al.*, 1973). Using $CH_3^{13}CH_2OH$, the only free radical spin trapped by 4-POBN was $CH_3\cdot CHOH$ (see Sovitj Pou, Carrol L. Ramos, Tim Gladwell, Erik Renks, Michelle Centra, David Young, Myron S. Cohen, and Gerald M. Rosen, "A kinetic approach to the selection of a sensitive spin trapping system for the detection of

hydroxyl radical," *Anal. Biochem.* **217** (1994): 76–83).

7. Disproportionation of nitroxides requires the presence of at least one hydrogen at the α-carbon position. There are exceptions to this general rule. For instance, unhindered bicyclic nitroxides exhibit remarkable stability, as a double bond at a bridge head would violate Bredt's rule (see Rose-Marie Dupeyre and André Rassat, "Nitroxides. XIX. Norpseudopelletierine-N-oxyl, a new stable, unhindered free radical," *J. Am. Chem. Soc.* **88** (1966): 3180–3181; Elmer J. Rauckman, Gerald M. Rosen, and Joy Cavagnaro, "Norcocaine nitroxide. A potential hepatotoxic metabolite of cocaine," *Mol. Pharmacol.* **21** (1982): 458–463).

Chapter 10

1. Based on the recorded EPR spectrum of DMPO–OH, the hyperfine splitting constant was originally determined to be 15.3 G, and published as such. More recent measurements have, however, found the value to be 14.9 G (see Anson S.W. Li, Michael S. Watson, Lisa D. Carlton, Garry R. Buettner, Anthony H. de Haas, and Colin F. Chignell, "A database for spin-trapping implemented on the IBM compatibles and Apple Macintoshes," 1989).

2. The initial step in the reaction of NO· with 2,5-dimethylhexadiene is the formation of a secondary free radical. In the case of the trans-isomer, initial rearrangement must proceed prior to ring closure. As the trans-isomer dominates the racemic mixture, alternative reactions have resulted in such a poor rate of nitroxide formation that it has been impossible to detect 2,2,5,5-tetramethyl-1-pyrrolinyloxyl by EPR spectroscopy using physiologic fluxes of NO·

generated during the metabolism of L-arginine by purified brain nitric oxide synthase (Sovitj Pou and Gerald M. Rosen, unpublished findings). In line with this observation, there has been much debate as to the nature of this addition reaction and whether NO·, NO$_2$·, or both free radicals are the reactive species with the conjugated diene. For contemporary papers on this controversy, see Antal Rockenbauer, and László Korecz, "Comment on conversion of nitric oxide into a nitroxide radical using 2,3-dimethylbutadiene and 2,5-dimethylhexadiene," *J. Chem. Soc., Chem. Commun.* (1994): 145; I. Gabr, R.P. Patel, M.C.R. Symons, and M.T. Wilson, "Novel reactions of nitric oxide in biological systems," *J. Chem. Soc., Chem. Commun.* (1995): 915–916; Issam Gabr, and Martyn C.R. Symons, "Reactions of conjugated dienes with nitrogen monoxide and dioxide," *J. Chem. Soc., Faraday Trans.* **92** (1996): 1767–1772; David R. Kelly, Simon Jones, John O. Adigum, Kevin S.V. Koh, and Simon K. Jackson, "The addition of nitric oxide to 2,5-dimethyl-hexa-2,4-diene gives nitrogen dioxide adducts," *Tetrahedon Lett.* **38** (1997): 1245–1248. It is for the reasons cited above that the fused cis-ringed diene 7,7,8,8-tetramethyl-*o*-quinodimethane and similarly related analogs are more reliable spin traps for NO·.

Chapter 11

1. For the purpose of this discussion, NO· has been shown to be an effective therapeutic agent, attenuating the effects associated with the reperfusion of an ischemic tissue. As presented in chapter 3, there is extensive debate as to whether ONOO$^-$, generated by the reaction of NO· with O$_2$·$^-$, is cytoprotective or cytotoxic.

Index